Lecture Notes in Computer Science 10404

More information about this series at http://www.springer.com/series/7407

Editors
Osvaldo Gervasi
University of Perugia
Perugia
Italy

Beniamino Murgante
University of Basilicata
Potenza
Italy

Sanjay Misra
Covenant University
Ota
Nigeria

Giuseppe Borruso
University of Trieste
Trieste
Italy

Carmelo M. Torre
Polytechnic University of Bari
Bari
Italy

Ana Maria A.C. Rocha
University of Minho
Braga
Portugal

David Taniar
Monash University
Clayton, VIC
Australia

Bernady O. Apduhan
Kyushu Sangyo University
Fukuoka
Japan

Elena Stankova
Saint Petersburg State University
Saint Petersburg
Russia

Alfredo Cuzzocrea
University of Trieste
Trieste
Italy

ISSN 0302-9743 ISSN 1611-3349 (electronic)
Lecture Notes in Computer Science
ISBN 978-3-319-62391-7 ISBN 978-3-319-62392-4 (eBook)
DOI 10.1007/978-3-319-62392-4

Library of Congress Control Number: 2017945283

LNCS Sublibrary: SL1 – Theoretical Computer Science and General Issues

Printed on acid-free paper

This Springer imprint is published by Springer Nature
The registered company is Springer International Publishing AG
The registered company address is: Gewerbestrasse 11, 6330 Cham, Switzerland

Osvaldo Gervasi · Beniamino Murgante
Sanjay Misra · Giuseppe Borruso
Carmelo M. Torre · Ana Maria A.C. Rocha
David Taniar · Bernady O. Apduhan
Elena Stankova · Alfredo Cuzzocrea (Eds.)

Computational Science and Its Applications – ICCSA 2017

17th International Conference
Trieste, Italy, July 3–6, 2017
Proceedings, Part I

Springer

Preface

These multiple volumes (LNCS volumes 10404, 10405, 10406, 10407, 10408, and 10409) consist of the peer-reviewed papers from the 2017 International Conference on Computational Science and Its Applications (ICCSA 2017) held in Trieste, Italy, during July 3–6, 2017.

ICCSA 2017 was a successful event in the ICCSA conference series, previously held in Beijing, China (2016), Banff, Canada (2015), Guimarães, Portugal (2014), Ho Chi Minh City, Vietnam (2013), Salvador, Brazil (2012), Santander, Spain (2011), Fukuoka, Japan (2010), Suwon, South Korea (2009), Perugia, Italy (2008), Kuala Lumpur, Malaysia (2007), Glasgow, UK (2006), Singapore (2005), Assisi, Italy (2004), Montreal, Canada (2003), (as ICCS) Amsterdam, The Netherlands (2002), and San Francisco, USA (2001).

Computational science is a main pillar of most present research as well as industrial and commercial activities and plays a unique role in exploiting ICT innovative technologies. The ICCSA conference series have been providing a venue to researchers and industry practitioners to discuss new ideas, to share complex problems and their solutions, and to shape new trends in computational science.

Apart from the general tracks, ICCSA 2017 also include 43 international workshops, in various areas of computational sciences, ranging from computational science technologies to specific areas of computational sciences, such as computer graphics and virtual reality. Furthermore, this year ICCSA 2017 hosted the XIV International Workshop on Quantum Reactive Scattering. The program also features three keynote speeches and four tutorials.

The success of the ICCSA conference series in general, and ICCSA 2017 in particular, is due to the support of many people: authors, presenters, participants, keynote speakers, session chairs, Organizing Committee members, student volunteers, Program Committee members, international Advisory Committee members, international liaison chairs, and various people in other roles. We would like to thank them all.

We would also like to thank Springer for their continuous support in publishing the ICCSA conference proceedings.

July 2017

Giuseppe Borruso
Osvaldo Gervasi
Bernady O. Apduhan

Welcome to Trieste

We were honored and happy to have organized this extraordinary edition of the conference, with so many interesting contributions and participants coming from more than 46 countries around the world!

Trieste is a medium-size Italian city lying on the north-eastern border between Italy and Slovenia. It has a population of nearly 200,000 inhabitants and faces the Adriatic Sea, surrounded by the Karst plateau.

It is quite an atypical Italian city, with its history being very much influenced by belonging for several centuries to the Austro-Hungarian empire and having been through several foreign occupations in history: by French, Venetians, and the Allied Forces after the Second World War. Such events left several footprints on the structure of the city, on its buildings, as well as on culture and society!

During its history, Trieste hosted people coming from different countries and regions, making it a cosmopolitan and open city. This was also helped by the presence of a commercial port that made it an important trade center from the 18th century on. Trieste is known today as a 'City of Science' or, more proudly, presenting itself as the 'City of Knowledge', thanks to the presence of several universities and research centers, all of them working at an international level, as well as of cultural institutions and traditions. The city has a high presence of researchers, more than 35 per 1,000 employed people, much higher than the European average of 6 employed researchers per 1,000 people.

The University of Trieste, the origin of such a system of scientific institutions, dates back to 1924, although its roots go back to the end of the 19th century under the Austro-Hungarian Empire. The university today employs nearly 1,500 teaching, research, technical, and administrative staff with a population of more than 16,000 students.

The university currently has 10 departments: Economics, Business, Mathematical, and Statistical Sciences; Engineering and Architecture; Humanities; Legal, Language, Interpreting, and Translation Studies; Mathematics and Geosciences; Medicine, Surgery, and Health Sciences; Life Sciences; Pharmaceutical and Chemical Sciences; Physics; Political and Social Sciences.

We trust the participants enjoyed the cultural and scientific offerings of Trieste and will keep a special memory of the event.

Giuseppe Borruso

Organization

ICCSA 2017 was organized by the University of Trieste (Italy), University of Perugia (Italy), Monash University (Australia), Kyushu Sangyo University (Japan), University of Basilicata (Italy), and University of Minho, (Portugal).

Honorary General Chairs

Antonio Laganà	University of Perugia, Italy
Norio Shiratori	Tohoku University, Japan
Kenneth C.J. Tan	Sardina Systems, Estonia

General Chairs

Giuseppe Borruso	University of Trieste, Italy
Osvaldo Gervasi	University of Perugia, Italy
Bernady O. Apduhan	Kyushu Sangyo University, Japan

Program Committee Chairs

Alfredo Cuzzocrea	University of Trieste, Italy
Beniamino Murgante	University of Basilicata, Italy
Ana Maria A.C. Rocha	University of Minho, Portugal
David Taniar	Monash University, Australia

International Advisory Committee

Jemal Abawajy	Deakin University, Australia
Dharma P. Agrawal	University of Cincinnati, USA
Marina L. Gavrilova	University of Calgary, Canada
Claudia Bauzer Medeiros	University of Campinas, Brazil
Manfred M. Fisher	Vienna University of Economics and Business, Austria
Yee Leung	Chinese University of Hong Kong, SAR China

International Liaison Chairs

Ana Carla P. Bitencourt	Universidade Federal do Reconcavo da Bahia, Brazil
Maria Irene Falcão	University of Minho, Portugal
Robert C.H. Hsu	Chung Hua University, Taiwan
Tai-Hoon Kim	Hannam University, Korea
Sanjay Misra	University of Minna, Nigeria
Takashi Naka	Kyushu Sangyo University, Japan

Rafael D.C. Santos	National Institute for Space Research, Brazil
Maribel Yasmina Santos	University of Minho, Portugal

Workshop and Session Organizing Chairs

Beniamino Murgante	University of Basilicata, Italy
Sanjay Misra	Covenant University, Nigeria
Jorge Gustavo Rocha	University of Minho, Portugal

Award Chair

Wenny Rahayu	La Trobe University, Australia

Publicity Committee Chair

Stefano Cozzini	Democritos Center, National Research Council, Italy
Elmer Dadios	De La Salle University, Philippines
Hong Quang Nguyen	International University (VNU-HCM), Vietnam
Daisuke Takahashi	Tsukuba University, Japan
Shangwang Wang	Beijing University of Posts and Telecommunications, China

Workshop Organizers

Agricultural and Environmental Big Data Analytics (AEDBA 2017)

Sandro Bimonte	IRSTEA, France
André Miralles	IRSTEA, France

Advances in Data Mining for Applications (AMDMA 2017)

Carlo Cattani	University of Tuscia, Italy
Majaz Moonis	University of Massachusettes Medical School, USA
Yeliz Karaca	IEEE, Computer Society Association

Advances Smart Mobility and Transportation (ASMAT 2017)

Mauro Mazzei	CNR, Italian National Research Council, Italy

Advances in Information Systems and Technologies for Emergency Preparedness and Risk Assessment and Mitigation (ASTER 2017)

Maurizio Pollino	ENEA, Italy
Marco Vona	University of Basilicata, Italy
Beniamino Murgante	University of Basilicata, Italy

Advances in Web-Based Learning (AWBL 2017)

Mustafa Murat Inceoglu	Ege University, Turkey
Birol Ciloglugil	Ege University, Turkey

Big Data Warehousing and Analytics (BIGGS 2017)

Maribel Yasmina Santos	University of Minho, Portugal
Monica Wachowicz	University of New Brunswick, Canada
Joao Moura Pires	NOVA de Lisboa University, Portugal
Rafael Santos	National Institute for Space Research, Brazil

Bio-inspired Computing and Applications (BIONCA 2017)

Nadia Nedjah	State University of Rio de Janeiro, Brazil
Luiza de Macedo Mourell	State University of Rio de Janeiro, Brazil

Computational and Applied Mathematics (CAM 2017)

M. Irene Falcao	University of Minho, Portugal
Fernando Miranda	University of Minho, Portugal

Computer-Aided Modeling, Simulation, and Analysis (CAMSA 2017)

Jie Shen	University of Michigan, USA and Jilin University, China
Hao Chenina	Shanghai University of Engineering Science, China
Chaochun Yuan	Jiangsu University, China

Computational and Applied Statistics (CAS 2017)

Ana Cristina Braga	University of Minho, Portugal

Computational Geometry and Security Applications (CGSA 2017)

Marina L. Gavrilova	University of Calgary, Canada

Central Italy 2016 Earthquake: Computational Tools and Data Analysis for Emergency Response, Community Support, and Reconstruction Planning (CIEQ 2017)

Alessandro Rasulo	Università degli Studi di Cassino e del Lazio Meridionale, Italy
Davide Lavorato	Università degli Studi di Roma Tre, Italy

Computational Methods for Business Analytics (CMBA 2017)

Telmo Pinto	University of Minho, Portugal
Claudio Alves	University of Minho, Portugal

Chemistry and Materials Sciences and Technologies (CMST 2017)

Antonio Laganà	University of Perugia, Italy
Noelia Faginas Lago	University of Perugia, Italy

Computational Optimization and Applications (COA 2017)

Ana Maria Rocha	University of Minho, Portugal
Humberto Rocha	University of Coimbra, Portugal

Cities, Technologies, and Planning (CTP 2017)

Giuseppe Borruso	University of Trieste, Italy
Beniamino Murgante	University of Basilicata, Italy

Data-Driven Modelling for Sustainability Assessment (DAMOST 2017)

Antonino Marvuglia	Luxembourg Institute of Science and Technology, LIST, Luxembourg
Mikhail Kanevski	University of Lausanne, Switzerland
Beniamino Murgante	University of Basilicata, Italy
Janusz Starczewski	Częstochowa University of Technology, Poland

Databases and Computerized Information Retrieval Systems (DCIRS 2017)

Sultan Alamri	College of Computing and Informatics, SEU, Saudi Arabia
Adil Fahad	Albaha University, Saudi Arabia
Abdullah Alamri	Jeddah University, Saudi Arabia

Data Science for Intelligent Decision Support (DS4IDS 2016)

Filipe Portela	University of Minho, Portugal
Manuel Filipe Santos	University of Minho, Portugal

Deep Cities: Intelligence and Interoperability (DEEP_CITY 2017)

Maurizio Pollino — ENEA, Italian National Agency for New Technologies, Energy and Sustainable Economic Development, Italy
Grazia Fattoruso — ENEA, Italian National Agency for New Technologies, Energy and Sustainable Economic Development, Italy

Emotion Recognition (EMORE 2017)

Valentina Franzoni — University of Rome La Sapienza, Italy
Alfredo Milani — University of Perugia, Italy

Future Computing Systems, Technologies, and Applications (FISTA 2017)

Bernady O. Apduhan — Kyushu Sangyo University, Japan
Rafael Santos — National Institute for Space Research, Brazil

Geographical Analysis, Urban Modeling, Spatial Statistics (Geo-and-Mod 2017)

Giuseppe Borruso — University of Trieste, Italy
Beniamino Murgante — University of Basilicata, Italy
Hartmut Asche — University of Potsdam, Germany

Geomatics and Remote Sensing Techniques for Resource Monitoring and Control (GRS-RMC 2017)

Eufemia Tarantino — Polytechnic of Bari, Italy
Rosa Lasaponara — Italian Research Council, IMAA-CNR, Italy
Antonio Novelli — Polytechnic of Bari, Italy

Interactively Presenting High-Quality Graphics in Cooperation with Various Computing Tools (IPHQG 2017)

Masataka Kaneko — Toho University, Japan
Setsuo Takato — Toho University, Japan
Satoshi Yamashita — Kisarazu National College of Technology, Italy

Web-Based Collective Evolutionary Systems: Models, Measures, Applications (IWCES 2017)

Alfredo Milani — University of Perugia, Italy
Rajdeep Nyogi — Institute of Technology, Roorkee, India
Valentina Franzoni — University of Rome La Sapienza, Italy

Computational Mathematics, and Statistics for Data Management and Software Engineering (IWCMSDMSE 2017)

M. Filomena Teodoro	Lisbon University and Portuguese Naval Academy, Portugal
Anacleto Correia	Portuguese Naval Academy, Portugal

Land Use Monitoring for Soil Consumption Reduction (LUMS 2017)

Carmelo M. Torre	Polytechnic of Bari, Italy
Beniamino Murgante	University of Basilicata, Italy
Alessandro Bonifazi	Polytechnic of Bari, Italy
Massimiliano Bencardino	University of Salerno, Italy

Mobile Communications (MC 2017)

Hyunseung Choo	Sungkyunkwan University, Korea

Mobile-Computing, Sensing, and Actuation - Fog Networking (MSA4FOG 2017)

Saad Qaisar	NUST School of Electrical Engineering and Computer Science, Pakistan
Moonseong Kim	Korean Intellectual Property Office, South Korea

Physiological and Affective Computing: Methods and Applications (PACMA 2017)

Robertas Damasevicius	Kaunas University of Technology, Lithuania
Christian Napoli	University of Catania, Italy
Marcin Wozniak	Silesian University of Technology, Poland

Quantum Mechanics: Computational Strategies and Applications (QMCSA 2017)

Mirco Ragni	Universidad Federal de Bahia, Brazil
Ana Carla Peixoto Bitencourt	Universidade Estadual de Feira de Santana, Brazil
Vincenzo Aquilanti	University of Perugia, Italy

Advances in Remote Sensing for Cultural Heritage (RS 2017)

Rosa Lasaponara	IRMMA, CNR, Italy
Nicola Masini	IBAM, CNR, Italy Zhengzhou Base, International Center on Space Technologies for Natural and Cultural Heritage, China

Scientific Computing Infrastructure (SCI 2017)

Elena Stankova	Saint Petersburg State University, Russia
Alexander Bodganov	Saint Petersburg State University, Russia
Vladimir Korkhov	Saint Petersburg State University, Russia

Software Engineering Processes and Applications (SEPA 2017)

Sanjay Misra	Covenant University, Nigeria

Sustainability Performance Assessment: Models, Approaches and Applications Toward Interdisciplinarity and Integrated Solutions (SPA 2017)

Francesco Scorza	University of Basilicata, Italy
Valentin Grecu	Lucia Blaga University on Sibiu, Romania
Jolanta Dvarioniene	Kaunas University, Lithuania
Sabrina Lai	Cagliari University, Italy

Software Quality (SQ 2017)

Sanjay Misra	Covenant University, Nigeria

Advances in Spatio-Temporal Analytics (ST-Analytics 2017)

Rafael Santos	Brazilian Space Research Agency, Brazil
Karine Reis Ferreira	Brazilian Space Research Agency, Brazil
Maribel Yasmina Santos	University of Minho, Portugal
Joao Moura Pires	New University of Lisbon, Portugal

Tools and Techniques in Software Development Processes (TTSDP 2017)

Sanjay Misra	Covenant University, Nigeria

Challenges, Trends, and Innovations in VGI (VGI 2017)

Claudia Ceppi	University of Basilicata, Italy
Beniamino Murgante	University of Basilicata, Italy
Lucia Tilio	University of Basilicata, Italy
Francesco Mancini	University of Modena and Reggio Emilia, Italy
Rodrigo Tapia-McClung	Centro de Investigación en Geografía y Geomática "Ing Jorge L. Tamayo", Mexico
Jorge Gustavo Rocha	University of Minho, Portugal

Virtual Reality and Applications (VRA 2017)

Osvaldo Gervasi	University of Perugia, Italy

Industrial Computational Applications (WICA 2017)

Eric Medvet	University of Trieste, Italy
Gianfranco Fenu	University of Trieste, Italy
Riccardo Ferrari	Delft University of Technology, The Netherlands

XIV International Workshop on Quantum Reactive Scattering (QRS 2017)

Niyazi Bulut	Fırat University, Turkey
Noelia Faginas Lago	University of Perugia, Italy
Andrea Lombardi	University of Perugia, Italy
Federico Palazzetti	University of Perugia, Italy

Program Committee

Jemal Abawajy	Deakin University, Australia
Kenny Adamson	University of Ulster, UK
Filipe Alvelos	University of Minho, Portugal
Paula Amaral	Universidade Nova de Lisboa, Portugal
Hartmut Asche	University of Potsdam, Germany
Md. Abul Kalam Azad	University of Minho, Portugal
Michela Bertolotto	University College Dublin, Ireland
Sandro Bimonte	CEMAGREF, TSCF, France
Rod Blais	University of Calgary, Canada
Ivan Blečić	University of Sassari, Italy
Giuseppe Borruso	University of Trieste, Italy
Yves Caniou	Lyon University, France
José A. Cardoso e Cunha	Universidade Nova de Lisboa, Portugal
Rui Cardoso	University of Beira Interior, Portugal
Leocadio G. Casado	University of Almeria, Spain
Carlo Cattani	University of Salerno, Italy

Dieter Kranzlmueller	LMU and LRZ Munich, Germany
Antonio Laganà	University of Perugia, Italy
Rosa Lasaponara	National Research Council, Italy
Maurizio Lazzari	National Research Council, Italy
Cheng Siong Lee	Monash University, Australia
Sangyoun Lee	Yonsei University, Korea
Jongchan Lee	Kunsan National University, Korea
Clement Leung	Hong Kong Baptist University, Hong Kong, SAR China
Chendong Li	University of Connecticut, USA
Gang Li	Deakin University, Australia
Ming Li	East China Normal University, China
Fang Liu	AMES Laboratories, USA
Xin Liu	University of Calgary, Canada
Savino Longo	University of Bari, Italy
Tinghuai Ma	NanJing University of Information Science and Technology, China
Sergio Maffioletti	University of Zurich, Switzerland
Ernesto Marcheggiani	Katholieke Universiteit Leuven, Belgium
Antonino Marvuglia	Research Centre Henri Tudor, Luxembourg
Nicola Masini	National Research Council, Italy
Nirvana Meratnia	University of Twente, The Netherlands
Alfredo Milani	University of Perugia, Italy
Sanjay Misra	Federal University of Technology Minna, Nigeria
Giuseppe Modica	University of Reggio Calabria, Italy
José Luis Montaña	University of Cantabria, Spain
Beniamino Murgante	University of Basilicata, Italy
Jiri Nedoma	Academy of Sciences of the Czech Republic, Czech Republic
Laszlo Neumann	University of Girona, Spain
Kok-Leong Ong	Deakin University, Australia
Belen Palop	Universidad de Valladolid, Spain
Marcin Paprzycki	Polish Academy of Sciences, Poland
Eric Pardede	La Trobe University, Australia
Kwangjin Park	Wonkwang University, Korea
Ana Isabel Pereira	Polytechnic Institute of Braganca, Portugal
Maurizio Pollino	Italian National Agency for New Technologies, Energy and Sustainable Economic Development, Italy
Alenka Poplin	University of Hamburg, Germany
Vidyasagar Potdar	Curtin University of Technology, Australia
David C. Prosperi	Florida Atlantic University, USA
Wenny Rahayu	La Trobe University, Australia
Jerzy Respondek	Silesian University of Technology Poland
Ana Maria A.C. Rocha	University of Minho, Portugal
Maria Clara Rocha	ESTES Coimbra, Portugal
Humberto Rocha	INESC-Coimbra, Portugal

Alexey Rodionov	Institute of Computational Mathematics and Mathematical Geophysics, Russia
Cristina S. Rodrigues	University of Minho, Portugal
Jon Rokne	University of Calgary, Canada
Octavio Roncero	CSIC, Spain
Maytham Safar	Kuwait University, Kuwait
Chiara Saracino	A.O. Ospedale Niguarda Ca' Granda - Milano, Italy
Haiduke Sarafian	The Pennsylvania State University, USA
Jie Shen	University of Michigan, USA
Qi Shi	Liverpool John Moores University, UK
Dale Shires	U.S. Army Research Laboratory, USA
Takuo Suganuma	Tohoku University, Japan
Sergio Tasso	University of Perugia, Italy
Ana Paula Teixeira	University of Tras-os-Montes and Alto Douro, Portugal
Senhorinha Teixeira	University of Minho, Portugal
Parimala Thulasiraman	University of Manitoba, Canada
Carmelo Torre	Polytechnic of Bari, Italy
Javier Martinez Torres	Centro Universitario de la Defensa Zaragoza, Spain
Giuseppe A. Trunfio	University of Sassari, Italy
Unal Ufuktepe	Izmir University of Economics, Turkey
Toshihiro Uchibayashi	Kyushu Sangyo University, Japan
Mario Valle	Swiss National Supercomputing Centre, Switzerland
Pablo Vanegas	University of Cuenca, Ecuador
Piero Giorgio Verdini	INFN Pisa and CERN, Italy
Marco Vizzari	University of Perugia, Italy
Koichi Wada	University of Tsukuba, Japan
Krzysztof Walkowiak	Wroclaw University of Technology, Poland
Zequn Wang	Intelligent Automation Inc., USA
Robert Weibel	University of Zurich, Switzerland
Roland Wismüller	Universität Siegen, Germany
Mudasser Wyne	SOET National University, USA
Chung-Huang Yang	National Kaohsiung Normal University, Taiwan
Xin-She Yang	National Physical Laboratory, UK
Salim Zabir	France Telecom Japan Co., Japan
Haifeng Zhao	University of California, Davis, USA
Kewen Zhao	University of Qiongzhou, China
Albert Y. Zomaya	University of Sydney, Australia

Additional Reviewers

A. Alwan Al-Juboori Ali	School of Computer Science and Technology, China
Aceto Lidia	University of Pisa, Italy
Acharjee Shukla	Dibrugarh University, India
Afreixo Vera	University of Aveiro, Portugal
Agra Agostinho	University of Aveiro, Portugal
Aguilar Antonio	University of Barcelona, Spain
Aguilar José Alfonso	Universidad Autónoma de Sinaloa, Mexico
Aicardi Irene	Politecnico di Torino, Italy
Alberti Margarita	University of Barcelona, Spain
Alberto Rui	University of Lisbon, Portugal
Ali Salman	University of Magna Graecia, Italy
Alvanides Seraphim	University at Newcastle, UK
Alvelos Filipe	Universidade do Minho, Portugal
Amato Alba	Seconda Università degli Studi di Napoli, Italy
Amorim Paulo	Instituto de Matemática da UFRJ (IM-UFRJ), Brazil
Anderson Roger	University of California Santa Cruz, USA
Andrianov Serge	Saint Petersburg State University, Russia
Andrienko Gennady	Fraunhofer-Institut für Intelligente Analyse- und Informationssysteme, Germany
Apduhan Bernady	Kyushu Sangyo University, Japan
Aquilanti Vincenzo	University of Perugia, Italy
Asche Hartmut	Potsdam University, Germany
Azam Samiul	United International University, Bangladesh
Azevedo Ana	Athabasca University, USA
Bae Ihn-Han	Catholic University of Daegu, South Korea
Balacco Gabriella	Polytechnic of Bari, Italy
Balena Pasquale	Polytechnic of Bari, Italy
Barroca Filho Itamir	Universidade Federal do Rio Grande do Norte, Brazil
Behera Ranjan Kumar	Indian Institute of Technology Patna, India
Belpassi Leonardo	National Research Council, Italy
Bentayeb Fadila	Université Lyon, France
Bernardino Raquel	Universidade da Beira Interiore, Portugal
Bertolotto Michela	University Collegue Dublin, UK
Bhatta Bijaya	Utkal University, India
Bimonte Sandro	IRSTEA, France
Blecic Ivan	University of Cagliari, Italy
Bo Carles	ICIQ, Spain
Bogdanov Alexander	Saint Petersburg State University, Russia
Bollini Letizia	University of Milano-Bicocca, Italy
Bonifazi Alessandro	Polytechnic of Bari, Italy
Bonnet Claude-Laurent	Université de Bordeaux, France
Borgogno Mondino Enrico Corrado	University of Turin, Italy
Borruso Giuseppe	University of Trieste, Italy

Bostenaru Maria	Ion Mincu University of Architecture and Urbanism, Romania
Boussaid Omar	Université Lyon 2, France
Braga Ana Cristina	University of Minho, Portugal
Braga Nuno	University of Minho, Portugal
Brasil Luciana	Instituto Federal Sao Paolo, Brazil
Cabral Pedro	Universidade NOVA de Lisboa, Portugal
Cacao Isabel	University of Aveiro, Portugal
Caiaffa Emanuela	Enea, Italy
Campagna Michele	University of Cagliari, Italy
Caniato Renhe Marcelo	Universidade Federal de Juiz de Fora, Brazil
Canora Filomena	University of Basilicata, Italy
Caradonna Grazia	Polytechnic of Bari, Italy
Cardoso Rui	Beira Interior University, Portugal
Caroti Gabriella	University of Pisa, Italy
Carravilla Maria Antonia	Universidade do Porto, Portugal
Cattani Carlo	University of Salerno, Italy
Cefalo Raffaela	University of Trieste, Italy
Ceppi Claudia	Polytechnic of Bari, Italy
Cerreta Maria	University Federico II of Naples, Italy
Chanet Jean-Pierre	UR TSCF Irstea, France
Chaturvedi Krishna Kumar	University of Delhi, India
Chiancone Andrea	University of Perugia, Italy
Choo Hyunseung	Sungkyunkwan University, South Korea
Ciabo Serena	University of l'Aquila, Italy
Coletti Cecilia	University of Chieti, Italy
Correia Aldina	Porto Polytechnic, Portugal
Correia Anacleto	CINAV, Portugal
Correia Elisete	University of Trás-Os-Montes e Alto Douro, Portugal
Correia Florbela Maria da Cruz Domingues	Instituto Politécnico de Viana do Castelo, Portugal
Cosido Oscar	University of Cantabria, Spain
Costa e Silva Eliana	University of Minho, Portugal
Costa Graça	Instituto Politécnico de Setúbal, Portugal
Costantini Alessandro	INFN, Italy
Crispim José	University of Minho, Portugal
Cuzzocrea Alfredo	University of Trieste, Italy
Danese Maria	IBAM, CNR, Italy
Daneshpajouh Shervin	University of Western Ontario, USA
De Fazio Dario	IMIP-CNR, Italy
De Runz Cyril	University of Reims Champagne-Ardenne, France
Deffuant Guillaume	Institut national de recherche en sciences et technologies pour l'environnement et l'agriculture, France
Degtyarev Alexander	Saint Petersburg State University, Russia
Devai Frank	London South Bank University, UK
Di Leo Margherita	JRC, European Commission, Belgium

Dias Joana	University of Coimbra, Portugal
Dilo Arta	University of Twente, The Netherlands
Dvarioniene Jolanta	Kaunas University of Technology, Lithuania
El-Zawawy Mohamed A.	Cairo University, Egypt
Escalona Maria-Jose	University of Seville, Spain
Faginas-Lago, Noelia	University of Perugia, Italy
Falcinelli Stefano	University of Perugia, Italy
Falcão M. Irene	University of Minho, Portugal
Faria Susana	University of Minho, Portugal
Fattoruso Grazia	ENEA, Italy
Fenu Gianfranco	University of Trieste, Italy
Fernandes Edite	University of Minho, Portugal
Fernandes Florbela	Escola Superior de Tecnologia e Gest ão de Bragancca, Portugal
Fernandes Rosario	USP/ESALQ, Brazil
Ferrari Riccardo	Delft University of Technology, The Netherlands
Figueiredo Manuel Carlos	University of Minho, Portugal
Florence Le Ber	ENGEES, France
Flouvat Frederic	University of New Caledonia, France
Fontes Dalila	Universidade do Porto, Portugal
Franzoni Valentina	University of Perugia, Italy
Freitas Adelaide de Fátima Baptista Valente	University of Aveiro, Portugal
Fusco Giovanni	Università di Bari, Italy
Gabrani Goldie	Tecpro Syst. Ltd., India
Gaido Luciano	INFN, Italy
Gallo Crescenzio	University of Foggia, Italy
Garaba Shungu	University of Connecticut, USA
Garau Chiara	University of Cagliari, Italy
Garcia Ernesto	University of the Basque Country, Spain
Gargano Ricardo	Universidade Brasilia, Brazil
Gavrilova Marina	University of Calgary, Canada
Gensel Jerome	IMAG, France
Gervasi Osvaldo	University of Perugia, Italy
Gioia Andrea	Polytechnic University of Bari, Italy
Giovinazzi Sonia	University of Canterbury, New Zealand
Gizzi Fabrizio	National Research Council, Italy
Gomes dos Anjos Eudisley	Universidade Federal da Paraíba, Brazil
Gonzaga de Oliveira Sanderson Lincohn	Universidade Federal de Lavras, Brazil
Gonçalves Arminda Manuela	University of Minho, Braga, Portugal
Gorbachev Yuriy	Geolink Technologies, Russia
Grecu Valentin	University of Sibiu, Romania
Gupta Brij	Cancer Biology Research Center, USA
Hagen-Zanker Alex	University of Surrey, UK

Hamaguchi Naoki	Tokyo Kyoiku University, Japan
Hanazumi Simone	University of Sao Paulo, Brazil
Hanzl Malgorzata	University of Lodz, Poland
Hayashi Masaki	University of Calgary, Canada
Hendrix Eligius M.T.	Operations Research and Logistics Group, The Netherlands
Henriques Carla	Inst. Politécnico de Viseu, Portugal
Herawan Tutut	State Polytechnic of Malang, Indonesia
Hsu Hui-Huang	National Chiao Tung University, Taiwan
Ienco Dino	La Maison de la télédétection de Montpellier, France
Iglesias Andres	Universidad de Cantabria, Spain
Imran Rabeea	NUST Islamabad, Pakistan
Inoue Kentaro	National Technical University of Athens, Greece
Josselin Didier	Université d'Avignon et des Pays de Vaucluse, France
Kaneko Masataka	Kisarazu National College of Technology, Japan
Kang Myoung-Ah	Blaise Pascal University, France
Karampiperis Pythagoras	National Center of Scientific Research, Athens, Greece
Kavouras Marinos	University of Athens, Greece
Kolingerova Ivana	University of West Bohemia, Czech Republic
Korkhov Vladimir	Saint Petersburg State University, Russia
Kotzinos Dimitrios	University of Cergy Pontoise, France
Kulabukhova Nataliia	Saint Petersburg State University, Russia
Kumar Dileep	SR Engineering College, India
Kumar Lov	National Institute of Technology, Rourkela, India
Kumar Pawan	Institute for Advanced Study, Princeton, USA
Laganà Antonio	University of Perugia, Italy
Lai Sabrina	Università di Cagliari, Italy
Lanza Viviana	Lombardy Regional Institute for Research, Italy
Lasala Piermichele	Università di Foggia, Italy
Laurent Anne	Laboratoire d'Informatique, de Robotique et de Microélectronique de Montpellier, France
Lavorato Davide	University of Rome, Italy
Le Duc Tai	Sungkyunkwan University, South Korea
Legatiuk Dmitrii	Bauhaus University, Germany
Li Ming	University of Waterloo, Canada
Lima Ana	University of São Paulo (UNIFESP), Brazil
Liu Xin	École polytechnique fédérale de Lausanne, Switzerland
Lombardi Andrea	University of Perugia, Italy
Lopes Cristina	Instituto Superior de Contabilidade e Administracao do Porto, Portugal
Lopes Maria João	Instituto Universitário de Lisboa, Portugal
Lourenço Vanda Marisa	Universidade NOVA de Lisboa, Portugal
Machado Jose	University of Minho, Portugal
Maeda Yoichi	Tokai University, Japan
Majcen Nineta	Euchems, Belgium
Malonek Helmuth	Universidade de Aveiro, Portugal

Mancini Francesco	University of Modena and Reggio Emilia, Italy
Mandanici Emanuele	Università di Bologna, Italy
Manganelli Benedetto	Università degli studi della Basilicata, Italy
Manso Callejo Miguel Angel	Universidad Politécnica de Madrid, Spain
Margalef Tomas	Autonomous University of Barcelona, Spain
Marques Jorge	University of Coimbra, Portugal
Martins Bruno	Universidade de Lisboa, Portugal
Marvuglia Antonino	Public Research Centre Henri Tudor, Luxembourg
Mateos Cristian	Universidad Nacional del Centro, Argentina
Mauro Giovanni	University of Trieste, Italy
McGuire Michael	Towson University, USA
Medvet Eric	University of Trieste, Italy
Milani Alfredo	University of Perugia, Italy
Millham Richard	Durban University of Technoloy, South Africa
Minghini Marco	Polytechnic University of Milan, Italy
Minhas Umar	University of Waterloo, Ontario, Canada
Miralles André	La Maison de la télédétection de Montpellier, France
Miranda Fernando	Universidade do Minho, Portugal
Misra Sanjay	Covenant University, Nigeria
Modica Giuseppe	Università Mediterranea di Reggio Calabria, Italy
Molaei Qelichi Mohamad	University of Tehran, Iran
Monteiro Ana Margarida	University of Coimbra, Portugal
Morano Pierluigi	Polytechnic University of Bari, Italy
Moura Ana	Universidade de Aveiro, Portugal
Moura Pires João	Universidade NOVA de Lisboa, Portugal
Mourão Maria	ESTG-IPVC, Portugal
Murgante Beniamino	University of Basilicata, Italy
Nagy Csaba	University of Szeged, Hungary
Nakamura Yasuyuki	Nagoya University, Japan
Natário Isabel Cristina Maciel	University Nova de Lisboa, Portugal
Nemmaoui Abderrahim	Universidad de Almeria (UAL), Spain
Nguyen Tien Dzung	Sungkyunkwan University, South Korea
Niyogi Rajdeep	Indian Institute of Technology Roorkee, India
Novelli Antonio	University of Bari, Italy
Oliveira Irene	University of Trás-Os-Montes e Alto Douro, Portugal
Oliveira José A.	Universidade do Minho, Portugal
Ottomanelli Michele	University of Bari, Italy
Ouchi Shunji	Shimonoseki City University, Japan
Ozturk Savas	Scientific and Technological Research Council of Turkey, Turkey
P. Costa M. Fernanda	Universidade do Minho, Portugal
Painho Marco	NOVA Information Management School, Portugal
Panetta J.B.	Tecnologia Geofísica Petróleo Brasileiro SA, PETROBRAS, Brazil

Pantazis Dimos	Otenet, Greece
Papa Enrica	University of Amsterdam, The Netherlands
Pardede Eric	La Trobe University, Australia
Parente Claudio	Università degli Studi di Napoli Parthenope, Italy
Pathan Al-Sakib Khan	Islamic University of Technology, Bangladesh
Paul Prantosh K.	EIILM University, Jorethang, Sikkim, India
Pengő Edit	University of Szeged, Hungary
Pereira Ana	IPB, Portugal
Pereira José Luís	Universidade do Minho, Portugal
Peschechera Giuseppe	Università di Bologna, Italy
Pham Quoc Trung	HCMC University of Technology, Vietnam
Piemonte Andreaa	University of Pisa, Italy
Pimentel Carina	Universidade de Aveiro, Portugal
Pinet Francois	IRSTEA, France
Pinto Livio	Polytechnic University of Milan, Italy
Pinto Telmo	Universidade do Minho, Portugal
Pinet Francois	IRSTEA, France
Poli Giuliano	Université Pierre et Marie Curie, France
Pollino Maurizio	ENEA, Italy
Portela Carlos Filipe	Universidade do Minho, Portugal
Prata Paula	Universidade Federal de Sergipe, Brazil
Previl Carlo	University of Quebec in Abitibi-Témiscamingue (UQAT), Canada
Prezioso Giuseppina	Università degli Studi di Napoli Parthenope, Italy
Pusatli Tolga	Cankaya University, Turkey
Quan Tho	Ho Chi Minh, University of Technology, Vietnam
Ragni Mirco	Universidade Estadual de Feira de Santana, Brazil
Rahman Nazreena	Biotechnology Research Centre, Malaysia
Rahman Wasiur	Technical University Darmstadt, Germany
Rashid Sidra	National University of Sciences and Technology (NUST) Islamabad, Pakistan
Rasulo Alessandro	Università degli studi di Cassino e del Lazio Meridionale, Italy
Raza Syed Muhammad	Sungkyunkwan University, South Korea
Reis Ferreira Gomes Karine	Instituto Nacional de Pesquisas Espaciais, Brazil
Requejo Cristina	Universidade de Aveiro, Portugal
Rocha Ana Maria	University of Minho, Portugal
Rocha Humberto	University of Coimbra, Portugal
Rocha Jorge	University of Minho, Portugal
Rodriguez Daniel	University of Berkeley, USA
Saeki Koichi	Graduate University for Advanced Studies, Japan
Samela Caterina	University of Basilicata, Italy
Sannicandro Valentina	Polytechnic of Bari, Italy
Santiago Júnior Valdivino	Instituto Nacional de Pesquisas Espaciais, Brazil
Sarafian Haiduke	Pennsylvania State University, USA

Santos Daniel	Universidade Federal de Minas Gerais, Portugal
Santos Dorabella	Instituto de Telecomunicações, Portugal
Santos Eulália	SAPO, Portugal
Santos Maribel Yasmina	Universidade de Minho, Portugal
Santos Rafael	University of Toronto, Canada
Santucci Valentinoi	University of Perugia, Italy
Sautot Lucil	MR TETIS, AgroParisTech, France
Scaioni Marco	Polytechnic University of Milan, Italy
Schernthanner Harald	University of Potsdam, Germany
Schneider Michel	ISIMA, France
Schoier Gabriella	University of Trieste, Italy
Scorza Francesco	University of Basilicata, Italy
Sebillo Monica	University of Salerno, Italy
Severino Ricardo Jose	Universidade de Minho, Portugal
Shakhov Vladimir	Russian Academy of Sciences (Siberian Branch), Russia
Sheeren David	Toulouse Institute of Technology, France
Shen Jie	University of Michigan, USA
Silva Elsa	INESC Tec, Porto, Portugal
Sipos Gergely	MTA SZTAKI Computer and Automation Research Institute, Hungary
Skarga-Bandurova Inna	Technological Institute of East Ukrainian National University, Ukraine
Skoković Dražen	University of Valencia, Spain
Skouteris Dimitrios	SNS, Italy
Soares Inês Soares Maria Joana	Universidade de Minho, Portugal
Soares Michel	Federal University of Sergipe, Brazil
Sokolovski Dmitri	Ikerbasque, Basque Foundation for Science, Spain
Sousa Lisete	Research, FCUL, CEAUL, Lisboa, Portugal
Stener Mauro	Università di Trieste, Italy
Sumida Yasuaki	Center for Digestive and Liver Diseases, Nara City Hospital, Japan
Suri Bharti	Guru Gobind Singh Indraprastha University, India
Sørensen Claus Aage Grøn	University of Aarhus, Denmark
Tajani Francesco	University of Rome, Italy
Takato Setsuo	Kisarazu National College of Technology, Japan
Tanaka Kazuaki	Hasanuddin University, Indonesia
Taniar David	Monash University, Australia
Tapia-McClung Rodrigo	The Center for Research in Geography and Geomatics, Mexico
Tarantino Eufemia	Polytechnic of Bari, Italy
Teixeira Ana Paula	Federal University of Ceará, Fortaleza, Brazil
Teixeira Senhorinha	Universidade do Minho, Portugal
Teodoro M. Filomena	Instituto Politécnico de Setúbal, Portugal
Thill Jean-Claude	University at Buffalo, USA
Thorat Pankaj	Sungkyunkwan University, South Korea

Tilio Lucia	University of Basilicata, Italy
Tomaz Graça	Instituto Politécnico da Guarda, Portugal
Torre Carmelo Maria	Polytechnic of Bari, Italy
Totaro Vincenzo	Polytechnic University of Bari, Italy
Tran Manh Hung	University of Danang, Vietnam
Tripathi Ashish	MNNIT Allahabad, India
Tripp Barba Carolina	Universidad Autónoma de Sinaloa, Mexico
Tut Zohra Fatema	University of Calgary, Canada
Upadhyay Ashish	Indian Institute of Public Health-Gandhinagar, India
Vallverdu Jordi	Autonomous University of Barcelona, Spain
Valuev Ilya	Russian Academy of Sciences, Russia
Varela Leonilde	University of Minho, Portugal
Varela Tania	Universidade de Lisboa, Portugal
Vasconcelos Paulo	Queensland University, Brisbane, Australia
Vasyunin Dmitry	University of Amsterdam, The Netherlands
Vella Flavio	University of Rome, Italy
Vijaykumar Nandamudi	INPE, Brazil
Vidacs Laszlo	University of Szeged, Hungary
Viqueira José R.R.	Agricultural University of Athens, Greece
Vizzari Marco	University of Perugia, Italy
Vohra Varun	Japan Advanced Institute of Science and Technology (JAIST), Japan
Voit Nikolay	Ulyanovsk State Technical University Ulyanovsk, Russia
Walkowiak Krzysztof	Wroclaw University of Technology, Poland
Wallace Richard J.	University College Cork, Ireland
Waluyo Agustinus Borgy	Monash University, Melbourne, Australia
Wanderley Fernando	FCT/UNL, Portugal
Wei Hoo Chong	Motorola, USA
Yamashita Satoshi	National Research Institute for Child Health and Development, Tokyo, Japan
Yamauchi Toshihiro	Okayama University, Japan
Yao Fenghui	Tennessee State University, USA
Yeoum Sanggil	Sungkyunkwan University, South Korea
Zaza Claudio	University of Foggia, Italy
Zeile Peter	Technische Universität Kaiserslautern, Germany
Zenha-Rela Mario	University of Coimbra, Portugal
Zoppi Corrado	Università di Cagliari, Italy
Zullo Francesco	University of l'Aquila, Italy
Zunino Alejandro	Universidad Nacional del Centro, Argentina
Žemlička Michal	Univerzita Karlova, Czech Republic
Živković Ljiljana	University of Belgrade, Serbia

Sponsoring Organizations

ICCSA 2017 would not have been possible without the tremendous support of many organizations and institutions, for which all organizers and participants of ICCSA 2017 express their sincere gratitude:

University of Trieste, Trieste, Italy
(http://www.units.it/)

University of Perugia, Italy
(http://www.unipg.it)

University of Basilicata, Italy
(http://www.unibas.it)

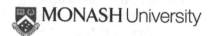

Monash University, Australia
(http://monash.edu)

KSU
九州産業大学
KYUSHU SANGYO UNIVERSITY

Kyushu Sangyo University, Japan
(www.kyusan-u.ac.jp)

Universidade do Minho
Escola de Engenharia

Universidade do Minho, Portugal
(http://www.uminho.pt)

Contents – Part I

General Tracks

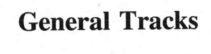

General Tracks

An Analysis of Reordering Algorithms to Reduce the Computational Cost of the Jacobi-Preconditioned CG Solver Using High-Precision Arithmetic

Sanderson L. Gonzaga de Oliveira[1(✉)], Guilherme Oliveira Chagas[2], and Júnior Assis Barreto Bernardes[1]

[1] Universidade Federal de Lavras, Lavras, Minas Gerais, Brazil
sanderson@dcc.ufla.br, jrassis@posgrad.ufla.br
[2] Instituto Nacional de Pesquisas Espaciais, São José dos Campos, São Paulo, Brazil
guilherme.o.chagas@gmail.com

Abstract. Several heuristics for bandwidth and profile reductions have been proposed since the 1960s. In systematic reviews, 133 heuristics applied to these problems have been found. The results of these heuristics have been analyzed so that, among them, 13 were selected in a manner that no simulation or comparison showed that these algorithms could be outperformed by any other algorithm in the publications analyzed, in terms of bandwidth or profile reductions and also considering the computational costs of the heuristics. Therefore, these 13 heuristics were selected as the most promising low-cost methods to solve these problems. Based on this experience, this article reports that in certain cases no heuristic for bandwidth or profile reduction can reduce the computational cost of the Jacobi-preconditioned Conjugate Gradient Method when using high-precision numerical computations.

Keywords: Bandwidth reduction · Profile reduction · Conjugate Gradient Method · Graph labeling · Reordering algorithms · Sparse matrices · Graph algorithm · High-precision arithmetic · Ordering · Sparse symmetric positive-definite linear systems · Combinatorial optimization · Heuristics

1 Introduction

In several scientific and engineering fields, such as finite element analysis, computational fluid mechanics, and structural engineering, a fundamental task is the resolution of large sparse linear systems with the form $Ax = b$, where A is an $n \times n$ sparse, symmetric, and positive-definite matrix, b is a vector of length n, and x is an unknown vector (which is sought) of length n. Generally, the highest computational cost of the simulation is required in the resolution of these

O. Gervasi et al. (Eds.): ICCSA 2017, Part I, LNCS 10404, pp. 3–19, 2017.
DOI: 10.1007/978-3-319-62392-4_1

linear systems. A substantial amount of memory and a high processing cost are necessary to store and to solve these large-scale linear systems. For the low-cost solution of large and sparse linear systems, a heuristic for bandwidth or profile reduction is often used so that the corresponding coefficient matrix A will have narrow bandwidth and small profile. Thus, heuristics for bandwidth and profile reductions are used to achieve low processing and storage costs for solving large sparse linear systems [14,17]. In particular, profile reduction is employed to reduce storage costs of applications that employ the skyline data structure [9] to represent large-scale matrices.

Let $A = [a_{ij}]$ be a symmetric sparse $n \times n$ matrix. The bandwidth of line i is $\beta_i(A) = i - min(j : (1 \leq j < i)\ a_{ij} \neq 0)$. Bandwidth of A is defined as $\beta(A) = \max[(1 \leq i \leq n)\ \beta_i(A)] = \max[(1 \leq i \leq n)\ (i - \min[j : (1 \leq j < i)] \mid a_{ij} \neq 0)]$. The profile of A is defined as $profile(A) = \sum_{i=1}^{n} \beta_i(A)$. The bandwidth and profile minimization problems are NP-hard [28,31]. Since these problems have associations with an extensive variety of other problems in scientific and engineering disciplines, several heuristics for bandwidth and profile reductions have been proposed for reordering the rows and columns of sparse matrices to solve the bandwidth and profile reduction problems.

A prominent algorithm for solving large-scale sparse linear systems is the Conjugate Gradient Method (CGM) [21,26]. Duff and Meurant [8] showed that a local ordering of the vertices of the corresponding graph of A can improve cache hit rates so that a computational cost reduction of the CGM is reached. Moreover, Burgess and Giles [3] and Das et al. [6] showed that such local ordering can be attained by using a heuristic for bandwidth reduction. Moreover, one should employ an ordering which does not lead to an increase of the number of iterations of the CGM when a preconditioner is applied [15].

In this work, we analyze cases where selected heuristics for bandwidth or profile reduction may not reduce the computational times of the Jacobi-preconditioned CGM (JPCGM). In previous publications [14,16], we showed preceding results and based on this experience [2,5,15,17], 13 heuristics were selected as the most promising methods in this field. Thus, the main objective of this work is to analyze the results of 13 potential state-of-the-art low-cost heuristics for bandwidth and profile reductions (that were selected from systematic reviews [2,5,15,17]) when executed to reduce the computational cost of the JPCGM using high-precision floating-point arithmetic.

Section 2 describes the systematic reviews accomplished to identify the potential best low-cost heuristics for bandwidth and profile reductions. Section 3 describes how the numerical experiments were conducted in this study. Section 4 shows the results. Finally, Sect. 5 addresses the conclusions.

2 Systematic Reviews

As described, since the bandwidth and profile reduction problems have connections with a wide range of other problems in scientific and engineering disciplines, a large number of heuristics for bandwidth and profile reductions has been proposed. In systematic reviews, 133 heuristics for bandwidth and/or profile reductions were identified [2,5,15,17], published between the 1960s and the present day, including a recent proposed heuristic for bandwidth and profile reductions [13]. From the analysis performed, respectively, seven and six heuristics for bandwidth and profile reductions were selected to be evaluated in this computational experiment as potentially being the best low-cost heuristics for bandwidth (Burgess-Lai [4], FNCHC [27], GGPS [38], VNS-band [30], hGPHH [24], CSS-band [23]) or profile (Snay [35], Sloan [34], Medeiros-Pimenta-Goldenberg (MPG) [29], NSloan [25], Sloan-MGPS [32]) reduction. The Reverse Cuthill-McKee method with starting pseudo-peripheral vertex given by the George-Liu algorithm (RCM-GL) [10] was selected in both systematic reviews of heuristics for bandwidth and profile reductions. In particular, the RCM-GL method [10] is contained in the Matlab software [36]. Therefore, from the 133 identified heuristics for bandwidth and profile reduction, 12 were selected to be evaluated in this computational experiment because no other simulation or comparison showed that these 12 heuristics could be superseded by any other heuristics in the analyzed papers, concerning bandwidth or profile reduction, when the computation costs of the given heuristic were also considered. Thus, these 12 heuristics could be deemed as the most promising low-cost heuristics to solve the problems.

The GPS heuristic [12] was not selected in these systematic reviews. In spite of this, it was also implemented and its results were compared with these 12 heuristics in this computational experiment because it is one of the most classic low-cost heuristics evaluated in the field for both bandwidth and profile reductions. Thus, 13 heuristics were implemented and/or evaluated in this work.

3 Description of the Tests, Implementation of the Heuristics, Testing, and Calibration

A 64-bit executable program of the VNS-band heuristic (which was kindly provided by one of the heuristic's authors) was used. This executable only runs with instances up to 500,000 vertices.

The FNCHC-heuristic source code was also kindly provided by one of the heuristic's authors. With this, the source code was converted and implemented in this present work using the C++ programming language.

The 11 other heuristics' authors were requested for the sources and/or executables of their algorithms. Some authors informed that they no longer had the source code or executable, some authors did not answer, and other authors explained that the programs could not be provided. Then, the 11 other heuristics were also implemented using the C++ programming language so that the computational costs of the heuristics could be properly compared [15]. Specifically, the g++ version 4.8.2 compiler was used.

The IEEE 754 double-precision binary floating-point arithmetic is composed of 11 bits of exponent (ranging between 10^{-307} and 10^{307}) and a matissa comprised of 53 bits, which describes approximately 16 decimal digits. Nowadays, this double-precision floating-point arithmetic is adequately accurate for most scientific computing applications. Nonetheless, for some scientific applications, the 64-bit IEEE arithmetic is no longer suitable for today's large-scale numerical simulations. Thus, some relevant scientific applications require high-precision floating-point computations. High-precision floating-point arithmetic is used in applications where the execution time of arithmetic is not a limiting factor, or where accurate results with many digits in the mantissa are needed. Some of these applications demand a significand of 64 bits or more to reach numerically useful results. These applications derive from numerous scientific applications, such as climate modeling, computational fluid dynamics (CFD) problems (e.g. vortex roll-up simulations), computational geometry, mesh generation, computational number theory, Coulomb N-body atomic system simulations, experimental mathematics, large-scale physical simulations performed on highly parallel supercomputers (e.g. studies of the fine structure constant of physics), and quantum theory [1]. Particularly, mesh generation, contour mapping, and other computational geometry applications substantially trust on highly precise arithmetic, mostly when the domain is the unit cube. The reason is that small numerical errors can induce geometrically questionable results. Such troubles are latent in the mathematics of the formulas commonly used in such computations and cannot be repaired without a considerable effort [1]. Specifically, in the applications mentioned, portions of the code normally contain numerically sensitive computations. When using double-precision floating-point arithmetic, these applications may return results with questionable precision, depending on the stopping criteria used. These imprecise results may in turn cause larger errors. On the other hand, it is normally cheaper and more reliable to use high-precision floating-point arithmetic to overcome these troubles [1]. Specifically, in this computational experiment, we used instances derived from meshes generated in discretizations of partial differential equations (that govern CFD problems) by finite volumes [19,20]. Hence, our numerical experiments will focus on high-precision floating-point arithmetic. We used the *GNU Multiple Precision Floating-point Computations with Correct-Rounding* (MFR) library with 256-bit (when using instances originating from discretizations of the Laplace equation) and 512-bit (when using instances contained in the University of Florida sparse matrix collection) precisions.

Many heuristics evaluated here are highly dependent on the starting vertex. Since Koohestani and Poli [24] did not explain which pseudo-peripheral vertex finder was used, the George-Liu algorithm [11] for computing a pseudo-peripheral vertex was used in this computational experiment. Hence, we will refer this heuristic as hGPHH-GL.

It was not our objective that the results of the C++ programming language versions of the heuristics supersede all the results of the original implementations. Our objective was to code reasonably efficient implementations of the heuristics evaluated to make it possible an adequate comparison of the results of the 13 heuristics. However, we tested and calibrated the C++ programming language versions of the heuristics implemented to compare our implementations with the codes used by the original proposers of the heuristics to ensure the codes we implemented were comparable to the algorithms that were originally proposed. We compared the results of the C++ programming language versions of the heuristics with the results presented in the original publications.

In particular, a previous publication [15] shows how the heuristics were implemented, tested, and calibrated. The C++ programming language implementations of the heuristics obtained similar results in bandwidth or profile reductions to the results presented in the original publications (see [15]).

Table 1 shows the characteristics of the five workstations used to perform the simulations. Particularly, the Ubuntu 14.04 LTS 64-bit operating system was used.

Table 1. Characteristics of the machines used to perform the simulations.

Machine	Processor unit: Intel®	Cache memory	Main memory (DDR3)	Linux kernel
M1	Core™ i3-2120 CPU 3.3 GHz	3 MB	8 GB 1.333 GHz	3.13.0-39-generic
M2	Xeon™ E5620 CPU 2.4 GHz	12 MB	24 GB 1.333 GHz	3.13.0-44-generic
M3	Core™ i5-3570 CPU 3.4 GHz	6 MB	8 GB 1.333 GHz	3.13.0-37-generic
M4	Core™ i7-4510U CPU 2.0 GHz	4 MB	8 GB 1.6 GHz	3.16.0-23-generic
M5	Core™ i7-4790K CPU 4.0 GHz	8 MB	12 GB 1.6 GHz	3.19.0-31-generic

Three sequential runs, with both a reordering algorithm and with the JPCGM, were carried out with each instance. In addition, for this experimental analysis of 13 low-cost heuristics for bandwidth and profile reductions, we followed the suggestions given by Johnson [22], aiming at reducing the computational cost of the JPCGM.

4 Numerical Experiments and Analysis

This section shows the results obtained in simulations using the JPCGM, executed after applying heuristics for bandwidth and profile reductions. Section 4.1 shows the results of the resolutions of linear systems arising from the discretization of the Laplace equation by finite volumes [19]. Section 4.2 shows the results of the resolutions of linear systems contained in the University of Florida sparse matrix collection [7].

Tables in this section show the dimension n of the respective coefficients matrix of the linear system (or the number of vertices of the graph associated with the coefficient matrix on it or the name of the instance contained in the University of Florida sparse matrix collection), the name of the reordering algorithms applied, the results with respect to profile and bandwidth reductions, the average results of the heuristics in relation to the computational cost, in seconds (s), and the memory requirements, in mebibytes (MiB). In addition, these tables show the number of iterations and the total computational cost, in seconds, of the JPCGM. Furthermore, in spite of the small number of executions for each heuristic in each instance, these tables show the standard deviation (σ) and coefficient of variation (C_v), referring to the total computational cost of the JPCGM. Additionally, these tables show "–" in the first row of a set of simulations performed with each instance. This means that no reordering algorithm was used. With this result, one can verify the speed-down of the JPCGM attained when using a heuristic for bandwidth or profile reduction, shown in the last columns of these tables. In the tables below, numbers in bold face are the best results (up to two occurrences) in the β, *profile*, t(s), and m.(MiB) columns. Figures in this section are presented as line charts for clarity.

4.1 Instances Originating from the Discretization of the Laplace Equation by Finite Volumes

This section shows the results of the resolutions of linear systems arising from the discretization of the Laplace equation by finite volumes [19]. These linear systems are divided into two datasets: seven and eight linear systems ranging from 7,322 to 277,118 and from 16,922 to 1,115,004 unknowns comprised of matrices with random order [see Fig. 1 and Tables 2 and 3 (with executions performed on the M1 machine)] and originally ordered using a sequence given by the Sierpiński-like curve [18,37] [see Fig. 2 and Tables 4 and 5 (with executions performed on the M2 machine)], respectively.

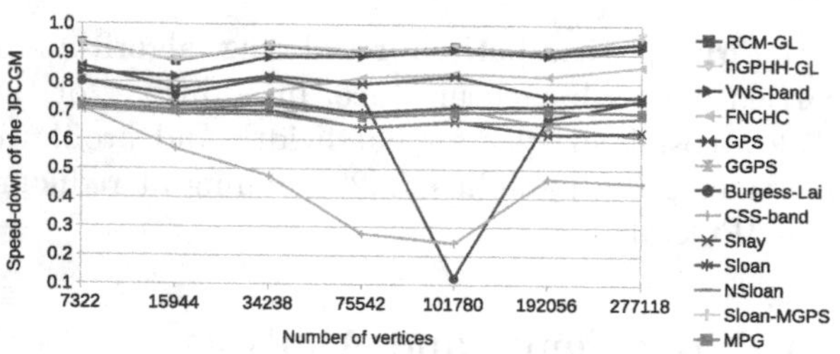

Fig. 1. Speed-downs of the JPCGM obtained using several heuristics for bandwidth and profile reductions applied to linear systems originating from the discretization of the Laplace equation by finite volumes and composed of matrices with random order (see Tables 2 and 3).

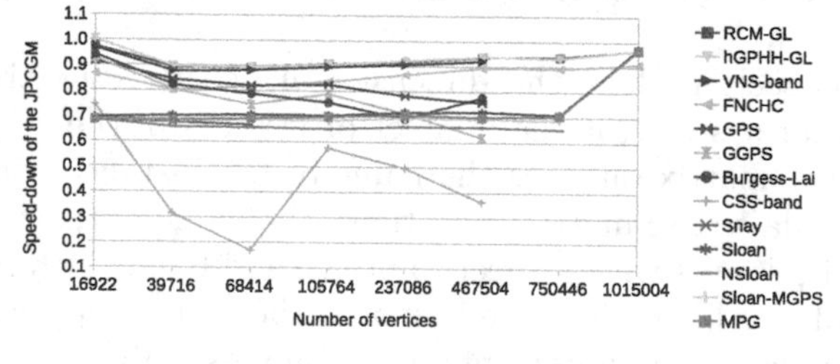

Fig. 2. Speed-downs of the JPCGM obtained using several heuristics for bandwidth and profile reductions applied to linear systems originating from the discretization of the Laplace equation by finite volumes and composed of matrices with a sequence given by the Sierpiński-like order (see Tables 4 and 5).

Tables 2, 3, 4 and 5 show that Sloan's heuristic almost always obtained the best profile results in these datasets. In addition, these tables show that the FNCHC heuristic achieved in general the best bandwidth results, but closely followed by the RCM-GL and hGPHH-GL heuristics, which presented much lower

Table 2. Resolution of three linear systems (derived from the discretization of the Laplace equation by finite volumes and composed of matrices with random order) using the JPCGM and vertices labeled by heuristics for bandwidth and profile reductions.

n	Heuristic	β	*profile*	Heuristic		JPCGM		σ	C_v (%)	Speed-down
				t(s)	m.(MiB)	iter.	t(s)			
7322	–	7248	16083808	–	–	498	**10**	0.02	0.16	–
	RCM-GL	80	396652	0.005	0.0	498	11	0.04	0.37	0.93
	hGPHH-GL	80	406461	0.006	0.0	498	11	0.03	0.29	0.93
	VNS-band	1599	966638	1.061	75.9	498	11	0.31	2.73	0.83
	FNCHC	**75**	444803	2.273	0.5	498	13	0.05	0.36	0.70
	GPS	78	404414	0.362	0.2	498	12	0.40	3.26	0.85
	GGPS	79	397534	0.415	1.3	498	13	0.04	0.30	0.81
	Burgess-Lai	152	407458	0.189	0.0	498	13	0.03	0.20	0.80
	CSS-band	7190	16103602	1.079	40.9	498	14	0.06	0.48	0.71
	Snay	1859	447914	0.240	0.3	498	14	0.10	0.03	0.72
	Sloan	391	375254	0.010	0.3	498	14	0.05	0.38	0.73
	NSloan	197	424476	**0.004**	0.3	498	15	0.01	0.07	0.72
	Sloan-MGPS	312	**374571**	0.030	0.2	498	14	0.02	0.14	0.73
	MPG	800	608671	0.010	0.2	498	15	0.03	0.17	0.72
15944	–	15902	76482022	–	–	745	**34**	0.21	0.63	–
	RCM-GL	121	1149442	0.020	0.0	745	39	0.04	0.10	0.87
	hGPHH-GL	124	1231692	0.020	0.0	745	39	0.29	0.76	0.88
	VNS-band	3916	5108940	1.190	196.7	745	40	0.49	1.23	0.82
	FNCHC	**113**	1321180	5.850	1.3	745	43	0.11	0.26	0.69
	GPS	118	1154030	1.580	0.5	745	42	1.38	3.31	0.78
	GGPS	118	1210195	3.550	2.8	745	43	0.07	0.17	0.73
	Burgess-Lai	212	1144254	1.090	0.0	745	44	0.07	0.16	0.75
	CSS-band	15749	77021429	8.560	192.7	745	51	0.33	0.65	0.57
	Snay	5862	1586436	1.020	0.5	745	47	0.06	0.13	0.70
	Sloan	484	**982693**	0.020	0.3	745	47	0.02	0.04	0.72
	NSloan	218	1222337	**0.010**	0.3	745	49	0.05	0.11	0.69
	Sloan-MGPS	481	1002661	0.100	0.3	745	47	0.14	0.30	0.71
	MPG	1277	1612396	0.030	0.3	745	48	0.20	0.41	0.70
34238	–	34059	357518296	–	–	1069	**105**	0.41	0.39	–
	RCM-GL	194	3411077	0.040	0.0	1069	114	0.90	0.79	0.93
	hGPHH-GL	192	3759478	0.040	0.0	1069	113	0.11	0.10	0.93
	VNS-band	2726	6767128	2.660	490.6	1069	116	1.23	1.06	0.89
	FNCHC	192	3913543	15.440	4.0	1069	122	0.38	0.31	0.77
	GPS	191	3545656	9.720	1.2	1069	118	0.59	0.50	0.82
	GGPS	**170**	3415253	19.690	5.2	1069	122	0.11	0.09	0.75
	Burgess-Lai	334	3282297	4.310	0.0	1069	125	0.29	0.23	0.81
	CSS-band	33923	359144453	67.280	910.7	1069	155	0.63	0.41	0.48
	Snay	21625	5150148	4.830	1.0	1069	145	0.53	0.36	0.70
	Sloan	917	**2578022**	0.060	0.8	1069	144	0.03	0.02	0.73
	NSloan	357	3608666	**0.030**	0.9	1069	153	0.14	0.09	0.69
	Sloan-MGPS	795	2671240	0.300	0.9	1069	146	0.07	0.05	0.72
	MPG	2322	3986576	0.060	0.9	1069	146	0.04	0.03	0.72

Table 3. Resolution of four linear systems (derived from the discretization of the Laplace equation by finite volumes and composed of matrices with random order) using the JPCGM and vertices labeled by heuristics for bandwidth and profile reductions.

n	Heuristic	β	*profile*	Heuristic		JPCGM		σ	C_v (%)	Speed-down
				t(s)	m.(MiB)	iter.	t(s)			
75542	–	75490	1744941733	–	–	1540	**328**	0.38	0.11	–
	RCM-GL	274	12086129	0.1	0	1540	362	0.62	0.17	0.91
	hGPHH-GL	277	13793938	0.1	0	1540	360	0.22	0.06	0.91
	VNS-band	21310	42564399	3.7	1198	1540	365	1.35	0.37	0.89
	FNCHC	**269**	13978910	40.8	7	1540	361	1.66	0.46	0.82
	GPS	272	12086603	41.5	4	1540	371	1.52	0.41	0.80
	GGPS	271	12405895	86.7	10	1540	391	0.44	0.11	0.69
	Burgess-Lai	460	11175444	42.5	0	1540	396	0.64	0.16	0.75
	CSS-band	74879	1747422045	692.6	3819	1540	497	1.33	0.27	0.28
	Snay	47789	28972039	41.7	2	1540	468	0.61	0.13	0.64
	Sloan	1521	**8981209**	0.2	1	1540	472	0.21	0.04	0.70
	NSloan	534	12805249	0.1	1	1540	508	0.04	0.01	0.65
	Sloan-MGPS	1236	9245713	1.0	1	1540	481	0.19	0.04	0.68
	MPG	4020	14107424	0.2	1	1540	481	0.06	0.01	0.68
101780	–	101583	3169282786	–	–	2173	**631**	0.36	0.06	–
	RCM-GL	405	21399542	0.1	0	2173	683	1.81	0.26	0.92
	hGPHH-GL	407	24041332	0.1	0	2173	684	0.30	0.04	0.92
	VNS-band	5207	25033097	5.6	1638	2173	685	2.07	0.30	0.91
	FNCHC	**391**	26974311	60.0	10	2173	699	1.41	0.20	0.83
	GPS	405	21399542	73.1	5	2173	694	4.57	0.66	0.82
	GGPS	400	21727818	153.2	16	2173	735	8.73	1.19	0.71
	Burgess-Lai	745	19394495	4385.7	0	2173	741	0.50	0.07	0.12
	CSS-band	101333	3160566736	1638.0	611	2173	944	2.03	0.22	0.24
	Snay	64553	45830895	61.1	3	2173	886	0.12	0.01	0.67
	Sloan	8845	**14909417**	0.3	2	2173	882	2.60	0.29	0.72
	NSloan	7602	21266761	0.1	2	2173	951	2.53	0.27	0.66
	Sloan-MGPS	8420	15400014	1.9	2	2173	902	2.78	0.31	0.70
	MPG	10502	24115880	0.2	2	2173	906	2.50	0.28	0.70
192056	–	191738	11329772559	–	–	2383	**1305**	0.72	0.06	–
	RCM-GL	360	42578191	**0.2**	0	2382	1437	3.39	0.24	0.91
	hGPHH-GL	364	48308977	0.3	0	2383	1429	0.12	0.01	0.91
	VNS-band	11142	99018771	16.5	3195	2383	1443	6.79	0.47	0.89
	FNCHC	**348**	48496246	114.8	21	2383	1469	1.65	0.11	0.82
	GPS	371	41541059	256.0	10	2383	1475	3.57	0.24	0.75
	GGPS	363	42925208	530.8	28	2383	1468	0.63	0.04	0.65
	Burgess-Lai	621	40149530	349.6	0	2383	1580	0.19	0.01	0.68
	CSS-band	191446	2737568773	793.5	1125	2383	1999	5.49	0.27	0.47
	Snay	112715	158031137	262.2	6	2384	1835	6.77	0.37	0.62
	Sloan	1963	**30916653**	0.7	4	2384	1815	0.49	0.03	0.72
	NSloan	750	44537494	**0.2**	4	2384	1968	0.25	0.01	0.66
	Sloan-MGPS	1759	31863871	4.2	4	2384	1858	1.06	0.06	0.70
	MPG	5366	47979879	0.5	4	2384	1853	0.37	0.02	0.70
277118	–	277019	23512579029	–	–	2771	**2236**	2.78	0.12	–
	RCM-GL	421	74726891	0.4	0	2771	2383	5.18	0.22	0.94
	hGPHH-GL	427	84714895	0.4	0	2771	2328	3.04	0.13	0.96
	VNS-band	12132	97666318	32.1	4618	2771	2397	6.11	0.26	0.92
	FNCHC	424	86076670	183.2	27	2771	2426	0.74	0.03	0.86
	GPS	**399**	72378558	510.3	16	2771	2459	12.28	0.50	0.75
	GGPS	420	75610158	1054.7	21	2771	2586	7.44	0.29	0.61
	Burgess-Lai	793	66880423	401.9	0	2771	2614	0.85	0.03	0.74
	CSS-band	276285	23509305627	1606.1	1680	2771	3314	26.97	0.81	0.45
	Snay	107539	310401674	516.6	10	2771	3032	0.75	0.02	0.63
	Sloan	2243	**55586226**	1.2	4	2771	3036	0.68	0.02	0.74
	NSloan	909	77343800	**0.3**	4	2771	3294	0.01	0.01	0.68
	Sloan-MGPS	2084	57032215	7.9	4	2771	3198	2.39	0.07	0.70
	MPG	7281	89227523	0.8	4	2771	3200	1.26	0.04	0.70

Table 4. Resolution of linear systems (ranging from 16,922 to 105,764 unknowns, derived from the discretization of the Laplace equation by finite volumes and composed of matrices ordered using a sequence determined by the Sierpiński-like curve) using the JPCGM and vertices labeled by heuristics for bandwidth and profile reductions.

n	Heuristic	β	*profile*	Heuristic		JPCGM		σ	C_v (%)	Speed-down
				t(s)	m.(MiB)	iter.	t(s)			
16922	–	16921	1710910	–	–	767	51	0.02	0.04	–
	RCM-GL	115	1252527	0.02	0.0	767	53	0.73	1.40	0.974
	hGPHH-GL	119	1321688	0.02	0.0	767	51	0.07	0.14	1.002
	VNS-band	4756	2393029	1.29	144.7	767	52	0.19	0.37	0.969
	FNCHC	**114**	1372628	8.21	2.0	767	51	0.04	0.09	0.865
	GPS	115	1252527	3.25	0.5	767	53	0.24	0.47	0.915
	GGPS	115	1321081	2.31	4.1	767	53	0.13	0.25	0.927
	Burgess-Lai	224	1235707	0.70	0.0	767	54	0.30	0.56	0.941
	CSS-band	16746	85797563	13.45	140.9	767	55	1.06	1.91	0.744
	Snay	6212	1508415	1.56	0.5	767	74	0.11	0.15	0.682
	Sloan	571	**1074251**	0.02	0.5	767	74	0.02	0.03	0.692
	NSloan	229	1336588	**0.01**	0.3	767	75	0.01	0.02	0.684
	Sloan-MGPS	462	1093326	0.12	0.3	767	75	0.05	0.06	0.685
	MPG	1231	1750944	0.03	0.5	767	75	0.08	0.10	0.683
39716	–	39715	6309342	–	–	1144	**188**	0.04	0.02	–
	RCM-GL	195	4376986	0.05	0.0	1144	210	0.52	0.25	0.894
	hGPHH-GL	192	4770829	0.05	0.0	1144	209	0.31	0.15	0.897
	VNS-band	5863	9979067	2.32	327.2	1144	212	1.11	0.52	0.877
	FNCHC	189	5021600	22.80	3.4	1144	211	1.00	0.47	0.803
	GPS	**180**	4464634	9.05	0.8	1144	213	0.65	0.31	0.845
	GGPS	194	4391324	19.90	3.1	1144	214	4.54	2.13	0.804
	Burgess-Lai	335	4156848	6.63	0.0	1144	221	0.15	0.07	0.824
	CSS-band	39346	480512986	341.56	537.9	1144	257	1.69	0.66	0.314
	Snay	22597	6548607	8.67	1.2	1144	269	0.30	0.11	0.677
	Sloan	830	**3342149**	0.08	1.0	1144	268	0.23	0.09	0.701
	NSloan	372	4634523	**0.03**	1.0	1144	286	0.11	0.04	0.656
	Sloan-MGPS	831	3461255	0.43	1.0	1144	275	0.24	0.09	0.682
	MPG	2358	5222214	0.08	1.0	1144	274	0.17	0.06	0.685
68414	–	68413	14882117	–	–	1514	**430**	0.03	0.01	–
	RCM-GL	238	9598308	0.08	0.0	1514	481	1.09	0.23	0.894
	hGPHH-GL	236	10705920	0.08	0.0	1514	481	0.39	0.08	0.894
	VNS-band	516	17030717	4.59	557.4	1514	483	0.93	0.19	0.882
	FNCHC	233	11325816	42.02	5.1	1514	485	1.66	0.34	0.817
	GPS	**225**	9751463	28.90	1.5	1514	495	0.29	0.06	0.821
	GGPS	233	9781546	68.94	6.2	1514	508	0.07	0.01	0.746
	Burgess-Lai	440	8910920	30.69	0.0	1514	515	0.43	0.08	0.788
	CSS-band	67862	1432183654	1962.01	2043.0	1514	591	3.97	0.67	0.168
	Snay	43837	21823074	36.43	2.1	1514	609	0.11	0.02	0.666
	Sloan	1284	**7093207**	0.17	1.0	1514	609	0.16	0.03	0.706
	NSloan	442	10128234	**0.06**	1.0	1514	656	0.10	0.01	0.656
	Sloan-MGPS	1082	7326579	0.96	1.0	1514	625	0.06	0.01	0.687
	MPG	2811	11460778	0.16	1.0	1514	623	0.44	0.07	0.690
105764	–	105763	29560801	–	–	1846	**816**	0.04	0.04	–
	RCM-GL	311	18180951	0.13	0.0	1846	899	2.09	0.23	0.907
	hGPHH-GL	309	20753083	0.14	0.0	1846	901	1.75	0.20	0.905
	VNS-band	2809	33762228	9.38	857.2	1846	901	0.36	0.04	0.896
	FNCHC	**289**	21067109	69.00	9.4	1846	905	2.95	0.33	0.837
	GPS	299	18336159	60.95	2.1	1846	927	0.80	0.09	0.826
	GGPS	306	18163269	136.02	7.5	1846	899	0.29	0.03	0.789
	Burgess-Lai	483	16959146	112.25	0.0	1846	968	0.40	0.04	0.755
	CSS-band	105406	3418070351	305.06	494.5	1846	1115	3.78	0.34	0.575
	Sloan	1756	**13247695**	0.32	2.2	1846	1159	0.35	0.03	0.703
	NSloan	602	19321158	**0.10**	1.2	1846	1255	0.76	0.06	0.650
	Sloan-MGPS	1512	13685106	1.88	1.9	1846	1191	0.15	0.01	0.684
	MPG	4447	20523176	0.27	1.2	1846	1167	2.19	0.19	0.699

Table 5. Resolution of four linear systems (derived from the discretization of the Laplace equation by finite volumes and composed of matrices originally ordered using a sequence determined by the Sierpiński-like curve) using the JPCGM and vertices labeled by heuristics for bandwidth and profile reductions.

n	Heuristic	β	*profile*	Heuristic		JPCGM			C_v (%)	Speed-down
				t(s)	m.(MiB)	iter.	t(s)	σ		
237086	–	237085	115804392	–	–	2611	**2629**	8.69	0.33	–
	RCM-GL	391	56621430	0.3	0	2612	2866	7.98	0.28	0.92
	hGPHH-GL	393	64411087	0.3	0	2612	2852	3.88	0.14	0.92
	VNS-band	1783	92303199	40.8	1912	2612	2858	5.97	0.21	0.91
	FNCHC	**388**	64592246	177.0	20	2612	2859	1.31	0.05	0.87
	GPS	392	56476790	392.3	13	2612	2965	5.43	0.18	0.78
	GGPS	389	57030510	820.9	16	2612	2869	11.37	0.40	0.71
	Burgess-Lai	718	52953332	698.3	0	2612	3100	0.90	0.03	0.69
	CSS-band	236418	2682971255	1657.8	1107	2612	3637	25.22	0.69	0.50
	Sloan	2044	**41300807**	1.0	7	2612	3661	3.25	0.09	0.72
	NSloan	812	59082821	**0.2**	6	2612	3981	2.10	0.05	0.66
	Sloan-MGPS	1898	42561396	6.2	6	2612	3777	1.33	0.04	0.70
	MPG	5707	64716730	0.7	6	2612	3718	1.27	0.03	0.71
467504	–	467503	382386929	–	–	3446	**6972**	2.65	0.04	–
	RCM-GL	448	130166482	0.6	0	3446	7434	3.43	0.05	0.94
	hGPHH-GL	445	149299971	0.6	0	3446	7423	6.17	0.08	0.94
	VNS-band	5001	227019725	154.2	4904	3446	7409	2.46	0.03	0.92
	FNCHC	449	156037685	371.9	40	3446	7416	6.92	0.09	0.90
	GPS	455	129684974	1593.1	27	3446	7618	2.46	0.03	0.76
	GGPS	**438**	132748941	3228.4	29	3446	8033	7.73	0.10	0.62
	Burgess-Lai	862	119436630	895.2	0	3446	8071	3.74	0.04	0.78
	CSS-band	466181	2526782462	9693.1	3644	3446	9490	6.99	0.07	0.36
	Sloan	2498	**95551358**	2.4	10	3449	9676	13.96	0.14	0.72
	NSloan	911	135054695	**0.5**	10	3449	10563	35.21	0.33	0.66
	Sloan-MGPS	2251	98187318	14.2	10	3449	10053	49.93	0.50	0.69
	MPG	7842	152632760	1.7	11	3449	9974	66.00	0.66	0.70
750446	–	750445	911516500	–	–	4246	**13660**	5.64	0.07	–
	RCM-GL	**461**	224589050	0.9	**0**	4245	14563	4.39	0.03	0.94
	hGPHH-GL	471	257304543	1.0	**0**	4246	14629	7.03	0.05	0.93
	FNCHC	715	304070803	615.5	60	4245	14676	0.23	0.01	0.89
	Sloan	2441	**164952184**	4.0	21	4232	19223	36.40	0.19	0.71
	NSloan	946	232760320	**0.8**	21	4232	20942	9.65	0.05	0.65
	Sloan-MGPS	2264	169448464	24.4	21	4232	19857	38.68	0.19	0.69
	MPG	7986	265072969	2.8	20	4232	19352	22.96	0.12	0.71
1015004*	–	1015003	1580908606	–	–	4557	**20025**	14.55	0.07	–
	RCM-GL	462	316593383	**1.0**	**0**	4557	20726	84.12	0.41	0.97
	hGPHH-GL	465	363030399	1.1	**0**	4557	20709	33.82	0.16	0.97
	FNCHC	**455**	365943729	819.9	88	4569	21174	10.56	0.05	0.91
	Sloan	2484	**233117499**	4.1	27	4557	20587	37.33	0.18	0.97
	MPG	8107	375103029	2.4	27	4557	20682	38.37	0.19	0.97

*Executions performed on the M3 machine.

computational costs. Nevertheless, no gain was attained regarding the speed-up of the JPCGM when using these heuristics. In particular, the FNCHC heuristic presented a much higher computational cost than the RCM-GL, Sloan's, MPG, NSloan, Sloan-MGPS, and hGPHH-GL heuristics.

A slight speed-up of the JPCGM applied to the linear system composed of 16,922 when using the hGPHH-GL heuristic (see Table 4) was reached, but this gain is marginal. Moreover, a speed-down of the JPCGM was obtained when using the other heuristics for bandwidth and profile reductions applied to the other linear systems.

Tables 4 and 5 do not show results of Snay's heuristic [35] applied to linear systems larger than 68,414 unknowns. Snay's heuristic obtained better results (related to reduce the JPCGM computational cost) than the results of the CSS-band [23] and NSloan [25] heuristics when applied to the linear systems composed of 39,716 and 68,414 unknowns. However, Snay's heuristic performed less favorably than the other heuristics when applied to the linear system comprised of 16,922 unknowns (see Table 4).

The GPS [12], Burgess-Lai [4], GGPS [38], and CSS-band [23] presented higher computational costs than the other heuristics [see t(s)(Heuristic) column in Tables 3 and 5]. Consequently, Table 5 does not show the results of these four heuristics applied to the linear systems composed of 750,446 and 1,015,004 unknowns, keeping in mind that the VNS-band execution program runs with instances up to 500,000 unknowns. Furthermore, Table 5 does not show the results of the NSloan [25] and Sloan-MGPS [32] heuristics applied to the linear system composed of 1,015,004 unknowns because these two heuristics performed less favorably than the five other heuristics when applied to linear systems contained in this dataset.

4.2 Instances Contained in the University of Florida Sparse Matrix Collection

Table 6 provides the characteristics of 11 linear systems (composed of symmetric and positive-definite matrices) contained in the University of Florida sparse matrix collection [7]. Tables 2, 3, 4 and 5 show that the RCM-GL [10], Sloan's [34], MPG [29], NSloan [25], Sloan-MGPS [32], and hGPHH-GL [24] heuristics presented much lower computational costs than the other heuristics evaluated in this computational experiment. Then, these six low-cost heuristics for bandwidth or profile reduction evaluated in this study were applied to the dataset presented in Table 6.

Table 6. Eleven linear systems (composed of symmetric and positive-definite matrices) contained in the University of Florida sparse matrix collection.

Instance	Size	β	profile	Density (%)	Description
nasa1824	1824	239	205547	1.18	Structure from NASA Langley
nasa2910	2910	859	525745	2.06	Structure from NASA Langley
sts4098	4098	3323	5217389	0.43	Finite element structural engineering matrix
nasa4707	4704	423	917562	0.47	Structure from NASA Langley
Pres_Poisson	14822	12583	9789525	0.33	Computational fluid dynamics problem
olafu	16146	593	4951980	0.39	Structure from NASA Langley
raefsky4	19779	11786	19611188	0.34	Buckling problem for container model
nasasrb	54870	893	20311330	0.09	Structure from NASA Langley
thermal1	82654	80916	175625317	0.01	Unstructured finite element steady-state thermal problem
2cubes_sphere	101492	100407	483241271	0.02	Finite element electromagnetics 2 cubes in a sphere
offshore	259789	237738	3588201815	0.01	3D finite element transient electric field diffusion

Tables 7 and 8 and Fig. 3 show the results of the resolutions of 11 linear systems contained in the University of Florida sparse matrix collection using the JPCGM and vertices labeled using heuristics for bandwidth and profile reductions. The hGPHH-GL heuristic obtained the best speed-up of the JPCGM when applied to the *Pres_Poisson* instance (see Table 7). On the other hand, speed-downs of the JPCGM were obtained when using these six heuristics for bandwidth and profile reductions when applied to the 10 other linear systems contained in the University of Florida sparse matrix collection that were used here.

Among the heuristics evaluated, the Sloan-MGPS and Sloan's (RCM-GL) heuristics obtained (almost always) the best profile (bandwidth) results when applied to the instances composed in this dataset. Nevertheless, speed-downs of the JPCGM were obtained when using these heuristics (except the simulation using the *Pres_Poisson* instance).

5 Conclusions

The results of 13 heuristics for bandwidth and profile reductions applied to reduce the computational cost of solving three datasets of linear systems using the Jacobi-preconditioned Conjugate Gradient Method in high-precision floating-point arithmetic are described in this paper. These heuristics were selected from systematic reviews [2,5,15,17].

In experiments using three datasets composed of large-scale linear systems, the hGPHH-GL heuristic performed best when applied to one linear system aiming at reducing the computational cost of the JPCGM (see Table 7). On the other hand, speed-downs of the JPCGM were obtained when applying these 13 heuristics for bandwidth and profile reductions to the other linear systems that were used in this computational experiment. Thus, the attained results show that in certain cases no heuristic for bandwidth or profile reduction can reduce the computational cost of the Jacobi-preconditioned Conjugate Gradient Method when using high-precision numerical computations.

Concerning the set of linear systems arising from the discretization of the Laplace equation by finite volumes comprised of matrices with random order, each vertex has exactly three adjacencies [19]. Probably because of this, relabeling the vertices did not improve cache hit rates.

Regarding the set of linear systems originating from the discretization of the Laplace equation by finite volumes and comprised of matrices originally ordered using a sequence given by the Sierpiński-like curve [19], a large number of cache misses may be occurred after applying heuristics for bandwidth and profile reductions. Probably, the reason is that a space-filling curve already provides an adequate memory-data locality so that a reordering algorithm is not useful in such cases. We applied these 13 heuristics in large-scale linear systems and cache memory is a relevant factor in the execution times of these simulations. Evidence from the experiments described in this paper does allow the assertion that a linear system should be studied carefully before using a heuristic for bandwidth

Table 7. Resolution of seven linear systems contained in the University of Florida sparse matrix collection using the JPCGM and vertices labeled using heuristics for bandwidth and profile reductions

Instance	Machine	Heuristic	β	*profile*	Heuristic		JPCGM		σ	C_v (%)	Speed-up/down
					t(s)	m.(MiB)	iter.	t(s)			
nasa1824	M1	–	**239**	205547	–	–	1350	**24**	0.14	0.58	–
		RCM-GL	282	229770	0.004	0.0	1350	25	0.10	0.38	0.96
		hGPHH-GL	293	291203	0.003	0.0	1350	25	0.14	0.56	0.98
		Sloan	1303	**186725**	0.005	0.0	1350	25	0.17	0.67	0.96
		NSloan	415	284963	**0.002**	0.0	1351	26	0.14	0.54	0.94
		Sloan-MGPS	1102	190128	0.012	0.0	**1347**	25	0.18	0.71	0.97
		MPG	1519	516936	0.010	0.0	1350	26	0.16	0.61	0.94
nasa2910	M1	–	**859**	525745	–	–	1846	**133**	0.32	0.24	–
		RCM-GL	875	522223	0.018	0.0	1846	143	1.02	0.72	0.93
		hGPHH-GL	869	1288759	0.016	0.0	1846	140	1.08	0.77	0.95
		Sloan	2015	**456322**	0.018	0.0	**1839**	138	0.15	0.11	0.97
		NSloan	1327	955899	**0.010**	0.0	1844	145	0.49	0.34	0.92
		Sloan-MGPS	2010	460149	0.016	0.0	1842	139	0.16	0.12	0.96
		MPG	2708	2288760	0.038	0.0	1842	147	0.02	0.01	0.91
sts4098	M4	–	3323	5217389	–	–	590	**20**	0.01	0.03	–
		RCM-GL	**1165**	2084237	0.009	**0.0**	590	22	0.07	0.30	0.87
		hGPHH-GL	1171	2981815	0.008	**0.0**	**588**	22	0.39	1.80	0.90
		Sloan	3195	518163	0.023	0.2	589	21	0.01	0.01	0.95
		NSloan	3020	2505064	**0.007**	0.2	589	22	0.01	0.05	0.90
		Sloan-MGPS	3351	**461998**	0.073	0.2	590	21	0.01	0.03	0.95
		MPG	3729	961548	0.017	0.2	**588**	21	0.08	0.38	0.93
nasa4704	M4		423	917562	–	–	4248	**190**	0.08	0.04	–
		RCM-GL	**419**	918658	0.009	0.0	4245	201	0.81	0.40	0.94
		hGPHH-GL	450	1079926	0.009	0.0	4244	202	4.37	2.16	0.94
		Sloan	3084	834354	0.024	0.0	4244	204	0.62	0.30	0.93
		NSloan	678	1076453	**0.005**	0.0	4247	210	2.46	1.17	0.90
		Sloan-MGPS	2753	**808577**	0.056	0.0	4246	203	1.20	0.59	0.93
		MPG	3680	2716364	0.074	0.0	4244	212	0.04	0.02	0.90
Pres_Poiss.	M4	–	12583	9789525	–	–	1009	309	1.27	0.41	–
		RCM-GL	**326**	3009635	0.060	**0.0**	1012	297	0.99	0.33	1.04
		hGPHH-GL	364	3130744	0.059	**0.0**	1009	**293**	0.46	0.16	**1.06**
		Sloan	642	**2827171**	0.066	0.3	1012	295	0.39	0.13	1.05
		NSloan	594	3951006	**0.044**	0.3	1012	328	0.38	0.12	0.94
		Sloan-MGPS	582	2834035	0.156	0.3	1009	294	0.94	0.32	1.05
		MPG	14168	26556694	3.845	0.3	1009	297	0.02	0.01	1.03
olafu	M3	–	593	4951980	–	–	16146	**6833**	5.10	0.07	–
		RCM-GL	**553**	5029301	0.146	**0.0**	16146	7253	32.13	0.44	0.94
		hGPHH-GL	573	5165776	0.132	**0.0**	16146	7189	10.30	0.14	0.95
		Sloan	6173	4768547	0.146	0.3	16146	7154	2.24	0.03	0.96
		NSloan	4760	7334811	**0.105**	0.3	16146	7290	0.58	0.01	0.94
		Sloan-MGPS	7775	**4489770**	0.171	0.3	16146	7219	3.00	0.04	0.95
		MPG	14467	29376748	4.268	0.3	16146	7491	9.28	0.12	0.91
raefsky4	M4	–	11786	19611188	–	–	11245	**5862**	1.05	0.02	–
		RCM-GL	**991**	12553981	0.130	**0.0**	**11157**	6293	3.03	0.05	0.93
		hGPHH-GL	1141	13120923	0.110	**0.0**	11182	6313	397.88	6.30	0.93
		Sloan	6550	8587731	0.180	0.3	11180	6200	29.26	0.47	0.95
		NSloan	2242	15308534	**0.070**	0.3	11246	6786	21.77	0.32	0.86
		Sloan-MGPS	8378	**7841072**	0.340	0.3	11245	6411	4.27	0.07	0.91
		MPG	18201	74604715	6.790	0.3	11248	6803	1.01	0.01	0.86

or profile reduction aiming at reducing the computational cost of the JPCGM (and probably when using a preconditioned CGM or other iterative linear system solver).

Table 8. Resolution of four linear systems contained in the University of Florida sparse matrix collection using the JPCGM and vertices labeled using several heuristics for bandwidth and profile reductions.

Instance	Machine	Heuristic	β	*profile*	Heuristic		JPCGM		σ	C_v (%)	Speed-down
					t(s)	m.(MiB)	iter.	t(s)			
nasasrb	M1	–	893	20311330	–	–	25326	**27902**	2.28	0.01	–
		RCM-GL	**586**	19448635	0.3	**0**	25327	30242	14.94	0.05	0.92
		hGPHH-GL	806	20545002	0.2	**0**	25326	28461	54.27	0.19	0.98
		Sloan	5063	19055047	0.3	1	25320	29893	147.67	0.49	0.93
		NSloan	4865	23564619	**0.1**	1	25326	31307	24.46	0.08	0.89
		Sloan-MGPS	4932	**18682599**	0.6	1	25320	29713	60.37	0.20	0.94
		MPG	45896	346820836	13.7	1	25326	32795	61.17	0.19	0.85
thermal1	M5	–	80916	175625317	–	–	1885	**456**	0.75	0.17	–
		RCM-GL	**220**	12017373	0.1	**0**	1885	562	1.21	0.22	0.81
		hGPHH-GL	240	12997244	0.1	**0**	1885	556	0.26	0.05	0.82
		Sloan	889	**10487409**	0.2	3	1885	537	0.19	0.04	0.84
		NSloan	429	13393908	0.1	3	1885	594	0.12	0.02	0.77
		SloanMGPS	661	10677120	0.7	3	1885	559	0.09	0.02	0.81
		MPG	16857	10958622	0.2	3	1885	529	0.65	0.12	0.86
2cubes_sph.	M5	–	100407	483241271	–	–	33	**14**	0.05	0.38	–
		RCM-GL	4709	268149672	0.3	**0**	33	18	0.05	0.29	0.78
		hGPHH-GL	**4693**	345191689	0.3	**0**	33	18	0.07	0.37	0.78
		Sloan	11186	**186478091**	14.7	2	33	17	0.04	0.23	0.43
		NSloan	9203	346819754	**0.2**	2	33	18	0.12	0.68	0.75
		SloanMGPS	13446	200449820	5.8	2	33	17	0.04	0.21	0.60
		MPG	95371	460302437	21.5	2	33	18	0.02	0.13	0.35
offshore	M5	–	237738	3588201815	–	–	1226	**1952**	1.73	0.09	–
		RCM-GL	**21035**	2634951939	0.7	**0**	1226	2494	0.30	0.01	0.78
		hGPHH-GL	23859	3897866179	0.7	**0**	1228	2519	9.04	0.36	0.78
		Sloan	121957	1837918281	72.1	5	1237	2417	10.46	0.43	0.78
		NSloan	102633	3264868562	0.7	5	1226	2638	11.08	0.42	0.74
		SloanMGPS	124658	**1510670726**	141.6	5	1226	2489	9.35	0.38	0.74
		MPG	253828	4262147507	260.5	5	1230	2438	1.25	0.05	0.72

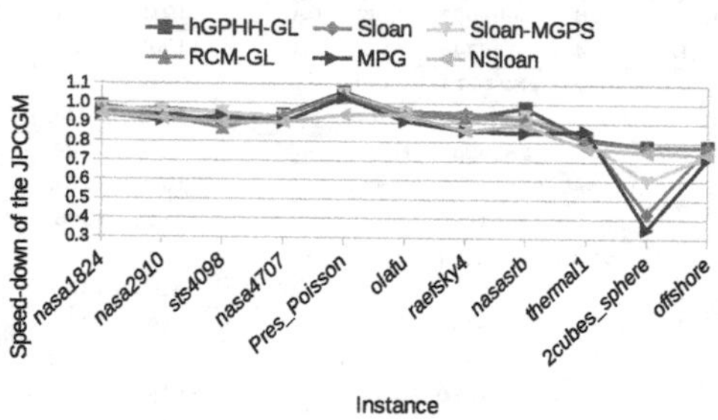

Fig. 3. Speed-downs of the JPCGM obtained using six heuristics for bandwidth and profile reductions applied to 11 linear systems contained in the University of Florida sparse matrix collection (see Tables 7 and 8).

As a continuation of this work, we intend to implement and evaluate the following preconditioners: Algebraic Multigrid, incomplete Cholesky factorization, threshold-based incomplete LU (ILUT), Successive Over-Relaxation (SOR), Symmetric SOR, and Gauss-Seidel. To provide more specific detail, we intend to study the effectiveness of the strategies when using incomplete or approximate factorization based preconditioners as well approximate inverse preconditioners. These techniques shall be used as preconditioners of the Conjugate Gradient Method and the Generalized Minimal Residual (GMRES) method [33] to evaluate their computational performance in conjunction with heuristics for bandwidth and profile reductions. Parallel strategies of the above algorithms will also be studied.

Extended (256-bit and 512-bit) precision was employed in this work. This reduces rounding errors. However, it increases the execution times by a large factor and it may not be performed when solving certain real problems. We intend to examine what occurs in double-precision arithmetic in future studies.

Acknowledgments. This work was undertaken with the support of the Fapemig - Fundação de Amparo à Pesquisa do Estado de Minas Gerais. The authors would like to thank respectively Prof. Dr. Dragan Urosevic, from the Mathematical Institute SANU, and Prof. Dr. Fei Xiao, from Beepi, for sending us the VNS-band executable programs, and the source code of the FNCHC heuristic. In addition, we would like to thank the reviewers for their valuable comments and suggestions.

References

1. Bailey, D.H.: High-precision floating-point arithmetic in scientific computation. Comput. Sci. Eng. **7**(3), 54–61 (2005)
2. Bernardes, J.A.B., Gonzaga de Oliveira, S.L.: A systematic review of heuristics for profile reduction of symmetric matrices. Procedia Comput. Sci. **51**, 221–230 (2015). (International Conference on Computational Science, ICCS)
3. Burgess, D.A., Giles, M.: Renumbering unstructured grids to improve the performance of codes on hierarchial memory machines. Adv. Eng. Softw. **28**(3), 189–201 (1997)
4. Burgess, I.W., Lai, P.K.F.: A new node renumbering algorithm for bandwidth reduction. Int. J. Numer. Methods Eng. **23**, 1693 1704 (1986)
5. Chagas, G.O., Gonzaga de Oliveira, S.L.: Metaheuristic-based heuristics for symmetric-matrix bandwidth reduction: a systematic review. Procedia Comput. Sci. (ICCS) **51**, 211–220 (2015)
6. Das, R., Mavriplis, D.J., Saltz, J.H., Gupta, S.K., Ponnusamy, R.: Design and implementation of a parallel unstructured Euler solver using software primitives. AIAA J. **32**(3), 489–496 (1994)
7. Davis, T.A., Hu, Y.: The University of Florida sparse matrix collection. ACM Trans. Math. Softw. **38**(1), 1:1–1:25 (2011)
8. Duff, I.S., Meurant, G.A.: The effect of ordering on preconditioned conjugate gradients. BIT Numer. Math. **29**(4), 635–657 (1989)
9. Felippa, C.A.: Solution of linear equations with skyline-stored symmetric matrix. Comput. Struct. **5**(1), 13–29 (1975)
10. George, A., Liu, J.W.: Computer Solution of Large Sparse Positive Definite Systems. Prentice-Hall, Englewood Cliffs (1981)
11. George, A., Liu, J.W.H.: An implementation of a pseudoperipheral node finder. ACM Trans. Math. Softw. **5**(3), 284–295 (1979)
12. Gibbs, N.E., Poole, W.G., Stockmeyer, P.K.: An algorithm for reducing the bandwidth and profile of a sparse matrix. SIAM J. Numer. Anal. **13**(2), 236–250 (1976)

13. Gonzaga de Oliveira, S.L., Abreu, A.A.A.M., Robaina, D., Kischinhevsky, M.: A new heuristic for bandwidth and profile reductions of matrices using a self-organizing map. In: Gervasi, O., et al. (eds.) ICCSA 2016. LNCS, vol. 9786, pp. 54–70. Springer, Cham (2016). doi:10.1007/978-3-319-42085-1_5
14. Gonzaga de Oliveira, S.L., Abreu, A.A.A.M., Robaina, D.T., Kischnhevsky, M.: An evaluation of four reordering algorithms to reduce the computational cost of the Jacobi-preconditioned conjugate gradient method using high-precision arithmetic. Int. J. Bus. Intell. Data Min. **12**(2), 190–209 (2017). http://dx.doi.org/10.1504/IJBIDM.2017.10004158
15. Gonzaga de Oliveira, S.L., Bernardes, J.A.B., Chagas, G.O.: An evaluation of low-cost heuristics for matrix bandwidth and profile reductions. Comput. Appl. Math. (2016). doi:10.1007/s40314-016-0394-9
16. Gonzaga de Oliveira, S.L., Bernardes, J.A.B., Chagas, G.O.: An evaluation of several heuristics for bandwidth and profile reductions to reduce the computational cost of the preconditioned conjugate gradient method. In: The XLVIII of the Brazilian Symposium of Operations Research (SBPO), Vitória, Brazil, September 2016
17. Gonzaga de Oliveira, S.L., Chagas, G.O.: A systematic review of heuristics for symmetric-matrix bandwidth reduction: methods not based on metaheuristics. In: The XLVII Brazilian Symposium of Operational Research (SBPO), Ipojuca-PE, Brazil, August 2015. Sobrapo
18. Gonzaga de Oliveira, S.L., Kischinhevsky, M.: Sierpiński curve for total ordering of a graph-based adaptive simplicial-mesh refinement for finite volume discretizations. In: Proceedings of the Brazilian National Conference on Computational and Applied Mathematics (CNMAC), pp. 581–585, Belém, Brazil (2008)
19. Gonzaga de Oliveira, S.L., Kischinhevsky, M., Tavares, J.M.R.S.: Novel graph-based adaptive triangular mesh refinement for finite-volume discretizations. Comput. Model. Eng. Sci. **95**(2), 119–141 (2013)
20. Gonzaga de Oliveira, S.L., Oliveira, F.S., Chagas, G.O.: A novel approach to the weighted laplacian formulation applied to 2D delaunay triangulations. In: Gervasi, O., Murgante, B., Misra, S., Gavrilova, M.L., Rocha, A.M.A.C., Torre, C., Taniar, D., Apduhan, B.O. (eds.) ICCSA 2015. LNCS, vol. 9155, pp. 502–515. Springer, Cham (2015). doi:10.1007/978-3-319-21404-7_37
21. Hestenes, M.R., Stiefel, E.: Methods of conjugate gradients for solving linear systems. J. Res. Natl. Bur. Stand. **49**(36), 409–436 (1952)
22. Johnson, D.: A theoretician's guide to the experimental analysis of algorithms. In: Goldwasser, M., Johnson, D.S., McGeoch, C.C., (eds.) Proceedings of the 5th and 6th DIMACS Implementation Challenges, Providence (2002)
23. Kaveh, A., Sharafi, P.: Ordering for bandwidth and profile minimization problems via charged system search algorithm. IJST-T Civ. Eng. **36**(2), 39–52 (2012)
24. Koohestani, B., Poli, R.: A hyper-heuristic approach to evolving algorithms for bandwidth reduction based on genetic programming. In: Bramer, M., Petridis, M., Nolle, L. (eds.) Research and Development in Intelligent Systems XXVIII, pp. 93–106. Springer, London (2011). doi:10.1007/978-1-4471-2318-7_7
25. Kumfert, G., Pothen, A.: Two improved algorithms for envelope and wavefront reduction. BIT Numer. Math. **37**(3), 559–590 (1997)
26. Lanczos, C.: Solutions of systems of linear equations by minimized iterations. J. Res. Natl. Bur. Stand. **49**(1), 33–53 (1952)
27. Lim, A., Rodrigues, B., Xiao, F.: A fast algorithm for bandwidth minimization. Int. J. Artif. Intell. Tools **3**, 537–544 (2007)
28. Lin, Y.X., Yuan, J.J.: Profile minimization problem for matrices and graphs. Acta Mathematicae Applicatae Sinica **10**(1), 107–122 (1994)

29. Medeiros, S.R.P., Pimenta, P.M., Goldenberg, P.: Algorithm for profile and wavefront reduction of sparse matrices with a symmetric structure. Eng. Comput. **10**(3), 257–266 (1993)
30. Mladenovic, N., Urosevic, D., Pérez-Brito, D., García-González, C.G.: Variable neighbourhood search for bandwidth reduction. Eur. J. Oper. Res. **200**, 14–27 (2010)
31. Papadimitriou, C.H.: The NP-completeness of bandwidth minimization problem. Comput. J. **16**, 177–192 (1976)
32. Reid, J.K., Scott, J.A.: Ordering symmetric sparse matrices for small profile and wavefront. Int. J. Numer. Methods Eng. **45**(12), 1737–1755 (1999)
33. Saad, Y., Schultz, M.H.: GMRES: a generalized minimal residual algorithm for solving nonsymmetric linear systems. SIAM J. Sci. Comput. **7**, 856–869 (1986)
34. Sloan, S.W.: A Fortran program for profile and wavefront reduction. Int. J. Numer. Methods Eng. **28**(11), 2651–2679 (1989)
35. Snay, R.A.: Reducing the profile of sparse symmetric matrices. Bulletin Geodésique **50**(4), 341–352 (1976)
36. The MathWorks, Inc.: MATLAB, 1994–2015. http://www.mathworks.com/products/matlab
37. Velho, L., Figueiredo, L.H., Gomes, J.: Hierarchical generalized triangle strips. Vis. Comput. **15**(1), 21–35 (1999)
38. Wang, Q., Guo, Y.C., Shi, X.W.: A generalized GPS algorithm for reducing the bandwidth and profile of a sparse matrix. Prog. Electromagn. Res. **90**, 121–136 (2009)

An Ensemble Similarity Model for Short Text Retrieval

Arifah Che Alhadi(✉), Aziz Deraman, Masita@Masila Abdul Jalil, Wan Nural Jawahir Wan Yussof, and Akashah Amin Mohamed

School of Informatics and Applied Mathematics, Universiti Malaysia Terengganu, 21030 Kuala Nerus, Terengganu, Malaysia
{arifah_hadi,a.d,masita,wannurwy}@umt.edu.my, akashahamin@gmail.com
http://www.umt.edu.my

Abstract. The rapid growth of World Wide Web has extended Information Retrieval related technology such as queries for information needs become more easily accessible. One such platform is online question answering (QA). Online community can posting questions and get direct response for their special information needs using various platforms. It creates large unorganized repositories of valuable knowledge resources. Effective QA retrieval is required to make these repositories accessible to fulfill users information requests quickly. The repositories might contained similar questions and answer to users newly asked question. This paper explores the similarity-based models for the QA system to rank search result candidates. We used Damerau-Levenshtein distance and cosine similarity model to obtain ranking scores between the question posted by the registered user and a similar candidate questions in repository. Empirical experimental results indicate that our proposed ensemble models are very encouraging and give a significantly better similarity value to improve search ranking results.

Keywords: Ensemble similarity model · Damerau-Levenshtein Distance · Cosine · Information retrieval

1 Introduction

The growth of online forum discussion or community-based question answering (CQA) sites is accompanied by a huge amount of potentially useful information that can be mined from their repositories. Online forum discussion is also used as a platform for distance learning where users can ask and discuss with the expert regarding their problems in education or working task. Online forums such as Stack Overflow, DreamInCode, MSDN, Tek-Tips and others have become popular platform used by computer users today.

Some users might also have the same problems and repeat a previously asked question without realizing it since there is no structure that organizes similar questions in an online forum discussion. All these sites are supported by experts whom are knowledgeable in providing answers on online forum. Question maybe

O. Gervasi et al. (Eds.): ICCSA 2017, Part I, LNCS 10404, pp. 20–29, 2017.
DOI: 10.1007/978-3-319-62392-4_2

left remain unanswered with many reasons like experts already answered the similar question and ask user to find themselves. This leads to the question starvation.

With a huge amount of questions, manual search is time consuming and impractical. Users turns to search engines to find the desired information. During this process, misspelled query terms is one of the main factors that affect the poor search result. Cucerzan et al., Duan et al. and Martins et al. mentioned that spelling errors frequently occurred in queries compared to written texts [6,8,18].

The widely used approach to identify similar text through lexical and semantic are based on string, corpus and knowledge [10]. There are also previous works that combines several techniques using ensemble or hybrid approach [3,16,19,20].

Even though ensemble or hybrid similarity model has shown significant contribution, it still suffers from the data-sparsity and noise in short texts such us the use of abbreviation, slang, misspelled, symbols and short form. This is due to the fact that current ensemble similarity model mostly rely on the third source (corpus or dictionary) and word-based model, hence insufficient and limited to deal with noise.

To overcome this issue, this research is proposed to analyze identified similarity model for short text retrieval by proposing an ensemble similarity model combining with edit distance model to handle misspell, as well as demonstrating the applicability of the proposed ensemble model.

In this paper, we work on measuring the similarity model between question and answer (QA) messages in online forum discussion using Damerau-Levenshtein distance (DLD) [7], and cosine similarity with *TF.IDF weights*. Our approach was based on works by [15,20] by using the same vector space model (VSM) and combined with edit distance similarity model.

2 Related Works

Extracting questions and answers from online forums are increasingly receiving academic attention. The major focus of previous research efforts on question answering is on Community-based Question Answering sites [5,22]. Cong et al. [5] used sequential pattern as feature to automatically distinguish between questions and non-questions in three different online forums (TripAdvisor, LonelyPlanet and BootsnAll). In addition, the classical features like 5W1H words and the presence of question marks were also used to detect the questions. They applied a graph based propagation method to identify and rank the candidate answers for the question in the same thread. Cosine similarity was used to measure the similarity between question answer pairs. The technique used was able to effectively extract question answer pairs and the average score for cosine similarity measured was 0.84.

Shtok et al. [22] reused the past resolved questions repository to answer the new question based on cosine similarity measure of question's titles. They calculate similarity score between two questions titles and bodies, entire title+body texts and entire text of each question and answer. However, based on our empirical analysis, using cosine similarity alone was not effective in finding similar question. This is due to the high dependency of the cosine model on the availability of exact terms in both questions.

Questions retrieval methods based on machine translation models was proposed by Jeon et al. [12] to expand queries and generate translation probabilities between words in similar questions pair. The translation is used to measure the likelihood that a candidate question is matches to the query. Nevertheless, such translation models are less sufficient on short texts queries.

The dictionary or corpus is also used to enrich short text to increase the similarity value for compared sentences [9,16,20]. This model relies on additional information from other source. For current growth of short text media platform, user tend to compact the text by using abbreviation, slangs, jargon, symbol or short form created by themselves [1,16] and might also lead to misspelled words. Then the use of additional corpus decrease the effectiveness in identifying the semantic or similarity of texts because the occurrence of spelling errors will reduce the similarity value due to unable to detect and match the words in corpus.

Some researcher exploited the strengths and weakness of hybrid or ensemble techniques by combining the various similarity measures [3,16,19,20]. Hybrid model based on semantic word embeddings and tf-idf information is used by [3] to reduce the impact of less informative terms and mentioned that the combination leads to a better model for semantic content within very short texts. [19] measures similarity for short queries from Web search logs. The results showed that lexical matching and probabilistic methods are good in finding semantically identical matches and interesting topically related matches, respectively. It was shown that the combination of lexical, stemmed, and probabilistic matches results were better than any method alone.

Meanwhile Noah et al. [20] explored the potential of word order similarity and semantic similarity as an ensemble method to classify semantic Malay sentences. The calculation of word order similarities relied on vector similarity measures (cosine) however have been proven less effective based on our empirical analysis. The use of open dictionary shows the dependency of third party source.

Spelling correction for search queries also gained attention in previous works [6,8,14,18]. An automatic spelling correction will improve the quality of search result retrieval. The major types of spelling error are typographical errors (insertion, deletion, substitution, transposition), word boundaries (concatenation, splitting) or unfamiliar new words [8,14]. To overcome the spelling errors, Hidden Markov model [14], Damerau-Levenshtein distance [2], Levenshtein distance [13], Markov n-gram transformation model [8] were applied.

3 Proposed Ensemble Similarity Model

In this paper an ensemble similarity model is proposed to analyze the similarity of questions in QA archive. To the best of our knowledge, there is no previous research work using our proposed similarity methods to analysis the QA for online forum discussion. A general overview of the processes is shown in Fig. 1.

The proposed method is a linear combination of VSM and edit-distance model similarity. For the VSM, we consider the weight of word occurrence and for edit distance, we utilize the minimum number of editions required to transform one string to the other.

We simplify the process by skipping the stemming and stop word removal. This pre-processing method was applied to reduce noise of textual data. However work by Gao et al. [9], Saif et al. [21], Hu et al. [11] and Martínez-Cámara et al. [17] claimed that stemming and word removal process negatively impacts the performance of short text analysis. Saif et al. [21] also mentioned that two major limitation of existing stop word list is too generic and outdated due to the new information and terms are continuously increasing.

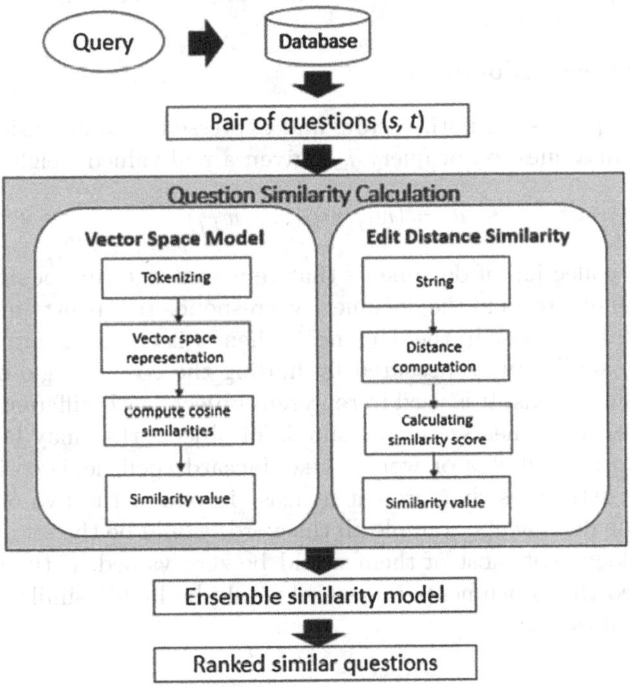

Fig. 1. Overview of the proposed model

A question in Stack Overflow, DreamInCode, MSDN and Yahoo! Answer contains title, question summary, and a body that have detailed explanation of the question. In most cases, when we refer to the question in this paper, it is actually referring to the question title. Jeon et al. [12] has demonstrated that using the question title for retrieval of similar questions is scored the highest in terms of effectiveness.

For each question, a similarity value is calculated for each similarity model. Using the VSM, question will be tokenized for breaking a question into words, phrases or symbols. Then the tokens will form a vector to represent each question and is organized in terms-document matrix. Weight of each word is then calculated based on the IDFs (inverse document frequencies). Finally the cosine similarity is computed to get the similarity value.

Whereas for edit distance model, string-string matrix will be computed for calculating the distance for each question pair. After receiving the similarity distance, we calculate the similarity score to get the similarity value. Finally the question similarity is derived by combining VSM and edit distance similarity value.

We will briefly describe the aforementioned processes of our proposed ensemble similarity questions detection in the following subsections.

3.1 Vector Space Model

Questions and queries are both vectors and expressed as t-dimensional vectors. Each term T, in a question or query j, is given a real valued weight w_{Tj}.

$$d_j = (w_{1j}, w_{2j}, \ldots, w_{Tj}) \tag{1}$$

Let m is a collection of documents that represent n term-document matrix. The weight (w) of a term in the document corresponds to an entry in the matrix. If the term does not exist in the document, then w is equal to zero.

The cosine similarity is measured by finding the cosine angle between two vectors of N dimensions. It is used to represent objects with different frequencies of its attributes. Documents are an example of objects that may have different frequencies of its attributes or words. Like Jaccard coefficient, cosine measure only considers attributes that present at least in one of the two objects being analyzed. In the documents example all the words would be the set of attributes, but for each document most of them would be zero valued. If the 0-0 matches were considered then documents in general would be highly similar. The cosine similarity is defined by:

$$cos(s,t) = \frac{s \cdot t}{\parallel s \parallel \parallel t \parallel} \tag{2}$$

The distance between the two vectors is an indication of the similarity of the two texts. The cosine of the angle between the two vectors is the most common distance measure. Assuming that we have a pair of questions, s and t of which:

- s : *Sipmle log java prgram*
- t : *Simple login java program*

where s is treated as the new query and t as the candidate question from the repository. The pairs representing spelling errors s paired with the correct spelling of the question t. Therefore, we will have a join set st = *[simple, sipmle, login, log, java, program, prgm]*. With the given of two vectors, we can measure the similarity of s and t by calculating their cosine product based on the term-document matrix. Thus the similarity value for both questions is 0.25.

$$\begin{array}{c} \\ t \\ s \end{array} \begin{array}{ccccccc} Simple & sipmle & login & log & java & program & prgm \\ \end{array}$$
$$\begin{matrix} t \\ s \end{matrix} \begin{pmatrix} 1 & 0 & 1 & 0 & 1 & 1 & 0 \\ 0 & 1 & 0 & 1 & 1 & 0 & 1 \end{pmatrix} \tag{3}$$

3.2 Edit-distance Similarity

DLD algorithm is used to calculate the similarity metrics which measuring the difference between two strings based on the number of changes that must be made to transform one strings to another resulting in a similarity or dissimilarity (distance) score [4]. Allowed changes are edition, insertion, deletion and transposition [7] - for example, the sequence (*siple*) can be converted to (*simple*) by one operations: transpose (*pm*).

Chen [4] also mentioned that similarity metric provides a floating point number indicating the level of similarity based on plain lexicographic match. Besides that, the DLD algorithm [7] also use to correct the spelling error as mentioned in [6,8,18], that spelling errors are most usually occurred in search queries.

Simply we can say that, the closeness of a match is measured in terms of the minimum number of operations necessary to convert the string into an exact match. We use the same question as mentioned earlier to calculate the DLD. For example, if the source string (in this case query input) is *sipmle* and the target string is *simple*, to transform *sipmle* to *simple*, we need to transpose *pm* to *simple*, thus, DLD will be 1. A small distance means the pairwise questions are very similar.

Figure 2 illustrates all the steps involved in calculating the DLD between *sipmle* and *simple*. To derive the DLD for s and t, the string s-string t matrix is constructed as shown in Fig. 2(b).

M_{ij} is a matrix of string s compared with string t. The recursion value of $M[i-1, j-1]$ is 0 if $s[i]$ and $t[j]$ are the same strings as shown in Fig. 2(c). Otherwise if $s[i] \neq t[j]$, the recursion value is computed using the following conditions:

$$1 + min(M[i-1, j-1],\ M[i-1, j],\ M[i, j-1]) \tag{4}$$

For example, the value of cell M[1,2] is the minimum value of the cell $M[0, 1]$, $M[0, 2]$ and $M[2, 1]$ plus 1 which given as follows:

$$M[1, 2] = 1 + min(0, 1, 2) \tag{5}$$

The illustration is shown in Fig. 2(d). Figure 2(e) shows transposing the character p and m which result in the string *simple*. Transposition is allowed only between such characters that are adjacent already in the original string. So, in the case of transposition if $s[i] = t[j-1]$ and $s[i-1] = t[j]$, then the value is $1 + (M[i-2, j-2])$ where

$$\begin{aligned} M[3, 3] &= 1 + m[3-2, 3-2] \\ &= 1 + m[1, 1] \end{aligned} \tag{6}$$

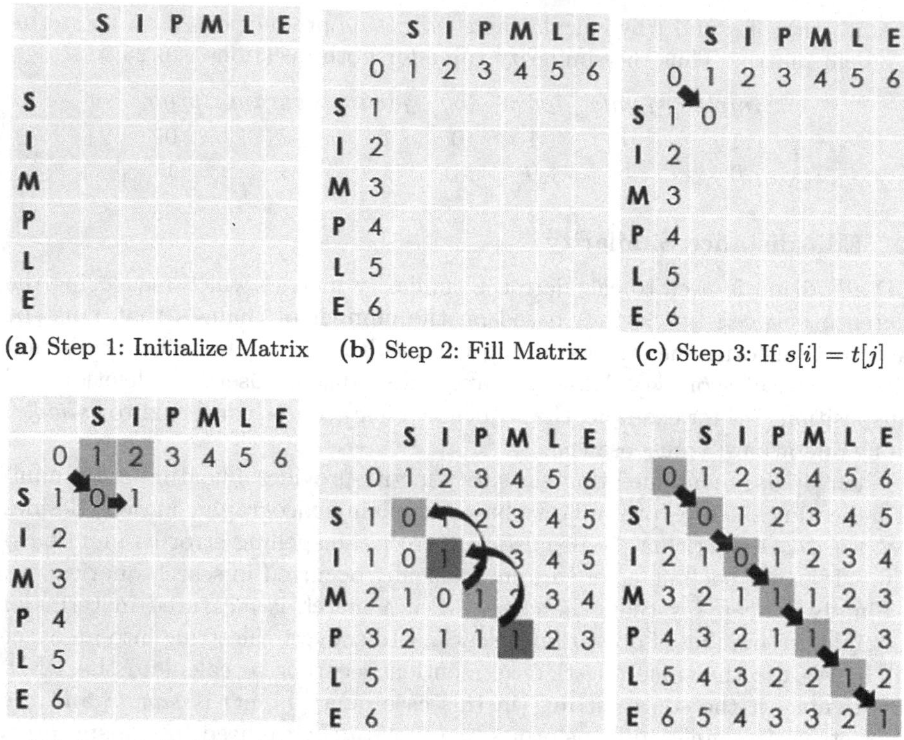

(a) Step 1: Initialize Matrix (b) Step 2: Fill Matrix (c) Step 3: If $s[i] = t[j]$

(d) Step 4: if $s[i] \neq t[j]$ (e) Step 5: If transposition (f) Step 6: DLD distance

Fig. 2. Steps involve in calculating DLD

After retrieving the distance of DLD, the similarity score is then calculated by using the following formula:

$$Sim_{DamerauLevenshtein}(s,t) = 1 - \frac{DamerauLevenshteinDist(s,t)}{max(|s|,|t|)} \tag{7}$$

where max is the length of the longest of the two given texts (s,t) and DLD is the Damerau-Levenshtein Distance. Thus the similarity value of s and t is 0.84. A large value means the strings are very similar.

3.3 Ensemble Similarity Measure

The ensemble similarity measure will calculate the overall similarity between two questions by a linear combinations as follows:

$$Sim_{overall} = \delta Sim_{word} + (1-\delta)Sim_{editDistance} \tag{8}$$

where δ is a constant value between 0 and 1, which decides the contribution of the involved similarity measure. We use 0.5 as constant value similar with [15,20]. Thus the ensemble similarity value of texts (s,t) is 0.55.

4 Preliminary Results and Discussion

We make several experiments to test the proposed methods to get the required results. The dataset used in this paper were collected from Stack Overflow with a few data with spelling error. Based upon the previous string-based similarity measures, we can derive ensemble similarity measures. Table 1 shows the results of cosine similarity, DLD and ensemble for pairwise analysis on new question or queries (s) with question in repository (t). We test them with different input queries to show the similarity value for both methods compared with our ensemble method.

If we look at the target query 1, *simple login java program*, for the third comparison, DLD (0.84) still gives higher value than cosine similarity measure (0.25) even there is a spelling error for the input question. The same results is also obtained for the fourth comparison which are DLD (0.92) and cosine (0.50). However, for the eighth comparison, the DLD still gives higher similarity percentage (0.16), compare with cosine which is 0. It shows that, DLD cannot work alone to rank the search result because based on our human interpretation, both questions are totally different in structure and meaning. To overcome this problem, we calculate the average between both methods to get similarity percentage and rank the search result.

Table 1. Pair of questions with its similarity value

Question	Questions compared	Cosine	DLD	Ensemble
Target query 1				
Simple login java program				
	1. Simple login java program	1	1	1
	2. Sipmle login java code	0.50	0.72	0.61
	3. Sipmle log java prgram	0.25	0.84	0.55
	4. Smple logon java program	0.50	0.92	0.71
	5. Simple login with java code	0.67	0.59	0.63
	6. Java GUI simple login back end mysql	0.57	0.42	0.50
	7. VERY simple user login system in Java	0.57	0.38	0.48
	8. Just any code	0	0.16	0.8
Target query 2				
Getting value from database				
	1. Getting value from database	1	1	1
	2. Getting specific value from Database	0.89	0.75	0.82
	3. Get value from detabase	0.50	0.81	0.66
	4. Get value from SQL database	0.67	0.70	0.69
	5. How get value from database	0.67	0.74	0.71
	6. Select one value from database	0.67	0.77	0.72
	7. Retrieve bit value from database	0.67	0.72	0.70
	8. Get value from data base	0.45	0.81	0.63

Then if the new question is similar to the existing question in the repositories, the results will be ranked according to the similarity measure. Thus, we can reduce the rate of unanswered questions by automatically giving the answers based on similar question answered in repositories.

5 Conclusion

In this paper, an ensemble similarity measure has been proposed in providing better search and retrieval of QA in forum discussion. The extracted QA pairs could be used to reduce unanswered question by giving the answer of new question which is similar to the past questions in repositories. The preliminary results also have shown the potential use of our ensemble similarity measure in overcome the weakness of both models by resolving the spelling error for the input query. Both techniques are useful to calculate the similarities of text but both does not consider any semantic in context. They literally calculate the similarities in physical words or letters without consider the meaning of words or the context of words in the question structure. However, the strength of both models also influence the achieving better similarity percentage. In future, human experts are needed to verify the results obtained by our proposed similarity method.

References

1. Anson, S., Watson, H., Wadhwa, K., Metz, K.: Analysing social media data for disaster preparedness: understanding the opportunities and barriers faced by humanitarian actors. Int. J. Disaster Risk Reduction **21**, 131–139 (2017)
2. Bard, G.V.: Spelling-error tolerant, order-independent pass-phrases via the damerau-levenshtein string-edit distance metric. In: Proceedings of the Fifth Australasian Symposium on ACSW Frontiers, ACSW 2007, vol. 68, pp. 117–124. Australian Computer Society Inc., Darlinghurst, Australia (2007)
3. Boom, C.D., Canneyt, S.V., Bohez, S., Demeester, T., Dhoedt, B.: Learning semantic similarity for very short texts. CoRR abs/1512.00765 (2015)
4. Chen, H.: String Metric and Word Similarity applied to Information Retrieval. Master's thesis, School of Computing. University of Eastern Findland (2012)
5. Cong, G., Wang, L., Lin, C.Y., Song, Y.I., Sun, Y.: Finding question-answer pairs from online forums. In: Proceedings of the 31st Annual International ACM SIGIR Conference on Research and Development in Information Retrieval, pp. 467–474, SIGIR 2008 (2008)
6. Cucerzan, S., Brill, E.: Spelling correction as an iterative process that exploits the collective knowledge of web users. In: Proceedings of EMNLP **4**, 293–300 (2004)
7. Damerau, F.J.: A technique for computer detection and correction of spelling errors. Commun. ACM **7**(3), 171–176 (1964)
8. Duan, H., Hsu, B.J.P.: Online spelling correction for query completion. In: Proceedings of the 20th International Conference on World Wide Web, pp. 117–126, WWW 2011, USA. ACM, New York (2011)
9. Gao, L., Zhou, S., Guan, J.: Effectively classifying short texts by structured sparse representation with dictionary filtering. Inf. Sci. **323**(C), 130–142 (2015)

10. Gomaa, W.H., Fahmy, A.A.: Article: a survey of text similarity approaches. Int. J. Comput. Appl. **68**(13), 13–18 (2013)
11. Hu, X., Tang, L., Tang, J., Liu, H.: Exploiting social relations for sentiment analysis in microblogging. In: Proceedings of the Sixth ACM International Conference on Web Search and Data Mining, pp. 537–546, WSDM 2013, USA. ACM, New York (2013)
12. Jeon, J., Croft, W.B., Lee, J.H.: Finding similar questions in large question and answer archives. In: Proceedings of the 14th ACM International Conference on Information and Knowledge Management, pp. 84–90, CIKM 2005, NY, USA. ACM, New York (2005)
13. Lhoussain, A.S., Hicham, G., Abdellah, Y.: Adaptating the levenshtein distance to contextual spelling correction. Int. J. Comput. Sci. Appl. **12**(1), 127–133 (2015)
14. Li, Y., Duan, H., Zhai, C.: A generalized hidden Markov model with discriminative training for query spelling correction. In: Proceedings of the 35th International ACM SIGIR Conference on Research and Development in Information Retrieval, pp. 611–620, SIGIR 2012, USA. ACM, New York (2012)
15. Li, Y., McLean, D., Bandar, Z.A., O'Shea, J.D., Crockett, K.: Sentence similarity based on semantic nets and corpus statistics. IEEE Trans. Knowl. Data Eng. **18**(8), 1138–1150 (2006)
16. Lochter, J.V., Zanetti, R.F., Reller, D., Almeida, T.A.: Short text opinion detection using ensemble of classifiers and semantic indexing. Expert Syst. Appl. **62**, 243–249 (2016)
17. Martínez-Cámara, E., Montejo-Ráez, A., Martín-Valdivia, M.T., Ureña López, L.A.: Sinai: machine learning and emotion of the crowd for sentiment analysis in microblogs. In: Second Joint Conference on Lexical and Computational Semantics (*SEM), vol. 2, Proceedings of the Seventh International Workshop on Semantic Evaluation, pp. 402–407, SemEval 2013. Association for Computational Linguistics, Atlanta, Georgia, USA, June 2013
18. Martins, B., Silva, M.J.: Spelling correction for search engine queries. In: Vicedo, J.L., Martínez-Barco, P., Muñoz, R., Saiz Noeda, M. (eds.) EsTAL 2004. LNCS (LNAI), vol. 3230, pp. 372–383. Springer, Heidelberg (2004). doi:10.1007/978-3-540-30228-5_33
19. Metzler, D., Dumais, S., Meek, C.: Similarity measures for short segments of text. In: Amati, G., Carpineto, C., Romano, G. (eds.) ECIR 2007. LNCS, vol. 4425, pp. 16–27. Springer, Heidelberg (2007). doi:10.1007/978-3-540-71496-5_5
20. Noah, S.A., Amruddin, A.Y., Omar, N.: Semantic similarity measures for malay sentences. In: Goh, D.H.-L., Cao, T.H., Sølvberg, I.T., Rasmussen, E. (eds.) ICADL 2007. LNCS, vol. 4822, pp. 117–126. Springer, Heidelberg (2007). doi:10.1007/978-3-540-77094-7_19
21. Saif, H., Fernandez, M., He, Y., Alani, H.: On stopwords, filtering and data sparsity for sentiment analysis of twitter. In: Proceedings of the Ninth International Conference on Language Resources and Evaluation (LREC 2014). European Language Resources Association (ELRA), Reykjavik, Iceland, May 2014
22. Shtok, A., Dror, G., Maarek, Y., Szpektor, I.: Learning from the past: answering new questions with past answers. In: Proceedings of the 21st International Conference on World Wide Web, pp. 759–768, WWW 2012 (2012)

Automatic Clustering and Prediction of Female Breast Contours

Haoyang Xie[1], Duan Li[1], Zhicai Yu[1], Yueqi Zhong[1,2(✉)], and Tayyab Naveed[1]

[1] College of Textiles, Donghua University, Shanghai 201620, China
haoyang.xie@gmail.com, zhyq@dhu.edu.cn
[2] Key Lab of Textile Science and Technology, Ministry of Education, Shanghai 201620, China

Abstract. The horizontal shape of breast is the key of shape categorization of female subjects. In this paper, Elliptic Fourier Analysis and two machine learning approaches (K-Means++ and Support Vector Machine) were used for the clustering and prediction of female breast. Female subjects were scanned by RGB-Depth camera (Microsoft Kinect). The breast contours and the under-breast contours were extracted via an anthropometric algorithm without manual intervention. Pearson Correlation Coefficient (PCC) was used to screen the breast candidate(s) for following shape clustering. Principal Component Analysis (PCA) was performed on the Elliptic Fourier Descriptors (EFDs), extracted during the Elliptic Fourier Analysis (EFA), followed by K-Means++ and SVM. K-Means++ was employed to determine the clustering number, meanwhile offered a credible labeled dataset for the subsequent Support Vector Machine (SVM). Finally, a prediction model was built through the SVM. The primary motivation for this research is to offer a quick reference tool for the designers of female bra. The proposed model was validated by reaching an accuracy of 90.5% for breast horizontal shape identification.

Keywords: Breast contour · Elliptic Fourier Analysis · Pearson Correlation Coefficient · Principal Component Analysis · K-Means++ · SVM

1 Introduction

Female breast shape analysis is an important research topic in fields including the design of bar [1], female garment industry [2], breast healthcare and shaping [3]. Researchers have done many related works from various perspectives. Early attempts of breast size classification use the girth of females' chest, under-bust and bust to parameterize the bra size, and introduce the alphabet A, B, C, and D to the breast classification system [4]. In order to utilize the three-dimensional information, Martin [5] takes into account the side view of breast curves and divides the breasts into four categories: flat, hemisphere, conic and goat shape. Chan et al. [6] point out the physical and physiological problems existed in design of bra including unfitted pressure of bra underwire, uncomfortable bra cups, inferior bra hooks and wrong-fitted bra shape and size. They also emphasize that the breast shape is one of the most significant aspect during the design process of female bra.

O. Gervasi et al. (Eds.): ICCSA 2017, Part I, LNCS 10404, pp. 30–42, 2017.
DOI: 10.1007/978-3-319-62392-4_3

In order to create a new bra sizing system, Zheng et al. [7] cluster the female breast shapes into eight clusters based on 3D body scanning data. They use the breast girth, the under-breast girth, and the breast depth ratio as the classification parameters. Morris et al. [8] propose a method for three-dimensional female breast calibration. They develop a range of 18 different cup shapes based on 50 subjects' breast shapes and dimensions. Dong and Zhang [9] use K-Means to divide the female breasts into four classes according to the size of bra. Nevertheless, the method they use to decide the number of clusters in K-Means procedure is not robust. Only the first four types of bra cups from their dataset are selected. Wang et al. [10] divide the breast shape of young women into five categories based on anthropometric measurement. However, their contact measurements require tedious manual operations. Chang [11] uses body scanner (TecMath) to collect three-dimensional data of female bodies and subdivides the female breast into nine categories through anthropometric measurement, but without a detailed illustration of the procedure they use to extract the features from 3D models.

In this paper, we focus our research on the horizontal shape of female breast, which is a significant component in breast shape analysis according to [7]. In our experiments, we used machine learning techniques to tackle the classification problem of female bosom contour. Technically, classification can be divided into two types: clustering and classification. Clustering is associated with unsupervised learning algorithm such as K-Means++. Classification is related to supervised learning algorithm like Support Vector Machine (SVM) [12]. The detailed explanation and corresponding results based on these approaches are provided in the following sections.

The remainder of this paper is organized as follows. Section 2 is an introduction to the experiment methods and algorithms. Section 3 is dedicated to demonstrate the experimental results and analysis. A brief conclusion is given in Sect. 4.

2 Methodology

As shown in Fig. 1, in order to complete the clustering and prediction of female breast contours, Elliptic Fourier Analysis (EFA) was adopted to parameterize the breast contours and the under-breast contours. Elliptic Fourier Descriptors (EFDs) extracted from these contours in EFA were considered as the original features. To determine the best candidate(s) for shape analysis, Pearson Correlation Coefficient (PCC) was used to assess the similarity of the fitting breast contour and the fitting under-breast contour, respectively. Principal Component Analysis (PCA) was performed to extract principal components (PCs), which can represent the original EFDs. These PCs were considered as shape features in the subsequent K-Means++ and SVM.

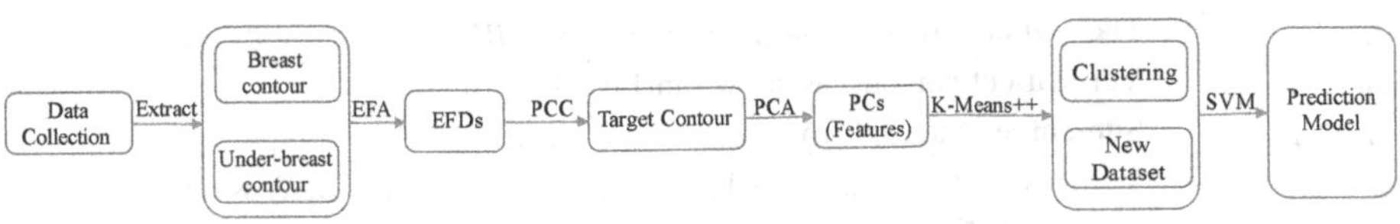

Fig. 1. Pipeline of the proposed methods.

2.1 Data Collection

In this study, 100 female subjects were scanned with the specified ‘A’ posture, as show in Fig. 2. With this ‘A’ pose, the armpits could be scanned clearly without significant loss of data.

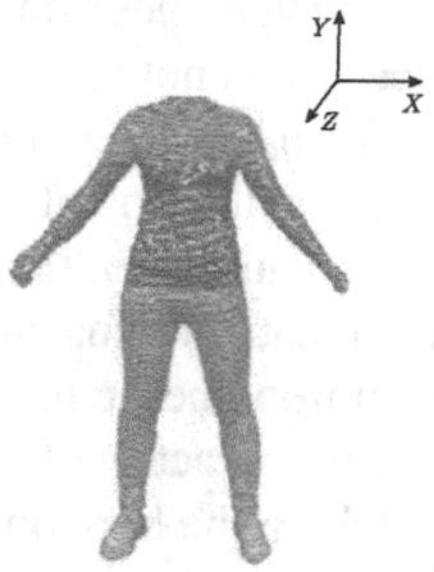

Fig. 2. Cut human body with ‘A’ pose.

After body scanning, the scanned human was cut horizontally at a fixed gauge (5 mm). Denoting the slices as $L : \{l_i | l_i \in L, i = 1, 2, \ldots, N\}$, where N is the number of total slices. The following notations can be defined for the convenience of explanation, as show in Table 1.

Table 1. Illustration of notations used in this algorithm

Notations	Illustration
$N(l_i)$	The number of closed-curves on slice l_i
$l_{i,j}$	*j-th* closed-curves on *i-th* slice l_i, where *j = 1, 2 or 3*
$V_{i,j,K}$	Vertices on $l_{i,j}$, $v_{i,j,k} \in V_{i,j,K}, k = 1, 2, \ldots, K$, where K is the number of vertices on $l_{i,j}$
$Z(V_{i,j,K})$	Z-coordinates of the vertices on $V_{i,j,K}$
$max(Z(V_{i,j,K}))$	The maximum Z-coordinate on $V_{i,j,K}$
BP	Breast point
S_{bp}	Breast contour
$l_{bp,j}$	The *j-th* closed-curve on the slice where *BP* locates. The *j-th* closed-curve on the slice where *BP* locates
UBP	Under-breast point
IP	Inflection point, the turning point of variation tendency of $Curv(v_k)$
$Curv(v_k)$	The curvature at the point v_k
$l_{ubp,j}$	The *j-th* closed-curve on the slice where *UBP* locates
l_{vert}	Vertical cut contour, as shown in Fig. 4
l_{vert_breast}	Sub-curve extracted from l_{vert}, as shown in Fig. 4
V_K	Vertices on l_{vert_breast}, $v_k \in V_K, k = 1, 2, \ldots, n, \ldots, k, \ldots, K$, where K is the number of vertices on l_{vert_breast}

Since breast circumference and under-breast circumference were two significant parameters for the design of bra and measurement of breast [7], we extracted these two contours from the scanned human body. The methods to extract these contours were described as follow:

Step 1. Extract bust point *BP* and breast contour S_{bp}.

Step 1.1. For *i-th* slice, l_i, if $N(l_i) = 3$, then find the vertex with maximum Z-coordinate in the second closed-curve $l_{i,2}$, as show in Fig. 3a and b.

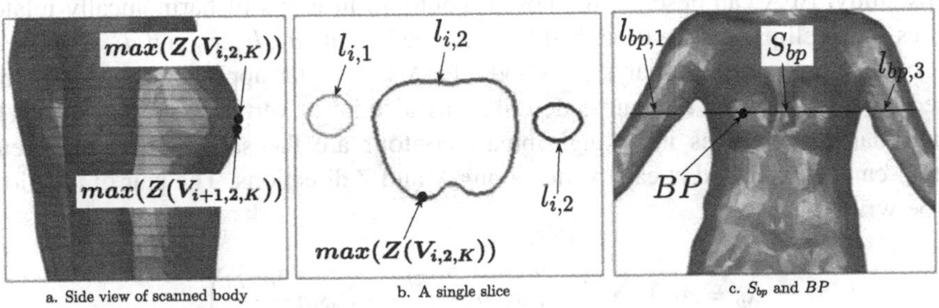

Fig. 3. Extract *BP* and S_{bp}; (a) side view of scanned body; (b) a slice with three closed-curve; (c) breast contour S_{bp}.

Step 1.2. Compare the Z-coordinates $Z(V_{i+1,\,2,K})$ and $Z(V_{i,2,K})$, if $max(Z(V_{i+1,\,2,K})) \leq max(Z(V_{i,2,K}))$, then $V_{i,2,K}$ is regarded as S_{bp}. Otherwise, repeat Step 1.1. The point with maximum Z-Coordinate in $V_{i,2,K}$ is *BP*, as shown in Fig. 3c.

Step 2. Extract *UBP* based on the variation of curvature. The human body is cut vertically at *BP*, as show in Fig. 4.

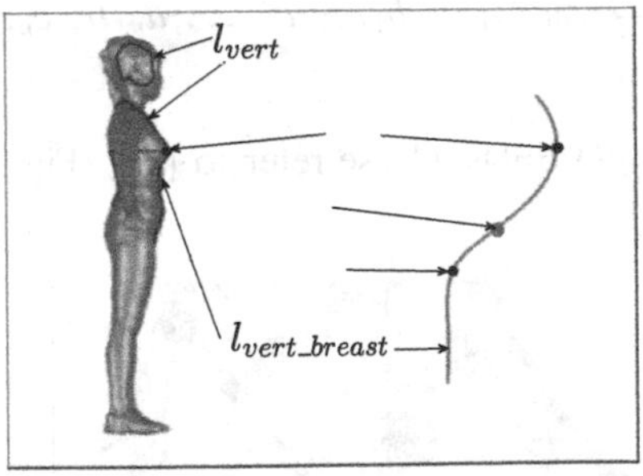

Fig. 4. Find *IP* and *UBP*.

Step 2.1. Start from *BP*, along the $-Y$ direction, we compare the curvature at v_n and v_{n+1}. If $Curv(v_{n+1}) > Curv(v_n)$, the v_n is *IP*, as show in Fig. 4.

Step 2.2. Continue this kind of comparison until $Curv(v_{k+1}) < Curv(v_k)$. The point v_k will be regarded as *UBP*.

Step 3. Extract the under-breast contour S_{ubp}. The horizontal slice at *UBP* consists of three closed curves since it will intersect with both arms and torso. The one with the maximum circumference will be regarded as the under-breast contour S_{ubp}, as shown in Fig. 4.

2.2 Elliptic Fourier Analysis

Extraction of EFDs. EFA was used to extract the original features of breast contours in this study. EFA can describe the bosom contours in terms of harmonically related ellipses. Each ellipse can be described by four coefficients a_n, b_n, c_n and d_n. Compared with other morphological analysis methods, EFA is a better approach to describe the closed curves. A bosom contour is described as a serial of vertices $v(x, y, z) \in V$. The *y*-coordinates of vertices in a single breast contour are the same. Thus, the breast contour can be expanded at each vertex along *X* and *Z* directions. The pair of functions can be written as:

$$x_p = A_0 + \sum_{n=1}^{\infty}\left(a_n cos\frac{2n\pi t_p}{T} + b_n sin\frac{2n\pi t_p}{T}\right) \tag{1}$$

$$z_p = C_0 + \sum_{n=1}^{\infty}\left(c_n cos\frac{2n\pi t_p}{T} + d_n sin\frac{2n\pi t_p}{T}\right) \tag{2}$$

where $t_p = \sum_{i=1}^{p} \Delta t_i$ is the length from the starting point to the *p*-th point, and the perimeter of the whole curve is denoted by *T*. n is the number of Fourier harmonics required to approximate the contour. a_n, b_n, c_n and d_n are the coefficients of Elliptic Fourier for *n-th* harmonic. A_0 and C_0 are the bias coefficients, corresponding to 0 frequency. The Elliptic Fourier Descriptors (EFDs) thus can be denoted as a row vector:

$$EFD = (A_0, C_0, a_1, b_1, c_1, d_1, \ldots, a_n, b_n, c_n, d_n). \tag{3}$$

For more details of EFA and EFDs, please refer to [13] (Fig. 5).

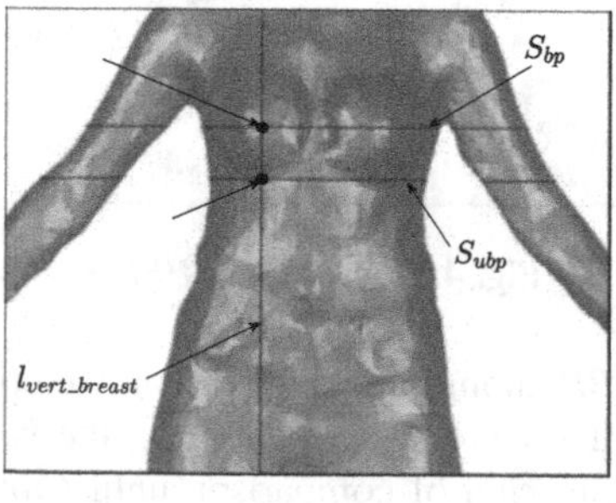

Fig. 5. Extract under-breast contour S_{ubp}.

Reconstruction Error Metrics. We select four parameters to measure the similarity of the original contour and fitting contour: (1) perimeter accuracy, (2) mean distance between original sampling points and fitting points, (3) standard deviation of distance between original sampling points and fitting points and (4) maximum distance between sampling points and reconstruction points.

The reason we chose perimeter as one of the error metrics is that the number of sampling point is closely related to the contour perimeter. Furthermore, the breast circumference is also a significant factor in garment industry. Taking O as the original/actual contour and F as the fitting contour. $P(O)$ and $P(F)$ is the perimeter of the original contour and the fitting contour, respectively. The accuracy A of perimeter can be denoted as

$$A = \frac{P(F)}{P(O)}. \tag{4}$$

The mean distance $\overline{D}$ reflects the deviation extent between the original shape and reconstruction shape.

$$\overline{\mathrm{D}} = \frac{1}{K}\sum\nolimits_{1}^{K} dist(O_i, F_i), i = 1, 2, \ldots, K \tag{5}$$

where K is the total number of the sampling points.

The standard deviation (σ_D) of distance reflects the measures of variability, i.e., $\sigma_D = std(D_i)$. The maximum distance between sampling points and fitting points is denoted as $D_{max} = Max\{dist(O_i, F_i), i = 1, 2, \ldots, K\}$.

2.3 Similarity Assessment

Before clustering, Pearson Correlation Coefficient (PCC) was used to assess the similarity among different samples in a cluster. Similarity is always defined in terms of how "close" the objects are in space, based on a distance function. The breast contours and under-breast contours were analyzed by PCC respectively. If the samples in a type of contours, under-breast contours for instance, are highly similar, it will be unable to provide sufficient contribution to the classification of breast shape. PCC is denoted as r and is calculated as:

$$r = \frac{\sum XY - \frac{\sum X \sum Y}{N}}{\sqrt{\left(\sum X^2 - \frac{(\sum X)^2}{N}\right)\left(\sum Y^2 - \frac{(\sum Y)^2}{N}\right)}} \tag{6}$$

where N is the number of samples, X and Y are the x and y coordinates of each sample. In our work, they are various features of a contour. The r value will locate between -1 and 1. The closer the absolute r to 1, the higher the similarity it is. Contrarily, the closer to 0, the less the similarity it is.

2.4 Principal Component Analysis

Technically, the more harmonics employed in EFA, the higher dimensions of EFDs. Nevertheless, the high dimensions can easily cause over-fitting [14]. In order to avoid over-fitting, PCA is adopt to reduce the dimensions. PCA is a widely used statistical data analysis tool. It tries to represent the feature of a subject with a set of independent variables rather than the original correlative variables. These independent variables are referred to as principal components (PCs).

PCs are linearly combination of the original EFDs and demonstrate the real source of variation between shapes. They are better as indicator variables compared to the original EFDs.

For each EFD, A_0 and C_0 are denoted as 0 frequency which are little effect on PCA. Thus, the input of PCA can be formatted as a $M \times 4N$ matrix,

$$E_{EFDs} = \begin{bmatrix} a_{11}b_{11}c_{11}d_{11} & \cdots & a_{1N}b_{1N}c_{1N}d_{1N} \\ \vdots & \ddots & \vdots \\ a_{M1}b_{M1}c_{M1}d_{M1} & \cdots & a_{MN}b_{MN}c_{MN}d_{MN} \end{bmatrix} \tag{7}$$

where M is the number of samples. $M = 100$ in our work. The algorithm of PCA is described briefly as follow.

Step 1. Calculate covariance matrix C_{EFDs} of the input E_{EFDs}.
Step 2. Compute the eigenvalues and eigenvectors of C_{EFDs}.
Step 3. The space spanned by the eigenvectors of the first k largest eigenvalues is denoted as a projection matrix $T_{4N \times k}$.
Step 4. $PC_s = E_{EFDs} \times T_{4n \times k}$, which reduce the dimensions from $4N$ to k where $k \leq 4N$.

2.5 K-Means++ Clustering

Traditional K-Means is a popular unsupervised learning algorithm and it has been widely used in many industrial and scientific fields [15, 16]. However, there are two major challenges: the first is that, the number of clusters must be given before clustering with prior knowledge. The second is that, the initial centers are selected randomly which always lead the algorithm to stuck at locally optimal solution accordingly.

In practice, the number of clusters K can be decided by visualizing the convergence process. In this paper, the following object function is used to determine the K value.

$$W_k = \sum\nolimits_k^K \sum\nolimits_i^N ||x_i - c_k||^2 \tag{8}$$

where N is the number of samples in each cluster, c is the cluster center, K is the proper number of clusters.

For the second challenge, we used Davide and Sergei's K-Means++ algorithm [17], whose principle idea is that the distance between cluster centers should be as far as possible. The algorithm is described briefly as follow:

Let $X \subset R^d$ be a dataset, $x \in X$.

Step 1. Select a cluster center S_1 from the input dataset randomly.
Step 2. For each sample x, calculate the distance $D(x)$ to its near cluster center. Euclidian distance was used in this our work.
Step 3. Select a new cluster center S_i, which maximizes the probability: $\frac{D(x')}{\sum D(x)}$, where $S_i = x' \in X$.
Step 4. Repeat step 2 and 3 until the K cluster centers are chosen.
Step 5. Implement the traditional K-Means algorithm according to the K initial cluster centers.

2.6 Support Vector Machine

SVM is a useful tool or technique of supervised learning for data classification and identification analysis. The primary objective of SVM is to find a hyper-plane that can divide the training dataset into two categories with maximum margin [18].

Given a serial of labeled training dataset $\{x_i, y_i\}_{i=1}^N, x_i \in R^n, y_i \in \{-1, 1\}$, where x_i represents the input data and y_i stands for output. The decision function is defined as $\omega^T x_i + b = 0$, where ω is the weight vector and b is the offset, which should satisfy the following condition,

$$y_i(\omega^T x_i + b) \geq 1 - \zeta \tag{9}$$

where ζ denotes the linearly separable slack variable. According to the normalization of minimum mean error through the decision, the following function can be deduced,

$$F(\omega, \zeta) = \frac{1}{2}\omega^T \omega + C\sum_{i=1}^N \zeta_i \tag{10}$$

where C is the penalty parameter in which to balance the complexity of algorithm and the proportion of error-classifying samples. The optimal solution can be transferred to optimization constraints according to the multiplier law of Lagrange,

$$\mathrm{Q}(a) = \sum_{i=1}^N a_i - \frac{1}{2}\sum_{i=1}^N \sum_{j=1}^N a_i a_j y_i y_j K(x_i, x_j) \tag{11}$$

where $\{a_i\}_{i=1}^N$ is the Lagrange multiplier. This function is subject to $\sum_{i=1}^N a_i y_i = 0, 0 \leq a_i \leq C, i = 1, 2, \ldots, N$. $K(x_i, x_j)$ is the kernel function, and it maps the original samples to a higher dimensional space. In this work, the Radial Basis Function (RBF) was selected as the kernel, a nonlinear map, which means that it can deal with non-linearly separable problem [19]. Notice that the PCs are employed as the features and they are linearly independence according to the principle of PCA. In addition, the number of hyper-parameters may influence the complexity of a SVM model. The number of hyper-parameters in RBF kernel, as Eq. 12, is less than other kernels like the polynomial kernel. Besides that, RBF is also easier to compute than others [20]. The RBF was denoted as

$$K(x_i, x_j) = \exp\left(-\gamma \left\| x_i - x_j \right\|\right)^2 \tag{12}$$

where γ is the kernel parameter.

3 Result and Discussion

3.1 Breast Contours Reconstruction

The breast contour could be reconstructed by EFA with various harmonics number N. Figure 6 demonstrated the various imitations with different harmonics of a breast contour.

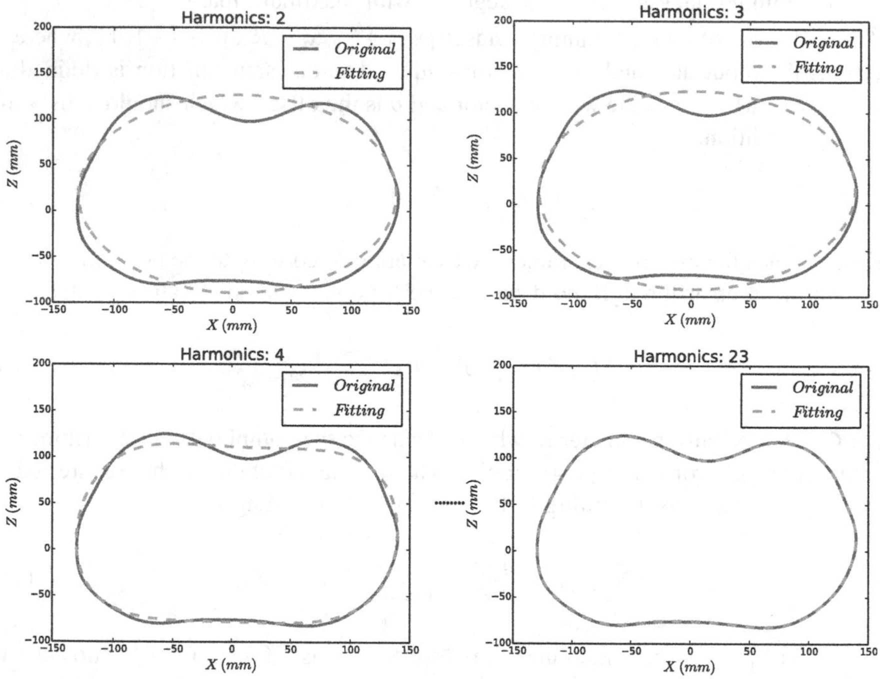

Fig. 6. The fitting results with different harmonic numbers.

Table 2 lists the harmonics and corresponding errors in our experiments. Based on our observation, the harmonics number $N = 23$ could provide sufficient accuracy for the whole breast contours in our dataset. The minimal perimeter accuracy (A), maximal $\overline{D}$, maximal D_{max} and maximal σ_D on the 100 fitting results of breast contours were 99.68%, 0.88, 0.99 and 0.56 respectively.

Table 2. Error metrics of different construction results.

Harmonics	Perimeter accuracy	Mean distance (mm)	Maximum distance (mm)	Standard deviation
2	89.67%	7.69	17.64	3.88
4	92.43%	3.22	8.13	1.94
6	95.43%	1.52	2.94	0.6
...	...	...	...	...
23	99.74%	0.48	0.99	0.18

The similar methods were applied to the under-breast contours, and the harmonics used to reconstruct the under-breast contours were truncated at $N = 18$.

3.2 Results of Similarity Assessment

PCC was applied to the fitting breast contours and the fitting under-breast contours to evaluate the similarity, respectively.

The average PCC of under-breast was 0.923, which stood for strong correlativity or highly similarity. The average PCC of breast contour was 0.236. The results of PCC proven that the under-breast contour was no significant contributions for clustering of horizontal shape even if it was an important factor for the design of bar and the breast measurement [6, 7, 21]. Therefore, we only considered the breast contours in the following test.

3.3 Results of PCA

The entire 100 breast contours' EFDs were subject to PCA without any pre-treatment. The first three, $k = 3$, PCs with maximum variance contributions were computed from the entire EFDs vectors (a matrix actually). PC1, PC2, and PC3, which demonstrated 63.29%, 19.06%, and 9.873% of the total variation respectively. The total sum (contribution) of them was up to 92.22%, as shown in Fig. 7.

In order to illustrate the influence of each PC on the breast contour, the mean and ±2 standard deviation (std) were used to compute a new EFD, which was used to restore three breast contours, as shown in Fig. 8. It was clear to see than PC1

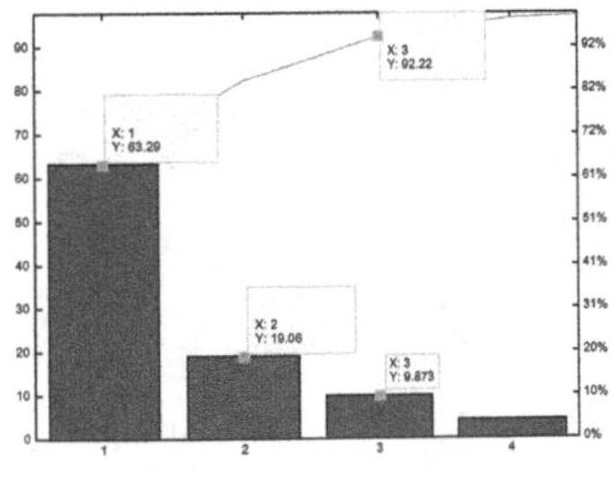

Fig. 7. Contributions of PCs.

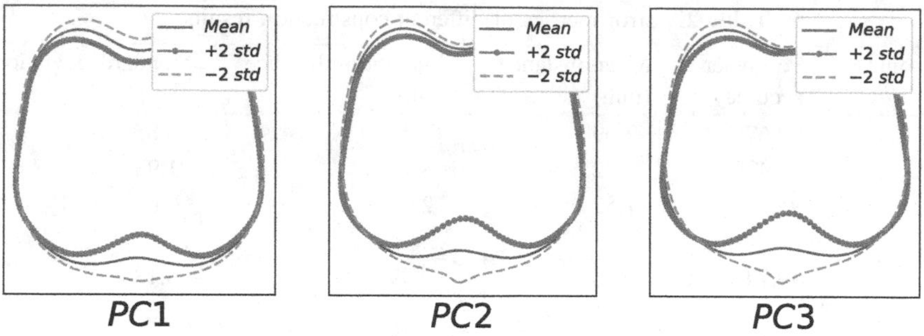

Fig. 8. Influences of PC1, PC2, and PC3 over breast contours.

demonstrated the degree of flatten. PC2 and PC3 illustrated the curvature variation of the breast curve.

3.4 Results of K-Means++ Clustering

Figure 9 illustrated the relationship between error, W_k, and the number of clusters, K. Obviously, the turning point indicated we can choose $K = 4$ as the optimal number of clusters. With this setting, the 100 female breast contours were divided into four categories through 27 iterations, and the number of samples in each group was 22, 20, 30 and 28, respectively. Figure 10 illustrated the difference among these four categories.

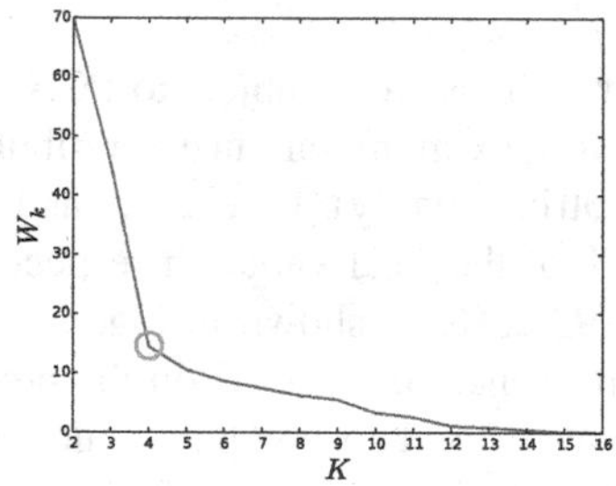

Fig. 9. Relationship between the error metric W_k and the number of clusters K.

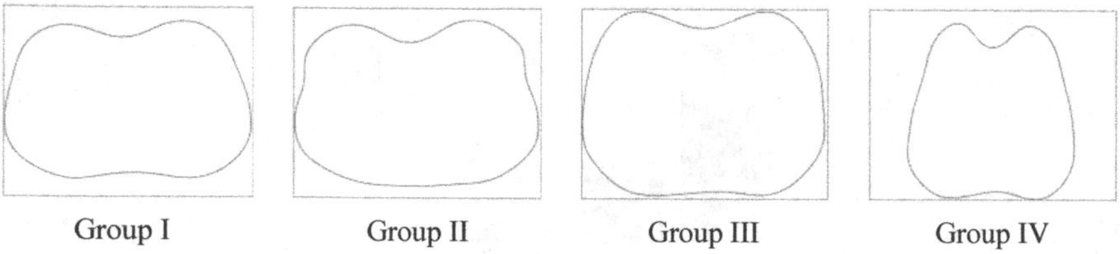

Fig. 10. The sample in each cluster with the minimum distance to its cluster center.

- Group I demonstrated a flatter or wider contour without obvious breast bulge.
- Group II demonstrated a flatter or wider contour with prominent breast bulge because of the concave sections on left and right sides.
- Group III demonstrated a well-rounded breast contour with natural bulge.
- Group IV represented a narrower breast but with obvious breast bulge.

The correlation analysis was also employed to validity the clustering performance. The correlation coefficients of the four groups were 0.98, 0.95, 0.96 and 0.99 respectively. All of them were higher than 0.95, which illustrated that the results of clustering were credible and valid, and thus could be used to build a credible labeled learning dataset.

3.5 SVM Prediction Model

Based on the results of K-Means++, a labeled dataset could be built from the existing unclassified dataset. We randomly selected 70% labeled data as the training set, and 30% as the test set.

The goal of SVM training was to pick a parameter pair (C, γ), where $C(C > 0)$ was the penalty explained in Eq. 10 and γ was the kernel parameter in Eq. 12, so that the SVM model could predict a new data accurately. In order to require the optimal parameter pair (C, γ), meanwhile avoiding over-fitting, we divided the training set into five subsets with equal size to implement 5-fold cross-validation. Each subset was tested using the prediction model trained based on the other four subsets sequentially. We employed grid-search during cross-validation. Various (C, γ) were tried and the specific pair with best cross-validation accuracy was select as the optimal parameters. Finally, γ was 0.5 and the penalty parameter C was 8.0.

According to the optimal pair (C, γ), the SVM prediction model could be setup and the accuracy on the test set achieved to 90.5%, which could be used as a reference to guide the recognition of female bosom contour shape.

4 Conclusion

In this paper, we described a novel system to deal with the clustering and identification of the horizontal shape of females' breast. During the whole procedures, four contributions were reached: firstly, the breast and under breast slices could be extract from a 3D model without manual intervention. Secondly, the under-breast contours proved to be no much contribution to the clustering of breast horizontal contours through PCC analysis. Thirdly, the female breast contour could be reconstructed by EFA accurately with 23 harmonics. Finally, four categories of female bosom contours were classified and the SVM prediction model achieved to 90.5% on the test set. Owing to the significance of horizontal shape to the design of female bra, the prediction model will provide references for relevant designers.

Acknowledgement. This work is supported by National Natural Science Foundation of China (Grant No. 61572124), Shanghai Natural Science Foundation (Grant No. 14ZR1401100), and the Fundamental Research Funds for the Central Universities (Grant No. CUSF-DH-D-2017006).

References

1. Zhou, J., Yu, W., Ng, S.P.: Studies of three-dimensional trajectories of breast movement for better bra design. Text. Res. J. **82**, 242–254 (2012)
2. Hirashima, T.: Garment with breast cups, such as brassiere, US (2010)
3. Blondeel, P., Hijjawi, J., Depypere, H., Roche, N., Van Landuyt, K.: Shaping the breast in aesthetic and reconstructive breast surgery: an easy three-step principle. Part II-breast reconstruction after total mastectomy. Plast. Reconstr. Surg. **123**, 794–805 (2009)
4. Bressler, K., Newman, K., Proctor, G.: A Century of Style: Lingerie. Quarto Publishing Plc., London (1998)
5. Martin, R., Altehenger, A., Saller, K.: Lehrbuch der Anthropologie. JSTOR (1968)
6. Chan, C.Y.C., Yu, W.W.M., Newton, E.: Evaluation and analysis of bra design. Des. J. **4**, 33–40 (2001)
7. Zheng, R., Yu, W., Fan, J.: Development of a new Chinese bra sizing system based on breast anthropometric measurements. Int. J. Ind. Ergon. **37**, 697–705 (2007)
8. Morris, D., Mee, J., Salt, H.: The calibration of female breast size by modeling. In: International Foundation of Fashion Technology Institutes Conference, Hong Kong (2002)
9. Dong, C., Zhang, D.: On the side shapes of women's breasts based on bra cups. J. Zhejiang Fash. Inst. Technol. **12**, 23–27 (2013)
10. Wang, F., Luo, Y., Wang, J.: Research on subdividing of female breast shape in Shanghai area. Prog. Text. Sci. Technol. 73–76 (2012)
11. Chang, L., Zhang, X., Qi, J.: Research on subdividing of female breast shapes based on 3-D body measurement. J. Text. Res. **27**, 21–24 (2006)
12. Pedregosa, F., Varoquaux, G., Gramfort, A., Michel, V., Thirion, B., Grisel, O., Blondel, M., Prettenhofer, P., Weiss, R., Dubourg, V., Vanderplas, J., Passos, A., Cournapeau, D., Brucher, M., Perrot, M., Duchesnay, E.: Scikit-learn: machine learning in Python. J. Mach. Learn. Res. **12**, 2825–2830 (2011)
13. Kuhl, F.P., Giardina, C.R.: Elliptical Fourier features of a closed contour. Comput. Graph. Image Process. **18**, 236–258 (1982)
14. Han, H.: Analyzing support vector machine overfitting on microarray data. In: Huang, D.S., Han, K., Gromiha, M. (eds.) ICIC 2014. LNCS, vol. 8590, pp. 148–156. Springer, Cham (2014). doi:10.1007/978-3-319-09330-7_19
15. Oyelade, O.J., Oladipupo, O.O., Obagbuwa, I.C.: Application of k-means clustering algorithm for prediction of students' academic performance. Comput. Sci. (2010)
16. Dhanachandra, N., Manglem, K., Chanu, Y.J.: Image segmentation using K-means clustering algorithm and subtractive clustering algorithm. Procedia Comput. Sci. **54**, 764–771 (2015)
17. Arthur, D., Vassilvitskii, S.: K-means ++: the advantages of careful seeding, pp. 1027–1035 (2007)
18. Cortes, C., Vapnik, V.: Support-vector networks. Mach. Learn. **20**, 273–297 (1995)
19. Guan, L., Xie, W., Pei, J.: Design of SVM based on radial basis function neural networks pre-partition, pp. 1480–1483 (2014)
20. Hsu, C.-W., Chang, C.-C., Lin, C.-J.: A practical guide to support vector classification (2003)
21. Chan, Y.C.: A conceptual model of intimate apparel design: an application to bra design (2010)

Parallel Ray Tracing for Underwater Acoustic Predictions

Rogério M. Calazan[1(✉)], Orlando C. Rodríguez[1], and Nadia Nedjah[2]

[1] LARSyS, University of Algarve, Campus de Gambelas, 8005-139 Faro, Portugal
{a53956,orodrig}@ualg.pt

[2] Department of Electronics Engineering and Telecommunications, Faculty of Engineering, State University of Rio de Janeiro, Maracanã, Rio de Janeiro 20550-013, Brazil

http://www.siplab.fct.ualg.pt, http://www.eng.uerj.br/~nadia

Abstract. Different applications of underwater acoustics frequently rely on the calculation of transmissions loss (TL), which is obtained from predictions of acoustic pressure provided by an underwater acoustic model. Such predictions are computationally intensive when dealing with three-dimensional environments. Parallel processing can be used to mitigate the computational burden and improve the performance of calculations, by splitting the computational workload into several tasks, which can be allocated on multiple processors to run concurrently. This paper addresses an Open MPI based parallel implementation of a three-dimensional ray tracing model for predictions of acoustic pressure. Data from a tank scale experiment, providing waveguide parameters and TL measurements, are used to test the accuracy of the ray model and the performance of the proposed parallel implementation. The corresponding speedup and efficiency are also discussed. In order to provide a complete reference runtimes and TL predictions from two additional underwater acoustic models are also considered.

Keywords: Parallel computing · Open MPI · Underwater acoustics · Ray tracing

1 Introduction

Ocean acoustic models are numerical tools, which provide a detailed description of sound propagation in the ocean waveguide. This is achieved through the computation of the pressure field transmitted by a set of acoustic sources and received on a set of hydrophones. Ocean acoustic models can be classified into different types, depending on the particular analytical approximation of the wave equation that the model implements numerically. Ray tracing models, for instance, are based on geometrical optics and address the solution of the wave equation using a high frequency approximation, which leads to the computation of wavefronts based on ray trajectories. In three-dimensional scenarios the underwater models are expected to deal with *out-of-plane* propagation, i.e.

O. Gervasi et al. (Eds.): ICCSA 2017, Part I, LNCS 10404, pp. 43–55, 2017.
DOI: 10.1007/978-3-319-62392-4_4

the models should be able to take into account the environmental variability in range, depth and azimuth, which is induced by a three-dimensional bathymetry or/and by a sound speed field [1–3].

The fundamental metric provided by an underwater model is the acoustic pressure, which is used to estimate the transmission loss (TL). TL by itself properly allows to measure the variation of a signal strength with distance [2], and represents a fundamental term in the sonar equations [4]; TL estimation is also fundamental for predictions of underwater noise [5] and source tracking [6], just to mention a few of many possible applications. However, the calculation of TL in a three-dimensional environment is computationally intensive and very time consuming.

Parallel processing is a strategy used to provide faster solutions to computationally complex problems by splitting the workload into sub-tasks, that would be allocated on multiple processors to run concurrently. In this sense, the distributed memory architecture is widely employed to achieve high performance computing. This architecture takes advantage of a Message Passing libraries to perform communication and synchronization such as, for instance, Open MPI [7].

This paper proposes a parallelization strategy of a ray tracing model, as well as its implementation using Open MPI to compute the acoustic pressure. In order to evaluate the proposed strategy and its implementation, data from a tank scale experiment are used to provide waveguide parameters and TL measurements to compare against predictions. The speedup and efficiency achieved are reported and discussed. Furthermore, two additional underwater acoustic models are used to compare the performance in terms of execution time and accuracy of TL predictions, against those obtained by the proposed implementation. The results show that the parallel implementation is able to improve the model performance significantly without compromising the accuracy of the predictions.

The remainder of this paper is organized as follows: Sect. 2 describes previous work in the field of parallel implementations of underwater acoustic models; Sect. 3 briefly describes the ray model; Sect. 4 describes the proposed strategy and its parallel implementation; Sect. 5 provides a description of the tank scale experiment; Sect. 6 discusses in detail the obtained results and compares them to those provided by the additional models; Sect. 7 presents the conclusions and points out directions for future work.

2 Previous Work

The discussion presented in [8] describes a parallel implementation of a parabolic equation based algorithm using MPI libraries, aimed at the analysis of 3D acoustic effects. The results indicate that for both idealized and realistic cases of underwater propagation the parallel implementation of the parabolic model reduced drastically the execution time. Alternatively, a GPU-based parallel implementation of a split-step Fourier parabolic equation is presented in [9], which shows that the GPU version could be ten times faster than a multi-core version using OpenMP. However, several idealized test cases considered to evaluate the implementations allowed to conclude that measurements of execution time of the multi-core were unreliable. The task of eigenray 3D search in the case of an irregular seabed using a parallel implementation based on OpenMP is discussed in [10]. The results are obtained for an idealized test case, and the implementation is reported to 3.76× faster than the sequential implementation, while preserving accuracy. However, the performance is evaluated using a reduced number of cores, and the solution scalability is limited to only one processor. Unlike the above discussed implementations the problem presented here will address the development of an efficient multi-core implementation, to be evaluated in terms of performance and accuracy against experimental data.

3 The Ray Tracing Model

The ray tracing model considered here is called *TRACEO3D*, and corresponds to a three-dimensional extension of the *TRACEO* ray model [11,12]. TRACEO3D is under current development at the Signal Processing Laboratory (SiPLAB) of the University of Algarve, and is able to predict acoustic pressure fields and particle velocity in environments of elaborate boundaries. The model produces ray, eigenray, amplitude, travel time information and can be able to take in account for out-of-plane propagation.

Generally speaking, TRACEO3D produces a prediction of the acoustic field in two steps: first, the set of Eikonal equations is solved in order to provide ray trajectories; second, ray trajectories are considered as the central axes of Gaussian beams, and the acoustic field at the position of a given hydrophone is computed as a coherent superposition of beam influences. The main steps of acoustic pressure calculations are summarized in Algorithm 1. It is important to remark that the total number of rays that are required to produce the prediction depends not only of the set of elevation angles, but also of the set of the azimuth angles. The computational time is therefore proportional to the total number of rays n, and to the hydrophone array size h. Furthermore, the choice of the parameter n is problem dependent because it is related to the amount of rays needed to sweep the 3D waveguide.

4 TRACEO3D on Distributed Memory Multi-core Processors

The Distributed Memory architecture is an efficient way to achieve high performance computing with multiprocessors, in which each processor has his own private physical address space [13]. The processors are interconnected via a high-speed network used for message exchange as illustrated in Fig. 1.

Algorithm 1. Sequential TRACEO3D version

load initial values
let ϕ = set of azimuth angles
let θ = set of elevation angles
let h = number of hydrophones
consider $n = \phi \times \theta$ number of rays
for $j := 1 \rightarrow n$ **do**
  launch ray_j with ϕ_j and θ_j
  solve the Eikonal equations
  compute the dynamic equations
  for $l := 1 \rightarrow h$ **do**
    compute the acoustic pressure for $hydrophone_l$
  end for;
end for;
return the acoustic pressure computed for all rays

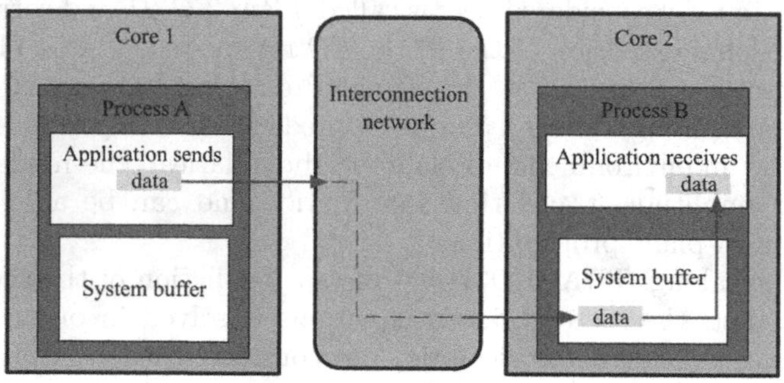

Fig. 1. Communication by message in a distributed memory multi-core architecture

4.1 MPI Basic Concepts

The *Message Passing Interface* (MPI) is a specification for a standard message library, which was defined in the MPI Forum [14]. The Open MPI [7] is a open source MPI implementation of the MPI specification, designed to be portable and efficient for high-performance architectures. A parallel process using MPI has his own private address space. When a process, say A needs to communicate with another process, say B, process A sends some data stored in its address space to the system buffer of process B, which can reside in distinct processor, as show in Fig. 1. The system buffer is a memory space reserved by the system to store incoming messages. This operation is cooperative and occurs only when process A performs a *send* operation and process B performs a *receive* operation.

There are blocking and no-blocking primitives for sending/receiving messages. In the case of blocking primitives, the processes involved in the operation would stop the program execution until the send/receive operation is completed. Otherwise, the program execution continues immediately after scheduling the communication. The MPI lists the communicating processes in groups, wherein the processes are identified by their classification within that group. This classification within the group is called *ranking*. Thus, a process in MPI is identified by a group number followed by its rank within this group. As a process may be part of more than one group, it may have different ranks. A group uses a specific communicator that describes the universe of communication between processes. The MPI_COMM_WORLD is the default communicator. It includes all processes defined by the user in a MPI application.

Algorithm 2 displays some of the main MPI primitives using FORTRAN. The command at line 5 defines and starts the environment required to run the MPI. The statement at line 6 identifies the process within a group of parallel processes. The function at line 7 returns the number of processes within a group. From that point on, each process runs in parallel as specified in the parallel region. Running processes may cooperate with each other via message passing. The routine at line 9 ends all MPI processes.

Algorithm 2. Example of MPI primitives implemented in Open MPI using FORTRAN

```
1: program example
2: include mpif.h
3:      integer rank, size
4:      /* Sequential region */
5:      call MPI_INIT()
6:      call MPI_COMM_RANK(MPI_COMM_WORLD,rank)
7:      call MPI_COMM_SIZE(MPI_COMM_WORLD,size)
8:           /* parallel region */
9:      call MPI_FINALIZE()
10:     /* Sequential region */
11: return
```

4.2 Parallel Implementation

The approach for the parallel Open MPI extension of TRACEO3D (hereafter called *TRACEO3Dompi*) is described in Algorithm 3. First, in line 2, a parallel environment is initialized for p distinct processes. After that each process loads the initial conditions, and therefore the algorithm proceeds with the division of the workload at ray level to balance the load of existing processes, as shown in line 8. The workload can be affected not only by the total number of rays but also by the ray path that it follows. When a given ray is launched, it follows his own trajectory and can be characterized by a given propagation time; computationally this means that each ray has his own execution time. The differences between such times can potentially lead to an unbalance within the process's workload, since one process may require more computational effort to trace one ray than another, thus increasing the overall execution time. On the other hand, rays with adjacent launching angles usually have similar trajectories, and consequently exhibit similar execution times. Thus, to avoid that the same process gets all the rays that might be more time consuming, adjacent rays are executed in different processes, selected according to the *rank* value, to balance the workload, as described in line 11. In line 12, each process starts to compute his respective set of rays. For each ray, the set of Eikonal equations is solved. After that, the computation of the dynamic equations is performed in line 15. Then, the contribution of the ray for the acoustic pressure in each hydrophone is calculated in line 17. When this is done for k rays each process sends a message to MASTER to compute all contributions, and return the final result. It is important to remark that there are only two moments when inter-process communication occurs: at the beginning (as shown in line 2), and at the end (as shown in lines 20–24).

Algorithm 3. Parallel TRACEO3D Open MPI Extension

1: **let** p = number of processes
2: **MPI_Init()**
3: **generate** *rank* for each process
4: **load** initial values
5: **let** ϕ = set of azimuth angles
6: **let** θ = set of elevation angles
7: **let** h = number of hydrophones
8: **consider** $n = (\phi \times \theta)/p$ number of rays per process
9: **do** $i = n \times rank$
10: **do** $k = i + n$
11: **select** the launch angles $\phi_{i..k}$ and $\theta_{i..k}$ according to rank
12: **for** $j := i \rightarrow k$ **do**
13:     **launch** ray_j with ϕ_j and θ_j
14:     **solve** the Eikonal Equations
15:     **compute** the Dynamic Equations
16:     **for** $l := 1 \rightarrow h$ **do**
17:         **compute** the acoustic pressure for $hydrophone_l$
18:     **end for**;
19: **end for**;
20: **if** $rank = Master$ **then**
21:     **receive** the computed acoustic pressure for *rank*
22: **else**
23:     **send** the computed acoustic pressure to *Master*
24: **end if**;
25: **MPI_Finalize()**
26: **return** the acoustic pressure computed for all set of processes

5 The Tank Scale Experiment

Experimental data acquired at an indoor shallow-water tank of the LMA-CNRS laboratory in Marseille was used to test the accuracy of predictions.

The experiment is described in detail in [3,15], thus a brief description is presented here. The inner tank dimensions were 10 m long, 3 m wide and 1 m deep. The source and the receiver were both aligned along the across-slope direction, as shown in Fig. 2. The bottom was filled with sand and a rake was used to produce a slope angle $\alpha \approx 4.5°$; sound speed in the water was considered constant and corresponded to 1488.2 m/s. The bottom parameters corresponded to c_p = 1655 m/s, ρ = 1.99 g/cm^3 and α_p = 0.5 dB/λ. The source was located at 8.3 m depth and bottom depth at the source position corresponded to 44.4 m. For modeling purposes the waveguide geometry is shown in Fig. 3, where the cross-slope range corresponds to 5 km. The ASP-H[1] data set is composed of time signals recorded at a fixed receiver depth, and source/receiver distances starting from r = 0.1 m until r = 5 m in increments of 0.005 m, providing a sufficiently fine representation of the acoustic field in terms of range. Three different receiver depths were considered, namely 10 mm, 19 mm and 26.9 mm, corresponding to data subsets referenced as ASP-H1, ASP-H2 and ASP-H3, respectively. Acoustic transmissions were performed for a wide range of frequencies. However, comparisons are presented only for data from the ASP-H1 subset with a highest frequency of 180.05 Hz; this is due to the fact that the higher the frequency the better the ray prediction. It is important to notice that a scale factor of 1000:1 is required to properly modify the frequencies and lengths of the experimental configuration. This implies that the following conversion of units is adopted: experimental frequencies in kHz become model frequencies in Hz, and experimental lengths in mm become model lengths in m. For instance, an experimental frequency of 180.05 kHz becomes a model frequency of 180.05 Hz, and an experimental distance of 10 mm becomes a model distance of 10 m. Sound speed remains unchanged, as well as compressional and shear attenuations.

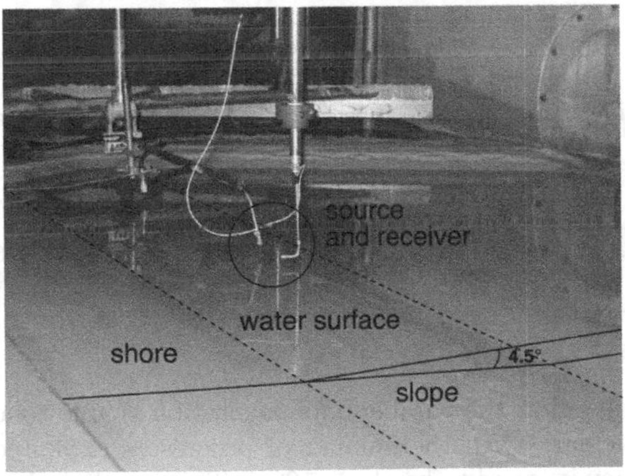

Fig. 2. Indoor shallow-water tank of the LMA-CNRS laboratory of Marseille [3]

[1] For *horizontal* measurements of cross-slope propagation.

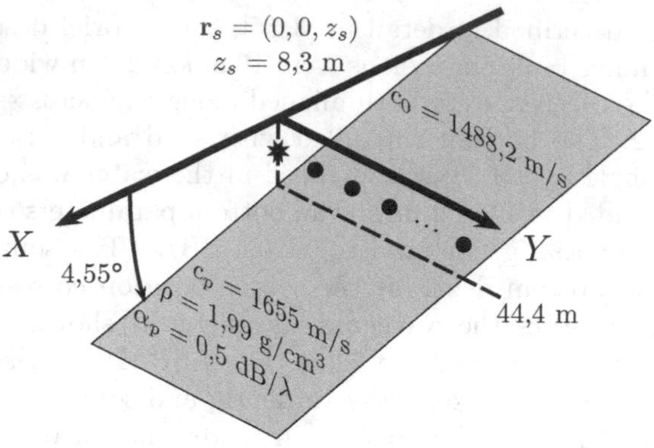

Fig. 3. Upslope environment parameterization for modeling

6 Results and Analysis

The set of waveguide parameters provided by the tank scale experiment were used as input for the models to compute predictions of TL. The models execution is performed by a computer server with two CPUs Xeon(R) E5-2420 of 1.90 GHz, where each CPU has 6 physical cores. In order to analyze the performance and scalability of the parallel implementation, the execution time of the sequential TRACEO3D version is compared to that of the TRACEO3Dompi using different number of processors. Table 1 presents the number of processors, followed by the achieved execution time and speedup, which are also shown in Fig. 4(a) and (b), respectively. Each MPI processes is mapped to one physical core. Figure 4(c) shows the efficiency of the parallel implementation, which is described as the speedup divided by the number of processes [16].

Table 1. Execution times and speedups for the sequential *vs.* parallel versions of TRACEO3D

#Processors	1	2	3	4	5	6	7	8	9	10	11	12
Runtime (s)	621.1	297.9	197.8	149.8	119.5	100.0	86.4	75.6	68.3	61.6	56.3	51.4
Speedup	1	2.1	3.1	4.1	5.1	6.2	7.2	8.2	9.1	10.1	11.0	12.1

The results show that the parallel implementation is able to achieve a linear speedup; in particular, when the number of processors exploited is between 2 and 8, the efficiency is found to be above 100%. This is believed to be due to the superlinearity effect of caches [16], as the parallel implementation scatters the items of several vectors, which introduces a lower memory use at each parallel process. Fig. 4(d) presents the root mean square error (RMSE) for TL predictions results between the sequential and parallel implementations. One can see that

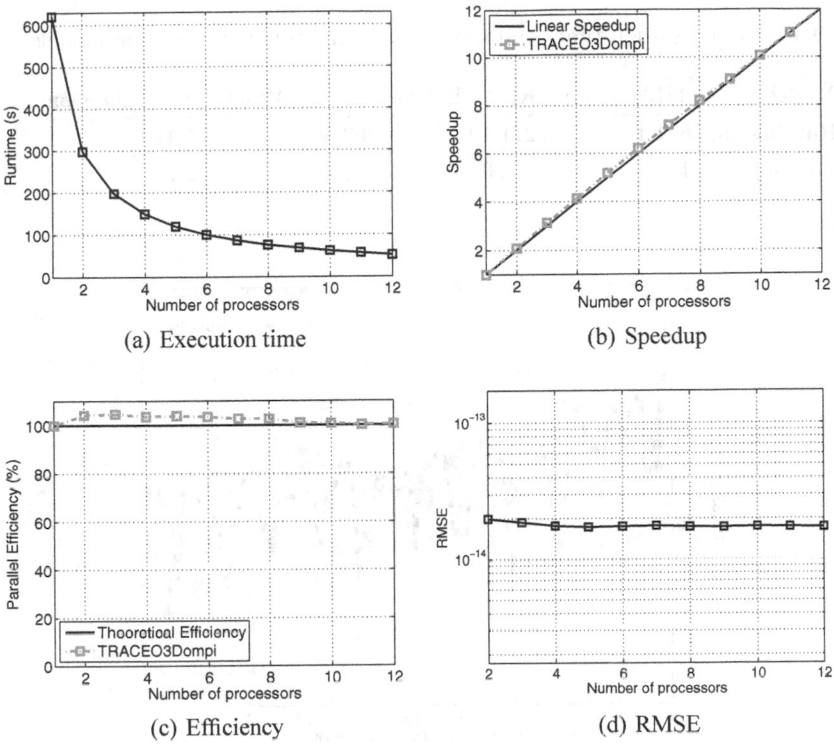

(a) Execution time (b) Speedup

(c) Efficiency (d) RMSE

Fig. 4. Results of sequential *vs.* parallel implementations

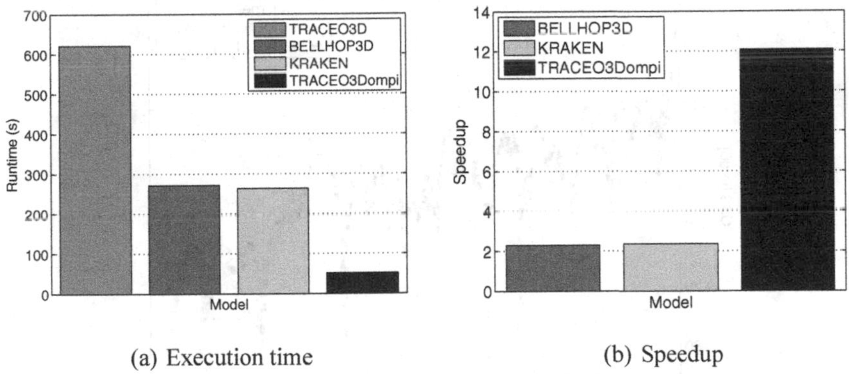

(a) Execution time (b) Speedup

Fig. 5. Execution time and speedup for model predictions of the tank scale experiment

the difference between the values is lower than 10^{-13} for any set of processes. Another issue is that the parallel efficiency is constrained to address each process to one physical core, and thus the parallel processes do not take advantage of

Table 2. Execution times and speedups for the three-dimensional models

Model	TRACEO3D	KRAKEN3D	BELLHOP3D	TRACEO3Dompi
Runtime (s)	621.1	271.10	263.05	51.41
Speedup	1	2.3	2.4	12.1

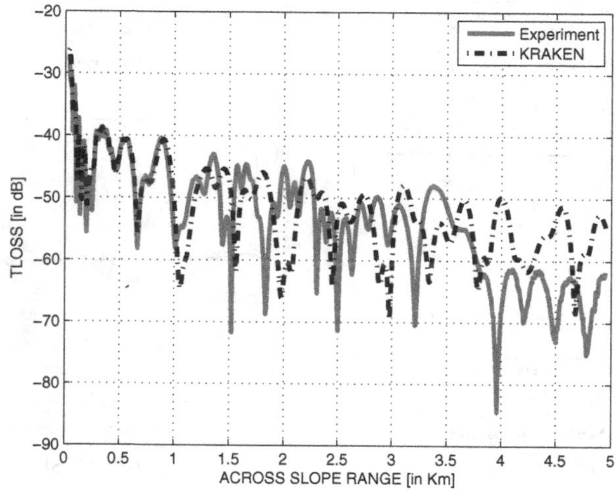

(a) KRAKEN

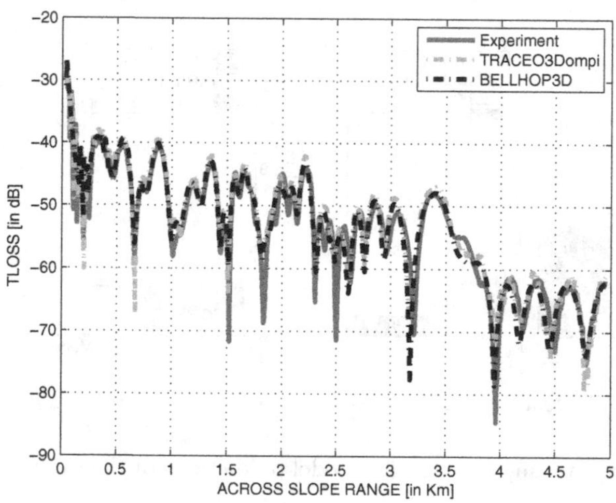

(b) TRACEO3Dompi

Fig. 6. Comparisons with experimental results for LMA CNRS H1 @ 180.05 Hz virtual cores. It is believed that the intensive computation makes use of the resources available at the physical core, disabling the advantage of simultaneous multithreading technology [13].

Additional analysis regarding the performance and accuracy of TRACEO3Dompi was performed using two other acoustic models: BELLHOP3D [17] (which is also based in ray theory and is a three-dimensional extension of Bellhop ray model [18]), and KRAKEN (which is a model based on normal mode theory [19,20]). Both models are able to provide TL predictions in three-dimensional environments[2]. Table 2 shows the execution times and speedups relative to TRACEO3D.

Figure 5(a) and (b) shows the results of execution time and speedup, respectively. It can be seen that the parallel implementation is up to 12× faster than the sequential version, and up to 5× faster that BELLHOP3D and KRAKEN. Comparisons against experimental data are shown in Fig. 6(a) for KRAKEN, while Fig. 6(b) shows the corresponding results for both BELLHOP3D and TRACEO3D. It can be seen that KRAKEN's prediction is only valid at the initial ranges[3], while TRACEO3D and BELLHOP3D predictions are able to follow accurately the experimental data.

7 Conclusions

This paper discussed the performance of a parallel implementation of the ray model TRACEO3D, using Open MPI, aiming at fast computations of acoustic pressure. The accuracy of the model and its parallel implementation were evaluated through comparisons with data from a tank scale experiment, which provides an ideal reference of complex three-dimensional propagation. Additional acoustic models were also used to benchmark the parallel version of TRACEO3D. The results show that the parallel implementation offers a linear speedup with respect to the number of processors used, as long as the workload becomes well distributed. Each parallel process addressed only one physical core because the simultaneous multithreading was not able to improve the overall performance. The implementation was found to be up to 12× faster than the sequential version without any loss of accuracy; additionally, when compared with other models the TRACEO3Dompi was found to be 5× faster as well, predicting the experimental curve with high accuracy.

Future work is expected to include an analysis of eigenray search, and of particle velocity calculations as part of the parallel extension. Moreover, it is intended to develop a parallel version of the ray model implemented in a graphic processing unit.

---

[2] KRAKEN and BELLHOP3D are part of the Acoustic Toolbox, which is available at the Ocean Acoustic Library [21].

[3] This is believed to happen because the wedge is too step for the adiabatic approximation used by KRAKEN to be valid.

Acknowledgments. This work received support from the Foreign Courses Program of CNPq and the Brazilian Navy. Thanks are due to the SiPLAB research team, LARSyS, FCT, University of Algarve. The authors are also deeply thankful to LMA-CNRS for allowing the use of the experimental tank data discussed in this work.

References

1. Tolstoy, A.: 3-D propagation issues and models. J. Comput. Acoust. **4**(03), 243–271 (1996)
2. Jensen, F.B., Kuperman, W.A., Porter, M.B., Schmidt, H.: Computational Ocean Acoustics, 2nd edn. Springer, New York (2011)
3. Sturm, F., Korakas, A.: Comparisons of laboratory scale measurements of three-dimensional acoustic propagation with solutions by a parabolic equation model. J. Acoust. Soc. Am. **133**(1), 108–118 (2013)
4. Hodges, R.P.: Underwater Acoustics: Analysis, Design and Performance of Sonar. Wiley, Hoboken (2011)
5. Soares, C., Zabel, F., Jesus, S.M.: A shipping noise prediction tool. In: OCEANS 2015, Genova (2015). doi:10.1109/OCEANS-Genova.2015.7271539
6. Felisberto, P., Rodriguez, O., Santos, P., Ey, E., Jesus, S.: Experimental results of underwater cooperative source localization using a single acoustic vector sensor. Sensors **13**(7), 8856–8878 (2013). doi:10.3390/s130708856
7. Open MPI: Open Source High Performance Computing, January 2017. http://www.open-mpi.org/
8. Castor, K., Sturm, F.: Investigation of 3D acoustical effects using a multiprocessing parabolic equation based algorithm. J. Comput. Acoust. **16**(02), 137–162 (2008)
9. Hursky, P., Porter, M.B.: Accelerating underwater acoustic propagation modeling using general purpose graphic processing units. In: OCEANS 2011 MTS/IEEE KONA, pp. 1–6 (2011)
10. Xing, C.X., Song, Y., Zhang, W., Meng, Q.X., Piao, S.C.: Parallel computing method of seeking 3D eigen-rays with an irregular seabed. In: 2013 IEEE/OES Acoustics in Underwater Geosciences Symposium, pp. 1–5 (2013)
11. Rodríguez, O.C.: The TRACEO ray tracing program. SiPLAB, University of Algarve (2011)
12. Rodríguez, O.C., Collis, J.M., Simpson, H.J., Ey, E., Schneiderwind, J., Felisberto, P.: Seismo-acoustic ray model benchmarking against experimental tank data. J. Acoust. Soc. Am. **132**(2), 709–717 (2012). http://dx.doi.org/10.1121/1.4734236
13. Patterson, D.A., Hennessy, J.L.: Computer Organization and Design: The Hardware/Software Interface. Morgan Kaufmann, Burlington (2013)
14. MPI Forum: Message Passing Interface (MPI) Forum Home Page, January 2017. http://www.mpi-forum.org/
15. Korakas, A., Sturm, F., Sessarego, J.P., Ferrand, D.: Scaled model experiment of long-range across-slope pulse propagation in a penetrable wedge. J. Acoust. Soc. Am. **126**(1), EL22–EL27 (2009)
16. Grama, A., Karypis, G., Kumar, V., Gupta, A.: Introduction to Parallel Computing, 2nd edn. Addison Wesley, Boston (2003)
17. Porter, M.B.: Out-of-plane effects in ocean acoustics. Technical report, DTIC Document (2013)
18. Porter, M.B.: Gaussian beam tracing for computing ocean acoustic fields. J. Acoust. Soc. Am. **82**(4), 1349 (1987). http://dx.doi.org/10.1121/1.395269

19. Porter, M.B.: A numerical method for ocean-acoustic normal modes. J. Acoust. Soc. Am. **76**(1), 244 (1984). http://dx.doi.org/10.1121/1.391101
20. Kuperman, W.A.: Rapid computation of acoustic fields in three-dimensional ocean environments. J. Acoust. Soc. Am. **89**(1), 125 (1991). http://dx.doi.org/10.1121/1.400518
21. HLS Research: Ocean acoustics library, January 2017. http://oalib.hlsresearch.com/

Influences of Flow Parameters on Pressure Drop in a Patient Specific Right Coronary Artery with Two Stenoses

Biyue Liu[1(✉)], Jie Zheng[2], Richard Bach[3], and Dalin Tang[4]

[1] Department of Mathematics, Monmouth University, West Long Branch, USA
bliu@monmouth.edu
[2] Mallinckrodt Institute of Radiology, Washington University, St. Louis, USA
zhengj@mir.wustl.edu
[3] Cardiovascular Division, Washington University, St. Louis, USA
rbach@wustl.edu
[4] Department of Mathematical Sciences, Worcester Polytechnic Institute, Worcester, USA
dtang@wpi.edu

Abstract. Blood pressure loss along the coronary arterial length and the local magnitude of the spatial wall pressure gradient (*WPG*) are important factors for atherosclerosis initiation and intimal hyperplasia development. The pressure drop coefficient (*CDP*) is defined as the ratio of mean trans-stenotic pressure drop to proximal dynamic pressure. It is a unique non-dimensional flow resistance parameter useful in clinical practice for evaluating hemodynamic impact of coronary stenosis. It is expected that patients with the same stenosis severity may be at different risk level due to their blood pressure situations. The aim of this study is to numerically examine the dependence of *CDP* and *WPG* on flow rate and blood viscosity using a patient-specific atherosclerotic right coronary artery model with two stenoses. Our simulation results indicate that the coronary model with a lower flow rate yields a greater *CDP* across a stenosis, while the model with a higher flow rate yields a greater pressure drop and a greater *WPG*. Increased blood viscosity results in a greater *CDP*. Quantitatively, *CDP* for each stenosis appears to be a linear function of blood viscosity and a decreasing quadratic function of flow rate. Simulations with varying size and location of the distal stenosis show that the influence of the distal stenosis on the *CDP* across the proximal stenosis is insignificant. In a right coronary artery segment with two moderate stenoses of the same size, the distal stenosis causes a larger drop of *CPD* than the proximal stenosis does. A distal stenosis located in a further downstream position causes a larger drop in the *CDP*.

Keywords: Pressure drop coefficient · Blood viscosity · Flow rate · Spatial wall pressure gradient

1 Introduction

Flowing blood imposes hemodynamic stress at the endothelial layer of the blood vessel, the blood-arterial wall interface. The stress development in the endothelial layer due to intravascular pressure and flow is a major contributing factor for the

O. Gervasi et al. (Eds.): ICCSA 2017, Part I, LNCS 10404, pp. 56–70, 2017.
DOI: 10.1007/978-3-319-62392-4_5

pathogenesis of atherosclerosis in the human arterial tree [1–4]. Strong clinical evidences suggest that blood pressure is a major determinant of vascular changes in the arterial system. Local low wall static pressure and local magnitude of the spatial wall pressure gradient (*WPG*) are strongly correlated to the wall thickening and the lumen cross-section narrowing [1–3, 5]. Giannoglou *et al.* [6] gave two main possible explanations about the static pressure distribution implication in atherosclerosis: (1) Atherogenic blood flow particles are probably forced to migrate towards low-pressure regions; (2) The wall pressure difference acting on the endothelium leads to a torque development which results in a redistribution of the initially accumulated atheromatous material within the subendothelial layer.

The magnitude of pressure drop has been clinically used to assess the severity of the lesion. The reduction in pressure drop as a result of angioplasty has been used to judge the success of an interventional procedure. The ratio of hyperemic flow in the presence of a stenosis to that in the absence of the stenosis was proposed and defined as the concept of fractional flow reserve (*FFR*) to assess the severity of stenosis [7–10]. In clinical practice, *FFR* of a coronary artery is the ratio of distal coronary pressure to aortic pressure. It is a pressure only prognostic marker in coronary disease. An *FFR* value of less than 0.75 indicates a functionally significant coronary artery stenosis. Currently, *FFR* is a very popular index of stenosis severity and is considered as a clinical standard for the determination of physiological significance of a stenosis. However, as discussed by Hoef *et al.* [11] and by Brosh *et al.* [12], *FFR* lacks the theoretical solid foundation to be a gold standard. It has some clinical and technical limitations in the evaluation of coronary disease severity in certain clinical situations. A non-dimensional index combining intracoronary pressure and flow may serve as an alternate parameter for the assessment of the physiologic significance of coronary artery disease.

Pressure drop coefficient (*CDP*) is defined as the ratio of mean trans-stenotic pressure drop to proximal dynamic pressure and is a unique non-dimensional flow resistance parameter. Banerjee *et al.* carried out *in-vivo* studies to validate the *CDP* for simultaneous evaluation of the severity of epicardial stenosis and microvascular dysfunction. They suggested that *CDP* has the advantage of higher resolving power for separating normal and diseased conditions of vessels. A lower value of *CDP* indicates a healthier artery while elevated values of *CDP* indicate abnormal stenosis. It may be useful in clinical practice for delineating the severity of epicardial disease under conditions of normal or abnormal microcirculation [13].

It is expected that patients with the same stenosis severity may be at different risk level due to their blood pressure situations. Thus it is important to understand how flow parameters affect *CDP* and *WPG* in order to properly assess the physiological significance of a stenosis on coronary blood flow. The objectives of the present work are to numerically analyze the dependences of *CDP* and *WPG* on blood viscosity and flow rate in stenotic right coronary arteries and to examine the effect of the distal stenosis on the *CDP* in coronary arteries with two stenoses. Three dimensional models with a patient specific geometry are developed to simulate the blood flow and computations are carried out under physiologic conditions.

2 Materials and Methods

In this study, blood flow in the coronary segment is assumed to be unsteady, laminar and incompressible with a density (ρ) of 1050 kg/m^3. The arterial wall is adopted as a rigid wall, comprised from non-elastic and impermeable materials. The flow is governed by the 3D Navier-Stokes equations with no external volume force. Blood is a complex suspension that demonstrates non-Newtonian rheological characteristics. Many non-Newtonian viscosity models have been used by researchers [5, 14–17]. The Carreau model is one of the most reputable models to simulate the blood flow through arteries. Cho and Kensey showed that the Carreau model is well adopted with experimental data. The blood in the present study is treated as a non-Newtonian fluid obeying the Carreau model [15].

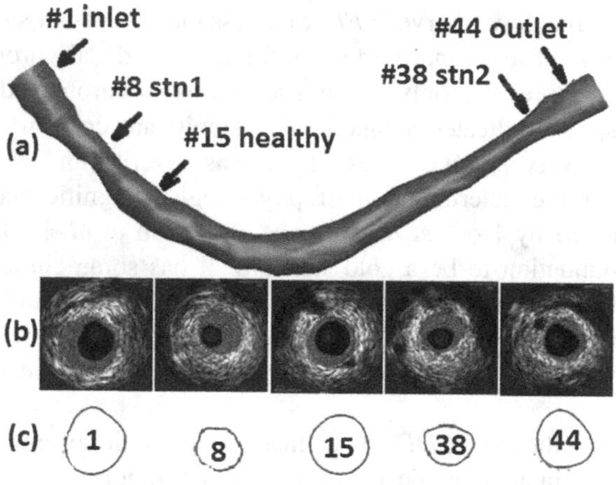

Fig. 1. Geometry of the stenotic artery with selected IVUS slices and corresponding contour curves of lumen boundaries

The geometry of a stenotic right coronary artery is reconstructed based on the lumen contour curves extracted from a 44-slice in-vivo 3D IVUS dataset casting of a patient and the angiographic image of the patient (see [18] for detailed information of the patient and the medical equipment used). Figure 1(a) shows the reconstructed stenotic right coronary artery segment of the patient. Figure 1(b) includes the selected IVUS slices at inlet (#1), proximal stenosis (#8), healthy region (#15), distal stenosis (#38), and outlet (#44), respectively. Figure 1(c) shows the corresponding contour curves of lumen boundary of the selected IVUS slices in Fig. 1(b). The lumen cross-section areas of the selected locations (as labeled in Fig. 1(a)) are 0.082377 cm^2, 0.029155 cm^2, 0.074325 cm^2, 0.035268 cm^2, and 0.067147 cm^2, respectively. The proximal stenosis near the inlet has a reduction of 66% approximately in lumen cross-section area and the distal stenosis near the outlet has a reduction of 47% approximately in lumen cross-section area. This stenotic right coronary artery segment has a length of 5.16 cm with a diameter of 0.32386 cm and 0.29239 cm at the inlet and the outlet, respectively. In the computer simulation, this coronary artery segment is extended at both the inlet and the outlet in length by 0.4 cm and 0.75 cm, respectively, to reduce the influence of the boundary conditions in the region of interest. Thus the stenotic right coronary artery segment as the geometry of the mathematical model has a total length of 6.31 cm. The extended parts at the inlet/outlet are not fully included in Fig. 1(a).

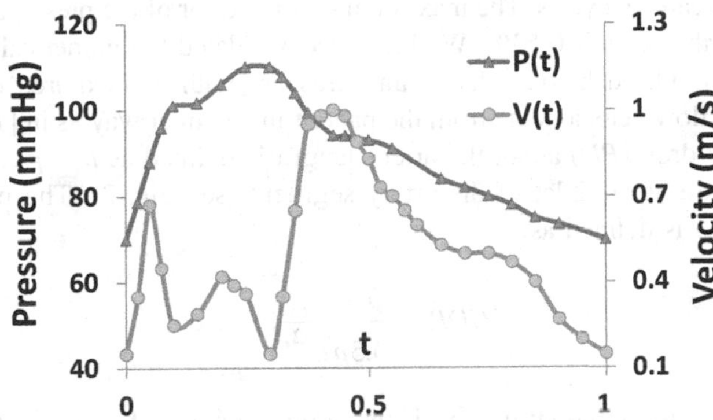

Fig. 2. Pulse waveforms of velocity and pressure

Figure 2 plots the velocity and pressure pulse waveforms extracted from the on-site blood pressure and flow velocity data of the patient. A normalized time length $t/t_p = 1$ is used for simplicity. The following boundary conditions are applied for Navier-Stokes equations: At the inlet boundary, a time dependent pressure with waveform P(t) and a no-viscous-stress condition are applied. At the outlet boundary, a fully developed velocity profile with pulse waveform V(t) is imposed as a normal outflow velocity. When the influence of flow rate on pressure drop is examined, a scalar is multiplied to the velocity function of the outlet boundary condition to yield different values of peak flow rate Q. When the effect of blood viscosity is studied, the peak flow rate Q is fixed as 220 ml/min while the infinity shear rate viscosity, η_∞, in the viscosity-shear rate relation [15] of the Carreau model is multiplied by a scalar m with $m = 0.7$, 1.0, 1.3, and 1.6, respectively. A non-slip boundary condition is imposed on the artery wall. The initial conditions for velocity components and pressure are obtained by solving the system of steady-state Navier-Stokes equations with the corresponding boundary conditions.

The geometry of a patient specific right coronary artery segment with two stenoses (see Fig. 1(a)) is created by a program written in MATLAB. It is then imported into a commercially available software package COMSOL5.2 as the computational domain. The momentum and continuity governing equations are numerically approximated using a finite element method over unstructured tetrahedral elements. The discretized system of equations is solved using the restarted generalized minimum residual (GMRES) iterative method with the number of iterations before restart specified as 50. The simulations are performed on HP Z640 Microsoft Windows Workstation.

The mesh is controlled by the following element size parameters: the maximum element size, the minimum element size, the maximum element growth rate, the curvature factor, and the resolution of narrow regions. To confirm that the numerical solutions are independent of the spatial mesh, the computations are repeated over the refined meshes with three different sizes: mesh1 with 1,780,784 elements, mesh2 with 2,355,505 elements and mesh3 with 4,696,734 elements. The maximum relative errors of the pressure for the flow with 220 ml/min peak flow rate over mesh2 and mesh3 compared to mesh1 are 0.57% and 1.16%, respectively. To ensure a truly periodic flow, computations are performed over four consecutive cardiac cycles. The maximum relative error of the pressure between the third and the forth cycles is 0.53%. We have also validated the numerical solutions by comparing the simulated blood velocity and pressure with the acquired on-site blood pressure and the flow velocity data from the patient in a similar way as in [17].

The pressure drop (*PD*) along the artery length is defined as $p - p_{in}$, where p_{in} is the blood pressure at the inlet of the artery segment (see Fig. 2). The pressure drop coefficient (*CDP*) is defined as

$$CDP = \frac{\widetilde{p} - \widetilde{p}_{in}}{0.5\rho \widetilde{\overline{U}}_{in}^2}, \tag{1}$$

where ρ is the density of blood; $\widetilde{p} - \widetilde{p}_{in}$ is the temporal mean of pressure drop from the inlet along the artery axial length; $\widetilde{\overline{U}}_{in}$ is the spatial and temporal mean blood velocity at the inlet of the artery. We also define a local index of pressure drop coefficient (CDP_i) for each stenosis in a similar way as in [19, 20].

$$CDP_i = \frac{-\Delta\widetilde{p}}{0.5\rho \widetilde{\overline{U}}_e^2}, \quad i = 1,2 \tag{2}$$

where $\Delta\widetilde{p} = \widetilde{p}_r - \widetilde{p}_e$, $\widetilde{p}_e$ and $\widetilde{p}_r$ represent the temporal mean pressure proximal and distal to each stenosis. The spatial wall pressure gradient (*WPG*) is defined as

$$WPG = \sqrt{\left(\frac{\partial p}{\partial x}\right)^2 + \left(\frac{\partial p}{\partial y}\right)^2 + \left(\frac{\partial p}{\partial z}\right)^2}, \tag{3}$$

where p is the wall static pressure. As stated by Giannoglou *et al*., the local magnitude of the spatial *WPG* is proposed to be an important initiating factor for atherosclerosis. The components of the *WPG* possibly have different effects on endothelial cells [6].

3 Results

3.1 Pressure Gradient-Flow Rate Relationship

To examine the effect of flow rate on the pressure drop, pressure drop coefficient and wall pressure gradient, computations are carried out by varying the peak flow rate Q (t = 0.425) from 100 ml/min to 280 ml/min. The infinity shear rate viscosity, η_∞, is fixed as 0.00345 $Pa{\cdot}s$. Figure 3(L) presents the plots of the slice-averaged CDP, PD_{mean} and WPG_{mean} in (L-a), (L-b), and (L-c), respectively, for Q = 100 ml/min, 160/ml/min, 220/ml/min, and 280 ml/min. The Reynolds numbers corresponding to these four different flow rates are 250, 398, 547 and 695. The Reynold number is defined as $Re = \widetilde{U}D\rho/\eta_{\infty}$, where $\widetilde{U}$ is the temporal mean velocity at the center of the inlet cross-section; D is the diameter of the inlet. Here PD_{mean} and WPG_{mean} are the temporal mean of PD and WPG in a cardiac cycle, respectively. A total of 128 points are picked approximately evenly spaced on the lumen boundary of each *IVUS* slice. The CDP, PD_{mean} and WPG_{mean} are averaged on the picked points of each slice, respectively. The horizontal axis is the axial length of the artery segment with $z = 0$ as the location of the first *IVUS* slice at the inlet. The neck of the proximal stenosis is at z = 0.725 cm. The distal stenosis has a long neck and it is located between z = 3.58 cm and z = 4.44 cm.

Figure 3(L-b) shows that the coronary model with a higher flow rate results in a greater pressure drop between the inlet and outlet, while the coronary model with a

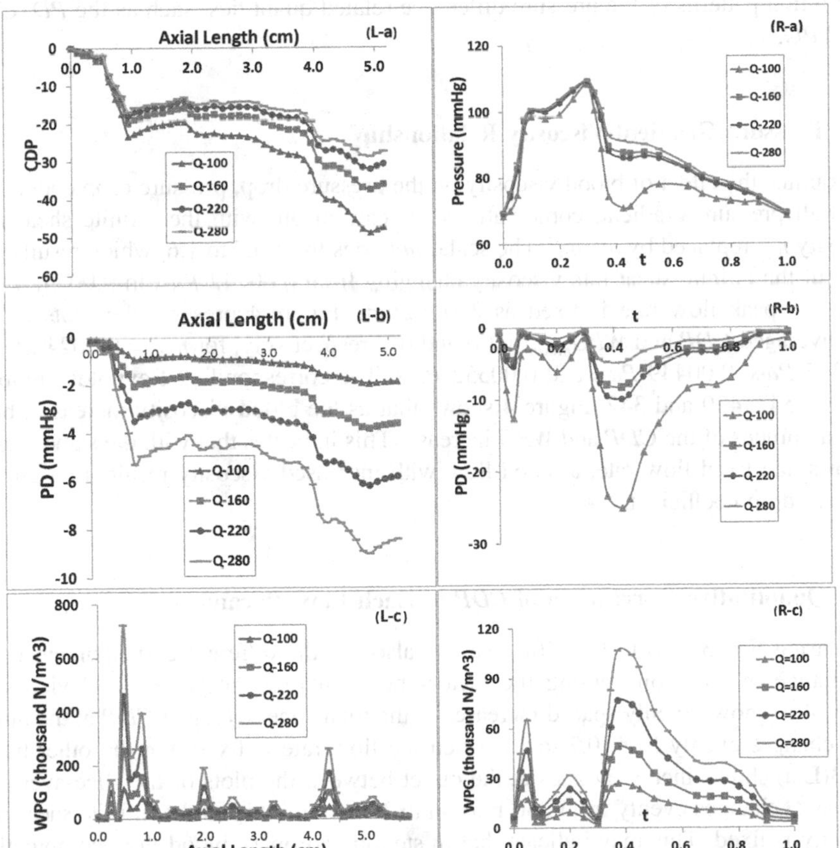

Fig. 3. Effect of flow rate. (L-a) Slice-averaged *CDP*, (L-b) Slice-averaged PD_{mean}, (L-c) Slice-averaged WPG_{mean}; (R-a) Pulse waveform of pressure (R-b) Pulse waveform of *PD* (R-c) Pulse waveform of *WPG* at the neck of the distal stenosis

lower flow rate yields a greater slice-averaged *CDP* (Fig. 3(L-a)). The plots in Fig. 3 (L-c) demonstrate that a higher flow rate leads to a greater *WPG*. The *WPG* reaches a peak value at the neck of each stenosis. Figures 3(R-a), (R-b) and (R-c) show the phasic waveforms of the pressure, *PD* and *WPG* at a point near the inner bend of the distal stenosis neck, respectively. From these plots we can see that *CDP* and *WPG* change markedly during a cardiac cycle. The pressure loss across a stenosis is more significant during diastole. The instantaneous maximum pressure drop occurs at the peak flow ($t = 0.425$) in a cardiac cycle. It is also interesting to compare the pulse waveform shapes of the plots in Fig. 3(R) with those of the velocity and pressure boundary conditions imposed at the inlet/outlet boundaries. We can observe that the shapes of the pressure phasic profiles in Fig. 3(R-a) are similar to that of the pressure plot in Fig. 2, while the shapes of the phasic profiles of the *PD* and *WPG* are similar to that of the velocity plot in Fig. 2. The waveform of the pressure drop coefficient is also similar to that of the velocity (plots are not included here). This indicates that the shape of the velocity pulse waveform of imposed boundary condition dominates the temporal distribution patterns of the pressure difference related quantities, such as the *PD*, *CDP* and *WPG*.

3.2 Pressure Gradient-Viscosity Relationship

To examine the effect of blood viscosity on the pressure drop, pressure drop coefficient and wall pressure gradient, computations are carried out with the infinite shear rate viscosity η_∞ replaced by $m \cdot \eta_\infty$. The scalar m varies from 0.7 to 1.6, which results in a value of the infinite shear rate viscosity changing from 0.00242 $Pa{\cdot}s$ to 0.00552 $Pa{\cdot}s$, while the peak flow rate is fixed as 220 ml/min. Figure 4 presents the plots of the slice-averaged *CDP* and WPG_{mean} in (a) and (b), respectively, for $\eta_\infty = 0.00242 Pa \cdot s$, 0.00345 $Pa{\cdot}s$, 0.00449 $Pa{\cdot}s$ and 0.00552 $Pa{\cdot}s$. The corresponding Reynolds numbers are 780, 547, 420 and 342. Figure 4 shows that as the blood viscosity increases, both the magnitudes of the *CDP* and *WPG* increase. This indicates that with the same size of stenosis and fixed flow rate, a blood flow with increased viscosity results in a greater pressure drop coefficient.

3.3 Quantitative Correlation of *CDP* to Each Flow Parameter

Comparing Fig. 3(L-a) to Fig. 4(a), we can also see the different correspondences of the change of *CDP* on varying the factors between flow rate and blood viscosity. Figure 4(a) shows evenly spaced increases in the total slice-averaged *CDP* at the outlet as m changes evenly from 0.7 to 1.6 when the flow rate is fixed. On the other hand, Fig. 3(L-a) shows increased gaps at the outlet between the plots of the slice-averaged *CDP* as Q changes evenly from 280 ml/min to 100 ml/min when the infinite shear rate viscosity is fixed. This may indicate that in stenotic coronary blood flow computation the flow rate has a stronger influence on the magnitude of *CDP* than the blood viscosity does.

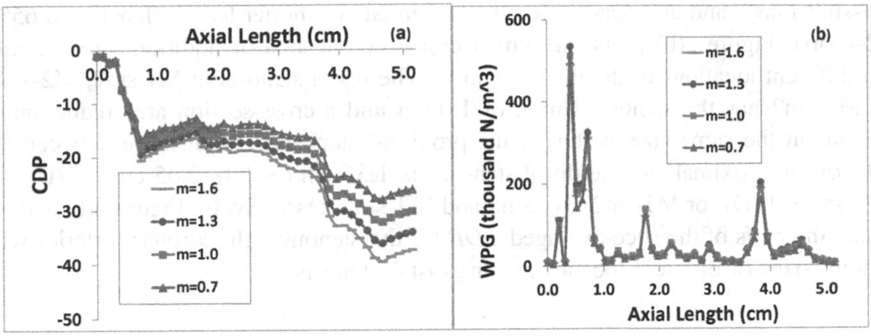

Fig. 4. Effect of blood viscosity. (a) Slice-averaged *CDP*, (b) Slice-averaged WPG_{mean}

Figure 5 further demonstrates the above mensioned diffirence on the correspondences of *CDP* to the changes of flow rate and blood viscosity. Figure 5 presents the plots of the CDP_i for each stenosis *vs* (a) the peak flow rate and (b) blood viscosity. From Fig. 5(a) we can see that for each stenosis CDP_i is approximately a decreasing and concave up parabolic function of the peak flow rate. The fitting functions quantitatively show that a reduction of flow rate at a lower value causes a greater increase of CDP_i for each stenosis. Figure 5(b) shows that the CDP_i for each stenosis is an increasing linear function of blood viscosity.

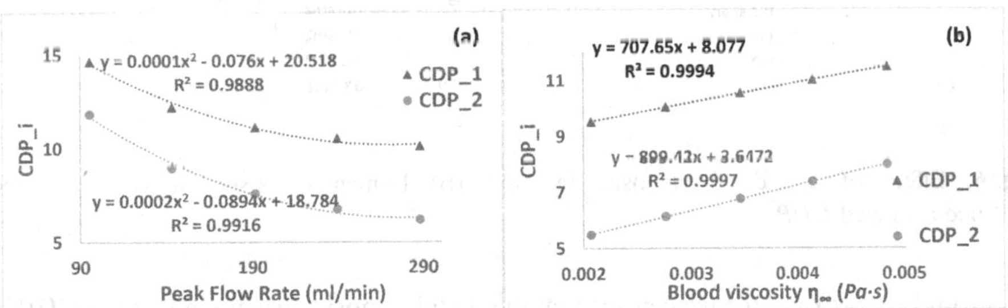

Fig. 5. Dependence of CDP_i for each stenosis (a) on flow rate, (b) on blood viscosity

3.4 Effect of the Distal Stenosis

To investigate the influence of the distal stenosis on the pressure drop coefficient in this patient specific right coronary artery with multiple stenoses, more blood simulations are performed in six additional artery geometries obtained by varying the size and location of the distal stenosis. Figure 6(a) plots the lumen cross-section area of four stenotic arteries: the original patient specific artery (labeled as 47%-stn2-patient), an artery with no reduction in lumen cross-section area for the distal stenosis (labeled as 0%-stn2), an artery with a reduction of 65% approximately in lumen cross-section area for the distal stenosis and the same length as that of the original patient's distal stenosis (labeled as

65%-stn2-long), and an artery with a distal stenosis in shorter length (labeled as 65%-stn2-short). Figure 6(b) plots the lumen cross-section area of addition three arteries with different locations of the distal stenosis. The distal stenosis in M1-stn2, M2-stn2, and M3-stn2 has the stenosis length of 1.0 cm and a cross-section area reduction of 65%, about the same size as that of the proximal stenosis. The distance between the necks of the proximal and the distal stenoses is 1.36 cm (~4D), 2.05 cm (~7D) and 3.47 cm (~11D) for M1-stn2, M2-stn2 and M3-stn2, respectively. Figure 6(c) and (d) present the plots of the slice-averaged *CDP* for the stenotic right coronary arteries with different sizes or different locations of the distal stenosis.

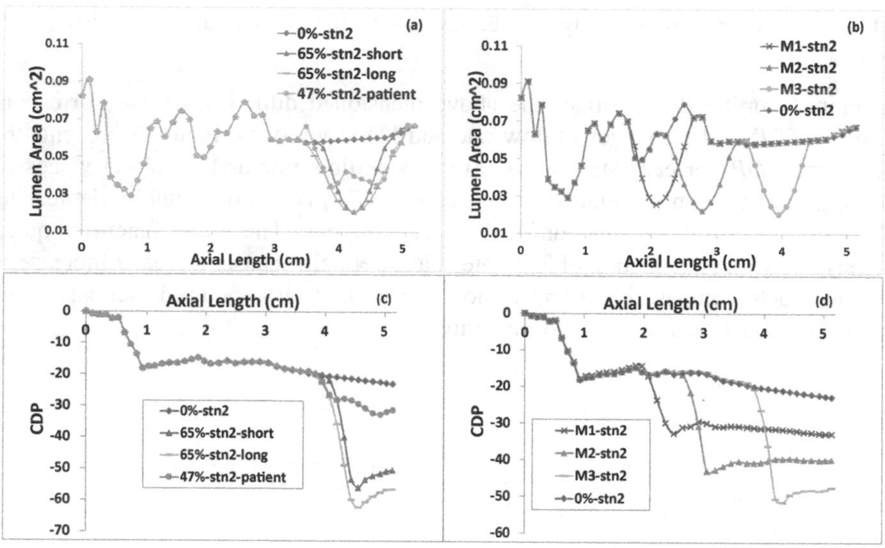

Fig. 6. Effect of the distal stenosis. (a) and (b) Lumen cross-section area, (c) and (d) Slice-averaged *CDP*

From Fig. 6(c) and (d) we can see that the distal stenosis does not affect the *CDP* in the upstream part of the artery. In this part the maximum relative errors of the slice averaged *CDP* for the three artery models plotted in Fig. 6(c) are less than 3% compared to that resulted from the real patient artery model. When the size of the area reduction of the distal stenosis changes from mild (47% of real patient) to moderate (65%), the *CDP* changes significantly in the downstream of the distal stenosis. In each of the two arteries with 65%-label in Fig. 6(c), the drop of the slice-averaged *CDP* across the distal stenosis is much larger than that across the proximal stenosis even though the percentages of the area reduction for both proximal and distal stenoses are approximately the same. This is also clearly shown quantitatively in Table 1, which lists the change of the slice-averaged *CDP* across each stenosis and the total change of the *CDP* from the inlet to the outlet for the seven arteries. Figure 6(c) and Table 1 also show that the change of the *CDP* across the distal stenosis varies notably from 31.25 to 37.71 as the length of the stenosis increases from 1.35 cm (65%-stn2-short) to 1.58 cm (65%-stn2-long). Figure 6(d) and the last three columns of Table 1 demonstrate clearly how the *CDP* across the distal stenosis is significantly affected by the location of the distal stenosis. The further downstream the distal stenosis locates, the larger the drop of the *CDP*.

Table 1. *CDP* change across the proximal stenosis, distal stenosis and total *CDP* change from the inlet to the outlet

	0%-stn2	47%-stn2-real patient	65%-stn2-short	65%-stn2-long	M1-stn2	M2-stn2	M3-stn2
1st stn	12.73	12.75	12.59	12.59	12.25	12.77	12.45
2nd stn	3.28	12.35	31.25	37.71	15.75	24.72	29.08
Total	22.94	31.31	50.42	56.49	32.95	39.97	47.64

4 Discussion

4.1 Effect of Flow Rate

In literature, many *in-vivo* and *in-vitro* researches were conducted to examine the relationship between pressure drop and flow velocity/rate [13, 20–30]. Various linear or quadratic fitting equations were obtained. Young *et al.* [21] and Gould *et al.* [22] demonstrated that the relation between pressure drop and flow velocity in stenosis can be described by $\Delta P = 0 + kv + Sv^2$. Here k $(mmHg \cdot s/cm)$ is the coefficient of pressure loss due to viscous friction. S $(mmHg \cdot s^2/cm^2)$ is the coefficient of pressure loss due to flow separation or localized turbulence. Takeda *et al.* investigated the effect of flow on pressure drop in 905 patients with all grades of aortic stenosis. They have shown that the mean pressure difference is linearly related to flow rate. They suggested that the slope of mean pressure difference against flow rate might be a better method of grading aortic stenosis since it does not rely on hemodynamic assumptions [25, 26]. Mario *et al.* assessed the feasibility of correlating stenosis severity with the slope of the instantaneous trans-stenotic pressure drop/velocity relationship [23, 24]. Marques *et al.* assessed the feasibility and reproducibility of the instantaneous diastolic flow velocity-pressure drop relationship. They found significant differences in this relationship among normal coronary arteries and arteries with intermediate and severe coronary stenosis. A steeper slope of this relationship corresponds to a severer stenosis [27]. However, literature is still lacking in a comparative simulation analysis to show how the key hydrodynamic parameters, the *CDP* and *WPG*, would differ for various flow rate.

In the present study we observe that the correspondence of *CDP* on flow rate is opposite to that of the pressure drop on flow rate. As flow rate decreases the magnitude of the slice-averaged *CDP* increases, that is, the coronary model with a lower flow rate yields a greater magnitude of *CDP*. Roy *et al.* [31] examined the effect of the guidewire on the pressure drop-flow relationship in moderate coronary artery stenosis using a 2D axisymmetric model. Their computation results show a decreasing trend of the values of the pressure drop coefficient as the flow rate increases from 50 ml/min to 170 ml/min. Mate *et al.* [32] used a large-scale instrumented model to study flows through isolated stenotic elements in large coronary arteries. The plots of Fig. 5 in [32] show that the flow with a greater Reynolds number results in a greater pressure drop coefficient. Our analysis from Fig. 3(L) qualitatively confirms that the observations made by Roy *et al.* [31] and by Mate *et al.* [32] in models of straight tube can also be observed in the blood flow model of a patient specific stenotic right coronary artery with curvature. We also quantitatively demonstrate the correlation of *CDP* index with flow rate. Since the elevated values of *CDP* indicate abnormal epicardial stenosis [13], our observation suggests that having the same size of lumen cross-section area reduction a patient with a lower coronary flow rate has a more hemodynamically significant coronary stenosis and thus may face a higher risk.

4.2 Effect of Blood Viscosity

It is known that blood viscosity is a determinant of capillary resistance to blood flow at both stenotic and arteriolar levels [33]. Increased blood viscosity plays an important role in the formation of atherosclerosis during inflammatory processes [34]. Rim *et al.* [33] conducted experiments using nine dogs and found that the increase in blood viscosity causes capillary resistance to rise, which decreases hyperemic coronary blood flow. Using statistical analysis, Indrianto *et al.* investigated the relationship of blood viscosity with the degree of stenosis among coronary heart disease patients. They concluded that blood viscosity increases the degree of coronary stenosis in coronary heart disease patients [34]. However, there was no publication that reported a quantitative analysis on the correspondence of *CDP* and *WPG* to blood viscosity in patient specific right coronary arteries with two stenoses. The computer simulation results from this study show that the slice-averaged *CDP* increases linearly as the blood viscosity increases when the flow rate is fixed. Furthermore, the increase in blood viscosity causes capillary resistance to rise and slows the blood flow [34] and thus increases the contact time between the atherogenic particles of the blood and the endothelium [35]. As discussed previously in Sect. 4.1, the blood flow with a lower flow rate yields a greater magnitude of CDP. Under these two folded effects a patient with higher blood viscosity may have a more functionally significant stenosis.

4.3 Influence of the Distal Stenosis

In arteries with serial stenoses, the blood flow pattern is complex. The effect of serial stenoses on coronary hemodynamics has drawn much attention. Many researchers have conducted the experimental and clinical investigations [30, 36] and the numerical simulations using 2D models of axisymmetric straight tube [29, 37, 38]. Talukder *et al.* did an experimental study of fluid dynamics in arteries with multiple mild stenoses. They showed that the total pressure drop across a series of stenoses increases linearly as the number of stenoses increases and the combined effect of multiple stenoses of mild severity can be critical [36]. D'Souza *et al.* assessed the effect of serial stenoses on the coronary diagnostic parameters using in-vitro experiments. They found that the pressure drop coefficient of the upstream stenosis varies insignificantly in the presence of a downstream stenosis with a varying degree of severity [30]. Lee *et al.* performed numerical simulations of turbulent flow through serial stenoses. They demonstrated that the maximum centerline velocity and disturbance intensity at the second stenosis are higher than those at the first stenosis when both stenoses have a 50% area reduction [37]. Bernad *et al.* carried out a computer simulation of blood flow in a patient specific right coronary artery with three stenoses. The middle stenosis has a 50% area reduction. Both proximal and distal stenoses have a reduction of 80% approximately in cross-section area. The numerical results listed in Table 3 of [29] show that the pressure difference across the distal stenosis is larger than that across the proximal stenosis.

Regarding the influence of the distal stenosis on the patient specific right coronary flow in this study, we have the following observations: (1) Table 1 shows insignificant differences in the change of the *CDP* across the proximal stenosis for all six models with different size, location and length of the distal stenosis. The distal stenosis only affects the distributions of the *CDP* in the region starting from the distal stenosis to the outlet, while the blood flow is not significantly affected in the upstream part of the artery; (2) The interstenotic distance between two serial stenoses plays an important role on the blood flow distribution in the downstream part of the stenotic artery. The last three columns in Table 1 show that the total change of the slice averaged *CDP* from the inlet to the outlet increases from 32.95 to 47.64 as the interstenotic distance between two stenoses increases from 4D to 11D. This indicates that a distal stenosis located in a further downstream position causes a larger drop in the *CDP*. (3) Comparing the 4th and 5th columns of Table 1, we can see that as the length of a stenosis increases, the change of *CDP* across the stenosis increases. (4) The distal stenosis has more influence on the *CDP* than the proximal stenosis does. The values listed in the last five columns of Table 1 show that the change of *CDP* across the distal stenosis is much larger than that across the proximal stenosis, even though both proximal and distal stenoses have a similar area reduction size (66% *vs* 65%) for each model. The results listed in Column 3 (real-patient artery) of Table 1 show that the changes of *CDP* across two stenoses are very close (12.75 *vs.* 12.35) while the area reductions of the proximal and distal stenoses are 66% and 47%, respectively. This is partially due to a stronger influence of the distal stenosis and partially due to a longer length of the distal stenosis.

Our observations in this study not only qualitatively confirm that some influences of the distal stenosis on the pressure drop coefficient in the axisymmetric straight tubes reported in literature are also valid in the patient specific right coronary arteries with curvature, but also show that the hemodynamics of blood flow in the right coronary arteries with multiple stenoses is more complex due to the additional influence from the curvature of bend. To our best knowledge this paper reports, for the first time, the effect of the distal stenosis on the *CDP* based on the blood simulations in the patient specific right coronary arteries by varying the size and location of stenosis. A further understanding of the insight of the interacting behavior of multiple stenoses will be important and useful to the clinical diagnosis and the decision for treatment of patients with multiple coronary stenoses.

5 Conclusions

The influences of hemodynamic flow parameters on the pressure drop coefficient and wall pressure gradient in a patient specific right coronary artery with two stenoses are investigated. This study shows that both *CDP* and *WPG* are strongly dependent on flow rate and blood viscosity. The CDP_i across each stenosis increases linearly as blood viscosity increases. As reported previously by many researchers, the pressure drop increases as flow rate increases. However, the pressure drop coefficient correlates negatively to flow rate. The CDP_i across each stenosis decreases quadratically as flow rate increases. The wall pressure gradient increases as the flow rate increases. The pressure drop coefficient and wall pressure gradient all proportionally increase as the blood viscosity increases. Based on our observations regarding the effects of the blood viscosity and flow rate on the pressure drop coefficient and wall pressure gradient, it maybe concluded that having the same size of a right coronary stenosis a patient with an increased blood viscosity and a lower flow rate has a more hemodynamically significant stenosis and is potentially at a higher risk. In a right coronary artery with two stenoses, the distal stenosis has an insignificant influence on the *CDP* and *WPG* across the proximal stenosis. When both proximal and distal stenoses are moderate, the distal stenosis results in a larger pressure drop and more disturbed blood flow than the proximal stenosis does. A distal stenosis located in a further downstream position causes a larger drop in the *CDP*.

Acknowledgement. This work is partially supported by a grant from the Simons Foundation (#210082 to Biyue Liu), a sabbatical grant from Monmouth University, NIH grant NIH/NIBIB R01 EB004759, and a Jiangsu Province Science and Technology Agency grant BE2016785.

References

1. Friedman, M.H., Hutchins, G.M., Bargeron, C.B., Deters, O.J., Mark, F.F.: Correlation between intimal thickness and fluid shear in human arteries. Atherosclerosis **39**, 425–436 (1981)
2. Glagov, S., Zarins, C.K., Giddens, D.P., Ku, D.N.: Mechanical factors in the pathogenesis, localization and evolution of atherosclerotic plaques. In: Camilleri, J.-P., Berry, C.L., Fiessinger, J.-N., Bariéty, J. (eds.) Diseases of the Arterial wall. Springer, Heideberg (1989). doi:10.1007/978-1-4471-1464-2_15
3. Salzar, R.S., Thubrikart, M.J., Eppink, R.T.: Pressure-induced mechanical stress in the carotid artery bifurcation: a possible correlation to atherosclerosis. J. Biomech. **28**, 1333–1340 (1995)
4. Samady, H., Eshtehardi, P., McDaniel, M.C., Suo, J., Dhawan, S.S., Maynard, C., Timmins, L.H., Quyyumi, A.A., Giddens, D.P.: Coronary artery wall shear stress is associated with progression and transformation of atherosclerotic plaque and arterial remodeling in patients with coronary artery disease. Circulation **124**, 779–788 (2011)
5. Liu, B., Zheng, J., Bach, R., Tang, D.: Correlations of coronary plaque wall thickness with wall pressure and wall pressure gradient: a representative case study. BioMed. Eng. OnLine **11**(43), 1–12 (2012)

6. Giannoglou, G.D., Soulis, J.V., Farmakis, T.M., Giannakoulas, G.A., Parcharidis, G.E., Louridas, G.E.: Wall pressure gradient in normal left coronary artery tree. Med. Eng. Phys. **27**, 455–464 (2005)
7. Young, D.F., Cholvin, N.R., Kirkeeide, R.L., Roth, A.C.: Hemodynamics of arterial stenosis at elevated flow rates. Circ. Res. **411**, 99–107 (1977)
8. Pijls, N.H.J., van Son, J.A., Kirkeeide, R.L., de Bruyne, B., Gould, K.L.: Experimental basis of determining maximum coronary, myocardial, and collateral blood flow by pressure measurements for assessing functional stenosis severity before and after percutaneous transluminal coronary angioplasty. Circulation **87**, 1354–1367 (1993)
9. Pijls, N.H.J., De Bruyne, B., Bech, G.J.W., Liistro, F., Heyndrickx, G.R., Bonnier, H.J.R.M., Koolen, J.J.: Coronary pressure measurement to assess the hemodynamic significance of serial stenoses within one coronary artery: validation in humans. Circulation **102**, 2371–2377 (2000)
10. Park, S.J., Ahn, J.M., Pijls, N.H.J., et al.: Validation of functional state of coronary tandem lesions using computational flow dynamics. Am. J. Cardiol. **110**, 1578–1584 (2012)
11. van de Hoef, T.P., Nolte, F., Rolandi, M.C., Piek, J.J., van den Wijngaard, J.P.H.M., Spaan, J.A.E., Siebes, M.: Coronary pressure-flow relations as basis for the understanding of coronary physiology. J. Mol. Cell. Cardiol. **52**, 786–793 (2012)
12. Brosh, D., Higano, S.T., Slepian, M.J., Miller, H.I., et al.: Pulse transmission coefficient: a novel nonhyperemic parameter for assessing the physiological significance of coronary artery stenoses. J. Am. Coll. Cardiol. **39**, 1012–1019 (2002)
13. Banerjee, R.K., Ashtekar, K.D., Effat, M.A., Helmy, T.A., Kim, E., Schneeberger, E.W., et al.: Concurrent assessment of epicardial coronary artery stenosis and microvascular dysfunction using diagnostic endpoints derived from fundamental fluid dynamics principles. J. Invasive Cardiol. **21**, 511–517 (2009)
14. Perktold, K., Peter, R.O., Resch, M.: Pulsatile non-Newtonian blood flow simulation through a bifurcation with an aneurysm. Biorheology **26**, 1011–1030 (1989)
15. Cho, Y.I., Kensey, K.R.: Effects of the non-Newtonian viscosity of blood on flows in a diseased arterial vessel. Part 1: steady flows. Biorheology **28**, 241–262 (1991)
16. Shibeshi, S.S., Collins, W.E.: The rheology of blood flow in a branched arterial system. Appl. Rheol. **15**, 398–405 (2005)
17. Liu, B., Zheng, J., Bach, R., Tang, D.: Influence of model boundary conditions on blood flow patterns in a patient specific stenotic right coronary artery. BioMed. Eng. Online, **14**(1), S6, 1–17 (2015)
18. Tang, D., Yang, C., Zheng, J., Bach, R., Wang, L., Muccigrosso, D., Billiar, K., Zhu, J., Ma, G., Maehara, A., Mintz, G.S., Fan, R.: Human coronary plaque wall thickness correlated positively with flow shear stress and negatively with plaque wall stress: an IVUS-based fluid-structure interaction multi-patient study. BioMed. Eng. OnLine **13**, 32 (2014)
19. Banerjee, R.K., Back, L.H., Back, M.R., Cho, Y.I.: Physiological flow simulation in residual human stenoses after coronary angioplasty. J. Biomech. Eng. **122**(4), 310–320 (2000)
20. Banerjee, R.K., Ashtekar, K.D., Helmy, T.A., Effat, M.A., Back, L.H., Khoury, S.F.: Hemodynamic diagnostics of epicardial coronary stenoses: in-vitro experimental and computational study. Biomed. Eng. Online **7**, 24 (2008)
21. Young, D.F., Cholvin, N.R., Roth, A.C.: Pressure drop across artificially induced stenoses in the femoral arteries of dogs. Circ. Res. **36**, 735–743 (1975)
22. Gould, K.L.: Pressure-flow characteristics of coronary stenoses in unsedated dogs at rest and during coronary vasodilation. Circ. Res. **43**, 242–253 (1978)
23. Serruys, P.W., di Mario, C., Meneveau, N., et al.: Intracoronary pressure and flow velocity with sensor-tip guidewires: a new methodologic approach for assessment of coronary hemodynamics before and after coronary interventions. Am. J. Cardiol. **71**, 41D–53D (1993)

24. di Mario, C., Gil, R., de Feyter, P.J., Schuurbiers, J.C., Serruys, P.W.: Utilization of translesional hemodynamics: comparison of pressure and flow methods in stenosis assessment in patients with coronary artery disease. Cathet. Cardiovasc. Diagn. **38**, 189–201 (1996)
25. Takeda, S., Rimington, H., Chambers, J.: The relation between transaortic pressure difference and flow during dobutamine stress echocardiography in patients with aortic stenosis. Heart **82**, 11–14 (1999)
26. Takeda, S., Rimington, H., Chambers, J.: Prediction of symptom-onset in aortic stenosis: a comparison of pressure drop/flow slope and haemodynamic measures at rest. Int. J. Cardiol. **81**, 131–137 (2001)
27. Marques, K.M., Spruijt, H.J., Boer, C., Westerhof, N., Visser, C.A., Visser, F.C.: The diastolic flow-pressure gradient relation in coronary stenoses in humans. J. Am. Coll. Cardiol. **39**, 1630–1636 (2002)
28. Bernad, E.S., Bernad, S.I., Craina, M.L.: Hemodynamic parameters measurements to assess severity of serial lesions in patient specific right coronary artery. Bio-Med. Mater. Eng. **24**(1), 323–334 (2014)
29. Bernad, S.I., Bernad, E.S., Totorean, A.F., Craina, M.L., Sargan, I.: Clinical important hemodynamic characteristics for serial stenosed coronary artery. Int. J. Des. Nat. Ecodyn. **10**, 97–113 (2015)
30. D'Souza, G.A., Peelukhana, S.V., Banerjee, R.K.: Diagnostic uncertainties during assessment of serial coronary stenoses: an in vitro study. J. Biomech. Eng. **136**, 021026 (2014)
31. Roy, A.S., Back, L.H., Banerjee, R.K.: Guidewire flow obstruction effect on pressure drop-flow relationship in moderate coronary artery stenosis. J. Biomech. **39**, 853–864 (2006)
32. Mates, R.E., Gupta, R.L., Bell, A.C., Klocke, F.J.: Fluid dynamics of coronary artery stenosis. Circ. Res. **42**, 152–162 (1978)
33. Rim, S.J., Leong-Poi, H., Lindner, J.R., Wei, K., Fisher, N.G., Kaul, S.: Decrease in coronary blood flow reserve during hyperlipidemia is secondary to an increase in blood viscosity. Circulation **104**, 2704–2709 (2001)
34. Indrianto, A.F., Samsuria, I.K., Kurniawan, K.D.: Blood viscosity increases the degree of coronary stenosis in coronary heart disease. Universa Medicina **34**, 168–176 (2015)
35. Giannoglou, G.D., Soulis, J.V., Farmakis, T.M., Farmakis, D.M., Louridas, G.E.: Haemodynamic factors and the important role of local low static pressure in coronary wall thickening. Int. J. Cardiol. **86**, 27–40 (2002)
36. Talukder, N., Karayannacos, P.E., Nerem, R.M., Vasko, J.S.: An experimental study of the fluid dynamics of multiple noncritical stenoses. J. Biomech. Eng. **99**, 74–82 (1977)
37. Lee, T.S., Liao, W., Low, H.T.: Numerical simulation of turbulent _ow through series stenoses. Int. J. Numer. Meth. Fluids **42**, 717–740 (2003)
38. Bertolotti, C., Qin, Z., Lamontagne, B., Durand, L.-G., Soulez, G., Cloutier, G.: Influence of multiple stenoses on echo-Doppler functional diagnosis of peripheral arterial disease: a numerical and experimental study. Ann. Biomed. Eng. **34**, 564–574 (2006)

Adaptive Sine Cosine Algorithm Integrated with Differential Evolution for Structural Damage Detection

Sujin Bureerat and Nantiwat Pholdee(✉)

Faculty of Engineering, Department of Mechanical Engineering,
Sustainable and Infrastructure Research and Development Center,
Khon Kaen University, Khon Kaen 40002, Thailand
nantiwat@kku.ac.th

Abstract. A sine cosine algorithm is one promising meta-heuristic recently proposed. In this work, the algorithm is extended to be self-adaptive and its main reproduction operators are integrated with the mutation operator of differential evolution. The new algorithm is called adaptive sine cosine algorithm integrated with differential evolution (ASCA-DE) and used to tackle the test problems for structural damage detection. The results reveal that the new algorithm outperforms a number of established meta-heuristics.

Keywords: Sine cosine algorithm · Differential evolution · Structural damage detection · Optimization · Meta-heuristics

1 Introduction

Structural health monitoring is an essential topic in the structural engineering field due to the requirement of reliability and safety use of the structure thought its life time. The key point of structural health monitoring is to identify the presence of structural damage, localising the damage and predicting its severity without destroying the structure or using nondestructive testing [1–3]. One of the most popular nondestructive techniques to identify damage location is to use static and/or dynamic testing data such as changing on strain data, structural deflection [4, 5], or modal data such as natural frequencies and mode shapes [1, 4–8].

Numerous work on damage detection based on changing of modal data has been reported worldwide [1, 4–16]. The main idea of this technique is to measure the modal data of an undamaged structure and use it as the baseline. When the modal data has changed, structural damage is supposed to occur. Identifying damage location can be achieved by applying a soft computing technique such as fuzzy logic systems [17, 18] and neural networks systems [17, 19–21]. For the latter, a large number of modal data of the damage and undamaged structures must be provided and used for training the system. Although much work has claimed that the technique works well, the requirement of a large number of training data and analysis time is an inevitable obstacle.

O. Gervasi et al. (Eds.): ICCSA 2017, Part I, LNCS 10404, pp. 71–86, 2017.
DOI: 10.1007/978-3-319-62392-4_6

Recently, the damage detection is traded as an optimization inverse problem. The main idea is to update mechanical properties of a mathematical model such as a finite element model until the modal data such as natural frequency of the model agree well with the testing data while the optimum parameters of the mechanical properties can be obtained by means of optimization [7, 9–14]. Over the last few decades, much research has successfully applied meta-heuristics (MHs) for this kind of problem, for example, genetic algorithm (GA) [9], differential evolution algorithm (DE) [16], ant colony optimization (ACO) [10], charged system search (ChSS) [11], etc. [7, 9–14, 16]. Although successful results have been reported, it is found that most of them choose to present only the best runs of the algorithms. This is not a proper way to investigate the use of MHs for real world applications. The methods must be run to solve a particular design problem many times and use the mean values of objective function as a performance indicator. Over a decade, over a thousand of MHs have been introduced based on a wide variety of search concepts and mechanisms. The algorithms can be regarded as a search method or an optimizer. The methods are simple to understand/use/code due to it being a soft computing technique. They can be used as an alternative to classical gradient-based optimization methods particularly for optimization problems in which sufficiently accurate function derivatives are not affordable. Moreover, they can be used for multiobjective or many-objective optimization more effectively than using the gradient-based optimizers since they can explore a Pareto optimal front within one optimization run. A genetic algorithm is arguable the best known and most used MH. Then, there can be differential evolution, particle swarm optimization and so on. Recently, there have been numerous MHs being published each year some of the new algorithms are a grey wolf optimizer [22], moth-fame optimization [23], the whale optimization algorithm [24] and a sine cosine algorithm [25]. Those algorithms are remained to be investigated when applying to practical design problem.

Therefore, this paper presents and extension of the sine cosine algorithm. An adaptive strategy is embedded into the new version while the mutation operator of differential evolution is integrated into the algorithm in order to further improve its performance. The new optimizer is then termed an adaptive sine cosine algorithm integrated with differential evolution (ASCA-DE). The optimizer is then implemented on several test problems for structural damage detection. Numerical results show that the proposed MH is superior to a number of established MHs found in the literature.

2 Formulation of a Damage Detection Optimization Problem

In this work, vibration based damage detection based on using natural frequencies is used for damage localization of truss structures. The main concept of using structural natural frequencies for damage detection of a truss structure is based on using a finite element model and the measured natural frequencies. When the natural frequencies and mode shapes are measured (usually the lowest n_{mode} natural frequencies), the finite element model is updated until the computed natural frequencies fit well with the measured ones. For the undamaged structure, natural frequencies can be calculated from a simple linear undamped free vibration finite element model which can be expressed as;

$$[\mathbf{K}]\{\phi_j\} - \omega_j^2[\mathbf{M}]\{\phi_j\} = 0 \tag{1}$$

where [**K**] is a structural stiffness matrix which can be expressed as the summation of element stiffness matrices [$\mathbf{k}_e$],

$$[\mathbf{K}] = \sum_{i=1}^{n_e} [\mathbf{k}_e] \tag{2}$$

where i is the ith element of the structure while n_e is the total number of elements. The matrix [**M**] is a structural mass matrix computed in similar fashion to the stiffness matrix. The variables ϕ_j and ω_j are the jth mode shape and its corresponding natural frequency, respectively. For the damaged structure, the stiffness matrix of the damaged element is assumed to be modified. The stiffness matrix of the damaged structure [$\mathbf{K}_d$] can be written as a percentage of damage in the elements as follows:

$$[\mathbf{K}_d] = \sum_{i=1}^{n_e} \frac{100 - p_i}{100} [\mathbf{k}_e] \tag{3}$$

where p_i is the percentage of damage on the ith element. The natural frequency of the damaged structure can be computed by solving Eq. (1) by replacing [**K**] with [$\mathbf{K}_d$].

The percentage of damage in the structural element (p_i) can be found by solving an optimization problem to minimise the root mean square error (RMSE) between natural frequencies measured from the damaged structure and natural frequencies computed by using the finite element model. The problem can be expressed as follow:

$$\text{Min: } f(\mathbf{x}) = \sqrt{\frac{\sum_{j=1}^{n_{mode}} \left(\omega_{j,damage} - \omega_{j,computed}\right)^2}{n_{mode}}} \tag{4}$$

where $\omega_{j,damage}$ and $\omega_{j,computed}$ are the structural natural frequency of mode j obtained from a damaged structure and that from solving (1) – (3). The design variables are those damage percentages of structural elements ($\mathbf{x} = \{p_1, \ldots, p_{nele}\}^T$) respectively. In this work, six vibration modes are used for calculation.

3 Test Problems with Trusses

Four truss damage detection optimization problems from two truss structures are used in this study. These are the test problems used in our previous studies [14]. Detail of the test problems are shown as follow:

3.1 Twenty-Five-Bar Truss

The structure is shown in Fig. 1 [10, 14]. The cross sections of all bar elements are set to be 6.4165 mm^2. Table 1 shown the material properties and simulated case study for this example. The data of natural frequencies of the undamaged and damaged 25-bar truss structures are shown in Table 2.

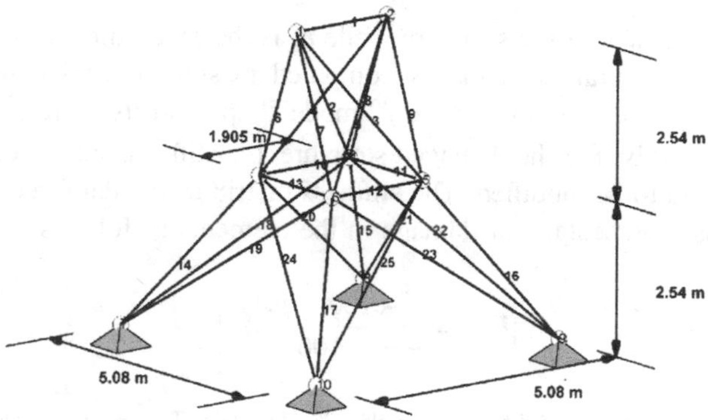

Fig. 1. Twenty-five bar truss

Table 1. Material properties and simulated case study for 25-bar truss.

Material density	7,850 kg/m^3
Modulus of elasticity	200 GPa
Simulated case study	Case I: 35% damage at element number 7
	Case II: 35% damage at element number 7 and 40% damage at element number 9

Table 2. Natural frequencies (Hz) of damaged and undamaged of 25 bar structure.

Mode	Undamaged	35% damage at element number 7	35% damage at element number 7 and 40% damage at element number 9
1	69.7818	69.1393	68.5203
2	72.8217	72.2006	71.3167
3	95.8756	95.3372	94.5625
4	120.1437	119.8852	119.6514
5	121.5017	121.4774	121.4253
6	125.0132	125.0130	125.0129

3.2 Seventy-Two-Bar Truss

The structure is shown in Fig. 2 [7, 14]. Four non-structural masses of 2270 kg are attached to the top nodes. The cross sections of all bar elements are set to be 0.0025 m^2. Table 3 shown the material properties and simulated case study for this example. The data of natural frequencies of the undamaged and damaged 72-bar truss structure are shown in Table 4.

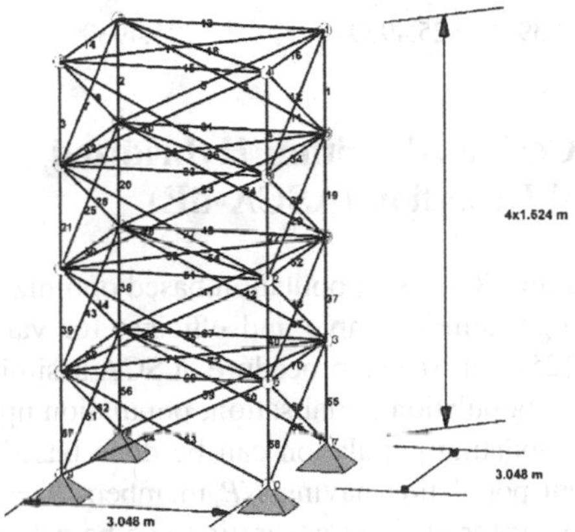

Fig. 2. Seventy-two bar truss

Table 3. Material properties and simulated case study for 72-bar truss.

Material density	2,770 kg/m^3
Modulus of elasticity	6.98×10^{10} Pa
Simulated case study	Case I: 15% damage at element number 55 (15% damage in element number 56, 57, or 58 results in the same set of natural frequencies)
	Case II: 10% damage at element number 4 and 15% damage at element number 58 (90, 180, and 270° rotation along the *z* axis lead to the same set of natural frequencies)

Table 4. Natural frequencies (Hz) of damaged and undamaged of 72 bar structure.

Mode	Undamaged	15% damage at element number 55	15% damage at element number 58 and 10% damage at element number 4
1	6.0455	5.9553	5.9530
2	6.0455	6.0455	6.0455
3	10.4764	10.4764	10.4764
4	18.2297	18.1448	18.0921
5	25.4939	25.4903	25.2437
6	25.4939	25.4939	25.4939

4 Adaptive Sine Cosine Algorithm Hybridized with Differential Evolution (ASCA-dE)

The Sine Cosine Algorithm (SCA) is a population based optimization method proposed by Mirjalili [25]. The algorithm is simple and efficient for various optimization test problems as reported in [25]. The search procedure of SCA is similar to other MH which contains three main steps; population initialisation, population updating and population selection. For the SCA, updating population can be done based on a sine and cosine function. Given a current population having *NP* members $\mathbf{X} = \{\mathbf{x}_1, \mathbf{x}_2, \ldots, \mathbf{x}_{NP}\}^T$, an element of a solution vector for the next generation can be calculated as follows:

$$x_{new,k} = \begin{cases} x_{old,k} + r_1 \sin(r_2)\left|r_3 x_{best,k} - x_{old,k}\right|, & if \quad r_4 < 0.5, \\ x_{old,k} + r_1 \cos(r_2)\left|r_3 x_{best,k} - x_{old,k}\right|, & otherwise \end{cases} \tag{5}$$

where $x_{\text{best,k}}$ is the kth matrix element of the current best solution. The variables r_2, r_3, and r_4 are random parameters in the ranges of [0, 2π], [0, 2] and [0, 1], respectively. The variable r_1 is an iterative adaption parameter,

$$r_1 = a - T\frac{a}{T_{\max}} \tag{6}$$

where a is a constant parameter while T is an iteration number. $T_{\max}$ is maximum number of iterations.

The search process of SCA start with generating an initial population at random, and then calculating their objective function values where the best solution is found. Then, the new population for the next generation is generated using Eq. (5) and the objective function values of its members are calculated. The current best will be compared with the best solution of the newly generated population and the better one is saved to the next generation. The process is repeated until a termination criterion is met. The computational steps of SCA are shown in Algorithm 1.

Algorithm 1 Sine Cosine Algorithm

Input: population size (N_p), number of generations (T_{max}), number of design variable (D)

Output: $\mathbf{x}_{best}$, f_{best}

Main algorithm

Initialise a population and set as the current population.
Find the best solution ($\mathbf{x}_{best}$)
For T=1 to T_{max}
    Calculate parameter r_1 using eq. (6)
    For l=1 to N_p
        For k = 1 to D
            Generate the parameter r_2, r_3 and r_4
            Update the kth element of the lth population ($\mathbf{x}_l$) using eq. (5)
        End For
    End For
    Calculate objective function values of the newly generated population and find the best ones ($\mathbf{x}_{best,new}$)
    Replace $\mathbf{x}_{best}$ by $\mathbf{x}_{best,new}$ if $f(\mathbf{x}_{best,new}) < f(\mathbf{x}_{best})$
End

For the proposed adaptive sine cosine algorithm with integration of DE mutation, the DE mutation operator as proposed in Pholdee and Bureerat [26] is integrated into the updating operation. The mutation equation is detailed as follow;

$$\mathbf{x}_{new} = \mathbf{x}_{best} + rand(-1, +1)F(\mathbf{x}_{r,1} + \mathbf{x}_{r,2} - \mathbf{x}_{r,3} - \mathbf{x}_{r,4}) \quad (7)$$

where $rand(-1, 1)$ gives either -1 or 1 with equal probability. F is a scaling factor while $\mathbf{x}_{r,1}$–$\mathbf{x}_{r,4}$ are four solutions randomly selected from the population.

At ASCA-DE updating operation, if a generated uniform random number in the interval [0,1] is lower than a probability value ($rand < P_{DE}$), the population will be updated using the SCA updating operation based on Eq. (5), otherwise, the population will be updated by DE mutation as detailed in Eq. (7).

The term of self-adaption of the proposed algorithm is accomplished in such way that the parameter r_2, r_3 and F are regenerated for each calculation based on the information from the previous iteration. For each calculation, the r_2 and r_3 are generated based on normally distributed random numbers with mean values, r_{2m} and r_{3m} respectively and standard deviation values, STD = 0.1 for both r_2 and r_3. The values of r_{2m} and r_{3m} are iteratively adapted based on the following equations:

$$r_{2m}(T+1) = 0.9r_{2m}(T) + 0.1mean(good_{r2m}), \quad (8)$$

and,

$$r_{3m}(T+1) = 0.9r_{3m}(T) + 0.1mean(good_{r3m}), \quad (9)$$

where $mean(good_{\mathrm{r2m}})$ and $mean(good_{\mathrm{r3m}})$ are the mean values of all values of r_2 and r_3 used in current iteration that lead to successful updates. The successful update means the created offspring is better than its parent from the previous iteration. In addition, for each calculation, the scaling factor F is generated by Cauchy distribution randomisation with the mean value F_{m} and STD value of 0.1 [27]. The F_{m} is iteratively adapted using the Lehmer mean [27] defined as follows:

$$F_m(T+1) = 0.9F_m(T) + 0.1\frac{sum(good_F^2)}{sum(good_F)} \quad (10)$$

where $good_F$ is a tray of all F used in the current iteration with successful updates.

The parameter P_{DE} is also regenerated in the similar fashion to r_2 and r_3 before updating a population. For an individual solution, the P_{DE} is generated by normal distribution randomising with the mean value of P_{DEm} and standard deviation of 0.1. P_{DEm} is iteratively adapted based on the following equation:

$$P_{DEm}(T+1) = 0.9P_{DEm}(T) + 0.1mean(good_{\mathrm{PDE}}), \quad (11)$$

where $good_{\mathrm{PDE}}$ means all P_{DE} values used in the current iteration with successful updates.

The search process of ASCA-DE start with initilaising a population, $r_{2\mathrm{m}}$, $r_{3\mathrm{m}}$, F_m and P_{DEm}. The $good_{\mathrm{r2m}}$, $good_{\mathrm{r3m}}$, $good_{\mathrm{F}}$ and $good_{\mathrm{PDE}}$ trays are empty initially. After having calculated objective function values, the current best solution will be obtained. To update a population, firstly, P_{DE} and a random number in [0, 1] are generated. If the generated random number is lower than P_{DE}, a scaling factor (F) is generated based on F_{m} and a new solution is created using Eq. (7), otherwise, a new solution is generated based on Eq. (5). For each calculation of Eq. (5), r_2 and r_3 are generated based on $r_{2\mathrm{m}}$ and $r_{3\mathrm{m}}$. If a newly generated solution is better than its parent, the new solution will be selected for the next generation while saving all used parameters P_{DE}, r_2, r_3 and F into the $good_{\mathrm{PDE}}$, $good_{\mathrm{r2m}}$, $good_{\mathrm{r3m}}$, and $good_{\mathrm{F}}$ trays, respectively. Then, update the $r_{2\mathrm{m}}$, $r_{3\mathrm{m}}$, F_m and P_{DEm} using Eqs. (8)–(11). The search process is repeated until a termination criterion is reached. The computational steps of ASCA-DE are shown in Algorithm 2.

Algorithm 2 ASCA-DE

Input: population size (N_p), number of generations (T_{max}), number of design variable (D)

Output: $\mathbf{x}_{best}$, f_{best}

Main algorithm

Initialise a population , r_{2m}, r_{3m}, F_m and P_{DEm}.
Find the best solution ($\mathbf{x}_{best}$)
For T=1 to T_{max}
    Calculate parameter r_1 using eq. (6)
    Empty $good_{r2m}$, $good_{r3m}$, $good_F$ and $good_{PDE}$
    For l=1 to N_p
        Generate P_{DE} by normal distribution random with mean values P_{DEm} and STD =0.1
        IF rand< P_{DE}
            Generate F by Cauchy distribution random with mean value F_m and STD = 0.1
            Updated a population using eq. (7)
        Else
            For k = 1 to D
                Generate the parameter r_2 and r_3 by normal distribution random with mean values r_{2m}, r_{3m}, and STD =0.1
                Random generate r_4 in rank [0, 1]
                Update the kth element of the lth population ($\mathbf{x}_l$) using eq. (5)
            End For
        End IF
        Calculate objective function values of the newly generated population
        IF $f(\mathbf{x}_{l,new}) < f(\mathbf{x}_{l,old})$
            Replace $\mathbf{x}_{l,old}$ by $\mathbf{x}_{l,new}$
            Add all generated r_2, r_3, F, and P_{DE}, into the $good_{r2m}$, $good_{r3m}$, $good_F$ and $good_{PDE}$ tray, respectively.
        End IF
    End For
    Find the best solution ($\mathbf{x}_{best}$)
    Update r_{2m}, r_{3m}, F_m, and P_{DEm} using eq. (8)-(11)
End

5 Numerical Experiment

The performance investigation of the proposed ASCA-DE for structural damage detection is carried out by employing the algorithm to solve the test problems in the previous section. ASCA-DE along with a number of MHs in the literature implemented to solve the test problems include (Note that the details of variables can be found in the original sources of each method) [14]:

- Differential evolution (DE) [28]: a DE/best/2/bin strategy was used. A scaling factor, and probability of choosing elements of mutant vectors (*CR*) are 0.5 and 0.8 respectively.
- Artificial bee colony algorithm (ABC) [29]: The number of food sources for employed bees is set to be $n_P/2$. A trial counter to discard a food source is 100.
- Real-code ant colony optimization (ACOR) [30]: The parameter settings are $q = 0.2$, and $\xi = 1$.
- Charged system search (ChSS) [31]: The number of solutions in the charge memory is $0.2 \times n_P$. The charged moving considering rate and the parameter *PAR* are set to be 0.75 and 0.5 respectively.
- League championship algorithm (LCA) [32]: The probability of success P_c and the decreasing rate to decrease P_c are set to be 0.9999 and 0.9995, respectively.
- Simulated annealing (SA) [33]: Starting and ending temperatures are 10 and 0.001 respectively. For each loop, n_{mode} candidates are created by mutating on the current best solution while other n_{mode} candidates are created from mutating the current parent. The best of those $2n_{mode}$ solutions are set as an offspring to be compared with the parent.
- Particle swarm optimization (PSO) [34]: The starting inertia weight, ending inertia weight, cognitive learning factor, and social learning factor are assigned as 0.5, 0.01, 0.5 and 0.5 respectively.
- Evolution strategies (ES) [35]: The algorithm uses a binary tournament selection operator and a simple mutation without the effect of rotation angles.
- Teaching-learning-based optimization (TLBO) [36]: Parameter settings are not required.
- Adaptive differential evolution (JADE) [27]: The parameters are self-adapted during an optimization process.
- Evolution strategy with covariance matrix adaptation (CMAES) [37]: The parameters are self-adapted during an optimization process.
- Sine Cosine Algorithm (SCA) (Algorithm 1) [25]: The constant a parameter is set to be 2.
- Adaptive Sine Cosine algorithm with integrating DE mutation (ASCDE) (Algorithm 2): The parameter a is set to be 2 while initial $r_{2\mathrm{m}}$, $r_{3\mathrm{m}}$, F_m and P_{DEm} are set to be 0.5.

Each optimizer is used to tackle each truss damage detection test problem for 30 optimization runs. The number of iterations (generations) is 300 for all case studies while the population size is set to be 30 and 50 for 25-bar and 72-bar trusses respectively. All methods will be terminated with two criteria: the maximum numbers of functions evaluatio as 20 × 300, 30 × 300 and 50 × 300 for the 25-bar and 72-bar trusses respectively, and the objective function value being less than or equal to 1×10^{-3}. The six lowest natural frequencies ($n_{mode} = 6$) are used to compute the objective function value. This number of selected frequencies is reasonable since it is practically easier to measure fewer lowest natural frequencies with sufficient accuracy.

6 Results and Discussions

After performing 30 optimization runs of all MHs on solving the four truss damage detection optimization problems, the results obtained from the various MHs are given in Tables 5, 6, 7 and 8. The mean of the objective function are used to indicate the search convergence of the algorithms in cases that the objective function threshold (1×10^{-3}) is not met during searching. Otherwise, the mean number of FEs is used as an indicator. The number of successful runs out of 30 runs is used to measure the search consistency. The algorithm that is terminated by the objective function threshold is obviously superior and any run being stopped with this criterion is considered a successful run.

Table 5. Results for 25 bar truss with 35% damage at element number 7

Optimizers	Mean objective function values	No. of successful runs from 30 runs	Mean of FEs
DE	0.0017	19	6019
ABC	0.0135	0	9000
ACOR	0.0089	0	9000
ChSS	0.1385	0	9000
LCA	0.9036	0	9000
SA	0.0089	0	9000
TLBO	0.0077	6	7772
CMAES	0.0033	0	9000
ES	0.0308	0	9000
PSO	8.3830	0	9000
JADE	0.0026	2	8953
SCA	0.0270	24	3262
ASCA-DE	0.0009	29	2835

6.1 Twenty-Five-Bar Truss

Table 5 shown the results of the 25-bar truss with 35% damage at element 7. The best performer based on the mean objective function values is ASCA-DE while the second best and the third best are DE and JADE respectively. When considering the number of successful runs, seven optimizers including DE, TLBO, JADE, SCA and ASCA-DE can detect the damage of the structure. The most efficient optimizer is ASCA-DE that can detect the damages of the structure for 29 times out of 30 runs with the average of 2835 function evaluations.

Table 6. Results for 25 bar truss with 35% damage at element number 7 and 40% damage at element number 9

Optimizers	Mean objective function values	No. of successful runs from 30 runs	Mean of FEs
DE	0.0096	27	5220
ABC	0.0326	0	9000
ACOR	0.0125	0	9000
ChSS	0.1590	0	9000
LCA	0.8080	0	9000
SA	0.0269	0	9000
TLBO	0.0405	1	8917
CMAES	0.0115	0	9000
ES	0.0356	0	9000
PSO	8.6012	0	9000
JADE	0.0042	6	8875
SCA	0.0930	0	9000
ASCA-DE	0.0032	26	5511

Table 7. Results for 72 bar truss with 15% damage at element number 55

Optimizers	Mean objective function values	No. of successful runs from 30 runs	Mean of FEs
DE	0.0087	14	12887
ABC	0.2184	0	15000
ACOR	0.0014	6	14831
ChSS	0.1727	0	15000
LCA	1.1499	0	15000
SA	0.0097	0	15000
TLBO	0.0035	27	5781
CMAES	0.0053	0	15000
ES	0.0010	29	9335
PSO	1.9146	0	15000
JADE	0.0019	1	15000
SCA	0.0070	23	4793
ASCA-DE	0.0008	30	1715

Results of the 25 bar truss with 35% damage at element 7 and 40% damage at the element number 9 are reported in Table 6. The best performer based on mean objective function values is ASCA-DE while the second best and the third best are JADE and DE respectably. When examining the number of successful runs, only two optimizers, DE and ASCA-DE can consistently detect the damage of the structure for 27 and 26 runs respectively while the average number of function evaluations to obtain the results are 5220 and 5511 respectively.

Table 8. Results for 72 bar truss with 15% damage at element number 58 and 10% damage at element number 4

Optimizers	Mean objective function values	No. of successful runs from 30 runs	Mean of FEs
DE	0.0127	7	13963
ABC	0.1591	0	15000
ACOR	0.0058	0	15000
ChSS	0.1348	0	15000
LCA	1.1049	0	15000
SA	0.0129	0	15000
TLBO	0.0045	7	13503
CMAES	0.0050	0	15000
ES	0.0023	2	14940
PSO	1.7726	0	15000
JADE	0.0031	0	15000
SCA	0.0260	2	14502
ASCA-DE	0.0035	21	9235

6.2 Seventy-Two-Bar Truss

Table 7 shows comparison results of the 72-bar truss with 15% damage at element 5. The best performer based on mean objective function values is ASCA-DE while the second best and the third best are ES and ACOR. When examining the number of successful runs, the most efficient method is ASCA-DE which can detect the damage of the structure for 30 times while the average numbers of function evaluations for the convergence results is only 1715.

Results of the 72 bar truss with 15% damage at element number 58 and 10% damage at element number 4 are given in Table 8. The best performer based on the mean of objective function values is ES while the second best and the third best are JADE and ASCA-DE respectively. The minimum objective function value is obtained by SCA. When considering the number of successful runs, only ASCA-DE can consistently detect the damage of the structure for 22 times from totally 30 optimization runs while the average number of function evaluations for the convergence results is 9235. Although ES and JADE given better mean objective function values, they fail to search for the damage location. ASCA-DE is said to be the most efficient optimizer for this case.

Overall, it was found that integrating DE mutation into and applying adaptive parameters to SCA lead to performance enhancement of the original SCA. The proposed ASCA-DE is the best performer on solving truss damage detection optimization problem. It is considered the most reliable method for this study.

7 Conclusions

Performance enhancement of a meta-heuristics called a sine cosine algorithm is proposed by integrating into it a mutation strategy of DE. Self-adaptive optimization parameters are employed to improve the search performance of the new algorithm. The proposed optimizer is implemented on solving a number of truss damage detection inverse problems. The results reveal that the new meta-heuristic is the best and most reliable method. Our future work is to investigate the new MH for solving other practical engineering design problems.

Acknowledgments. The authors are grateful for the support from the Thailand Research Fund (TRF).

References

1. Sinou, J.J.: A review of damage detection and health monitoring of mechanical systems from changes in the measurement of linear and non-linear vibrations. In: Sapri, R.C. (ed.) Mechanical Vibrations: Measurement, Effects and Control, pp. 643–702. Nova Science Publishers, Inc., Hauppauge (2009)
2. Chen, H., Shi, X., He, Q., Mao, J.H., Liu, Y., Kang, H., Shen, J.: A multiresolution investigation on fatigue damage of aluminum alloys at micrometer level, Int. J. Damage. Mech. **26** (2017). doi:10.1177/1056789517693411
3. Shen, J., Mao, J., Boileau, J., Chow, C.L.: Material damage estimated via linking micro/macroscale defects to macroscopic mechanical properties. Int. J. Damage Mech **23**, 537–566 (2014)
4. Wang, X., Hu, N., Fukunaga, H., Yao, Z.: Structural damage identification using static test data and changes in frequencies. Eng. Struct. **23**, 610–621 (2001). doi:10.1016/S0141-0296(00)00086-9
5. Gerist, S., Maheri, M.R.: Multi-stage approach for structural damage detection problem using basis pursuit and particle swarm optimization. J. Sound Vib. **384**, 210–226 (2016). doi:10.1016/j.jsv.2016.08.024
6. Koh, B.H., Dyke, S.J.: Structural health monitoring for flexible bridge structures using correlation and sensitivity of modal data. Comput. Struct. **85**, 117–130 (2007). doi:10.1016/j.compstruc.2006.09.005
7. Kaveh, A., Zolghadr, A.: An improved CSS for damage detection of truss structures using changes in natural frequencies and mode shapes. Adv. Eng. Softw. **80**, 93–100 (2015). doi:10.1016/j.advengsoft.2014.09.010
8. Laier, J.E., Villalba, J.D.: Ensuring reliable damage detection based on the computation of the optimal quantity of required modal data. Comput. Struct. **147**, 117–125 (2015). doi:10.1016/j.compstruc.2014.09.020
9. Chou, J.H., Ghaboussi, J.: Genetic algorithm in structural damage detection. Comput. Struct. **79**, 1335–1353 (2001). doi:10.1016/S0045-7949(01)00027-X
10. Majumdar, A., Maiti, D.K., Maity, D.: Damage assessment of truss structures from changes in natural frequencies using ant colony optimization. Appl. Math. Comput. **218**, 9759–9772 (2012). doi:10.1016/j.amc.2012.03.031

11. Tabrizian, Z., Amiri G.G., Beigy, M.H.A.: Charged system search algorithm utilized for structural damage detection. Shock Vib. **2014**, Article ID 194753, 13 p. (2014). doi:10.1155/2014/194753
12. Xu, H., Ding, Z., Lu, Z., Liu, J.: Structural damage detection based on Chaotic Artificial Bee Colony algorithm. Struct. Eng. Mech. **55**(6), 1223–1239 (2015). doi:10.12989/sem.2015.55.6.1223
13. Ding, Z.H., Huang, M., Lu, Z.R.: Structural damage detection using artificial bee colony algorithm with hybrid search strategy. Swarm Evol. Comput. **28**, 1–13 (2016). doi:10.1016/j.swevo.2015.10.010
14. Pholdee, N., Bureerat, S.: Structural health monitoring through meta-heuristics – comparative performance study. Adv. Comput. Des. **1**, 315–327 (2016). doi:10.12989/acd.2016.1.4.315
15. Pal, J., Banerjee, S.: A combined modal strain energy and particle swarm optimization for health monitoring of structures. J. Civil Struct. Health Monit. **5**, 353–363 (2015). doi:10.1007/s13349-015-0106-y
16. Casciati, S.: Stiffness identification and damage localization via differential evolution algorithms. Struct. Control Health Monit. **15**, 436–449 (2008). doi:10.1002/stc.236
17. Agarwalla, D.K., Dash, A.K., Bhuyan, S.K., Nayak, P.S.K.: Damage detection of fixed-fixed beam: a fuzzy neuro hybrid system based approach. In: Panigrahi, B.K., Suganthan, P.N., Das, S. (eds.) SEMCCO 2014. LNCS, vol. 8947, pp. 363–372. Springer, Cham (2015). doi:10.1007/978-3-319-20294-5_32
18. Jiao, Y.B., Liu, H.B., Cheng, Y.C., Gong, Y.F.: Damage identification of bridge based on chebyshev polynomial fitting and fuzzy logic without considering baseline model parameters. Shock Vib. **2015**, Article ID 187956, 10 p. (2015). doi:10.1155/2015/187956
19. Pan, D.-G., Lei, S.-S., Wu, S.-C.: Two-stage damage detection method using the artificial neural networks and genetic algorithms. In: Zhu, R., Zhang, Y., Liu, B., Liu, C. (eds.) ICICA 2010. LNCS, vol. 6377, pp. 325–332. Springer, Heidelberg (2010). doi:10.1007/978-3-642-16167-4_42
20. Abdeljaber, O., Avci, O.: Nonparametric structural damage detection algorithm for ambient vibration response: utilizing artificial neural networks and self-organizing maps. J. Archit. Eng. **22**, 04016004 (2016). doi:10.1061/(ASCE)AE.1943-5568.0000205
21. Sidibe, Y., Druaux, F., Lefebvre, D., Maze, G., Léon, F.: Signal processing and Gaussian neural networks for the edge and damage detection in immersed metal plate-like structures. Artif. Intell. Rev. **46**, 289–305 (2016). doi:10.1007/s10462-016-9464-z
22. Mirjalili, S., Mirjalili, S.M., Lewis, A.: Grey wolf optimizer. Adv. Eng. Softw. **69**, 46–61 (2014). doi:10.1016/j.advengsoft.2013.12.007
23. Mirjalili, S.: Moth-flame optimization algorithm: a novel nature-inspired heuristic paradigm. Knowl.-Based Syst. **89**, 228–249 (2015). doi:10.1016/j.knosys.2015.07.006
24. Mirjalili, S., Lewis, A.: The whale optimization algorithm. Adv. Eng. Softw. **95**, 51–67 (2016). doi:10.1016/j.advengsoft.2016.01.008
25. Mirjalili, S.: SCA: a sine cosine algorithm for solving optimization problems. Knowl.-Based Syst. **96**, 120–133 (2016). doi:10.1016/j.knosys.2015.12.022
26. Bureerat, S., Pholdee, N.: Optimal truss sizing using an adaptive differential evolution algorithm. J. Comput. Civil Eng. **30**, 04015019 (2015). doi:10.1061/(ASCE)CP.1943-5487.0000487
27. Zhang, J., Sanderson, A.C.: JADE: adaptive differential evolution with optional external archive. IEEE T. Evolut. Comput. **13**, 945–958 (2009). doi:10.1109/TEVC.2009.2014613
28. Storn, R., Price, K.: Differential evolution – a simple and efficient heuristic for global optimization over continuous spaces. J. Global Optim. **11**, 341–359 (1997). doi:10.1023/A:1008202821328

29. Karaboga, D., Basturk, B.: A powerful and efficient algorithm for numerical function optimization: artificial bee colony (ABC) algorithm. J. Global Optim. **39**, 459–471 (2007). doi:10.1007/s10898-007-9149-x
30. Socha, K., Dorigo, M.: Ant colony optimization for continuous domains. Eur. J. Oper. Res. **185**, 1155–1173 (2008). doi:10.1016/j.ejor.2006.06.046
31. Kaveh, A., Talatahari, S.: A novel heuristic optimization method: charged system search. Acta Mech. **213**, 267–289 (2010). doi:10.1007/s00707-009-0270-4
32. Husseinzadeh, K.A.: An efficient algorithm for constrained global optimization and application to mechanical engineering design: League championship algorithm (LCA). Comput. Aided Design. **43**, 1769–1792 (2011). doi:10.1016/j.cad.2011.07.003
33. Bureerat, S., Limtragool, J.: Structural topology optimization using simulated annealing with multiresolution design variables. Finite Elem. Anal. Des. **44**, 738–747 (2008). doi:10.1016/j.finel.2008.04.002
34. Venter, G., Sobieszczanski-Sobieski, J.: Particle swarm optimization. AIAA J. **41**, 1583–1589 (2003). doi:10.2514/2.2111
35. Back, T.: Evolutionary Algorithms in Theory and Practice. Oxford University Press, Oxford (1996)
36. Rao, R.V., Savsani, V.J., Vakharia, D.P.: Teaching–learning-based optimization: a novel method for constrained mechanical design optimization problems. Comput. Aided Des. **43**, 303–315 (2011). doi:10.1016/j.cad.2010.12.015
37. Hansen, N., Muller, S.D., Koumoutsakos, P.: Reducing the time complexity of the derandomized evolution strategy with covariance matrix adaptation (CMA-ES). Evol. Comput. **11**, 1–18 (2003). doi:10.1162/106365603321828970

Resource Production in StarCraft Based on Local Search and Planning

Thiago F. Naves(✉) and Carlos R. Lopes

Faculty of Computing, Federal University of Uberlândia, Uberlândia, Brazil
{tfnaves,crlopes}@ufu.br

Abstract. This paper describes a approach for resource production in real-time strategy games (RTS Games). RTS games is a research area that presents interesting challenges for planning of concurrent actions, satisfaction of preconditions and resource management. Rather than working with fixed goals for resource production, we aim at achieving goals that maximize real-time resource production. The approach uses the Simulated Annealing (SA) algorithm as a search tool for deriving resource production goals. The authors have also developed a planning system that works in conjunction with SA to operate properly in a real-time environment. Analysis of performance compared to human and bot players corroborate in confirming the efficiency of our approach and the claims we have made.

Keywords: Real-time strategy games · Actions · Resources · Planning · Search

1 Introduction

Determining which resources should be built into a game is one of the main challenges in RTS Games. Along with the choice of resources, it is necessary to automate the planning of actions that build resources, satisfy constraints between actions and manage parallelism of actions. All these planning issues must be made under uncertainty, since the opponent is hidden over fog of war. These features are commonly found in problems that are relevant to the artificial intelligence planning area. Based on this information, we have explored ideas from the automated planning area for determining resource production goals as well as a plan of actions that achieves them.

The resources in RTS games originate from many sources of raw materials, basic construction, military units and civilization. The resources are used in attack and defense against enemies. In fact, there is an initial stage of the game in which players spend time aiming at developing the maximum amount of resources. During this time there are no direct confrontations and players who achieve better quality in resource production have advantage in the confrontation phase. Thus, resource production is a vital task for success in the game.

We have developed an approach that maximizes resource production based on goals established at runtime. Subsequently, it is possible to determine a plan

O. Gervasi et al. (Eds.): ICCSA 2017, Part I, LNCS 10404, pp. 87–102, 2017.
DOI: 10.1007/978-3-319-62392-4_7

of actions and their scheduling that leads the game from an initial resource state to a goal state that contains the new resources produced to improve the army. The plan of actions contains all the necessary actions to achieve the goal. A goal is described by the resources that this can achieve. Actually, as it is described later, the plan of actions by itself describes the resources or goals that should be produced in order to maximize the armed force of a player.

For this, the Simulated Annealing algorithm (SA) [1] is used along with planners and checkers that have been developed so that the approach is able to operate within the RTS games domain. The SA performs a search process, where an initial plan of actions goes through several changes in order to maximize the quality of the resource production and increase the force of the army that will be generated.

The domain used in our work is StarCraft, which is considered a game with a high number of constraints in planning tasks [6]. The work has been developed to solve a common deficiency in research involving RTS games, which is the choice of which resource production goals to strive for. For instance, [2,5,6] present planning and scheduling algorithms to determine which actions should be carried out to reach a resource production goal from an initial resource state. However, the establishment of a goal is not based on any explicit criterion that assures quality in its planning. Most probably the specification of another goal might bring better results. Different from previous work our approach does not take into account a fixed goal for resource production.

In our approach an initial plan of actions is generated. In order to reach this, a planner system called POPlan has been developed, which is able to plan and schedule the necessary actions to achieve an initial amount of resources, which may change at runtime. The initial plan is the basis for exploring new plans in order to find the one that has the greatest number of resources. In conjunction, a consistency checker called SHELRChecker was developed. SHELRChecker validates and manages the changes made during the search for new plans of actions. By doing this we claim that action planning based on a fixed goal is not necessary, but what is important is a specification of a policy that maximizes resource production.

The algorithms shown in this paper can also be used to build tools to assist players, such as an interface to help in choosing a goal and which resources should be built at given moments of the match. This work is a result of our efforts in developing action planning algorithms for resource production in RTS games. In this paper we focus on describing the planning system. Results of experiments and performance compared to human players and bot players that compete in StarCraft competitions corroborate in confirming the efficiency of our approach and the claims made in this work.

The remainder of the paper is organized as follows. Section 2 presents the characterization of the problem. Section 3 describes related work. In Sect. 4 there is a brief description of the use of SA in the approach proposed here. The description of POPlan and its operation is in Sect. 5. One finds in Sect. 6 a description of the consistency checker SHELRChecker. The experiments and discussion of the results are found in Sect. 7 and finally the conclusion in Sect. 8.

2 Characterization of the Problem

RTS games are strategy games in which military decisions and actions occur in a real-time environment [3]. To execute any action you must ensure that its predecessor resources are available. The predecessor word can be understood as a precondition to perform an action and the creation of the resource as an effect of its execution. Resources can be placed into one of the following categories: *Require*, *Borrow*, *Produce* and *Consume*. The resources are based on StarCraft game, which has three different character classes: Zerg, Protoss and Terran. This work focuses on the resources of the class Terran.

The planning problem in RTS games is to find a sequence of actions that leads the game to a goal state that achieves a certain amount of resources. This process must be efficient. In general, the search for efficiency is related to the time that is spent in the execution of the plan of action (makespan). However, in our approach we seek actions that increase the amount of resources that raise the army power of a player. This is achieved by introducing changes into a given plan of action in order to increase the resources to be produced, especially the military units, without violating the preconditions between the actions that will be used to build these resources.

In our approach each plan of action has a time limit, army points, feasible and unfeasible actions. The time limit is the deadline of the plan of action, i.e., the maximum makespan value that a plan has to perform their actions. The time limit works as a due date for any plan of action, which cannot be exceeded. The army points are used to evaluate the strength of the plan in relation to its military power. Each resource has a value that defines its ability to fight the opponent. The sum of this value for each resource present in a plan corresponds to its army points. Feasible actions in a plan are those actions that can be carried out, i.e., they have all their preconditions satisfied within the current planning. Unfeasible actions are those that cannot be performed in the plan and are waiting for its preconditions to be satisfied at some stage of planning. Unfeasible actions do not consume planning resources and do not contribute to the evaluation of the army points of the plan.

As an example of operations in a plan, suppose that a time limit of 170 s was imposed to find a goal and the initial plan was set to perform *1 Firebat*. In this case, the completed plan of action has the following actions: *(8 collect-minerals, 1 collect-gas, 1 build-refinery, 1 build-barrack, 1 build-academy, 1 build-firebat)* with makespan equals to 160 and 16 army points. There are several operations that could be made in this plan and that would leave unfeasible actions. For example, exchange one of the actions *collect-minerals* that is a precondition of *Barracks* with any other action would leave the resource unfeasible. In consequence the resources *Firebat* and *Academy* would also be unfeasible, since these have *Barracks* as one of their preconditions. In this way, the plan would have 0 army points.

Some specific combinations of operations can make the plan increase its resource production and consequently its army strength. For example, suppose that a *build-marine* action is inserted in place of the *collect-gas* action. The new action is feasible in the plan, because it uses the *Minerals* left by the action *build-firebat*, which became unfeasible due to lack of *Gas*. Now, if in the next two operations the actions of *collect-minerals* and *collect-gas* are inserted in the plan, the resource *Firebat* once again becomes feasible. This happens because the action of minerals would be executed by a *Scv* resource which is idle and the scheduling process allocates it as early as possible in the plan. The gas action would be performed at the same time as a result of the scheduling process. Thus, the final goal now has 28 army points without exceeding the time limit, still having the same makespan of 160 s. The state of resources when starting a match in StarCraft is *4 Scv, 1 CommandCenter*.

3 Related Work

There is little research in maximizing goals of resource production in RTS games. Some of the reasons are: the complexity involved in searching and managing the state space, planning and decision making under uncertainty and the resource management under real-time constraints.

Our approach is based on the work developed by [8]. It also uses SA to explore the state space of the StrarCraft. But in this case to balance the different classes in the game by checking the similarity of plans in the each class. In our work Simulated Annealing is used to determine a goal to be achieved that maximizes resource production, given the current state of resources in the game and a time limit for the completion of actions.

In [5] a linear planner was developed to generate a plan of action given an initial state and a goal for production of resources, which is defined without the use of any explicit criterion. The generated plan is scheduled in order to reduce the makespan. Our work now uses a planner system with scheduling. Our approach deals with dynamic goals for resource production. This is necessary in order to figure out a goal that maximizes the strength of the army. To the contrary, [5] works with a goal with fixed and unchangeable resources to be achieved.

Work developed by [2] has the same objective as that pursued by [5]. Their approaches differ in the use of the planning and scheduling algorithms. [2] developed their approach using Partial Order Planning and SLA* for scheduling. [2] also works with a goal with fixed and unchangeable resources to be achieved and cannot be adapted to our problem.

The work of [6] presents an approach based on the FF [11] algorithm which uses enhanced hill-climbing and a delete-relaxation based heuristic on to develop a plan of action that achieves a specific goal in the shortest possible time (minimum makespan). The goals to be used are expert goals, recovered from a database created from analyses of games already played. In this approach, heuristics and efficiency strategies to reduce the search space and achieve better results. In this work the amount of goals and the number of produced resources are limited those goals that are at the database, at some point of the game a more appropriate goal maybe cannot be considered, by not being at the database or not produce the correct amount of resources. The approach proposed in this paper aims to select and build goals through plans of actions without resource constraints and various possibilities of productions.

The works of [14,18] uses an approach with Case Based Reasoning (CBR) and Goal Driven Autonomy (GDA). CBR is used to select expert goals from a database and GDA allows a goal to be discarded during its execution and a new goal is chosen that takes into account the game situation. This approach has similarities with that proposed in this article, especially in relation to planning goals within the game, however this approach also has a limit on the number of goals that are considered, which are present in a database.

The work of [12] also uses expert goals to select a goal in RTS games. In this, an analysis is made on already played matches and goals are scored according to characteristics of current state of the game, such as, the amount of remaining units and amount of downed enemies. To select a goal within the game is used genetic algorithm (GA), which is responsible for selecting a goal according to its score and characteristics that better satisfy the current objective of the player. The strategy of choosing goals from characteristics of the state of the game is interesting, but the approach proposed here considers a wider range of possible goals and seeks to find this without the help of prior information or expert goals.

We also surveyed some already existing planners that could be applied. Approaches described in [7,10,13] were considered for our problem at hand, but both have a different approach and would have to be modified to adapt to our goal. In short, most of the approaches surveyed have different objectives and the techniques used are not efficient for the domain that we are exploring.

4 Simulated Annealing in the Maximization of Resource Production

Simulated Annealing is a meta-heuristic which belongs to the class of local search algorithms [1]. SA was chosen due to the robustness and possibility of it being combined with other techniques. It was also the algorithm that achieved the best results during the initial tests that were completed to develop this approach. SA takes as input a possible solution to the problem, in our case a plan of action that was developed by POPlan, which is going to be described in the next section. The mechanism that generates neighbors will perform operations of exchange and replacement of actions in the plan. It should be noted that these operations generate new plans of action. The following will briefly describe some of the main features of the algorithm and the adjustments made for use in the RTS games environment. More details of the functioning of the SA are available in [16].

In SA a possible solution is evaluated through the use of an objective function. In our approach the objective function is responsible for counting the army points that exist in each plan of action. Another attribute present in resources could be used to evaluate the plan, but thinking about the game as a whole the attack force is a good value to be maximized. In this way it is possible to find out which states have the highest attack strength. A state that has a better evaluation than the current state becomes the current state. Even if the new plan generated is not chosen as the current solution there is still a probability of its acceptance.

There are several ways to generate a new neighbor plan from the current solution in SA. The operation used in the approach is the following: we choose randomly between select two actions within the plan and switch their positions, or replace an action for another randomly chosen. This mechanism yields nice results. When a neighbor plan is generated, algorithm SHELRChecker is used to manage and evaluate the changes made in the goal.

5 POPlan

The POPlan is a system that uses the concept of partial order planning (POP) [15] to generate a plan of action and perform the scheduling task during this process. Two reasons explain why a partial order planner is required. The first reason is the fact that SA need a plan as input data that enables the search for solutions with parallelism of actions. Another reason has to do with finding an integrated solution that reduces time during the process of planning and scheduling.

The system architecture is composed of two levels. At one level we have the planner component responsible for plan generation. At another level we have the scheduling of the actions that make the generated plan. These two levels interweave their activities and are depicted in Fig. 1. With this architecture, the system has the concept of strong coupling [9] where planning and scheduling problems are reduced to a uniform representation. In Fig. 1 the domain element represents the environment of RTS games where the planning is conducted, the restrictions represent the preconditions and precedences of the actions of a specific RTS game and the processes of planning and scheduling use the previous information in their executions. The restrictions satisfaction process is made defining the actions and causal links between them and the resources available, which are used to build a certain resource production goal. During the scheduling it is checked if any of the links established in the previous process became invalid, i.e., if any link has become inconsistent. For example, if during the process of creating links an action has to have a link to a *Minerals* resource, it means that this action will use it at a certain time. However, if this *Minerals* is used by another action that will be scheduled first it may cause inconsistency. It is necessary to reconstruct this link with other *Minerals* resource available.

With the POPlan architecture if any inconsistency is found in the scheduling step it is possible to check and satisfy it in the planning stage that is being performed intercalated with that. Thus, it is not necessary to interrupt the process for such a verification. In this coupling, the planner is essential for successfully achieving this representation, because the causal links defined between the actions in the planning stage have the temporal order of constraints among the actions, which also assists in the production and use of resources. With that, the scheduling step only figures the best time for the action to be performed, leaving planning tasks to be used in the next stage.

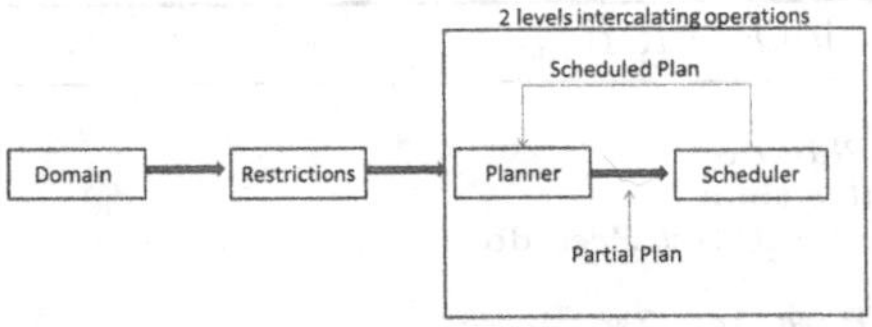

Fig. 1. Representation of the strong coupling intercalating POP and scheduling.

Algorithm 1. POPlan($Plan$, R_{goal}, T_{limit}))

```
1: Plan ← LinkBuild(Plan, R_goal)
2: for each Action Act ∈ Plan do
3:    if Act.unity = false then
4:       times ← Quantity(Act)
5:    end if
6:    while i < times do
7:       for each Action ActRs ∈ Plan do
8:          if ActRs = Act.rbase then
9:             Schedule(Plan, ActRs, Act)
10:         end if
11:      end for
12:      contr ← Constructs(Plan, Act)
13:      if contr = false then
14:         Exit()
15:      end if
16:      i ← i + 1
17:   end while
18: end for
19: return Plan
```

Algorithm 1 describes the pseudo-code of the POPlan.

The POPlan receives as parameters the list where the actions that will make the plan will be inserted (*Plan*), the resource that is the current goal (R_{goal}) and the time limit for completion of the plan of action T_{limit}. At the beginning of the operation, the planner calls the method (*LinkBuild*()) (line 1), responsible for defining all the actions that will make the plan and its respective causal links.

Using the concept of POP, all actions that make up the plan have causal links. It is possible to separate those links into two types. The first is the precedence link, where an action that has such a link is necessary for the execution of another action that receives the second type of link, which is the condition link. For example, an action *collect-minerals* with a link to an action *build-factory* means that the action of collecting mineral precedes the execution of the action that builds a *Factory*. This action in turn has a link indicating that the action of collecting minerals is a condition for its execution. Thus, when an action has some of its attributes changed, it becomes easier to find out which other actions should change due to this occurrence. Each action with a link is located in lists, precedence links is in the list *link* and condition link in *linkReb* list (in Algorithm 2).

Algorithm 2 receives as parameters the plan of action (*Plan*) and the resource production goal (R_{goal}). Algorithm 2 shows the pseudo-code of the LinkBuild method.

Algorithm 2. LinkBuild(*Plan*, R_{goal})

```
1: Plan ← Extract(R_goal)
2: for each Action Act ∈ Plan do
3:    if Act.finished = false then
4:       for each Resource Rsc ∈ Act.Pred do
5:          Action Actn
6:          exist ← CheckPred(Act, Rsc, Actn)
7:          if exist = false then
8:             Act.linkReb.push(Actn)
9:             Actn.link.push(Act)
10:            Plan.push(Actn)
11:            Plan ← LinkBuild(Plan, R_goal)
12:         else
13:            Act.linkReb.push(Actn)
14:            Actn.link.push(Act)
15:         end if
16:      end for
17:      Act.finished = true
18:   end if
19: end for
20: return Plan
```

Algorithm 2 inserts the first action in the plan and from its preconditions produces the remaining necessary actions. When an action is selected, it is checked to see if all its preconditions have been satisfied on the plan through the attribute exist (line 3). This checking is necessary due to the fact that when an action is placed in the plan to satisfy certain preconditions, the preconditions of this action that have just been entered are immediately checked (line 11). So it is possible that an action of the plan that will be checked has had its preconditions already satisfied.

Between lines 3 and 4 Algorithm 2 checks for each action of the plan to see if its precondition resources are present in the plan or must be inserted into it. This check is made by *CheckPred*() (line 6), which checks whether there is an action in the plan that can satisfy the precondition or if it is necessary to insert a new action. If a new action is necessary, the value false is assigned to the variable *exist* and the algorithm creates the links between the action and its precondition, putting the new action in the plan and calls itself to enter the preconditions of the new action (lines 7 and 11). If the action that satisfies the precondition already exists, then only the necessary links between these actions are created (lines 12–14).

The function *CheckPred*() (line 6) receives the variable *Actn* as a parameter and assigns it to either the new action that will be inserted into the plan or the action that is already there and can satisfy the necessary precondition. The algorithm validates the attribute, which exists for each action and that already has all its preconditions satisfied. At the end, the plan containing the actions and the links between them is returned to the POPlan to continue its execution. The plan that Algorithm 2 builds and returns to POPlan contains only the actions necessary to reach the resource goal production and the respective links between them. The starting and ending times of each action will be defined later by the scheduler, which does not need to perform any specific planning task, since the actions are already configured.

The POPlan after setting up the plan with the necessary actions and links through *LinkBuild*(), starts to go through the actions of the plan. Algorithm 1 between lines 2 and 5 verifies to see if the action is not the type of production unit, i.e., whether the action is the mineral or gas type, but only resources that are not of unit type. The function *Quantity*() (line 4) is called to check how many times the action should be executed. For example, if the action *build-minerals* has a link to a *build-academy* that requires 150 *Minerals*, three executions of the same action are necessary because each execution generates 50 *Minerals*.

After setting the starting time of the action through the scheduling process, the planner will now finish setting its attributes, again using the planning stage. The method *Constructs*() (line 12) is responsible for this task, by setting the starting and ending execution time of the action, inserting new actions of renewable resources if necessary and, modifying some link action if necessary due to the scheduling that can change these links. This procedure also checks there is any threat held in these (according to the concept of POP), and checks to see if the plan time limit has not been exceeded. The final plan that is built by the planner, contains the actions that reach a certain amount of resources.

6 SHELRChecker the Consistency Checker

With the search for goals toward plans of actions that have quality in army points and scheduled actions, verification and validation of new plans becomes even more important. The scheduled consistency checker, SHELRChecker, is used in this step. Algorithm 3 shows its pseudo-code.

Algorithm 3. SHELRChecker(*Plan*, *opt*, T_{limit})

```
Invib(ActsInv, opt)
for each Action Act ∈ ActsInv do
    if Act.feasi = true then
        for each Action Actl ∈ Act.link do
            UndoL(Actl, ActsInv)
        end for
        for each Action Actl ∈ Act.Clink do
            ReleaL(Actl, ActsInv)
        end for
    end if
end for
Rearrange(Plan, ActsInv, opt)
while cont = true do
    for each Action Act ∈ ActsInv do
        feasible ← Gather(Plan, Act)
        if feasible = true then
            for each Action ActRs ∈ Plan do
                if ActRs = Act.rBase then
                    Schedule(Plan, ActRs, Act)
                    cont ← Constructs(Plan, Act)
                end if
            end for
        end if
    end for
end while
return Plan
```

The Algorithm 3, receives as parameters the current solution of the SA (*Plan*), the operation that will be performed in the plan *opt* and the time limit of the solution to be generated (T_{limit}). Initially the method *Invib*() is called (line 1). It is responsible for checking what actions will be unfeasible due to the operation. The method places the first actions in the unfeasible action list *ActsInv*, which is necessary for the algorithm to find out which others will also become unfeasible.

The algorithm between lines 2 and 11 finds the other actions that will be unfeasible. For each one that still has the attribute (*feasi*) valid (line 3), i.e., every action that has just been added to the list of unfeasible, the algorithm calls the methods responsible for removing its links and finding out which other actions will also be unfeasible. The method (*UndoL*()) (line 5) is called to put each resource that receives an action link that was unfeasible (*Act*) within the list of unfeasible actions. This is carried out as (*Act*) which is a precondition of such actions, which will no longer run. For example, if an action *build-refinery* becomes unfeasible this means that the *Refinery* does not exist in the game anymore, so all actions *collect-gas* receiving a link of this resource will also become unfeasible.

The method *ReleaL*() (line 8) is called just after *UndoL*(), it is responsible for releasing the actions that were serving as a precondition for the action that became unfeasible, thus eliminating the links from other actions that came from it. This frees the resources used by the action to be used by other actions. After the list of unfeasible actions has been completely filled, the algorithm calls the *Rearrange*() (line 12). It performs the operation on the plan so that it becomes a new neighbor plan. From line 12 onwards, the algorithm focuses on what actions can become feasible again, configures its attributes and schedules the action. The function *Gather* (line 15) is used with every unfeasible action. It seeks the preconditions of the action that is being checked *Act*, to see if the available actions can be used to execute it.

When the preconditions of an action are satisfied within the plan, the algorithm schedules the action and changes its attributes, as they become feasible. A search for base resources of the action (line 17) is made to define which among those available would execute the action. When a base resource is found the scheduling algorithm is called *Schedule*() (line 19). The algorithm is the same as used in the planner. It performs the schedule for each feasible action. Finally, the function *Constructs*() (line 20) is called to change the attributes of the action that became feasible again. It sets up the execution times of actions, builds all the links between them and their preconditions, removes the action from the list of unfeasible and checks to see if the time limit has not been reached. If this time has been reached the verifier stops the process returning the plan with the last action that became feasible as new solution for SA.

7 Experiments and Discussion of Results

The experiments were conducted by using the StarCraft game environment. It was possible through the use of the BWAPI (BroodWar API) [4], which allows us to test the developed algorithms within the game and collect the results. We compare our results against human players and a bot from competitions. We ran our experiments using a computer equipped with an Intel Core i7 1.73 GHz with 8 GB RAM on Windows 7 system.

The procedures used in the experiments are: $SA(S)$, $S(SS)$, *Player*(E) and BT. $SA(S)$ refers to our approach, with SA plus POPlan and SHELRChecker. $SA(SS)$ also refers to our approach in conjunction with the subgoals strategy, which will be presented during the experiments. *Player*(E) is a human player experience level in the game. To be classified as an experienced player a 5-year experience with StarCraft was considered. BT is the bot called BTHAI [17], who participated in the 2014 AIIDE StarCraft Competition[1].

In all tests, the used approaches tried to achieve the best goal with resource production, i.e., a goal focused on produce resources that allow a player to advance against the enemy bases and overcome it in a confrontation. These goals produce various types of resources that increase the amount of army points. Produce resources with ability to fight enemys leads to the production of other high level types of resources present in the game, since its are required to enable more powerful resources. In fact, a feature observed in various players when played matches of the game is the trend to produce offensive resources along the match and only produce other types of resources when these are needed to enable the coming of new resources that will be used to attack and resist to the enemy's army. Thus, the approaches will be compared according the amount of resources that can produce within a time limit. Different time limits will be imposed on each test and approaches should try to achieve the best goal without exceeding this time limit (Table 1).

Table 1. Results of Experiment 1

Time limit	Player(E)	SA(S)	Makespan of SA(S)	Runtime of **SA(S)**
150 s	32	32	148 s	28.8 *s*
250 s	95	**98**	250 s	42.6 *s*
350 s	**151**	132	350 s	63.2 *s*
450 s	**207**	188	599 s	96.8 *s*
550 s	**254**	210	547 s	129.2 *s*
650 s	**369**	298	645 s	164.5 *s*
750 s	**437**	307	746 s	199.1 *s*

In Experiment 1, we conduct a performance test between the $SA(S)$ and *Player*(E). *Player*(E) was instructed to use as a strategy goals that produce resources he would use to form a base to confront and defend against enemy. For each time limit, we considered five executions and collected the average

[1] StarCraft Competition is a competition between Starcraft bots, this competition is an event that is part of AIIDE (AAAI Conference on Artificial Intelligence and Interactive Digital Entertainment).

values for army points, makespan, and runtime. *SA*(*S*) was able to defeat the experienced player in one occasion, tying in another one and losing in five others. One can see that *SA*(*S*) achieves its best results when searching for goals with average time intervals between 150 and 350 s. This happens because in these time intervals the number of actions that the approach needs to manage is not as large as those at longer intervals, which helps the algorithm to achieve better performance (Table 2).

Table 2. Results of Experiment 2

SA(S)			SA(SS)		
Army points	Makespan	Runtime	Army points	Makespan	Runtime
242	596 s	148.6 s	**299**	599 s	100.1 *s*
245	594 s	149.9 s	**303**	600 s	101.4 *s*
241	592 s	148.5 s	**297**	600 s	98.8 *s*
242	590 s	147.1 s	**302**	597 s	99.4 *s*
243	593 s	151.5 s	**297**	600 s	98.2 *s*
243	596 s	148.8 s	**312**	598 s	97.7 *s*
242	598 s	150.9 s	**301**	600 s	100.1 *s*
245	588 s	149.4 s	**301**	598 s	98.3 *s*
237	594 s	149.7 s	**297**	600 s	99.6 *s*
240	593 s	151.2 s	**307**	598 s	97.1 *s*

In order to use our approach under real-time constraints we used a strategy that takes into account the fact the best results have been achieved considering time intervals that range from 150 to 300 s. This strategy consists of decomposing a resource production goal to be achieved in time intervals superior to 300 s into smaller resource production subgoals to be achieved in time intervals in which the algorithm can find the best results. As soon as a subgoal is achieved its plan of actions is carried out. While executing the actions that satisfies a subgoal, the algorithm keeps working in order to figure out a plan of action that satisfies the next subgoal, and so on. For example, suppose a 600 s time limit for resource production. In this case, the algorithm splits the resource production goal for this time interval into three resource production subgoals considering a 200 s time limit for each one. Whenever a goal is divided into subgoals, these have time intervals between 150 and 200 s.

Experiment 2 demonstrates the performance of the subgoals strategy. In this experiment, the *SA*(*S*) and *SA*(*SS*) (that uses the subgoals strategy) were executed ten times to find goals with time limits of 600 s. For comparison, the left column of the table contains the algorithm with its normal execution, where on the right hand column of the table, the strategy of dividing the main goal into subgoals was used. The algorithm divided the goal of 600 s into 3 subgoals of 200 s.

Through an analysis of the results in Experiment 2, the algorithm using subgoals in most cases found better solutions in a shorter runtime. This is due to subgoals containing smaller quantity of actions, which facilitates algorithm operations during the management of actions that become unfeasible due to changes in the plan. This feature also allows the algorithm to perform more modifications in the plans. In this way, the algorithm can exploit goals with better quality and with more army points. By using a single goal, the number of actions in the plan is high and causes the generated solutions to become complex to manage. In the subgoals strategy, when a subgoal finishes its execution we have as a result resources which are used in achieving the next subgoal.

Table 3. Results of Experiment 3

Player(E)		SA(SS)	
Army points	Number of military resources	Army points	Number of military resources
181	9	174	9
182	8	**184**	12
180	9	172	9
184	10	178	11
181	9	**182**	11
178	8	**184**	12
182	10	173	11
184	10	174	9
184	10	148	6
183	9	177	10

Table 3 presents the results of Experiment 3. A comparison is made again using *Player*(*E*), only this time with *SA*(*SS*). Ten executions with the time limit of 400 s were made. The *Player*(*E*) obtained seven wins over *SA*(*SS*) with respect to the army points. The *SA*(*SS*) goals were able to produce more resources than the *Player*(*E*) in most experiments. This occurred at times when the algorithm was able to produce more military resources than the human player, due to the approach of using subgoals strategy. Again, this approach generates solutions that have more resources with lower production cost. This happens because *SA*(*SS*) schedules their actions in order to produce more resources. Due to space limitations, the description of the scheduling algorithm is going to appear in an upcoming paper. *Player*(*E*) frequently concentrated on producing more powerful resources, but production was realized on a small scale due to the time limit imposed.

In Experiment 4 a comparison between the *SA*(*SS*) and *BT* bot player is made. This was chosen because it is an agent with open code and also by the use of the resources of Terran class in Starcraft, the same used by our approach.

To accomplish this experiment tests were performed to determine the maximum time that *BT* was able to produce resources without using their resources on direct attacks within the game. This was considered the time that both approaches would have to perform their resource production. At the end, the quality of both in terms of army points was evaluated. *BT* was chosen for having an offensive resource production strategy, similar to that described at the beginning of this section. During each experiments execution the bot concentrated only on achieving goals that increase its ability to confront and be confronted by enemies. The time was set in 270 s and ten executions were carried out with this time.

With the results described in Table 4, $SA(SS)$ was able to overcome the *BT* in most of the executions. The source code of *BT* has not been changed since this could alter its characteristics and resource selection strategies. In most of the executions *BT* concentrated in producing Firebat (Terran resource) units, resource which has a good cost benefit. However, this strategy has a maximum amount of resources that can be built in a time interval and thus the values of army points achieved by *BT* were often similar. $SA(SS)$ does a search for the best combination of resources that maximize production and raise the power of the army produced. During program execution, $SA(SS)$ produces different combination of resources within the time interval and the best one is chosen as a goal along with its plan of actions.

Table 4. Results of Experiment 4

SA(SS)	BT	
Army points	Army points	Execution
109	85	1
85	**89**	2
106	84	3
109	85	4
104	82	5
106	85	6
108	89	7
109	84	8
106	85	9
108	82	10

Experiment 5 shows performance analysis of $SA(SS)$. Table 5 describes the average results of 15 runs with three different time intervals. For goals with an interval of 250 s, the amount of near-optimum solutions is large. This result is highlighted by the average army points (93) whose value is close to the optimal value. At intervals of 400 s good results were obtained. In the 600 s intervals, a

Table 5. Results of Experiment 5 using SA(SS)

Time limit	250 s	400 s	600 s
Optimum value of army points	100	184	312
Number of runs over 95% of optimum	84%	41%	37%
Number of runs over 90% of optimum	6%	49%	44%
Average value of ***runtime***	32.2 s	64.6 s	99.1 s
Average value of army point	93	167	281
Number of runs	15	15	15

drop in performance begins to be presented. The results point to the effectiveness of this approach in the search for goals in shorter time intervals, in which the algorithm can obtain best results. This feature also reinforces the use of the subgoals strategy.

8 Conclusion and Future Work

In the previous sections we described our approach to resource production that aims to find goals with quality in RTS games. The approach uses Simulated Annealing, which from an initial plan of action figures a new plan that maximizes the production of resources within a given time limit.

Experiments allowed us to conclude that our approach was able to compete with an experienced human player and beat a mid-level player, in terms of quality in the production of resources. With respect to performance, experiments that make use of smaller time limits produce results close to the optimal value and in some cases the optimal plan of actions.

As a future work we are going to investigate whether other local search algorithms might improve the results already achieved. We also want to investigate other strategies that can be used to reduce the complexity in the management of actions during the generation of new plans. In the experiments we compare the performance of our approach with just one bot player (BT). Others AI players are going to be used for comparison. In the near future our goal is the management of resources to be used in tactics for for attack and defense in conjunction with the approach already developed in order to build a complete player agent.

Acknowledgments. This research is supported in part by the Coordination for the Improvement of Higher Education Personnel (CAPES), Research Foundation of the State of Minas Gerais (FAPEMIG) and Faculty of Computing (FACOM) from Federal University of Uberlândia (UFU).

References

1. Aarts, E., Korst, J.: Simulated Annealing and Boltzmann Machines: A Stochastic Approach to Combinatorial Optimisation and Neural Computing. Wiley, New York (1989)

2. Branquinho, A., Lopes, C.R., Naves, T.F.: Using search and learning for production of resources in RTS games. In: The 2011 International Conference on Artificial Intelligence (2011)
3. Buro, M.: Call for AI research in RTS games. American Association for Artificial Intelligence (AAAI) (2004)
4. BWAPI: BWAPI - an API for interacting with starcraft: broodwar. http://code.google.com/p/bwapi/. Accessed 09 Apr 2015
5. Chan, H., Fern, A., Ray, S., Wilson, N., Ventura, C.: Online planning for resource production in real-time strategy games. In: ICAPS (2007)
6. Churchil, D., Buro, M.: Build order optimization in starcraft. In: AI and Interactive Digital Entertainment Conference, AIIDE 2011 (2011)
7. Do, M.B., Kambhampati, S.: Sapa: a scalable multi-objective metric temporal planner. J. AI Res. **20**, 63–75 (2003)
8. Fayard, T.: Using a planner to balance real time strategy video game. In: Workshop on Planning in Games, ICAPS 2007 (2007)
9. Fink, E.: Changes of Problem Representation: Theory and Experiments. Springer, Heidelberg (2003)
10. Gerevini, A., Saetti, A., Serina, I.: Planning through stochastic local search and temporal action graphs in LPG. J. Artif. Intell. Res. **20**, 239–290 (2003)
11. Hoffmann, J., Nebel, B.: FF: the fast-forward planning system. J. Artif. Intell. Res., 253–302 (2001)
12. Jay, Y., Nick, H.: Evolutionary learning of goal priorities in a real-time strategy game. In: Proceedings of the Eighth AAAI Conference on Artificial Intelligence and Interactive Digital Entertainment (2012)
13. Kecici, S., Talay, S.S.: TLplan-C: an extended temporal planner for modeling continuous change. In: The International Conference on Automated Planning and Scheduling (ICAPS) (2010)
14. Klenk, M., Molineaux, M., Aha, D.: Goal-driven autonomy for responding to unexpected events in strategy simulations. Comput. Intell., 187–203 (2013)
15. Minton, S., Bresina, J., Drummond, M.: Total-order and partial-order planning: a comparative analysis. J. Artif. Intell. Res. **2**, 227–262 (1994)
16. Naves, T.F., Lopes, C.R.: Maximization of the resource production in RTS games through stochastic search and planning. In: IEEE International Conference on Systems, Man and Cybernetics - SMC, pp. 2241–2246 (2012)
17. Ontanon, S., Synnaeve, G., Uriarte, A., Richoux, F., Churchill, D., Preuss, M.: A survey of real-time strategy game AI research and competition in starcraft. IEEE Trans. Comput. Intell. AI Games (2013)
18. Weber, B.G., Mawhorter, P., Mateas, M., Jhala, A.: Reactive planning idioms for multi-scale game AI. In: Proceedings of IEEE CIG (2010)

Inductive Synthesis of the Models of Biological Systems According to Clinical Trials

Vasily Osipov[1], Mikhail Lushnov[2], Elena Stankova[3], Alexander Vodyaho[4], and Nataly Zukova[4,5(✉)]

[1] St. Petersburg Institute for Informatics and Automation of the Russian Academy of Sciences (SPIIRAS), St. Petersburg, Russia

[2] St. Petersburg Almazov Cardiological Center, St. Petersburg, Russia
lushnov_ms@almazovcentre.ru

[3] St. Petersburg State University, St. Petersburg, Russia
e.stankova@spbu.ru

[4] St. Petersburg State Electrotechnical University, St. Petersburg, Russia
{aivodyaho,nazhukova}@mail.ru

[5] St. Petersburg National Research University of Information Technologies, Mechanics and Optics, St. Petersburg, Russia

Abstract. In the article an approach to solving the problem of inductive synthesis of the models of biological systems according to clinical trials is suggested. Suggested approach to inductive synthesis of biological models on the base of results of clinical trials allows essentially decrease computational complexity of this problem. Formalization of biological models in the form of graph of parameters allows use well developed mathematical apparatus of theory of graphs, which suggest effective methods of models transformation. Nowadays suggested approach is used in Almazov Cardiological Center for automatic medical data processing.

Keywords: Models of biological systems · Inductive synthesis of the models

1 Introduction

Nowadays one of the important problems in the domain of computer simulation of human biological systems is synthesis of models on the base of the results of clinical trials. Existence of such models which are correlated with age, gender and other patient features can be helpful for current healing and prophylactic activities of doctors. Nowadays essential efforts are directed to investigation of dependences of processes in human organism. In order to receive more adequate human biological models it is necessary to mine unknown dependences between entities of the model. It is well known that destruction of such couples indicates presence of misbalance in organism and results the patient's condition worsening [1, 2].

From the computational point of view the problem of mining of these dependences is rather sophisticated one. Each subsystem can have hundreds of parameters and number of results of measurement of these parameters may reach many hundreds. It is necessary to mention that number of possible variants of links of a set of M parameters is equal to the sum of their different combinations.

O. Gervasi et al. (Eds.): ICCSA 2017, Part I, LNCS 10404, pp. 103–115, 2017.
DOI: 10.1007/978-3-319-62392-4_8

On one hand, if more dependences are taken into account for model building, the model becomes more adequate; on the other hand, with increasing of the number of dependences, which are taken into account, the process of model syntheses becomes more complex.

In order to decrease complexity of the problem one can use methods based on reducing the space of analyzed parameters. Procedures of merging and cutting of loosely coupled parameters are known [3]. One can use an approach based on selection of so called concentrators and linear parameters. Concentrator parameters give generalized presentation of one parameter or set of linear parameters, which are correlated with this concentrator [4]. In cooperation with correlation and regression analysis this approach can give satisfactory results, but it can not decrease the complexity up to the level which is enough to built high quality models of biological systems.

One of the perspective approaches to solving this complex problem is usage of "divide and conquest" approach. The development of this approach for solving the problem of inductive synthesis of the models of biological systems according to clinical trials is described in this paper.

There is an important problem, which is not discussed in this article. It is a problem of generated model verification. There are a lot of known approaches to solving of this problem. One can use, for example, the approach, described in [5, 6] or any other.

2 Problem Definition

Let us assume, that results of clinical trails of liquids discharged by an organism which characterize a biological system are known. It can be a periphery blood system, acid-base blood status, urine parameters, etc. Results of clinical trials can be presented for example as a matrix

$$\mathbf{P} = \begin{bmatrix} p_{11}\, p_{12} \cdots\cdots\cdots p_{1N} \\ p_{21}\, p_{22} \cdots\cdots\cdots p_{2N} \\ \cdots\cdots\cdots\cdots\cdots\cdots \\ p_{M1}\, p_{M2} \cdots\cdots p_{MN} \end{bmatrix}$$

In general case such system can be described by a graph of parameters (vertexes) and dependences between arcs. The example of such graph is shown in Fig. 1.

Each value of the parameter in the system can be expressed by a generalized function, value of which depends upon coupled parameters. For example, i-*th* value of the first by order parameter p_{1i} according to the graph presented in Fig. 1 can be presented as

$$p_{1i} = \phi_{1i}(\phi_{12i}(p_{2i}), \phi_{13i}(p_{3i}), \phi_{14i}(p_{4i}), \phi_{15i}(p_{5i}), \phi_{16i}(p_{6i})); i = \overline{1, N},$$

where $\varphi_{1i}(.)$ - a partial function which couples the first parameter with other parameters; $\varphi_{12i}(.)$–$\varphi_{16i}(.)$ - more partial functions which couple values of the first parameter with separate values of other parameters and defining the power of these couples; $p_{2i}, p_{3i}, p_{4i}, p_{5i}, p_{6i}$ - values of parameters from which depends the value of the first

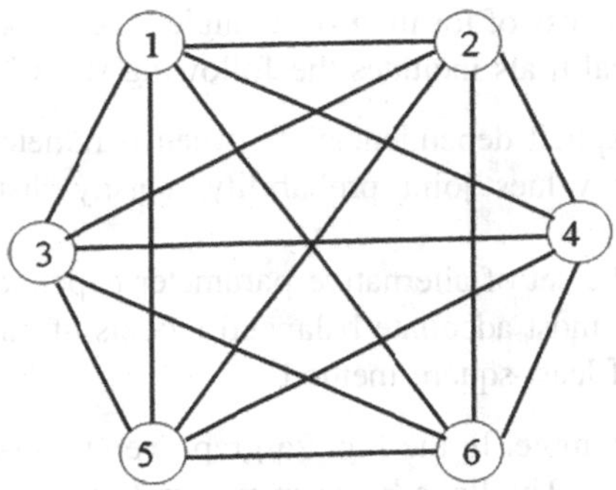

Fig. 1. The example of the graph of coupled parameters

parameter; N – is the number of values of each parameter, received as a results of clinical trials.

The inductive synthesis S is used for defining unique functions $\varphi_j(.)$ for all M parameters on the base of possible $\varphi_{ji}(.)$

$$S(\varphi_{ji}(.);\ i - \overline{1, N}) \to \varphi_j(.);\ j = \overline{1, M}.$$

For solving this problem at big values of M and N it is necessary to use special methods of decreasing the problem complexity. In the next section generalized algorithm of solving this problem with the help of decomposition approach is discussed.

3 Algorithm of Synthesis

Suggested algorithm of synthesis allows build hierarchical typified dependency graph of the parameters for biological systems with attributed vertexes and attributed arcs. For separate parameters one can define directions of the arcs. All vertexes (parameters) can be divided into 3 groups: free (not connected with other), linear and concentrator vertexes. One can associate correlation, functional and logical dependencies with arcs. Vertex attributes define type and form of dependency presentation which couples concentrator vertex with linear vertexes. Arc attributes are optional. If these attributes are used, they show the weight of linear parameter in the dependency reflected by concentrator vertex. Suggested algorithm is base on the following procedures. Let us assume that there is one level of hierarchy. In order to define linear parameters and concentrator parameters on initial step it is necessary to define explicit couples which form the skeleton of single level dependency model. Specification of the model is realized by means of fulfillment of this skeleton model. The procedure of skeleton fulfillment includes a number of procedures of analyses of different types of dependences and their base models. Each model can be specified iteratively. One can define values of linear parameters as a sum of other parameters multiplied by undefined coefficient. In nonlinear models parameters are expressed as nonlinear functions of other parameters. The process of forming of a single level dependency model (graph) from the results of clinical trials includes the following steps.

Step 1. Mining of explicit dependencies between parameters on the base of estimation of parameter values joint probability density distribution and skeleton forming.
Step 2. Forming of the set of alternative parameter dependency from each other.
Step 3. Choosing the most adequate balanced models of parameters dependencies by means of usage of least-square method.

Let us consider an example. In the Fig. 2a graph vertexes correspond to the initial set of analyzed parameters. The links between parameters are absent. As the result of analysis one concentrator parameter p_0^c, which is linked with five linear parameters (Fig. 2b) is received. Figure 2c shows defined logical dependencies.

Usage of the algorithm of logical dependencies mining allows find two additional concentrator parameters $(p_1^c \text{ and } p_2^c)$ and allows define dependencies between 10 parameters.

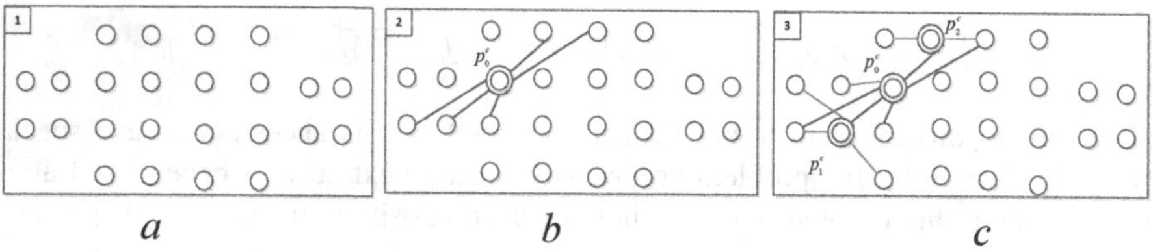

Fig. 2. The example of single level parameters dependencies model synthesis

When a hierarchical graph is build for biological system modeling, the following approach can be used. Hierarchical graph forming is based on usage of decomposition and aggregation operations. Decomposition assumes dividing of initial set of parameters into a number of parameter subsets. Inside each subset dependencies are defined. Aggregation assumes defining of dependencies between concentrator parameters of the neighboring levels. Both measured values of parameters and number of hierarchy levels are conceded as initial data. Attributed hierarchical graph of biological system parameter dependencies can be conceded as the result of synthesis. For realization of the synthesis procedure one can use the following sequence of steps.

Step 1. Decomposition of the parameter set into a number of subsets with the help of cluster analysis (logical parameter decomposition).
Step 2. Selection of the concentrator parameters and the linear parameters for all logical parameter sets of the current level. After it the level of hierarchy is increased by one.
Step 3. If the level of hierarchy is lower then maximum then go to step 1.

Step 4. Estimation of dependencies between concentrator parameters for different levels of hierarchy. Establishment links between neighboring levels parameters. This is vertical links in the hierarchical graph.

The example of such synthesis is shown in Fig. 3. For building the first level of hierarchy on the base of initial set of parameters (Fig. 3a) cluster analysis algorithms are used.

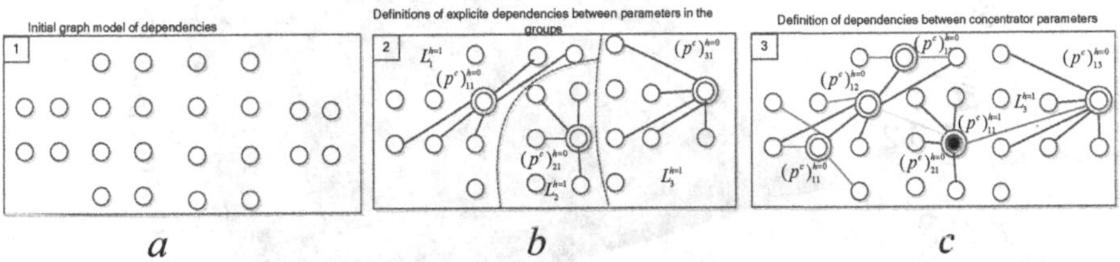

Fig. 3. The example of the synthesis of hierarchical model of parameters dependencies

As a result three sets of parameters are formed, which are shown in Fig. 3b. For each set concentrator parameters are defined. Established logical and functional links between concentrator parameters for selected sets are shown in the Fig. 3c.

Parameter $(p^c)_{11}^{h=1}$ is concentrator parameter of the second level of hierarchy. Links $(p^c)_{11}^{h=1}$–$(p^c)_{12}^{h=0}$ and $(p^c)_{11}^{h=1}$–$(p^c)_{13}^{h=0}$ are vertical links.

4 Separate Rules Peculiarities

Let us discuss the peculiarities of the synthesis rules which allow define dependencies between parameters of biological systems. The rule set is defined as a result of processing of arrays of experimental data in cooperation with medical staff. Main rules are the following: (i) rules for defining dependencies between parameters on the base of the computation of joint probability density function of two random values; (ii) rules for defining correlation dependencies between parameters; (iii) rules for defining functional and logical links.

Mining of dependences between values of parameters can be realized by means of computation of joint probability density function of two random variables in a following way.

Let the analyzed vector of parameters $\mathbf{P} = (p_1, \ldots p_i, \ldots p_j \ldots p_M)^T$ includes M parameters.

The procedure of the analysis includes M steps. One parameter from the set is excluded on each step. Then on j-*th* step the set of parameters can be presented as $P = \{p_j\} \cup \{P \backslash p_j\}$, excluded parameter p_j one can consider as a potential candidate to be the concentrator parameter, all the rest parameters are candidates to be the linear

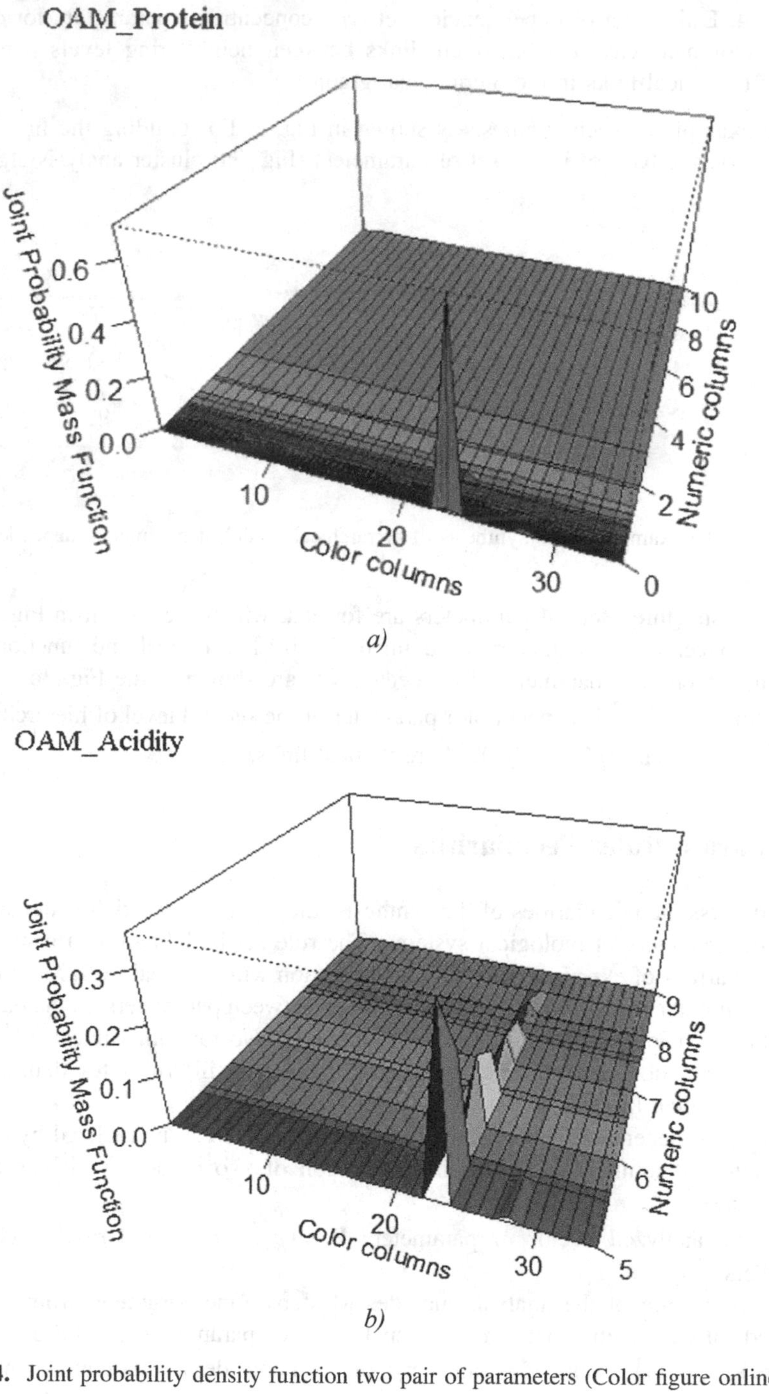

Fig. 4. Joint probability density function two pair of parameters (Color figure online)

parameters. Calculations of joint probability of p_j and parameters $P \backslash p_j$ are to be made. For paired analysis of quality parameters it is necessary to calculate basis which is build on the base of the attribute values. The bases for quantized parameters are either attribute values or intervals of values. For estimation of presence of the parameter dependences criterion χ^2 is used.

The example of joint probability density function two pair of parameters is shown in the Fig. 4. The parameters "Color" and "OAM_Protein" have explicit dependence (Fig. 4a), for parameters "Color" and "OAM_Acidity" the dependence is doubtful.

For finding the correlation dependencies the following mechanism is used. On the input there is a matrix of result of analysis. The pair correlation coefficients with the help of known formula must be calculated [7]. As a result the correlation matrix is formed. With the help of this matrix all parameters are divided into sets depending upon values of the correlation coefficients.

In order to mine functional dependencies it is necessary to receive estimations of presence linear and nonlinear regression dependencies between parameters with the help of the least-square method. For defining values of parameters one can also use Levenberg-Marquardt method [8]. Logical dependencies can be found with the help of the algorithms of search of maximum and closed partial data sets in the table of trial results [9].

5 Results of Experiments

These results were received on the base of analyses of the results of patients' physiological systems parameters measurements. Data were received from the Almazov Cardiological Center (Saint-Petersburg, Russia, http://www.almazovcentre.ru/). In the center medical information system «qMS» produced by SP. ARM (Saint-Petersburg, Russia, http://www.sparm.com/) is deployed. In its data base hundreds thousands urine analysis received by means of automatic laboratory devices usage are stored. Besides it the data base contains medical reports of urine analysis that describe verbally urine properties.

Doctors can make requests to the data base for receiving additional information about patient's condition taking into account all available verbal and numerical information about urine properties.

During the experiment more then 13000 results of urine trials of patients who were treated in the Center from 2014 were investigated. The patients had cardiovascular diagnosis (ICD-10: I20-I22). The number of analyzed parameters for each patient was about 50.

The experiment included following steps.

Step 1. Data preprocessing. While data preprocessing, main attention was paid to processing of potential concentrator parameters values. The majority of quantity parameters were transformed into quality parameters with the help of sampling algorithms. In order to form the set of quality values a domain vocabulary (domain model) was used, the methods of frequency estimations of parameters values were used also. The text Mining [10, 11] algorithms were used for selections of the most important parameters. In particular methods of inverse document frequency (IDF) and singular value decomposition (SVD) [11] were used.

Step 2. Model building. On the base of data about each patient, single level dependency model was build. For defining links correlation-regression analysis methods realized in the packet STATISTICA (StatSoft/Dell, https://www.statsoft.com) were used.

Step 3. Dependency model estimation. Patients were grouped according with the diagnosis. Each group was investigated separately. With the help of usage of the operation of graph (model) merge, generalized graphs in accordance with diagnosis were built. In the process of merging the weight was assigned to each vertex. The weight is defined by the number of patients for which this parameter was identified as concentrator parameter. For the arcs the weights were assigned in a similar way.

It was discovered that distribution of a set of concentrator parameters and linear parameters does not depend upon the diagnosis. The distribution for these parameters was normal for all diagnosis. Results for two parameters are shown in Fig. 5. Parameter "OAM_Oksalati" (Fig. 5a) does not depend upon the diagnosis, parameter "OAM_Urati" is different.

The distributions analysis shows that the difference between graphical dependencies model is not more then 5%. One concentrator parameter describes from 3 to 10 linear parameters and 7 parameters by median. Estimations are calculated after deleting of 10% of noise from left and right from the center of mass of distributions, i.e. 10% of concentrator parameters, which include minimum and maximum number of links, which are not taken into account.

Quality estimation of the dependency model for physiological system was done. The example of solving the exact problem of analysis of such systems on the base of parameters of liquids discharged by organism is described below.

The problem was to study the structure of color clinical semantic gamma of the text variable "Color" and its correlation with quantity parameters of urine in order to use this verbal information in the process of clinical decision making. The main problem was that laboratory technicians do not have a tradition of usage colorimeters when formulating verbal conclusions on the color of urine. For defining the term "Color" they use many different words, such as light yellow, yellow, stramineous, straw-color, yellowish-brownish, reddish-brownish, light-brown, bloody, yellowish-green with a brown tinge, intensive yellow, brown, orange, greenish-yellowish with a brownish tinge, yellowish-pink, dark brown, amber, pink with blood, colorless, pink, milk white, lemon yellow, beer color, etc. Totally 54 terms indicating different colors were included into vocabulary and synonyms were defined. During analysis, inverse document frequencies were calculated. It is assumed that in a medical context selected terms are the most semantically focused. This transformation reflects word images (frequencies) and total frequency of its appearance in the parameters set to be analyzed. As a rule, it is called inverse document frequency (IDF). After it, in order to select medical terms, singular value decomposition (SVD) was performed.

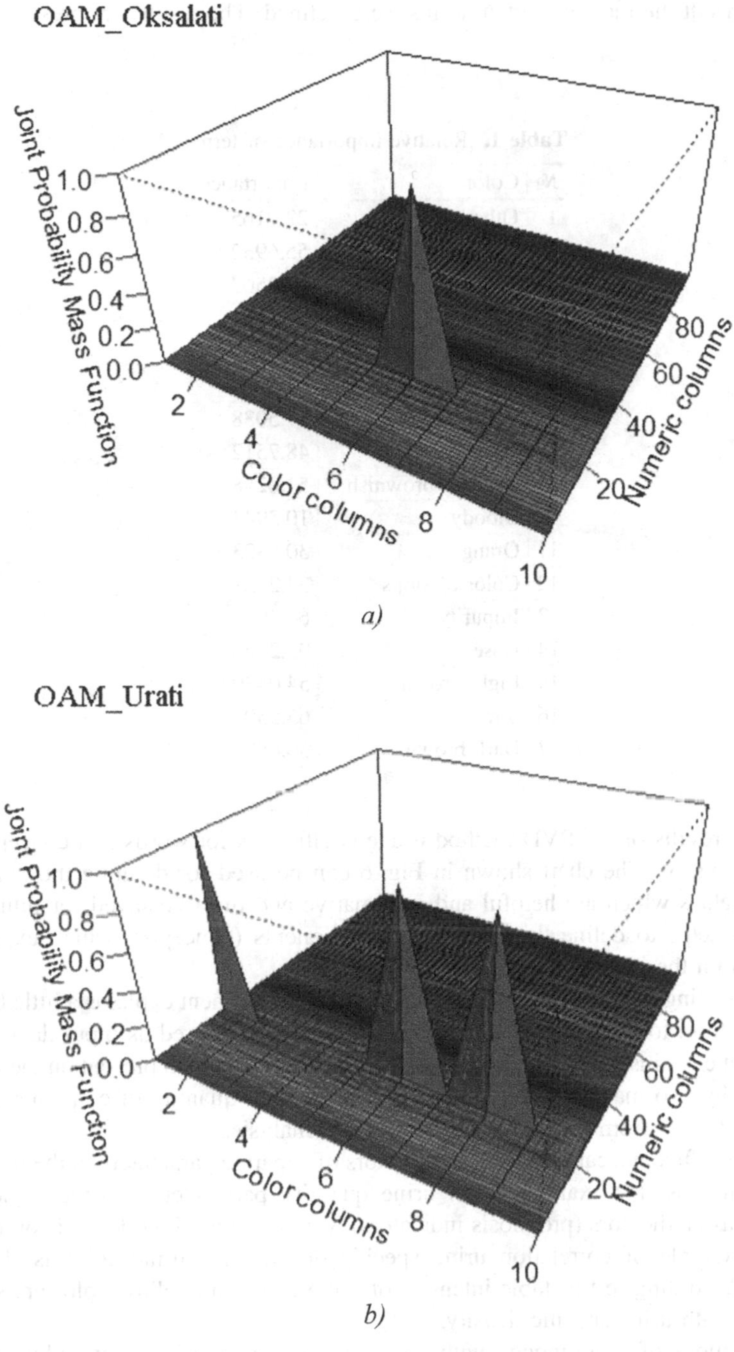

Fig. 5. Distribution of concentrator parameters and linear parameters for subsystems (Color figure online)

As a result the most important terms were defined. The results are presented in the Table 1.

Table 1. Relative importance of terms

№	Color	Importance
1	Other	22.9165
2	Colorless	55.6932
3	Brownish	47.3502
4	Yellowish-brownish	19.1181
5	Yellowish-greenish	79.0909
6	Yellowish-pinkish	18.1297
7	Greenish-yellow	17.5038
8	Intensive	48.7512
9	Reddish-brownish	51.0248
10	Bloody	10.3943
11	Orange	30.5323
12	Color of slops	34.2251
13	Impurity	64.3964
14	Rose	93.2726
15	Light-brown	54.0430
16	Straw	65.5507
17	Dark brown	50.6809

As the results of the SVD method usage coefficients for words and concepts were received (Fig. 6). The chart shown in Fig. 6 can be used for defining the number of singular values which are helpful and informative and must be saved for future analysis. This helps to define the number of components (concepts), which explain the dispersion on the input.

Investigating the chart one can say that the first component explains a little bit more then 16% from total dispersion for 19 words, which were used as input data. Second component explains about 8%, etc. So 24% of dispersion, which present on the input, is explained by two main components. So, not more then quantity of components (concepts) on the left from this point are helpful for analysis.

Word coefficients can be used as predictors of separate parameters in the process of links definitions, for example, with urine quantity parameters or other systems of organism as predicators (prognosis indicators), which are used while decision making.

The example of correlation urine specific density and urine color is shown in Table 2. According to this table intensity of urine color and yellow color are strongly correlated with urine specific density.

Correlations of urobilinogen with yellow-orange, greenish-yellow colors are presented in Tables 3 and 4 respectively. Results in the Table 3 are evident facts, but they are received from the text analysis.

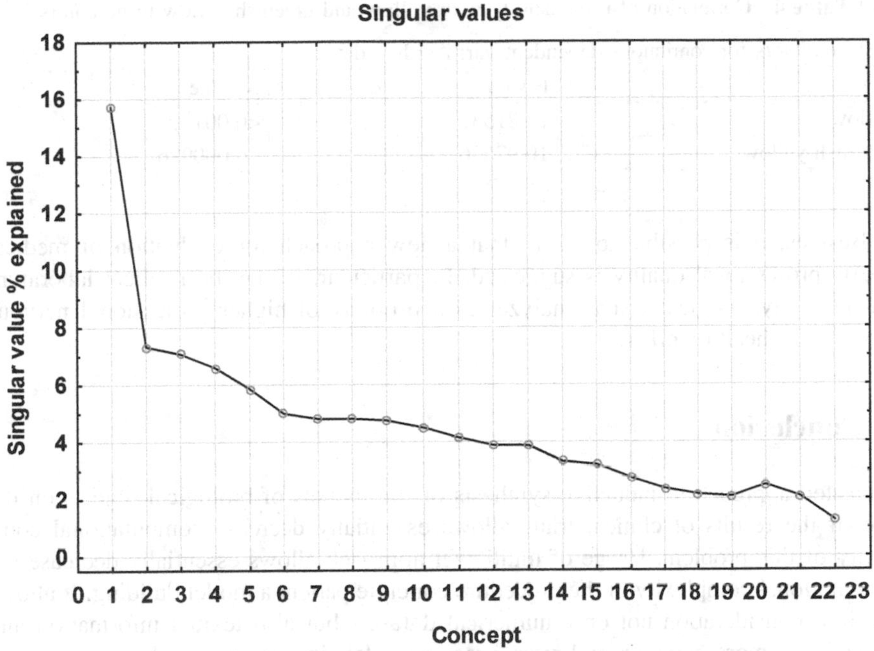

Fig. 6. Concept terms chart

Table 2. Correlation of urine specific density and urine color

Best predictors for continuous dependent var: specific density		
	F-value	p-value
Intensity	111.2486	<0.001
Yellow	14.8404	<0.001

Table 3. Correlation of urobilinogen with yellow-orange, greenish and other urine colors

Best predictors for continuous dependent var: urobilinogen		
	F-value	p-value
Yellow	114.6666	<0.001
Orange	47.4889	<0.001
Other	6.5016	0.010793
Greenish-yellow	6.4359	0.011199

So, the suggested approach based on analysis of urine color has shown its correctness and utility for solving real tasks of clinic practice. Such approach can be used for automatic processing of medical conclusions and other medical documents.

Table 4. Correlation of urine acidity with yellow and greenish-yellow urine colors

Best predictors for continuous dependent var: Acidity_device		
	F-value	p-value
Yellow	27.81532	<0.001
Greenish-yellow	10.97656	0.000926

Besides, it is possible to affirm that a new approach for evaluation of medical experts professional quality is suggested. In particular, it may be medical laboratory and their physiological visual analyzer as a property of higher professional nervous activity of a health worker.

6 Conclusion

Suggested approach to inductive synthesis of the models of biological objects on the base of the results of clinical trials allows essentially decrease computational complexity of this problem. Usage of multilevel approach allows essentially decrease the computational complexity of the procedure concrete patient a model building. It allows take into consideration not only numerical datasets but also textual information and allows build more complex and more precise models in run time mode.

Nowadays suggested approach is used in Almazov Cardiological Center for automatic medical data processing. The models for patients treated in the center in 2014 were built and evaluated. The overall volume of processed data was about 4 GB. Formalization of biological models in the form of graph of parameters allows use well developed mathematical apparatus of theory of graphs, which suggests effective methods of transformation of such models. In particular it is possible to build graph morphism and optimize computational operation on the sets of linked elements.

Usage of suggested algorithms of model building is not limited by the medicine domain. They can be used in other subject domains such as biology and sociology, where it is necessary to build in run time models of the objects, taking into account not only numerical but also textual information from experts, when volumes of information to be processed is big or very big.

Results of experiments show that suggested algorithms can be effectively used for processing tens of gigabytes of information. This is quite enough for solving real problems at least in medical domain. If it would be necessary to process bigger volumes of information it can be done with the help of 3 mechanisms: (i) usage of parallel algorithms on clusters; (ii) usage of n-layer algorithms instead of two layer algorithms; (iii) usage of parallel n-layer algorithms.

Acknowledgement. This work was partially financially supported by Government of Russian Federation, Grant 074-U01

References

1. Lushnov, A., Lushnov, M.: Medical Information Systems: Multidimensional Analyses of Medical and Ecology Data. Helicon Plus, SpB (2013)
2. Kupershtoh, V., Mirkin, B., Trofimov, V.: The sum of internal links as indicator of classification results quality. Avtomatika i Telemekhanika **3**, 133–141 (1976). (in Russian)
3. Kriete, A., Eils, R.: Computational Systems Biology. Elsevier Academic Press, Cambridge (2006)
4. Akbari, Z., Unland, R.: Automated determination of the input parameter of DBSCAN based on outlier detection. In: Iliadis, L., Maglogiannis, I. (eds.) AIAI 2016. IFIP AICT, vol. 475, pp. 280–291. Springer, Cham (2016). doi:10.1007/978-3-319-44944-9_24
5. Stankova, E.N., Balakshiy, A.V., Petrov, D.A., Shorov, A.V., Korkhov, V.V.: Using technologies of OLAP and machine learning for validation of the numerical models of convective clouds. In: Gervasi, O., et al. (eds.) ICCSA 2016, Part III. LNCS, vol. 9788, pp. 463–472. Springer, Cham (2016). doi:10.1007/978-3-319-42111-7_36
6. Raba, N.O., Stankova, E.N.: On the problem of numerical modeling of dangerous convective phenomena: possibilities of real-time forecast with the help of multi-core processors. In: Murgante, B., Gervasi, O., Iglesias, A., Taniar, D., Apduhan, B.O. (eds.) ICCSA 2011. LNCS, vol. 6786, pp. 633–642. Springer, Heidelberg (2011). doi:10.1007/978-3-642-21934-4_51
7. Freedman, D.: Statistical Models: Theory and Practice, 2nd edn. Cambridge University Press, Cambridge (2009)
8. Gill, P.R., Murray, W., Wright, M.H.: The Levenberg-Marquardt method. In: Practical Optimization, pp. 136–137. Academic Press, London (1981)
9. Zaki, M.J., Meira Jr., W.: Data Mining and Analysis: Fundamental Concepts and Algorithms. Cambridge University Press, Cambridge (2014)
10. Feldman, R., Sanger, J.: The Text Mining Handbook: Advanced Approaches in Analyzing Unstructured Data. Cambridge University Press, Cambridge (2006)
11. Leskovec, J., Rajaraman, A., Ullman, J.: Mining Massive Datasets. Stanford University, Stanford (2014)

Solving Sparse Differential Riccati Equations on Hybrid CPU-GPU Platforms

Peter Benner[1], Ernesto Dufrechou[2], Pablo Ezzatti[2], Hermann Mena[3], Enrique S. Quintana-Ortí[4], and Alfredo Remón[1](✉)

[1] Max Planck Institute for Dynamics of Complex Technical Systems, Magdeburg, Germany
{benner,remon}@mpi-magdeburg.mpg.de
[2] Instituto de Computación, Universidad de la República, 11300 Montevideo, Uruguay
{edufrechou,pezzatti}@fing.edu.uy
[3] School of Mathematical Sciences and Information Technology, Yachay Tech, San Miguel de Urcuquí, Ecuador
mena@yachaytech.edu.ec
[4] Dep. de Ingeniería y Ciencia de la Computación, Universidad Jaime I, Castellón, Spain
quintana@icc.uji.es

Abstract. The numerical treatment of the linear-quadratic optimal control problem requires the solution of Riccati equations. In particular, the differential Riccati equations (DRE) is a key operation for the computation of the optimal control in the finite-time horizon case. In this work, we focus on large-scale problems governed by partial differential equations (PDEs) where, in order to apply a feedback control strategy, it is necessary to solve a large-scale DRE resulting from a spatial semi-discretization. To tackle this problem, we introduce an efficient implementation of the implicit Euler method and linearly implicit Euler method on hybrid CPU-GPU platforms for solving differential Riccati equations arising in a finite-time horizon linear-quadratic control problems. Numerical experiments validate our approach.

Keywords: Differential Riccati equation (DRE) · Implicit Euler · Linearly implicit Euler · LQR · Feedback control

1 Introduction

The linear-quadratic optimal control problem is a constrained optimization problem. The dynamics constraints are given by the state equation

$$\dot{x}(t) = Ax(t) + Bu(t), \quad t > 0, \quad x(0) = x_0, \tag{1}$$

where $A \in \mathbb{R}^{n \times n}$ is the state matrix, $B \in \mathbb{R}^{n \times m}$ is the input matrix, and the cost function to be minimized is given by

$$J(x_0, u) := \int_0^{t_f} (x(t)^T Qx(t) + u(t)^T Ru(t))dt + x(t_f)^T Gx(t_f). \tag{2}$$

O. Gervasi et al. (Eds.): ICCSA 2017, Part I, LNCS 10404, pp. 116–132, 2017.
DOI: 10.1007/978-3-319-62392-4_9

If Q is positive semidefinite, R is positive definite, and the time horizon is finite (i.e., $t_f < \infty$), the minimum and consequently optimal control is given in feedback form by: $u_*(t) = -K(t)^T x(t)$, where the feedback $K(t)$ is a matrix-valued function defined as: $K(t) = X_*(t)BR^{-1}$, and $X_*(t)$ is the unique non-negative self-adjoint solution of the differential Riccati equation (DRE) [1]:

$$\begin{aligned} \dot{X}(t) &= -(Q + A^T X(t) + X(t)A - X(t)BR^{-1}B^T X(t)), \\ X(t_f) &= G. \end{aligned} \tag{3}$$

In this paper, we consider optimal control of partial differential equations (PDEs) [2,3]. Large-scale DREs appear in these problems because the dimension of A reflects the number of states resulting from the spatial discretization. Here by large-scale we refer to n being of order 10^3 to 10^6. Moreover, the coefficient matrices of the resulting DRE often exhibit a special structure. Specifically, in many practical applications [4], A is sparse, while Q is of low (numerical) rank which implies that it can be approximated by a low rank factor (LRF). In practice, for large-scale applications it is not possible to construct explicitly $X(t_k) := X_k$, i.e., the matrices evaluated at each point of the mesh t_k, because these are in general dense. However, X_k usually has a low rank as well, and therefore it can be accurately approximated by a LRF $Z_k \in \mathbb{R}^{n \times z_k}$ such that $X_k \approx Z_k Z_k^T$ with $z_k \ll n$. This property has been deeply studied and is frequent in applications [5–8]. The DRE is perhaps the most exhaustively addressed nonlinear matrix differential equation as it arises in optimal control, optimal filtering, H_∞ control of linear-time varying systems, and differential games, among others [9–11]. There is a large variety of approaches in the literature to compute the solution of the DRE [12–15]. Most of the methods consider stiff DREs as due to the fact that fast and slow modes are present. For our hybrid CPU-GPU platforms we follow the algorithms developed in [16]. In [16,17], the authors consider the solution of large-scale DREs arising in optimal control problems for parabolic partial differential equations via numerical methods for ordinary differential equations such as, the Rosenbrock method and the backward differentiation formula (BDF). Although the performance of these methods is competitive, their computational cost is quite high for large-scale problems, asking for the application of high performance computing (HPC) techniques to accelerate their execution. We point out that a new approach based on splitting methods has been recently proposed [18,19]. The numerical implementation of this approach relies on the computation of the matrix exponential, and therefore we will not consider it here.

In this work, we propose a parallel solver for (3) that exploits the massive hardware concurrency of graphics processors (GPUs) to tackle the computational complexity of these problems. In recent years, there has been a rapid evolution in the number and computational power of the cores integrated into general-purpose CPUs, as well as an increasing adoption of GPUs as hardware accelerators in scientific computing. A large number of efforts have demonstrated the remarkable speed-up that hybrid CPU-GPU systems provide for the solution of dense and sparse linear algebra problems. In particular, such hardware platforms have been previously employed for the numerical solution of DREs, with dense coefficient matrix A, in [20]. In this paper we turn our attention towards the more challenging sparse case. Due to the large-scale nature of the problem, we will focus on BDF and Rosenbrock methods of order one.

The remainder of the paper is organized as follows. In Sect. 2 we describe the algorithms for solving the DRE via BDF and Rosenbrock methods offering a few implementation details. In Sect. 3, we present our hybrid CPU-GPU implementation of the methods. In Sect. 4 we report results for a collection of experiments that validate the efficiency of our approach. Finally, a discussion of conclusions and future work closes the paper in Sect. 5.

2 Differential Riccati Equations

In order to obtain the solution of the DRE (3), we can integrate the DRE forward in time. In the following we describe the numerical methods for time-varying symmetric DREs of the form

$$\begin{aligned} \dot{X}(t) &= Q(t) + X(t)A(t) + A^T(t)X(t) - X(t)S(t)X(t) = F(t, X(t)),\\ X(t_0) &= X_0, \end{aligned} \tag{4}$$

where $t \in [t_0, t_f]$ and $Q(t), A(t), S(t), X(t) \in \mathbb{R}^{n\times n}$. We assume that the coefficient matrices are piece-wise continuous, locally bounded, matrix-valued functions, which ensures the existence and uniqueness of the solution of (4); see, e.g., [9, Theorem 4.1.6]. In this paper we will focus on time-invariant systems. We briefly revisit the BDF and Rosenbrock methods of order one for DREs, i.e., the implicit and linearly implicit Euler methods, [16].

BDF Methods. Consider $X_{k+1} \approx X(t_{k+1}), Q_{k+1} := Q(t_{k+1}), A_{k+1} := A(t_{k+1})$, and $S_{k+1} := S(t_{k+1})$. The application of a BDF method of order one yields the algebraic Riccati equation (ARE) for X_{k+1}

$$\begin{aligned} (hQ_{k+1} - X_k) + (hA_{k+1} - \tfrac{1}{2}I)^T X_{k+1} + X_{k+1}(hA_{k+1} - \tfrac{1}{2}I)&\\ -X_{k+1}(hS_{k+1})X_{k+1} = 0,& \end{aligned} \tag{5}$$

where I stands for the identity matrix (of appropriate order), h is the time-step size, and thus, $t_{k+1} = h + t_k$. Assuming that

$$\begin{aligned} Q_k &= C_k^T C_k, \qquad & C_k &\in \mathbb{R}^{p\times n},\\ S_k &= B_k B_k^T, & B_k &\in \mathbb{R}^{n\times m},\\ X_k &= Z_k Z_k^T, & Z_k &\in \mathbb{R}^{n\times z_k}, \end{aligned}$$

the ARE (5) can be then re-written as

$$\begin{aligned} \hat{C}_{k+1}^T\hat{C}_{k+1} + \hat{A}_{k+1}^T Z_{k+1}Z_{k+1}^T + Z_{k+1}Z_{k+1}^T\hat{A}_{k+1}&\\ - Z_{k+1}Z_{k+1}^T\hat{B}_{k+1}\hat{B}_{k+1}^T Z_{k+1}Z_{k+1}^T = 0,& \end{aligned} \tag{6}$$

where $\hat{A}_{k+1} = hA_{k+1} - \frac{1}{2}I, \hat{B}_{k+1} = \sqrt{h}B_{k+1}, \hat{C}_{k+1}^T = \sqrt{h}C_{k+1}^T - Z_kZ_k^T$.

In this way, we obtain exactly the linear implicit Euler method. To solve AREs, we use the Newton's method as a one-step iteration. This results in the need to solve one Lyapunov equation at each step.

Remark 1. For higher order BDF methods, due to the fact that some parameters of the method are negative (see e.g., [21]), a direct implementation requires complex arithmetic. However, an LDL^T decomposition [22] of the right-hand side of the related Lyapunov equation overcomes this problem. This approach was recently proposed in [23]. In this work we do not consider this option as we focus on the method of order one.

Rosenbrock Methods. These algorithms can also be written in terms of the LRF of the solution. Concretely, the one-stage Rosenbrock method applied as a matrix-valued algorithm to autonomous DREs of the form (4) can be written as

$$\begin{aligned} \bar{A}_k^T K_1 + K_1 \bar{A}_k &= -F(X_k) - hF_{t_k}, \\ X_{k+1} &= X_k + K_1, \end{aligned} \tag{7}$$

where $\bar{A}_k = A_k - S_k X_k - \frac{1}{2h} I$, and $F_k := F(t_k, X_k), F$ is defined as in (4). Moreover, the Eq. (7) can be reformulated such that the next iterate is computed directly from the Lyapunov equation

$$\bar{A}_k^T X_{k+1} + X_{k+1} \bar{A}_k = -Q - X_k S X_k - \frac{1}{h} X_k. \tag{8}$$

The right-hand side of (8) is simpler to evaluate than that in (7), and consequently the implementation of (8) is more efficient [17]. If we assume,

$$\begin{aligned} Q &= C^T C, & C &\in \mathbb{R}^{p \times n}, \\ S &= B B^T, & B &\in \mathbb{R}^{n \times m}, \\ X_k &= Z_k Z_k^T, & Z_k &\in \mathbb{R}^{n \times z_k}, \end{aligned}$$

with $p, m, z_k \ll n$, and set $N_k = [\, C^T \;\, Z_k(Z_k^T B) \;\, \sqrt{h^{-1}} Z_k \,]$, the Lyapunov equation (8) results in

$$\bar{A}_k^T X_{k+1} + X_{k+1} \bar{A}_k = -N_k N_k^T, \tag{9}$$

where $\bar{A}_k = A - B(Z_k(Z_k^T B))^T - \frac{1}{2h} I$. Observing that $rank(N_k) \le p + m + z_k \ll n$, we can use the modified version of the *alternating directions implicit* (ADI) iteration to solve (9). The application of the low rank Cholesky factor ADI (LRCF-ADI) iteration to (9) yields LRFs $Z_{k+1} \in \mathbb{R}^{n \times z_{k+1}}$, with $z_{k+1} \ll n$, such that $X_{k+1} \approx Z_{k+1} Z_{k+1}^T$.

Lyapunov Equation Solver. Efficient numerical methods to tackle large and sparse Lyapunov equations have been proposed in recent years. A common approach is the LRCF-ADI iteration [24–26]. In our implementation we use recent advances of ADI based solvers [25,27]. The low rank representation of the solution relies on the decay of the singular values. Methods based on extended and rational Krylov subspace have also proved to be practical alternatives [28]; however, we do not consider these approaches in this work. A state-of-the-art survey of the methods is presented in [29].

To close this section, we briefly describe the ADI iteration for the Lyapunov equation:

$$F^T Y + YF = -WW^T, \tag{10}$$

where F is c-stable [1], and $W \in \mathbb{R}^{n \times n_w}$. The ADI iteration applied to (10) boils down to

$$\begin{aligned}(F^T + p_j I)Y_{j-\frac{1}{2}} &= -WW^T - Y_{j-1}(F - p_j I),\\ (F^T + \overline{p_j} I)Y_j^T &= -WW^T - Y_{j-\frac{1}{2}}(F - \overline{p_j} I);\end{aligned}$$

where $\overline{p}_j$ denotes the complex conjugate of $p_j \in \mathbb{C}$ for $j = 1, \ldots, k$ and $j - \frac{1}{2}$ represent an intermediate evaluation, see [30].

3 A Hybrid CPU-GPU Implementation

The BDF and the Rosenbrock methods of order one require the solution of a Lyapunov equation in their inner "loop". Thus, both methods rely on the same building block and, therefore, their implementations are similar. In this section we focus on the BDF methods of order one, but part of the implementation issues are relevant for the Rosenbrock method as well. Although the code is written in MATLAB, it relies on external high performance libraries to perform the most expensive operations from the point of view of execution time. Hence, the result is an efficient and easy-to-use implementation for multicore platforms. Additionally, we modified the Lyapunov solver, the most expensive and critical kernel in the method, such that most of its suboperations are offloaded to the GPU. Thus, the implementation of the BDF method presented in this paper exploits the computational resources in a CPU-GPU platform.

As argued earlier, from a computational viewpoint, the performance bottleneck lies in the sparse Lyapunov equations, which are solved using a modified version of LYAPACK [31]. In turn, the performance of the underlying Lyapunov solver strongly depends on the actual implementation of two key operations: the solution of linear systems of the form $Ax = b$, where x, b are vectors and A is sparse; and the solution of shifted linear systems of the form $(A + p_i I)X = B$, where A is a sparse matrix, X, B are matrices with a small number of columns, and p_i is the i-th parameter from the ADI iteration. The flexible design of LYAPACK allows the user to select (or provide) the implementation of the linear system solver (and other basic operations) that best suits the problem data and the underlying hardware. This allows to conveniently exploit properties such as the sparsity pattern of the coefficient matrix, its condition number, or its dimension. In this study, we develop an iterative solver, based on the Bi-Conjugate Gradient Stabilized (BICGSTAB) method, especially designed to leverage the computing capabilities of the GPU. This algorithm is suitable to tackle sparse linear systems of large dimension, and the use of a GPU renders an important acceleration factor. We also incorporate a tridiagonal solver based in [32] for problems where A presents this particular structure.

Algorithm 1. x =BiCGSTAB(A, σ, b)

1: $\hat{r}_0 = r_0 = b - Ax_0$
2: $\rho_0 = \alpha = \omega_0 = 1$
3: $v_0 = p_0 = 0$
4: **for** $i = 1$ **to** ... **do**
5: $\quad \rho_i = (\hat{r}_0, r_{i-1})$ (DOT)
6: $\quad \beta = (\rho_i/\rho_{i-1})(\alpha/\omega_{i-1})$
7: $\quad p_i = r_{i-1} + \beta(p_{i-1} - \omega_{i-1}v_{i-1})$ (AXPY+SCAL+AXPY)
8: $\quad v_i = Ap_i$ SpMV
9: $\quad$ **if** $\sigma \neq 0$ **then** $v_i = v_i + \sigma p_i$ (AXPY)
10: $\quad \alpha = \rho_i/(\hat{r}_0, v_i)$ (DOT)
11: $\quad s = r_{i-1} - \alpha v_i$ (AXPY)
12: $\quad t = As$ SpMV
13: $\quad$ **if** $\sigma \neq 0$ **then** $t = t + \sigma s$ (AXPY)
14: $\quad \omega_i = (t, s)/(t, t)$ (DOT+NRM2)
15: $\quad x_i = x_{i-1} + \alpha p_{i-1} - \omega_i s$ (AXPY+AXPY)
16: $\quad$ **if** x_i is accurate enough **then** stop
17: $\quad r_i = s - \omega_i t$ (AXPY)
18: **end for**

The GPU implementation of the BiCGStab algorithm is based on that proposed by Naumov [33]. A description of BiCGStab is given in Algorithm 1 where, next to each operation, we name the CUBLAS routine that supports it (inside brackets). The most time-consuming type of operation in this method is the sparse matrix-vector product (SpMV). Furthermore, we note that this kernel is invoked twice at each iterative step of the algorithm (lines 8 and 12). Therefore, it is crucial for the efficient execution of the whole solver.

In order to efficiently compute the SpMV of the shifted matrices $(A + \sigma I) \cdot y$ present in the ADI iteration, on a GPU, we rewrite this operation as $(Ay) + (\sigma Iy) = Ay + \sigma y$, i.e., a SpMV and a vector scaling. To accelerate the computation of the SpMV, A is stored in CSR format [34], which facilitates the row-wise access to the matrix and therefore, it is appropriate for the computation of SpMV products. Both operations are provided as part of Nvidia libraries. In particular, the SpMV product is implemented in the `csrmv` routine included in CUSPARSE, while the vector scaling is implemented in routine `axpy` of CUBLAS.

The remaining operations in Algorithm 1 are performed via calls to the corresponding CUBLAS kernels. As stated previously, our implementation only addresses the time-invariant case. This allows us to transfer the coefficient matrix A to the GPU only once, before the computation starts. The rest of the data transfers between the host and the accelerator are: (1) transfer the right-hand side vector of each linear system to the GPU; and (2) retrieve the solution vector back to the CPU. The time required by these data transfers, when compared to the cost of the computations, is negligible.

The number of columns of the LRF Z_j, grows as the computation proceeds since n_w new columns are appended at each step. This implies an increment in the computational cost of the following steps of the algorithm, which might turn the cost of the algorithm unaffordable. Given the LRF of the solution X, this problem can be alleviated by means of a column compression algorithm. Concretely, we employ the RRQR factorization in [35].

4 Experimental Evaluation

This section presents the numerical evaluation of our versions of the BDF and Rosenbrock solvers in terms of accuracy and execution time.

All the experiments in this section were performed on a server equipped with recent CPU+GPU technology from INTEL and NVIDIA; Specifically, an INTEL I7-4770 processor with 4 cores at 3.4 GHz and 16 GB of RAM connected to a NVIDIA K40C with 2,880 CUDA cores and 12 GB of RAM. Additionally, we use CentOS Rel. 6.4, gcc v4.4.7, INTEL MKL 11.1, MATLAB Version R2015a and CUDA/CUSPARSE/CUBLAS 7.0. Double-precision floating-point arithmetic was employed in all cases.

The default number of shifts was employed in all cases ($l = 10$). The values returned by the selection procedure in LYAPACK required only real arithmetic (and, therefore, Basic Matrix Operations, or BMOs, that operate with real data). To provide a fair comparison, all our timings include the cost of data transfers between CPU and GPU for the CUDA-enabled routines. Furthermore, the original version of the LYAPACK routine is executed on the INTEL CPU using one thread per physical core. The stopping criterion of the BiCGStab iteration was determined such that the accuracy achieved in the solution was in the same order of magnitude as that obtained with the direct LU-based method.

4.1 Test Cases

To evaluate the proposal we use three different families of test cases:

- CHOI-LAUB. This model is extracted from [36], and is based on the following symmetric DRE of size $n \times n$

$$\begin{aligned} \dot{X}(t) &= k^2 I_n - X^2(t), \, t_0 \leq t \leq T, \\ X(t_0) &= X_0, \end{aligned}$$

  where k is a parameter that controls the stiffness of the equation. When X_0 is diagonalizable, i.e., $X_0 = S \Sigma S^{-1}$ and $\Sigma = \text{diag}\{\sigma_i\}_{i=0\ldots n}$, the solution of this equation can be expressed as

$$X(t) = S \, \text{diag} \left(\frac{k \sinh kt + \sigma_i \cosh kt}{\cosh kt + \frac{\sigma_i}{k} \sinh kt} \right)_{i=0,\ldots,n} S^{-1}$$

  In our experiments, $k = 100, X_0 = I_n$ with $n = 2,000$ and $0 \leq t \leq 1$.

- POISSON. This case models a 2-D distribution of temperature. The state matrix A is obtained from the discretization of the Poisson's equation with the 5-point operator on an $N \times N$ mesh, resulting in a block tridiagonal sparse matrix of order N^2. The input matrix is constructed as $B = \left[e_1^T, 0_{1\times N(N-1)}\right]^T$, with $e_1 \in \mathbb{R}^N$ denoting the first column of the identity matrix. We evaluate five instances of this problem resulting from the use of different discretization meshes: $N =$ 50, 100, 150, 200 and 250.
- HIGHW. This example corresponds to a mathematical model of position and velocity control for a string of high speed vehicles. The model is know as smart highway and is described in [37, Example 3.1]. Here we use less columns in matrix B as the target problems we want to focus on, LQR problems governed by PDEs, usually present this form. Four instances of this problem are evaluated, $n =$ 524,289, 786,433, 1,048,577 and 1,572,865.

4.2 Numerical Results

The first experiment aims to asses the accuracy of the two methods, and the impact of reducing the time-step h on the accuracy of the solution. Thus, we first test our solvers on a matrix equation with a known analytic solution, i.e., the CHOI-LAUB test case, in the interval $[a, b] = [0, 1]$.

For the experiments conducted in this section, we employed the original contents of LYAPACK to solve the Lyapunov equations involved in the BDF and the Rosenbrock methods. This version is based on the LU-based solver provided by the `lu` function in MATLAB. Internally, when A is sparse, MATLAB relies on UMFPACK [38] to perform this factorization and solve the sparse triangular systems. Table 1 shows the difference between the exact solution and the solution obtained with non-tuned implementations of the methods, in terms of the spectral norm and the Frobenius norm, in $t = 1$. A similar experiment performed with the GPU-enabled version obtained similar results. Thus, the optimizations included did not imply any loose of accuracy.

Table 1. Difference between the exact and the computed solutions for an instance of dimension $n = 2,000$ of the CHOI-LAUB Benchmark using different timesteps h in both methods.

Method	h	$\|\cdot\|_2$	$\|\cdot\|_F$	$\|X_{0.01} - X_h\|_F \,/\, \|X_{0.01}\|_F$
BDF	0.10	7.105427e−14	3.177644e−12	7.1054e−16
	0.05	4.263256e−14	1.906586e−12	4.2633e−16
	0.01	0	0	–
Ros	0.10	1.162429e−07	5.198539e−06	1.1624e−09
	0.05	4.263256e−14	1.906586e−12	4.2633e−16
	0.01	0	0	–

Considering the results summarized in Table 1, we note that the BDF method produces better approximations to the solution, which is expected since the BDF requires more computations. The results also show that decreasing the time-step improves the accuracy of both methods, but the impact is more relevant for the Rosenbrock method; therefore, the Rosenbrock alternative requires a finer discretization than the BDF method to attain a similar level of accuracy. Moreover, the solutions for each time t, should converge to the exact solution if h becomes sufficiently small. Therefore, if we take the solution obtained with the smallest time-step as the reference, the difference between the reference solution and a solution computed with a larger time-step should become smaller with h as well. In this line, Table 1 also includes the norm of the relative difference between the solution obtained using the larger time-steps (i.e., $h = 0.1$ and $h = 0.05$) and the solution with $h = 0.01$.

For the POISSON test case, we evaluate the five instances described in Sect. 4.1, on the interval $[a, b] = [0, 1]$ and three different time steps, $h = 0.1, 0.05$, and 0.01. Unlike for the Choi-Laub example, an analytic solution of the equation is not available for this benchmark, so we use the solution computed with $h = 0.01$ as the reference and compare it with the solutions obtained with other values of h. These results are displayed in Table 2. For all the evaluated instances, the difference with respect to the reference solution decreases when the value of h is reduced. As it was observed in the Choi-Laub experiment, this decrease is more important for the Rosenbrock method.

The dimension of the instances of HIGHW turns impossible an analogous evaluation since constructing the solution explicitly from the factors would imply a prohibitive storage cost, and an out-of-core approach to calculate the norms using only the factors of the solution demands an important execution time. For this reason we do not perform the accuracy evaluation of this benchmark.

Table 2. Comparison between the solution computed with $h = 0.01$ and the ones computed with a larger timestep for the POISSON Benchmark.

n	h	$\|X_{0.01} - X_h\|_F \,/\, \|X_{0.01}\|_F$	
		BDF	Ros
2,500	0.10	0.003743	0.016413
	0.05	0.001742	0.006460
10,000	0.10	0.003742	0.017895
	0.05	0.001740	0.007074
22,500	0.10	0.003743	0.018406
	0.05	0.001742	0.007302
40,000	0.10	0.003744	0.018649
	0.05	0.001742	0.007404
62,500	0.10	0.003670	0.018837
	0.05	0.001436	0.007509

4.3 Performance Evaluation

We next evaluate the performance of our implementations. In this experiment we solve the DRE in the same interval and use the same time-steps that were selected for the evaluation in the previous subsection. In order to asses the quality and performance of the new CPU-GPU solver, we include an implementation of the linear system solver based on the BiCGStab method (on the CPU) provided by MATLAB. This will serve as a reference for our GPU version of the BiCGStab solver. All execution times in this section are expressed in seconds.

Table 3 shows an important reduction of the execution time with respect to the original version of LYAPACK, which grows with the problem dimension. The high execution times of the original LYAPACK version were expected, due to the considerable computational cost of the LYAPACK solver. Concretely, this algorithm computes a new set of shift parameters. In practice, each new shift parameter implies the computation of an LU factorization (of the corresponding shifted matrix). In contrast, the performance of the pure CPU version with the BiCGStab solver does not show much improvement. As the dimension of the instances is rather small, it is likely that the SpMV kernel, which is the main operation of the BiCGStab solver, cannot fully exploit the massive parallelism offered by the GPU.

Table 4 reports the performance of the Rosenbrock method for Benchmark POISSON. Here we also present the execution time of the different stages of

Table 3. Execution times (in seconds) of the BDF method for Benchmark POISSON.

h	n	t_{lu_cpu}	t_{bicg_cpu}	t_{bicg_gpu}
0.1	2,500	15.70	13.10	22.89
	10,000	144.09	43.39	45.39
	22,500	637.04	87.43	80.89
	40,000	2,373.10	161.29	123.33
	62,500	6,373.02	265.51	175.59
0.05	2,500	29.20	21.10	34.87
	10,000	275.60	64.43	70.96
	22,500	1,488.74	140.09	126.07
	40,000	5,282.45	277.40	198.23
	62,500	14,510.51	423.34	291.87
0.01	2,500	130.16	77.39	121.30
	10,000	1,490.71	231.57	257.36
	22,500	9,573.87	555.18	464.61
	40,000	32,742.11	1,107.31	852.06
	62,500	70,631.85	2,401.73	1,679.60

Table 4. Execution times (in seconds) of the different stages of Rosenbrock method for Benchmark POISSON with $[a, b] = [0, 1]$ and different values of h.

h	n	*solver*	params	*pre.adi*	*lrcfadi*	*rrqr*	Total
0.10	2,500	lyapack_{orig}	0.46	1.79	1.25	2.02	5.53
		bicgstab_{cpu}	0.22	0.00	4.53	1.96	6.72
		bicgstab_{gpu}	0.23	0.00	6.99	1.97	9.20
	10,000	lyapack_{orig}	1.39	22.29	11.01	6.83	41.54
		bicgstab_{cpu}	1.36	0.00	12.70	7.30	21.38
		bicgstab_{gpu}	1.38	0.00	8.15	7.41	16.95
	22,500	lyapack_{orig}	4.32	107.58	39.82	11.83	163.57
		bicgstab_{cpu}	4.24	0.00	24.28	13.21	41.76
		bicgstab_{gpu}	4.25	0.00	9.85	12.57	26.69
	40,000	lyapack_{orig}	9.90	423.07	94.87	18.38	546.29
		bicgstab_{cpu}	9.72	0.00	41.74	16.11	67.64
		bicgstab_{gpu}	9.67	0.00	12.30	16.40	38.44
	62,500	lyapack_{orig}	17.43	1,199.53	205.71	22.27	1,445.03
		bicgstab_{cpu}	17.56	0.00	67.56	23.25	108.47
		bicgstab_{gpu}	17.88	0.00	14.84	22.71	55.54
0.05	2,500	lyapack_{orig}	0.41	3.51	2.06	3.28	9.27
		bicgstab_{cpu}	0.42	0.00	6.48	3.29	10.19
		bicgstab_{gpu}	0.43	0.00	9.92	3.30	13.66
	10,000	lyapack_{orig}	2.49	42.48	19.50	9.69	74.18
		bicgstab_{cpu}	2.46	0.00	19.08	10.10	31.66
		bicgstab_{gpu}	2.47	0.00	11.50	9.87	23.87
	22,500	lyapack_{orig}	7.74	255.43	70.98	17.43	351.62
		bicgstab_{cpu}	7.81	0.00	35.68	17.34	60.87
		bicgstab_{gpu}	7.65	0.00	14.11	17.22	39.02
	40,000	lyapack_{orig}	18.04	965.67	202.31	24.27	1,210.39
		bicgstab_{cpu}	18.08	0.00	64.26	25.23	107.67
		bicgstab_{gpu}	18.01	0.00	19.11	25.03	62.25
	62,500	lyapack_{orig}	32.31	2,640.04	507.23	36.15	3,215.92
		bicgstab_{cpu}	32.83	0.00	102.88	38.90	174.80
		bicgstab_{gpu}	32.35	0.00	23.13	37.97	93.65
0.01	2,500	lyapack_{orig}	1.91	16.08	8.84	10.92	37.80
		bicgstab_{cpu}	1.93	0.01	20.95	10.85	33.78
		bicgstab_{gpu}	2.08	0.01	30.38	10.92	43.42
	10,000	lyapack_{orig}	11.38	236.52	78.42	31.29	357.72
		bicgstab_{cpu}	11.40	0.02	58.60	32.03	102.13
		bicgstab_{gpu}	11.42	0.02	34.09	30.69	76.30
	22,500	lyapack_{orig}	34.96	1,612.62	353.13	52.35	2,053.25
		bicgstab_{cpu}	34.21	0.02	113.86	51.81	200.06
		bicgstab_{gpu}	35.13	0.02	43.87	53.00	132.17
	40,000	lyapack_{orig}	99.99	5,560.96	1,132.78	86.93	6,881.13
		bicgstab_{cpu}	101.48	0.02	207.47	88.36	397.81
		bicgstab_{gpu}	100.47	0.02	61.38	91.72	254.07
	62,500	lyapack_{orig}	229.29	12,443.45	2,335.54	140.60	15,149.64
		bicgstab_{cpu}	230.34	0.02	360.95	135.29	727.34
		bicgstab_{gpu}	230.24	0.02	79.77	136.55	447.33

the method. The column *solver* indicates which version of the code is executed; column *params* shows the time needed to calculate the shift parameters of the LRCF-ADI algorithm; *pre. adi* refers to the preparation stage of the shifted linear system solver (which in the case of the original LYAPACK involves the LU factorization of all the shifted matrices); and *rrqr* is the time consumed by the column compression technique applied to the partial low rank solution of the equation. The performance gain with respect to the original version of LYAPACK is significant, almost 34× for the largest instance, but the improvement with respect to the CPU version of BICGSTAB is rather modest. However, the GPU is only employed for the solution of the Lyapunov equation, and this operation is up to 4× faster in the GPU-based variant. We also note that RRQR becomes the most time consuming stage of the accelerated version of the method when the problem dimension is large. We plan to address the acceleration of the RRQR algorithm in future work.

We next address the HIGHW test case. As the main matrix of this Benchmark presents a tridiagonal structure, we enhanced the Lyapunov solver in LYAPACK with a GPU version of a tridiagonal system solver. The pure-CPU variant employs the MATLAB default tridiagonal solver. Table 5 presents the execution times for the HIGHW case. The results show a runtime reduction close to 40% for all the evaluated instances. Table 6 exhibits the results for the Rosenbrock method applied to the HIGHW problem, decoupling the execution times of all the stages in the method. Although the reduction of the total runtime obtained by the GPU-based solver is below 20%, the column compression (*rrqr*) phase

Table 5. Execution times (in seconds) of the BDF method for Benchmark HIGHW with $[a, b] = [0, 1]$.

h	n	$time_{cpu}$	$time_{gpu}$
0.10	524,289	197	126
	786,433	292	183
	1,048,577	413	238
	1,572,865	628	401
0.05	524,289	370	232
	786,433	569	340
	1,048,577	781	458
	1,572,865	1,175	716
0.01	524,289	1,596	990
	786,433	2,396	1,485
	1,048,577	3,342	1,980
	1,572,865	5,024	3,249

is again the most time-consuming phase of the method, and this operation is fully computed in the CPU. All the GPU-enabled stages of the method report time savings. The computation of the shift parameters for the Lyapunov solver is about 20% faster, while the Lyapunov solver based on the LRCF-ADI method is 3× faster for the largest instance tested (Figs. 1 and 2).

Table 6. Execution times (in seconds) of the different stages of Rosenbrock method for Benchmark HIGHW with $[a, b] = [0, 1]$.

h	n	*solver*	params	*pre.adi*	*lrcfadi*	*rrqr*	Total
0.10	524,289	CPU	19.35	0.15	22.47	67.12	109.18
		GPU	15.50	0.20	10.77	55.19	81.75
	786,433	CPU	29.54	0.30	35.04	86.64	151.64
		GPU	24.49	0.34	16.71	95.26	136.95
	1,048,577	CPU	43.33	0.40	48.09	133.55	225.56
		GPU	29.32	0.41	21.48	140.52	191.90
	1,572,865	CPU	58.44	0.62	71.10	188.61	319.05
		GPU	43.16	0.61	32.46	168.06	244.54
0.05	524,289	CPU	36.33	0.30	42.36	110.11	189.25
		GPU	27.23	0.34	20.49	108.63	156.85
	786,433	CPU	56.58	0.58	68.75	182.24	308.40
		GPU	45.13	0.58	32.22	180.67	258.86
	1,048,577	CPU	75.58	0.78	91.28	237.75	405.74
		GPU	64.99	0.75	41.16	261.34	368.58
	1,572,865	CPU	111.80	1.23	136.41	394.47	644.41
		GPU	81.04	1.17	62.57	381.99	527.27
0.01	524,289	CPU	176.97	1.47	209.85	601.04	990.06
		GPU	128.36	1.49	97.84	567.53	795.97
	786,433	CPU	271.07	2.52	337.75	830.68	1,443.16
		GPU	196.85	2.89	158.13	917.36	1,276.48
	1,048,577	CPU	364.25	4.00	443.61	1,177.29	1,990.85
		GPU	293.67	3.83	202.50	1,221.99	1,723.67
	1,572,865	CPU	543.33	5.83	676.20	1,647.30	2,875.13
		GPU	438.62	5.70	312.91	1,802.18	2,561.89

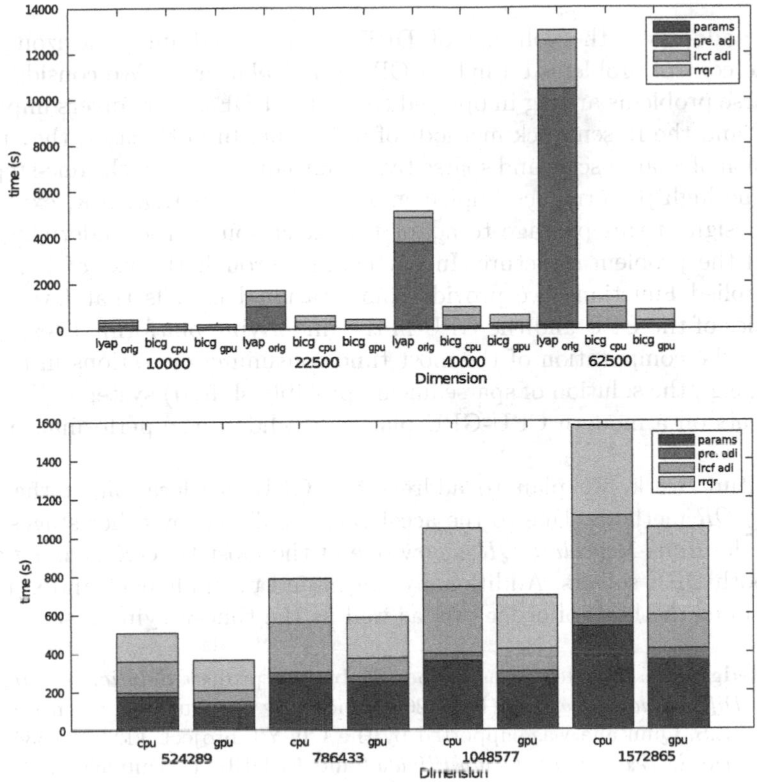

Fig. 1. Execution times (in seconds) of the different stages of Rosenbrock method for Benchmark POISSON (top) and HIGHW (bottom) with $[a, b] = [0, 10]$ and $h = 0.1$.

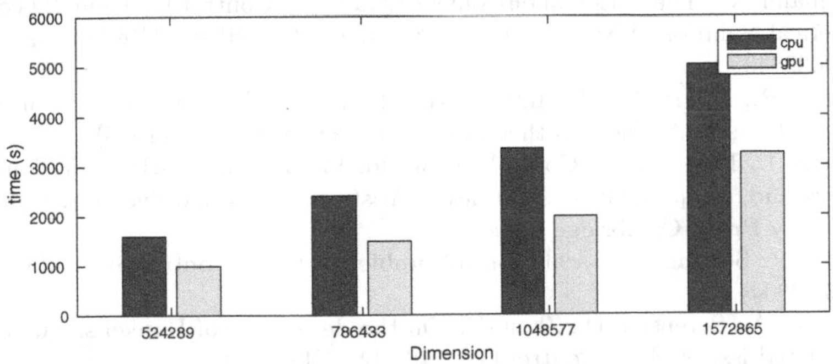

Fig. 2. Execution times of the BDF method for Benchmark HIGHW with $[a, b] = [0, 1]$ and $h = 0.01$.

5 Conclusions

We have addressed the solution of DREs and related finite horizon linear-quadratic control problems on hybrid CPU-GPU platforms. We consider large-scale sparse problems arising in optimal control of PDEs. Our solvers implement the BDF and the Rosenbrock methods of order one. In both cases, they require the solution of a large-scale and sparse Lyapunov equation at each time-step evaluated. The high performance implementations based on LYAPACK, exploit the flexible design of this package to adapt the execution to the underlying hardware and the problem structure. In particular, through the use of LYAPACK's User Supplied Functions, we provide computational kernels that leverage the capabilities of the CPU and the GPU in the underlying platform. These kernels accelerate the computation of the most time-consuming operations in both the methods, e.g., the solution of sparse linear (possibly shifted) systems. Numerical experiments on a modern CPU-GPU platform validate the performance of our solvers.

As future work, we plan to address the GPU acceleration of the *Rank-Revealing QR* methods. Due to the acceleration suffered by other stages of the method, the *Rank-Revealing QR* is now one of the most time-consuming operation in both DRE solvers. Additionally, we intend to implement and study the corresponding methods of order two, as well as the time-varying case.

Acknowledgment. H. Mena was supported by the project *Solution of large-scale Lyapunov Differential Equations* (P 27926) founded by the Austrian Science Foundation FWF. E.S. Quintana was supported by the CICYT project TIN2014-53495-R of the *Ministerio de Economía y Competitividad* and FEDER. E. Dufrechou, P. Ezzatti and A. Remón were supported by the EHFARS project funded by the German Ministry of Education and Research BMBF.

References

1. Mehrmann, V.: The Autonomous Linear Quadratic Control Problem: Theory and Numerical Solution. LNCIS, vol. 163. Springer, Heidelberg (1991). doi:10.1007/BFb0039443
2. Curtain, R., Zwart, H.: An Introduction to Infinite-Dimensional Linear System Theory. Texts in Applied Mathematics. Springer, New York (1995)
3. Lasiecka, I., Triggiani, R.: Control Theory for Partial Differential Equations: Continuous and Approximation Theories I: Abstract Parabolic Systems. Cambridge University Press, Cambridge (2000)
4. Benner, P.: Solving large-scale control problems. IEEE Control Syst. Mag. **14**(1), 44–59 (2004)
5. Antoulas, A., Sorensen, D., Zhou, Y.: On the decay rate of Hankel singular values and related issues. Syst. Control Lett. **46**, 323–342 (2000)
6. Grasedyck, L.: Existence of a low rank or H-matrix approximant to the solution of a Sylvester equation. Numer. Lin. Algebra Appl. **11**, 371–389 (2004)
7. Penzl, T.: Eigenvalue decay bounds for solutions of Lyapunov equations: the symmetric case. Syst. Control Lett. **40**, 139–144 (2000)

8. Sabino, J.: Solution of large-scale Lyapunov equations via the block modified smith method. Ph.D. thesis, Rice University, Houston, Texas (2007)
9. Abou-Kandil, H., Freiling, G., Ionescu, V., Jank, G.: Matrix Riccati Equations in Control and Systems Theory. Birkhäuser, Basel (2003)
10. Ichikawa, A., Katayama, H.: Remarks on the time-varying H_∞ Riccati equations. Syst. Control Lett. **37**(5), 335–345 (1999)
11. Petersen, I., Ugrinovskii, V., Savkin, A.V.: Robust Control Design Using H^∞ Methods. Springer, London (2000)
12. Choi, C., Laub, A.: Efficient matrix-valued algorithms for solving stiff Riccati differential equations. IEEE Trans. Automat. Control **35**, 770–776 (1990)
13. Davison, E., Maki, M.: The numerical solution of the matrix Riccati differential equation. IEEE Trans. Automat. Control **18**, 71–73 (1973)
14. Dieci, L.: Numerical integration of the differential Riccati equation and some related issues. SIAM J. Numer. Anal. **29**(3), 781–815 (1992)
15. Kenney, C., Leipnik, R.: Numerical integration of the differential matrix Riccati equation. IEEE Trans. Automat. Control **AC–30**, 962–970 (1985)
16. Benner, P., Mena, H.: Numerical solution of the infinite-dimensional LQR-problem and the associated differential Riccati equations. J. Numer. Math. (2016, online)
17. Benner, P., Mena, H.: Rosenbrock methods for solving differential Riccati equations. IEEE Trans. Autom. Control **58**, 2950–2957 (2013)
18. Hansen, E., Stillfjord, T.: Convergence analysis for splitting of the abstract differential Riccati equation. SIAM J. Numer. Anal. **52**, 3128–3139 (2014)
19. Stillfjord, T.: Low-rank second-order splitting of large-scale differential Riccati equations. IEEE Trans. Autom. Control **10**, 2791–2796 (2015)
20. Benner, P., Ezzatti, P., Mena, H., Quintana-Ortí, E.S., Remón, A.: Solving matrix equations on multi-core and many-core architectures. Algorithms **6**, 857–870 (2013)
21. Ascher, U., Petzold, L.: Computer Methods for Ordinary Differential Equations and Differential-Algebraic Equations. SIAM, Philadelphia (1998)
22. Golub, G., Loan, C.V.: Matrix Computations, 3rd edn. Johns Hopkins University Press, Baltimore (1996)
23. Lang, N., Mena, H., Saak, J.: On the benefits of the LDL factorization for large-scale differential matrix equation solvers. Linear Algebra Appl. **480**, 44–71 (2015)
24. Li, J., White, J.: Low rank solution of Lyapunov equations. SIAM J. Matrix Anal. Appl. **24**(1), 260–280 (2002)
25. Benner, P., Li, J., Penzl, T.: Numerical solution of large Lyapunov equations, Riccati equations, and linear-quadratic control problems. Numer. Linear Algebra Appl. **15**, 755–777 (2008)
26. Penzl, T.: Numerische Lösung großer Lyapunov-Gleichungen. Ph.D. thesis, Technische Universität Chemnitz (1998)
27. Benner, P., Kürschner, P., Saak, J.: Efficient handling of complex shift parameters in the low-rank cholesky factor ADI method. Numer. Algorithms **62**, 225–251 (2012)
28. Simoncini, V.: A new iterative method for solving large-scale Lyapunov matrix equations. SIAM J. Sci. Comput. **29**, 1268–1288 (2007)
29. Benner, P., Saak, J.: Numerical solution of large and sparse continuous time algebraic matrix Riccati and Lyapunov equations: a state of the art survey. GAMM Mitt. **36**, 32–52 (2013)
30. Wachspress, E.: Iterative solution of the Lyapunov matrix equation. Appl. Math. Lett. **107**, 87–90 (1988)
31. Dufrechou, E., Ezzatti, P., Quintana-Ortí, E.S., Remón, A.: Accelerating the Lyapack library using GPUs. J. Supercomput. **65**, 1114–1124 (2013)

32. Alfaro, P., Igounet, P., Ezzatti, P.: A study on the implementation of tridiagonal systems solvers using a GPU. In: 30th International Conference of the Chilean Computer Science Society (SCCC 2011), pp. 219–227. IEEE (2011)
33. Naumov, M.: Incomplete-LU and Cholesky preconditioned. Iterative methods using CUSPARSE and CUBLAS. Nvidia white paper (2011)
34. Saak, J.: Effiziente numerische Lösung eines Optimalsteuerungsproblems für die Abkühlung von Stahlprofilen. Diplomarbeit, Fachbereich 3/Mathematik und Informatik, Universität Bremen, D-28334 Bremen (2003)
35. Saak, J.: RRQR-MEX a MATLAB mex-interface for the rank revealing QR factorization. https://www2.mpi-magdeburg.mpg.de/mpcsc/mitarbeiter/saak/Software/rrqr.php. Accessed 07 July 2015
36. Choi, C., Laub, A.: Constructing Riccati differential equations with known analytic solutions for numerical experiments. IEEE Trans. Automat. Control **35**, 437–439 (1990)
37. Abels, J., Benner, P.: CAREX - a collection of benchmark examples for continuous-time algebraic Riccati equations. Technical report, SLICOT working note 1999–2014 (1999). http://www.slicot.org
38. Davis, T.: Umfpack version 4.4 user guide. Department of Computer and Information Science and Engineering, University of Florida, Gainesville, FL (2005)

A Hybrid CPU-GPU Scatter Search for Large-Sized Generalized Assignment Problems

Danilo S. Souza[1(✉)], Haroldo G. Santos[1], Igor M. Coelho[2], and Janniele A.S. Araujo[1]

[1] Federal University of Ouro Preto, Ouro Preto, MG, Brazil
danilo.gdc@gmail.com, haroldo@iceb.ufop.br, janniele@decsi.ufop.br
[2] State University of Rio de Janeiro, Rio de Janeiro, RJ, Brazil
igor.machado@gmail.com

Abstract. In the Generalized Assignment Problem, tasks must be allocated to machines with limited resources, in order to minimize processing costs. This problem has several industrial applications and often appears as substructure of other combinatorial optimization problems. By harnessing the massive computational power of Graphics Processing Units in a Scatter Search metaheuristic framework, we propose a method that efficiently generates a solution pool using a Tabu list criteria and an Ejection Chain mechanism. Common characteristics are extracted from the pool and solutions are combined by exploring a restricted search space, as a Binary Programming model. Classic instances vary from 100–1600 jobs and 5–80 agents, but due to the big amount of optimal and near-optimal solutions found by our method, we propose novel large-sized instances up to 9000 jobs and 600 agents. Results indicate that the method is competitive with state-of-the-art algorithms in literature.

Keywords: Scatter search · Tabu search · Ejection chain · Binary programming · Generalized Assignment Problem

1 Introduction

The Generalized Assignment Problem (GAP), originally proposed by Ross and Soland [24], is defined as finding an assignment of minimal cost given an amount jobs and agents, such that one job can be allocated to a unique agent satisfying the capacity constraints. GAP is an NP-hard problem [25], with applications in various industrial problems.

Formally, the GAP can be defined as: given n the number of jobs and m the number of agents, let us consider the set of jobs $J = \{1, ..., n\}$ and the set of agents $I = \{1, ..., m\}$, aiming to determine a minimum cost allocation in order to assign each job to exactly one agent, satisfying the resource constraints of each

O. Gervasi et al. (Eds.): ICCSA 2017, Part I, LNCS 10404, pp. 133–147, 2017.
DOI: 10.1007/978-3-319-62392-4_10

agent. By allocating job j to agent i, it is assigned a cost d_{ji} and a consumption of resource r_{ji}, such that the total amount of resources available to agent i is denoted by b_i. The binary decision variable x_{ji} receives 1 if the job j is allocated to the agent i, and 0 otherwise. Then, a mathematical model of the GAP can be formulated as follows:

$$Minimize \sum_{j \in J} \sum_{i \in I} d_{ji} x_{ji} \tag{1}$$

Subject to

$$\sum_{j \in J} r_{ji} x_{ji} \leq b_i \quad \forall i \in I \tag{2}$$

$$\sum_{i \in I} x_{ji} = 1 \quad \forall j \in J \tag{3}$$

$$x_{ji} \in \{0, 1\} \quad \forall i \in I, j \in J \tag{4}$$

Several approaches have been proposed to tackle this problem, including exact and heuristic methods. Works proposed in the literature using heuristic algorithms include Genetic Algorithms [3,7,28], Simulated Annealing [20] and Tabu Search [6,13,15,20]. Also, some exact algorithms have been proposed: the branch-and-price algorithm by Savelsbergh [26], the Cutting Plane algorithm by Avella et al. [1], and Branch-and-Cut by Nauss [18], where optimal solutions were found for many instances with up to 200 jobs and 20 agents.

Chu and Beasley [3] present a heuristic for the generalized assignment problem based on the genetic algorithm. In addition to standard GA procedures, the proposed heuristic presents a non-binary representation that automatically satisfies job assignment constraints, a fitness-unfitness pair evaluation and a heuristic operator that helps to improve the cost and feasibility of the solution. The computational results show that the proposed GA heuristic is capable of finding optimal solutions for several small size problems.

In the paper proposed by Posta et al. [22] an exact algorithm to solve the generalized allocation problem is implemented. The problem treated as optimization is reformulated in a sequence of decision problems, and variable-fixing rules are used to solve these effectively. Decision problems are solved by a Depth-First Lagrangian Branch-and-Bound method, and the variable-fixing rules to prune the search tree improve the performance of the method. These rules are based on Lagrangian reduced costs which are calculated using a dynamic programming algorithm. The approach of successively solving a decision problem for each generated subproblem is very fast in practice, since they start from a good initial limit. The results show that the proposed method is able to find optimal solutions for large instances.

For more details on solution methods for GAP, Cattrysse and Van Wassenhove [2] and Öncan [19] presented extensive reviews of applications and the existing exact and heuristic algorithms.

The use of optimization techniques to obtain close to optimal solutions to NP-Hard problems often demands high processing times, especially when dealing with large problems. However, high-performance computational techniques can be applied to reduce computational time and improve algorithm performance in solving the problem. Recently, the use of Graphics Processing Unit (GPU) emerged in order to accelerate the calculations in many application areas with low cost computational equipment. Successful implementations in GPU can obtain a speed up of 100 times better than a sequential implementation in CPU [14]. However, due to specific details of the GPU, such as the Single Instruction Multiple Threads (SIMT) paradigm, classic algorithms usually need to be entirely rewritten in order to explore the massive amount of processing cores and abundant device memory bandwidth.

This paper describes a hybrid metaheuristic inspired by Scatter Search metaheuristic implemented to run on a heterogeneous CPU-GPU platform to produce high quality solutions the GAP. The proposed hybrid method includes a parallel implementation of the Tabu Search (TS) metaheuristic combined with a Binary Programming Pool Search (BPPS). The use of the TS combined with scatter search is proposed due to its promising background for solving other hard optimization problems and its very simple structure, which makes is very suitable for GPU based implementation. Our method includes a fast GPU based implementation of Ejection Chains, so that a large neighborhood is explored in parallel in the GPU at each TS iteration. The execution of the TS heuristic in GPU allows the generation of several solutions in reduced computational time, thus facilitating the composition of a pool of elite solutions used as reference set for BPPS method. The BPPS enables a rapid search process by using rules to fix decision variables using most common characteristics of the solution pool and rapidly reduce the search space of the problem.

The reminder of the paper is organized as follows. Section 2 discusses the design and implementation of proposed hybrid method. Section 3 details computational experiments designed for evaluating and analyzes experiment results. Section 4 reports the findings and future work to improve the method.

2 Proposed Method

This section presents the proposed approach to approach the Generalized Assignment Problem. The proposed method is a hybrid approach composed of an adaptation of the scatter search using tabu search techniques and a binary programming pool search.

The Scatter search (SS) is an evolutionary metaheuristic that was first introduced in Glover [8] with objective to explore a solution space from the evolution of a a set of good solutions called the Reference Set [23]. For a general introduction to SS, we refer the reader to [8,16,17]. The Algorithm 1 shows the main structure of our method.

Algorithm 1. Proposed method

1: Construct an initial population P with size of $|P|$ using the Greedy randomized method.
2: Build the reference set using TS in GPU to improve the solutions in step 1
3: Select the solution with more characteristics similar to the other solutions (denominated in this work degree of similarity).
4: Uses the BPPS from the solutions selected in step 3 to perform Solution Improvement, Reference Set Update, Subset Generation and Solution Combination.

In the first phase of the approach a Greedy Randomized algorithm is used to generate a set of distinct initial solutions. To improve solutions, the tabu search approach is implemented using the parallelism in GPU, where the method is executed several times in parallel (the current generation GPUs typically have thousands of processing cores) generating several solutions, where the best solutions are stored in a pool (Reference Set).

Using the pool it is possible to calculate the degree of similarity from each solution to each other, and the solution with the highest degree is used as input to the next phase. For each decision variable each input solution in the pool, a value h_{ji} computes the number of times that the variable x_{ji} appears with value 1 in solutions of the pool of solutions.

The Binary Program Pool Search proposes the fixation of the variables with greater values of h_{ji}, in order to reduce the search space for the BPPS and to propagate the most common features for new solutions, and this process is repeated by gradually increasing the search space whenever it is not possible to obtain an improvement in the solution.

The methods used are presented in details in the following subsections.

2.1 Greedy Randomized Method

A greedy randomized method is used to generate the initial solutions. The proposed method considers a weight to prioritize the allocations that will allow the reduction of the final cost. This weight is calculated in relation to the cost c_{ji} and consumption of resources r_{ji}: $W_{ji} = w_1 c_{ji} + w_2 r_{ji}$. The allocation is performed by the job, that is, given a job j it will be allocated to the agent i with the lower weight W_{ji} which satisfies the capacity constraint. Task are chosen sequentially.

A pseudocode of the method is presented in the Algorithm 2.

In the greedy method, initially the weight W_{ji} is calculated for each job j to be allocated in an agent i. The iterator k is responsible for going through each job, so that it is allocated by exactly one agent. The variable a receives the lowest weight agent W_{ki}, soon after it is verified if the job k has not yet been allocated and if it is allocated to agent a the capacity constraints are met, if so, the assignment of the job k to the agent a is performed. Otherwise an infinite value is assigned to the weight W_{ka} so that agent a will not be chosen later to be execute the job k. The process is repeated until all tasks are allocated to exactly one agent. It is worth mentioning that depending on the values assigned

Algorithm 2. Greedy method

Input: w_1, w_2 ▷ Weight assigned to the cost and the resource.
c_{ji}, r_{ji} ▷ Cost and resources of allocating job j to agent i
b_i ▷ Available capacity of agent i
Output: x ▷ Solution Initial

$W_{ji} \leftarrow w_1 c_{ji} + w_2 r_{ji} \; \forall j \in N, i \in M$
for $k := 1$ **to** n **do** ▷ $n = |N|$
  for $z := 1$ **to** m **do** ▷ $m = |M|$
    $a \leftarrow \underset{(i)\in M}{\textbf{Argmin}} \{W_{ki}\}$
    if $\sum_{i \in I} x_{ki} = 0$ **and** $r_{ka} + \sum_{j \in J} x_{ja} r_{ja} \leq b_a$ **then**
      $x_{ka} \leftarrow 1$
    else
      $W_{ka} \leftarrow \infty$
    end if
  end for
end for
Return x

to the weights w_1 and w_2 as well as the resources r_{ji} and costs c_{ji}, the proposed method can generate infeasible solutions. In case of infeasibility, the method is reinitialized using values other than w_1 and w_2, until a feasible solution is found. The weight w_1 was determined by a random real value contained in the interval $[0.5, 1.5]$, thus generating different solutions at each iteration. For the weight w_2 the value was assigned by the instance group, in groups C and D, w_2 will be the value of $w_1 + 1$, for the instances of group E, the value of w_2 is $w_1 + 19$. For the instances of group G, the value of w_1 by a random real value contained in the interval $[0.5, 1.5]$ and the value of w_2 is $w_1 - 0.1$.

2.2 Tabu Search

Tabu Search (TS) is a metaheuristic based on local search or neighborhood search that admits worsening solutions, proposed by Glover [9,10] to solve complex problems of combinatorial optimization. The tabu search is an iterative procedure that aims to explore the solution space of the problem in order to make successive moves from one solution x to another solution x' in its neighborhood $N(X)$. However, only this search engine is not enough to escape from local optimum, since there may be a return to a previously generated solution. To avoid this, the algorithm uses the concept of tabu list.

The tabu list (adaptive memory) may be short-term memory or long-term memory. The short-term memory restricts the search by prohibiting certain movements allowing you to overcome great locations, prevent cycling and direct the search to unexplored regions. The use of the long-term memory provides the return of the search for regions considered promising. For this is used a measure of the frequency of attributes of the best solutions found during the search called elite solutions [11].

Each iteration of the Tabu Search algorithm consists of main tasks: neighborhood generation, neighborhood evaluation, choosing the best solution within a neighborhood and updating the tabu list, and finally moving to the new solution [4]. In paper, we propose the parallel implementation with the use of GPU for the phases generation neighborhood and evaluation neighborhood, provided that the permutations forming the neighborhood are all stored in memory. The other phases will run on CPU on a single thread. It requires data transfer between GPU and CPU in every iteration.

The use of the GPU is performed by means of implementations of *kernels* that are executed in a defined set of threads. Theoretically, all threads run the same code *simultaneously* and have a unique identifier. In practice, threads that belong to a same *warp* of 32 threads will execute the same code (this value can vary in some GPU microarchitectures). The threads are divided into groups called blocks, which will be executed by different Stream Multiprocessors (SMs). The maximum dimension for the number of threads and blocks is limited by the GPU, but it can be configured by the user in order to optimize the execution time of each *kernel* [27]. In other words, the same method can be executed by several threads and these in turn grouped in several blocks. It is noteworthy that between threads of the same block it is possible to share data through the "shared memory".

The neighborhood is generated from the ejection chain method presented in the Subsect. 2.2, where each thread generates a certain value of ejection chains and the best chain will be stored in the shared memory and thus propagated to the next phase. Each block of the GPU architecture consists of several threads, a single thread from each block will be responsible for evaluating the best generated chains stored in shared memory, updating the solution with the best chain for that block. Each block will execute a local search procedure, and in the end it will generate a reference set (Solution Poll) with the best solutions generated by each block. But long-term memory is shared by all blocks. The short-term memory is used in the Ejection Chain method, on the other hand, the long-term memory is applied as a tie-break criterion in the evaluation of the solutions.

Ejection Chain. Ejection chain methods are variable depth methods that generate sequences of simple moves, in order to achieve more complex compound moves. In other words, an ejection chain of L levels consists of successive operations performed on a set of elements, where the lth operation changes the state of one or more solution elements (that are said to be ejected in operation $l+1$). The states change as well as the ejection chain advances, and these depend on the cumulative effect of the previous steps [12].

In this paper, the number of operations performed on the solution s is limited to a size k.

The moves are based on the size k, such that:

- for $k = 1$, a randomly selected job j will be allocated to a new agent w, this new allocation must respect the capacity of the agent w, thus corresponding a *shift* movement;

- for $k = 2$, two jobs j_1 and j_2 are randomly chosen, where i_1 and i_2 are the agents allocated to perform the jobs j_1 and j_2 respectively. The movement corresponds to a swap movement, where the job j_1 will be allocated to agent i_2 if possible, and the job j_2 to agent i_1;
- for $k \geq 3$, the ejection chain corresponds to a sequence of exchanges for different agents and jobs, where jobs are randomly selected $j_1, j_2, ..., j_k$ and their respective agents called $i_1, i_2, ..., i_k$. The move must always obey the capacity constraints of the problem, so each job j_k, with the exception of the last, will be allocated to agent i_{k+1}, that is, job j_1 will be allocated to agent i_2, job j_2 will be allocated to agent i_3, successively until the job $j_{(k-1)}$ be assigned to agent i_k, and finally job j_k will be allocated to the agent i_1. The process is assembled iteratively in order to verify chains of size 2 to k, returning the chain that will have the best improvement for the solution (which will contribute to a decrease in the objective function). Therefore, the returned chain can have size contained between 2 and k.

One of the main differences in the approach proposed by Yagiura et al. [29] is that all the movements used in our algorithm must be feasible. Another difference is that in the work of Yagiura et al. [29], is that the probability of choosing the jobs is not the same for all, thus favoring a specific group of jobs, and in our algorithm the probabilities are equal.

A pseudocode of the ejection chain method is presented in the Algorithm 3. In this method, we represent an allocation $x_{ji} = 1$ as a function $\phi(j) = i$.

2.3 Reference Set (Solution Pool)

As previously stated, the reference set is generated from the execution of the Tabu Search in GPU. The Reference Set is used in later phases to calculate the degree of similarity between the solutions as well as to determine which features will be propagated to for the next solutions. The number of solutions stored in Solution Pool was determined by the maximum number of blocks allowed by the GPU architecture.

2.4 Binary Programming Pool Search

The proposed Binary Programming Pool Search (BPPS) is based on Relaxation Induced Neighborhood Search (RINS) [5] and Ellipsoidal Cuts [21], using characteristics contained in the solution pool.

Initially, a solution of the solution pool is chosen as input to the BP formulation, and the solution with the greater value of objective function is chosen. To determine common characteristics in the solution pool a residency Matrix h_{ji} is calculated, where for each variable x_{ji} the value h_{ji} contains the number of times that $x_{ji} = 1$ in the solutions contained in the pool of solutions, that is, given x_{ji} is equal to 1 in w solutions of the pool, the value of h_{ji} will be set to w.

Algorithm 3. Ejection chain method

Input: x ▷ Solution, $x_{ji} = 1$ if job j is allocated to agent i
$\phi()$ ▷ Function representing the allocation
$RAND()$ ▷ Function returning random element from set
r_{ji}, c_{ji} ▷ Resource and cost to allocate job j to agent i
b_i, u_i ▷ Available capacity and resource for agent i
k ▷ Size of the ejection chain
δ_k ▷ Cost change in solution for chain length k
Δ ▷ Cost difference between solutions
N, M ▷ Sets with n jobs and m agents
L ▷ Set with jobs contained in the tabu list
Output: C ▷ Vector of ordered pairs to store the ejection chain

if $k = 1$ **then**
  $j_1 \leftarrow RAND(N \backslash L)$; $i_1 \leftarrow RAND(M)$;
  if $u_{i_1} + r_{j_1 i_1} \leq b_{i_1}$ **then**
    $C \leftarrow C + (j_1, i_1)$; $L \leftarrow L \cup \{j_1\}$;
  end if
end if
if $k \geq 2$ **then**
  $\Delta \leftarrow 0$; $\delta_{best} \leftarrow \infty$; $size_{best} \leftarrow 0$;
  $j_1 \leftarrow RAND(N \backslash L)$; $i_1 \leftarrow \phi(j_1)$; $N \leftarrow N \backslash \{j_1\}$;
  for $a := 2$ **to** k **do**
    $j_a \leftarrow RAND(N \backslash L)$, $i_a \leftarrow \phi(j_a)$;
    $T \leftarrow (N \backslash L) \backslash \{j_a\}$; ▷ Unused jobs in eject chain step
    while $u_{i_a} - r_{j_a i_a} + r_{j_{a-1} i_a} > b_{i_a} \wedge |T| \neq 0$ **do**
      $j_a \leftarrow \min(N \backslash L)$; $i_a \leftarrow \phi(j_a)$; $T \leftarrow T \backslash \{j_a\}$;
    end while
    if $|T| = 0$ **then**
      $k \leftarrow size_{best}$;
    else
      $\Delta \leftarrow c_{j_{a-1} i_a} - c_{j_a i_a}$;
    end if
    if $u_{i_1} - r_{j_1 i_1} + r_{j_a i_1} > b_{i_1}$ **then**
      $\delta_a \leftarrow c_{j_a i_1} - r_{j_1 i_1}$;
    else
      $\delta_a \leftarrow \infty$;
    end if
    if $\Delta + \delta_a < \delta_{best}$ **then**
      $\delta_{best} \leftarrow \Delta + \delta_a$; $size_{best} \leftarrow a$;
    end if
    $N \leftarrow N \backslash \{j_a\}$;
  end for
  for $a := 2$ **to** $size_{best}$ **do**
    $C \leftarrow (j_{(a-1)}, i_a)$;
  end for
  $C \leftarrow (size_{best}, i_1)$;
end if
Return C

With the initial solution and the residency matrix h_{ji} of decision variables, the BPPS uses these parameters to perform the ellipsoidal cuts (thus reducing the solution space). At each iteration the constraint 5 is added to the model, where the value of α_{ji} is calculated from the normalization of the data contained in the matrix h_{ji}. The β is the amount of slack to define the ellipsoidal neighborhood and it is defined as parameter and incremented every iteration in case there is no improvement in the solution.

$$\sum_{j \in N, i \in M | x_{ji}=1} \alpha_{ji}.(1 - x_{ji}) \leq \beta \tag{5}$$

The constraint presented in the Eq. 6 limits the value of the objective function by the value of the incumbent solution, and is updated at each iteration.

$$\sum_{j \in N, i \in M} x_{ji}.c_{ji} \leq F_o^{'} \tag{6}$$

Each execution of the solver is limited to 2 situations, if there is improvement in the incumbent solution the number of 500 node exploited without improvement is used as criterion, otherwise a time limit of 10 s is used. A pseudocode for the formulation is presented in the Algorithm 4.

Algorithm 4. BP pool search

Input: x ▷ Solution, $x_{ji} = 1$ if job j is allocated to agent i
h_{ji} ▷ Residency Matrix
N ▷ Set with n jobs
M ▷ Set with m agents
Parameters: β ▷ Slack to define the ellipsoidal neighborhood
timeLimit ▷ Time limit for BPPS
Output: $x^{'}$ ▷ Best solution found

$t \leftarrow \beta$
while $time \leq timeLimitl \wedge k \leq |N|$ **do**
  $m_1 = \textbf{MAX}(h_{ji})\ \forall i \in N, j \in M |\ x_{ji} = 1$;
  $m_2 = \textbf{MIN}(h_{ji})\ \forall i \in N, j \in M |\ x_{ji} = 1$;
  $\alpha_{ji} \leftarrow (h_{ji} - m_2)/(m_2 - m_1)\ \forall i \in N, j \in M$;
  $F_o^{'} \leftarrow f(x)$;
  $x^{'} \leftarrow RunSolver(x, \alpha_{ji}, F_o^{'}, \beta, timeLimit - time)$;
  if $f(x^{'}) < f(x)$ **then**
    $h_{ji} \leftarrow h_{ji} + 1, \forall x_{ji} = 1$;
    $\beta \leftarrow t$;
  else
    $\beta \leftarrow \beta + 1$;
  end if
  $x \leftarrow x^{'}$
end while
Return $x^{'}$

---

3 Computational Results

In this section, we evaluate the performance of the proposed hybrid GPU heuristic based on Scatter Search, named GSS. Our main experimental platform is composed of a CPU Intel i7 3.40 GHz, with 24 GB of memory, and a GPU GeForce GTX 780, with 2304 cores and 2 GB of global memory. It is worth to mention that twelve core of CPU was used for phase of the BPPS. The TS metaheuristic was implemented in C with CUDA SDK 7.5. The BPPS was coded in C++ and AMPL language, solved by the Gurobi Solver version 7.0.2.

The instances used for the experiments are subdivided four types of benchmark instances called types C, D, and E proposed for Chu and Beasley [3] and Laguna et al. [15]. Type G instances were proposed in that work to deal with much larger problems, with 10 times more agents and jobs (all instances are freely available in the author's website). Instances of these types are generated as follows:

- **Type C:** r_{ji} are random integers from uniform interval $[5, 25]$, d_{ji} are random integers from [10, 50], and $b_i = 0.8 * (\sum_{j \in J} r_{ji})/m$.
- **Type D:** r_{ji} are random integers from $[1, 100]$, $d_{ji} = 111 - r_{ji} + e_1$, where e_1 are random integers from $[-10, 10]$, and $b_i = 0.8 * (\sum_{j \in J} r_{ji})/m$.
- **Type E:** $r_{ji} = 1 - 10 * ln\ e_2$, where e_2 are random numbers from $(0, 1]$, $d_{ji} = 1000/aij - 10 * e_3$, where e_3 are random numbers from $[0, 1]$, and $b_i = 0.8 * (\sum_{j \in J} r_{ji})/m$.
- **Type G:** r_{ji} are random integers from $[1, 100]$, $d_{ji} = 111 - r_{ji} + e_4$, where e_4 are random integers from $[-10, 10]$, and $b_i = 1.75 * (\sum_{j \in J} r_{ji})/m$.

We tested the following three sets of problem instances.

- **MEDIUM:** Total of 18 instances of types C, D, and E with n up to 200. This set has combinations of 100 and 200 jobs with 5, 10 and 20 agents.
- **LARGE:** Total of 27 instances of types C, D, and E with n up to 1600. This set has combinations of 400, 900 and 1600 jobs with 10, 15, 20, 30, 40, 60 and 80 agents.
- **VERY LARGE:** Total of 12 instances of type G with n up to 9000, all of which were generated by us[1]. This set has combinations of 1000, 2000, 4000 and 9000 jobs with 50, 100, 200, 300, 400 and 600 agents.

As a parameter of GTS metaheuristic, size of tabu list was defined from an offline calibration, where some list sizes were tested $(5, 10, 15, 20, (0.5, 0.75, 1, 1.25, 1.4, 1.5, 1.6) * (|N|/(max(k) + 1)))$, but the one that presented the best result was $1.25 * (|N|/(max(k) + 1))$, and the stop condition was set for the maximum time of 15 s without improvement. The number of ejection chains generated per thread was defined for medium and large instances by performing

[1] The proposed instances are available on the website: https://drive.google.com/open?id=0B20uFG9WVmWMM2t1MlIxNG5aNW8.

tests with the values of 1, 5, 10, 15 and 20, thus obtaining a better performance in relation to *cost-time* with the value 5. For very large instances, the value of ejection chain per thread was defined with the value 100. Size k of the ejection chain was calibrated with values 1, 5, 10, 15 and 20 obtaining better performance for chain of size 1 and 10. The β parameter of the BPPS was set to 2 for medium and large instances and $0.02 * |N|$ for very large instances, given tests performed with values 1, 2, 3, 4, 5, 6, 8, 10, 11, $0.01 * |N|$, $0.02 * |N|$ and $0.002 * |N|$.

Table 1 presents a comparison between the state-of-the-art algorithm in literature for the GAP, named TS [29], and the proposed GSS and GTS algorithms, used the medium and large instances. The methods were executed 10 times and the best solutions values are presented and compared to a Lower Bound of each problem instance.The GTS consists in the first phase of the GSS, where a Tabu Search is performed inside the GPU, in order to populate a reference set including good quality solutions. The second phase of GSS consists of the BPPS model running over an exact solver, which tries to prove the optimality of the current best solution, including new solutions found in the process back to the Scatter Search reference set.

When the average gaps are compared, the GSS is able to achieve lower values for instance groups C and E (0.03%, 0.23% and 0.02%) when compared to the TS (0.04%, 0.20% and 0.04%). The GSS is able to find 21 best known solutions and 14 ties (in bold values) from 45 instances, also proving the optimality of 15 solutions (solutions marked with an asterisk). In fact, since the gaps are quite low when computed over a Lower Bound for the GAP, it is much likely that most of the current best known solutions are already optimal (although not proven yet). The instance group D is still quite challenging for the GSS, consuming a bigger share of the computational times, while providing solutions with bigger gaps.

In Table 2 the results of computational experiments for instances of type G are presented. Our algorithm GSS obtained solutions with gap between 0.6% and 9.4% and medium gap of 3.5%. The method GTS can generate quality solutions for the reference set, obtaining an average gap of about 6.8%. Due to the size of the instances the computational time was increased. Where the BPPS method was set a runtime of 10 min. For four of the twelve instances of type G the method BPPS can not improve the solutions found by GTS, generally occurring in larger instances.

Finally, the GPU configuration consisted of 12 blocks (in order to maximize device occupancy), each one performing an independent Tabu Search and contributing to a single solution in the pool (the reference set in Scatter Search consisted of 12 solutions). This configuration achieved an acceleration of 3 to 8 times, when compared to a pure CPU implementation of the GTS module.

Table 1. Computational experiments for classic datasets

Inst.	Lower Bound	GSS (GTS + BPPS) Best	Gap	Time	GTS Best	Gap	Time	TS Best	Gap	Time
C.100.5	1930	**1931***	0.05	26.3	**1931**	0.05	23.9	**1931**	0.05	0.6
C.100.10	1400	**1402***	0.14	42.5	1404	0.29	32.8	**1402**	0.14	3.0
C.100.20	1242	**1243***	0.08	48.7	1252	0.81	37.9	**1243**	0.08	21.6
C.200.5	3455	**3456***	0.03	31.1	3475	0.58	21.3	**3456**	0.03	3.7
C.200.10	2804	**2806***	0.07	68.7	2818	0.50	30.7	**2806**	0.07	100.5
C.200.20	2391	**2391***	0.00	123.6	2408	0.71	54.4	2392	0.04	137.4
C.400.10	5596	**5597***	0.02	83.7	5629	0.59	63.8	**5597**	0.02	105.8
C.400.20	4781	**4782***	0.02	133.5	4827	0.96	49.0	**4782**	0.02	130.4
C.400.40	4244	**4244***	0.00	316.9	4279	0.82	62.0	**4244**	0.00	157.6
C.900.15	11339	**11341**	0.02	744.2	11431	0.81	71.1	**11341**	0.02	759.6
C.900.30	9982	**9983**	0.01	332.7	10087	1.05	13.3	9985	0.03	720.0
C.900.60	9325	**9327**	0.02	917.6	9425	1.07	150.6	9328	0.03	704.7
C.1600.20	18802	**18803**	0.01	831.4	18948	0.78	138.1	**18803**	0.01	691.6
C.1600.40	17144	**17146**	0.01	873.6	17340	1.14	143.1	17147	0.02	2103.7
C.1600.80	16284	**16288**	0.02	894.8	16456	1.06	203.2	16291	0.04	4317.2
C Average			**0.03**	**364.61**		**0.75**	**73.0**		**0.04**	**663.8**
D.100.5	6350	**6353***	0.05	151.3	6449	1.56	33.2	6357	0.11	62.9
D.100.10	6342	**6355**	0.20	510.0	6541	3.14	24.1	6358	0.25	107.2
D.100.20	6177	6229	0.84	209.8	6345	2.72	76.3	**6221**	0.71	111.0
D.200.5	12741	**12745**	0.03	691.3	12889	1.16	93.7	12746	0.04	95.5
D.200.10	12426	**12437**	0.09	1203.3	12688	2.11	116.2	12446	0.16	129.2
D.200.20	12230	**12267**	0.30	1155.2	12601	3.03	57.9	12284	0.44	120.7
D.400.10	24959	**24969**	0.04	616.0	25457	2.00	220.3	24974	0.06	16.1
D.400.20	24561	24621	0.24	500.6	25354	3.23	201.7	**24614**	0.22	81.3
D.400.40	24350	24506	0.53	416.6	25081	3.00	114.9	**24463**	0.46	165.2
D.900.15	55403	**55430**	0.05	1585.6	56888	2.68	218.2	55435	0.06	112.9
D.900.30	54833	54970	0.25	873.5	56713	3.43	271.0	**54910**	0.14	234.4
D.900.60	54551	54782	0.42	1438.9	56089	2.82	237.8	**54666**	0.21	833.8
D.1600.20	97823	97887	0.07	2527.7	100067	2.29	419.9	**97870**	0.05	143.0
D.1600.40	97105	97260	0.16	1938.7	99965	2.95	322.2	**97177**	0.07	1294.0
D.1600.80	97034	97316	0.29	587.1	99836	2.89	281.2	**97109**	0.08	4795.4
D Average			**0.23**	**960.3**		**2.60**	**179.2**		**0.20**	**553.5**
E.100.5	12673	**12681***	0.06	67.6	12868	1.54	58.4	12682	0.07	39.9
E.100.10	11568	**11577***	0.08	132.4	11879	2.69	45.8	**11577**	0.08	31.6
E.100.20	8431	**8436**	0.06	140.1	8938	6.01	59.1	8443	0.14	90.4
E.200.5	24927	**24930***	0.01	85.1	25228	1.21	78.4	**24930**	0.01	20.0
E.200.10	23302	**23307***	0.02	157.8	23518	0.93	95.9	**23307**	0.02	34.1
E.200.20	22377	**22379***	0.01	115.4	22814	1.95	24.0	22391	0.06	209.3
E.400.10	45745	**45746**	0.00	72.9	46135	0.85	15.8	**45746**	0.00	260.7
E.400.20	44876	**44877**	0.00	143.0	45436	1.25	20.9	44882	0.01	212.9
E.400.40	44557	**44570**	0.03	202.9	45570	2.27	95.4	44589	0.07	217.7
E.900.15	102420	102426	0.01	103.9	103643	1.19	18.0	**102423**	0.00	157.4
E.900.30	100426	**100431**	0.00	501.9	101354	0.92	16.8	100442	0.02	758.9
E.900.60	100144	**100171**	0.03	644.0	104007	3.86	42.6	100185	0.04	483.7
E.1600.20	180642	180654	0.01	616.6	182784	1.19	16.0	**180647**	0.00	1683.2
E.1600.40	178293	**178307**	0.01	620.0	179571	0.72	16.5	178311	0.01	2214.7
E.1600.80	176816	**176846**	0.02	624.6	179740	1.65	19.5	176866	0.03	2473.1
E Average			**0.02**	**281.9**		**1.88**	**41.5**		**0.04**	**592.5**
Average			**0.09**	**535.60**		**1.74**	**97.9**		**0.09**	**603.2**

*Indicates that an optimal solution was found (and proven) by the BPPS module

Table 2. Computational experiments for proposed large dataset

Inst.	*Lower*	GSS (GTS + BPPS)			GTS		
	Bound	Best	Gap	Time	Best	Gap	Time
G - 1000 - 50	13484	**13587**	0.8	678.6	14569	8.0	78.6
G - 1000 - 100	12802	**13258**	3.6	702.2	14400	12.5	102.2
G - 1000 - 200	12636	**13504**	6.9	696.0	14968	18.5	96.0
G - 2000 - 50	26794	**26942**	0.6	670.7	28456	6.2	70.7
G - 2000 - 100	26068	**26784**	2.7	675.6	27674	6.2	75.6
G - 2000 - 200	25094	**27457**	9.4	741.5	27554	9.8	141.5
G - 4000 - 100	50873	**51691**	1.6	714.1	52783	3.8	114.1
G - 4000 - 200	50345	**52519**	4.3	866.5	**52519**	4.3	266.5
G - 4000 - 400	50219	**53508**	6.5	1173.2	**53508**	6.5	573.2
G - 9000 - 150	113399	**115424**	1.8	851.8	116173	2.4	251.8
G - 9000 - 300	113176	**115248**	1.8	1124.1	**115248**	1.8	524.1
G - 9000 - 600	113238	**115406**	1.9	1219.6	**115406**	1.9	619.6
G Average			**3.5**	**842.8**		**6.8**	**242.8**

4 Conclusions

This work presented a hybrid GPU heuristic inspired by Scatter Search metaheuristic in order to deal with the Generalized Assignment Problem, or GAP. The proposed algorithm (named GSS), consisted of two integrated phases, where in the first phase named GTS, 12 independent Tabu Searches were performed in a GPU to form a pool of 12 high quality solutions. This pool was later passed to the BPPS module, consisting of an exact solver for a Binary Programming model considering a limited version of the original problem. The proposed technique managed to find 21 best known solutions, when compared to the state-of-the-art algorithm for the GAP, also proving the optimality of 15 solutions. The average gaps are smaller than literature, when compared to a lower bound of the problem. The 12 new instances be proposed for this challenging and classical problem, incorporating characteristics of the hardest problems in D group, named of G group. Where it was possible to verify the efficiency of the proposed method when dealing with very large instances.

Although the achieved gaps are quite low for most instances, we propose the following future improvements for the GSS algorithm. Many low level optimizations can be applied in GTS regarding the GPU performance, such as: reducing number of registers, increasing shared memory usage, storage of read-only data in specific constant memories, better kernel launching parameters that increase memory and compute throughput, while reducing the general occupancy.

Acknowledgements. The authors would like to thank Brazilian funding agencies CAPES, CNPq, FAPEMIG and FAPERJ for supporting the current work.

References

1. Avella, P., Boccia, M., Vasilyev, I.: A computational study of exact knapsack separation for the generalized assignment problem. Comput. Optim. Appl. **45**(3), 543–555 (2010). Springer, Heidelberg
2. Cattrysse, D.G., Van Wassenhove, L.N.: A survey of algorithms for the generalized assignment problem. Eur. J. Oper. Res. **60**(3), 260–272 (1992). Elsevier, Amsterdam
3. Chu, P.C., Beasley, J.E.: A genetic algorithm for the generalised assignment problem. Comput. Oper. Res. **24**(1), 17–23 (1997). Elsevier, Amsterdam
4. Czapiński, M., Barnes, S.: Tabu Search with two approaches to parallel flowshop evaluation on CUDA platform. J. Parallel Distrib. Comput. **71**(6), 802–811 (2011). Elsevier, Amsterdam
5. Danna, E., Rothberg, E., Pape, C.L.: Exploring relaxation induced neighborhoods to improve MIP solutions. Math. Program. **102**(1), 71–90 (2005). Springer, Heidelberg
6. Diaz, J.A., Fernández, E.: A Tabu search heuristic for the generalized assignment problem. Eur. J. Oper. Res. **132**(1), 22–38 (2001). Elsevier, Amsterdam
7. Feltl, H., Raidl, G.R.: An improved hybrid genetic algorithm for the generalized assignment problem. In: Proceedings of the 2004 ACM Symposium on Applied Computing, pp. 990–995. ACM, New York (2004)
8. Glover, F.: Heuristics for integer programming using surrogate constraints. Decis. Sci. **8**(1), 156–166 (1977). Wiley Online Library, New Jersey
9. Glover, F.: Tabu search-part I. ORSA J. Comput. **1**(3), 190–206 (1989). INFORMS, Catonsville
10. Glover, F.: Tabu search-part II. ORSA J. Comput. **2**(1), 4–32 (1990). INFORMS, Catonsville
11. Glover, F., Laguna, M.: Tabu Search. Springer, New York (2013)
12. Glover, F., Rego, C.: Ejection chain and filter-and-fan methods in combinatorial optimization. 4OR: Q. J. Oper. Res. **4**(4), 263–296 (2006). Springer, Heidelberg
13. Higgins, A.J.: A dynamic Tabu search for large-scale generalised assignment problems. Comput. Oper. Res. **28**(10), 1039–1048 (2001). Elsevier, Amsterdam
14. Kirk, D.B., Wen-Mei, W.H.: Programming Massively Parallel Processors: A Hands-on Approach, vol. 2, pp. 10–14. Morgan Kaufmann, San Francisco (2012)
15. Laguna, M., Kelly, J.P., Gonzlez-Velarde, J., Glover, F.: Tabu search for the multilevel generalized assignment problem. Eur. J. Oper. Res. **82**(1), 176–189 (1995). Elsevier, Amsterdam
16. Martí, R., Duarte, A., Laguna, M.: Advanced scatter search for the max-cut problem. INFORMS J. Comput. **21**(1), 26–38 (2009). INFORMS, Catonsville
17. Martí, R., Laguna, M., Glover, F.: Principles of scatter search. Eur. J. Oper. Res. **169**(2), 359–372 (2006). Elsevier, Amsterdam
18. Nauss, R.M.: Solving the generalized assignment problem: an optimizing and heuristic approach. INFORMS J. Comput. **15**(3), 249–266 (2003). INFORMS, Catonsville
19. Öncan, T.: A survey of the generalized assignment problem and its applications. INFOR: Inf. Syst. Oper. Res. **45**(3), 123–141 (2007). Canadian Operational Research Society, Ottawa
20. Osman, I.H.: Heuristics for the generalised assignment problem: simulated annealing and Tabu search approaches. Oper.-Res.-Spektrum **17**(4), 211–225 (1995). Springer, Heidelberg

21. Pigatti, A., De Aragao, M.P., Uchoa, E.: Stabilized branch-and-cut-and-price for the generalized assignment problem. Electron. Notes Discret. Math. **19**, 389–395 (2005). Optimization Online, North-Holland
22. Posta, M., Ferland, J.A., Michelon, P.: An exact method with variable fixing for solving the generalized assignment problem. Comput. Optim. Appl. **52**(3), 629–644 (2012). Springer, Heidelberg
23. Resende, M.G., Ribeiro, C.C., Glover, F., Martí, R.: Scatter search and path-relinking: fundamentals, advances, and applications. In: Gendreau, M., Potvin, J.-Y. (eds.) Handbook of metaheuristics, pp. 87–107. Springer, Heidelberg (2010). doi:10.1007/978-1-4419-1665-5_4
24. Ross, G.T., Soland, R.M.: A branch and bound algorithm for the generalized assignment problem. Math. Program. **8**(1), 91–103 (1975). Springer, Heidelberg
25. Sahni, S., Gonzalez, T.: P-complete approximation problems. J. ACM (JACM) **23**(3), 555–565 (1976). ACM, New York
26. Savelsbergh, M.: A branch-and-price algorithm for the generalized assignment problem. Oper. Res. **45**(6), 831–841 (1997). INFORMS, Catonsville
27. Sulewski, D., Edelkamp, S., Kissmann, P.: Exploiting the computational power of the graphics card: optimal state space planning on the GPU. In: 21st International Conference on Automated Planning and Scheduling - ICAPS, Freiburg (2011)
28. Wilson, J.M.: A genetic algorithm for the generalised assignment problem. J. Oper. Res. Soc. **48**(8), 804–809 (1997). Elsevier, Amsterdam
29. Yagiura, M., Ibaraki, T., Glover, F.: A path relinking approach with ejection chains for the generalized assignment problem. Eur. J. Oper. Res. **169**(2), 548–569 (2006)

Vector Field Second Order Derivative Approximation and Geometrical Characteristics

Michal Smolik(✉) and Vaclav Skala

Department of Computer Science and Engineering, Faculty of Applied Sciences, University of West Bohemia, Pilsen, Czech Republic
{smolik, skala}@kiv.zcu.cz

Abstract. Vector field is mostly linearly approximated for the purpose of classification and description. This approximation gives us only basic information of the vector field. We will show how to approximate the vector field with second order derivatives, i.e. Hessian and Jacobian matrices. This approximation gives us much more detailed description of the vector field. Moreover, we will show the similarity of this approximation with conic section formula.

Keywords: Vector field · Critical point · Geometry · Conic section · Hessian matrix

1 Introduction

The visualization of vector field topology is a problem that arises naturally when studying the qualitative structure of flows that are tangential to some surface. The knowledge of the data in a single point would be of little help when the goal is to obtain knowledge and understanding of the whole vector field. The individual numbers can be of little interest. It is the connection between them, which is important.

Helman and Hesselink [6] introduced the concept of the topology of a planar vector field to the visualization community. They extracted critical points and classified them into sources, sinks and saddles, and integrated certain stream lines called separatrices from the saddles in the directions of the eigenvectors of the Jacobian matrix. Later, topological methods have been extended to higher order critical points [14], boundary switch points [10], and closed separatrices [21]. In addition, topological methods using classification have been applied to simplify [15, 16], smooth [20], compress [1, 7], compare [11] and design vector fields.

The published research methods use for classification of critical points and vector field description only linear approximation of the vector field. None of it uses an approximation with second order partial derivatives, i.e. Hessian matrix. This approximation gives a more detailed description of the vector field around a critical point and can be used for a more detailed classification. Use of the approximation with Hessian matrix will be described in this paper.

O. Gervasi et al. (Eds.): ICCSA 2017, Part I, LNCS 10404, pp. 148–158, 2017.
DOI: 10.1007/978-3-319-62392-4_11

2 Vector Field Approximation

Vector fields [18] on surfaces [17] are important objects, which appear frequently in scientific simulation in CFD (Computational Fluid Dynamics) [2, 12] or modelling by FEM (Finite Element Method). To be visualized [5, 8], such vector fields are usually linearly approximated for the sake of simplicity and performance considerations. Other possible approximations are [3, 4, 9].

The vector field can be easily analyzed when having an approximation of the vector field near some location point. The important places to be analyzed are so called critical points. Analyzing the vector field behavior near these points gives us the information about the characteristic of the vector field.

2.1 Critical Point

Critical points ($\boldsymbol{x}_0$) of the vector field are points at which the magnitude of the vector vanishes

$$\frac{d\boldsymbol{x}}{dt} = \boldsymbol{v}(\boldsymbol{x}) = \boldsymbol{0}, \tag{1}$$

i.e. all components are equal to zero

$$\begin{bmatrix} \frac{dx}{dt} \\ \frac{dy}{dt} \end{bmatrix} = \begin{bmatrix} 0 \\ 0 \end{bmatrix}. \tag{2}$$

A critical point is said to be isolated, or simple, if the vector field is non-vanishing in an open neighborhood around the critical point. Thus for all surrounding points $\boldsymbol{x}_\varepsilon$ of the critical point $\boldsymbol{x}_0$ the Eq. (1) does not apply, i.e.

$$\frac{d\boldsymbol{x}_\varepsilon}{dt} \neq \boldsymbol{0}. \tag{3}$$

At critical points, the direction of the field line is indeterminate, and they are the only points in the vector field where field lines can intersect (asymptotically). The terms singular point, null point, neutral point or equilibrium point are also frequently used to describe critical points.

These points are important because together with the nearby surrounding vectors, they have more information encoded in them than any such group in the vector field, regarding the total behavior of the field.

2.2 Linearization of Vector Field

Critical points can be characterized according to the behavior of nearby tangent curves. We can use a particular set of these curves to define a skeleton that characterizes the global behavior of all other tangent curves in the vector field. An important feature of differential equations is that it is often possible to determine the local stability of a critical point by approximating the system by a linear system. These approximations are aimed at studying the local behavior of a system, where the nonlinear effects are expected to be small. To locally approximate a system, the Taylor series expansion must be utilized locally to find the relation between $\boldsymbol{v}$ and position $\boldsymbol{x}$, supposing the flow $\boldsymbol{v}$ to be sufficiently smooth and differentiable. In such case, the expansion of $\boldsymbol{v}$ around the critical points $\boldsymbol{x}_0$ is

$$\boldsymbol{v}(\boldsymbol{x}) = \boldsymbol{v}(\boldsymbol{x}_0) + \frac{\partial \boldsymbol{v}}{\partial \boldsymbol{x}}(\boldsymbol{x} - \boldsymbol{x}_0). \tag{4}$$

As $\boldsymbol{v}(\boldsymbol{x}_0)$ is according to (1) equal zero for critical points, we can rewrite Eq. (4) using matrix notation

$$\begin{bmatrix} v_x \\ v_y \end{bmatrix} = \begin{bmatrix} \frac{\partial v_x}{\partial x} & \frac{\partial v_x}{\partial y} \\ \frac{\partial v_y}{\partial x} & \frac{\partial v_y}{\partial y} \end{bmatrix} \cdot \begin{bmatrix} x - x_0 \\ y - y_0 \end{bmatrix} \tag{5}$$

$$\boldsymbol{v} = \boldsymbol{J} \cdot (\boldsymbol{x} - \boldsymbol{x}_0), \tag{6}$$

where $\boldsymbol{J}$ is called Jacobian matrix and characterizes the vector field behavior around a critical point $\boldsymbol{x}_0$.

2.3 Approximation Using Hessian Matrix

Vector fields are approximated using only linear approximation to determine the local behavior of the vector field. However, linearization gives as basic classification of the critical points and about the flow around them, the approximation using second order derivatives will give us some more information.

The approximation of vector field around a critical point using the second order derivative must be written for each vector component (v_x and v_y) separately, see the following equations

$$v_x = \begin{bmatrix} \frac{\partial v_x}{\partial x} \\ \frac{\partial v_x}{\partial y} \end{bmatrix}^T \cdot \begin{bmatrix} \Delta x \\ \Delta y \end{bmatrix} + \frac{1}{2} \begin{bmatrix} \Delta x \\ \Delta y \end{bmatrix}^T \cdot \begin{bmatrix} \frac{\partial^2 v_x}{\partial x^2} & \frac{\partial^2 v_x}{\partial x \partial y} \\ \frac{\partial^2 v_x}{\partial y \partial x} & \frac{\partial^2 v_x}{\partial y^2} \end{bmatrix} \cdot \begin{bmatrix} \Delta x \\ \Delta y \end{bmatrix} \tag{7}$$

$$v_y = \begin{bmatrix} \frac{\partial v_y}{\partial x} \\ \frac{\partial v_y}{\partial y} \end{bmatrix}^T \cdot \begin{bmatrix} \Delta x \\ \Delta y \end{bmatrix} + \frac{1}{2} \begin{bmatrix} \Delta x \\ \Delta y \end{bmatrix}^T \cdot \begin{bmatrix} \frac{\partial^2 v_y}{\partial x^2} & \frac{\partial^2 v_y}{\partial x \partial y} \\ \frac{\partial^2 v_y}{\partial y \partial x} & \frac{\partial^2 v_y}{\partial y^2} \end{bmatrix} \cdot \begin{bmatrix} \Delta x \\ \Delta y \end{bmatrix}, \tag{8}$$

where $\Delta x = x - x_0$ and $\Delta y = y - y_0$. These two equations can be written in matrix notation as well

$$v_x = \boldsymbol{J}_x \cdot (\boldsymbol{x} - \boldsymbol{x}_0) + \frac{1}{2}(\boldsymbol{x} - \boldsymbol{x}_0)^T \cdot \boldsymbol{H}_x \cdot (\boldsymbol{x} - \boldsymbol{x}_0) \tag{9}$$

$$v_y = \boldsymbol{J}_y \cdot (\boldsymbol{x} - \boldsymbol{x}_0) + \frac{1}{2}(\boldsymbol{x} - \boldsymbol{x}_0)^T \cdot \boldsymbol{H}_y \cdot (\boldsymbol{x} - \boldsymbol{x}_0), \quad (10)$$

where $\boldsymbol{H}_x$ and $\boldsymbol{H}_y$ are Hessian matrices, $\boldsymbol{J}_x$ is the first row of Jacobian matrix and $\boldsymbol{J}_y$ is the second row of Jacobian matrix.

The Hessian matrix is a square matrix of second-order partial derivatives of a scalar-valued function, or scalar field. It describes the local curvature of a function of many variables.

Approximation of vector field using (7) and (8) gives us more detailed description than approximation of vector field using (5), see Fig. 1. The approximation in Fig. 1 (right) gives us the same information like in Fig. 1 (left), although we can see the curvature of the two main axis for the saddle.

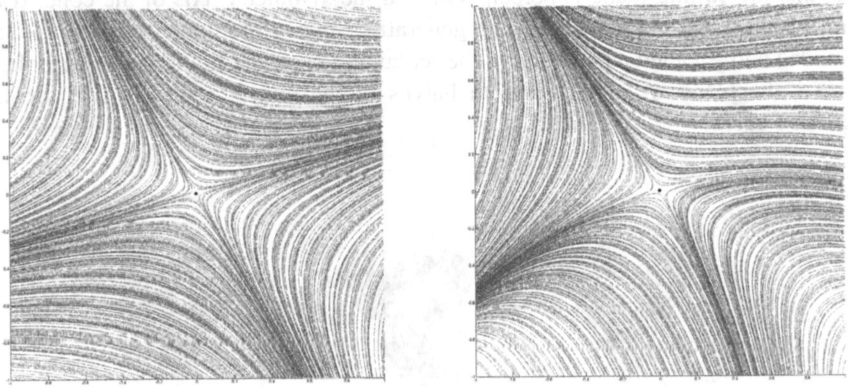

Fig. 1. Comparison between the phase portraits for the vector field approximated using linear approximation (left) and using second order derivative (right).

Equations (7) and (8) can be rewritten in different formulas as follows

$$v_x = \frac{1}{2}[\Delta x \quad \Delta y \quad 1] \cdot \begin{bmatrix} \frac{\partial^2 v_x}{\partial x^2} & \frac{\partial^2 v_x}{\partial x \partial y} & \frac{\partial v_x}{\partial x} \\ \frac{\partial^2 v_x}{\partial y \partial x} & \frac{\partial^2 v_x}{\partial y^2} & \frac{\partial v_x}{\partial y} \\ \frac{\partial v_x}{\partial x} & \frac{\partial v_x}{\partial y} & 0 \end{bmatrix} \cdot \begin{bmatrix} \Delta x \\ \Delta y \\ 1 \end{bmatrix}, \quad (11)$$

$$v_y = \frac{1}{2}[\Delta x \quad \Delta y \quad 1] \cdot \begin{bmatrix} \frac{\partial^2 v_y}{\partial x^2} & \frac{\partial^2 v_y}{\partial x \partial y} & \frac{\partial v_y}{\partial x} \\ \frac{\partial^2 v_y}{\partial y \partial x} & \frac{\partial^2 v_y}{\partial y^2} & \frac{\partial v_y}{\partial y} \\ \frac{\partial v_y}{\partial x} & \frac{\partial v_y}{\partial y} & 0 \end{bmatrix} \cdot \begin{bmatrix} \Delta x \\ \Delta y \\ 1 \end{bmatrix}. \quad (12)$$

These two equations have some geometrical background. When v_x and v_y are equal zero, each equation describes some conic section.

Approximation of the vector field using Hessian matrix, i.e. using second order derivatives, is a bit more computationally expensive than the standard linear approximation but gives us more detailed description of the vector field as will be seen in the following chapters.

Conic Section

A conic is the curve obtained as the intersection of a plane, called the cutting plane, with a double cone, see Fig. 2. Planes that pass through the vertex of the cone will intersect the cone in a point, a line or a pair of intersecting lines. These are called degenerate conics and some authors do not consider them to be conics at all.

There are three types of non-degenerated conics, the ellipse, parabola, and hyperbola, see Fig. 2. The circle is a special kind of ellipse. The circle and the ellipse arise when the intersection of the cone and plane is a closed curve. The circle is obtained when the cutting plane is parallel to the plane of the generating circle of the cone, this means that the cutting plane is perpendicular to the symmetry axis of the cone. If the cutting plane is parallel to exactly one generating line of the cone, then the conic is unbounded and is called a parabola. In the remaining case, the figure is a hyperbola. In this case, the plane will intersect both halves of the cone, producing two separate unbounded curves.

Fig. 2. Types of conic sections, i.e. parabola, circle and ellipse, and hyperbola.

A conic section is described by the following implicit equation

$$\begin{bmatrix} x & y & 1 \end{bmatrix} \cdot \begin{bmatrix} a_{11} & a_{12} & a_{13} \\ a_{21} & a_{22} & a_{23} \\ a_{31} & a_{32} & a_{33} \end{bmatrix} \cdot \begin{bmatrix} x \\ y \\ 1 \end{bmatrix} = 0. \tag{13}$$

where $a_{ij}i, j \in \{1, 2, 3\}$ are coefficients of conic section. Depending on these values, we can classify the types of conic sections. To do that, we need to compute two determinants

$$\Omega = \begin{vmatrix} a_{11} & a_{12} & a_{13} \\ a_{21} & a_{22} & a_{23} \\ a_{31} & a_{32} & a_{33} \end{vmatrix} \tag{14}$$

$$\omega = \begin{vmatrix} a_{11} & a_{12} \\ a_{21} & a_{22} \end{vmatrix}. \tag{15}$$

When knowing determinants Ω and ω we can easily classify the type of conic section using the following table

Table 1. Classification of conic section.

	$\omega \neq 0$		$\omega = 0$
$\Omega \neq 0$	$\omega > 0$	$\omega < 0$	Parabola
	Ellipse	Hyperbola	
$\Omega = 0$	Pair of intersecting lines		Pair of parallel lines

Equations (11) and (12) are the same as (13) when $v_x = 0$ and $v_y = 0$ and therefore they geometrically represent conic sections (Table 1).

3 Classification of Critical Points

There exist a finite set of fundamentally different critical points, defined by the number of inflow and outflow directions, spiraling structures etc., and combinations of these. Since the set is finite, each critical point can be classified. Such a classification defines the field completely in a close neighborhood around the critical point. By knowing the location and classification of critical points in a vector field, the topology of the field is known in small areas around these. Assuming a smooth transition between these areas, one can construct a simplified model of the whole vector field. Such a simplified representation is useful, for instance, in compressing vector field data into simpler building blocks [13].

The critical points are classified based on the vector field around that point. The information derived from the classification of critical points aids the information selection process when it comes to visualizing the field. By choosing seed points for field lines based on the topology of critical points, field lines encoding important information is ensured. A more advanced approach is to connect critical points, and use the connecting lines and surfaces to separate areas of different flow topology [1, 19].

3.1 Standard Classification Using a Linear Approximation

The fact that a linear model can be used to study the behavior of a nonlinear system near a critical point is a powerful one [1]. We can use the Jacobian matrix to characterize the vector field and the behavior of nearby tangent curves, for nondegenerate critical point.

The eigenvalues and eigenvectors of Jacobian matrix are very important for vector field classification and description (see Fig. 3). A real eigenvector of the Jacobian matrix defines a direction such that if we move slightly from the critical point in that direction, the field is parallel to the direction we moved. Thus, at the critical point, the real eigenvectors are tangent to the trajectories that end on the point. The sign of the corresponding eigenvalue determines whether the trajectory is outgoing (repelling) or incoming (attracting) at the critical point. The imaginary part of an eigenvalue denotes circulation about the point.

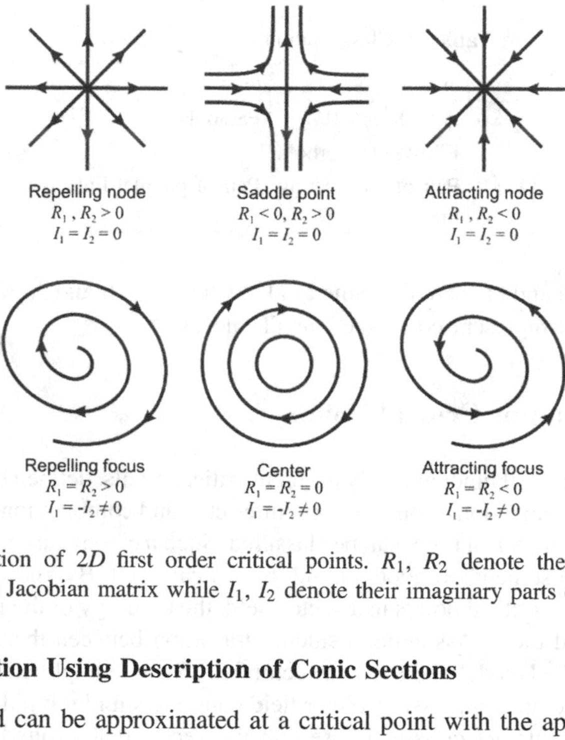

Fig. 3. Classification of 2*D* first order critical points. R_1, R_2 denote the real parts of the eigenvalues of the Jacobian matrix while I_1, I_2 denote their imaginary parts (from [1]).

3.2 Classification Using Description of Conic Sections

Each vector field can be approximated at a critical point with the approximation that uses the second order derivatives, i.e. Hessian matrix. One such example of approximated vector field around a critical point $\boldsymbol{x}_0 = [0,0]^T$ can be

$$v_x = \frac{1}{2}[\Delta x \quad \Delta y \quad 1] \cdot \begin{bmatrix} -1 & 1 & 1 \\ 1 & -1 & 2 \\ 1 & 2 & 0 \end{bmatrix} \cdot \begin{bmatrix} \Delta x \\ \Delta y \\ 1 \end{bmatrix}, \tag{16}$$

$$v_y = \frac{1}{2}[\Delta x \quad \Delta y \quad 1] \cdot \begin{bmatrix} 0 & 0 & -1 \\ 0 & 0 & 1.5 \\ -1 & 1.5 & 0 \end{bmatrix} \cdot \begin{bmatrix} \Delta x \\ \Delta y \\ 1 \end{bmatrix}. \tag{17}$$

Equation (16) represents for $v_x = 0$ a parabola and (17) for $v_y = 0$ a line. This approximated vector field can be seen in Fig. 4.

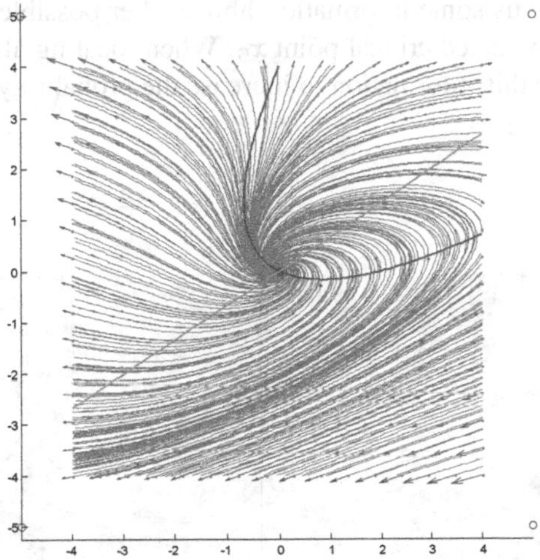

Fig. 4. Vector field approximated as (16) and (17). The zero iso-lines are a line and a parabola.

Now, we showed conic sections that have only one intersection point at $[0,0]^T$. Two conic sections can have up to four intersections. Each intersection defines a critical point. Therefore, we can approximate a vector field around one critical point and some more critical points in the neighborhood will be included in this approximation.

Vector fields around a focus critical point can be for some real vector field approximated for example as

$$\begin{aligned} v_x &= \frac{1}{2}[\Delta x \quad \Delta y \quad 1] \cdot \begin{bmatrix} 1 & -3 & 1 \\ -3 & 1 & 2 \\ 1 & 2 & 0 \end{bmatrix} \cdot \begin{bmatrix} \Delta x \\ \Delta y \\ 1 \end{bmatrix} \\ v_y &= \frac{1}{2}[\Delta x \quad \Delta y \quad 1] \cdot \begin{bmatrix} 0 & 0 & -1 \\ 0 & 0 & 1.5 \\ -1 & 1.5 & 0 \end{bmatrix} \cdot \begin{bmatrix} \Delta x \\ \Delta y \\ 1 \end{bmatrix} \end{aligned} \quad (18)$$

$$\begin{aligned} v_x &= \frac{1}{2}[\Delta x \quad \Delta y \quad 1] \cdot \begin{bmatrix} -0.5 & 0.5 & 1 \\ 0.5 & -0.5 & 2 \\ 1 & 2 & 0 \end{bmatrix} \cdot \begin{bmatrix} \Delta x \\ \Delta y \\ 1 \end{bmatrix} \\ v_y &= \frac{1}{2}[\Delta x \quad \Delta y \quad 1] \cdot \begin{bmatrix} -1 & 1 & -1 \\ 1 & -1 & 1.5 \\ -1 & 1.5 & 0 \end{bmatrix} \cdot \begin{bmatrix} \Delta x \\ \Delta y \\ 1 \end{bmatrix} \end{aligned} \quad (19)$$

This both approximations of vector fields describe behavior around a focus critical point at $[0, 0]^T$. Both of them contain one more critical point, which is a saddle critical point. These saddle critical points do not have to be real critical points of the approximated vector field, but they can be present in the vector field. Therefore, this approximation can give us some information about other possible critical points in the neighborhood of approximated critical point $\boldsymbol{x}_0$. When locating all critical points in the vector field, we can use this information to increase the probability of finding all critical points (Fig. 5).

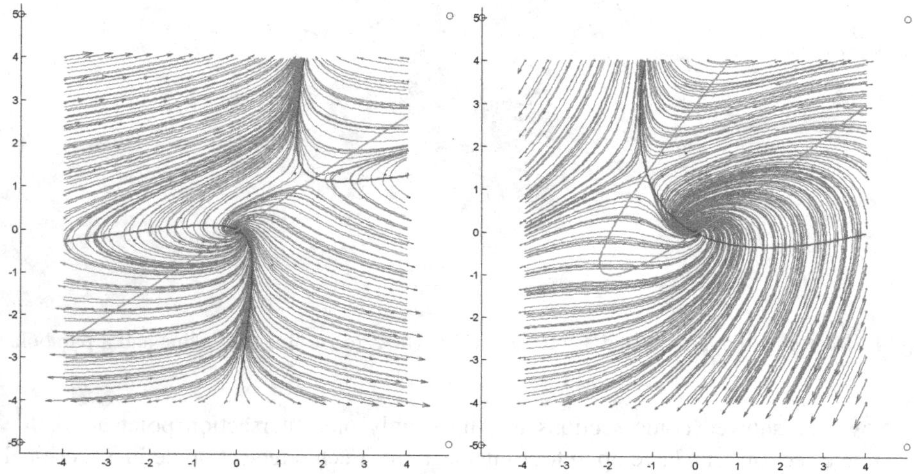

Fig. 5. Vector field approximated as (18) (left) and (19) (right). The zero iso-lines are a line and a hyperbola (left), or two parabolas (right).

The maximal number of two conic sections intersection points is four. In the next example, we will show it. Let us have a vector field, which can be approximated at point $\boldsymbol{x}_0$ for example as

$$\begin{aligned} v_x &= \frac{1}{2}\begin{bmatrix}\Delta x & \Delta y & 1\end{bmatrix} \cdot \begin{bmatrix} -0.25 & 0 & 1 \\ 0 & -1 & 2 \\ 1 & 2 & 0 \end{bmatrix} \cdot \begin{bmatrix}\Delta x \\ \Delta y \\ 1\end{bmatrix} \\ v_y &= \frac{1}{2}\begin{bmatrix}\Delta x & \Delta y & 1\end{bmatrix} \cdot \begin{bmatrix} -1 & 1 & -1 \\ 1 & -1 & 1.5 \\ -1 & 1.5 & 0 \end{bmatrix} \cdot \begin{bmatrix}\Delta x \\ \Delta y \\ 1\end{bmatrix} \end{aligned} \tag{20}$$

$$v_x = \frac{1}{2}[\Delta x \quad \Delta y \quad 1] \cdot \begin{bmatrix} 1 & 1 & 1 \\ 1 & 2 & 2 \\ 1 & 2 & 0 \end{bmatrix} \cdot \begin{bmatrix} \Delta x \\ \Delta y \\ 1 \end{bmatrix}$$
$$v_y = \frac{1}{2}[\Delta x \quad \Delta y \quad 1] \cdot \begin{bmatrix} 1 & -1 & -1 \\ -1 & 1.5 & 1.5 \\ -1 & 1.5 & 0 \end{bmatrix} \cdot \begin{bmatrix} \Delta x \\ \Delta y \\ 1 \end{bmatrix} \quad (21)$$

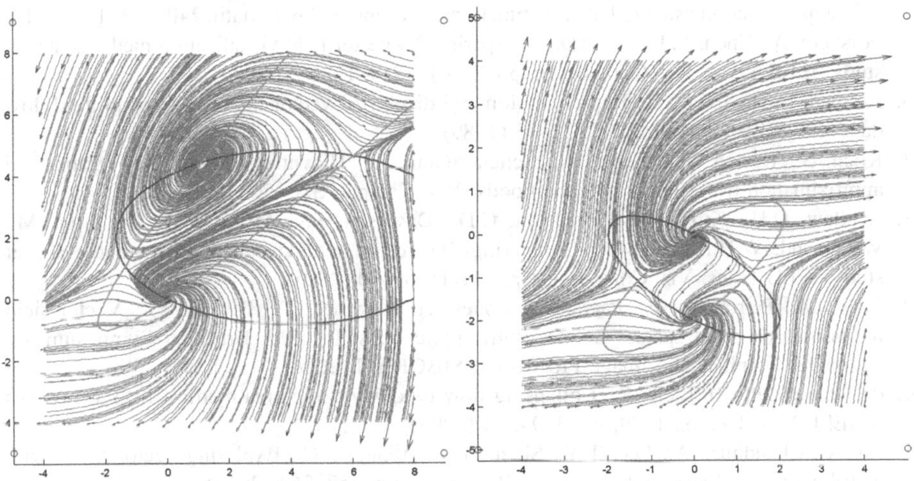

Fig. 6. Vector field approximated as (20) (left) and (21) (right). The zero iso-lines are a parabola and an ellipse (left), or two ellipses (right).

These two approximations (20) and (21) of vector fields are visualized in Fig. 6. It can be seen, that each approximation contains four critical points, i.e. one critical point where the vector field was approximated and three more critical points.

4 Conclusion

A new vector field critical points description using the second order derivatives approximation is described. The approximation can be rewritten in a matrix form of a conic section formula. We proved, that approximation using Hessian matrix, rather than only Jacobian matrix, gives us better representation of a vector field and it can help with localization of critical points in a vector field.

Acknowledgments. The authors would like to thank their colleagues at the University of West Bohemia, Plzen, for their comments and suggestions, their valuable comments and hints provided. The research was supported by projects Czech Science Foundation (GACR) No. 17-05534S and partly by SGS 2016-013.

References

1. Agranovsky, A., Camp, D., Joy, K.I., Childs, H.: Subsampling-based compression and flow visualization. In: IS&T/SPIE Electronic Imaging, International Society for Optics and Photonics (2015)
2. Balduzzi, F., Bianchini, A., Maleci, R., Ferrara, G., Ferrari, L.: Critical issues in the CFD simulation of Darrieus wind turbines. Renew. Energy **85**, 419–435 (2016)
3. Benbourhim, M.N., Bouhamidi, A.: Approximation of vectors fields by thin plate splines with tension. J. Approx. Theory **136**(2), 198–229 (2005)
4. Cabrera, D.A.C., González-Casanova, P., Gout, C., Juárez, L.H., Reséndizd, L.R.: Vector field approximation using radial basis functions. J. Comput. Appl. Math. **240**, 163–173 (2013)
5. Forsberg, A., Chen, J., Laidlaw, D.: Comparing 3D vector field visualization methods: a user study. IEEE Trans. Vis. Comput. Graph. **15**(6), 1219–1226 (2009)
6. Helman, J., Hesselink, L.: Representation and display of vector field topology in fluid flow data sets. IEEE Comput. **22**(8), 27–36 (1989)
7. Koch, S., Kasten, J., Wiebel, A., Scheuermann, G., Hlawitschka, M.: 2D Vector field approximation using linear neighborhoods. Vis. Comput. **32**, 1563–1578 (2015)
8. Laidlaw, D.H., Kirby, R.M., Jackson, C.D., Davidson, J.S., Miller, T.S., Da Silva, M., Warrenand, W.H., Tarr, M.J.: Comparing 2D vector field visualization methods: a user study. IEEE Trans. Vis. Comput. Graph. **11**(1), 59–70 (2005)
9. Lage, M., Petronetto, F., Paiva, A., Lopes, H., Lewiner, T., Tavares, G.: Vector field reconstruction from sparse samples with applications. In: 19th Brazilian Symposium on Computer Graphics and Image Processing, SIBGRAPI (2006)
10. de Leeuw, W., van Liere, R.: Collapsing flow topology using area metrics. In: Proceedings of IEEE Visualization 1999, pp. 349–354 (1999)
11. Lu, K., Chaudhuri, A., Lee, T.Y., Shen, H.W., Wong, P.C.: Exploring vector fields with distribution-based streamline analysis. PacificVis, pp. 257–264 (2013)
12. Peng, C., Teng, Y., Hwang, B., Guo, Z., Wang, L.P.: Implementation issues and benchmarking of lattice Boltzmann method for moving rigid particle simulations in a viscous flow. Comput. Math. Appl. **72**(2), 349–374 (2016)
13. Philippou, P.A., Strickland, R.N.: Vector field analysis and synthesis using threedimensional phase portraits. Graph. Models Image Process. **59**(6), 446–462 (1997)
14. Scheuermann, G., Krüger, H., Menzel, M., Rockwood, A.: Visualizing non-linear vector field topology. IEEE Trans. Vis. Comput. Graph. **4**(2), 109–116 (1998)
15. Skraba, P., Rosen, P., Wang, B., Chen, G., Bhatia, H., Pascucci, V.: Critical point cancellation in 3D vector fields: robustness and discussion. IEEE Trans. Vis. Comput. Graph. (2016)
16. Skraba, P., Wang, B., Chen, G., Rosen, P.: 2D vector field simplification based on robustness. In: Pacific Visualization Symposium (PacificVis), IEEE, pp. 49–56 (2014)
17. Smolik, M., Skala, V.: Spherical RBF vector field interpolation: experimental study. SAMI 2017, pp. 431–434, Slovakia (2017)
18. Smolik, M., Skala, V.: Vector field interpolation with radial basis functions. SIGRAD 2016, pp. 15–21, Sweden (2016)
19. Weinkauf, T., Theisel, H., Shi, K., Hege, H.-C., Seidel, H.-P.: Extracting higher order critical points and topological simplification of 3D vector fields. In: Proceedings of IEEE Visualization 2005, pp. 559–566, Minneapolis, U.S.A. (2005)
20. Westermann, R., Johnson, C., Ertl, T.: Topology-preserving smoothing of vector fields. IEEE Trans.Vis. Comput. Graph **7**(3), 222–229 (2001)
21. Wischgoll, T., Scheuermann, G.: Detection and visualization of closed streamlines in planar flows. IEEE Trans. Vis. Comput. Graph. **7**(2), 165–172 (2001)

Exhaustive Analysis for the Effects of a Feedback Regulation on the Bi-Stability in Cellular Signaling Systems

Chinasa Sueyoshi[1] and Takashi Naka[2(✉)]

[1] Graduate School of Information Science, Kyushu Sangyo University, 2-3-1, Matsukadai, Higashi-ku, Fukuoka-shi, Fukuoka, Japan
k15djk01@st.kyusan-u.ac.jp

[2] Faculty of Science and Engineering, Kyushu Sangyo University, 2-3-1, Matsukadai, Higashi-ku, Fukuoka-shi, Fukuoka, Japan
naka@is.kyusan-u.ac.jp

Abstract. Cellular signaling systems regulate biochemical reactions operating in cells for various functions. The regulatory mechanisms have been recently studied intensively since the malfunction of the regulation is thought to be one of the substantial causes of cancer formation. However, it is rather difficult to develop the theoretical framework for investigation of the regulatory mechanisms due to their complexity and nonlinearity. In this study, more general approach is proposed for elucidation of emergence of the bi-stability in cellular signaling systems by construction of mathematical models for a class of cellular signaling systems and the exhaustive simulation analysis over the variation of network architectures and the values of parameters. The model system is formulated as regulatory network in which every node represents an activation-inactivation cyclic reaction for respective constituent enzyme of the network and the regulatory interactions between the reactions are depicted by arcs between nodes. The emergence of the stable equilibrium point in steady states of the network is analyzed with the Michaelis-Menten reaction scheme as the reaction mechanism in each cyclic reaction. The analysis is performed for all variations of the regulatory networks comprised of two nodes, three nodes, and four nodes with a single feedback regulation loop. The ratios and the aspects of the emergence of the stable equilibrium points are analyzed over the exhaustive combinations of the parameter values for each node with the common Michaelis constant for the regulatory networks. It is revealed that the shorter feedback length is favorable for bi-stability. Furthermore, the bi-stability and the oscillation is more likely to develop in the case of low value of the Michaelis constant than in the case of high value, implying that the condition of the higher saturation levels, which induces stronger nonlinearity. In addition to these results, the analysis for the parameter regions yielding the bi-stability and the oscillation are presented.

Keywords: Cellular signaling systems · Regulatory networks · Cyclic reaction · Michaelis-Menten reaction mechanism · Bi-stability · Oscillation

O. Gervasi et al. (Eds.): ICCSA 2017, Part I, LNCS 10404, pp. 159–173, 2017.
DOI: 10.1007/978-3-319-62392-4_12

1 Introduction

Cellular signaling systems regulate biochemical reactions operating in cells for various functions such as cell differentiation, cell proliferation, and homeostasis. Cellular signal transduction systems have been studied intensively from the recent viewpoint that their disorder is thought to be one of the causes for cancer formation since the systems are known to regulate biochemical reactions in cells. Figure 1 shows MAPK (Mitogen-activated Protein Kinase) cascade which is one of the typical cellular signaling systems and has been studied intensively elucidating the various regulatory characteristics for activities of living cells, such as, the switch-like responses, bi-stability, oscillations, robustness, and so on. [1–6]. The cellular signaling systems are comprised of enzymatic reactions, such as, phosphorylation-dephosphorylation cyclic reactions which is primal components of MAPK cascade as seen in Fig. 1. Cyclic reaction seems to be primal reaction system for other cellular signaling system, such as, the Rac1, PAK, and RhoA signaling network [7].

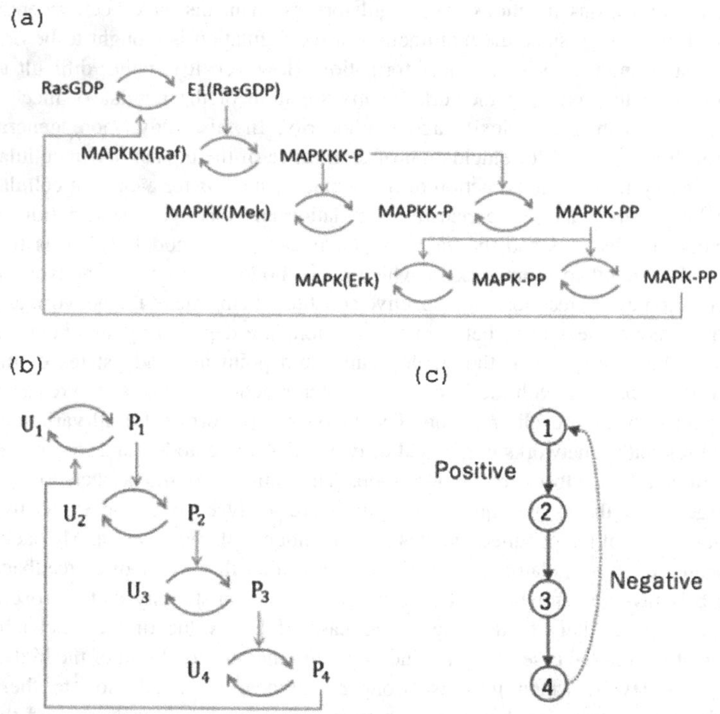

Fig. 1. Regulatory network for MAPK cascade. (a): The MAPK cascade is composed of several cyclic reactions and has a feedback regulation from MAPK-PP. (b): Simplified version of the MAPK cascade. (c): Regulatory network representing the MAPK cascade.

Many studies have employed simulation analysis with the given values for the parameters in the systems because of the difficulty developing the analytical method for the systems due to their non-linearity. In this study, more general approach is proposed for elucidation of emergence of the bi-stability in cellular signaling systems by construction of mathematical models for a class of cellular signaling systems and the exhaustive simulation analysis over the variation of network architectures and the values of parameters. The model system is formulated as regulatory network in which every node represents an activation-inactivation cyclic reaction for respective constituent enzyme of the network and the regulatory interactions between the reactions are depicted by arcs between nodes. The emergence of the stable equilibrium point in steady states of the network is analyzed with the Michaelis-Menten reaction scheme as the reaction mechanism in each cyclic reaction. Since stable equilibrium points are convergent states of the relaxation process in dynamic changes due to random noises, and seem to correspond to the distinct biochemical states, such as, normal states or malfunctional states, it is biologically significant to analyze the aspects of bi-stability in cellular signaling systems. Similar approaches have been taken in several studies [5, 8–13]. Kuwahara et al. applied the similar approach using rather simpler regulatory networks to elucidate the effects of network architectures on the stochastic characteristics [8]. Ma et al. use the same regulatory networks of three nodes, but focused on the biochemical adaptation mechanisms for living cells [9].

2 Formulation of Signaling Systems as Regulatory Networks

We formulate cellular signaling systems as regulatory networks. Every node represents an activation-inactivation cyclic reaction for respective constituent enzyme of the network The regulatory interactions between the reactions are depicted by arcs between nodes. The activated enzyme in a node acts on another node as activating enzyme which is called positive regulation or activating enzyme of reverse path which is called negative regulation. The MAPK cascade is comprised of several cyclic reactions and has a feedback regulation from MAPK-PP as shown in Fig. 1 (a). Figure 1 (b) is a simplified version of the MAPK cascade. Figure 1(c) shows the regulatory network representation for the MAPK cascade which is used in this study.

We employ the Michaelis-Menten mechanism as the reaction mechanisms in the cyclic reactions which is formulated as Eqs. (1)–(5). To be precise, Michaelis-Menten approximation equation is adopted as a reaction mechanism. Therefore, the enzyme substrate complex does not appear in the reaction rate equations. Figure 2 shows a cyclic reaction in which a node i is regulated positively by a node j and negatively by a node k. T_i is the total concentration of the enzyme P_i. R_i is the relative concentration, and L_i is the normalized Michaelis constant. In addition to the Michaelis constant, the normalized rate equation has a parameter K_i which is the ratio of maximum inactivation velocity to maximum activation velocity.

$$\dot{P}_i = \frac{a_i P_j U_i}{M_i + U_i} - \frac{d_i P_k P_i}{M_i + P_i} \tag{1}$$

$$T_i = U_i + P_i \tag{2}$$

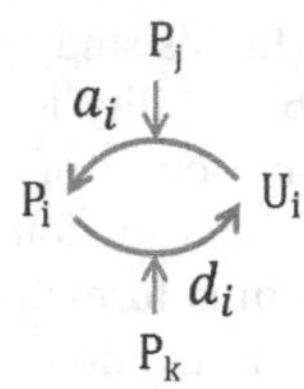

Fig. 2. Cyclic reaction. U_i is the inactive form of the enzyme at node *i*. P_i is the active form of the enzyme at node *i*. P_j is the enzyme catalyzing activation, P_k is the enzyme catalyzing inactivation. The rate constant for the activation and the inactivation are denoted by a_i and d_i, respectively.

$$R_i = \frac{P_i}{T_i}, L_i = \frac{M_i}{T_i} \tag{3}$$

$$\frac{T_i}{a_i T_j}\dot{R}_i = \frac{R_j(1 - R_i)}{L_i + 1 - R_i} - \frac{K_i R_k R_i}{L_i + R_i} \tag{4}$$

$$K_i = \frac{d_i T_k}{a_i T_j} \tag{5}$$

The value of R_j is set to be unity in the case of no positive regulation for node *i*, and the value of R_k is set to be unity in the case of no negative regulation.

We analyze the aspects of emergence of multi-stability for all variations of two-node regulatory networks, three-node regulatory networks, and four-node regulatory networks with a single feedback regulation loop as show in Fig. 3. Solid arrows and dashed arrows depict the positive and negative regulations, respectively. Figure 3I represents the regulatory network for the MAPK cascade with a negative feedback regulation shown in Fig. 1.

3 Parameter Values and Appearance Ratio of Bi-Stability for the Analysis

The exhaustive analysis is performed with the variation of parameter values in appropriate range to cover the assumed values for reactions of MAPK cascade in other studies [14–18]. We assume that the value of normalized Michaelis constant *L* is common for all nodes to reduce the computational costs. The value of *L* is changed over the discrete values of $2^{-5}, 2^{-4}, \ldots, 2^{5}$, that is, 11 different values. K_is are sets to be independent values for each node, and have 11 different values as same as *L*. Therefore, the total number of parameter combinations of K_is is 121 (= 11^2), 1331 (= 11^3), and 14641 (= 11^4) for two-node networks, three-node networks, and four-node networks, respectively.

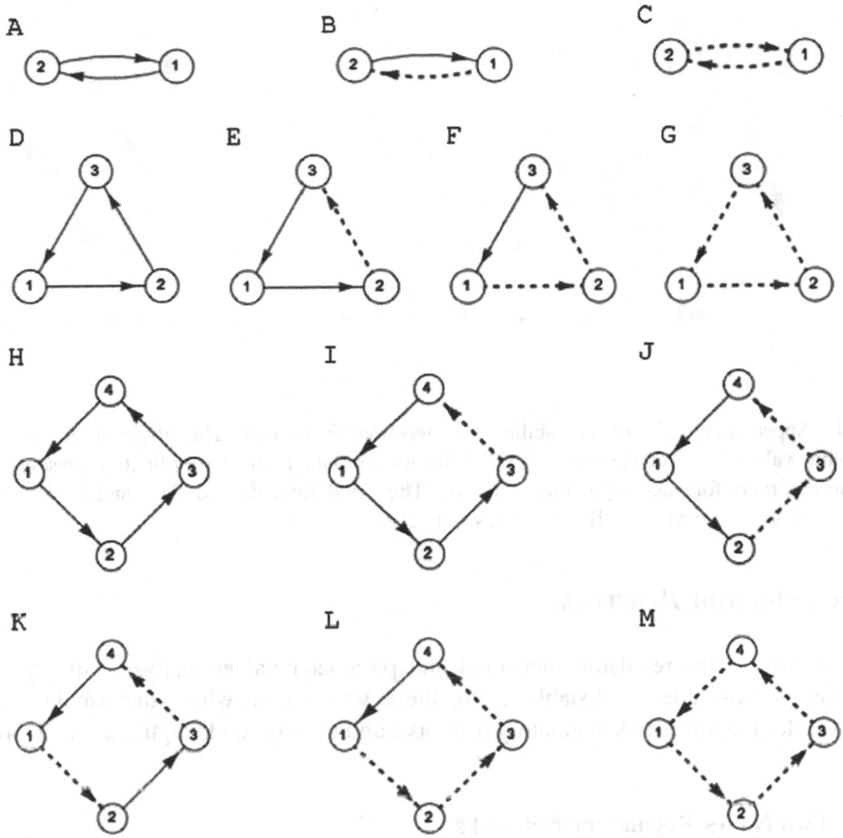

Fig. 3. Regulatory networks with one feedback regulation for two-nodes, three-nodes, and four-nodes. Solid lines and broken lines denotes positive and negative regulation, respectively. The number in each node correspond to the axis label in graphs of parameter space, such as, Fig. 5 for two-node networks, Figs. 8 and 11 for three-node networks.

Bi-stability is assessed by the standard stability theory [19]. At first, a set of algebraic equations derived by setting the right hand side of Eq. (4) to be zero is solved to obtain the equilibrium points of the system. Then, the eigenvalues at those points are calculated for the Jacobian matrix of a set of equations. If the real parts of all eigenvalues are negative, the point is thought to be the stable equilibrium point. If more than two stable equilibrium points exist, then the system is multi-stable. We define appearance ratio of bi-stability as the percentage of the total number of combinations for the parameter (K_is) values yielding the bi-stability to the total number of examined combinations for parameter values, which is used to evaluate the aspect of appearance of bi-stability quantitatively.

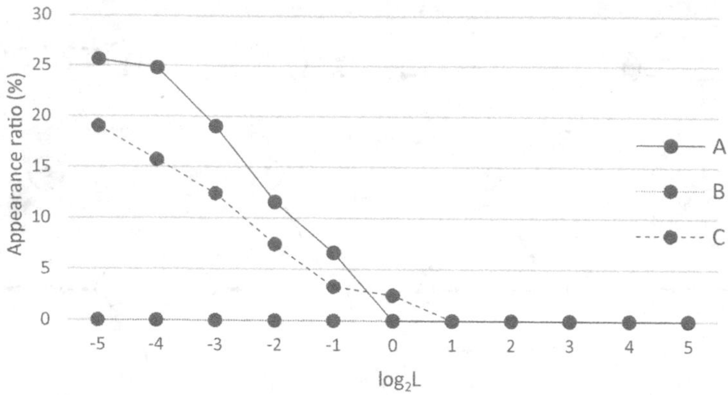

Fig. 4. Appearance ratio of bi-stability for two-node networks. The abscissa depicts the logarithm values of 2 for the normalized Michaelis constant L and the ordinate represents the appearance ratio for each regulatory network. The solid line, dashed line, and broken line correspond to the network A, B, and C, respectively.

4 Results and Discussion

With respect to the regulatory networks and parameter values analyzed, the system becomes mono-stable or bi-stable. Also, there was a case where the oscillation is observed for the three-node regulatory networks and the four-node regulatory networks.

4.1 Two Nodes Regulatory Networks

Two-node regulatory network has become either of mono-stable or bi-stable. Figure 4 shows the appearance ratio of bi-stability for individual regulatory networks. Bi-stability appears in the network A comprised of double positive regulations and the network C comprised of double negative regulations as shown in Fig. 3, and the network A has higher appearance ratio of bi-stability than the network C. we can also see that smaller normalized Michaelis constant L yields higher appearance ratio of bi-stability. The highest appearance ratio was 25.6% and 19.0% in the case of the smallest value for normalized Michaelis constant L in the network A and the network C, respectively.

Figure 5 shows the bi-stable points in the parameter space. The circle marks are for the network A and the triangle marks are for the network C. As seen in these graphs, bi-stability appears where both of two K_i values are small for the network A comprised of double positive regulations corresponding to the region where the activity of activation is higher than that of the inactivation. Bi-stability appears where both of two K_i values are large for the network C comprised of double negative regulations correspond to the region where the activity of inactivation is higher than that of the activation. It is suggested that the parameter values which enhance the feedback regulations are favorable for emergence of bi-stability. Furthermore, the bi-stable region with smaller

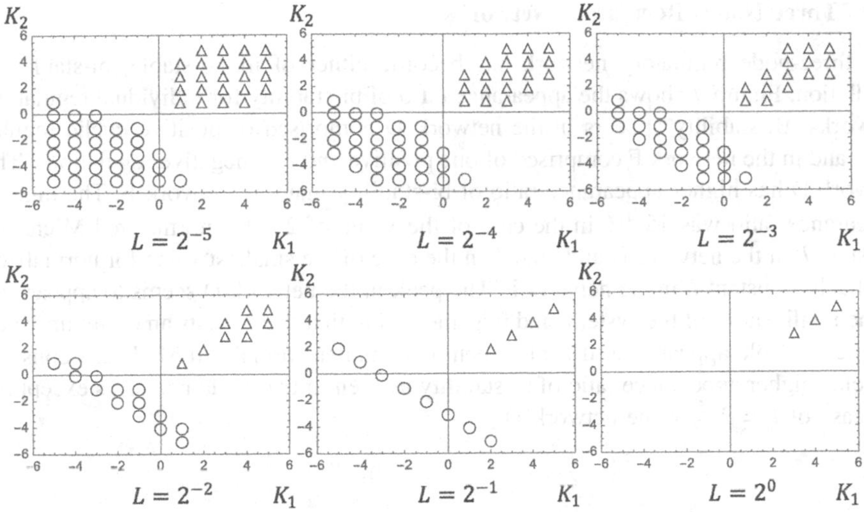

Fig. 5. Distributions of bi-stable points in the parameter space for two-node networks. L denotes the values of the normalized Michaelis constant. The abscissa and the ordinate depict the logarithm values of 2 for K_1 and K_2, respectively. The circle marks and the triangle marks indicate the bi-stable points for the network A and C, respectively.

normalized Michaelis constant L includes one with larger normalized Michaelis constant L in the network C.

Figure 6 shows the effects of Michaelis constant L on values of stable equilibrium points for each network. The distances between pair of equilibrium points are depicted by a circle marks. It is seen that the distance between two points decreases as the Michaelis constant L increases. Then, the bifurcation occurs at around L of 1/2 and 1 for the network A and the network C, respectively.

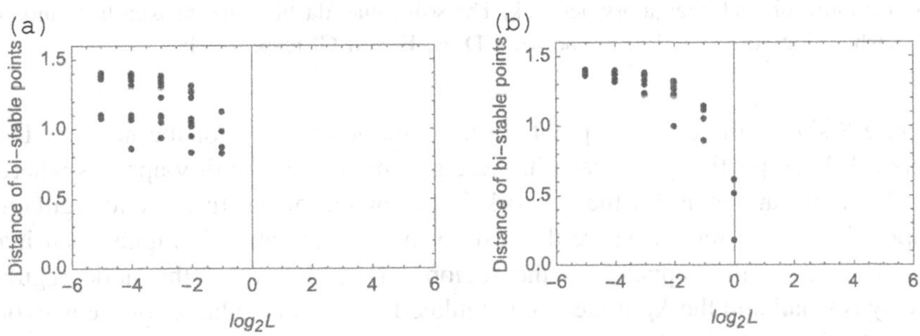

Fig. 6. The effects of Michaelis constant on values of stable equilibrium points for each node in network A (a) and network C (b). Each circle mark corresponds to pair of equilibrium points in a bi-stable state. The abscissa depicts the logarithm values of 2 for the normalized Michaelis constant L and the ordinate represents the distance between two stable equilibrium points.

4.2 Three Nodes Regulatory Networks

The three-node regulatory network has become either of mono-stable, bi-stable, or oscillation. Figure 7 shows the appearance ratio of bi-stability for individual regulatory networks. Bi-stability appears in the network D comprised of positive cyclic regulations and in the network F comprised of one positive and two negative regulations. The network D has higher appearance ratio of bi-stability than the network F. The highest appearance ratio was 15.4% in the case of the value of 2^{-4} for normalized Michaelis constant L in the network D and 10.4% in the case of the smallest value for normalized Michaelis constant L in the network F. The peak in the network D seems to appear due to the nonlinearity of the system, and it is interesting that there exists an optimum value for the bi-stable appearance. It is also seen that smaller normalized Michaelis constant L yields higher appearance ratio of bi-stability as seen in two-node networks except for the case of $L = 2^{-5}$ in the network D.

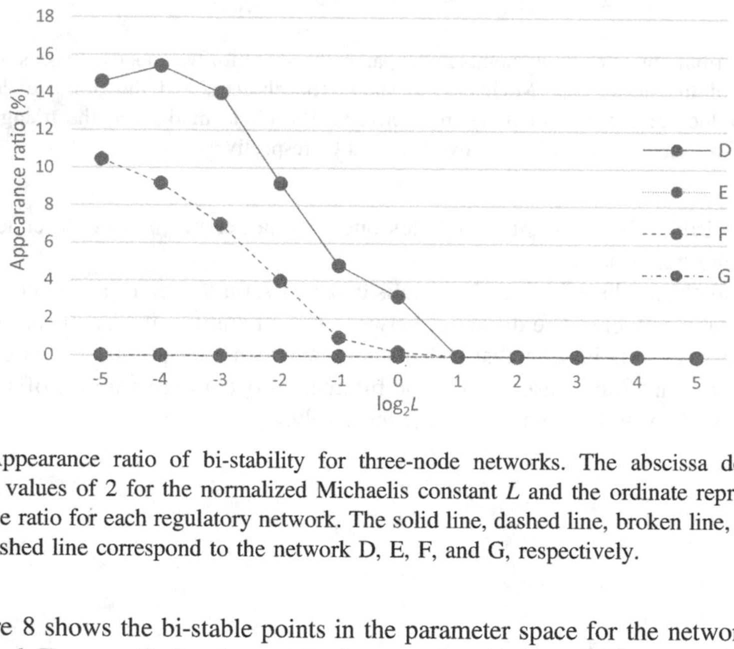

Fig. 7. Appearance ratio of bi-stability for three-node networks. The abscissa depicts the logarithm values of 2 for the normalized Michaelis constant L and the ordinate represents the appearance ratio for each regulatory network. The solid line, dashed line, broken line, and chain double dashed line correspond to the network D, E, F, and G, respectively.

Figure 8 shows the bi-stable points in the parameter space for the network D and the network F, respectively. As seen in these graphs, the bi-stability appears where all of two K_i values are small for the network D comprised of positive cyclic regulations, corresponding to the region where the activity of the activation is higher than inactivation. The bi-stability appears in the region where the K_i of the node regulated positively is small and the K_i of the node regulated negatively is large for the network F comprised of one positive and two negative regulations. As seen in the two-node networks, it is suggested that the parameter values which enhance the feedback regulations are favorable for emergence of bi-stability. Furthermore, the bi-stable region with smaller normalized Michaelis constant includes one with larger normalized Michaelis constant in the network F.

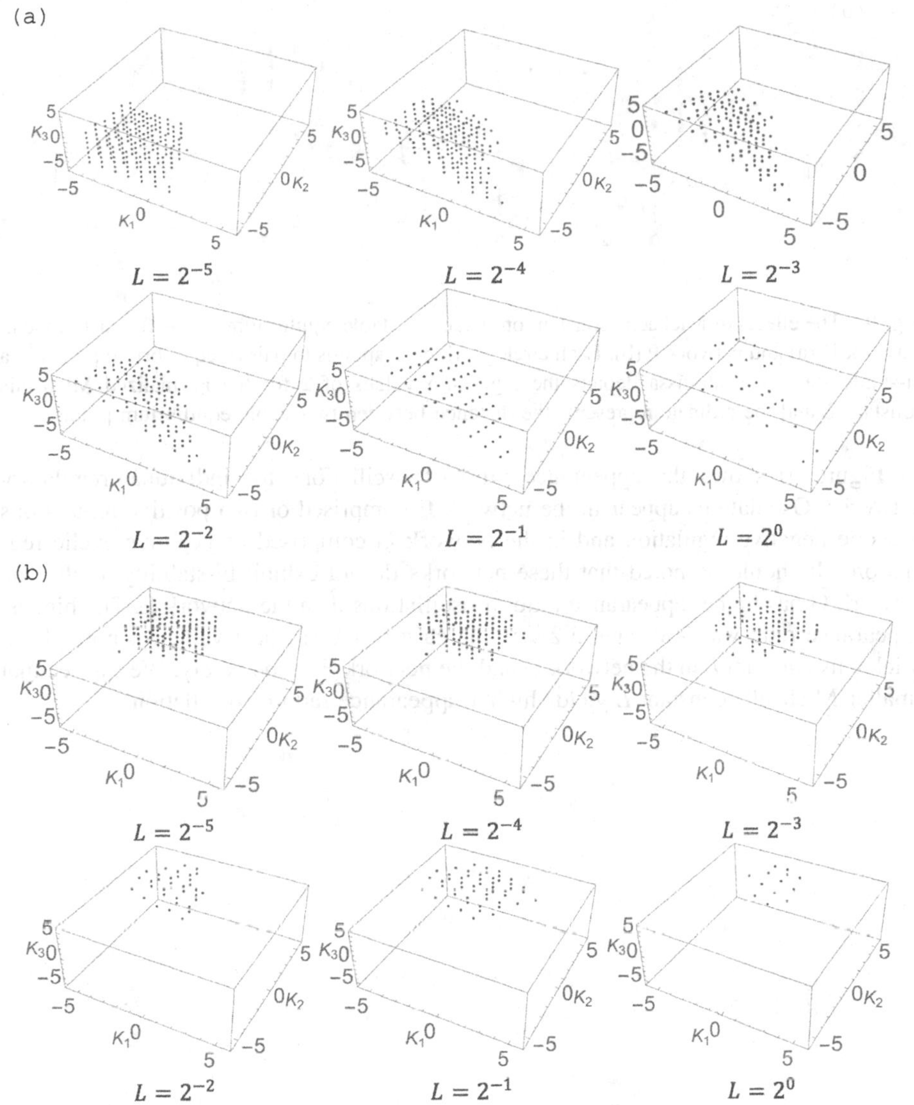

Fig. 8. Distributions of bi-stable equilibrium points in the parameter space for three-node network D (a) and F (b). L denotes the values of the normalized Michaelis constant. The abscissa, the ordinate, and the horizontal axis depict the logarithm values of 2 for K_1,K_2, and K_3, respectively.

Figure 9 shows the effects of Michaelis constant L on the values of stable equilibrium points for each network. The distances between pair of equilibrium points are depicted by a circle marks. It is seen that the distance between two points decreases as the normalized Michaelis constant L increases as seen in two-node networks. Then, the bifurcation of the system occurs at L of 1 for both of the network D and the network F.

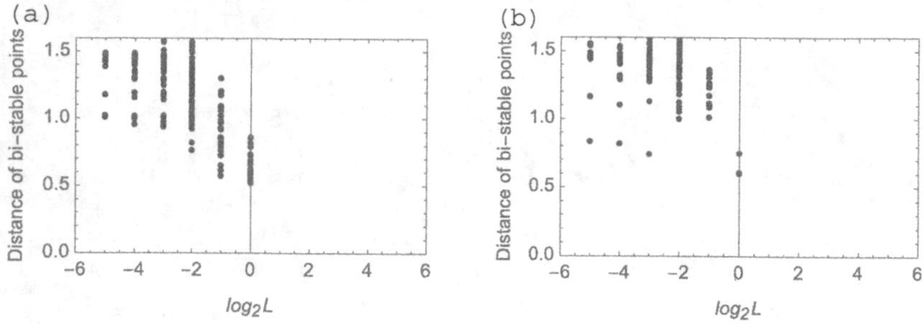

Fig. 9. The effects of Michaelis constant on values of stable equilibrium points for each node in network D (a) and network F (b). Each circle mark corresponds to pair of equilibrium points in a bi-stable state. The abscissa depicts the logarithm values of 2 for the normalized Michaelis constant L and the ordinate represents the distance between two stable equilibrium points.

Figure 10 shows the appearance ratio of oscillations for individual regulatory networks. Oscillations appear in the network E comprised of two positive regulations and one negative regulation and in the network G comprised of negative cyclic regulations. It should be noted that these networks do not exhibit bi-stability at all. The network G has higher appearance ratio of oscillations than the network E. The highest appearance ratio was 7.8% and 5.2% in the case of the smallest value for normalized Michaelis constant L in the network G and the network E, respectively. We can see that smaller Michaelis constant L yields higher appearance ratio of oscillation.

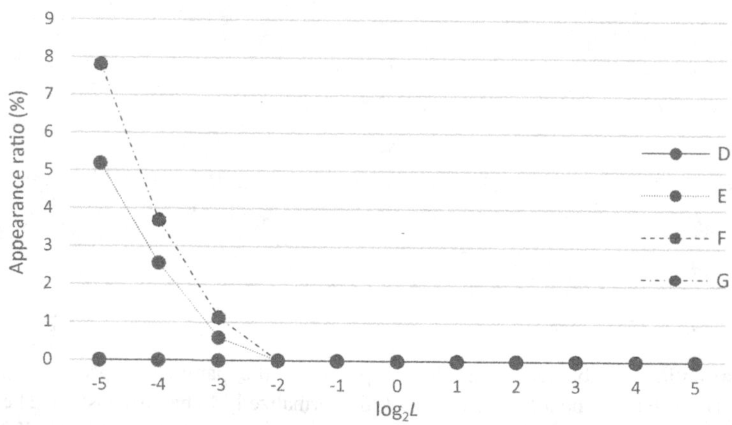

Fig. 10. Appearance ratio of oscillation for three-node networks. The abscissa depicts the logarithm values of 2 for the normalized Michaelis constant L and the ordinate represents the appearance ratio for each regulatory network. The solid line, dashed line, broken line, and chain double dashed line correspond to the network D, E, F, and G, respectively.

Figure 11 shows the oscillation points in the parameter space for the network E and the network G, respectively. As seen in these graphs, the oscillation appears where all of two K_i values are large for the network G comprised of negative cyclic regulations,

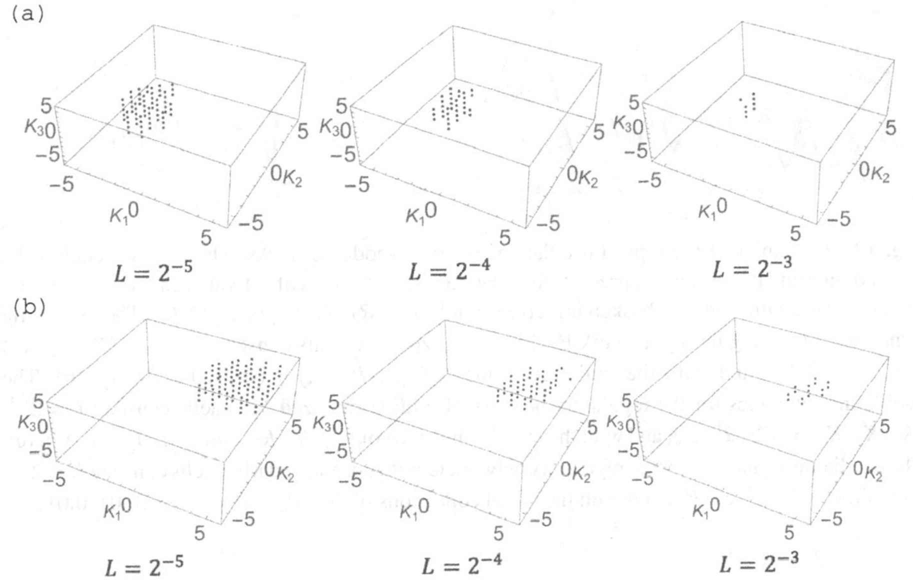

Fig. 11. Distributions of oscillation points in the parameter space for three-node network E (a) and G (b). L denotes the values of the normalized Michaelis constant. The abscissa, the ordinate, and the horizontal axis depict the logarithm values of 2 for K_1,K_2, and K_3, respectively.

corresponding to the region where the activity of the inactivation is higher than activation. And oscillation appears in the region where the K_i of the node regulated positively is small and the K_i of the node regulated negatively is large for the network E comprised of two positive and one negative regulations. As seen in emergence of bi-stability, it is suggested that the parameter values which enhance the feedback regulations are favorable for emergence of oscillations. Furthermore, the oscillation region with smaller normalized Michaelis constant includes one with larger normalized Michaelis constant in the network G. Figure 12 demonstrates the typical oscillation dynamics.

4.3 Four Nodes Regulatory Networks

The four-node regulatory network has become either of mono-stable, bi-stable, or oscillation. Figure 13 shows the appearance ratio of bi-stability for individual regulatory networks. Bi-stability appears in the network H comprised of positive cyclic regulations and in the network M comprised of negative cyclic regulations, and network J and K. The highest appearance ratio was 10.0% in the case of the value of 2^{-3} for normalized Michaelis constant L in the network H. Highest appearance ratios for the network J, K, and M are 5.8%, 5.5%, and 4.6% in the case of the smallest value for normalized Michaelis constant L, respectively. It should be noted that there exists a peak in the network H as seen in the three-node network D. It is also seen that smaller normalized Michaelis constant L yields higher appearance ratio of bi-stability as seen in two-node and three-node networks except for the network H. Bifurcations seems to occur at around the value of 2 for normalized Michaelis constant L.

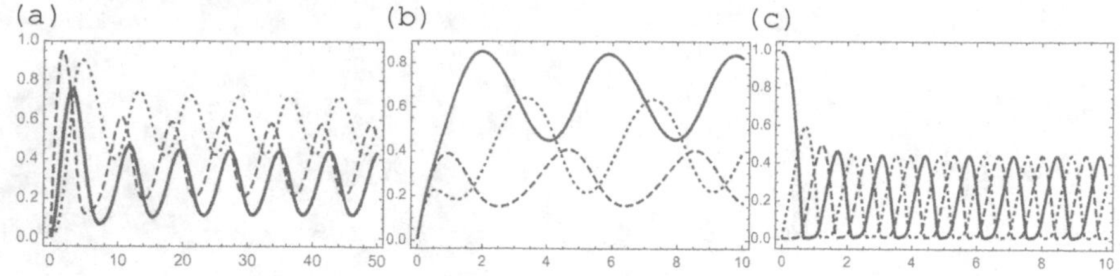

Fig. 12. Dynamics of the typical oscillations for three-node networks. The abscissa depicts the elapsed time and the ordinate represents the relative concentrations of activated enzymes. The solid line, the dashed line, and the broken line correspond to R_1, R_2, R_3, respectively. (a): The oscillating dynamics for the regulatory network E with normalized Michaelis constant $L = 2^{-3}$, $(K_1, K_2, K_3) = (2^{-1}, 2^{-2}, 2^{1})$, and with the initial conditions of $(R_1, R_2, R_3) = (0.01, 0.01, 0.01)$. (b): The oscillating dynamics for the regulatory network G with normalized Michaelis constant $L = 2^{-3}$, $(K_1, K_2, K_3) = (2^{1}, 2^{1}, 2^{2})$, and with the initial conditions of $(R_1, R_2, R_3) = (0.01, 0.01, 0.01)$. (c): The oscillating dynamics for the regulatory network G with normalized Michaelis constant $L = 2^{-5}$, $(K_1, K_2, K_3) = (2^{3}, 2^{3}, 2^{3})$, and with the initial conditions of $(R_1, R_2, R_3) = (0.99, 0.01, 0.01)$.

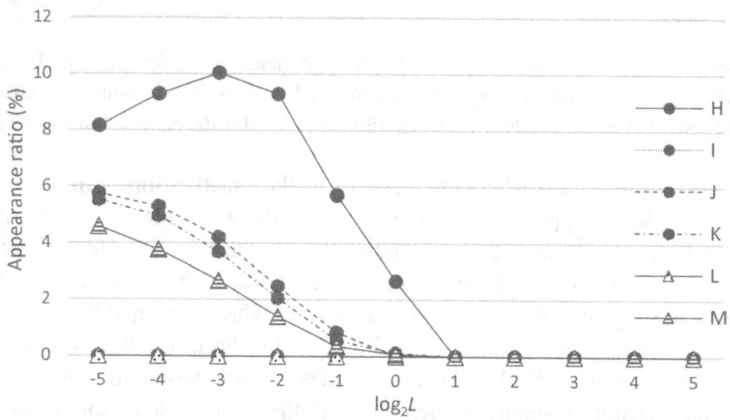

Fig. 13. Appearance ratio of bi-stability for four-node networks. The abscissa depicts the logarithm values of 2 for the normalized Michaelis constant L and the ordinate represents the appearance ratio for each regulatory network. The solid line, the dashed line, the broken line, the chain double dashed line, the solid line with triangle marks hatched with dots, and the solid line with triangle marks hatched with horizontal lines correspond to the network H, I, J, K, L, and M, respectively.

Figure 14 shows the appearance ratio of oscillation for individual regulatory networks. Oscillation appears in the network I comprised of three positive regulations and one negative regulation and in the network L comprised of one positive regulation and three negative regulations. It should be noted that these networks do not exhibit bi-stability at all. The highest appearance ratios are 5.6% and 4.4% in the case of the smallest value for normalized Michaelis constant L in the network I and L, respectively. It is also seen that smaller normalized Michaelis constant L yields higher appearance ratio of oscillations as seen in three-node networks. Bifurcations seems to occur at around the value of 2^{-1} for normalized Michaelis constant L.

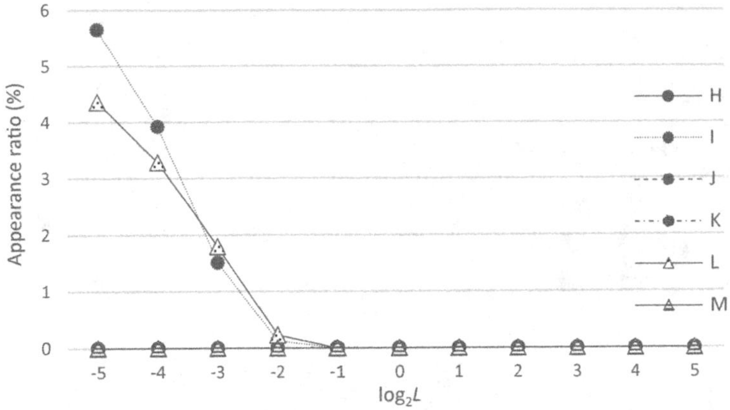

Fig. 14. Appearance ratio of oscillations for four-node networks. The abscissa depicts the logarithm values of 2 for the normalized Michaelis constant L and the ordinate represents the appearance ratio for each regulatory network. The solid line, the dashed line, the broken line, the chain double dashed line, the solid line with triangle marks hatched with dots, and the solid line with triangle marks hatched with horizontal lines correspond to the network H, I, J, K, L, and M, respectively.

4.4 Effects of the Length of Feedback Regulations on the Appearance of Bi-Stability and Oscillations

Figures 15 and 16 show the average appearance ratio of bi-stability and oscillation, respectively. The average appearance ratio is the mean value of appearance ratios for all regulatory networks comprised of the same number of nodes, that is, two-nodes, three-nodes, and four-nodes. It is seen that the average appearance ratio both of

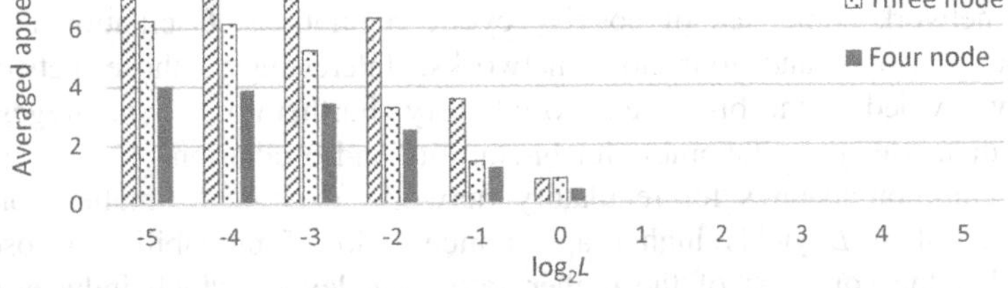

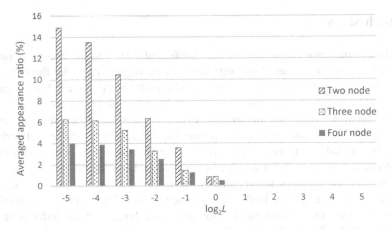

Fig. 15. Averaged appearance ratio of bi-stability. The abscissa depicts the logarithm values of 2 for the normalized Michaelis constant L and the ordinate represents the averaged appearance ratio over the regulatory networks. The striped bar, the dotted bar, and the gray bar correspond to two-node network, three-node network, and four-node network, respectively.

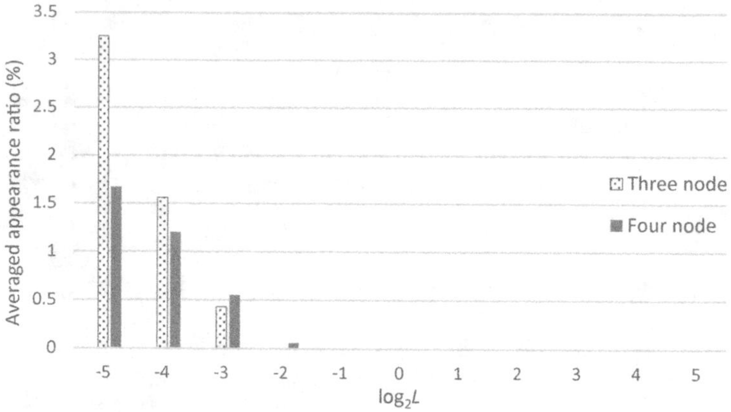

Fig. 16. Averaged appearance ratio of oscillation. The abscissa depicts the logarithm values of 2 for the normalized Michaelis constant L and the ordinate represents the averaged appearance ratio over the regulatory networks. The dotted bar, and the gray bar correspond to three-node network and four-node network, respectively.

bi-stability and oscillation are lower as the number of nodes in networks increase, but with two exceptions. The average appearance ratio of oscillations for four-node networks is slightly higher than one for three-node network at the value of 2^{-3} for normalized Michaelis constant L. And oscillation occurs only for four-node networks at the value of 2^{-2} for normalized Michaelis constant L. Therefore, it is suggested that shorter distance of feedback mechanisms is favorable for emergence of bi-stability and oscillations. In addition, it is seen that smaller normalized Michaelis constant L yields higher appearance ratio of bi-stability and oscillation as seen in individual networks.

5 Conclusions

The highest appearance ratios are 25.6%, 15.4%, and 10.0% in the two-node network A, three-node network D, and four-node network H, respectively. All these networks are comprised of positive cyclic regulations. It should be noted that the MAPK cascade with a positive feedback regulation as seen in Ferrel et at. [1] is correspond to the four-node network comprised of positive cyclic regulations. Oscillation appears in three node networks and four node networks. Interestingly, those networks are exclusively divided to the bi-stable or oscillatory. Furthermore, it is suggested that shorter feedback length is favorable for bi-stability and oscillation.

As in common features for regulatory networks examined, smaller normalized Michaelis constant L yields higher appearance ratio of bi-stability or oscillation implying that the condition of the higher saturation levels, which induces stronger nonlinearity. Regarding to the parameter spaces, bi-stability or oscillation appears in the region where the K_i is small in the case that the node is regulated positively and the K_i is large in the case that the node is regulated negatively, suggesting that the stronger regulation makes the tendency of bi-stability or oscillation higher.

References

1. Ferrell Jr., J.E., Machleder, E.M.: The biochemical basis of an all-or-none cell fate switch in Xenopus oocytes. Science **280,** 895–898 (1998)
2. Jeschke, M., Baumgartner, S., et al.: Determinants of cell-to-cell variability in protein kinase signaling. PLoS Comput. Biol. **9**(12), e1003357 (2013)
3. Kholodenko, B.N.: Cell-signalling dynamics in time and space. Nat. Rev. Mol. Cell Biol. **7**(3), 165–176 (2006)
4. Mai, Z., Liu, H.: Random parameter sampling of a generic three-tier MAPK cascade model reveals major factors affecting its versatile dynamics. PLoS One **8**(1), e54441 (2013)
5. Qiao, L., Nachbar, R.B., et al.: Bistability and oscillations in the Huang-Ferrell model of MAPK signaling. PLoS Comput. Biol. **3**(9), 1819–1826 (2007)
6. Volinsky, N., Kholodenko, B.N.: Complexity of receptor tyrosine kinase signal processing. Cold Spring Harb. Perspect. Biol. **5**(8), a009043 (2013)
7. Byrne, K.M., Monsefi, N., et al.: Bistability in the Rac1, PAK, and RhoA signaling network drives actin cytoskeleton dynamics and cell motility switches. Cell Syst. **2**, 38–48 (2016)
8. Kuwahara, H., Gao, X.: Stochastic effects as a force to increase the complexity of signaling networks. Sci. Rep. **3**, 2297 (2013)
9. Ma, W., Trusina, A., et al.: Defining network topologies that can achieve biochemical adaptation. Cell **138**(4), 760–773 (2009)
10. Mobashir, M., Madhusudhan, T., et al.: Negative interactions and feedback regulations are required for transient cellular response. Sci. Rep. **4**, 3718 (2014)
11. Ramakrishnan, N., Bhalla, U.S.: Memory switches in chemical reaction space. PLoS Comput. Biol. **4**(7), e1000122 (2008)
12. Shah, N.A., Sarkar, C.A.: Robust network topologies for generating switch-like cellular responses. PLoS Comput. Biol. **7**(6), e1002085 (2011)
13. Tsai, T.Y., Choi, Y.S., et al.: Robust, tunable biological oscillations from interlinked positive and negative feedback loops. Science **321**(5885), 126–129 (2008)
14. Huang, C.F., Ferrell Jr., J.E.,: Ultrasensitivity in the mitogen-activated protein kinase cascade. In: Proceedings of the National Academy of Science of the United States of America, vol. 93, pp. 10078–10083 (1996)
15. Brightman, F.A., Fell, D.A.: Differential feedback regulation of the MAPK cascade underlies the quantitative differences in EGF and NGF signalling in PC12 cells. FEBS Lett. **482**, 169–174 (2000)
16. Levchenko, A., Bruck, J., et al.: Scaffold proteins may biphasically affect the levels of mitogen-activated protein kinase signaling and reduce its threshold properties. Proc. Natl. Acad. Sci. U.S.A. **97**(11), 5818–5823 (2000)
17. Schoeberl, B., Eichler-Jonsson, C., et al.: Computational modeling of the dynamics of the MAP kinase cascade activated by surface and internalized EGF receptors. Nat. Biotechnol. **20**, 370–375 (2002)
18. Hatakeyama, M., Kimura, S., et al.: A computational model on the modulation of mitogen-activated protein kinase (MAPK) and Akt pathways in heregulin-induced ErbB signalling. Biochem. J. **373**(2), 451–463 (2003)
19. Heinrich, R., Schuster, S.: Fundamentals of Biochemical Modeling the Regulation of Cellular Systems. Chapman & Hall, New York (1996)

Fault Classification on Transmission Lines Using KNN-DTW

Bruno G. Costa[2], Jean Carlos Arouche Freire[1,2(✉)], Hamilton S. Cavalcante[2], Marcia Homci[2], Adriana R.G. Castro[1], Raimundo Viegas[2], Bianchi S. Meiguins[2,3], and Jefferson M. Morais[2,3]

[1] Faculty of Electrical Engineering, Federal University of Pará, City Belém 66075-110, Brazil
jeanarouche@ufpa.br
[2] Faculty of Computing, Federal University of Pará, City Belém 66075-110, Brazil
[3] Information Visualization Lab - Faculty of Computing, Federal University of Pará, City Belém 66075-110, Brazil

Abstract. To maintain the quality of electricity is necessary to know the main disturbances in the electrical power system, an investigation into signal behavior is presented in this research through the short circuit fault type classification in transmission lines. The analysis of the database UFPAFaults using the KNN algorithm with a change in the calculation of similarity allowed the classifier to execute multivariate time series. On the other hand, the DTW calculation dispenses preprocessing steps as front ends adopted in several papers and presents satisfactory results in the classification of these faults. The comparison of this classifier with Frame Based Sequence Classification architecture, shows the relevance of direct classification of faults using KNN-DTW.

Keywords: Quality of electricity · Power system · Short circuit · Fault classification.

1 Introduction

The increasing demand for electrical energy and the increase in consumption of electro-electronic equipment that is more vulnerable to electrical disturbances increases the need for higher quality of electrical energy (QEE). This reality drives the electric power systems (EPS) (Fig. 1) to have an acceptable configuration of physical, operational, monitoring and control infrastructure with more consistently mechanisms that assist avoid and reduce electrical disturbances in power transmission lines, here called short-circuit type faults [1].

A mechanism widely used by EPSs to detect electrical occurrences in power transmission lines is the protection devices. An example of such instruments is the relays, in which their role is to interrupt and diagnose in the shortest possible time the transmission of energy in the face of a short-circuit fault. It is an

O. Gervasi et al. (Eds.): ICCSA 2017, Part I, LNCS 10404, pp. 174–187, 2017.
DOI: 10.1007/978-3-319-62392-4_13

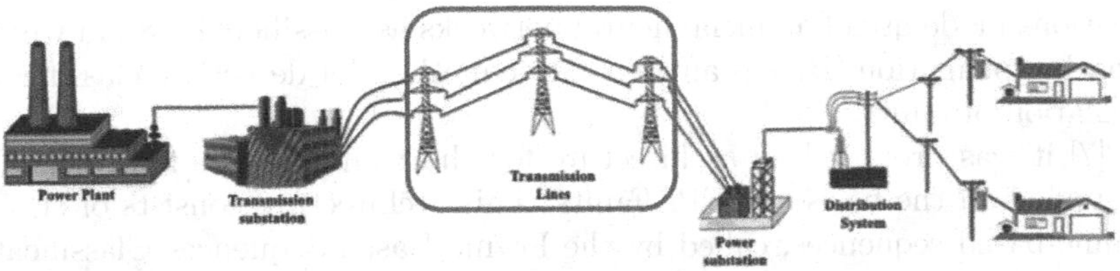

Fig. 1. Representation of the Electric Power System (EPS).

alternative to automate control and monitoring in EPSs, helping in the decision making at the operational level in the event of an electrical disturbance [2].

A relevant factor in this context is that electric power companies adopt oscillography equipment to maintain the QEE, where they use some with a monitoring function that have simplistic algorithms for obtaining and retaining information about electrical disturbances in EPSs. Other elements, such as intelligent electronic devices (IEDs) and digital relays, they provide more sophisticated and more specific algorithms for studying and analyzing short-circuit type faults in EPSs.

In relation to the study and analysis of short-circuit type faults in EPSs, many studies have explored the classification of these faults using front ends (Processing tool that convert sample), since they are multivariate temporal time series that present different sizes, making it impossible to recognize patterns directly by the conventional classifiers, it is soon evident the need for a pre-processing to organize the data, consequently bringing a greater computational cost and delaying the time to assist in decision making in the EPSs operating sector.

Because of that, it would be valid to explore mechanisms and methodologies that reduce computational cost in time series classification multivariate without the use of front ends for data preprocessing. Thus, most of the literature currently found with a fault classification approach, wavefront multiresolution frontends [3] are used, but due to the different possibilities of implementation it is important to test the variations of parameters and coefficients reported.

In the research of [4] an algorithm was proposed for transmission line fault classification based on discrete wavelet transform and in [2] different techniques of artificial neural networks were evaluated for the detection and classification of transmission line faults of energy. In the study of [5], the wavelet transform were used to synchronize the voltage and current samples with respect to time and through the comparison of thresholds of the fault coefficients it is possible to classify them and they are can located with techniques of artificial neural networks.

[6] proposed an algorithm immune to the impedance for real-time application, which also uses wavelet transform and has input as synchronized current values, a technique that extracts characteristics of current signals, based on harmonics generated due to faults in the transmission lines of energy Three-phase. Other

applications made use of artificial neural networks as classifiers in [8], in which it used back propagation (BP) as an alternative method for detection, classification and isolation of faults.

In [7] it was proposed an architecture for short circuit type fault classification, applied on the basis of UFPAFaults. This architecture consists of classifying frame-based sequences, called by the Frame Based Sequences Classification (FBSC) author, and promises to be a flexible architecture that has possibilities to achieve good results with low computational cost. The main idea behind this processing is to segment the input sequence into a fixed-size vector called a frame and repeatedly invoke a classifier to process each frame. The user can choose the front end, the classifier and the combination rules, so the parameters must be rigorously evaluated to avoid biased conclusions.

In view of the above, the main objective of this work is to propose a methodology that directly classifies a multivariate time series without the use of front ends. The work case study was the classification of short-circuit type faults in power transmission lines, because it was a type of multivariate time series. For this, the KNN-DTW algorithm was used, in which it adopted the DTW (Dynamic Time Warping) instead of the Euclidean distance to calculate the distances between neighbors closest to the KNN classifier that allows the classification of multivariate sequences without the use of pre-processing (Front end), so can know which class belongs to the fault. It was then compared with FBSC-based architecture proposed by [7].

This paper is organized as follows. Section 1 presents the introduction of the search. Section 2 shows the UFPAFaults database and explains what are time series of assorted sizes. In Sect. 3 the details of the algorithm KNN-DTW are given. In Sect. 4 a comparison with FBSC architecture is presented, and Sect. 5 shows the results. Section 6 contains the conclusions and perspectives of fault classification using KNN-DTW.

2 UFPAFaults Database

For a better understanding of the study it is necessary to have prior knowledge about the database used in the research experiments, so it will detail some peculiarities of this base which stores a basic cause of electrical power quality events, short circuit. A database called UFPAFaults was developed since 2005 by the Research Group of the Laboratory of Signal Processing (LaSP) of the Federal University of Pará (UFPA), being a database of public domain[1].

The absence of a database available and rich in the cause of the events of interest, motivated the obtaining and composition of this base. Short circuits were generated using simulators via one of the best tools for performing analysis and digital simulations of transient phenomena in power systems [7].

The software Alternative Transient Program (ATP) was used to obtain the database simulating an electrical power system, invoked repeatedly by AmazonTP, software developed in Java used to simulate only short-circuit events,

[1] Properly labeled and found in https://github.com/bruno1307/ufpafaults-knn-dtw.

storing the simulated faults for a three-phase system of transmission of electric power. Since ATP is a program written in Fortran most of its users use the ATP-Draw graphical interface to elaborate a circuit to be simulated, in Fig. 2 the block is presented for the simulation of a fault, with the SW elements representing the switches and R the resistances of the ATPDraw software.

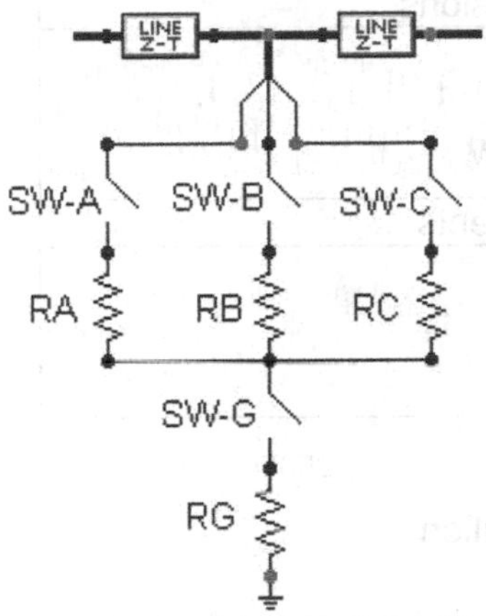

TIPO FALTA	SW-A	SW-B	SW-C	SW-G
AG	fechada	aberta	aberta	fechada
BG	aberta	fechada	aberta	fechada
CG	aberta	aberta	fechada	fechada
AB	fechada	fechada	aberta	aberta
AC	fechada	aberta	fechada	aberta
BC	aberta	fechada	fechada	aberta
ABC	fechada	fechada	fechada	aberta
ABG	fechada	fechada	fechada	fechada
ACG	fechada	aberta	fechada	fechada
BCG	aberta	fechada	fechada	fechada
ABCG	fechada	fechada	fechada	fechada

Fig. 2. Simulation of a fault with ATPDraw software

The phases are represented by A, B, C, and an "AB" fault is identified when the short circuit occurs between phases A and B. By considering the possibility of a short circuit with the ground phase (G), ten possible Causes are found: AG, BG, CG, AB, AC, BC, ABC, ABG, ACG and BCG.

Currently the UFPAFaults base is in the fifth version, composed of 27,500 simulations, organized in five sets of 100, 200, ..., 1000 faults each. The Amount is sufficient for the robust training of some classifiers, allowing to obtain curves of sample complexity, facilitating analysis of the classifiers performance.

The voltage and current waveforms generated by the ATP simulations had a sampling period of 0.25 μs (ATP, deltaT = 2.5E−5), which corresponds to a sampling frequency fs = 40k Hz. Figure 3 shows the voltage and current behavior in the sample frequency range from 1,200 to 28,500 Hz. With the deformation of the waveforms, it is possible to identify that the short circuit occurred between phases A and B as an AB foul. By analyzing the sample range, it is clear that each short circuit has a different duration of time for each fault. Consequently the database is composed of a multivariate sequence with respect to time. In other words the base UFPAFaults is composed of data from time series multivariate with variable duration. This is the main characteristic of the data which it is based upon. For all simulations, the 40 kHz signal was low-pass filtered and decimated by 20 create a signal with fs = 2 kHz.

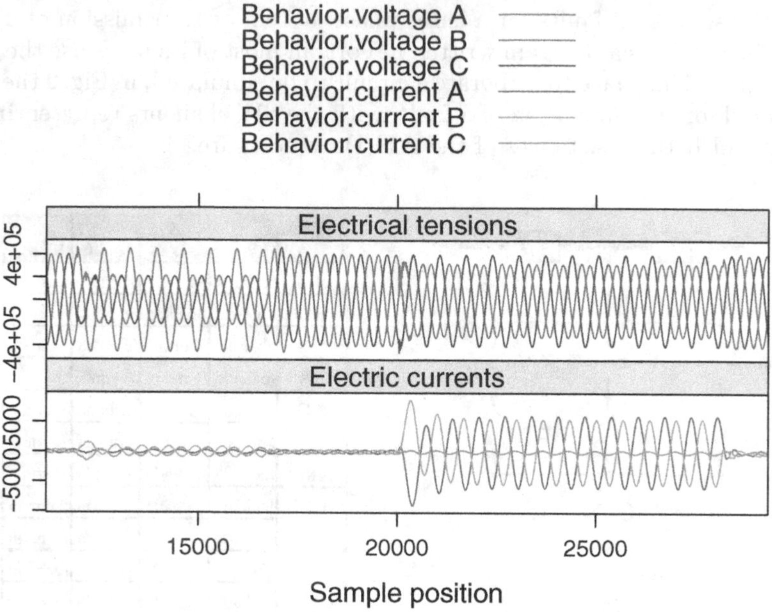

Fig. 3. Waveforms representing voltages and currents of fault AB.

3 The KNN-DTW Algorithm

The experiments of the study used the UFPAFaults database to submit to a pattern recognition technique that uses the training base as a template to classify a new event. This particularity of KNN algorithm training will be an advantage over computational cost if the database used is consistent and normalized. In order to recognize the class of a new sample, it is necessary that the nearest K-neighbors algorithm performs a scan on all the samples of the training base, calculating the distance between the unknown sample and the others that compose the classification model, This measurement of similarity is commonly found with the calculation of the Euclidean distance, but in this research the distance is calculated with dynamic temporal alignment DTW, in Fig. 4, extracted in [9], It is possible to visualize that the euclidean distance compares the two sequences in the same time interval (linear), while the DTW distance compares according to similar form of the samples.

Knowing all distances it is necessary to determine the amount of samples closest to K, and the highest frequency among the nearest K neighbors will define the class of the new sample [10]. In choosing the value of K, it is not interesting to assign even value because it increases the chances of a tie between the quantities of the highest frequency of k, making the decision of the classifier impossible. A large value of K causes a high computational cost, since the base used is robust. In this research the value 1 was assigned for K, and the classification found

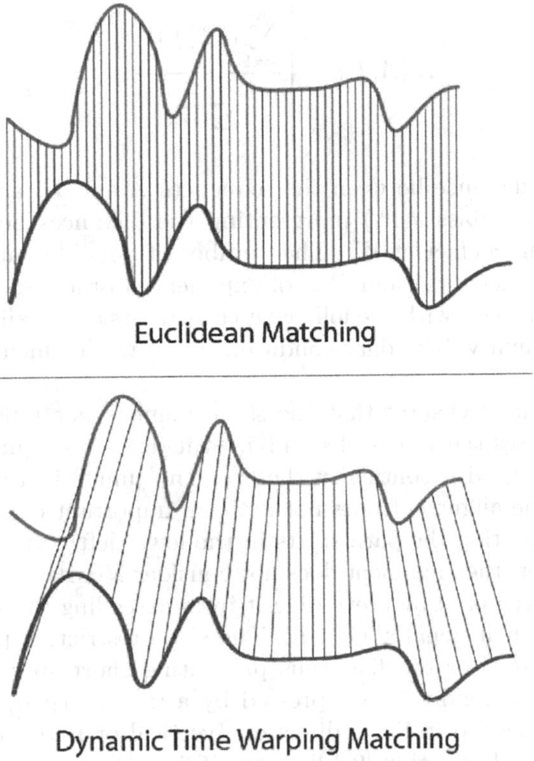

Fig. 4. Comparison of euclidean distance measurements and DTW.

presented a reliable result with low error rate. This classifier used extensively in the area of pattern recognition, proved to be efficient in class detection of the organized database UFPAFaults [11].

3.1 DTW Calculation

The algorithm that has become popular with voice recognition is now widely used in several areas that deal with time series [12]. The objective is to align and compare two multivariate time series, using a dynamic programming algorithm DTW as a measure of KNN similarity. The choice of the algorithm was given by the ability to measure the distance between two sequences of different lengths relative to the time axis, then the DTW dynamic time alignment [13] will be known, considering the vectors with the following characteristics:

$$A = a_1, a_2, \ldots, a_i, \ldots, a_n \tag{1}$$

$$B = b_1, b_2, \ldots, b_j, \ldots, b_m \tag{2}$$

In order to find a sequence equivalent to the two vectors, a matrix with the sequences A and B is constructed, and submitted to an alignment function. The first step is the normalization of the time series with the following equation:

$$D(A,B) = \left[\frac{\sum_{s=1}^{k} d(P_s).W_s}{\sum_{s=1}^{k} W_s} \right] \tag{3}$$

The shortest distance between the sequences A and B, are calculated as a function of the variables $d(P_s)$ representing the distances between i_s and j_s, and the weighting coefficient W_s. The number of possible paths in the matrix formed by the sequences A and B is of exponential order, so it is necessary to reduce the search space with the following constraints of the alignment function: monotocity, continuity, boundary condition, window alignment and constraints of inclination.

The monotonicity ensures that the same point is not repeated, because it remains or increases the values of i and j, so it does not return the same position of the matrix, with continuity there are no jumps in the path sequence. It ensures that the alignment does not omit an important characteristic. While the boundary condition the path starts in the lower left corner and ends in the upper right corner, the alignment does not consider a partial sequence. Window alignment limits the path by preventing it from exceeding the width of the window, ensuring near-diagonal alignment. The slope restriction prevents the path from being too steep or too flat, thus preventing short steps from coinciding with long ones, the condition is expressed by a ratio of (p/q) where (p) is the number of steps desired in Same direction (vertical or horizontal) and q is the time in the diagonal direction [9,14]. Figure 5 is a classic representation of the final result of the DTW algorithm and shows the sequences (A) and (B), the alignment matrix, the constraints and the sequence of interest found, details at http://www.psb.Ugent.be/cbd/papers/gentxwarper/DTWalgorit.

4 Comparison KNN-DTW with FBSC Architecture

A flexible architecture called FBSC was used for frame-based sequence classification, promising to achieve good results with low computational cost to solve the fault classification problem of the UFPAFaults database. The idea behind this architecture is to segment the input sequence into fixed-frame vectors (frames) and repeatedly invoke a conventional classifier to process each frame.

Architecture FBSC makes use of front ends, a pre-processing phase that organizes the data into a fixed-size array to be processed by a conventional classifier. Some front ends and the concatenation of these generate a high dimensionality data set, increasing the computational cost of the classifiers. In order to reduce high dimensionality, a selection of parameters was applied before the pattern recognition stage [15,16]. In the FBSC architecture the user chooses the front end, the classifier and the combination rule, this allows a great degree of freedom in the design of the classifier and must be rigorously evaluated to avoid biased conclusion.

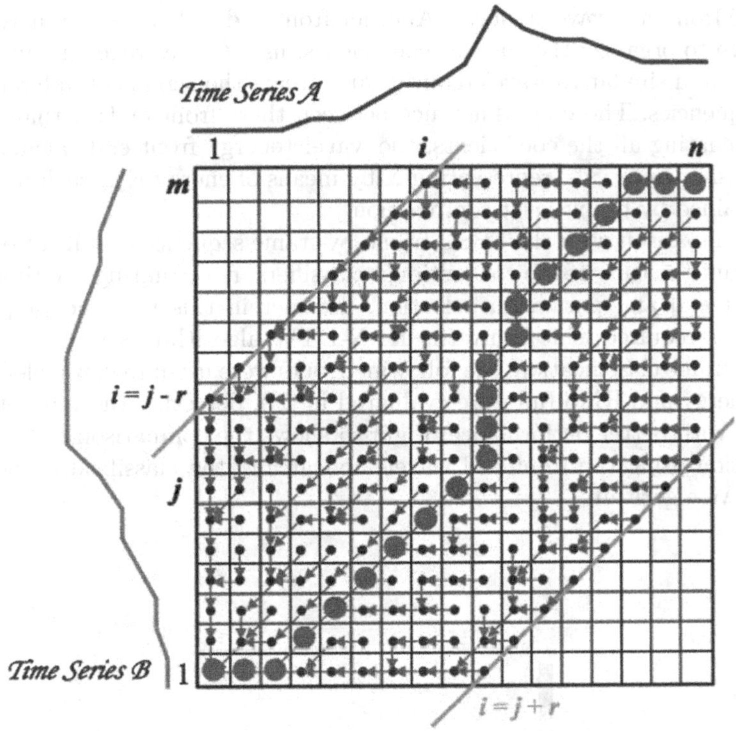

Fig. 5. Sequence found after the DTW calculation.

4.1 Front Ends: Raw, RMS e-wavelets

Any sample that represents a fault, does not contain sufficient characteristics that allows the decision using a classification algorithm. For this reason, a front end has the function of converting the samples into parameters, generating a sequence that allows reasonable decisions through the conventional classifiers.

The front end raw is the simplest, because its output parameters correspond to values of the original sample, without any other processing that organizes the samples into a Z matrix, where classification will take place. Another well-used front end is the RMS, the data organization process allowing to obtain a rough estimate of the fundamental frequency amplitude of the waveform. This front end consists in calculating the windowed RMS value for each of the waveforms.

The front ends of wavelets are found in most of the literature, as it allows a large number of implementations. On the other hand, special care must be taken in replicating a multi-resolution front end, in order to avoid jobs with unfeasible results.

In this work, front end wavelet concatenates all coefficients and organizes into a matrix Z, taking into account the coefficients that have different sampling frequencies, forming a table with organized coefficients. This process is called the front end waveletconcat. Another front end called waveletenergy is an alternative to organize the wavelet coefficients, uses the average energy of each coefficient and similar to waveletconcat, and treats the signals of different sampling frequencies. The main difference between these front ends is that instead of concatenating all the coefficients, the waveletenergy front end calculates the energy of short intervals, representing X by means of energy E in each frequency band obtained by the wavelet composition.

The comparison with the FBSC frame-by-frame sequence classification takes into account the time of the conventional classifiers, not counting the time spent with front end and parameter selection, however in this work no preprocessing stage is required, since only the KNN-DTW algorithm is required. In the frame-by-frame classification, the following fronts were used: raw, waveletconcat, waveletenergy and RMS (all were evaluated in [7]), detecting the waveletconcat front end with better performance. Figure 6 shows the comparison of the FBSC classification using the front end waveletconcat and the classification with the KNN-DTW algorithm.

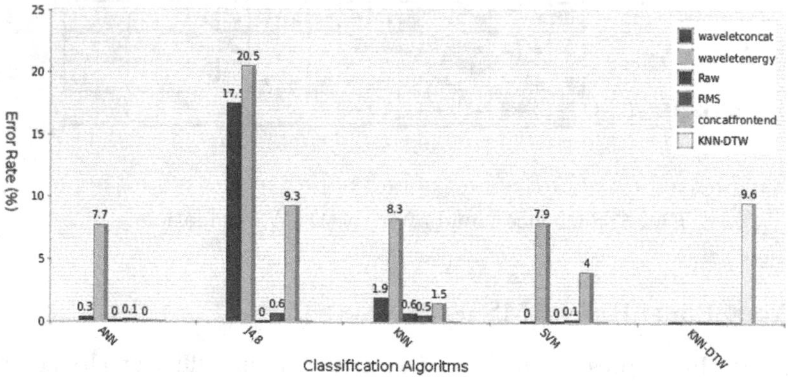

Fig. 6. The best result of the FBSC and KNN-DTW architecture.

5 Analysis of Results

The KNN-DTW algorithm implemented in Java was duly tested and validated, received as input to the UFPAFaults database, and observed the behavior of the error rate curve for the following training samples 100, 200, 300, ..., 1000 Faults, used to enable the KNN-DTW algorithm to classify 1000 test samples. In Fig. 7, the error rate starts with a value above 30% and decreases when the number of training samples is increased, initially a marked decrease in the curve is observed until reaching the mark of 300 samples and error rate less than 15%, Then the curve continues to fall gradually and reached 6.3% error rate with the total of 1000 sorted samples.

In Fig. 7, 100 training samples are submitted to initiate the analysis of the error rate behavior, which shows the initial value above 71%, a decrease is observed when the number of training samples increases. Initially, there is a sharp drop in the next event that ranks 200 faults, reaching 33.6% with the largest drop in error rate, with 300 samples and error rate improves considerably and falls below 15%. Then the curve continues to drop gradually, exhibiting fluctuations in the results and reaching the best 8.7% mark with 900 training samples, used to classify 1000 test samples without noise.

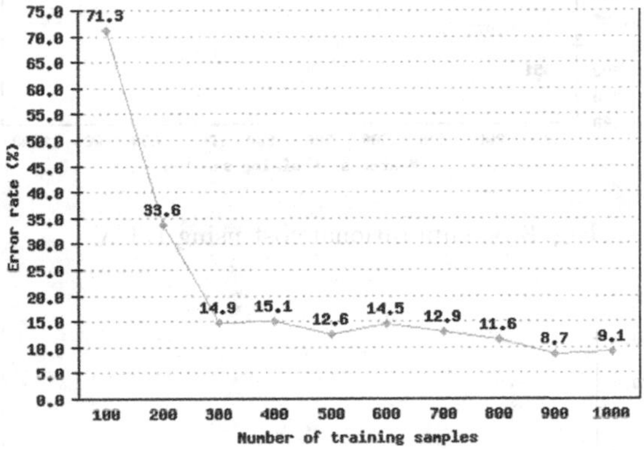

Fig. 7. Error rate using DTW.

The computational cost using dynamic temporal alignment DTW in Fig. 7 showed a growth in the time axis proportional to the increase of samples 100, 200, 300, ..., 1000, respectively.

5.1 Result with Gaussian White Noise

To simulate the various undesirable interferences present in the transmission lines, Gaussian white noise was added on the test base. This signal is a steady random ergonomic process with mean zero and constant power spectral density for all frequencies, hence called "white noise", analogous to white light having all visible wavelengths [17]. A Sample of the Gaussian process has a Gaussian probability distribution for its amplitude. The white noise Gaussian inserted in the experiment, proved that the KNN-DTW algorithm is robust for short-circuit type fault classification in power transmission lines (Fig. 8).

The error rates of the fault classification with different noise levels are shown in Fig. 9, the highest noise level inserted in the fixed test deck with 1000 samples, corresponds to 10 decibels (db) and goes up to 50 db at intervals of 10 by 10 db. Initially an error rate is 80% displayed and drops sharply to near 15 db and with a decrease in noise level, tending to stabilize around 9% from 30 db.

The classification with different levels of noise, showed a higher cost in relation to the time when compared to the classification without noise. Figure 10

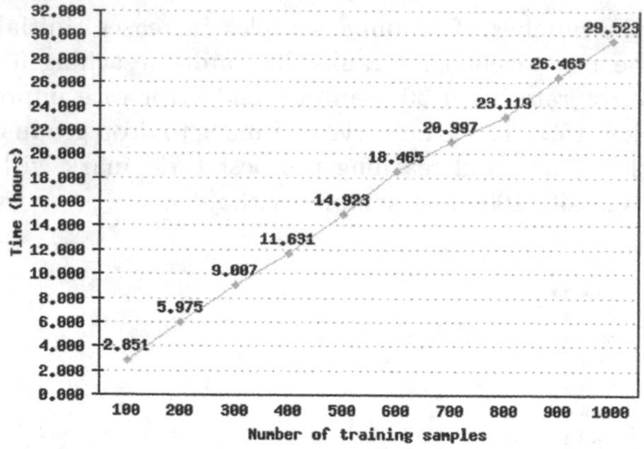

Fig. 8. Computational cost using DTW.

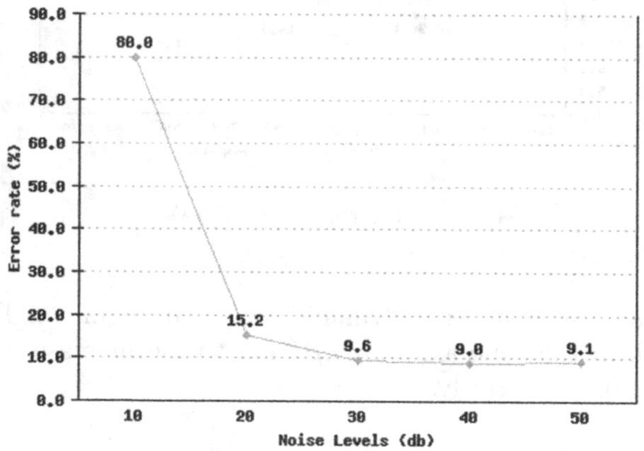

Fig. 9. Classification with different levels of noise.

shows a ratio of time in hours to variable noise rates. The highest noise level took 30 h and 45 min, while the noise processing that took less time was the 40 db rate with 30 h and 4 min.

The results of the classification of faults with noise in power transmission lines using the KNN-DTW algorithm is compared in Fig. 11 with FBSC architecture, which used front ends such as waveletconcat, waveletenergy, raw, RMS and concatfrontend. After this preprocessing phase, it was classified with conventional KNN and noise was also inserted with the interval of 10 to 50 db.

As can be seen in the results of Fig. 11, the KNN-DTW noise classification presented acceptable accuracy and reliability without loss of performance compared to FBSC frame-by-frame classification using front ends for pre-processing of data for the use of conventional classifiers.

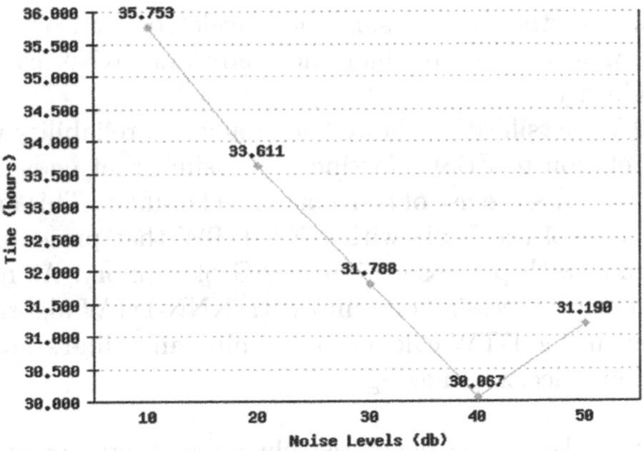

Fig. 10. Time to rate different noise levels.

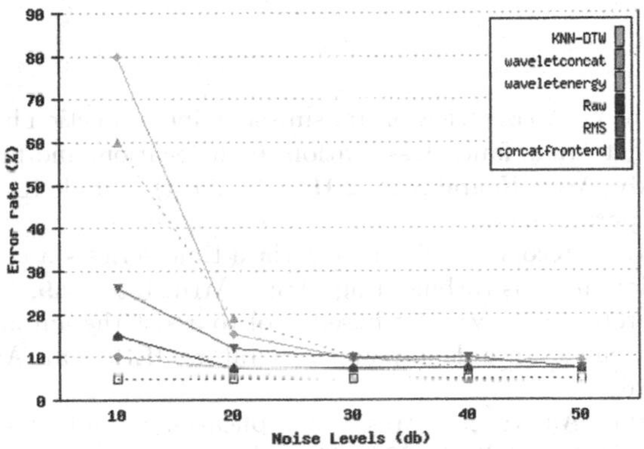

Fig. 11. Classification of KNN-DTW and FBSC using convectional KNN.

6 Conclusion

We investigated a problem that compromises the quality of electric power, proposing a solution without using the processing phases that precedes the classification of faults, in which techniques were applied on a database of short circuit in transmission lines. The classical algorithm KNN was used, with a simple change in its similarity measure, allowing the classification of an event with multivariate characteristics of variable duration. The efficiency of the results was proven by comparing with an architecture developed to classify the faults of the UFPAFaults database.

The experiment of this work overcame the classification of the FBSC architecture using the front end RMS, when it was frame-by-frame architecture that used the front end raw only, the neural networks ANN (Artificial Neural Network) and SVM (Support Vector Machine) were better. Results of the KNN-DTW classification compared to FBSC architecture using front end waveletenergy, only the

ANN classifier was better. While using the waveletconcat front end the KNN-DTW algorithm was better only than the traditional KNN classifier with no change in its similarity.

The KNN-DTW classification showed accuracy and reliability without loss of performance in relation to FBSC classification using front ends and parameter selection, which precedes the use of conventional classifiers. This article proposes a direct classification of the faults with KNN-DTW that presented results generated in a machine with processor i7 and 16G of memory. As future work we can propose the cloud processing of faults with KNN-DTW algorithm and evaluate modifications in the DTW calculation, aiming an online system that allows real-time operational decision making.

Acknowledgment. The authors thank Eletrobras-Eletronorte for the technical support and Coordination of Personal Improvement of Higher Education of the Ministry of Education and Culture of Braisil for research.

References

1. Yadav, A., Dash, Y.: An overview of transmission line protection by artificial neural network: fault detection, fault classification, fault location, and fault direction discrimination. Adv. Artif. Neural Syst. **2014**, 20 (2014). Article ID 230382, Hindawi Publishing Corporation
2. Fontes, C.: Pattern recognition in multivariate time series - a case study applied to fault detection in a gas turbine. Eng. Appl. Artif. Intell. **49**, 10–18 (2015)
3. Pereira, S., Moreto, M.: A wavelet based tool to assist the automated analysis of waveform disturbances records in power generators. IEEE Lat. Am. Trans. **14**(8), 3621–3629 (2016)
4. Patel, V., Chistian, A.: Wavelet transform application to fault classification. Indian J. Appl. Res. **5**(2) (2015). ISSN: 2249-555X
5. Shaik, A., Pulipaka, R.: A new wavelet based fault detection, classification and location in transmission lines. Electr. Power Energy Syst. **64**, 35–40 (2014)
6. Reddy, M., Rajesh, D., Mohanta, D.: Robust transmission line fault classification using wavelet multi-resolution analysis. Comput. Electr. Eng. Comput. Electr. Eng. **39**, 1219–1247 (2013)
7. Morais, J.: A framework for evaluating automatic classification of underlying causes of disturbances and its application to short-circuit faults. IEEE (2011)
8. Tayeb, E.B.M., Rhirn, O.A.A.A.: Transmission line faults detection, classification and location using artificial neural network. IEEE (2012). Copyright Notice: 978-1-4673-6008-11
9. Yurtman, A., Barshan, B.: Detection and evaluation of physical therapy exercises by dynamic time warping using wearable motion sensor units. In: Gelenbe, E., Lent, R. (eds.) Information Sciences and Systems 2013. Lecture Notes in Electrical Engineering, vol. 264, pp. 305–314. Springer, Cham (2013). doi:10.1007/978-3-319-01604-7_30
10. Tang, B., He, H.: ENN: extended nearest neighbor method for pattern recognition. Res. Front. **10**(3) (2015). IEEE, Print ISSN: 1556-603X
11. Campos, G., ZimekEmail, A., Sander, J.: On the evaluation of unsupervised outlier detection: measures, datasets, and an empirical study. Data Min. Knowl. Discov. **30**(4), 891–927 (2016). doi:10.1007/s10618-015-0444-8

12. Giorgino, T.: Computing and visualizing dynamic time warping alignments in R: the dtw package. J. Stat. Softw. **31**, 1–24 (2009)
13. Geler, Z., Kurbalija, V., Radovanović, M., Ivanović, M.: Impact of the Sakoe-Chiba band on the DTW time series distance measure for kNN classification. In: Buchmann, R., Kifor, C.V., Yu, J. (eds.) KSEM 2014. LNCS, vol. 8793, pp. 105–114. Springer, Cham (2014). doi:10.1007/978-3-319-12096-6_10
14. Silva, D., Batista, G.: Speeding up all-pairwise dynamic time warping matrix calculation. In: Proceedings of the 2016 SIAM International Conference on Data Mining, eISBN: 978-1-61197-434-8 Book Code: PRDT16
15. Homci, M., Chagas, P., Miranda, B., Freire, J., Viégas, R., Pires, Y., Meiguins, B., Morais, J.: A new strategy based on feature selection for fault classification in transmission lines. In: Montes-y-Gómez, M., Escalante, H.J., Segura, A., Murillo, J.D. (eds.) IBERAMIA 2016. LNCS (LNAI), vol. 10022, pp. 376–387. Springer, Cham (2016). doi:10.1007/978-3-319-47955-2_31
16. Valenti, A., Giuffrida, S., Linguanti, F.: Decision trees analysis in a low tension real estate market: the case of Troina (Italy). In: Gervasi, O., Murgante, B., Misra, S., Gavrilova, M.L., Rocha, A.M.A.C., Torre, C., Taniar, D., Apduhan, B.O. (eds.) ICCSA 2015. LNCS, vol. 9157, pp. 237–252. Springer, Cham (2015). doi:10.1007/978-3-319-21470-2_17
17. Marmarelis, V.Z.: Gaussian White Noise, Nonlinear Dynamic Modeling of Physiological Systems (2012). Online ISBN: 9780471679370

Intelligent Twitter Data Analysis Based on Nonnegative Matrix Factorizations

Gabriella Casalino[1](✉), Ciro Castiello[1], Nicoletta Del Buono[2], and Corrado Mencar[1]

[1] Department of Informatics, University of Bari Aldo Moro, 70125 Bari, Italy
{gabriella.casalino,ciro.castiello,corrado.mencar}@uniba.it
[2] Department of Mathematics, University of Bari Aldo Moro, 70125 Bari, Italy
nicoletta.delbuono@uniba.it

Abstract. In this paper we face the problem of intelligently analyze Twitter data. We propose a novel workflow based on Nonnegative Matrix Factorization (NMF) to collect, organize and analyze Twitter data. The proposed workflow firstly fetches tweets from Twitter (according to some search criteria) and processes them using text mining techniques; then it is able to extract latent features from tweets by using NMF, and finally it clusters tweets and extracts human-interpretable topics. We report some preliminary experiments demonstrating the effectiveness of the proposed workflow as a tool for Intelligent Data Analysis (IDA), indeed it is able to extract and visualize interpretable topics from some newly collected Twitter datasets, that are automatically grouped together according to these topics. Furthermore, we numerically investigate the influence of different initializations mechanisms for NMF algorithms on the factorization results when very sparse Twitter data are considered. The numerical comparisons confirm that NMF algorithms can be used as clustering method in place of the well known k-means.

Keywords: Nonnegative matrix factorization · Intelligent Data Analysis · Twitter data · Clustering · Topic extraction

1 Introduction

Twitter[1] is a very popular social network which allows users to share short SMS-like messages called tweets. Tweets are a source of latest commentaries, personal opinions on various topics, comments on life events: the list of different ways to use Twitter could be really long. It has been estimated that there are 500 millions of tweets per day[2]. Clearly, human capabilities are unsuitable to process this big amount of data, hence automatic mechanisms are indispensable tools to extract useful information and knowledge. Keyword extraction mechanisms are

[1] twitter.com.
[2] According to Twitter CEO Dick Costolo in October 2012. (http://www.telegraph.co.uk/technology/twitter/9945505/Twitter-in-numbers.html).

O. Gervasi et al. (Eds.): ICCSA 2017, Part I, LNCS 10404, pp. 188–202, 2017.
DOI: 10.1007/978-3-319-62392-4_14

devoted to automatically identify a set of terms that best describe the subject of a document. Keyword extraction plays an important role in many tasks in which documents written in natural language are involved. Extracted keywords can be used, for instance, to build indexes for a document collection or to groups documents into homogeneous clusters. On the other hand, topic extraction enables intelligent document analysis since it is possible to classify documents according to semantic categories.

Building an accurate topic extraction mechanism for Twitter domain could bring some advantages in identifying and characterize communities [1], understanding political inclinations [2] or discovering user behavior [3], just to cite some examples. However before being analyzed with any automatic learning mechanisms, social data (such as tweets) need to be collected, preprocessed and then translated from their unstructured form into a more useful format. In this context, the vector space model for text data, which represents documents by columns and words by rows of the so called term-document matrix (where each element weights the corresponding term for the selected document) provides a useful way to transform unstructured Twitter data into structured data. The matrix representation of tweets can be easily processed by automatic learning mechanisms in order to extract topics from data.

In order to process tweets, we rely on dimensionality reduction mechanisms, which project the term-document matrix X into a lower dimensional space whose basis are topics, proved to be very advantageous in text mining and topic extraction contexts [4]. The well known Latent Semantic Analysis based on the Singular Value Decomposition (SVD) of X, however, presents some disadvantages: the presence of negative values. Negative values cannot be easily interpreted so that the topics discovered via SVD do not have a clear meaning. On the other hand, Nonnegative Matrix Factorizations (NMF) are powerful methods recently proposed to uncover latent low-dimensional structures intrinsic in high-dimensional data and provide a nonnegative, part-based, representation of data [5,6]. Nonnegativity enhances meaningful interpretations of mined information and distinguishes NMF from other traditional dimensionality reduction algorithms, such as SVD, motivating their success in several areas such as bioinformatics, pattern recognition and document clustering [7–10]. In the context of Twitter data analysis, NMF have been used to analyze Twitter networks and obtain trends on Twitter [11], to learn topic from term correlation data derived from short texts [12], for emotion detection from text written in Indonesian language [13]. A peculiar on-line NMF framework was also proposed to modeling the evolution of topics so as to aid a fast discovery of emerging themes in streaming social media content [14]; NMF proved to be faster than classical k-means algorithm and provided more easily interpretable results when mining Twitter data from World Cup Tweets [15].

In this paper, we present a Twitter-analysis workflow based on NMF. The proposed workflow is thought to collect data from Twitter using specific search criteria and to appropriately organize collected data in a term-tweet matrix X. Once this data matrix is derived, the core of the process is triggered to factorize the matrix through NMF as the product of two nonnegative matrices W and H, such that WH approximates X. This core process is therefore devoted to find k vectors (the columns of W) that are linearly independent in the vector space spanned by the collected tweets (columns of X). These vectors can be used to reveal the latent structure of the data. In particular, due to the nonnegativity property of NMF, the derived factorization finds an immediate and intuitive interpretation as topics underlying tweets [7,16]. The factorization results can then be exploited to cluster original tweets and to visualize topics through specific word cloud tools. The initialization phase is crucial in NMF and can lead to different matrix decompositions [17], therefore we included different initialization mechanisms in the proposed process. Furthermore, the proposed workflow allows the user to select different NMF algorithms and to inject a-priori knowledge in the factorization process. The aim of the paper is twofold; firstly it intends to present an experimental workflow devoted to standardize the technical steps to be performed when NMF are applied to pattern discovery from twitter datasets. Secondly, it would point out the advantages of NMF in the context of Intelligent Data Analysis stressing how nonnegativity of matrix factors provides meaningful interpretations of mined information and distinguishes NMF from other traditional clustering algorithms, such as k-means.

The paper is organized as follows. In Sect. 2 we describe in some details the main steps assembling the proposed workflow. In Sect. 2.3 we discuss how NMF are involved in the considered scenario and the benefits deriving from the adoption of these methods in topic extraction and cluster analysis of tweets. In Sect. 3 we illustrate some preliminary experiments to empirically demonstrate the effectiveness of the proposed workflow to extract interpretable topics from some newly collected Twitter datasets. Moreover, we report the results obtained using different initialization mechanisms for NMF algorithms together with some discussions on how an initialization method is critical for the quality of the final results of NMF decomposition. In the final section we sketch the future work oriented to address some open problems.

2 Twitter Data Analysis Workflow

The workflow for tweet data analysis is illustrated in Fig. 1. Three sequential phases comprise all the activities of the proposed process. These phases are:

1. **Dataset Creation** is devoted to the collection, preprocessing and the reorganization of tweets in a structured matrix form;
2. **NMF Decomposition** is the core of the process as it is responsible of the factorization of the data matrix obtained as output of the first phase;
3. **Data Analysis** exploits the latent factors computed by NMF to carry out topic extraction and tweet clustering.

A detailed description of the tasks involved in each single phase is reported in the following subsections.

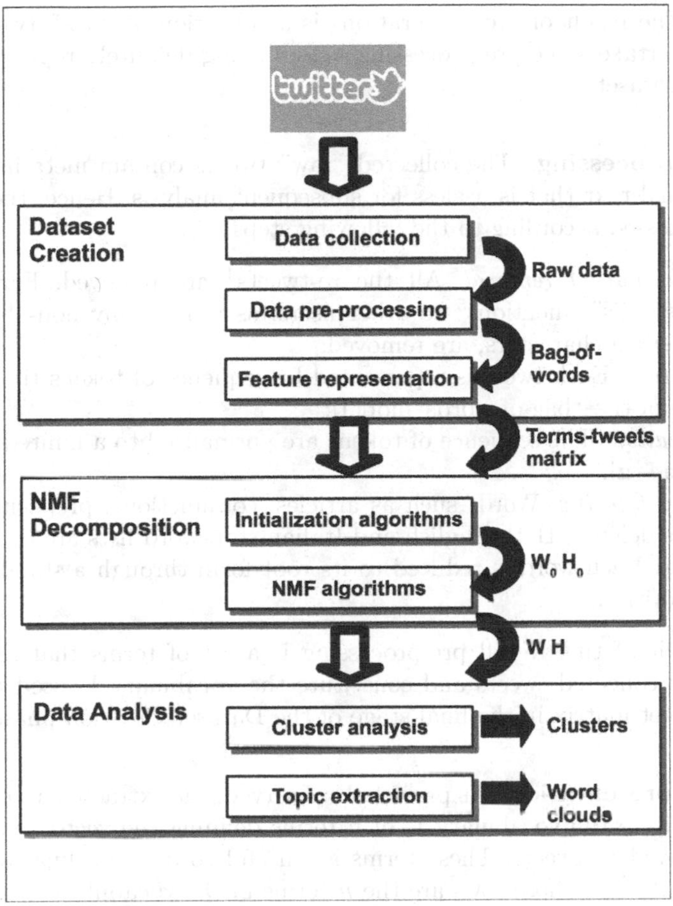

Fig. 1. Workflow of Twitter data analysis based on NMF

2.1 Dataset Creation

The Dataset Creation phase deals with collecting tweets, pre-processing them and representing the extracted dataset as a numerical matrix. This matrix relates each tweet with a collection of terms belonging to an automatically extracted vocabulary. The output of the Dataset Creation phase is a term-tweet nonnegative real matrix of proper dimensions. The main tasks performed in this phase are described as follow.

Data Collection. A number of tweets can be extracted from Twitter through the provided API[3] (Application Program Interface). Tweet extraction is based on some search criteria; particularly, the tweets are searched by some user-defined

[3] https://dev.twitter.com/apps.

keywords. The result of these operations is a collection of "raw" tweets which have to undertake some pre-processing before being definitely represented as a structured dataset.

Data Pre-processing. The collected "raw" tweets contain meta-information and additional text that is useless for subsequent analysis. Hence, these tweets are pre-processed according to the following steps.

1. *Meta-information removal.* All the re-tweets[4] are removed. Furthermore, URLs, "emojis", mentions[5] to other users, as well as any non-alphabetical and numerical characters, are removed;
2. *Tokenization.* Each tweet is represented by sequence of tokens (i.e. words in the sense of the "bag-of-words" model);
3. *Normalization.* The sequence of tokens are normalized to a limited character set, i.e. $[\mathtt{a}-\mathtt{z}]$;
4. *Stop-word filtering.* Words such as articles, conjunctions, prepositions, pronouns are deleted. Both English and Italian stop-word lists are considered;
5. *Stemming.* Each word is reduced to its root form through a standard stemming algorithm.

The result of the overall pre-processing is a set of terms that are able to describe the collected tweets and constitute the vocabulary V used the derive the term-tweet matrix in the final stage of the Dataset Creation phase.

Matrix Representation. As previously observed, the extracted and processed words constitute the vocabulary V of n terms defining the vector space model of the retrieved m tweets. These terms are useful to derive a final term-tweet matrix $X \in \mathbb{R}_+^{n \times m}$, whose rows are the n terms in the vocabulary V and whose columns are the m tweets. Each element x_{ij} represents the "weight" of the i-th term in describing the j-th tweet and is computed using classical tf-idf weighting function used in Information Retrieval. It should be pointed out that tweets are limited to 140 characters, hence the number of terms associated to each tweet is usually very small (at most 60 terms *circa*). On the other hand, the size of the vocabulary V (number of different terms) is very high. As a consequence, the matrix X is very sparse.

2.2 NMF Decomposition

The core of the proposed workflow is the application of Nonnegative Matrix Factorizations (NMF) to the term-tweet matrix X, extracted at the end of the first phase. NMF decomposes $X \in \mathbb{R}_+^{n \times m}$ in two matrices: a *term-topic* matrix $W \in \mathbb{R}_+^{n \times k}$ and a *topic-tweet* matrix $H \in \mathbb{R}_+^{k \times m}$ such that $X \approx WH$. The value k, which must satisfy $k < \min(m, n)$, is the rank of the factorization and defines

[4] Tweets that a user received in her stream and shared to her followers.

[5] Text beginning with the symbol '@' followed by any unique user name.

the number of latent topics hidden in X. Each collected tweet (columns X_j of the term-tweet matrix X) can be written as a linear combination of columns w_i of the matrix W as

$$X_j \approx h_{1j}w_1 + h_{2j}w_2 + \ldots + h_{kj}w_k,$$

the element h_{ij} representing the projection of the j-th tweet vector in the direction of the i-th topic vector w_i. As a consequence, the columns of W can be interpreted as latent topics defining the semantics of the tweet vector space, while the elements of H represent the degree to which each tweet belongs to each topic.

From the computational viewpoint, NMF can be carried out through a number of algorithms [7,18]. Some of these algorithms have been considered in the proposed process and can be selected by the user. Particularly, we considered the following NMF algorithms:

- Multiplicative *NMF* algorithm, based on the Euclidean distance which is considered the baseline method for NMF [5,19];
- Alternating Nonnegative Least Squares Projected Gradient (*ALS*) [20];
- Sparse Nonnegative Matrix Factorization (*SNMF*), which is able to control the sparsity of the factors W and H [21];
- Nonsmooth Nonnegative Matrix Factorization (*NSNMF*), which is able to extract highly localized patterns in data, forcing the global sparseness of the factors W and H [22].

Beside different NMF algorithms forming the core of proposed workflow, some initialization mechanisms for the matrices W and H are adopted too. In fact, all NMF algorithms are iterative mechanisms, hence they require some initial matrices as a starting point. Moreover, since different initialization algorithms can lead to different solutions of NMF, a correct initialization is critical for the quality of the final results of NMF decomposition. As a consequence, a throughout experimental analysis is required to choose the correct initialization scheme for the problem at hand.

Among several initialization mechanisms proposed in literature [8], we selected three different random initialization algorithms (which require low computational costs, but have the drawback of generating poor informative initial matrices), namely RANDOM, RANDOM_C and RANDOM_VCOL initialization [23], as well as NNDSVD initialization [24]), which is considered one of the best initialization mechanisms, though it is more computationally expensive than random methods.

2.3 Data Analysis

The Data Analysis phase is dedicated to revealing the hidden topics from tweets. This can be done by exploiting the factors W and H obtained at the end of the NMF Decomposition phase. Particularly, the approach adopted for the extraction of the hidden topics is illustrated in Fig. 2. After topics have been extracted and interpreted, tweets can be clustered accordingly. Furthermore, this phase allows to effectively visualize each cluster and to easily display its semantics using appropriate tools.

Topic Extraction. As previously highlighted, the columns of the matrix factor W stand for the hidden topics embedded into the vector space describing the tweets (after the pre-processing phase). In particular, a column w_k in W associates a weight w_{ki} to each term in V: the higher this weight, the more important the term in defining the hidden topic. To allow an easier interpretation of hidden topics, for each column of W the topmost r terms (in order of their weight) can be selected.

The example reported in Fig. 2 illustrates how to extract the hidden topics using the matrices W and H, by electing the topmost three terms from each column of W.

It is important to highlight that the topics are automatically discovered by analyzing the original tweets; in fact they usually are not known in advance but are learned from data.

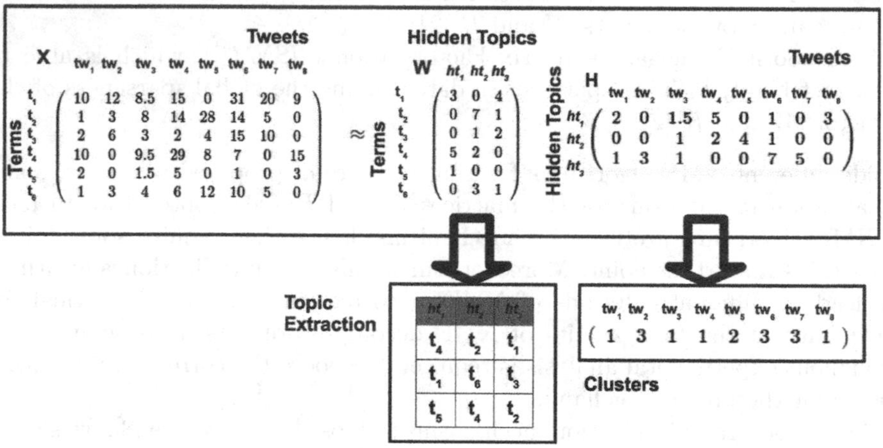

Fig. 2. Data analysis with NMF.

Cluster Analysis with NMF. Each tweet exhibits multiple topics with different relevance. Through NMF it is possible to suggest the importance of each topic in each tweet. The encoding matrix H maps the hidden topics (rows of H) with the tweets (columns of H), and elements h_{ij} indicate the importance (weight) that the i-th topic has in the j-th tweet. In the example in Fig. 2, the first tweet tw_1 is about the hidden topics ht_1 and ht_3 with weights 2, and 1, respectively, while it does not refer to the topic ht_2.

Each tweet is represented as a vector in the new space spanned by the column vectors w_j. Thus NMF can be used for hard document clustering by assigning the tweets to the nearest basis in the space [25,26]. This is equivalent to assigning each tweet to the topic with the highest weight in the column of H. Since NMF and spherical k-means have been proved to be equivalent [27], each hidden topic in W can be seen as the centroid of a cluster. However, due to the non-negativity of NMF (and differently from the k-means) each centroid is also interpretable. In the example (reported in Fig. 2, the tweets tw_1, tw_3, tw_4, tw_8 are assigned to the first cluster (that is about the terms t_4, t_1 and t_5), the tweet tw_5 to the second cluster and the tweets tw_6 tw_5 to the third cluster.

Finally, to help the cluster visualization, the word-cloud representation can be adopted. This mechanism allows to show the selected words with different font sizes: the more a word is important in a tweet, the bigger and bolder it appears in the word cloud.

2.4 Implementation Details

The proposed workflow has been implemented using the Python 3.x language, using the following libraries:

- TWEEPY[6] is used to provide direct access to the public stream of tweets, which can be downloaded according to some search criteria;
- NLTK[7] is used to perform tweet pre-processing;
- SCIKITLEARN[8] is used to compute the tf-idf weights and the clustering measures;
- NIMFA[9] is used for NMF initialization and decomposition.

3 Preliminary Experiments

In this section some preliminary experiments are reported to illustrate the effectiveness of the proposed workflow in analyzing Twitter data and in extracting interpretable topics from some newly collected Twitter datasets. Particularly, the aim of the experiments is to numerically compare the influence of different NMF initializations and algorithms on the clustering results and on the semantic meaning of the topics extracted from the collected Twitter data. The experiments have been performed by using three different datasets (described in Subsect. 3.1); initialization mechanisms for NMF have been compared and the obtained results are described in Subsect. 3.3 together with quantitative and qualitative evaluation of the obtained clusters.

---

6 http://docs.tweepy.org/en/v3.5.0/.
7 http://www.nltk.org/py-modindex.html.
8 http://scikitlearn.org.
9 http://nimfa.biolab.si/.

3.1 Datasets

Three datasets of Italian tweets were created using the *Dataset Creation* phase of the process (as described in Sect. 2.1). Four distinct groups of tweets were acquired using as search criterion the presence –in the tweet text– of four Italian keywords for each group. Table 1 reports the used keywords together with some statistics on the data. It should be noted that the selected keywords have a very general meaning: we made this choice to better investigate the capability of NMF in topic extraction.

Table 1. Summary of the three datasets.

Dataset	Keyword 1	Keyword 2	Keyword 3	Keyword 4	#terms	#tweets
1	RELIGIONE (religion)	TECNOLOGIA (technology)	SCUOLA (school)	AMORE (love)	4219	2272
2	AMORE (love)	SPORT (sport)	VIAGGIO (travel)	MUSICA (music)	2840	995
3	AMORE (love)	SCUOLA(school)	CLIMA (climate)	CIBO (food)	4350	2312

3.2 Factorization Rank

The choice of the factorization rank is crucial for the results. In fact the number of clusters, and the hidden topics is given by the rank value k. It is an input parameter for the problem. In the preliminary experiments, we set the factorization rank at $k = 4$ as the number of the keywords used to acquire the tweets. In this way we are able to compare the hidden topics, automatically extracted by the NMF algorithms, with the semantic categories expressed by the keywords used for retrieving tweets.

3.3 Results

Numerical results on different initialization algorithms have been compared first. Then initial matrices have been used to run different NMF algorithms as requested in the NMF decomposition phase, in order to derive matrices W and H. The obtained results have been used in the Data analysis phase to extract hidden topics and to finally group tweets accordingly.

Initialization Algorithm Comparisons. The initialization algorithms have been compared in terms of initial approximation error and required time in order to verify whether inexpensive, but less informed algorithms, lead to acceptable results, so that they could be used in place of more informed but computationally expensive algorithms. Particularly,

- The *Initial Error* evaluates the error obtained by approximating the original matrix X with the initial pair W_0, H_0 obtained by the initialization algorithm. This error is computed by means of the Frobenius norm applied to the difference between the original matrix and its initial approximation:

$$\|X - W_0H_0\|_F^2$$

Table 2. Comparisons of the performance of initialization algorithms.

	Dataset 1		Dataset 2		Dataset 3	
Init. algorithm	Init. err	Ex. time	Init. err	Ex. time	Init. err	Ex. time
RANDOM	2493.50	0.016	1662.94	0.006	2619.00	0.034
RANDOM_C	0.49995	1.535	0.49994	0.592	0.49995	2.008
RANDOM_VCOL	0.49997	0.597	0.49996	0.082	0.49996	0.417
NNDSVD	0.47650	64.97	0.47636	3.291	0.41622	37.08

- The *Execution Time* measures the time (in seconds) needed by the initialization algorithm to construct the initial matrices[10].

Table 2 reports the performance of the initialization algorithms on each dataset. As expected, the simplest random generation of the initial matrices is the fastest but less accurate method. On the contrary, NNDSVD is the slowest, but shows the minimum initial error. The computational time of NNDSVD depends polynomially on the matrix dimensions, as witnessed by the remarkable differences in execution time with between datasets of different dimensions (namely, datasets 1 and 2 vs. dataset 3). This could be a problem when dealing with big amounts of data. On the other hand, the semi-informed initialization algorithms RANDOM_C and RANDOM_VCOL give initial error values that are comparable with those given by NNDSVD, but with a significant reduction of the computation time. In this case, the differences among the datasets are limited.

NMF Comparison. After evaluating the initialization performance, the analysis of the influence of initialization algorithms on the NMF results has been considered. Comparisons are reported in terms of:

- *Final Error*, which evaluates the approximation error of the final matrices W and H in reconstructing the original matrix X (again, by using Frobenius norm on the difference matrix $X - WH$);
- *Execution Time*, measuring the time (in seconds) the algorithm required to obtain a solution;
- The *Iterations Number*, reporting the number of iterations required by the algorithm to reach at least one stopping criterion (maximum number of iterates or error reduction under a threshold).

Different behaviors of any pair (initialization, NMF algorithm) have been observed over the three datasets. We show in detail the numerical results of the experiments on the first dataset (Table 3). Standard NMF and NSNMF are almost insensitive to the initialization mechanism. On the other hand, ALS is strongly sensitive to the initial matrices and it converges to the best results with

[10] Experiments have been run on a machine equipped with an Intel i5-480M 2.6 GHz CPU with 8 GiB of RAM.

Table 3. Performance of the NMF algorithms initialized with different strategies applied to Dataset 1. (We use the exponential notation with base 10 for the final error.)

	NMF			ALS			NSNMF			SNMF		
Init.	Fin. err.	Time	It.	Fin. err.	Time	It.	Fin. err.	Time	It.	Fin. err.	Time	It.
RANDOM	2.1e+3	0.027	100	1.5e+0	0.032	13	2.9e+4	0.025	100	1.0e−07	0.038	95
RANDOM_C	2.1e+3	0.028	100	4.2e−3	0.059	100	2.9e+4	0.047	100	9.0e−08	0.078	100
RANDOM_VCOL	2.1e+3	0.043	100	1.5e−05	0.048	100	2.9e+4	0.043	100	1.8e−07	0.094	59
NNDSVD	2.1e+3	0.052	100	8.7e−05	0.057	74	2.9e+4	0.062	100	6.7e−08	0.104	100

RANDOM_VCOL. The use of NNDSVD guarantees the best results in terms of final error (with the exception of ALS, but even in this case the final error is very small), but at the expenses of a high execution time to perform initialization. Finally, we observe that ALS and SNMF yield the most accurate approximations of the initial matrix, because they show early convergence (within the prescribed limit of 100 iterations) and a final error that is extremely small, especially when compared with NMF and NSNMF.

Cluster Results. Very accurate solutions may not be the most significant in terms of grouping tweets according to their topics. In order to evaluate this ability, we evaluate the quality of cluster analysis. Clustering results have been evaluated in terms of the *V-measure*, an external cluster evaluation measure based on entropy [28]. External cluster measures compare the obtained labeling with the *a-priori* known classes. In particular, the V-measure is defined by the harmonic mean between *homogeneity* and *completeness* as follows:

$$v = 2 * \frac{homogeneity * completeness}{homogeneity + completeness}$$

where homogeneity quantifies how much clusters contain data points belonging to the same class, while completeness quantifies how much data points belonging to the same class are grouped in the same cluster. Both values range in $[0, 1]$.

Table 4 reports the clustering performances on the three datasets. We added the results computed by standard k-means as a term of comparison. On the overall, we observe that k-means performs badly on the first dataset but it is comparable with NMF on the other datasets. On the other hand, NSNMF gives very poor results in terms of V-measure, when initialized with random methods. This results confirm that NMF can be used as a clustering algorithm in place of k-means. Moreover, differently from k-means, NMF is able to detect interpretable centroids, used to define the topic of each cluster.

Topic Extraction. In order to appreciate the semantics of the hidden topics returned by NMF, we have represented each topic with the topmost 10 terms (in order of their weight in the corresponding column of W). We used word-clouds

Table 4. Cluster performance of NMF and initialization algorithms in terms of V-measure.

Init.-NMF alg	Dataset 1	Dataset 2	Dataset 3
Random-NMF	0.755	0.540	0.668
Random-ALS	0.795	0.570	0.719
Random-NSNMF	0.066	0.043	0.094
Random-SNMF	0.688	0.705	0.718
Random_c-NMF	0.696	0.715	0.726
Random_c-ALS	0.694	0.702	0.722
Random_c-NSNMF	0.048	0.011	0.081
Random_c-SNMF	0.690	0.705	0.720
Random_vcol-NMF	0.680	0.688	0.714
Random_vcol-ALS	0.703	0.703	0.755
Random_vcol-NSNMF	0.060	0.030	0.099
Random_vcol-SNMF	0.688	0.709	0.723
NNDSVD-NMF	0.699	0.710	0.742
NNDSVD-ALS	0.678	0.705	0.745
NNDSVD-NSNMF	0.638	0.635	0.578
NNDSVD-SNMF	0.692	0.712	0.731
k-means	0.433	0.806	0.721

for a visual representation of these topics. Figure 3 shows the four hidden topics extracted from the first dataset by the NSNMF algorithm initialized with the NNDSVD. The tweets are grouped in four clusters (in accordance with the rank value $k = 4$) and depicted in the respective frames of the picture as word-clouds of the ten terms with highest weight. As it can be observed, the main terms in each frame are exactly the Italian (stemmed) keywords used to acquire the tweets (Love, School, Religion, Technology). NMF has been able to correctly capture the hidden meaning in the tweets. Furthermore, the terms appearing in each word-cloud are semantically correlated. As an example, the term RELIGION is grouped together with the Italian words *terror*, *Bruxelles*, *Islam* and the tag *stopislam*. Of course, these words do not define the concept of *religion*, but Twitter data are strictly related to the temporal instants they are acquired, so that they reflect the current events and the respective people thoughts and feelings. The numerical experiments were conducted after the terrorist attacks in Bruxelles (on March 22th, 2016), thus explaining why those words have been grouped together. We observe a similar result with the keywords TECNOLGIA and SCUOLA. In particular, the terms connected to TECNOLGIA are also related to the terrorists' facts; in fact in that days the possibility of accessing to confidential information contained in the terrorists' phones has been discussing. That is why the terms "Iphone" and "Apple" have a big weight (i.e. bold font and big size)

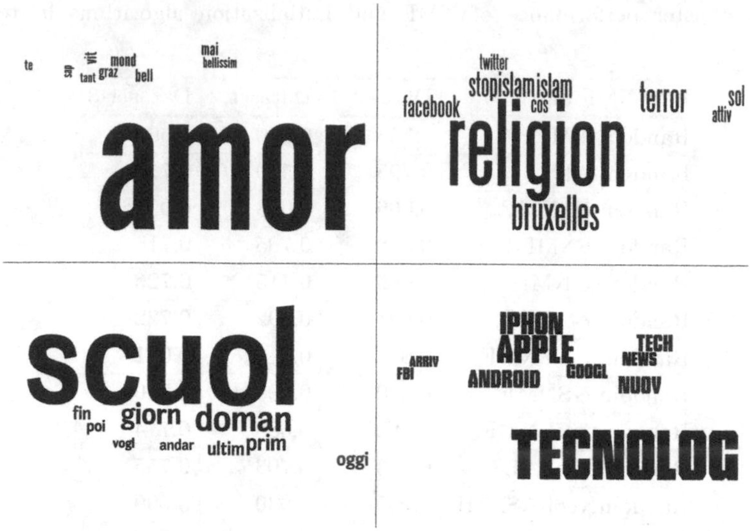

Fig. 3. Hidden topics obtained with Dataset 1, NNDSVD initialization and NSNMF algorithm.

in the word cloud, but also "FBI", though to a lesser account. Finally, the terms related to returning to school have been grouped with the keyword SCUOLA, because in the days tweets have been collected, the students were coming back to school after Easter holidays.

In conclusion, we observed that NMF algorithms are able to detect localized patterns in sparse matrices as the term-tweet matrix is. In particular, NSNMF it is able to preserve this sparsity in the factorization process giving more interpretable bases than the other algorithms.

4 Final Remarks

In this paper we have proposed a workflow to intelligently analyze a particular kind of textual data: the Twitter data, which are characterized by a small number of terms belonging to a large vocabulary. After retrieving tweets according to some search criteria, the proposed workflow transforms the selected tweets in a structured matrix form, which is suitable for NMF decomposition. Finally, tweets are clustered in groups related to their hidden topics. In this work we verified the effectiveness and efficiency of the proposed techniques on three datasets, comparing the results obtained by different combinations of initialization and NMF algorithms. The proposed experimental workflow is mainly devoted to standardize the technical steps one has to perform when nonnegative matrix factorizations are applied to pattern discovery from twitter datasets. Beside different NMF algorithms forming the core of proposed workflow, also some initialization mechanisms are considered in order to allow the user to chose starting matrices for NMF algorithms. In fact, a correct initialization is critical for the quality of the final results of NMF decomposition in an Intelligent Data Analysis context.

Furthermore, we have compared the NMF cluster results with the well known k-means clustering algorithm, showing that NMF give comparable results with a better interpretability that is evidenced by the word cloud representation used to visualize the hidden topics discovered in data.

Future work will be addressed to codify and test different NMF algorithms and other initialization methods. Moreover, we intend to investigate the appropriate choice of the factorization rank k which is connected to the number of clusters NMF are able to extract.

Acknowledgements. This work has been supported in part by the GNCS (*Gruppo Nazionale per il Calcolo Scientifico*) of Istituto Nazionale di Alta Matematica Francesco Severi, P.le Aldo Moro, Roma, Italy.

References

1. Gupta, A., Joshi, A., Kumaraguru, P.: Identifying and characterizing user communities on Twitter during crisis events. In: Proceedings of the 2012 Workshop on Data-Driven User Behavioral Modelling and Mining from Social Media, DUB-MMSM 2012, pp. 23–26. ACM, New York (2012)
2. Wong, F.M.F., Tan, C.W., Sen, S., Chiang, M.: Quantifying political leaning from tweets, retweets, and retweeters. IEEE Trans. Knowl. Data Eng. **28**(8), 2158–2172 (2016)
3. Jin, L., Chen, Y., Wang, T., Hui, P., Vasilakos, A.V.: Understanding user behavior in online social networks: a survey. IEEE Commun. Mag. **51**(9), 144–150 (2013)
4. Aggarwal, C.C., Zhai, C.: Mining Text Data. Springer Science & Business Media, New York (2012)
5. Lee, D.D., Seung, H.S.: Learning the parts of objects by non-negative matrix factorization. Nature **401**(6755), 788–791 (1999)
6. Gillis, N.: The why and how of nonnegative matrix factorization. In: Signoretto, M., Suykens, J.A.K., Argyriou, A. (eds.) Regularization, Optimization, Kernels, and Support Vector Machines. Machine Learning and Pattern Recognition Series. Chapman and Hall/CRC, Boca Raton (2014)
7. Casalino, G., Del Buono, N., Mencar, C.: Nonnegative matrix factorizations for intelligent data analysis. In: Naik, G.R. (ed.) Non-negative Matrix Factorization Techniques. SCT, pp. 49–74. Springer, Heidelberg (2016). doi:10.1007/978-3-662-48331-2_2
8. Casalino, G., Del Buono, N., Minervini, M.: Nonnegative matrix factorizations performing object detection and localization. Appl. Comp. Intell. Soft Comput. **2012**, 15:1–15:19 (2012)
9. Cichocki, A., Zdunek, R., Phan, A.H., Amari, S.: Nonnegative Matrix and Tensor Factorizations: Applications to Exploratory Multi-way Data Analysis and Blind Source Separation. Wiley, Hoboken (2009)
10. Del Buono, N., Esposito, F., Fumarola, F., Boccarelli, A., Coluccia, M.: Breast cancer's microarray data: pattern discovery using nonnegative matrix factorizations. In: Pardalos, P.M., Conca, P., Giuffrida, G., Nicosia, G. (eds.) MOD 2016. LNCS, vol. 10122, pp. 281–292. Springer, Cham (2016). doi:10.1007/978-3-319-51469-7_24
11. Kim, Y.-H., Seo, S., Ha, Y.-H., Lim, S., Yoon, Y.: Two applications of clustering techniques to Twitter: community detection and issue extraction. Discret. Dyn. Nat. Soc. **2013**, 8 (2013)

12. Yan, X., Guo, J., Liu, S., Cheng, X., Wang, Y.: Learning topics in short texts by non-negative matrix factorization on term correlation matrix. In: Proceedings of the SIAM International Conference on Data Mining SIAM 2013, pp. 749–757 (2013)
13. Arifin, A.Z., Sari, Y.A., Ratnasari, E.K., Mutrofinn, S.: Emotion detection of tweets in Indonesian language using non-negative matrix factorization. Int. J. Intell. Syst. Appl. **6**(9), 8 (2014)
14. Saha, A., Sindhwani, V.: Learning evolving and emerging topics in social media: a dynamic NMF approach with temporal regularization. In: Proceedings of the Fifth ACM International Conference on Web Search and Data Mining, WSDM 2012, pp. 693–702. ACM, New York (2012)
15. Godfrey, D., Johns, C., Sadek, C., Meyer, C., Race, S.: A case study in text mining: interpreting Twitter data from world cup tweets (2014)
16. Alonso, J.M., Castiello, C., Mencar, C.: Interpretability of fuzzy systems: current research trends and prospects. In: Kacprzyk, J., Pedrycz, W. (eds.) Springer Handbook of Computational Intelligence, pp. 219–237. Springer, Heidelberg (2015). doi:10.1007/978-3-662-43505-2_14
17. Casalino, G., Del Buono, N., Mencar, C.: Subtractive clustering for seeding non-negative matrix factorizations. Inf. Sci. **257**, 369–387 (2014)
18. Berry, M., Browne, M., Langville, A., Pauca, P., Plemmons, R.: Algorithms and applications for approximate nonnegative matrix factorization. Comput. Stat. Data Anal. **52**(1), 155–173 (2007)
19. Lee, D.D., Seung, H.S.: Algorithms for non-negative matrix factorization. In: Leen, T.K., Dietterich, T.G., Tresp, V. (eds.) Advances in Neural Information Processing Systems 13, pp. 556–562. MIT Press, Cambridge (2001)
20. Lin, C.-J.: Projected gradient methods for nonnegative matrix factorization. Neural Comput. **19**(10), 2756–2779 (2007)
21. Kim, H., Park, H.: Sparse non-negative matrix factorizations via alternating non-negativity-constrained least squares for microarray data analysis. Bioinformatics **23**(12), 1495–1502 (2007)
22. Pascual-Montano, A., Carazo, J.M., Kochi, K., Lehmann, D., Pascual-Marqui, R.D.: Nonsmooth nonnegative matrix factorization (nsNMF). IEEE Trans. Pattern Anal. Mach. Intell. **28**(3), 403–415 (2006)
23. Albright, R., Cox, J., Duling, D., Langville, A., Meyer, C.: Algorithms, initializations, and convergence for the nonnegative matrix factorization. Technical report, NCSU Technical Report Math 81706 (2006)
24. Boutsidis, C., Gallopoulos, E.: SVD based initialization: a head start for nonnegative matrix factorization. Pattern Recogn. **41**, 1350–1362 (2008)
25. Xu, W., Liu, X., Gong, Y.: Document clustering based on non-negative matrix factorization. In: Proceedings of the 26th Annual International ACM SIGIR Conference on Research and Development in Information Retrieval, SIGIR 2003, pp. 267–273. ACM, New York (2003)
26. Shahnaz, F., Berry, M.W., Pauca, V.P., Plemmons, R.J.: Document clustering using nonnegative matrix factorization. Inf. Process. Manag. **42**(2), 373–386 (2006)
27. Ding, C., He, X., Simon, H.D.: On the equivalence of nonnegative matrix factorization and k-means - spectral clustering. In: Proceedings of the SIAM Data Mining Conference, pp. 606–610. SIAM (2005)
28. Rosenberg, A., Hirschberg, J.: V-measure: a conditional entropy-based external cluster evaluation measure. In: Proceedings of the 2007 Joint Conference on Empirical Methods in Natural Language Processing and Computational Natural Language Learning (EMNLP-CoNLL), pp. 410–420 (2007)

Q-matrix Extraction from Real Response Data Using Nonnegative Matrix Factorizations

Gabriella Casalino[1(✉)], Ciro Castiello[1], Nicoletta Del Buono[2], Flavia Esposito[2], and Corrado Mencar[1]

[1] Department of Informatics, University of Bari Aldo Moro, 70125 Bari, Italy
gabriella.casalino@uniba.it

[2] Department of Mathematics, University of Bari Aldo Moro, 70125 Bari, Italy

Abstract. In this paper we illustrate the use of Nonnegative Matrix Factorization (NMF) to analyze real data derived from an e-learning context. NMF is a matrix decomposition method which extracts latent information from data in such a way that it can be easily interpreted by humans. Particularly, the NMF of a score matrix can automatically generate the so called Q-matrix. In an e-learning scenario, the Q-matrix describes the abilities to be acquired by students to correctly answer evaluation exams. An example on real response data illustrates the effectiveness of this factorization method as a tool for EDM.

Keywords: Nonnegative Matrix Factorization · Educational Data Mining · Q-matrix · Skill interpretation

1 Introduction

Searching and extracting useful information from large datasets is a complex process which is essential in several applications. Data collected in real-world contexts are often characterized by intrinsic inaccuracy. That is due to a number of reasons which could be related to the very methods employed to gather information. Sometimes inappropriate measurement instruments or subjective judgments are involved in such processes. Additionally, the hidden relationships among data may be too complex to be expressed in readable form, especially when the latent interactions of the features characterizing a dataset contribute to produce ambiguous and overlapping pieces of information.

During the usual teaching activities large amounts of data can be produced, deriving from student examinations, questionnaire evaluations, contributes in intelligent tutoring systems. Manually collecting explicit and useful information from this data is not an easy task: human capabilities are unsuitable to process big amounts of data and discover the latent factors which mainly influence student learning. Therefore, automatic tools are indispensable to investigate the cognitive processes and to realize some kind of innovative learning mechanisms calibrated on student performance.

O. Gervasi et al. (Eds.): ICCSA 2017, Part I, LNCS 10404, pp. 203–216, 2017.
DOI: 10.1007/978-3-319-62392-4_15

Educational Data Mining (EDM) is an emerging research topic concerning the analysis of automatic techniques for knowledge extraction from data in e-learning contexts [25]. EDM methods aim at collecting, archiving and analyzing data related to learning mechanisms and student evaluations. The goal of these methods is to reveal conceptual categories which are not directly observable, such as inclinations, interests, personalities, and cognitive capabilities [26].

In this paper we adopt the Nonnegative Matrix Factorization (NMF) for extracting useful knowledge from educational data. NMF is a constrained optimization mechanism for reducing data dimensionality which differentiates from other well-known decomposition techniques (such as Singular Value Decomposition or Principal Component Analysis) since it allows to extract from data a kind of latent information which is more intuitive and interpretable [1,21,22]. When NMF is applied to educational data mining problems, these pieces of information represent the building blocks of a cognitive learning model that can be expressed in terms of the so-called question matrix (Q-matrix). This matrix captures the hidden relationships existing between the questions proposed to the students for their assessment and the skills they should have acquired in order to produce correct answers [27]. In particular, our investigation concerns the application of NMF to a real-world response dataset related to tests administered to undergraduate students approaching their university careers. We are interested in verifying how the factors computed by NMF contribute to improve interpretability in a such a sensible context.

The paper is organized as follows. In the following section we provide a brief introduction to the cognitive learning model we are interested to investigate. In particular, we discuss how a factorization mechanism can be involved in the considered scenario, showing the benefits deriving from the adoption of the NMF technique. In Sect. 3 we report the results obtained by applying a specific implementation of the NMF algorithm (namely, the constraint alternating least-square NMF) to a dataset of student responses. In the final section we sketch the future work oriented to address some open problems.

2 Factorization Mechanism in EDM

Classical Test Theory (CTT) [14] and Item Response Theory (IRT) [20] stand as two major influencing theories when appropriate methods for analyzing e-learning data must be selected. On the one hand, CTT describes a theoretical framework based on a simple linear model where the manifest test score depends on the sum of two unobservable variables, i.e. the true score and the error score. On the other hand, IRT evaluates the performance exhibited by a student while facing a test question as a function of the latent skills related to the cognitive efforts applied for answering that question. Both theories assume that the responses to questionnaires are indicative of latent factors and skills, which cannot be directly observable, but can be implicitly measured and extracted from the collected evaluation data. In this sense, the challenge is to develop mechanisms for automatic extraction of such hidden features.

The definition of latent factors depends on the representation adopted to describe the data and on the specific operators exploited to manage them. Typically, data available in e-learning scenarios can be represented as a real-value matrix $R \in \mathbb{R}_+^{m \times n}$ (called "score matrix"), recording the scores obtained by n students participating to a test composed by m questions. Each question is usually referred to as an "item". The elements R_{ij} in the score matrix are nonnegative values indicating the score obtained by the jth student on the ith item. As a simple case, such values can be limited to the binary digits "1" and "0" corresponding to "correct" and "incorrect" responses, respectively.

We assume that the latent factors involved in a cognitive scenario can be represented as nonnegative vectors of skills: the students' abilities can be defined as linear combinations of these skills. Formally, we suppose that the score matrix R can be decomposed in the following form:

$$R = QS + E, \tag{1}$$

where $Q \in \mathbb{R}^{m \times k}$ is the item-skill matrix encoding the relationship between m items and k latent skills [27]; $S \in \mathbb{R}^{k \times n}$ is the skill-student matrix describing the capabilities of n students with respect to k latent skills, $k \in \mathbb{N}$ (with $k \leq \min(m, n)$) is the number of latent skills; and $E \in \mathbb{R}^{m \times n}$ is a real-value matrix encoding a white noise inevitably connected with the score measurement process.

The Q-matrix is a nonnegative (possibly binary) matrix, with values ranging in an interval. Each element in Q quantifies how much a particular skill influences the possibility to correctly face an item. Depending on the adopted formalism, the Q-matrix can be interpreted in different ways. In a *conjunctive* interpretation, all the skill values in a row are necessary to successfully face the corresponding item; in a *disjunctive* interpretation any skill is sufficient to fulfill such a role; in a *compensatory* (or *additive*) interpretation the chance to correctly face a selected item increases with the number of skills acquired by a student.

As an example, the following binary Q-matrix can be considered, involving four items and three skills:

$$Q = \begin{pmatrix} 0\,1\,1 \\ 0\,0\,1 \\ 1\,0\,0 \\ 1\,0\,1 \end{pmatrix} \tag{2}$$

Each row of Q corresponds to an item i_l $(l = 1, \ldots, 4)$ and each column corresponds to a skill s_j $(j = 1, \ldots, 3)$. As previously asserted, the interpretation of the Q matrix may be different. With reference to item i_1, a conjunctive interpretation assumes that both skills s_2 and s_3 are required to correctly answer i_1; a disjunctive interpretation assumes that at least one of the skills s_2, s_3 is required to correctly answer i_1; a compensatory interpretation assumes that skills s_2 and s_3 are required in order to provide a greater possibility to correctly answer i_1. It should be pointed out that the cognitive model represented in (1) relies on standard matrix multiplicative operator: the elements of R are computed as a linear combination of the skill values. Such an additive operator finds its proper correspondence with a compensatory model: the more a student is endowed with pertinent skills, the greater her ability is to correctly answer an item.

In the linear model (1), skills can be thought as abstract data vectors useful to understand a large dataset. Since skills are latent and their effective characterization is unknown, the construction of a Q-matrix is a nontrivial process. However, due to relationship $k \leq \min(m, n)$, a dimensionality reduction must be performed (starting from the analysis of the original data). Therefore, the construction of Q can be carried out by resorting to some dimensionality reduction techniques, such as matrix factorization, as we are going to discuss in the next section.

2.1 NMF for Q-matrix Approximation

Generally, given a data matrix $R \in \mathbb{R}_+^{m \times n}$ whose elements are nonnegative, it is possible to provide a Nonnegative Matrix Factorization (NMF) of R in terms of additive combination of nonnegative factors representing realistic building blocks for reconstructing the original data [11,18]. Formally, we can write:

$$R \approx WH, \tag{3}$$

with $W \in \mathbb{R}_+^{m \times k}$ and $H \in \mathbb{R}_+^{k \times n}$. The value k is the rank of the factorization representing the number of building blocks useful to synthetically describe the original matrix. In general, k is chosen so that $(n + m)k < nm$.

NMF has been successfully applied in EDM contexts. Particularly, it proved to be able to extract a Q-matrix moving from the analysis of real data concerning student responses [8–10,24]. The matrices W and H derived through the NMF approach can be regarded as the item-skill and the skill-student matrices involved in the model (1), respectively. In this sense, the columns of the matrix W represent the skills, and the rows of the matrix H represent the degree of acquisition of a particular skill by the students. The nonnegativity property automatically preserved by NMF factors is particularly useful in this contexts. In fact, it is straightforward to observe that negative values in W and H would make no sense: in psychometric testing, for example, it is rather impractical to refer to items which are "negatively" associated with the latent skills being tested. Similarly, it would be unreasonable to refer to students who demonstrate "negative" capabilities for some skill. Finally, the capability of NMF to automatically derive a Q-matrix (in terms of the factor matrix W) must be related to a compensatory interpretation of the cognitive model represented in (1). That is due to the additive operator involved in the NMF definition.

To compute the NMF of R, we must take into account some quality measures to evaluate how effective the product WH is in approximating R. Therefore, we consider the Frobenius norm of the difference matrix $R - WH$ which must be minimized by identifying the nonnegative matrices W and H with rank k:

$$\min_{W \geq 0, H \geq 0} \|R - WH\|_F^2, \tag{4}$$

where $\|\cdot\|_F$ denotes the Frobenius norm on matrices. The factorization (3) can be computed by solving the minimization (4) with non-negativity constraint. This optimization problem is NP-hard and non convex in both W and H, that are unknown variables. However, it is convex with respect to each variable taken separately; this makes possible to split the original minimization problem into two sub-problems (first H is fixed to estimate W, then W is fixed to estimate H). Such an approach is carried on by exploiting alternating optimization techniques, i.e. iterative methods suitable for solving this kind of problems. A number of alternating optimization techniques may be considered (recent reviews of such techniques are reported in [7,13]): in this work we refer to a particular version of the Alternating Least Squares (ALS) algorithm [3]. Our choice is justified by some advantages deriving from the adoption of ALS, including: the positions of the zero elements in W and H are not fixed (in contrast with what happens with other methods); a single matrix requires to be initialized to start the computation; a fast rate of convergence to a local minimum solution is obtained. Moreover, since we adopt the Constraint Alternating Least Square (CALS) algorithm [16,17], the sparsity on both factors W and H is enforced: the additional sparsity constraint produces factors which reveal to be more interpretable in an e-learning scenario. The CALS algorithm is designed in order to solve a modified optimization problem where (4) is penalized by adding two terms which guarantee further sparsity patterns for both matrices W and H. In particular, the algorithm used in this paper is designed to solve the following optimization problem:

$$\min_{W,H}(\|R - WH\|_F^2 + \alpha\|H\|_F^2 + \beta\sum_i \|W_{i:}\|_1^2), \tag{5}$$

being $W_{i:}$ the i row of W and $\|\cdot\|_1$ the 1-norm to be evaluated on vectors; the constants α, β are user defined parameters adopted to balance the trade-off between the approximation accuracy and the sparseness of W and H. The main steps performed by the CALS algorithm are reported in Algorithm 1, where the involved stopping criterion requires (5) to fall under a specified threshold value.

As can be observed, Algorithm 1 requires the specification of the rank value k which must be someway estimated, as we are going to assess in the following section.

2.2 Rank Estimation

The choice of the rank parameter k is critical in NMF and only empirical approaches are known to select it. In e-learning contexts, k defines the number of latent skills. This parameter is usually assigned as "a-priori" knowledge by the domain expert; recently the application of an SVD-based approach to infer an appropriate value of k has been explored [2]. Here, we adopted an approach which is able to automatically select k in a given interval of values [5].

It should be observed that the choice of k determines the number of latent factors standing as the guiding directions for grouping the samples inside the data matrix. However, for each established k value such a grouping may be

Algorithm 1. Constraint Alternating Least Square NMF (CALS)

```
Require: $R \in \mathbb{R}_+^{m\times n}$ {score matrix}
Require: $k < \min(m, n)$ {rank value}
Require: parameters $\alpha, \beta \in (0, 1)$
Require: $W \in \mathbb{R}_+^{m\times k}$ {$W$ is initialized as a random matrix; }
while stopping criterion is not met do
    solve the matrix equation $(W^\top W + \sqrt{\alpha} I_k)H = W^\top R$ w.r.t. $H$ {being $I_k$ the identity matrix}
    set all negative elements in $H$ to 0;{projection step for $H$}
    solve the matrix equation $W(HH^\top + \sqrt{\beta}\mathbf{1}_{k\times k}) = RH^\top$ w.r.t. $W$ {being $\mathbf{1}_{k\times k}$ the $k \times k$ matrix of ones}
    set all negative elements in $W$ to 0; {projection step for $W$}
end while
```

varying. To assess the stability of the clustering process associated to the NMF mechanism, the cophenetic correlation coefficient (CCC) can be considered [4]. At the same time, the choice of k affects the quality of WH in terms of its capability to approximate R. More precisely, the trend of the approximation error can be represented by a curve determined by the residual sum of squares (RSS) [15]. In this way, we evaluate the CCC and RSS values for each k in the given interval of values: k is finally set to the minimum value determining either the point which foreruns the lowest CCC value or the inflection point of the RSS curve.

3 *Q*-matrix Extraction from Real Response Data

The numerical test illustrated in this section aims at automatically extracting a Q-matrix from real response data using NMF. The extracted Q-matrix has been interpreted in the compensatory sense and the experimental session has been practically conducted resorting to a NMF package available in the R software environment [12]. Moreover, the parameters needed by CALS were set to $\alpha = -1$ and $\beta = 0.01$.

3.1 Dataset Description and Data Preprocessing

We considered a dataset composed by the responses provided by 410 students to 24 items. The population of students underwent an entrance exam for enrolling in a Bachelor degree course in Computer Science. The items administered to the students during the exam have been extracted from a repository where each included item pertains to one of six different subjects: Logics, Reasoning Competence, Geometry, Equations and Inequalities, Basic Number Properties, and Elementary Algebra. The available data correspond to the students' responses and were recorded as a binary score matrix: elements with value equal to "1" indicate correct answers to items; elements with value equal to "0" indicate wrong responses to items. However, since the students were allowed not to provide any

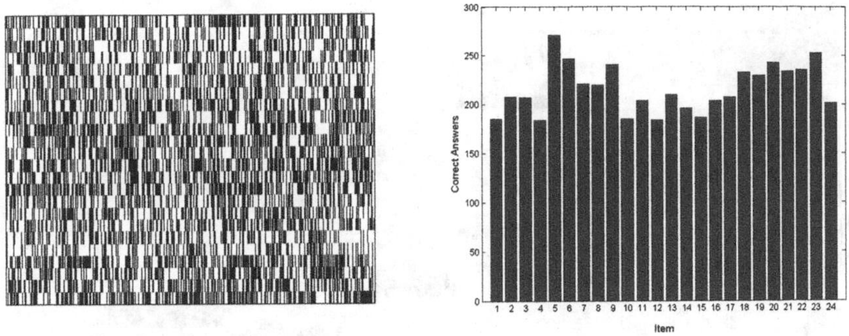

(a) (b)

Fig. 1. (a) Score-matrix with the results of 391 students (columns) to 24 questions (rows). White color indicates correct answers, black color indicates wrong answers. (b) Distribution of correct answers per item.

answer to some items, the resulting dataset comprises missing values which have been treated during a preliminary pre-processing stage. Particularly, the Little's test has been applied to the data at hand, providing evidence that the missing values belong to a specific category commonly referred to as "missing completely at random" values [19]. Such a result, together with the limited occurrence of missing values, suggested to simply tackle this problem by imposing the deletion method.

In this way, the obtained score matrix $R \in \mathbb{R}^{24\times391}$ contains the responses given by 391 students to 24 questions (items). A final re-arrangement of the items was accomplished in order to group them according to their inherent subjects. In this way, the entire set of 24 items can be intended as a sequence of 6 subsets composed by 4 items pertaining to: Logics (items 1–4), Reasoning Competence (items 5–8), Geometry (items 9–12), Equations and Inequalities (items 13–16), Basic Number Properties (items 17–20), and Elementary Algebra (items 21–24). Figure 1(a) reports the heatmap representation of the matrix R ("1" values are reported in white, "0" values are reported in black), while Fig. 1(b) depicts the distribution per item of the correct answers provided by the students.

Since the test was not originally developed on the basis of a predefined cognitive diagnosis model, the skills measured by the questionnaire are unknown. The conducted data analysis aimed at determining the existence of some basic set of skills measured by the test and their potential relationship with the a-priori topic labels assigned to each item. Particularly, the NMF analysis is oriented to identify the item relational structure and to extract some meaningful skills reflecting specific mathematical abilities connected with preassigned item topics. The information extracted from the NMF analysis can be used as a template to better understand the student learning behavior, in order to remedy specific failures in the students' competence.

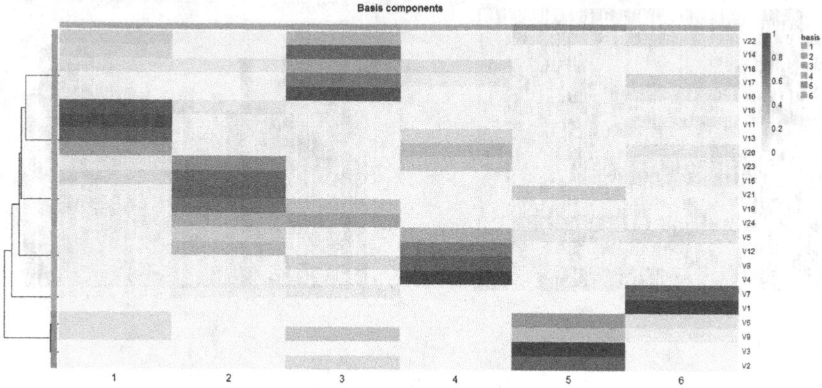

(a) Matrix W

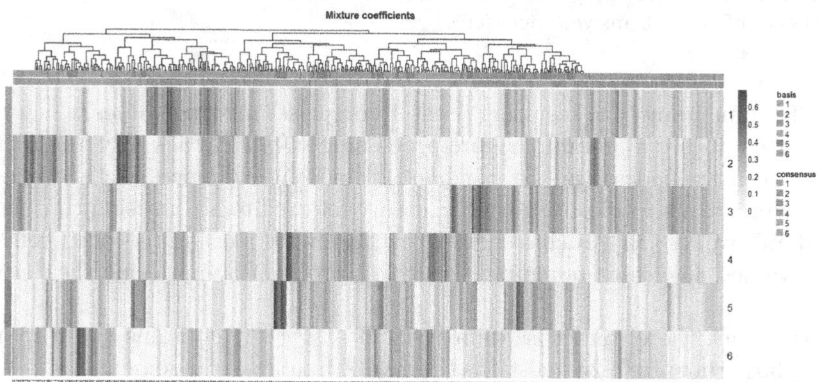

(b) Matrix H

Fig. 2. Heatmap representation of Q-matrix W (a) and skill-student matrix H (b) obtained by applying the CALS algorithm with rank value $k = 6$ (Color figure online)

3.2 *Q*-matrix Generated by the NMF Algorithm

We carried on a twofold experimentation, based on the choice for the k value. Our first attempt was to a-priori set $k = 6$, related to the different subjects involved in the questionnaire. In this sense, we intend to verify a correspondence between the latent skills to be derived by NMF analysis and the pre-defined topics under examination.

By applying the NMF analysis with $k = 6$, we obtained a factorization of the original score matrix R in terms of a Q-matrix $W \in \mathbb{R}_+^{24\times 6}$ and a coefficient matrix $H \in \mathbb{R}_+^{6\times 391}$. In order to provide a visualization of such a result, Fig. 2 illustrates the heatmaps related to the obtained factors[1]. The color shade of each

[1] The actual values of W and H have been normalized to allow the heatmap representation.

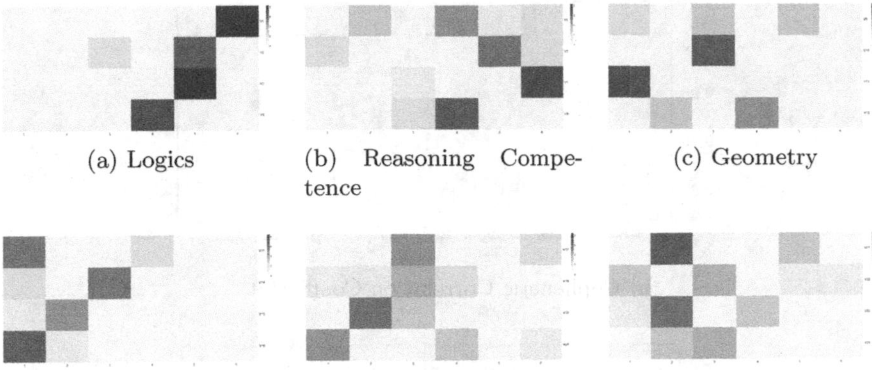

(a) Logics
(b) Reasoning Competence
(c) Geometry
(d) Equations and Inequalities
(e) Basic Number Properties
(f) Elementary Algebra

Fig. 3. Heatmap representations of the sub-matrices of W related to the 6 different item subjects.

cell indicates the weight of the skills in characterizing the reported information. Particularly, darker shades in the W matrix reveal a greater influence of the skill to tackle a particular item; darker shades in the H matrix reveal a greater degree of competence for a student in a specific skill. Moreover, in Fig. 2(a) the rows of W have been rearranged in order to provide a more readable representation: items pertaining to similar latent skills are reported in adjacent positions, while distancing those characterized by different skills. An analogous re-ordering has been applied on the columns of H in Fig. 2(b). Such a rearrangement process has been carried out through a hierarchical clustering of row/column values with average linkage (the material procedure was provided by the R software environment).

It can be observed that the items reported in the W matrix appear to cluster around the 6 latent skills derived by the NMF analysis. Even though such a partition does not coincide with the 6 original subjects involved in the questionnaire, a deeper analysis of matrix W highlights how the items pertaining to Logics and Reasoning Competence are characterized by a specific subset of skills (namely, skills 4, 5, and 6). On the other hand, the remaining items are characterized by the other derived skills (namely, skills 1, 2, and 3). Such a dichotomy emerges also from the analysis of Fig. 3, separately depicting the heatmaps related to the items of the different subjects (actually, the derived matrix W has been split in the Figs. 3(a–f)). It can be observed that the clouds composed by the darker shades lean towards the right-side of the heatmaps in case (a–b), while leaning towards the left-side of the heatmaps in the other cases.

This kind of results may suggest an affinity of items pertaining to Logics and Reasoning Competence: this is in agreement with the semantic related to those particular subjects. In fact, those kind of questions are presumably tackled by arguing some line of reasoning. On the other hand, the items pertaining to the

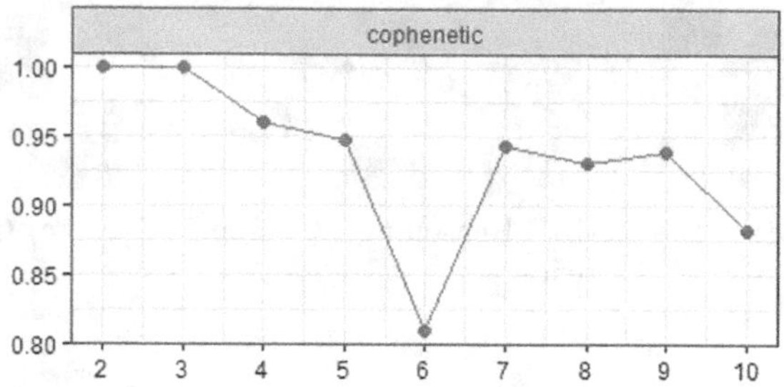

(a) Cophenetic Correlation Coefficient

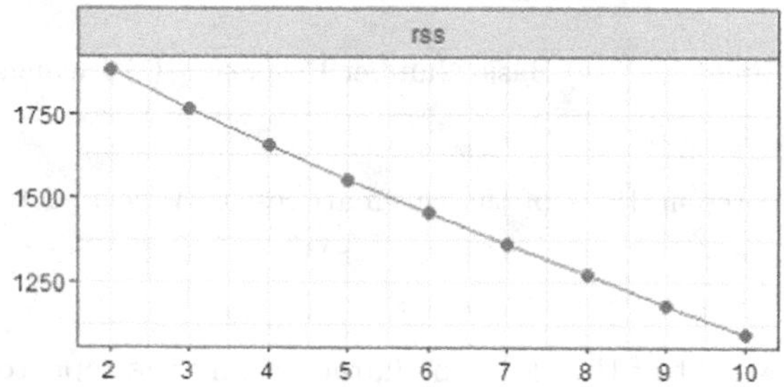

(b) Residual Sum of Squares

Fig. 4. Behavior of the CCC in the interval [2, 10]. A decreasing behavior followed by a slow drop off can be observed in the sub-interval [5, 7].

(a) k=2

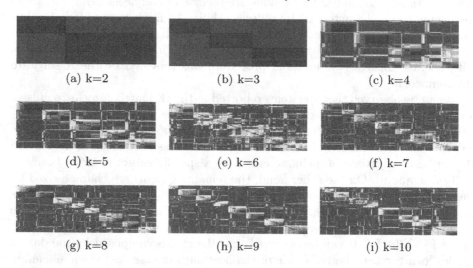

(b) k=3

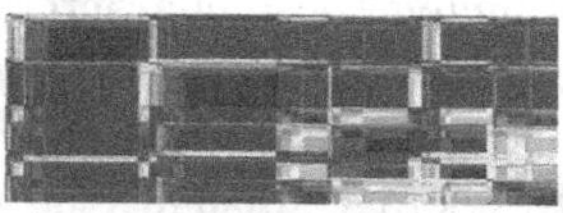

(c) k=4

(d) k=5

(e) k=6

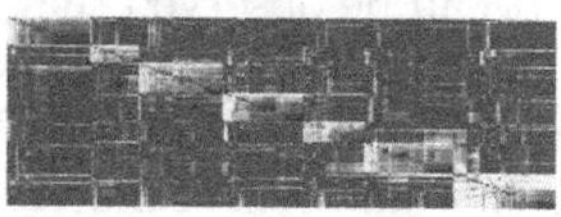

(f) k=7

(g) k=8

(h) k=9

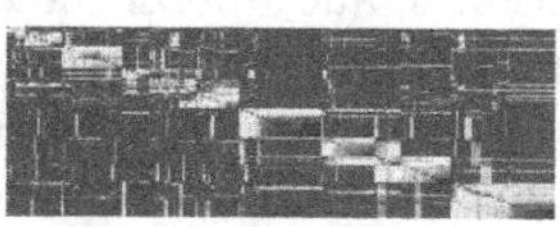

(i) k=10

Fig. 5. Consensus matrices computed for each rank value k in the interval [2, 10].

other subjects are usually tackled by applying some learned rules. In this sense, an affinity can be traced among them, even if it appears to be somewhat fuzzier, as shown by the shape of the corresponding cloud shades.

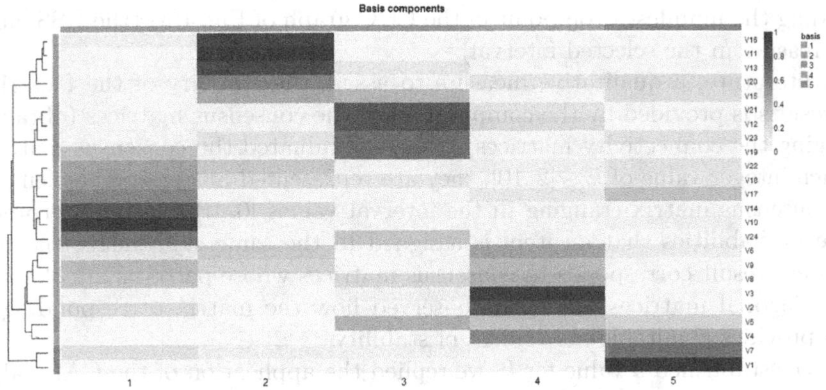

(a) Matrix W

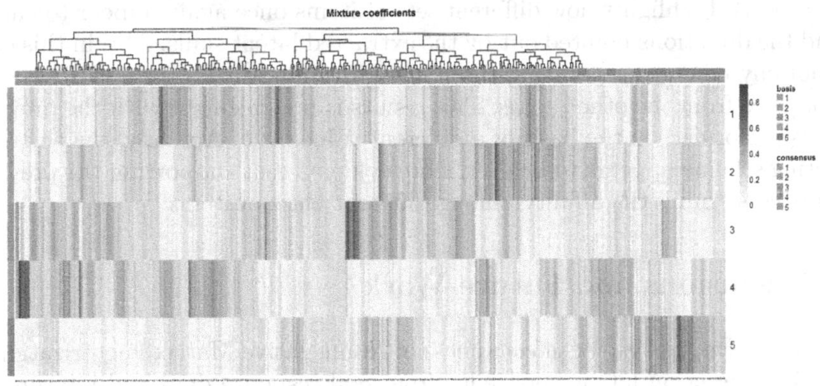

(b) Matrix H

Fig. 6. Heatmap representation of Q-matrix W and skill-student matrix H obtained using CALS algorithm with rank value $k = 5$

As a second attempt to conduct the NMF analysis on the available data, we performed an estimation for the k value. As previously discussed, such a value should be selected as an integer falling into a specific interval which in our case turns out to be [2, 24] (spanning the range defined by the item cardinality). However, we restricted our analysis to the reduced range [2, 10] which appears to be a suitable compromise both to reduce the computational burden required to complete the experimentation and to ensure the interpretability of the final results (in terms of the number of extracted latent skills).

For each integer value inside the considered interval, 50 runs of the CALS algorithm have been performed, all of them initialized with a random nonnegative matrix $W_0 \in \mathbb{R}^{24 \times k}$ (alternative initialization methods can be adopted: see [6]). Figure 4 illustrates the trends of the CCC and RSS curves. The automatic process of rank estimation provided the value $k = 5$, corresponding to the value preceding the manifest drop point in the CCC graph of Fig. 4(a) (the RSS curve is decreasing in the selected interval).

In literature, a qualitative measure to assess the stability of the factorization results is provided by the computation of the consensus matrices (obtained averaging the connectivity matrices) [23]. We evaluated the consensus matrices for each integer value of $k \in [2, 10]$: they are represented in Fig. 5. The elements of a consensus matrix (ranging in the interval values $[0, 1]$) can be interpreted as the probabilities that an item is assigned to the same skill; hence the best consistent result corresponds to consensus matrices whose patterns are closer to block diagonal matrices. It can be observed how the matrix corresponding to $k = 5$ provides a suitable assessment of stability.

Once established a value for k, we replied the application of the CALS algorithm to the score matrix R. Figure 6 illustrates the heatmaps related to the obtained factors $W \in \mathbb{R}_+^{24\times5}$ and $H \in \mathbb{R}_+^{5\times391}$. The rearrangement of the rows composing W highlights how different sets of items once again appear to cluster around the directions pointed out by the extracted latent skills. Also in this case, a dichotomy emerges separating the items pertaining to Logics and Reasoning Competence from the other items. This results is in agreement with the previous analysis performed on the basis of a different value for k. Actually, the emerging clustering scheme appears to be even more evident, thus supporting the previous remarks concerning the affinity among different class of items.

4 Conclusions and Future Work

This paper investigates the effectiveness of Nonnegative Matrix Factorization as a tool for extracting latent knowledge from real response data and interpreting it in terms of Q-matrix: this particular representation of knowledge can be useful to understand and direct student learning. A real score matrix containing the responses given by 391 students to 24 questions (items) has been analyzed using a specific constraint NMF algorithm, namely the CALS algorithm, and different sets of latent skills were extracted. Two experimental sessions were conducted: the first one aimed at employing the CALS algorithm with an a-priori defined number of skills, while the second one performed an NMF analysis of the score matrix to automatically extract the proper number of latent skills as suggested by the adopted extraction strategy. Both the experimental sessions have been discussed in order to investigate the extracted latent factors and determine the potential relationship with the a-priori topic subjects characterizing each item included in the questionnaire. Even if none of the patterns corresponding to the extracted skills fits the a-priori established groups of items, our analysis implies that the derived latent skills suggest that the items inherently hint at the existence of some overlapping latent abilities. Moreover, a dichotomy emerges which separates the items pertaining to Logics and Reasoning Competence (which must be faced by conducting some line of reasoning) from the remaining items pertaining to other subjects (which are usually tackled by applying some learned rules).

Further investigation is required to deepen the understanding of how extracted knowledge influences the student learning behavior and how it can be adopted to remedy specific failures in the students' competence. Moreover, future work will address to study the complexity and the scalability of the method when big data are analyzed, test the effectiveness of other NMF algorithms in EDM contexts, and to design different mechanisms estimating the proper number of latent skills for each score matrix.

Acknowledgements. This work has been supported in part by the GNCS (*Gruppo Nazionale per il Calcolo Scientifico*) of Istituto Nazionale di Alta Matematica Francesco Severi, P.le Aldo Moro, Roma, Italy.

References

1. Alonso, J.M., Castiello, C., Mencar, C.: Interpretability of fuzzy systems: current research trends and prospects. In: Kacprzyk, J., Pedrycz, W. (eds.) Springer Handbook of Computational Intelligence, pp. 219–237. Springer, Heidelberg (2015). doi:10.1007/978-3-662-43505-2_14
2. Beheshti, B., Desmarais, M.C., Naceur, R.: Methods to find the number of latent skills. In: Proceedings of the 5th International Conference on Educational Data Mining, EDM 2012, pp. 81–86 (2012)
3. Berry, M.W., Browne, M., Langville, A.N., Pauca, V.P., Plemmons, R.J.: Algorithms and applications for approximate nonnegative matrix factorization. Comput. Stat. Data Anal. **52**, 155–173 (2007)
4. Brunet, J.P., Tamayo, P., Golub, T.R., Mesirov, J.P.: Metagenes and molecular pattern discovery using matrix factorization. Proc. Nat. Acad. Sci. **101**(12), 4164–4169 (2004)
5. Del Buono, N., Esposito, F., Fumarola, F., Boccarelli, A., Coluccia, M.: Breast cancer's microarray data: pattern discovery using nonnegative matrix factorizations. In: Pardalos, P.M., Conca, P., Giuffrida, G., Nicosia, G. (eds.) MOD 2016. LNCS, vol. 10122, pp. 281–292. Springer, Cham (2016). doi:10.1007/978-3-319-51469-7_24
6. Casalino, G., Del Buono, N., Mencar, C.: Subtractive clustering for seeding non-negative matrix factorizations. Inf. Sci. **257**, 369–387 (2014)
7. Casalino, G., Del Buono, N., Mencar, C.: Non negative matrix factorizations for intelligent data analysis. In: Naik, G.R. (ed.) Non-negative Matrix Factorization Techniques: Advances in Theory and Applications. SCT, pp. 49–74. Springer, Heidelberg (2016). doi:10.1007/978-3-662-48331-2_2
8. Desmarais, M.C.: Conditions for effectively deriving a q-matrix from data with non-negative matrix factorization (2011)
9. Desmarais, M.C., Beheshti, B., Naceur, R.: Item to skills mapping: deriving a conjunctive q-matrix from data. In: Cerri, S.A., Clancey, W.J., Papadourakis, G., Panourgia, K. (eds.) ITS 2012. LNCS, vol. 7315, pp. 454–463. Springer, Heidelberg (2012). doi:10.1007/978-3-642-30950-2_58
10. Desmarais, M.C., Naceur, R.: A matrix factorization method for mapping items to skills and for enhancing expert-based q-matrices. In: Lane, H.C., Yacef, K., Mostow, J., Pavlik, P. (eds.) AIED 2013. LNCS, vol. 7926, pp. 441–450. Springer, Heidelberg (2013). doi:10.1007/978-3-642-39112-5_45

11. Donoho, D., Stodden, V.: When does non-negative matrix factorization give a correct decomposition into parts? In: Thrun, S., Saul, L., Schölkopf, B. (eds.) Advances in Neural Information Processing Systems 16. MIT Press, Cambridge (2004)
12. Gaujoux, R., Seoighe, C.: A flexible R package for nonnegative matrix factorization. BMC Bioinform. **11**(1), 1 (2010)
13. Gillis, N.: The Why and How of Nonnegative Matrix Factorization. Machine Learning and Pattern Recognition Series. Chapman and Hall/CRC, Boca Raton (2014). pp. 257–291
14. Gulliksen, H.: Theory of Mental Tests. Lawrence Erlbaum, Hillsdale (1950)
15. Hutchins, L.N., Murphy, S.M., Singh, P., Graber, J.H.: Position-dependent motif characterization using non-negative matrix factorization. Bioinformatics **24**, 2684–2690 (2008)
16. Kim, H., Park, H.: Sparse non-negative matrix factorizations via alternating non-negativity-constrained least squares for microarray data analysis. Bioinformatics **23**(12), 1495 (2007). http://dx.doi.org/10.1093/bioinformatics/btm134
17. Kim, H., Park, H.: Nonnegative matrix factorization based on alternating nonnegativity constrained least squares and active set method. SIAM J. Matrix Anal. Appl. **30**(2), 713–730 (2008)
18. Lee, D.D., Seung, H.S.: Algorithms for non-negative matrix factorization. In: Proceedings of the Advances in Neural Information Processing Systems Conference, vol. 13, pp. 556–562. MIT Press (2000)
19. Little, R.J.A.: A test of missing completely at random for multivariate data with missing values. J. Am. Stat. Assoc. **83**(404), 1198–1202 (1988)
20. Lord, F.: A theory of test scores. Psychometrika Monogr. **7** (1952)
21. Mencar, C., Castiello, C., Fanelli, A.M.: Fuzzy user profiling in e-learning contexts. In: Lovrek, I., Howlett, R.J., Jain, L.C. (eds.) KES 2008. LNCS, vol. 5178, pp. 230–237. Springer, Heidelberg (2008). doi:10.1007/978-3-540-85565-1_29
22. Mencar, C., Torsello, M., Dell'Agnello, D., Castellano, G., Castiello, C.: Modeling user preferences through adaptive fuzzy profiles. In: ISDA 2009–9th International Conference on Intelligent Systems Design and Applications, pp. 1031–1036 (2009)
23. Monti, S., Tamayo, P., Mesirov, J., Golub, T.: Consensus clustering: a resampling-based method for class discovery and visualization of gene expression microarray data. Mach. Learn. **52**(1), 91–118 (2003)
24. Oeda, S., Yamanishi, K.: Extracting time-evolving latent skills from examination time series. In: EDM2013, pp. 340–341 (2013)
25. Romero, C., Ventura, S.: Data mining in education. WIREs Data Min. Knowl. Discov. **3**, 12–27 (2013)
26. Silva, C., Fonseca, J.: Educational data mining: a literature review. In: Rocha, Á., Serrhini, M., Felgueiras, C. (eds.) Europe and MENA Cooperation Advances in Information and Communication Technologies. AISC, vol. 520, pp. 87–94. Springer, Cham (2017). doi:10.1007/978-3-319-46568-5_9
27. Tatsuoka, K.K.: Rule space: an approach for dealing with misconceptions based on item response theory. J. Educ. Measur. **20**(4), 345–354 (1983)

Building Networks for Image Segmentation Using Particle Competition and Cooperation

Fabricio Breve(✉)

São Paulo State University (UNESP), Rio Claro, SP 13506-900, Brazil
fabricio@rc.unesp.br
http://www.rc.unesp.br/igce/demac/fbreve/

Abstract. Particle competition and cooperation (PCC) is a graph-based semi-supervised learning approach. When PCC is applied to interactive image segmentation tasks, pixels are converted into network nodes, and each node is connected to its k-nearest neighbors, according to the distance between a set of features extracted from the image. Building a proper network to feed PCC is crucial to achieve good segmentation results. However, some features may be more important than others to identify the segments, depending on the characteristics of the image to be segmented. In this paper, an index to evaluate candidate networks is proposed. Thus, building the network becomes a problem of optimizing some feature weights based on the proposed index. Computer simulations are performed on some real-world images from the Microsoft GrabCut database, and the segmentation results related in this paper show the effectiveness of the proposed method.

Keywords: Particle competition and cooperation · Image segmentation · Complex networks

1 Introduction

Image Segmentation is the process of dividing an image into multiple parts, separating foreground from background, identifying objects, or other relevant information [31]. This is one of the hardest tasks in image processing [23] and completely automatic segmentation is still a big challenge, with existing methods being domain dependent. Therefore, interactive image segmentation, partially supervised by an specialist, became an interesting approach in the last decades [1–4,20,21,24,25,28–30,33].

Many interactive image segmentation approaches are based on semi-supervised learning (SSL), category of machine learning which is usually applied to problems where unlabeled data is abundant, but the process of labeling them is expensive and/or time-consuming, usually requiring intense work of human specialists [19,34]. SSL techniques employ both labeled and unlabeled data in their training process, overcoming the limitations of supervised and unsupervised learning, in which only labeled or unlabeled data is used for training, respectively. Regarding

O. Gervasi et al. (Eds.): ICCSA 2017, Part I, LNCS 10404, pp. 217–231, 2017.
DOI: 10.1007/978-3-319-62392-4_16

the interactive segmentation task, SSL techniques spread labels provided by the user for some pixels to the unlabeled pixels, based on their similarity.

Particle competition and cooperation (PCC) [12] is a graph-based SSL approach, which employs particles walking on a network represented by an undirected and unweighted graph. Nodes represent the data elements and the particles represent the problem classes. Particles from the same class cooperate with each other and compete against particles representing different classes for the possession of the network nodes.

Many graph-based SSL techniques are similar and share the same regularization framework [34]. They usually spread the labels globally, while PCC employs a local propagation approach, through the walking particles. Therefore, its computational cost is close to linear ($O(N)$) in the iterative step, while many other state-of-the-art methods have cubic computational complexity ($O(N^3)$).

PCC was already applied to some important machine learning tasks, such as overlapped community detection [10,11], learning with label noise [9,17,18], learning with concept drift [8,16], active learning [6,13,14], and interactive image segmentation [5,7].

In the interactive segmentation task, PCC is applied to a network built from the image to be segmented. Each pixel is represented by a network node. Edges are created between nodes corresponding to similar pixels. Then, particles representing the labeled pixels walk through the network trying to dominate most of the unlabeled pixels, spreading their label and trying to avoid invasion from enemy particles representing other classes in the nodes they already possess. In the end of the iterative process, the particles territory frontiers are expected to coincide with the frontiers among different image segments [5].

In the network formation stage, the edges between nodes are created based on the similarity between the corresponding pixels, according to the Euclidean distance between their features, which are extracted from the image. A large amount of features may be extracted from each pixel. These include RGB (red, green, and blue) components, intensity, hue, and saturation. Other features take pixel location and neighborhood into account. Given an image, each feature may have more or less discriminative capacity regarding the classes of interest. Therefore, it is important to weight each feature according to its discriminative capacity, so the PCC algorithm segmentation capacity is also increased.

Unfortunately, defining these weights is a difficult task. The methods proposed so far work well in some images, but fail in others. In [15], four automatic feature weight adjustment methods were proposed based on feature values (mean, standard deviation, histogram) for each class in the labeled pixels. They were applied to three images from the Microsoft GrabCut database [30]. Three of the methods were able to increase PCC segmentation accuracy in at least one image, but none of them increased accuracy in all of the three images.

In this paper, a new method to automatically define feature weights is proposed. It is based on an index, which is extracted from candidate networks built with all the features and their candidate weights. This approach has some advantages over the methods that consider only individual features. For instance, individual features may not be good discriminators, but combined they may have a higher discriminative capacity. A candidate network is built considering the combination of all features and their respective weights. Therefore, the proposed index, extracted from the candidate networks, may be used to evaluate if a given set of weights leads to a proper network to be used by PCC.

In this sense, finding a good set of weights is just a matter of optimizing the weights based on the index extracted from the candidate network built with them. In this paper, a genetic algorithm [22,27] is used to optimize the weights, with the proposed index used as the fitness function to be maximized.

Computer simulations are performed using some real-world images extracted from the Microsoft GrabCut database. The PCC method is applied to both a network built with feature weights optimized by the proposed method and a network built with non-weighted features, used as baseline. The segmentation accuracy is calculated on both resulting images, comparing them to the *ground truth* images labeled by human specialists. The results show the efficacy of the proposed method.

The remaining of this paper is organized as follows. Section 2 presents the particle competition and cooperation model. In Sect. 3, the proposed method is explained. Section 4 presents the experiments used to validate the method. In Sect. 5, the computer simulation results are presented and discussed. Finally, some conclusions are drawn on Sect. 6.

2 Image Segmentation Using Particle Competition and Cooperation

In this section, the semi-supervised particle competition and cooperation approach for interactive image segmentation is presented. The reader can find more complete expositions in [5,12].

Overall, PCC may be applied to image segmentation tasks by converting each image pixel into a network node, represented by an undirected and unweighted graph. Edges among nodes are created between similar pixels, according to the Euclidean distance between the pixel features. Then, a particle is created for each labeled node, i.e., nodes representing labeled pixels. Particles representing the same class belong to the same team, they cooperate with their teammates to dominate unlabeled nodes, at the same time that they compete against particles from other teams. As the system runs, particles walk through the network following a random-greedy rule.

Each node has a set of domination levels, each level belonging to a team. When a particle visits a node, it raises its team domination level on that node, at the same time that it lowers the other teams domination levels. Each particle has a strength level, which changes according to its team domination level on the node its visiting. Each team of particles also has a table to store the distances between all the nodes it has visited and the closest labeled node of its class. These distance tables are dynamically updated as the particles walk. At the end of the iterative process, each pixel will be labeled by the team that has the highest domination level on its corresponding node.

A large amount of features may be extracted from each pixel x_i. In this paper, 23 features are considered: (1) the pixel row location; (2) the pixel column location; (3) the red (R) component of the pixel; (4) the green (G) component of the pixel; (5) the blue (B) component of the pixel; (6) the hue (H) component of the pixel; (7) the saturation (S) component of the pixel; (8) the value (V) component of the pixel; (9) the ExR component; (10) the ExG component; (11) the ExB component; (12) the average of R on the pixel and its neighbors (MR); (13) the average of G on the pixel and its neighbors (MG); (14) the average of B on the pixel and its neighbors (MB); (15) the standard deviation of the R on the pixel and its neighbors (SDR); (16) the standard deviation of G on the pixel and its neighbors (SDG); (17) the standard deviation of B on the pixel and its neighbors (SDB); (18) the average of H on the pixel and its neighbors (MH); (19) the average of S on the pixel and its neighbors (MS); (20) the average of V on the pixel and its neighbors (MV); (21) the standard deviation of H on the pixel and its neighbors (SDH); (22) the standard deviation of S on the pixel and its neighbors (SDS); (23) the standard deviation of V on the pixel and its neighbors (SDV).

For all measures considering the pixel neighbors, an 8-connected neighborhood is used, except on the borders where no wraparound is applied. All components are normalized to have mean 0 and standard deviation 1. They may also be scaled by a vector of weights λ in order to emphasize/deemphasize each feature during the network generation. ExR, ExG, and ExB components are obtained from the RGB components using the method described in [26]. The HSV components are obtained from the RGB components using the method described in [32].

The network is represented by the undirected and unweighted graph $\mathbf{G} = (\mathbf{V}, \mathbf{E})$, where $\mathbf{V} = \{v_1, v_2, \ldots, v_N\}$ is the set of nodes, and $\mathbf{E}$ is the set of edges (v_i, v_j). Each node v_i corresponds to the pixel x_i. Two nodes v_i and v_j are connected if v_j is among the k-nearest neighbors of v_i, or vice-versa, considering the Euclidean distance between the features of x_i and x_j. Otherwise, v_i and v_j are disconnected.

For each node $v_i \in \{v_1, v_2, \ldots, v_L\}$, corresponding to a labeled pixel $x_i \in \mathfrak{X}_L$, a particle ρ_i is generated and its initial position is defined as v_i. Each particle ρ_j has a variable $\rho_j^\omega(t) \in [0, 1]$ to store its strength, which defines how much it impacts the node it is visiting. The initial strength is always set to maximum, $\rho_j^\omega(0) = 1$.

Each team of particles has a distance table, shared by all the particles belonging to the team. It is defined as $\mathbf{d_c(t)} = d_c^1(t), \ldots, d_c^N(t)\}$. Each element $d_c^i(t) \in [0 \quad N-1]$ stores the distance between each node v_i and the closest labeled node of the class c. Particles initially know only that the distance to any labeled node of its class is zero ($d_c^i = 0$ if $y(x_i) = c$). All other distances are adjusted to the maximum possible value ($d_c^i = n - 1$ if $y(x_i) \neq c$) and they are updated dynamically as the particles walk.

Each node v_i has a dominance vector $\mathbf{v_i^\omega(t)} = \{v_i^{\omega_1}(t), v_i^{\omega_2}(t), \ldots, v_i^{\omega_C}(t)\}$, where each element $v_i^{\omega_c}(t) \in [0,1]$ corresponds to the domination level of the team c over the node v_i. The sum of all domination levels in a node is always constant:

$$\sum_{c=1}^{C} v_i^{\omega_c} = 1. \tag{1}$$

Nodes corresponding to labeled pixels have constant domination levels, and they are always adjusted to maximum for the corresponding team and zero for the others. On the other hand, nodes that correspond to unlabeled pixels have variable dominance levels, initially equal for all teams, but varying as they are visited by particles. Therefore, for each node v_i, the dominance vector $\mathbf{v_i^\omega}$ is defined by:

$$v_i^{\omega_c}(0) = \begin{cases} 1 \text{ if } x_i \text{ is labeled and } y(x_i) = c \\ 0 \text{ if } x_i \text{ is labeled and } y(x_i) \neq c \\ \frac{1}{C} \text{ if } x_i \text{ is unlabeled} \end{cases}. \tag{2}$$

When a particle ρ_j visits an unlabeled node v_i, domination levels are adjusted as follows:

$$v_i^{\omega_c}(t+1) = \begin{cases} \max\{0, v_i^{\omega_c}(t) - \frac{0{,}1\rho_j^\omega(t)}{C-1}\} \\ \quad \text{if } c \neq \rho_j^c \\ v_i^{\omega_c}(t) + \sum_{r \neq c} v_i^{\omega_r}(t) - v_i^{\omega_r}(t+1) \\ \quad \text{if } c = \rho_j^c \end{cases}, \tag{3}$$

where ρ_j^c represents the class label of particle ρ_j. Each particle ρ_j will change the node its visiting v_i by increasing the domination level of its class on it ($v_i^{\omega_c}$, $c = \rho_j^c$) at the same time that it decreases the domination levels of other classes ($v_i^{\omega_c}$, $c \neq \rho_j^c$)). Since nodes corresponding to labeled pixels have constant domination levels, (3) is not applied to them.

The particle strength changes according to the domination level of its class in the node it is visiting. Thus, at each iteration, a particle strength is updated as follows: $\rho_j^\omega(t) = v_i^{\omega_c}(t)$, where v_i is the node being visited, and $c = \rho_j^c$.

When a node v_i is being visited, the particle updates its class distance table as follows:

$$d_c^i(t+1) = \begin{cases} d_c^q(t) + 1 \text{ if } d_c^q(t) + 1 < d_c^i(t) \\ d_c^i(t) \quad \text{otherwise} \end{cases}, \tag{4}$$

where $d_c^q(t)$ is the distance from the previous visited node to the closest labeled node of the particle class, and $d_c^i(t)$ is the current distance from the node being visited to the closest labeled node of the particle class. Notice that particles have no knowledge of the graph connection patterns. They are only aware of which are the neighbors of the node they are visiting. Unknown distances are discovered dynamically as the particles walk and distances are updated as particles naturally find shorter paths to the nodes.

At each iteration, each particle ρ_j chooses a node v_i among the neighbors of its current node to visit. The probability of choosing a node v_i is given by: (a) the particle class domination on it, $v_i^{\omega_c}$, and (b) the inverse of its distance to the closest labeled node from the particle class, d_c^i, as follows:

$$p(v_i|\rho_j) = \frac{W_{qi}}{2\sum_{\mu=1}^{n} W_{q\mu}} + \frac{W_{qi}v_i^{\omega_c}(1+d_c^i)^{-2}}{2\sum_{\mu=1}^{n} W_{q\mu}v_\mu^{\omega_c}(1+d_c^\mu)^{-2}}, \tag{5}$$

where q is the index of the node being visited by particle ρ_j, c is the class label of particle ρ_j, $W_{qi} = 1$ if there is an edge between the current node and the node v_i, and $W_{qi} = 0$ otherwise. A particle will stay on the visited node only if, after applying (3), its class domination level is the largest on that node; otherwise, the particle is expelled and it goes back to the node it was before, staying there until the next iteration.

The stop criterion is defined as follows. Periodically, the highest domination level on each node is taken and their mean is calculated ($\langle v_i^{\omega_m}\rangle$, $m = \arg\max_c v_i^{\omega_c}$). This value usually has a quick increase in the first iterations, then it stabilizes at a high level and it starts oscillating slightly. At this moment, for each node v_i, if $v_i^{\omega_c} > 0.9$, then the class c is assigned to the corresponding pixel ($y(x_i) = c$). The remaining nodes (if any) will be labeled at a second phase.

The second phase is a quick iterative process, where each unlabeled pixel x_i adjusts its corresponding $\mathbf{v_i^\omega}$ as follows:

$$\mathbf{v_i^\omega}(t+1) = \frac{1}{a}\sum_{j\in\eta} \mathbf{v_j^\omega}(t)\,\mathrm{dist}(x_i, x_j), \tag{6}$$

where η is the subset of a adjacents pixels of x_i. $a = 8$, except in the borders where no wraparound is applied. $\mathrm{dist}(x_i, x_j)$ is the function that returns the Euclidean distance between features x_i e x_j, weighted by λ. Therefore, each unlabeled pixel receives contributions of the neighboring pixels, which are proportional to their similarity. The second phase ends when $\langle v_i^{\omega_m}\rangle$ stabilizes. Now unlabeled pixels finally receive their labels, $y(x_i) = \arg\max_c v_i^{\omega_c}(t)$.

3 Building Networks for PCC

As explained in Sect. 2, pixel features may be scaled by a vector of weights λ in order to emphasize/deemphasize each feature to the upcoming network generation step. Increased segmentation accuracy by PCC is expected with a proper choice of weights. Therefore, it is desirable to find methods to automatically define λ.

In [5], λ was optimized using a genetic algorithm [22,27], but the segmentation accuracy, measured comparing the algorithm output with ground truth images segmented by humans, was used as the fitness function. This approach was acceptable as proof of concept, but in real-world segmentation tasks ground truth images are not available. Thus, in [15], four methods were proposed to automatically adjust λ based on the data distribution for each feature and each class, with only the user labeled pixels considered. That approach led to some mixed results, three of the four methods were able to increase PCC segmentation accuracy, when compared to the results achieved without weighting the features, in at least one of the three tested images. But none of the four methods increased PCC segmentation accuracy for all the three tested images, which were extracted from the Microsoft GrabCut database [30].

In this paper, a different approach is proposed. Instead of evaluating individual features before the network construction, networks are built with some candidate values for λ. The resulting candidate networks are evaluated using a proposed network index. Therefore, finding a good λ becomes an optimization problem, where the proposed network index is maximized.

This approach has the advantage of considering all the features together, already weighted by the candidate λ. Therefore, individual features, which are not good discriminators alone and would be deemphasized in previous approaches, may be combined to produce a proper network for PCC segmentation.

The proposed network index ϕ, to be maximized, is calculated by analyzing the edges between labeled nodes in the candidate network. It is defined as follows:

$$\phi = \frac{z_i}{z_t}, \tag{7}$$

where z_i is the amount of edges between two labeled nodes representing the same class, and z_t is the total amount of edges between any labeled nodes, no matter which class they belong. Thus, ϕ is higher as the proportion of edges between nodes of the same class increases. Candidate networks with fewer edges between nodes representing different classes are desirable, since this is a clue that different classes data are well-separated in that network, making the PCC job easier.

Notice that theoretically $0 \leq \phi \leq 1$, but $\phi \approx 1$ in most practical situations, so the difference in ϕ for networks built with different λ may be very small. Therefore, an improved index α is defined as:

$$\alpha = \left(\frac{z_i}{z_t}\right)^{\sigma}, \tag{8}$$

where

$$\sigma = \frac{\ln(0.5)}{\ln(\varPhi)}, \tag{9}$$

with $\varPhi$ as the result of (7) when it receives a network built without any feature weighting, i.e., the same as if $\lambda = \{1, 1, \ldots, 1\}$. Notice that $0 \leq \alpha \leq 1$ with the differences in α for different choices of λ being much easier to notice then in σ. $\alpha < 0.5$ means that the choice of λ is probably bad and may lead to PCC accuracy worse than when it is applied to the features without any weighting. $\alpha > 0.5$ means the choice of λ is probably effective. The higher α is, more appropriate the network is expected to be. So, α is maximized to find a proper network to feed PCC.

Figure 1a shows a example of a candidate network. Suppose that it was built without any feature weighting. There are 27 nodes, 8 of them belong to the "blue" class, 8 of them belong to the "orange" class, and the remaining 11 nodes are unlabeled. There are 15 edges (colored green) connecting nodes from the same class and 5 edges (colored red) connecting nodes from different classes. Therefore, by applying (7), $\phi = \frac{15}{20} = 0.75$. Then, $\sigma = 2.4094$. The index α for the same network will be $\alpha = \left(\frac{15}{20}\right)^{2.4094} = 0.5$.

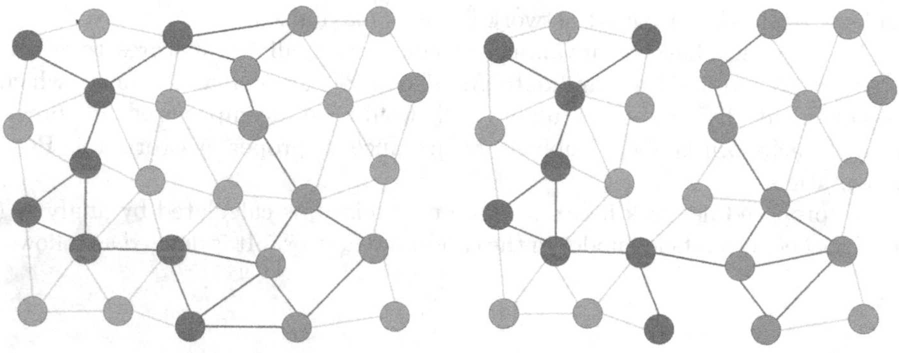

Fig. 1. Examples of candidate networks with 27 nodes. Labeled nodes are colored in blue and orange. Unlabeled nodes are colored gray. (a) 15 edges between nodes of the same class are represented in green, while 5 edges between nodes of different classes are represented in red. (b) 16 edges between nodes of the same class are represented in green, while a single edge between nodes of different classes is represented in red. (Color figure online)

Now, suppose that, during the optimization process, the network represented in Fig. 1b is built given a candidate λ. By applying (8), we have $\alpha = \left(\frac{16}{17}\right)^{2.4094} = 0.8641$. The higher α means that this network have higher class separability and it would probably allow PCC to achieve a higher classification accuracy then the network on Fig. 1a.

4 Experiments

In order to validate the proposed technique, three images were selected from the Microsoft GrabCut database [30]. The selected images, their *trimaps* providing seed regions, and the *ground truth* images are shown on Fig. 2. In the *trimaps*, black (0) represents the background, which is ignored; dark gray (64) is the labeled background; light gray (128) is the unlabeled region, which labels will be estimated by the proposed method; and white (255) is the labeled foreground.

In the first experiment, networks were built for each image without any weighting and with different values for the parameter k. PCC was applied to

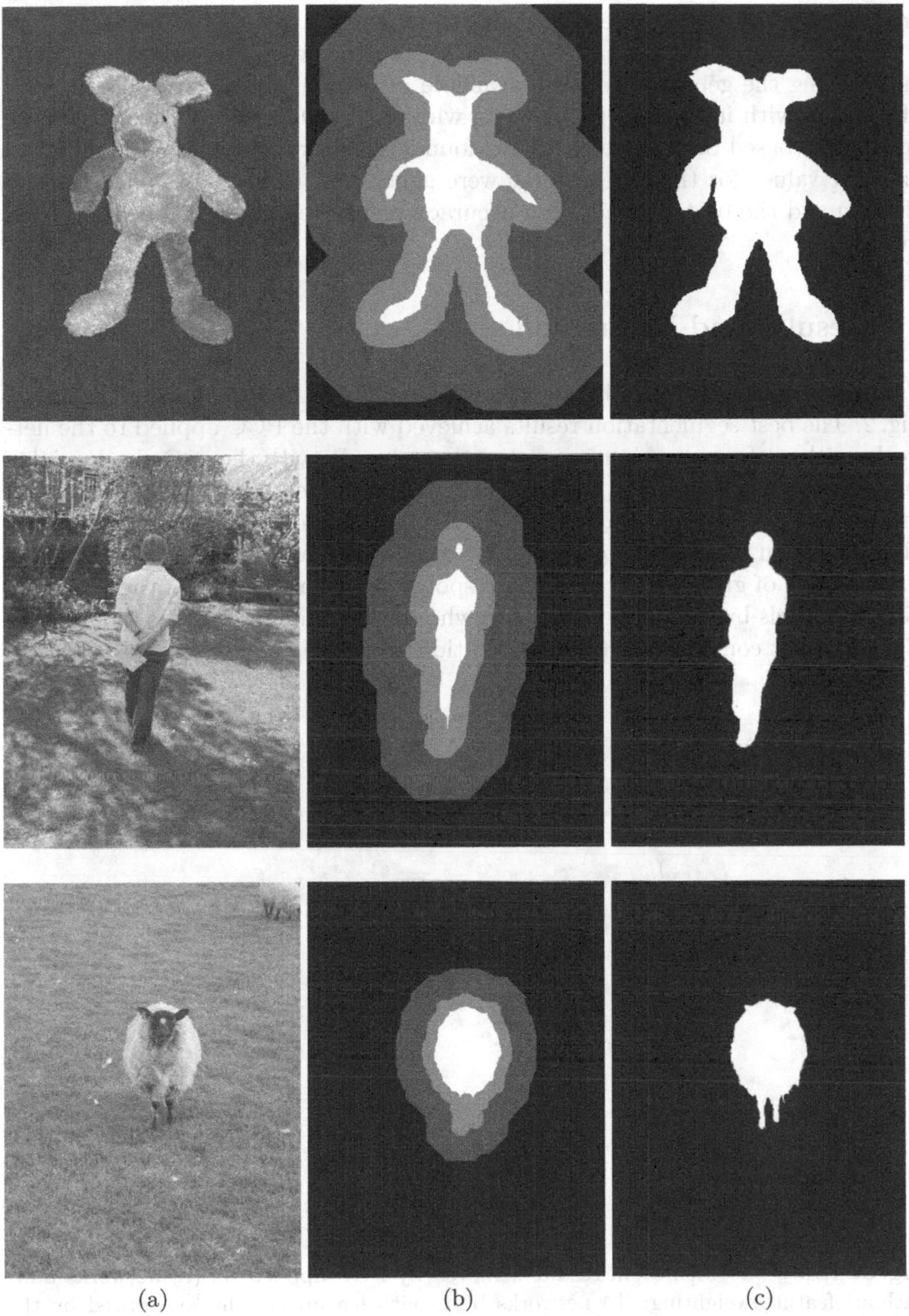

(a) (b) (c)

Fig. 2. (a) Original images from the GrabCut dataset, (b) the *trimaps* providing the seed regions, and (c) the original *ground truth* images.

each of them and the best segmentation accuracy result was taken for each image. These results are used as the baseline.

In the second experiment, for each image, the weight vector λ was optimized using the genetic algorithm available in Global Optimization Toolbox of MATLAB, with its default parameters, while $k = 100$ was kept fixed. Once the optimal λ (based on the index σ) was found, networks with the optimal λ and different values for the parameter k were generated. PCC was applied to each of them and the best segmentation accuracy result was taken for each image as well.

5 Results and Discussion

The experiments described in Sect. 4 were applied on the three images shown on Fig. 2. The best segmentation results achieved with the PCC applied to the networks without feature weighting and to the networks with the optimized weights are shown on Figs. 3, 4, and 5. Error rates are computed as the fraction between the amount of incorrectly classified pixels and the total amount of unlabeled pixels (light gray on *trimaps* images). Notice that *ground truth* images have a thin contour of gray pixels, which corresponds to uncertainty, i.e., they received different labels by the different persons who did the manual classification. These pixels are not computed in the classification error.

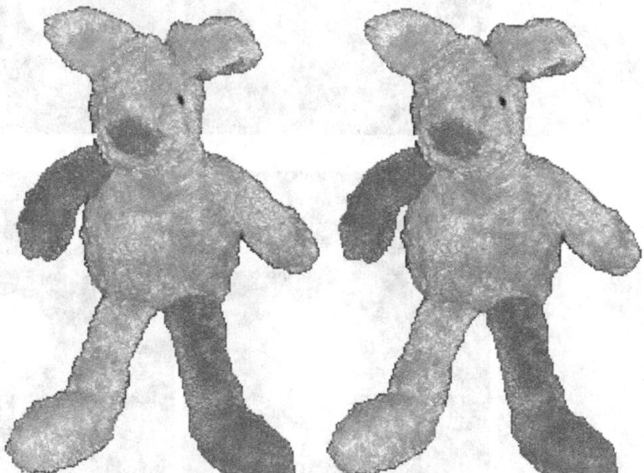

(a) Error: 1.89% (b) Error: 1.86%

Fig. 3. Teddy - Segmentation results achieved by PCC applied to: (a) networks built without feature weighting; (b) networks built with feature weights optimized by the proposed method

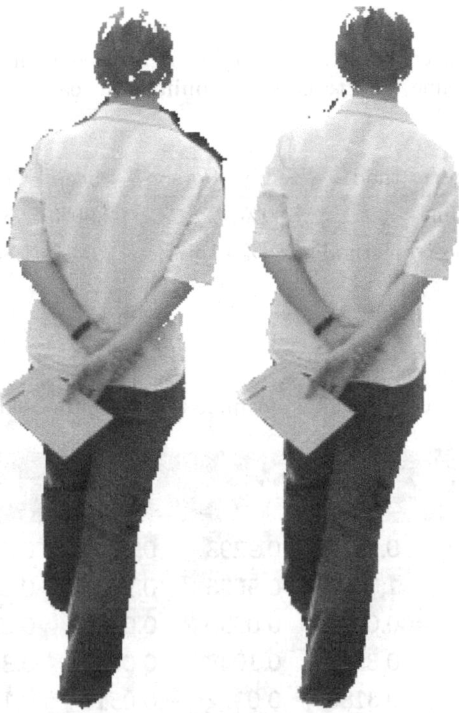

(a) Error: 2.81% (b) Error: 1.67%

Fig. 4. Person7 - Segmentation results achieved by PCC applied to: (a) networks built without feature weighting; (b) networks built with feature weights optimized by the proposed method

Segmentation error rates are also summarized on Table 1. By analyzing the results we notice that the feature weight optimization using the proposed method lead to lower segmentation error rates on the three tested images, showing its effectiveness.

The optimized indexes σ found for each image were 1.0 in all scenarios (32-bit float precision), which means they would probably improve the segmentation results, as they actually did. The networks generated for "teddy" easily reached $\sigma = 1.0$, as more than half of the random selected weights would lead to $\sigma = 1.0$. This explains why the first random generated weights (first individual) were returned by the genetic algorithm. On the other hand, "person7" and "sheep" took 40 and 164 generations, respectively, to finally reach $\sigma = 1.0$. Each generation has 200 individuals. The optimized features weights (λ) are shown on Table 2.

In the selected images, the row and the column of the pixels clearly are the most important features. Though the other features got lower weights in mean, the proper weights for each image were important to provide the decrease in classification error.

Table 1. Segmentation error rates when PCC is applied to networks built without feature weighting (baseline) and to networks built with feature weights optimized by the proposed method

Image/method	Teddy	Person7	Sheep	Mean
Baseline	1.89%	2.81%	2.90%	2.53%
Proposed method	1.86%	1.67%	2.04%	1.86%

Table 2. Feature weights optimized by the proposed method

Image / Feature	teddy	person7	sheep	Mean
Row	0.5377	0.9293	0.9908	0.8193
Col	1.0000	0.9686	0.9901	0.9862
R	0.0000	0.0550	0.0080	0.0210
G	0.8622	0.1048	0.0700	0.3457
B	0.3188	0.0372	0.0512	0.1357
H	0.0000	0.0476	0.0287	0.0254
S	0.0000	0.0186	0.0562	0.0249
V	0.3426	0.0977	0.0697	0.1700
ExR	1.0000	0.0732	0.0049	0.3594
ExB	1.0000	0.2085	0.0146	0.4077
ExG	0.0000	0.1051	0.1173	0.0741
MR	1.0000	0.0734	0.0237	0.3657
MG	0.7254	0.0674	0.0486	0.2805
MB	0.0000	0.0419	0.0408	0.0276
SDR	0.7147	0.1788	0.0145	0.3027
SDG	0.0000	0.0380	0.0042	0.0141
SDB	0.0000	0.0161	0.0377	0.0180
MH	1.0000	0.0363	0.2545	0.4303
MS	1.0000	0.1754	0.2584	0.4779
MV	1.0000	0.1079	0.0301	0.3794
SDH	0.6715	0.0098	0.1917	0.2910
SDS	0.0000	0.0239	0.1267	0.0502
SDV	0.7172	0.0787	0.0270	0.2743

(a) Error: 2.90% (b) Error: 2.04%

Fig. 5. Sheep - Segmentation results achieved by PCC applied to: (a) networks built without feature weighting; (b) networks built with feature weights optimized by the proposed method

6 Conclusion

In this paper, a new approach to build networks representing image pixels was proposed. The networks are used in the image segmentation task, using the semi-supervised learning method known as particle competition and cooperation (PCC). The approach consists in optimizing a proposed index which is calculated for each candidate network. The optimization process automatically calculates weights for the features which are extracted from the image to be segmented.

Computer simulations with some real-world images show that the proposed method is effective in improving segmentation accuracy, lowering pixel classification error. As future work, the method will be applied on more images and using more features, searching for some pattern on the images and the corresponding optimized weights. The index may also be improved to provide even better networks to feed PCC and further increase segmentation accuracy. The optimized feature weights might be used on similar images. Features with low weight might be excluded to improve execution time and segmentation accuracy. Finally, the method may be applied to images with less labeled pixels, like "scribbles" instead of "trimaps", since PCC is a semi-supervised method and does not require so many labeled data points.

Acknowledgment. The author would like to thank the São Paulo Research Foundation - FAPESP (grant #2016/05669-4) and the National Counsel of Technological and Scientific Development - CNPq (grant #475717/2013-9) for the financial support.

References

1. Artan, Y.: Interactive image segmentation using machine learning techniques. In: 2011 Canadian Conference on Computer and Robot Vision (CRV), pp. 264–269, May 2011
2. Artan, Y., Yetik, I.: Improved random walker algorithm for image segmentation. In: 2010 IEEE Southwest Symposium on Image Analysis Interpretation (SSIAI), pp. 89–92, May 2010
3. Blake, A., Rother, C., Brown, M., Perez, P., Torr, P.: Interactive image segmentation using an adaptive GMMRF model. In: Pajdla, T., Matas, J. (eds.) ECCV 2004. LNCS, vol. 3021, pp. 428–441. Springer, Heidelberg (2004). doi:10.1007/978-3-540-24670-1_33
4. Boykov, Y., Jolly, M.P.: Interactive graph cuts for optimal boundary & region segmentation of objects in ND images. In: Proceedings of the Eighth IEEE International Conference on Computer Vision, ICCV 2001, vol. 1, pp. 105–112 (2001)
5. Breve, F., Quiles, M.G., Zhao, L.: Interactive image segmentation using particle competition and cooperation. In: 2015 International Joint Conference on Neural Networks (IJCNN), pp. 1–8, July 2015
6. Breve, F.: Active semi-supervised learning using particle competition and cooperation in networks. In: The 2013 International Joint Conference on Neural Networks (IJCNN), pp. 1–6, August 2013
7. Breve, F., Quiles, M.G., Zhao, L.: Interactive image segmentation of non-contiguous classes using particle competition and cooperation. In: Gervasi, O., Murgante, B., Misra, S., Gavrilova, M.L., Rocha, A.M.A.C., Torre, C., Taniar, D., Apduhan, B.O. (eds.) ICCSA 2015. LNCS, vol. 9155, pp. 203–216. Springer, Cham (2015). doi:10.1007/978-3-319-21404-7_15
8. Breve, F., Zhao, L.: Particle competition and cooperation in networks for semi-supervised learning with concept drift. In: The 2012 International Joint Conference on Neural Networks (IJCNN), pp. 1–6, June 2012
9. Breve, F., Zhao, L.: Particle competition and cooperation to prevent error propagation from mislabeled data in semi-supervised learning. In: 2012 Brazilian Symposium on Neural Networks (SBRN), pp. 79–84, October 2012
10. Breve, F., Zhao, L.: Fuzzy community structure detection by particle competition and cooperation. Soft Comput. **17**(4), 659–673 (2013). http://dx.doi.org/10.1007/s00500-012-0924-3
11. Breve, F., Zhao, L., Quiles, M., Pedrycz, W., Liu, J.: Particle competition and cooperation for uncovering network overlap community structure. In: Liu, D., Zhang, H., Polycarpou, M., Alippi, C., He, H. (eds.) ISNN 2011. LNCS, vol. 6677, pp. 426–433. Springer, Heidelberg (2011). doi:10.1007/978-3-642-21111-9_48
12. Breve, F., Zhao, L., Quiles, M., Pedrycz, W., Liu, J.: Particle competition and cooperation in networks for semi-supervised learning. IEEE Trans. Knowl. Data Eng. **24**(9), 1686–1698 (2012)
13. Breve, F.A.: Combined active and semi-supervised learning using particle walking temporal dynamics. In: 2013 BRICS Congress on Computational Intelligence and 11th Brazilian Congress on Computational Intelligence (BRICS-CCI CBIC), pp. 15–20, September 2013
14. Breve, F.A.: Query rules study on active semi-supervised learning using particle competition and cooperation. In: Anais do Encontro Nacional de Inteligncia Artificial e Computacional (ENIAC), So Carlos, pp. 134–140 (2014)

15. Breve, F.A.: Auto feature weight for interactive image segmentation using particle competition and cooperation. In: Proceedings - XI Workshop de Viso Computacional WVC 2015, pp. 164–169 (2015)
16. Breve, F.A., Zhao, L.: Semi-supervised learning with concept drift using particle dynamics applied to network intrusion detection data. In: 2013 BRICS Congress on Computational Intelligence and 11th Brazilian Congress on Computational Intelligence (BRICS-CCI CBIC), pp. 335–340, September 2013
17. Breve, F.A., Zhao, L., Quiles, M.G.: Semi-supervised learning from imperfect data through particle cooperation and competition. In: The 2010 International Joint Conference on Neural Networks (IJCNN), pp. 1–8, July 2010
18. Breve, F.A., Zhao, L., Quiles, M.G.: Particle competition and cooperation for semi-supervised learning with label noise. Neurocomputing **160**, 63–72 (2015)
19. Chapelle, O., Schölkopf, B., Zien, A. (eds.): Semi-Supervised Learning. Adaptive Computation and Machine Learning. The MIT Press, Cambridge (2006)
20. Ding, L., Yilmaz, A.: Interactive image segmentation using probabilistic hypergraphs. Pattern Recogn. **43**(5), 1863–1873 (2010). http://www.sciencedirect.com/science/article/pii/S0031320309004440
21. Ducournau, A., Bretto, A.: Random walks in directed hypergraphs and application to semi-supervised image segmentation. Comput. Vis. Image Underst. **120**, 91–102 (2014). http://www.sciencedirect.com/science/article/pii/S1077314213002038
22. Goldberg, D.E.: Genetic Algorithms in Search, Optimization and Machine Learning, 1st edn. Addison-Wesley Longman Publishing Co. Inc., Boston (1989)
23. Gonzalez, R.C., Woods, R.E.: Digital Image Processing, 3rd edn. Prentice-Hall Inc., Upper Saddle River (2008)
24. Grady, L.: Random walks for image segmentation. IEEE Trans. Pattern Anal. Mach. Intell. **28**(11), 1768–1783 (2006)
25. Li, J., Bioucas-Dias, J., Plaza, A.: Semisupervised hyperspectral image segmentation using multinomial logistic regression with active learning. IEEE Trans. Geosci. Remote Sens. **48**(11), 4085–4098 (2010)
26. Lichman, M.: UCI machine learning repository (2013). http://archive.ics.uci.edu/ml
27. Mitchell, M.: An Introduction to Genetic Algorithms. MIT Press, Cambridge (1998)
28. Paiva, A., Tasdizen, T.: Fast semi-supervised image segmentation by novelty selection. In: 2010 IEEE International Conference on Acoustics Speech and Signal Processing (ICASSP), pp. 1054–1057, March 2010
29. Protiere, A., Sapiro, G.: Interactive image segmentation via adaptive weighted distances. IEEE Trans. Image Process. **16**(4), 1046–1057 (2007)
30. Rother, C., Kolmogorov, V., Blake, A.: "GrabCut": interactive foreground extraction using iterated graph cuts. ACM Trans. Graph. **23**(3), 309–314 (2004). http://doi.acm.org/10.1145/1015706.1015720 http://doi.acm.org/10.1145/1015706.1015720
31. Shapiro, L., Stockman, G.: Computer Vision. Prentice Hall, Upper Saddle River (2001)
32. Smith, A.R.: Color gamut transform pairs. ACM Siggraph Comput. Graph. **12**, 12–19 (1978). ACM
33. Xu, J., Chen, X., Huang, X.: Interactive image segmentation by semi-supervised learning ensemble. In: International Symposium on Knowledge Acquisition and Modeling, KAM 2008, pp. 645–648, December 2008
34. Zhu, X.: Semi-supervised learning literature survey. Technical report 1530, Computer Sciences, University of Wisconsin-Madison (2005)

QueueWe: An IoT-Based Solution for Queue Monitoring

Gibeon S. Aquino Jr., Cícero A. Silva(✉), Itamir M.B. Filho, Dênis R.S. Pinheiro, Paulo H.Q. Lopes, Cephas A.S. Barreto, Anderson P.N. Silva, Renan O. Silva, Thalyson L.G. Souza, and Tyrone M. Damasceno

Department of Informatics and Applied Mathematics, Federal University of Rio Grande Do Norte, Campus Universitário, Lagoa Nova, PO Box 1524, Natal, RN, Brazil
gibeon@info.ufrn.br, cicero@ppgsc.ufrn.br, itamir.filho@imd.ufrn.br, denis.rspinheiro@gmail.com, paulo.hq.lopes@gmail.com, cephasax@gmail.com, andersonpablo@hotmail.com.br, renan.osilva@hotmail.com, thalysonluiz@gmail.com, tyronedamasceno@gmail.com
http://www.dimap.ufrn.br

Abstract. Internet of Things allows people's everyday objects to be connected to the Internet and to each other, which gives them the ability to communicate between themselves and with the end users. This allows such objects to contribute to the improvement of the accomplishment of tasks such as: environment monitoring, people's health monitoring, natural resources management, and many other activities. This way, this work designs, implements and evaluates a solution for queue monitoring. This solution used as its first use scenario the University Restaurant (UR) of the Universidade Federal do Rio Grande do Norte (Federal University of Rio Grande do Norte - UFRN). The goal was to inform the restaurant's users the best times to go to the UR in order to avoid queues, consequently making better use of their time. To achieve this goal, a prototype of the solution was developed involving sensors, connectivity, a mobile application and an IoT platform.

Keywords: Internet of Things (IoT) · Sensors · Platform · Raspberry Pi · Android

1 Introduction

Nowadays, the use of smart devices and other types of sensors has been widely disseminated, mainly because they have a low cost, low power consumption, high information processing power and high connectivity capacity [1]. Thus, devices that are connected to the Internet surpassed the number of people in the world in 2011; by the year of 2013, 9 billion devices were connected to the large network [4]. In addition, it is estimated that this number reaches 25 billion in 2020 [3].

O. Gervasi et al. (Eds.): ICCSA 2017, Part I, LNCS 10404, pp. 232–246, 2017.
DOI: 10.1007/978-3-319-62392-4_17

This way, it is possible to note the fast dissemination of these devices and the fact that it is foreseeable that in the near future they will be even more present in people's daily lives. In this context, the Internet of Things (IoT) emerges, which represents a paradigm in which people's everyday objects are equipped to have the ability to communicate with each other and with users, which makes them part of the Internet [9]. Moreover, in this paradigm, the things connected to the network collaborate in the accomplishment of important tasks that will bring numerous benefits to everyone's lives [5].

Many places (banks, supermarkets, shops, etc.) face issues related to the existence of large queues at certain times of their operation, which results in loss of time for their users and, consequently, in dissatisfaction with the service provided. On the other hand, at times the flow of people in these environments is very low and they are underutilized. Thus, it is possible to identify the opportunity to develop solutions that can help users decide on the best time to attend such places. In particular, this problem is part of the daily life of UFRN's University Restaurant, which causes annoyances to students, teachers and administrative technicians who usually attend the place.

This way, the main goal of this article is to design, implement and evaluate a queue monitoring solution using UFRN's UR as the first test scenario. Therefore, the proposed solution intends to inform the UR's users of the times when there is a less flow of people in such environment, thus helping them to decide what is the best time to eat their meals.

The rest of this article is organized as follows: Section 2 presents UFRN's University Restaurant. Section 3 describes the proposed queue monitoring solution, which was named QueueWe. Section 4, in its turn, shows details related to the implementation of the solution and discusses the evaluation process. Finally, Sect. 5 discusses the conclusions and proposes future works for this study.

2 UFRN's University Restaurant

By analyzing some data gathered from UFRN's Superintendência de Informática (Informatics Superintendence - SINFO), it is possible to note the gradual increase in demand for the services offered by the University Restaurant. In 2016, 740,237 meals were served. This number reflects the use of the services provided by the Restaurant located in UFRN's main campus in the period of 01/02/2016 to 12/18/2016 and represents an increase of 13.24% over the same period in the previous year.

Currently, the University Restaurant operates in three periods of time a day, offering breakfast, lunch, and dinner to University staff, students, and administrative technicians. These times are [8]:

- Breakfast from 6:30 A.M. to 9:15 A.M. On Saturdays, Sundays and holidays from 7:30 A.M. to 8:30 A.M.;
- Lunch from 10:30 A.M. to 2:30 P.M. On Saturdays, Sundays and holidays from 11 A.M. to 1 P.M.;
- Dinner from 5 P.M. to 7 P.M. On Saturdays, Sundays and holidays from 5 P.M. to 6 P.M.

An eminent fact is that, with the expansion of UFRN's main campus, the Institution has won hundreds of new students. Each semester, new students join the University, making the demand for the services provided by the UR grow.

With this growing demand, the agglomeration of users at the entrance of the building is one of the most disturbing factors, since they usually have to face huge queues, as shown in Fig. 1.

Fig. 1. Queues outside the UR.

In addition, with the problem of agglomeration of people at the entrance of the University Restaurant, it is necessary to create a solution that serves those who usually enjoy the UR's service. Taking this into consideration, the idea to develop a system that facilitates the organization and monitoring of people in the UR's queue emerged. This way, the user would have real-time access to the status of the queue through an application made available via mobile device. With this tool in hand, the user has the possibility to choose the best time to go to the UR in order to avoid facing large queues. In addition, the UR's management could also monitor the queue and take actions to increase and improve the service in order to avoid uncontrolled queue increase. Besides that, it would also be possible to analyze historical data using analytics techniques [7], which could assist the management to continue to operate in order to maintain the expected demand.

3 QueueWe: IoT-Based Queue Monitoring Solution

Currently, users of the University Restaurant do not have a mechanism that enables them to find out the queue size at a given time without having to go to the venue. This makes it difficult for them to choose the best time to go there. Unable to get the information in advance, many end up going to the Restaurant, finding huge queues and consequently contributing to further increase the number of people waiting to be served, which causes them frustration and disruptions.

Considering all the reported problems related to the UR's queues, it was thought to develop a solution that would measure and show the restaurant queue's approximate size to its users in real time. The idea was that they could view this information from their mobile devices, such as smartphones. This way, university students would be able to assess the best time to go to the UR so as to make better use of their time and avoid possible annoyances.

One challenge for accomplishing this is that the restaurant's queue usually does not follow a well-defined logical pattern since there are no physical structures in place to guide it. In order for the project to work, we decided to build a structure that would make the queue pattern more organized so that it would follow a unique path, making the process of counting the people in the queue easier. Figure 2 illustrates the proposal.

Along the queue, several sensors are installed to detect presence. They communicate with the Kaa Project[1], the IoT platform chosen for the project, which acts as a middleware connecting other two parts of the solution: the intelligence application and the mobile devices, as shown in Fig. 3. The sensors use Kaa

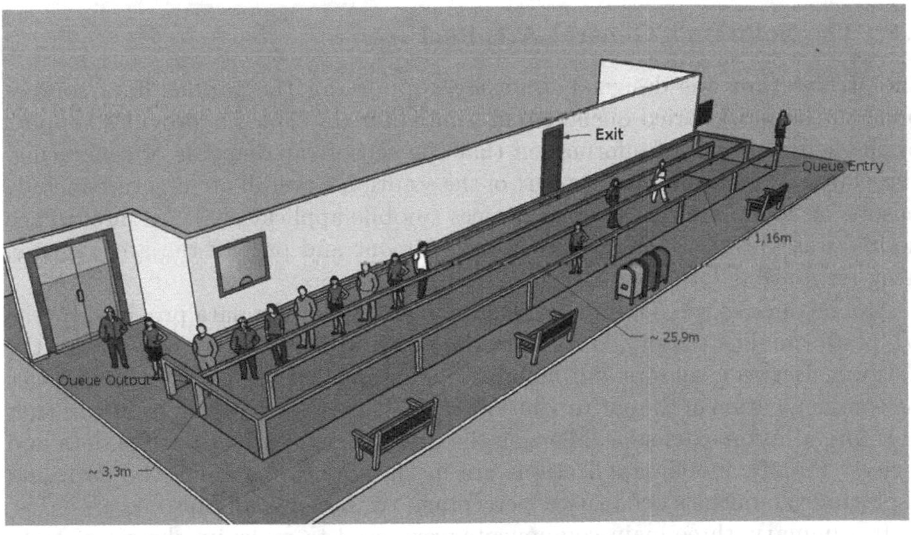

Fig. 2. UR's queue design.

[1] http://www.kaaproject.org/.

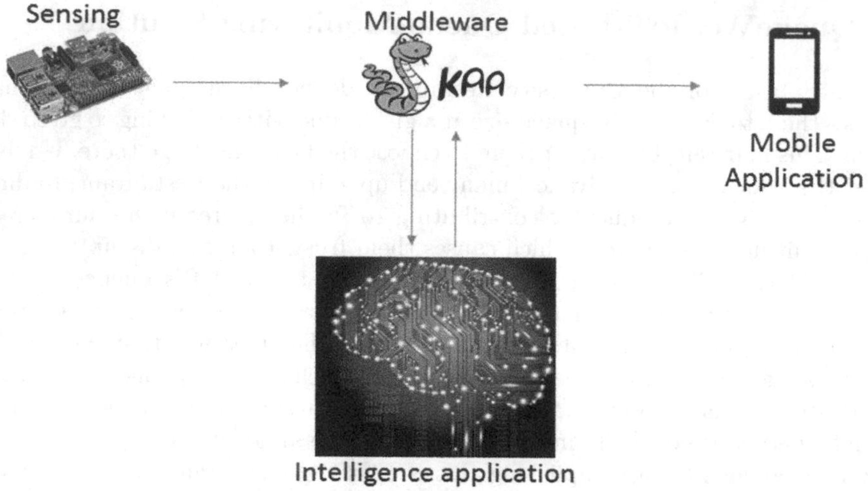

Fig. 3. Parts that make up the solution and their relationships.

Project to send information about people's presence to the intelligence application, which is responsible for collecting the data and estimating the queue size based on the readings performed by the sensors. The intelligence application communicates back to the Kaa Project sending information on the estimated queue size so that the Kaa Project can retransmit it to the mobile applications of users who have registered to receive information about the UR's occupation.

3.1 The Solution's General Architecture

The architecture was designed to ensure support for the various data sources found in the most varied queue monitoring contexts. Thus, it provides support for the heterogeneity of information that the sensors are capable of generating. Therefore, components that are part of the solution's overall architecture include sensors, an IoT platform, and end devices (mobile applications). An intelligence module was also designed as both a complement and part of the architecture, being responsible for processing data from the sensors.

Mobile devices will act exclusively as consumers for the data provided by the IoT platform. The sensors, in its turn, are responsible for providing data to the platform. Between sensors and mobile applications, the IoT platform manages the communication of the entire solution. It provides resources and protocols that make the solution extensible, allowing the integration of other types of data and devices. Finally, mobile applications are in charge of consuming the data and displaying the queue's occupation percentage to the application's users.

In summary, three main components were used to make up the set of technologies involved in the operation of the QueueWe solution: Raspberry Pi with UltraSound Sensors, Kaa Project platform, and mobile application. In addition, the intelligence module was used to process the data.

3.2 Technologies Involved

Among the components mentioned in the architecture, it was necessary to make decisions about which technologies had to be used to support and meet the project's initial requirements. Regarding the IoT platform, the Platform Kaa Project [2] was chosen because it is mentioned and used in several articles and because it has characteristics that are considered important for the project [6]. It was also identified that Kaa Project uses an alternative communication method, providing a Software Development Kit (SDK) for the application to communicate with the platform in an abstract way. This criterion was initially identified as positive for the development of the project, but some productivity issues were found with its use.

Kaa Project is opensource, so there is a community of developers who work for project growth. The community is very active, since all the doubts that we found in the development of QueueWe were solved in the own Kaa Project forum in a few hours. The average time to update documentation errors is one day, also the project evolution is fast, as they provide new versions of the platform in short times. Another reason is to allow the installation of a local machine with network configurations geared towards the project. The project sought to work with a platform that uses a data format that is consolidated and accepted among the existing ones [6]. Thus, JSON stands out as the standard data format to be used. Also, as an initial requirement for the project, it was necessary to work with a platform that provides good documentation quality and Kaa Project provided a complete documentation, allowing us to faster understand how it works [6].

The mobile application (application for end users) was developed on the Android platform, which was chosen due to its great popularity. Since the target audience of the QueueWe project is mostly made of students who use the UR, it is notable that Android impacts more potential users. Another reason to use Android is the smaller number of difficulties found in the beginning of the development process. It does not require large financial expenses to create an application and the number of software requirements is also low since you only need a computer with minimal settings to run Android Studio.

Regarding the **Raspberry Pi with presence sensors**, there was a variety of sensors that could be used in queue monitoring. However, since we are dealing with a specific region (linear) and not a radial area (beam), the strategy for using sensors was to assign checkpoints along the queue, identifying the points in which there are people. The UR's space is wide and open, so using a presence sensor, for example, could provide inaccurate data, leading to presenting users with wrong information. In the chosen strategy, using a presence sensor would lead to a situation where people close to the queue but outside it would activate the sensor, giving the impression that there were more people in line than it actually existed. Thus, among the existing sensors with specific region characteristics, precision tests were performed on UltraSound Sensors, proving them to have good indications of use. According to the tests, the maximum distance with acceptable accuracy was greater than the width of a queue, therefore its use seemed fair. Further details of its use are described in Sect. 4.2.

Finally, it was necessary to create an Intelligence Module for the project. It was identified, throughout the development of the work, that Kaa Project does not provide a data treatment/data analysis functionality in the version used (0.8.x). Thus, the system's intelligence module is responsible for synchronizing, identifying, organizing and converting the data into useful information for the Android application's users. It was developed in Java and installed on the same machine where the Kaa Project is configured. It receives data from the Raspberry Pi that are organized in a range and thus builds a snapshot of the queue. With this, it is possible to calculate and determine which points of the queue are occupied, which can give the user a final value that indicates the queue's occupation percentage.

By knowing the technologies that make up the project and its needs, it is easier to understand the operation of the entire solution.

3.3 The Solution's Operation

The solution proposed in this work works as shown in Fig. 4. As it is possible to see, QueueWe follows a set of five steps:

1. A certain group of sensors performs readings regarding the presence of objects in front of them. The data from these readings are routed to the devices (Raspberry Pi) and pooled until all of the sensors connected to the device have sent their readings;
2. This data, along with the identification of the device and its sensors, are sent to the IoT platform (Kaa Project);

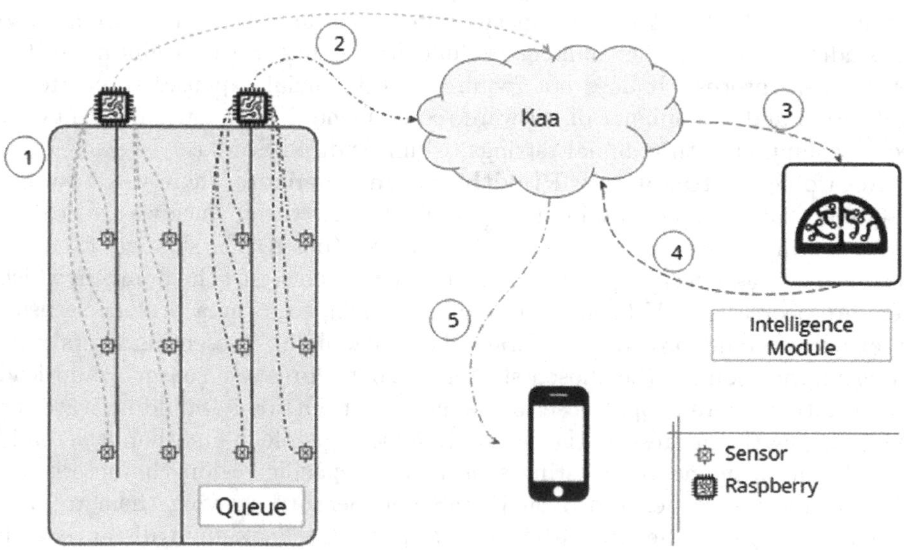

Fig. 4. The solution's operation.

3. Soon after, the IoT platform sends the received data to the Intelligence Module and it performs the necessary treatments and calculations so that the queue's occupation is correctly informed;
4. Then, the Intelligence Module sends the queue's occupation value to the IoT platform;
5. Finally, the IoT platform sends the queue's occupation percentage to the mobile application. The data is processed so that it is displayed to the user in a graphic format.

4 The Solution's Implementation and Evaluation

In this section, the proof of concept implemented in this study is described.

4.1 Proofs of Concept

Aiming to use technologies that make the development of this proposal easier, a survey of the IoT platforms available to carry out the treatment of information received from sensors and the communication with the Android client application was performed.

Table 1. Comparison of IoT Platforms

	AWS	Azure	Carriots	Fiware	Kaa	OpenIoT	ThingSpeak
Open source	No	No	No	Yes	Yes	Yes	No
Free	No	No	No	Yes	Yes	Yes	No
Local installation	No	No	No	Yes	Yes	Yes	No
Programming language	C, Node.Js	.NET, Java, C, Node.Js	Rest API	Java	Java	Java	Java, C, Node.js
Support	Active	Active	Active	Active	Active	Inactive	Active
Interoperability	Yes	Yes	Yes	Yes	Yes	Yes	Yes

After the survey, the Kaa Project platform was chosen because it is open source, it is implemented in a well-known and used language (Java), it has good support and documentation and it is free for development, as seen in Table 1.

Although the Fiware also meets many of the requirements, it was discarded for having a complex module implementation and for having high hardware requirements for the proposed implementation.

Regarding the sensors, UltraSound Sensors were chosen. In order to connect these sensors, the Raspberry Pi platform was used since it has its own operating system, as well as input and output components that make the development of the proposed solution easier. The tests performed aimed to identify the sensors' accuracy. The sensors underwent tests simulating different distances between them and the obstacles, which were able to identify an ideal value for the distances. With this value, it is possible to define the maximum width for the queue.

Table 2 presents the data obtained by varying the distance between the sensor and the barrier, which, in the UR queue's case, is a person; as well as analyzing how high the sensor would work better: if on the floor, 75 cm or 100 cm from the ground. It was possible to notice that the sensor can obtain good results in distances up to two meters; for longer distances, the results become inconsistent and the tests often generate errors.

Table 2. UltraSound sensor's accuracy

Distance (cm)/Height	100 cm	75 cm	Floor
60	57,7	77,9	79,5
120	116,3	125,5	135,4
180	174,8	184,5	200,1
240	~230	~190	~200
300	0	0	0

Upon reaching the 240 cm test, 50% of the data had reception failures; the rest of them were able to score approximately 230 cm. With these data, we can observe that the ideal positioning distance between the sensors and the obstacle is no greater than 200 cm and that the height in which the sensors were when they were able to get closer to informing the exact distance to the obstacle was 100 cm.

Based on the observations resulting from these tests, the queue's structure can be designed according to Fig. 2, with the sensors positioned at points that would improve the accuracy of the readings and consequently the accuracy of the solution as a whole.

4.2 Implementation

UltraSound Sensor. The set of UltraSound Sensors is responsible for maintaining sensing in the physical structure of the queue's handrails. Each sensor is positioned to represent the occupation range in a space. This is previously determined in accordance with the characteristics of the environment to be monitored, which, in this case, is the University Restaurant.

Individually, each sensor must behave like a flag that indicates the presence of users within the space where it is located. As a solution centered on viability, the HC-SR04 model was chosen, which has determinant characteristics and proven efficiency for distances between 2 cm up to 2 m, being able to capture presence in a controlled radius of 15° from the sound source when positioned at 1 m in height, which is satisfactory for the queue's structure. Each sensor is statically arranged in the queue so that it can represent an estimated range for a given number of people, thus forming the sensor range.

For the operation of each sensor, it is necessary to have 5 V working at 15 mA (on the sensor, Vcc port). Therefore, a single Raspberry Pi model B can be capable of delivering the energy requirement of up to 600 mA, making it possible to use the same power supply for several sensors. The same proposal is valid for grounding (on the sensor, GND port). Therefore, two ports of the Raspberry Pi device can respond to all sensors in the range.

All UltraSound Sensors should receive a request (on the sensor, trigger port) in the form of 8 40Khz pulses. This pulse triggers each of the elements in the sensor range so that they can independently hear some feedback. It is worth noting that for the queue's structure the pulse is expected to return positive for human presence only if it can hear the trigger request at a distance lower than the width between the handrails. In this case, each UltraSound Sensor occupies a previously controlled position in the Raspberry Pi. For the other positive values that are higher than the distance of the handrail (i.e., above the distance between handrails) and also for the negative values, the sensor considers that there is no human presence detected for the calculation of the queue's occupancy estimation.

Raspberry Pi. Due to the characteristics of the built-in UltraSound Sensor model, the Raspberry Pi acts as a mediator between the information collected by each UltraSound Sensor and the IoT platform. This equipment potentially has the following responsibilities:

- supply power to the sensors;
- send measurement requests to the sensor (trigger pulse);
- individually collect the response from each sensor (echo);
- calculate the distance between the intercepted object and the UltraSound Sensor considering the time difference between the request and the response;
- and, finally, send the collected data to the Kaa Project Platform.

This data is organized in a structure that contains the references for the microcontroller and each of its sensors. It is up to the RaspBerry Pi to send the collected data to the IoT platform where, in the future, it should be validated and the queue's occupancy values should be indicated.

Thus, the microcontroller is responsible for providing connectivity between the sensor range and the Kaa Project platform, as well as for providing the technical requirements for the operation of the sensors. For the prototype, the model B Raspberry Pi 2 was used with an ARM Cortex-A7 quad-core processor and 1 GHz of RAM memory, powered by a dual voltage source with 5 V output voltage and a maximum output current of 2A.

Kaa Project. It is a platform that is still in development, as well as an open source middleware supported by CyberVision Inc. The purpose of its use is to make possible the communication between devices that communicate in different ways, making the system interoperable. It has peculiar characteristics such as the generation of a Software Development Kit (SDK).

Kaa Project's SDKs are embedded in the device and implement real-time bi-directional data exchange with the server. These kits provide all the methods responsible for performing communication between the platform and other devices without having to thoroughly understand the communication data, for example, port connection, IP and transfer mode.

Middleware administration can be performed in two ways: through a User Interface (UI) or a REST API. By using these modes, it is possible to create users with different types of permissions, define profiles and configure communication modes.

These communication modes can be separated into three different types: Notification, Log, and Events. They are used for different purposes and follow particular data flows. For device management, the Notification is commonly used.

The Kaa Project notification feature allows the transfer of any data between the server and the endpoints. The structure of the data that is transported by the notifications is defined by the notification scheme configured in the Kaa Project server and embedded in the Kaa Project endpoints [2].

For storing data in the cloud, it is necessary to use the Log communication mode. It allows to extract data from devices and send them to the cloud. The data are stored in storage systems that are already used in scientific and technological environments, such as Cassandra, MongoDB and Oracle [2]. It is possible to note, therefore, that between these two types of Kaa Project communication there is no possibility of performing point-to-point communication between the devices, that is, to send data captured by a sensor, pass it through the server and deliver it to a mobile device. The only possible way to accomplish this is using Events, which was done in this solution.

The Kaa Project Events subsystem allows the generation of sensor data in real time by manipulating these events in a Kaa Project server and sending them to the mobile application that belongs to the same user [2]. The Kaa Project events structure is determined by an Event Class Schema (ECF).

In the QueueWe project, the ECFs were the *SensorEventCF* and the *QueueInfoEventCF*, which are respectively used to perform communication between the sensors and the Kaa Project and from the Kaa Project to mobile devices. It was also necessary to create two classes of events, one for each communication interval: *SensorEvent* and *QueueEvent*. Both classes of events are classified as the Record type, with the variable name message and String type. As described in Sect. 3.3, the data is written through the Raspberry Pi Application in the *SensorEvent* and later sent to the Kaa Project. Upon arriving at the Kaa Project, the Intelligence Module listens to this action, processes the data and writes it in the *QueueEvent*, thus sending it to anyone who listens to the *QueueEvent*, that is, Android applications.

As mentioned, the events communicate only among the same users. Therefore, it was also necessary to create different types of users to make the use of the application possible. There are four types of users: the platform's general administrator (*Administrator*); the application's administrator (*Tenant Administrator*); a user responsible for configuring the *EventClass* (*Tenant Developer*); and a user that was not used in our application nor described in the Kaa Project documentation (*Tenant User*). A single user checker, the *smartru.verifier*, was then created to be used in the entire application.

With this configuration, it was possible to generate the SDKs to be embedded in their individual applications. Thus, the SDK was generated for the Raspberry Pi Application, for the Intelligence Module, and for the Android Application.

Intelligence Module. The Intelligence Module is responsible for knowing the queue's configuration and form according to the arrangement of the sensors in the real environment and the calculation of the queue's occupancy percentage according to the physical layout chosen for the sensors.

For the correct functioning of the Intelligence Module, it was necessary to disregard the old idea of what a queue is and define the domain involved as follows: a queue is the static organization of several devices, which periodically sends readings from each device to the platform in the cloud (Kaa Project). The devices, in turn, are composed of several sensors that are responsible for performing the readings for a certain position in the queue. The need to formalize the concept of a queue for the application started with the goal of having a completely adaptable and reconfigurable solution, a characteristic that is under the exclusive responsibility of the Intelligence Module.

The Intelligence Module works by receiving readings from each device (Raspberry Pi) and observing the development of a queue occupation "snapshot" at a given time. If all the devices have sent their readings, the "snapshot" is considered complete and a queue occupation matrix is stored in a buffer. When a given number of queue occupancy matrices is obtained, the application performs parameterized validity calculations of the readings for each sensor and summarizes a final value for queue occupation.

It is necessary to mention that all of these steps are needed given the sensors' vulnerability to exception cases such as people who do not move along with the queue, sensor that stopped working, etc. With this, the Intelligence Module must deal with missing, noisy or invalid data and function properly even in these situations.

After the process described above, the queue's occupancy percentage calculated is sent back to the IoT platform (Kaa Project) for proper forwarding.

Android. The Android application is in charge of displaying queue size information to users. This information will only load when the application is being used, thus avoiding unnecessary processing and data consumption.

Kaa Project is in charge of sending queue information to all applications that are connected at that time. It does this using its SDK, in other words, through the *QueueInfoEventClassFamily* event family.

The application displays, in percentage, the size of the UR's queue. The percentage refers to the maximum capacity supported by the queue's physical structure. The values are displayed in graphics, which are accompanied by texts that indicate whether the queue is too full or not. Figure 5 illustrates this, showing different scenarios for queue status.

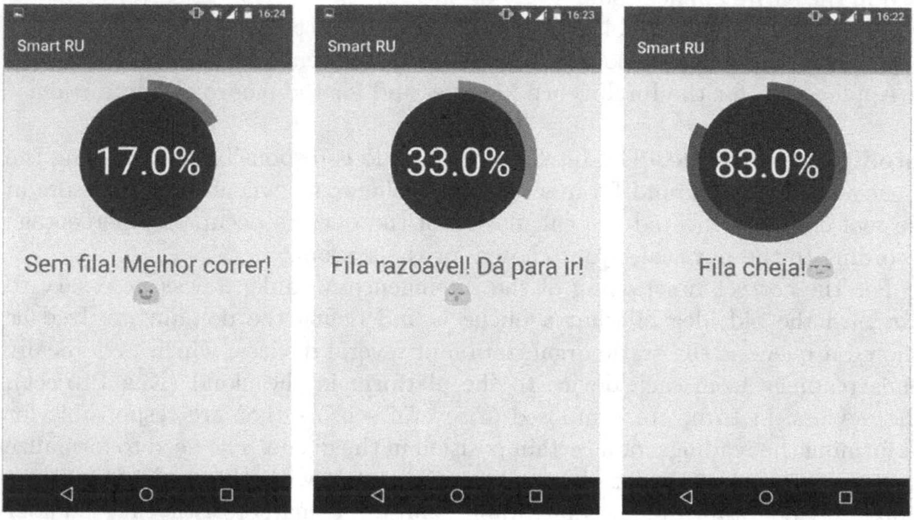

Fig. 5. UR's application with different queue statuses.

Figure 6 represents a real photograph of the prototype. In it, we can observe a basic example of how the queue structure will be when the system is implemented, as well as the positioning of the sensors, which can be seen in the Fig. 7.

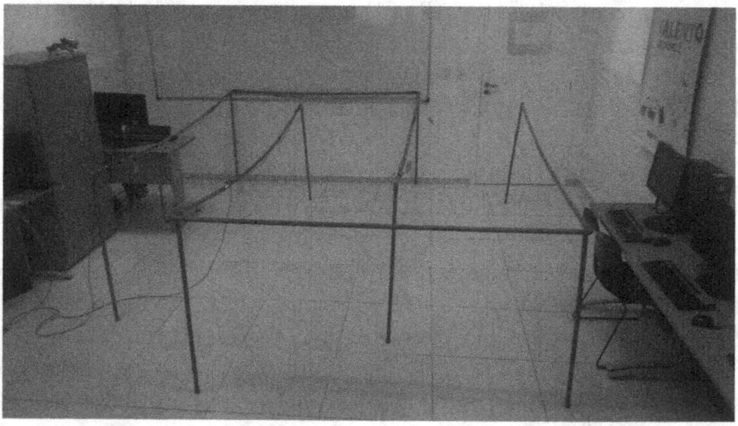

Fig. 6. Prototype photography

Fig. 7. UltraSound Sensor used in the prototype

5 Conclusions and Future Works

This article presents a queuing monitoring solution based on the IoT infrastructure, which uses as components UltraSound Sensors, microcontroller, IoT platform, intelligence module and an Android mobile application. The implementation of this solution was a challenge given the characteristics of the chosen IoT platform which, at the time of development, was in its initial version; and the specificity of the UltraSound Sensors considering resistance to dust and water, since they must be robust because they are exposed to an external environment. Regarding the challenge of using the IoT platform (Kaa Project), it was necessary to create the Intelligence Module to obtain information from the presence data captured in the queue. This module interprets the captured data and reports in real time the queue's current status.

As future works for this solution, we intend to carry out the evaluation of the implementation results and how it is helping to promote comfort for UFRN's University Restaurant users. We intend to perform a comparative analysis of our solution with others that have the same purpose. We also intend to use infrared sensors and RFID to calculate the occupation inside the UR. Finally, we intend to use images obtained from cameras as an alternative solution to using sensors. In this approach, we intend to use image processing techniques to estimate the number of people in the queue and inside the University Restaurant.

Acknowledgments. The authors would like to thank the Instituto Nacional de Engenharia de Software (National Institute of Software Engineering - INES), UFRN's Superintendência de Informática (Informatics Superintendence - SINFO) and Instituto Metrópole Digital (Digital Metropolis Institute - IMD), especially its IT Department (DTI-IMD) and the Inova Metrópole (Inova Metropolis) for the support provided at all stages of this work.

References

1. Cheng, L., Wu, C., Zhang, Y., Wu, H., Li, M., Maple, C.: A survey of localization in wireless sensor network. Int. J. Distrib. Sens. Netw. **8**(12), 1–12 (2012)
2. CyberVision. Kaa project. https://www.kaaproject.org/. Accessed 20 Dec 2016
3. Gartner. Gartner says 4.9 billion connected “things” will be in use in 2015. http://www.gartner.com/newsroom/id/2905717. Accessed 20 Dec 2016
4. Gubbi, J., Buyya, R., Marusic, S., Palaniswami, M.: Internet of things (IOT): a vision, architectural elements, and future directions. Future Gener. Comput. Syst. **29**(7), 1645–1660 (2013)
5. Maia, P., Batista, T., Cavalcante, E., Baffa, A., Delicato, F.C., Pires, P.F., Zomaya, A.: A web platform for interconnecting body sensors and improving health care. Proced. Comput. Sci. **40**, 135–142 (2014)
6. Silva, E.C.G.F., Oliveira, M.I.S., Oliveira, E., Gama, K.S., Lóscio, B.F.: Um survey sobre plataformas de mediação de dados para internet das coisas. In: Congresso da Sociedade Brasileira de Computação (CSBC). CSBC (2015)
7. Sun, G.-D., Ying-Cai, W., Liang, R.-H., Liu, S.-X.: A survey of visual analytics techniques and applications: state-of-the-art research and future challenges. J. Comput. Sci. Technol. **28**(5), 852–867 (2013)
8. RU UFRN. Ru em números. http://www.ru.ufrn.br/#funcionamento. Accessed 20 Dec 2016
9. Zanella, A., Bui, N., Castellani, A., Vangelista, L., Zorzi, M.: Internet of things for smart cities. IEEE Int. Things J. **1**(1), 22–32 (2014)

Inter-building Routing Approach for Indoor Environment

Tiara Annisa Dionti[1(✉)], Kiki Maulana Adhinugraha[1], and Sultan Mofareh Alamri[2]

[1] School of Computing, Telkom University, Bandung, Indonesia
tiaraannisadionti@students.telkomuniversity.ac.id, kikimaulana@telkomuniversity.ac.id

[2] College of Computing and Informatics, Saudi Electronic University, Dammam, Saudi Arabia
salamri@seu.edu.sa

Abstract. Routing system has been implemented in the outdoor routing and the indoor routing. There are significant differences that make indoor routing is more complex than outdoor routing, which is the outdoor routing implements two dimensional spaces, while at the indoor routing allows the routing of the three dimensional spaces that represent multi-level building. This research concern about the prototype development of the inter-building routing system. The construction of this prototype needs to consider both outdoor and indoor routing. Shortest path algorithms could be implemented after the construction of three dimensional spaces spatial data structure in order to inform the users about the shortest route between two points in indoor spaces.

Keywords: Spatial · Inter-building rout · Three dimensional spaces · Graph · Shortest path algorithm

1 Introduction

In recent years, a navigation system or outdoor routing systems like Google Maps is become very beneficial [1]. This navigation system is also implemented in a smaller area, like mapping on indoor spaces [2]. Even the needs of the construction of this indoor routing system is increasing continuously, especially for the public point of interest like malls, airports, offices, school, etc. [3–7]. With the indoor routing system, someone will be facilitated in finding any rooms since most individuals spend their lives in indoor environments [8–10]. However, the majority of users still use manual mapping system by displaying a room map plan in the building.

There are significant differences that make indoor routing more complex than outdoor routing, which is the outdoor routing generally implemented in two-dimensional spaces, while at the indoor routing allows the routing of the three-dimensional spaces that represent high rise building [11]. A large number of

O. Gervasi et al. (Eds.): ICCSA 2017, Part I, LNCS 10404, pp. 247–260, 2017.
DOI: 10.1007/978-3-319-62392-4_18

applications focus only on the region (not the exact coordinate location) [12,13]. A building may have numbers of rooms and numbers of corridors [14]. Each room allows for a variety of doors that connect the room with another spaces. And the building can be a multi-level building that has stairs, lifts, or elevators to move from one level to another. The entire spaces must have identified by labels as well as connectivity between the spaces.

The aim of this research is to develop the prototype of Inter-Building route system. We combine the outdoor routing system with indoor routing system. There are several kinds of liaisons between buildings like corridors, highways, tunnels, flyovers, etc. Three dimensional spaces could be a solution to build this system. We define and categorize the indoor distances between indoor uncertain objects [15]. This data structure will identify an object accurately by storing geographic data that are represented to undirected graph form which has three-dimensional attributes x, y, and z, where x is the longitude, y is the latitude, and z represents the height level of the points. For further research, this routing approach could be used to implement any kind of indoor indexing method. We intend to develop a new indoor indexing method and compare with others for the performance check.

This paper is organized as follow. Section 2 describes the data structure and how to build the prototype. The next section is about how to implement shortest path algorithms in the indoor routing system in three dimensional spaces. Section 4 presents the testing and analysis result of the system performance. Section 5 are the conclusion and future work of this research.

2 Inter-building Data Structure

Inter-building routing system is a routing system implementation in the buildings. The main problem in inter-building routing system is how to construct the data structure that represent the indoor spaces (Fig. 1).

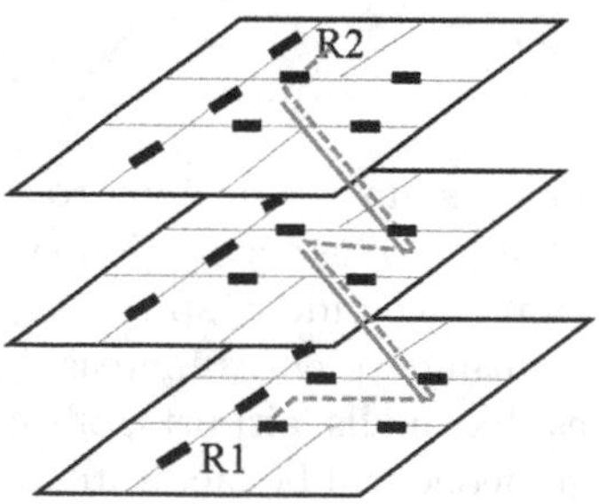

Fig. 1. Illustration of distance between two rooms.

By considering the illustration, it would be easy for human to find out the path to go from R1 to R2. But for the computer, it needs the exact data structure that can represent the rooms, the corridors, and also the stairs. The undirected graph can be implemented.

2.1 Three Dimensional Spaces

Consider the example in Fig. 2. That is the picture of how the three dimensional spaces could be implement in representing the buildings. As we can see here, a building could be not just in one level, but it also could be a multi-level building. Each nodes of rooms, stairs, corridors will have its own three dimensional attribute. x attribute is the longitude coordinate in decimal, y attribute is the latitude coordinate also in decimal, and z attribute is the height or floor level of the building which is in ordinal. The value of z could be negative that presents the underground building.

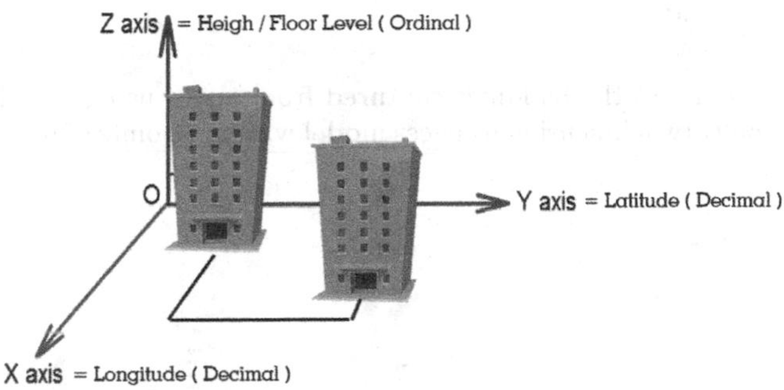

Fig. 2. Illustration of three dimensional spaces for inter-building routing.

And in Table 1, it is the idea of how to represent the indoor routing system using the undirected graph.

Table 1. The concept of indoor spaces modeling

Domain concept	Modeling concept
Room	A node
Door	An edge
Corridor	One or more nodes with one or more edges
Stair	One or more nodes with one or more edges
Elevator	One node with several edges
Pathway	One or more nodes with several edges

In this research, the case study is the area of Telkom University, Bandung, Indonesia especially the D, E, and F buildings in School of Computing, Telkom University. There are many kinds of buildings connector, such as: corridors, flyovers, tunnels, etc. In this case, there are two kinds of corridors that connect the buildings, the open corridors and the close corridors (Figs. 3 and 4).

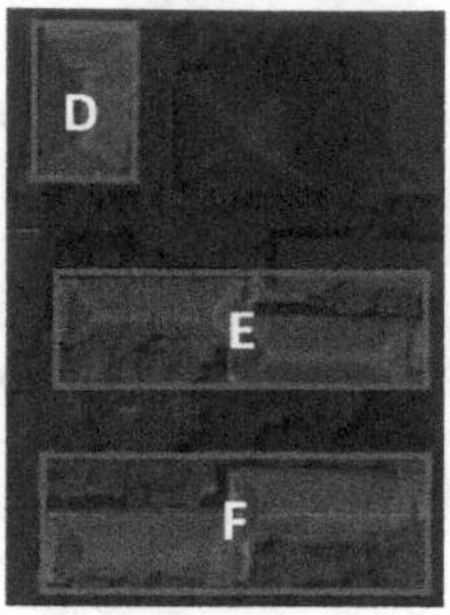

(a)

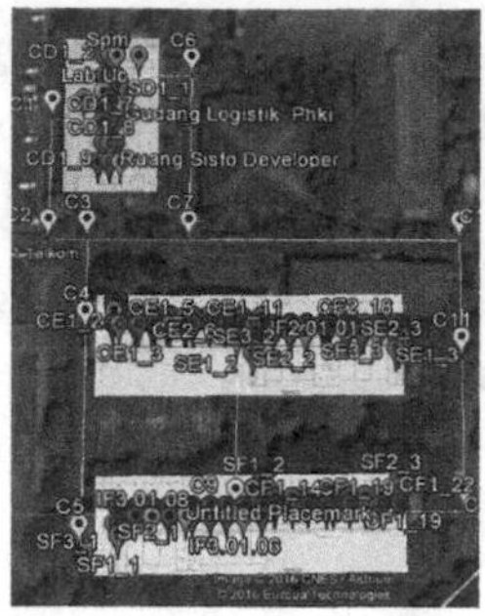

(b)

Fig. 3. (a) The picture of the buildings captured from above using Google Earth. (b) The overlaid of multi-two dimensional spaces model with the points of rooms, corridors, and stairs.

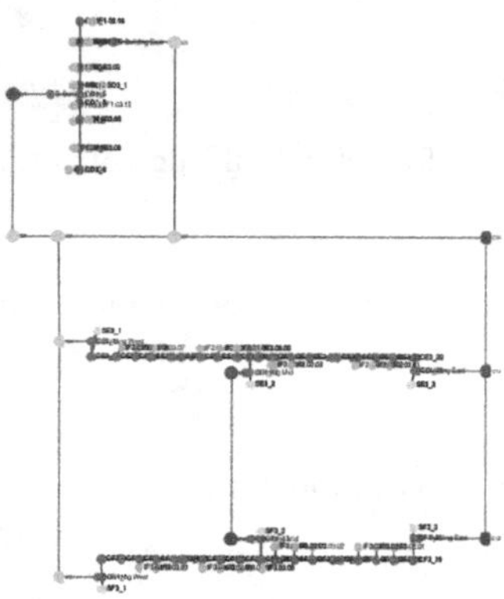

(a)

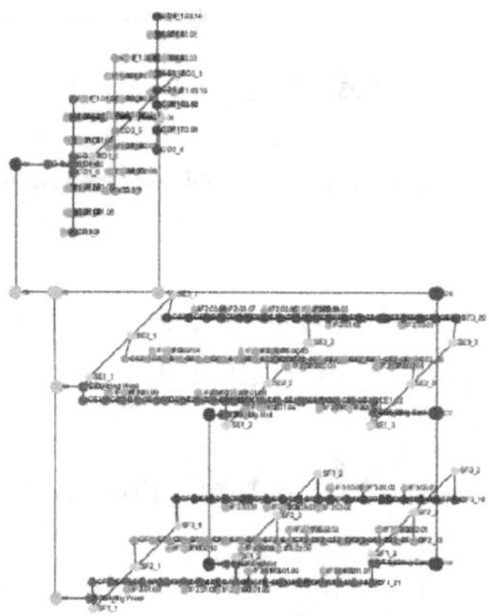

(b)

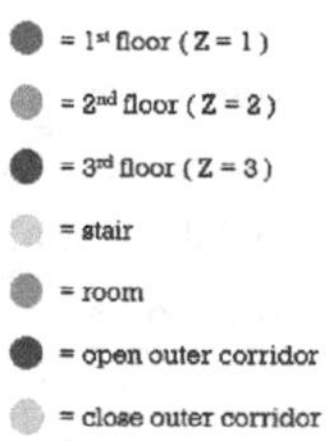

(c)

Fig. 4. (a) The overlaid of multi-two dimensional spaces undirected graph. (b) The same graph after coordinate shifting process. (c) The graph's legend.

3 Shortest Path Calculation

The routing system method by applying two dimensional spaces should also applicable to three dimensional spaces. Consider the main process of the system in Algorithm below[1].

Algorithm 1. Algorithm for indoor routing system

Input: Starting Point and Destination Point in
Output: The shortest route and the closed shortest route out
Initialisation :
```
Load datasets
Input Starting Point and Destination Point
Select the Shortest Path Algorithm
if (closed shortest path available) then
    return the shortest route and the closed shortest route
else
    return the shortest route
end if
System performance check
```

The system will be able to receive two inputs, the first input is the location of starting point and the second input is the destination point. The system would then output the shortest route from those points using the shortest algorithm. There are two kinds of shortest route, first is the shortest route that will consider all nodes in the graph, second is the closed shortest route that will eliminate all the open space nodes like the outdoor corridor that have no roof over it.

The small example of graph as shown in Fig. 5 is the graph that will be used it this section to be implement in each algorithm. Assume that the source node is node A and the destination node is node F.

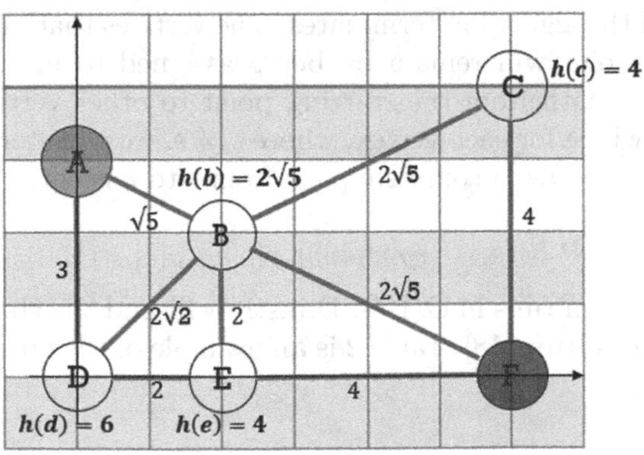

Fig. 5. Smaller graph example

[1] Datasets and source code available at http://bit.ly/2ql47JQ.

```
Dijkstra(G, w, s) G-Graph, w-weight, s-source
    for each vertex in Vertex[G]
        dist[v] ← ∞
        dist[s] ← 0
S ← empty set
while Vertex[G] is not an empty set
    u ← Extract_Min(Vertex[G])
    S ← S union {u}
    for each edge (u,v) outgoing from u
        if dist[u] + weight(u,v)< dist[v]
            dist[v] ← dist[u] + weight(u,v)
```

Fig. 6. Dijkstra's algorithm pseudocode

3.1 Dijkstra's Algorithm

Figure 6 is the pseudocode.

Dijkstras algorithm is an iterative procedure of dynamic programming functional equation associated with the shortest path problem given that the arc lengths are non-negative [16]. Dijkstras algorithm will do n iterations until all vertices have been visited. From the list of unvisited vertices we have to choose the vertex which has the minimum value at its label. We will choose a starting point s first. After that, we will consider all neighbors of this vertex. For each unvisited neighbor we will consider a new length, which is equal to the sum of the labels value at the initial vertex $v(d[v])$ and the length of edge l that connects them. If the resulting value is less than the value at the label, then we have to change the value in that label with the newly obtained value [17].

$$d[neighbors] = min(d[neighbors], d[v] + l) \tag{1}$$

After considering all of the neighbors, we will assign the initial vertex as visited ($u[v] = true$). After repeating this step n times, all vertices of the graph will be visited and the algorithm terminates. The vertices that are not connected with the starting point will remain by being assigned to infinity. In order to restore the shortest path from the starting point to other vertices, we need to identify array $p[]$, where for each vertex, where $v \neq s$, we will store the number of vertex $p[v]$. In other words, a complete path from s to v is equal to the following statement [17].

$$P = (s, \cdots, p[p[p[v]]], p[p[v]], p[v], v) \tag{2}$$

Dijkstra's algorithm runs in $O(V^2)$. Dijkstra will find the shortest path from a single source to all vertex [18]. Table 2 is an example of the output of Dijkstra's algorithm.

3.2 Floyd-Warshall Algorithm

Figure 7 is the pseudocode of Floyd-Warshall algorithm. Consider the graph G that contains n vertices. Notation d_{ijk} means the shortest path from i to j that

Table 2. Dijkstra's algorithm result

Step	Visited	A	B	C	D	E	F
0		0	∞	∞	∞	∞	∞
1	A	0	$\sqrt{5}$	∞	3	∞	∞
2	A, B	0	$\sqrt{5}$	$3\sqrt{5}$	**3**	$2+\sqrt{5}$	$3\sqrt{5}$
3	A, B, D	0	$\sqrt{5}$	$3\sqrt{5}$	3	$2+\sqrt{5}$	$3\sqrt{5}$
4	A, B, D, E	0	$\sqrt{5}$	$3\sqrt{5}$	3	$2+\sqrt{5}$	$3\sqrt{5}$
5	A, B, D, E, F	0	$\sqrt{5}$	$3\sqrt{5}$	3	$2+\sqrt{5}$	$3\sqrt{5}$
6	A, B, E, D, F, C	**0**	$\sqrt{5}$	$3\sqrt{5}$	**3**	$2+\sqrt{5}$	$3\sqrt{5}$

passes through vertex k. If there is exists edge between vertices i and j, it will be equal to d_{ij0}, otherwise it can assigned as infinity. The value of d_{ijk} will be equal to d_{ijk-1} if the shortest path from i to j does not pass through the vertex k. But if the shortest path from i to j passes through the vertex k then first it goes from i to k, after that goes from k to j. In this case the value of d_{ijk} will be equal to $d_{ikk-1} + d_{kjk-1}$. And in order to determine the shortest path we need to find the minimum of these two statements [17].

$$d_{ij0} = the\ length\ of\ edge\ between\ vertices\ i\ and\ j \tag{3}$$

$$d_{ijk} = min(d_{ijk-1}, d_{ikk-1} + d_{kjk-1}) \tag{4}$$

```
The Floyd-Warshall Algorithm
Floyd-Warshall (G,w,s) G-Graph, w-weight, s-source
init path[ ][ ];
for k = 1 to n
    for each (i, j) in (1 ... n)
        path[i][j] = min(path[i][j],
        path[i][k]+path[k][j]);
end
```

Fig. 7. Floyd-Warshall algorithm pseudocode

Floyd-Warshall algorithm has three times looping of the vertex which mean it runs in $O(V^3)$. Floyd-Warshall will find the shortest path for all pairs shortest path [19] as shown in Table 3 as the result of this case.

3.3 Bellman-Ford Algorithm

Bellman-Ford algorithm acknowledges the edges with negative weights. However the graph we have in this research does not contain any cycles of negative weights.

Table 3. Floyd-Warshall algorithm result

	A	B	C	D	E	F
A	0	2.236068	6.708204	3	4.236068	6.708204
B	2.236068	0	4.472136	2.828427	2	4.472136
C	6.708204	4.472136	0	7.300563	6.472136	4
D	3	2.828427	7.300563	0	2	6
E	4.236068	2	6.472136	2	0	4
F	6.708204	4.472136	4	6	4	0

Consider we have array $d[]$, it will store the minimal length from the starting point s to other vertices. The algorithm consists of several phases, where in each phase it needs to minimize the value of all edges by replacing $d[b]$ where b are vertices of the graph to following statement $d[a]+c; a$ where c is the corresponding edge that connects them. The length of all shortest paths requires $n1$ phases, but the value of elements of the array will remain by being assigned to infinity for those vertices of a graph that are unreachable, [17].

Figure 8 is the pseudocode of Bellman-Ford algorithm.

```
Bellman-Ford(G,w,s) G-Graph, w-weight, s-source
for each vertex in Vertex[G]
    dist[v] ← ∞
    dist[s] ← 0
for i ← all vertices in G do
    for each edge (u,v) in Edge[G] do
        Relax (u,v,w)
    for each edge (u,v) in Edge[G] do
        if dist[v] > dist[u] + weight(u,v) then return false
    return true
```

Fig. 8. Bellman-Ford algorithm pseudocode

Bellman-Ford algorithm runs in $O(VE)$. Just like Dijkstra's algorithm, Bellman-Ford will find the shortest path from a single source node to all nodes [20]. The output of this algorithm for this case is shown in Table 4.

Table 4. Bellman-Ford algorithm result

	A	B	C	D	E	F
A	0	2.236068	6.708204	3	4.236068	6.708204

3.4 A* Algorithm

The A* use a heuristic value to be considered in the algorithm. A heuristic value is a value that assume as the best length from a vertex to the goal. The formula itself is expressed as:

$$f(n) = g(n) + h(n) \tag{5}$$

$f(n)$ is the estimate of the best solution that goes through n, $g(n)$ is the actual cheapest cost of arriving at from the start node to n and $h(n)$ is the heuristic estimate of the cost to the goal from n. The time complexity is increased depending on the number of nodes and edges in the graph [18]. Figure 9 is the pseudocode.

```
The AStar Algorithm
A*(G,s,w) G-Graph, w-weight, s-source
var closed = the empty set
var q = make_queue(path(start))
while q is not empty
    var p = remove_first(q)
    var x = the last node of p
    if x in closed
        continue
    if x = goal
        return p
        add x to closed
        foreach y in successors(p)
            enqueue(q, y)
return failure
```

Fig. 9. A* algorithm pseudocode

A* runs in $O(V)$. A* find the shortest path for a single source node to a single destination node as shown in Fig. 10 as the result of this case.

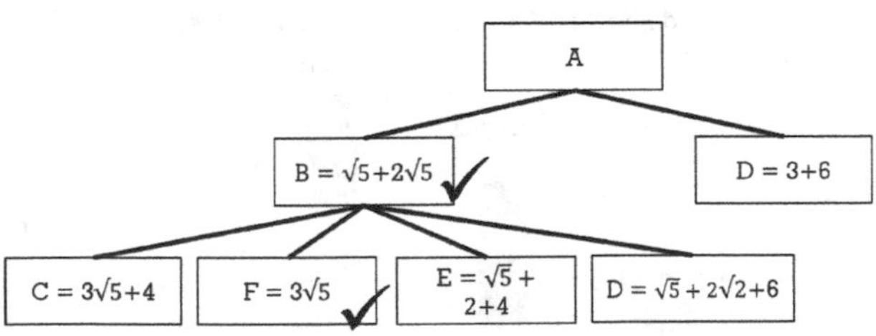

Fig. 10. A* algorithm output

4 System Testing and Analyzing

The result of this testing and analyzing process will show how good the performance of the system is. There are two aspects that will be the concern in this step, they are execution time and the correctness of the outputs. We use Java programming to implement the algorithms. The specification of the device is Windows 7 Professional 64 bit Intel Core i7-3770 CPU with 4 GB RAM. Also we use NetBeans IDE 8.1 with standard Java platform JDK 8.

4.1 Time Execution Testing Result

Here are the result of time execution testing. For the testing sample, the shortest path between node IF1.01.01 to node IF3.03.07 were chosen as a sample. The testing do the searching process five times for each algorithm. All time values are in seconds. Consider the chart in Fig. 11 that shows the comparison in shortest path calculation which has 272 nodes to be calculated. Figure 12 shows the comparison in closed shortest path calculation which has 266 nodes to be calculated. Figure 13 shows the comparison in both shortest path and closed shortest path calculation since the system will give the result of both in once it runs. From

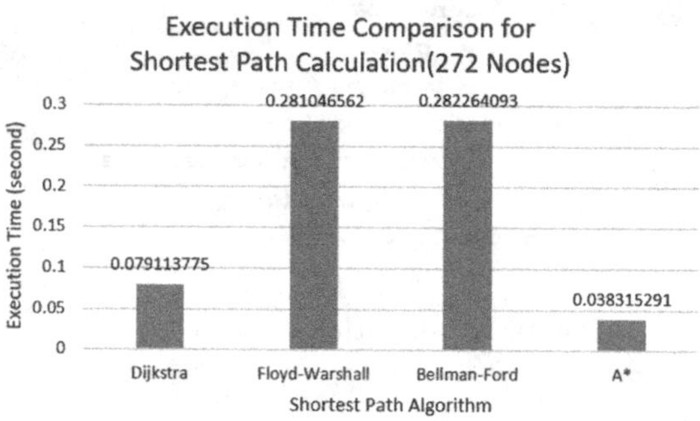

Fig. 11. Execution time comparison for shortest path calculation

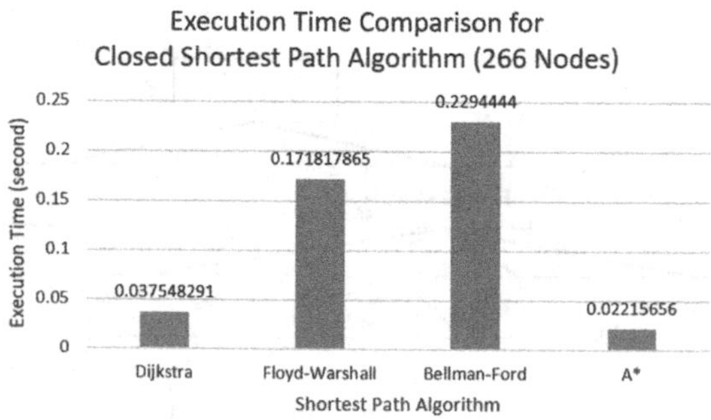

Fig. 12. Execution time comparison for closed shortest path

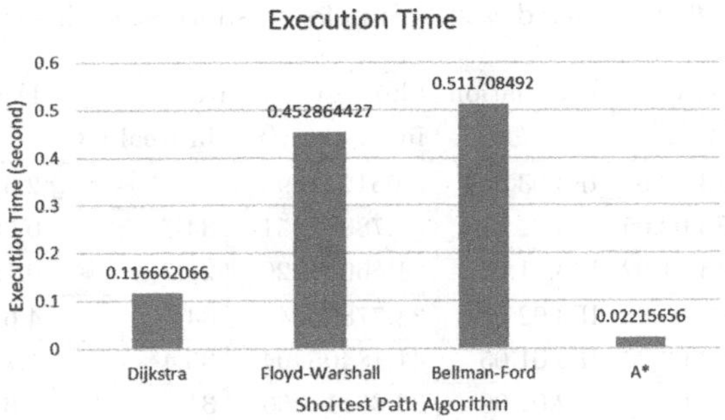

Fig. 13. Execution time comparison in system

those three charts we can see that A* always give us the lower execution time than Dijkstra's algorithm. It could be concluded that A* is more suitable to be implemented in this Indoor Routing System case.

4.2 Distance Result Testing

Here are the result of the system's accuracy testing. This testing will show the deviation between the result of system calculation and the real distance measurements as shown in Tables 5 and 6. All the distances are in meters.

Figures 14 and 15 shows the fluctuation of the testing deviation result. From the chart, we can see that the deviation result is quiet fluctuate. Because of that, we can use the formula of sample standard deviation denoted by s. By using this

Table 5. Distance deviation in shortest path search

No.	Scenario	Source	Destination	Shortest distance		Deviation (|S - R|)
				in System (S)	in Real (R)	
1	Same	IF3.03.04	IF3.03.05	8.661221996	6	2.661221996
2	Building,	IF3.02.05	IF3.02.01	34.78388051	34.7	0.083880508
3	Same Z	IF3.01.02	IF3.01.03	21.56047226	21.7	0.139527743
4	Same	IF3.03.03	IF3.02.05	38.7783564	34.1	4.6783564
5	Building,	IF3.03.03	IF3.01.05	41.18405404	35.4	5.784054043
6	Different Z	IF2.01.10	IF2.02.09	36.66114963	31	5.661149626
7	Different	IF2.01.05	IF3.01.05	69.19729964	69.3	0.102700358
8	Building,	IF2.01.10	IF3.01.08	102.1953267	94	8.195326717
9	Same Z	IF2.01.10	IF1.01.01	120.0752443	114.2	5.875244347
10	Different	IF3.01.05	IF2.02.05	78.71397663	79.6	0.886023374
11	Building,	IF2.02.01	IF3.01.02	120.9040472	113.1	7.804047212
12	Different Z	IF3.03.04	IF2.01.05	87.21359285	87.6	0.386407153
AVERAGE						3.521495

Table 6. Distance deviation in closed shortest path search

No.	Scenario	Source	Destination	Shortest distance		Deviation (|**S** - **R**|)
				in System (S)	in Real (R)	
1	Same Building, Same Z	IF3.03.04	IF3.03.05	8.661221996	6	2.661221996
2		IF3.02.05	IF3.02.01	34.78388051	34.7	0.083880508
3		IF3.01.02	IF3.01.03	21.56047226	21.7	0.139527743
4	Same Building Different Z	IF3.03.03	IF3.02.05	38.7783564	34.1	4.6783564
5		IF3.03.03	IF3.01.05	41.18405404	35.4	5.784054043
6		IF2.01.10	IF2.02.09	36.66114963	31	5.661149626
7	Different Building, Same Z	IF2.01.05	IF3.01.05	168.010905	168.8	0.789095019
8		IF2.01.10	IF3.01.08	102.1953267	94	8.195326717
9		IF2.01.10	IF1.01.01	140.9867135	135.5	5.48671352
10	Different Building, Different Z	IF3.01.05	IF2.02.05	177.5121245	177.8	0.287875482
11		IF2.02.01	IF3.01.02	234.2925391	234.6	0.307460939
12		IF3.03.04	IF2.01.05	187.8782596	184.8	3.07825964
AVERAGE						3.096076803

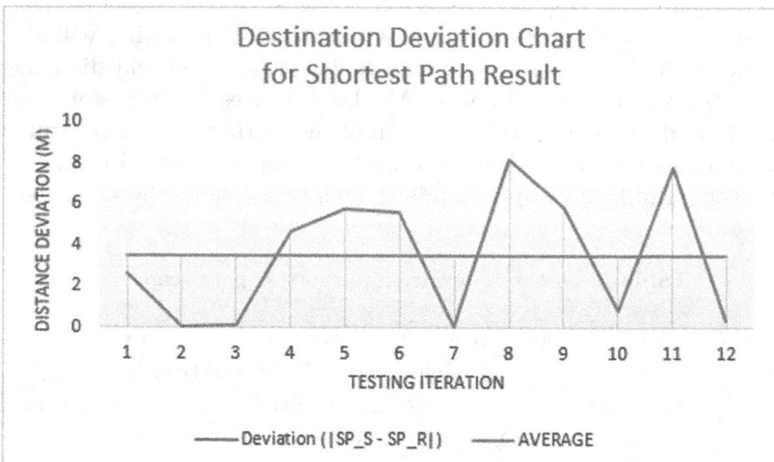

Fig. 14. Distance deviation testing result for shortest path

formula, we get the standard deviation by the sample data distribution that is **3.142778377203** meters. The standard deviation formula is equal to the following statement.

$$s = \sqrt{\frac{1}{N-1} \sum_{i=1}^{N} (x_i - \bar{x})} \tag{6}$$

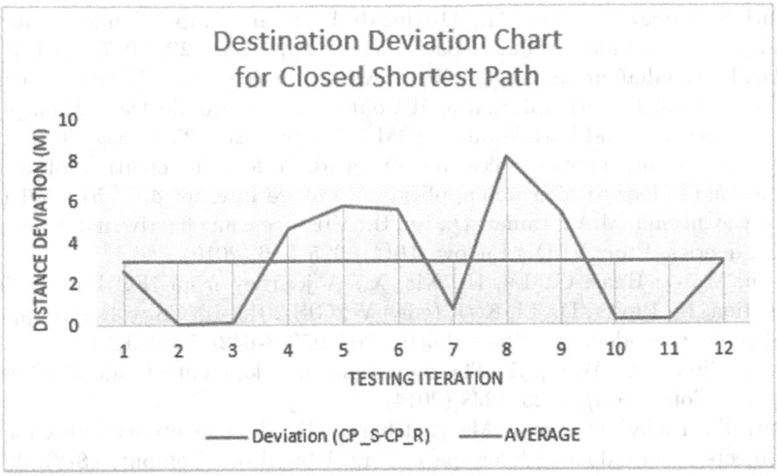

Fig. 15. Distance deviation testing result for close shortest path

5 Conclusion

Indoor Routing System could be implemented using three dimensional spaces. The most suitable Shortest Path Algorithm to the three dimensional spaces data between Dijkstra and A* Algorithms is A* Algorithm that give the lowest value of execution time. This System still lack of resulting the exact distance after it compared with the distance in real life. There are a few reason of this drawbacks such as the process in getting the nodes coordinate is still not accurate enough and for buildings that have a bit length of stairs to connect to the outer space, it assumed as one height level, so there will be a little bit difference between the calculation result and the measurement result. In this case, the calculation result could be called as an approximate distance. The shortest path through the specific spaces could be obtained by eliminating the unwanted nodes (In this case, we eliminate the open space nodes to get the closed shortest path). There are a lot of suggestions from the authors about the continuation of this research. More research needed to obtain the exact longitude and latitude coordinate of the points. It would be better if the graph could be viewed with the 3D blueprints of the buildings. This system could be implemented in more spacious area to become an inter-area rout system with more kinds of liaison.

References

1. Gotlib, D., Gnat, M., Marciniak, J.: The research on cartographical indoor presentation and indoor route modeling for navigation applications. In: 2012 International Conference on Indoor Positioning and Indoor Navigation (IPIN), pp. 1–7. IEEE (2012)
2. Afyouni, I., Cyril, R., Christophe, C.: Spatial models for context-aware indoor navigation systems: a survey. J. Spat. Inf. Sci. **1**(4), 85–123 (2012)

3. Alamri, S., Taniar, D., Safar, M., Al-Khalidi, H.: Spatiotemporal indexing for moving objects in an indoor cellular space. Neurocomputing **122**, 70–78 (2013)
4. Dudas, P.M., Ghafourian, M., Onalin, H.A.K.: Ontology and algorithm for indoor routing. In: 2009 Tenth International Conference on Mobile Data Management: Systems, Services and Middleware, MDM 2009, pp. 720–725. IEEE (2009)
5. Goetz, M.: Using crowdsourced indoor geodata for the creation of a three-dimensional indoor routing web application. Future Internet **4**(2), 575–591 (2012)
6. Shao, Z., Cheema, M.A., Taniar, D., Lu, H.: VIP-Tree: an effective index for indoor spatial queries. Proc. VLDB Endow. **10**(4), 325–336 (2016)
7. Boysen, M., de Haas, C., Lu, H., Xie, X.: A journey from IFC Files to indoor navigation. In: Pfoser, D., Li, K.-J. (eds.) W2GIS 2013. LNCS, vol. 8470, pp. 148–165. Springer, Heidelberg (2014). doi:10.1007/978-3-642-55334-9_10
8. Han, L., Zhang, T., Wang, Z.: The design and development of indoor 3D routing system. J. Softw. **9**(5), 1223–1228 (2014)
9. Alamri, S., Taniar, D., Safar, M., Al-Khalidi, H.: A connectivity index for moving objects in an indoor cellular space. Pers. Ubiquitous Comput. **18**(2), 287–301 (2014)
10. Alamri, S., Taniar, D., Safar, M.: Indexing moving objects in indoor cellular space. In: 2012 15th International Conference on Network-Based Information Systems (NBiS), pp. 38–44. IEEE (2012)
11. Güting, R.H.: An introduction to spatial database systems. VLDB J. Int. J. Very Large Data Bases **3**(4), 357–399 (1994)
12. Alamri, S., Taniar, D., Safarb, M., Al-Khalidi, H.: Tracking moving objects using topographical indexing. Concurr. Comput.: Pract. Exp. **27**(8), 1951–1965 (2015)
13. Alamri, S., Taniar, D., Safar, M.: A taxonomy for moving object queries in spatial databases. Future Gener. Comput. Syst. **37**, 232–242 (2014)
14. Goetz, M., Zipf, A.: Formal definition of a user-adaptive and length-optimal routing graph for complex indoor environments. Geo-Spat. Inf. Sci. **14**(2), 119–128 (2011)
15. Xie, X., Lu, H., Pedersen, T.B.: Distance-aware join for indoor moving objects. IEEE Trans. Knowl. Data Eng. **27**(2), 428–442 (2015)
16. Shehzad, F., Shah, M.A.A.: Evaluation of shortest paths in road network. Pak. J. Commer. Soc. Sci. **3**, 67–79 (2009)
17. Magzhan, K., Jani, H.M.: A review and evaluations of shortest path algorithms. Int. J. Sci. Technol. Res. **2**(6), 99–104 (2013)
18. Sathyaraj, B.M., Jain, L.C., Finn, A., Drake, S.: Multiple uavs path planning algorithms: a comparative study. Fuzzy Optim. Decis. Making **7**(3), 257–267 (2008)
19. Hougardy, S.: The floyd-warshall algorithm on graphs with negative cycles. Inf. Process. Lett. **110**(8–9), 279–281 (2010)
20. Goldberg, V.A., Radzik, T.: A heuristic improvement of the bellman-ford algorithm. Appl. Math. Lett. **6**(3), 3–6 (1993)

A CFD Study of Wind Loads on High Aspect Ratio Ground-Mounted Solar Panels

Giovanni Paolo Reina and Giuliano De Stefano(✉)

Dipartimento di Ingegneria Industriale e dell'Informazione,
Università della Campania, Aversa, Italy
{giovannipaolo.reina,giuliano.destefano}@unicampania.it

Abstract. Computational fluid dynamics is used to study the wind loads on a high aspect ratio ground-mounted solar panel. Reynolds-averaged Navier-Stokes simulations are performed using a commercial finite volume-based code with two different numerical approaches. First, the entire panel is directly simulated in a three-dimensional domain. Then, a small portion of the panel is considered, by imposing periodic boundary conditions in the spanwise homogeneous direction. The comparison shows a good match between the results obtained with the two different models, in terms of pressure coefficient and aerodynamic loads. The main consequence is a considerable reduction of the computational costs when using the reduced model.

Keywords: Computational fluid dynamics · Solar panel · Wind loads

1 Introduction

In recent years, solar power has been strongly emerging as the first source of alternative energy for industrial applications. The main reasons lie in the simplicity of the electricity production process, the possibility of reducing global warming and air pollution and the continuously decreasing price of photovoltaic (PV) panels.

However, the efficiency of PV cells represents a challenging issue, since the maximum value reached by the large-area commercial cells is about 24% [1]. In the industrial field, a widespread solution to the problem of low efficiency is the use of solar farms, which are large-scale PV power stations capable of generating large quantities of electricity. In these plants, even thousands of ground-mounted solar panels are connected to form systems of length equal to tens of meters.

One of the most important goals of the engineering research on PV systems is the analysis of the aerodynamic loads acting on both the panels and their support structures. However, from the experimental point of view, the study of this ground-mounted systems is very complex, since the characteristic spatial scales that are involved require the realization of very small models for boundary layer wind tunnel tests.

Among the others, important experimental findings were obtained by Stathopoulos et al. [2], who studied the aerodynamic loads on a stand-alone

O. Gervasi et al. (Eds.): ICCSA 2017, Part I, LNCS 10404, pp. 261–272, 2017.
DOI: 10.1007/978-3-319-62392-4_19

PV panel for different configurations and inclinations. The pressure field on the upper and lower surfaces of a single panel for different wind directions was investigated by Abiola-Ogedengbe et al. [3], while the effect of lateral and longitudinal spacing between panels on the wind loading of a ground-mounted solar array was studied by Warsido et al. [4].

Also due to the difficulties encountered in wind tunnel testing, the Computational Fluid Dynamics (CFD) analysis of PV ground-mounted systems has recently become an important predictive tool. For instance, Aly and Bitsuamlak [5] investigated the sensitivity of wind loads on the geometric scale of a stand-alone panel. Jubayer and Hangan [6] carried out unsteady Reynolds-averaged Navier-Stokes (URANS) simulations in order to investigate the wind load and the flow field features for a ground-mounted stand-alone PV system immersed in the atmospheric boundary layer (ABL), with a variable wind direction.

The main goal of this work is the CFD study of the wind loads on high aspect ratio ground-mounted solar panels that are commonly used in solar farms. Two different numerical approaches are followed. First, the entire panel is directly simulated in a three-dimensional domain. The second approach consists in the simulation of a small portion of the panel by imposing periodic boundary conditions in the spanwise homogeneous direction.

2 Computational Domain and Grid

The geometric model investigated in this work corresponds to a high aspect ratio ground-mounted solar panel. The PV system, which is typical of solar farms, consists of 36 panels arranged in a single row. The dimensions of each row are 1.2 m in length, 2 m in width, and 0.007 m in thickness, so the overall dimensions of the whole system are 43.2 m (L) and 2 m (C). The panel is supported by means of six columns, modeled as bars 1 m high with square cross section, equally spaced by 7.2 m. The analyses are conducted at two inclination angles of the PV panel with respect to the horizontal wind direction: $\theta = -25°$ and $\theta = 25°$.

The gaps between the single panels in the real geometry are neglected in the realization of the CFD model, since their effect is considered to be negligible, as suggested by Wu [7]. The dimensions of the final computational domain are 32 C (length), 6.3 C (height) and 21.6 C (width). The distances of the boundaries from the panel are given in Fig. 1. The entire model is created with similar characteristics to the geometry described in [6], which is chosen as main reference for the validation of the present approach.

A sub-domain with a length of 3.6 C is obtained from the full model, as depicted in Fig. 2, in which the geometric scale is enlarged with respect to the previous figure. The two side faces, obtained by cutting vertically the previous domain, and therefore also the panel, have an equal distance of 1.8 C from the center of one column along the z-axis direction. This model is used in the hypothesis of infinite panel, imposing periodicity to the two side surfaces in the CFD solver.

An unstructured mesh is generated throughout the fluid domain in both models with a minimum cell size of 1×10^{-3} m. The only exception is in areas

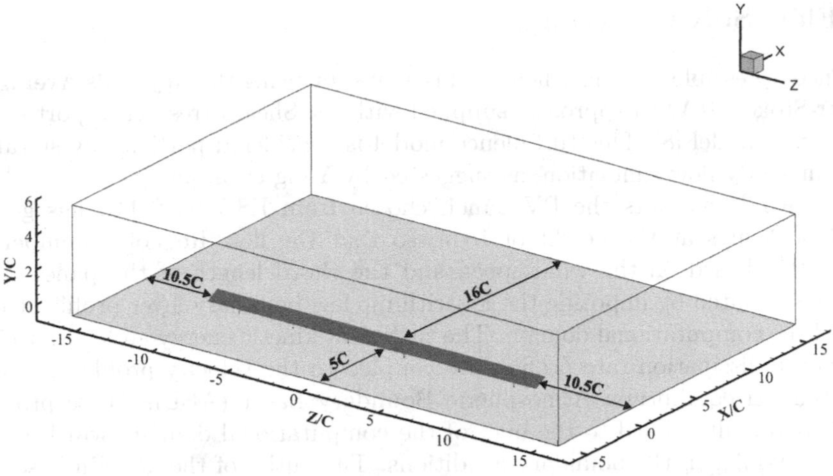

Fig. 1. Computational domain for the complete model of the PV panel

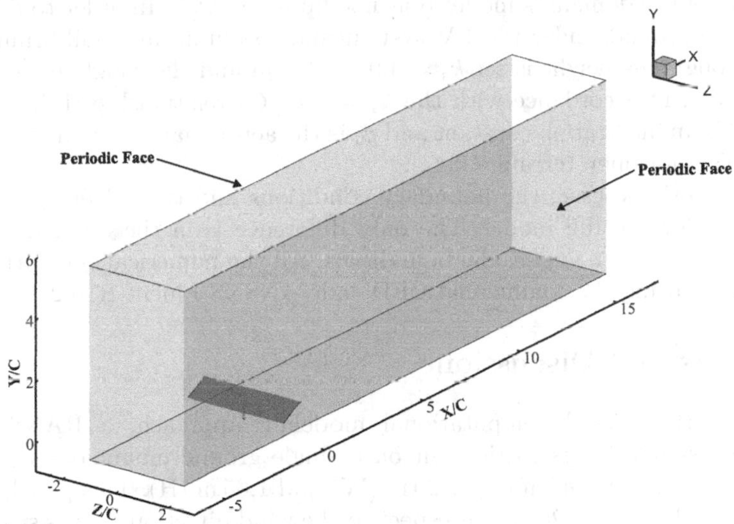

Fig. 2. Computational domain for the periodic model of the PV panel

close to the solid walls (panel surfaces, columns and ground), where 20 layers of hexahedral cells are inflated in order to obtain a value of $y^+ \leq 1$, so as to better calculate the boundary layer. The overall numerical grid involves about 12 million computational cells for the full model, and about 2 million cells for the periodic one. In order to impose the periodicity of the model, in the second domain the two side faces are matched. This way, the nodes on each side are coincident with those of the other side, so as to obtain a global mesh that can be virtually repeated infinite times along the longitudinal direction.

3 Flow Solver Settings

The incompressible turbulent flow field is solved by using the Reynolds Averaged Navier-Stokes (RANS) approach, supplied with the Shear Stress Transport $k-\omega$ turbulence model [8]. This turbulence model is considered particularly suitable for blunt body flow simulation, as suggested by Yang et al. [9].

The wind that hits the PV panel, chosen from ESDU [10,11], has got a speed of 26 m/s at the height of 10 m, so that the flow Reynolds number is 3.56×10^6, based on the wind speed and the chord length of the panel. The wind is simulated by imposing the logarithmic law boundary layer profile at the inlet of the computational domain. The turbulent kinetic energy and the specific turbulence dissipation rate profiles are coupled to the velocity profile, in order to obtain an equilibrium Atmospheric Boundary Layer (ABL). These profiles are therefore all applied to the inlet of the computational domain, which is set as velocity-inlet in the boundary conditions. The outlet of the domain is set as pressure-outlet, the sides and upper boundaries as symmetry. Regarding solid walls, the panel and the columns surfaces are all set as no-slip smooth walls, while the bottom of the domain is modeled as no-slip rough wall. In order to correctly simulate the ground under the PV system and to obtain an equilibrium ABL flow, the roughness height is set $k_s = 4.9 \times 10^{-6}$ m and the roughness constant $C_s = 6 \times 10^4$, in accordance with the $k_s = Ey_0/C_s$ relationship [12], in which $E = 9.793$ is an integration constant and y_0 is the aerodynamic roughness length, that is 0.03 m for open terrain [10,11].

For the periodic case, the boundary conditions and the solver settings are the same as for the full model. The only difference is in the two sides of the domain, which are set as periodic boundaries. All the numerical simulations are performed by using the commercial CFD code ANSYS Fluent R16.2.

4 Results and Discussion

To validate the overall computational modeling approach, a RANS three-dimensional simulation is carried out on a single ground-mounted solar panel, which dimensions are 2.48 m (B), 7.29 m (W) and 1.65 m (H). This panel, shown in Fig. 3, is inclined by $-25°$ with respect to the wind direction. It was studied in two previous works by Jubayer and Hangan [6] and Abiola-Ogedengbe et al. [3], both numerically and experimentally. The pressure coefficient along the mid-line of the panel surface determined from the present simulation is compared with the previous results in Fig. 4. The diagram shows a good agreement for the upper surface of the panel with both reference data, while the curve of the pressure coefficient on the lower surface is more similar to the wind tunnel result than the numerical one.

On the other hand, in Table 1, the aerodynamic coefficients are compared with that ones obtained by Jubayer [6]. It can be seen that the drag and the lift coefficients have a percentage error that is fully acceptable.

After this preliminary validation, the analyses on the present models are performed at two angles of inclination, that are $\theta = -25°$ (shown in Fig. 5)

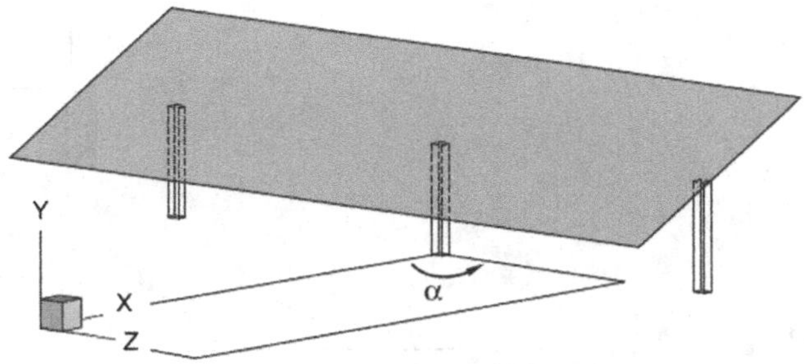

Fig. 3. Numerical model of the PV panel (from Jubayer and Hangan [6])

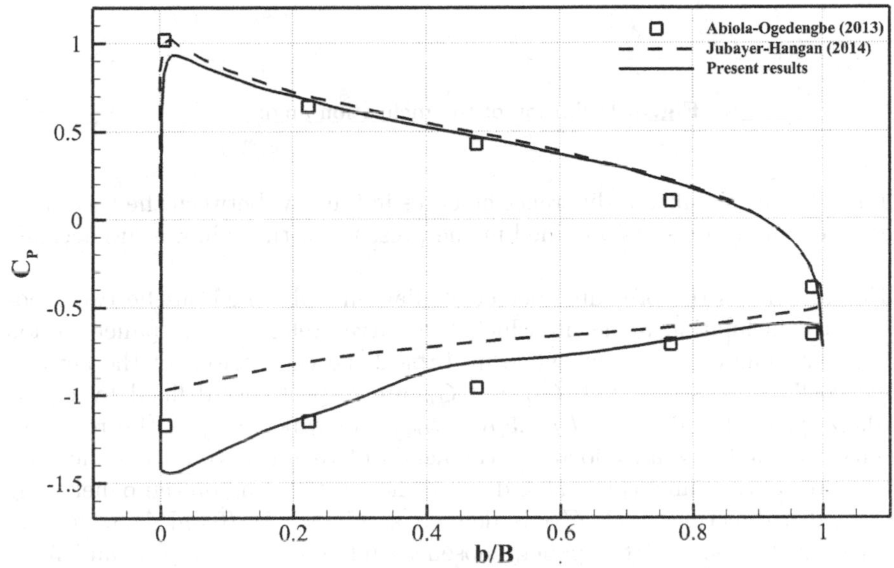

Fig. 4. Comparison of C_P for the validation case

Table 1. Aerodynamic coefficients comparison with the CFD study of Jubayer and Hangan [6]

	C_D	C_L	C_M
Present results	0.56	−1.20	0.11
Previous results	0.57	−1.24	-
Error	−1.7%	−3.2%	-

and $\theta = 25°$ with respect to the wind direction. Each calculation is made on a parallel workstation by employing 2 Intel Xeon 2.20 GHz processors and it took about 12 h for the full models and about 1 h and a half for the periodic ones. The simulations are carried out with and without considering the supporting

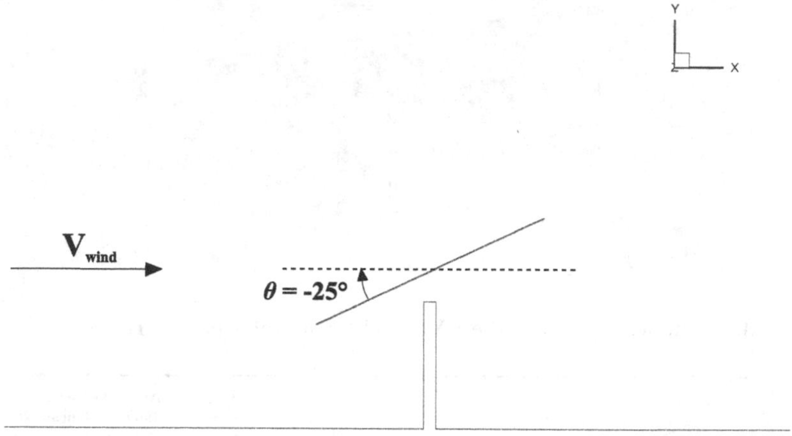

Fig. 5. Definition of the inclination angle θ

columns, but no significant difference emerges in the flow between the two cases. Therefore, only the results obtained in the presence of the columns are actually given.

Figure 6 shows the pressure coefficient diagrams obtained on the two models at each inclination angle, in which the curves refers to the panel section at the centerline of a column, while in Table 2 the comparison of the aerodynamic coefficients is reported. C_D and C_L are respectively defined by $C_D = D/(0.5\rho V_{ref}^2 S_{ref})$ and $C_L = L/(0.5\rho V_{ref}^2 S_{ref})$, in which V_{ref} is the reference velocity, that is, the wind velocity at the height of 10 m (m/s), S_{ref} is the front surface area of the panel (m^2), D and L are the forces acting on the panel along the x and y directions (N). C_M is defined as $C_M = M/(0.5\rho V_{ref}^2 S_{ref} C)$, in which C is the chord of the panel, chosen as reference length (m), and M is the aerodynamic moment about the z-axis (Nm), calculated with respect to the center of the panel.

The pressure distribution is practically the same for both inclination angles. Regarding the aerodynamic coefficients, as expected, the angle variation causes a change of sign of both C_L and C_M. Furthermore, the absolute values of the C_L and C_D are comparable for the two inclination angles, owing to the symmetry of the problem. In Figs. 7 and 8 the distributions of C_P on the upper surface for $\theta = -25°$ and for the lower surfaces for $\theta = 25°$ are shown. A good match emerges from the comparison between the periodic model and the central part of the full one. The pressure coefficient tends to decrease at the lateral side of the full model, which is shown only for a half in Fig. 7. This is due to the existing three-dimensionality of the flow.

These results demonstrate that the aerodynamic behavior of a solar panel of considerable length can be derived from the study of a part of it, characterized by a more limited length, assuming that this is repeated in a periodic

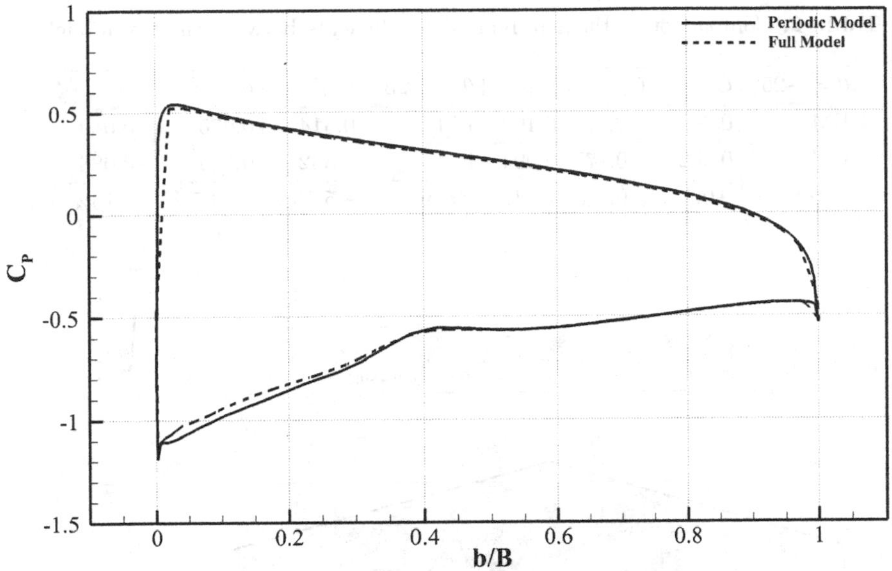

(a)

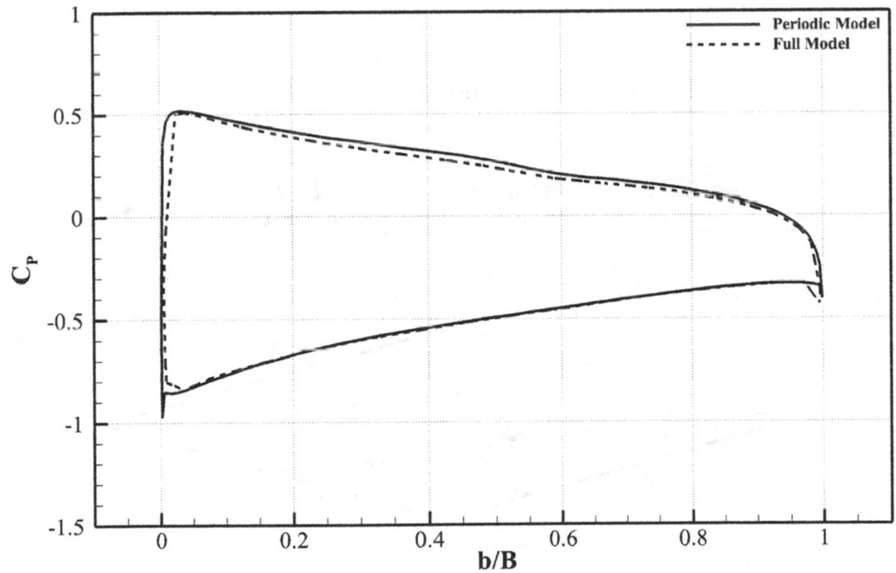

(b)

Fig. 6. Comparison of C_P on the panel section at the centerline of a supporting column between the full model and the periodic one at $\theta = -25°$ (a), $25°$ (b)

Table 2. Comparison of the aerodynamic coefficients between the two models

$\theta = -25°$	C_D	C_L	C_M	$\theta = -25°$	C_D	C_L	C_M
FM	0.360	−0.774	0.102	FM	0.314	0.675	−0.089
PM	0.362	−0.779	0.094	PM	0.332	0.716	−0.088
Error	−0.5%	−0.6%	7.84%	Error	−5.73%	−6.07%	−1.13%

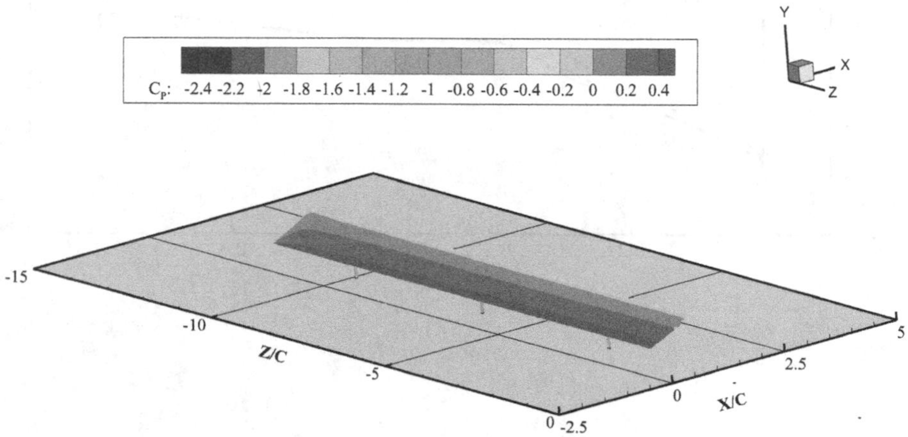

(a)

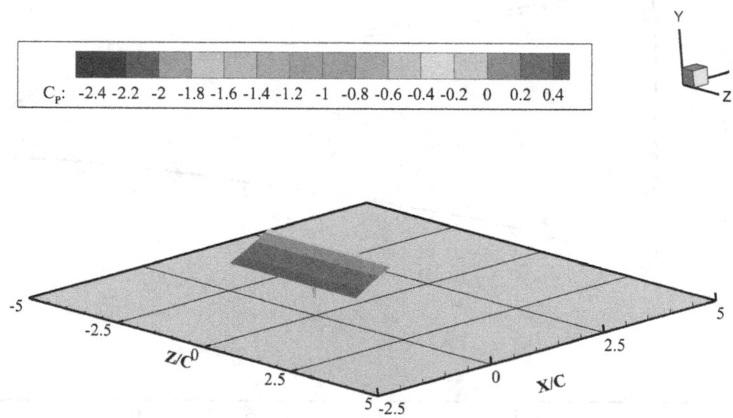

(b)

Fig. 7. Contours of C_P on the panel for $\theta = -25°$

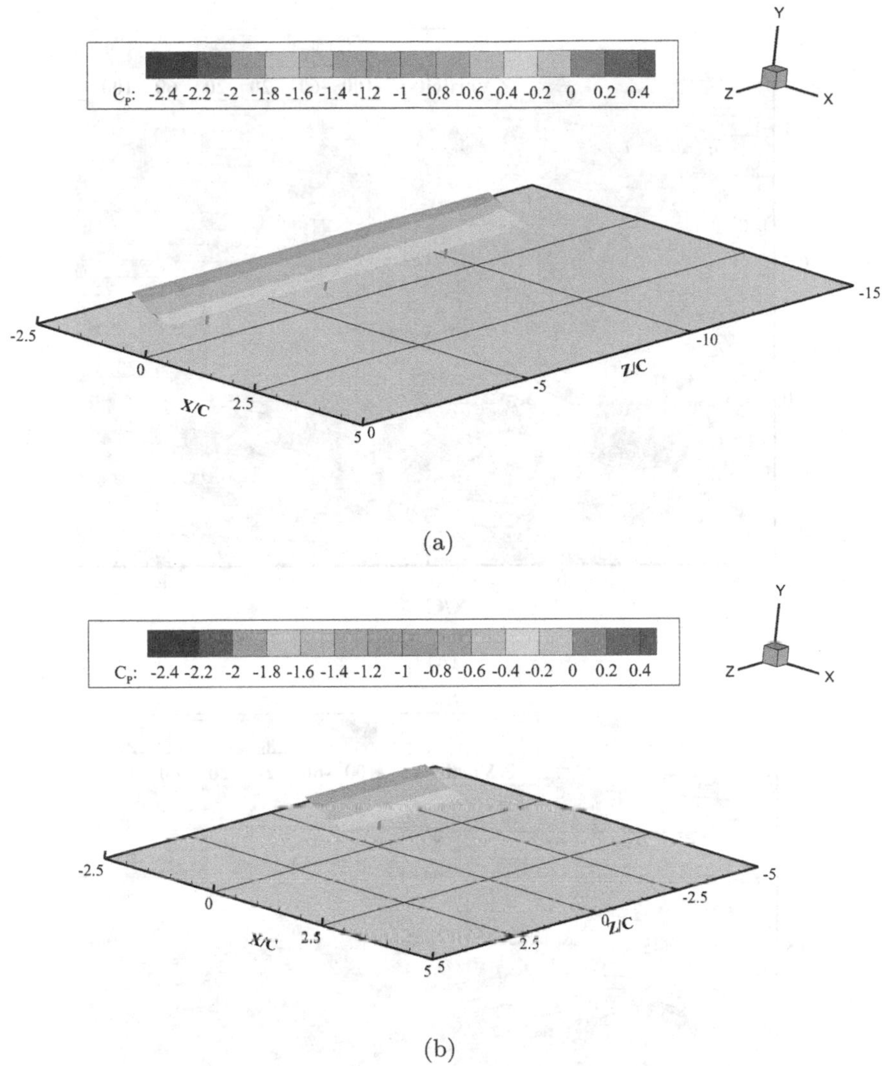

Fig. 8. Contours of C_P on the panel for $\theta = 25°$

manner along the longitudinal direction. The main consequence is a considerable reduction of the computational cost and of the simulation time for a single CFD analysis. Lastly, to get a better view of the flow around the panel, the streamlines superimposed with the contour of vorticity in the middle section of the support column are shown in Fig. 9.

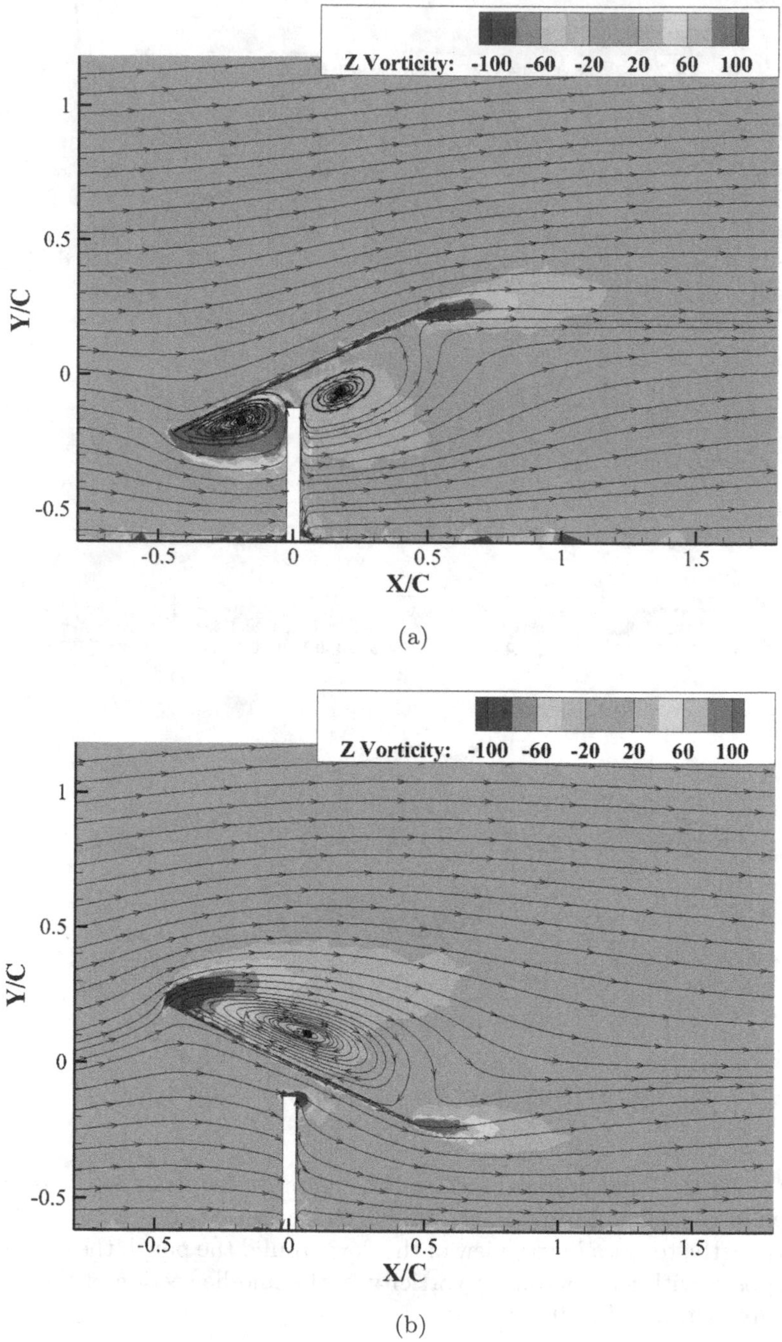

Fig. 9. Sectional flow streamlines superimposed with the vorticity contours on the periodic model at $\theta = -25°$ (a), $25°$ (b)

5 Conclusions

In this work a CFD study of wind loads on high aspect ratio ground-mounted photovoltaic panels was performed by following two different approaches. The entire panel was directly simulated in a full three-dimensional domain and a reduced model was obtained by considering a small portion of the panel, by imposing periodic boundary conditions in the homogeneous spanwise direction.

The PV panel was analyzed for two different inclination angles with respect to the wind direction. The simulations were carried out using a Reynolds-Averaged Navier Stokes (RANS) numerical approach supplied with the Shear Stress Transport $k-\omega$ turbulence model with a hybrid mesh scheme. The wind flow at the inlet of the computational domain was simulated by imposing the logarithmic law boundary layer profile, in order to obtain an equilibrium Atmospheric Boundary Layer (ABL).

The results obtained on the panel portion showed a good correspondence when compared to the full panel, both in terms of pressure coefficient, calculated at the centerline section of the two models, and the aerodynamic coefficients. It was demonstrated that, by means of CFD, it is possible to determine the wind loads on a panel of high aspect ratio from the simulation of a small portion of it, assuming that this is repeated periodically along the longitudinal direction.

The main advantage is a considerable saving in terms of computational time, since each calculation on the periodic model was achieved in approximately 1/8 of the time required for the simulation on the full model.

References

1. Blakers, A., Zin, N., McIntosh, K., Fong, K.: Efficiency improvements of photovoltaic panels using a sun-tracking system. Energy Procedia **33**, 1–10 (2013)
2. Stathopoulos, T., Zisis, I., Xypnitou, E.: Local and overall wind pressure and force coefficients for solar panels. J. Wind Eng. Ind. Aerodyn. **125**, 195–206 (2014)
3. Abiola-Ogedengbe, A., Hangan, H., Siddiqui, K.: Experimental investigation of wind effects on a standalone photovoltaic (PV) module. Renew. Energy **78**, 657–665 (2015)
4. Warsido, W., Bitsuamlak, G., Barata, J.: Influence of spacing parameters on the wind loading of solar array. J. Fluid Struct. **48**, 295–315 (2014)
5. Aly, A., Bitsuamlak, G.: Aerodynamics of ground-mounted solar panels: test model scale effects. J. Wind Eng. Ind. Aerodyn. **123**, 250–260 (2013)
6. Jubayer, C., Hangan, H.: Numerical simulation of wind effects on a stand-alone ground mounted photovoltaic (PV) system. J. Wind Eng. Ind. Aerodyn. **134**, 56–64 (2014)
7. Wu, Z., Gong, B., Wang, Z., Li, Z., Zang, C.: An experimental and numerical study of the gap effect on wind load on heliostat. Renew. Energy **35**, 797–806 (2010)
8. Menter, F.: Two-equation eddy-viscosity turbulence models for engineering applications. Am. Inst. Aeronaut. Astronaut. J. **32**, 1598–1605 (1994)
9. Yang, Y., Gu, M., Jin, X.: New inflow boundary conditions for modeling the neutral equilibrium atmospheric boundary layer in SST k-model. In: Proceedings of the Seventh Asia Pacific Conference on Wind Engineering, Taipei, Taiwan November, pp. 8–12 (2009)

10. ESDU: Strong winds in the atmospheric boundary layer. part 1: Mean hourly wind speeds. Engineering Science Data Unit Number 82026 (1982)
11. ESDU: Strong winds in the atmospheric boundary layer. part 2: Discreet gust speeds. Engineering Science Data Unit Number 82045 (1983)
12. Blocken, B., Stathopoulos, T., Carmeliet, J.: Cfd simulation of the atmospheric boundary layer: wall function problems. Atmos. Environ. **41**, 238–252 (2007)

A Comparative Study of a GUI-Aided Formal Specification Construction Approach

Fumiko Nagoya[1(✉)] and Shaoying Liu[2]

[1] College of Commerce, Nihon University, Tokyo, Japan
nagoya.fumiko@nihon-u.ac.jp

[2] Faculty of Computer and Information Sciences, Hosei University, Tokyo, Japan
sliu@hosei.ac.jp

Abstract. Formal specification techniques still remain a challenge for applying formal methods in practice how to reduce unnecessary changes during and after writing formal specifications. We proposed a GUI-aided approach to constructing formal specifications, but its effectiveness has not been evaluated. This paper describes an experiment we have conducted to systematically evaluate the GUI-aided approach by comparing it with the existing refinement approach for constructing a formal specification based on an informal requirements specification. We chose a travel reservation system as the target system for the experiment, and developed half of its functions using the GUI-aided approach and the other using the existing refinement approach. The comparative experiment shows how we analyzed the data collected during the development and presents the findings. The result indicates that the GUI-aided approach is superior to the existing refinement approach due to focus on adequate requirement acquisitions in the early phase of software design.

1 Introduction

Graphical user interface (GUI) design has been seen as an important factor in developments of highly interactive software systems, whether the interfaces and system behaviors meet to users' perceptions. Many definitions and disciplines of User eXperience (UX) [1,2] have been suggested by human-computer interaction community. In industry, some prototyping tools [3,4] are available for constructing visual models to give a framework of an interactive web system or service. The visual models may help to evaluate a website layout at an early stage of development, but it is not enough to ensure the quality of the interactive software systems. Such models do not provide exactly what services or functions in interactive software systems to implement and how to verify whether the final product executes properly in accordance with the users' expectations.

Formal methods are powerful techniques for software developments, and comprise of formal specification and verification techniques. Formal specifications can clarify user requirements by mathematical notations, and formal verifications support to ensure correct behaviors. Some of formal specification techniques applied

O. Gervasi et al. (Eds.): ICCSA 2017, Part I, LNCS 10404, pp. 273–283, 2017.
DOI: 10.1007/978-3-319-62392-4_20

in practice for deleting inconsistency and ambiguity in specifications. Honestly speaking, they are useful for describing required functions in rigorous ways, but not suitable for designing GUIs. In addition, if a GUI changes during and after writing a formal specification, it will raise development cost considerably.

To address the problems, our research group proposed a *GUI-aided approach* for constructing formal specifications [5,6]. The *GUI-aided approach* uses GUI models derived from the structured informal requirements specification that clarifies required functions, input and output data items, and their constraints for a system. And then, the initial informal specification is improved on the result of observations and discussions through the GUI models. Finally, a formal specification is constructed on the basis of the improved informal specification for the design of the system. This approach applies rapid prototyping techniques [7] for eliciting end-users' requirements in the early stage of development. However, at the same moment, it contains a disadvantage of rapid prototyping developments: "*it may increase the complexity of the system as the scope of the system may expand beyond original plans*" [8]. To evaluate this point, we conducted a comparison between the *GUI-aided approach* and the existing refinement approach that constructs a formal specification based on an informal requirements specification. In a development of a travel reservation system, we developed half of functions using the *GUI-aided approach* and the other using the existing refinement approach. In order to facilitate the construction of a formal specification in a comprehensible manner, we use the Structured Object-Oriented Formal Language (SOFL) [9] as the specification language. It provides well-designed requirement analysis [10], and systematic inspection [11] and testing [12] in practical software developments.

The rest of this paper is organized as follows. Section 2 briefly introduces a refinement approach for constructing a formal specification in SOFL. Section 3 explains an outline of the *GUI-aided approach*. Then, Sect. 4 describes a comparative study in accordance with our *GUI-aided approach* and the existing refinement approach. Section 5 reviews the related work. Lastly, in Sect. 6, we give conclusions and point out future research directions.

2 Refinement Approach in SOFL

SOFL integrates Vienna Development Method Specification Language (VDM-SL) [13], Data Flow Diagram [14], and Petri Nets [15] for setting up practical textual notations and graphical notations. The formal textual notations give precise definitions of data and operations by simple propositional logics, basic set theory and predicates for describing formal specifications. The graphical notations give understandings of overall architecture, associations with functions, and hierarchic structure of the system.

Developing software systems with SOFL begins to document user's requirements informally based on communications between software developers and the end-users. Figure 1 illustrates the refinement approach in SOFL. It shows an outline of constructing a formal specification from an informal requirements specification. An informal specification in SOFL is composed of three parts: **functions**,

data resources and **constraints** as the left-hand side of Fig. 1. Informal specifications equal informal documentations to be written in a natural language in order to summarize findings from the communications. The developers precisely describe the functions to be implemented by the system, the data resources to be used, and the necessary constraints on both functions and resources.

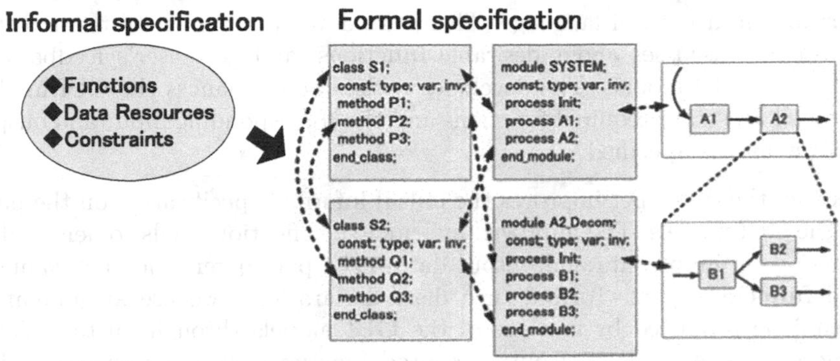

Fig. 1. Refinement approach in SOFL

The right-hand side of Fig. 1 shows a formal specification. A formal specification is composed of classes, modules, and hierarchical Condition Data Flow Diagrams (CDFDs). A class contains a set of variables and a set of methods. The value of each variable represents an attribute of the corresponding object, and each method provides an operation that can change or access to the state of the object. A module is a functional abstraction: it consists of a module name, constant declarations, type declarations, variable declarations, an invariant section, and a list of process names. A process defines an operation for transforming into outputs from inputs: it is composed of a process name, input and output ports, pre-condition, and post-condition. The pre-condition describes a constraint on the input data flows before the execution of the process, while the post-condition provides a constraint on the output data flows after the execution. These classes and modules represent an architecture of the entire system. A CDFD is a diagram for representing functional behaviors, by each box, for a process; each directed line, for an input or output data flow; and each box with a number, for a data store.

The refinement approach in SOFL is systematically designed, but it is impossible to keep away from any change of specifications during and after writing them due to the lack of sufficient requirement analysis. Model based developments are well known to improve development productivity in engineering. Models make developers shift of focus on the early phases in development processes. We will explain the *GUI-aided approach* for constructing formal specifications, and how to reduce unnecessary changes on the later phases of development processes.

3 A GUI-aided Approach

In accordance with our *GUI-aided approach*, a construction of a formal specification includes three steps as follows:

- First, a developer derives GUI models from an initial informal specification. The informal specification defines **functions**, **data resources**, and **constraints** in a natural language. Then, the developer demonstrates it to the end-users, discusses about desirable functions, and gets user's feedback to refine the GUI models. The developer revises and enhances the GUI models until all necessary required functions and the corresponding input and output data items are specified.
- Second, the developer improves the initial informal specification on the basis of the GUI models. The improved informal specification needs to denote declarations of input data and output data, each parameter, and each value in the **functions** part. To find such data declarations, we use an animation technique produced by motions of the GUI models through button-related functions. The motions of button-related functions guide a common understanding about system behaviors of the target system and it is effective to elicit users' real requirements.
- Finally, a formal specification is constructed based on the improved informal specification. A hierarchical structure of operations is derived from the **functions** part in the improved informal specification. Then, a sequence of modules and processes is specified in a hierarchical fashion. Also, he/she draws CDFDs for each module with an emphasis on contract with the corresponding module.

The next section shows an example of the *GUI-aided approach* for constructing of a formal specification step by step.

4 A Comparative Study

We conducted a comparative study of constructing a formal specification to measure cost-effectiveness between the *GUI-aided approach* and the existing refinement approach for a development of functionality and interactive features service. In this study, one designer, with eight end-users supports, engaged in describing an informal specification, developing GUI models for half of functions, and defining the half of formal specification by the *GUI-aided approach* and the other by the ordinary refinement approach in SOFL. One might assume that it would be better to have applied the *GUI-aided approach* and the refined approach in all functions, and then compare the results for each function. We took our experimental methodology because we have only one subject who could conduct the experiment. If the designer was required to carry out the comparison for the same application using the two different techniques, it would create a threat to the validity of the experiment result because the use of one technique will definitely impact on the result of applying another technique to the same functions of the same system. Our methodology is not ideal, but it allows us to independently observe the effect of applying each technique on the functions that possess similar feature and nature.

The end-users required the new service they want, reviewed the informal specification and GUI-models, and provided their feedback to improve the informal specification. A program was implemented independently based on each formal specification in different approach, and verified by functional testing whether it meets the user requirements. Also, the designer tracked the actual time that was spent for each development step to evaluate the approaches. We will explain the informal specification, GUI-models, formal specification, the test result, and spending times for the development.

4.1 Informal Specification

The travel reservation system makes it easy to find a good hotel from partnership hotels around the world, find flights for one-way or round-trip including international flights, and find fun things to do as day trips & excursions from international travel agents. Moreover, booking the hotels, the air tickets, and the activities are available. The system includes management functions by the user to show the reservation statuses and cancel them. An informal specification of the travel reservation system is specified as follows:

Functions:
- Find hotels
- Book hotels
- Find flights
- Book flights
- Find day trips & excursions
- Book day trips & excursions
- Confirm and cancel reservations

– Data Resources:
  - hotels: hotel ID, hotel name, rank, location, credit card availability, room list, best price list.
  - rooms: room ID, hotel ID, room type, maximum occupancy, vacancy list.
  - air tickets: flight number, airline name, departure date, departure time, departure place, arrival date, arrival time, arrival place, seat class, credit card availability, seat list.

  ⋮

– Constraints:
  - Hotel ID should be unique.
  - A set of flight number and departure date should be unique.
  - All search conditions of the hotels should include locations.

- All search conditions of the flights should include the number of passengers.

⋮

GUI models for half of functions, such as **Find hotels**, **Book hotels**, **Find flights**, **Book flights**, were created in our comparative study. In contrast, the other functions in the informal specification were translated into a formal specification directly.

4.2 GUI Models

In our comparative study, eleven GUI models for representing four functions were implemented under the Eclipse environment using Swing in Java. These models incorporate buttons, menu bar, or menu items to represent event handling functions, but each handler does not need to be associated with any database.

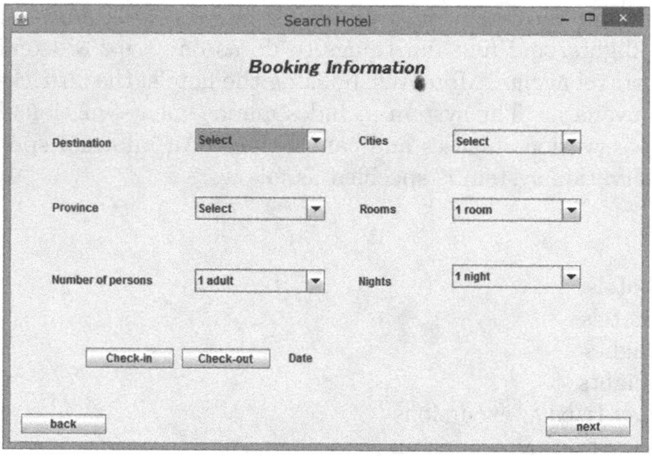

Fig. 2. The GUI model for finding hotels

For instance, Fig. 2 displays a look and feel for the function of **Find hotels**. The GUI model requires input data: travel destination, number of persons, number of rooms, number of hotel stays, and check-in/check-out dates. After inputting them and pushing "next" button, an accommodation list will appear in a next GUI model. The designer demonstrated it in front of the end-users, discussed with the clients about desirable functions, refined the model on the basis of users' feedback. In our comparative study, the designer expected only showing available rooms for the requested period is enough for users. However, the end-users wanted to add the optional function that all accommodations are displayed, including fully booked hotels for the requested dates. Not only that, they wanted to sort the accommodation list by price rather than hotel rank.

The GUI models facilitate more deepening of understanding about functional behaviors and the corresponding input and output data in operation levels. In this way, the informal specification was improved to clarify a hierarchical structure of operations and add data declarations for input and output data.

4.3 Formal Specification

Half of the functions in the travel reservation system are improved through a prototype analysis by GUI models, and a formal specification was created as follows:

```
module Travel_System/FindingHotels_Decom
type
    roomData = composed of
        rID : roomID
        hotel : hotelID
        rank : string
        capacity : nat
        vacancy : boolean
        amount : nat
 end;
......
var
    rooms: seq of roomData; ......
process Show _unlimit (hotelIDs:hotelIDs, amount:nat, date:Calendar) d:hotelIDs
ext rd rooms
pre  true
post forall[i:inds(hotelIDs)]| exists[j:inds(hotelIDs(i).roomIDs)]|
    rooms(hotelIDs(i), roomIDs(j)).capacity >= humanNum and
    rooms(hotelIDs(i), roomIDs(j)).amount(date) <= amount) and
    d = conc(~d, hotelIDs(j))
end_process;
```

As we explained in Sect. 2, a module defines a module name, constant declarations, type declarations, variable declarations, an invariant section, and a sequence of processes. The module *FindingHotels_Decom* relates with the GUI model given in the Fig. 2.

The process of *Show_unlimit* declares input parameter: a list of accommodations, amount, date, output parameter: a new list of available accommodations, external variable: rooms, and their types respectively. The pre-condition defines no specific constraints as **true**. The post-condition denotes that if an accommodation fits within a budget, less than maximum occupancy of a room, and a reservation is also possible during the requested period, then the accommodation's ID adds a list of the available accommodations. Figure 3 represents the functional behaviors of the module of *FindingHotels_Decom*.

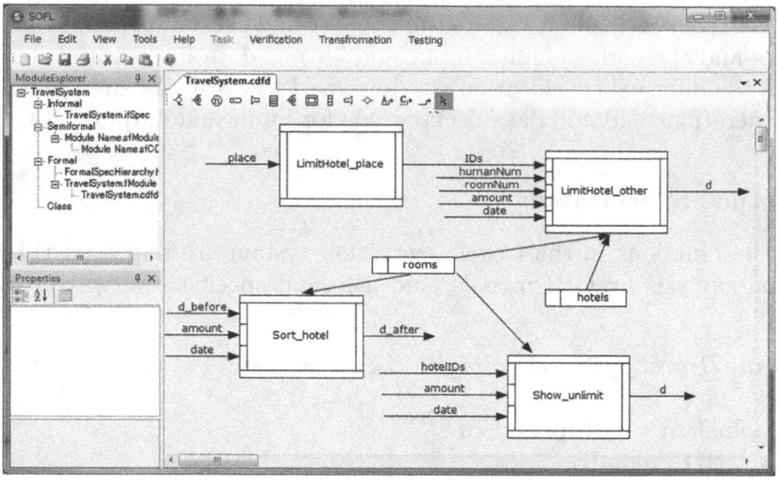

Fig. 3. The CDFD for finding hotels

At the same time, the rest of the three functions in the travel reservation system required to write a formal specification without GUI models for comparison. We translated into a formal specification from the informal specification about the functions: **Find day trips & excursions, Book day trips & excursions, Confirm and cancel reservations**, then drew the CDFDs for each module.

4.4 Results

In order to evaluate the effectiveness between the *GUI-aided approach* and the existing refinement approach, we implemented a Java program using each half of formal specification, generated test cases, and executed them. A programmer translated from the processes specified in the formal specifications into a class or a set of classes in the Java program with the intention of identifying a relationship between the formal specification and the program. Next, functional scenarios derived from the formal specifications for generation of the test cases. We applied a specification based testing [12] to verify the correctness of program by generating test cases from the formal specification. Then, we verified whether the program covers functional scenarios defined in the specification including failure paths cases. Table 1 shows the test result that how many bugs found by testing in the source codes. It is not surprising that the failures in functional testing by the *GUI-aided approach* is few, since the approach used a superior model for analyzing functional behaviors.

Also, we recorded each time for the following activities: constructing the informal specification, GUI modeling & improved informal specification, constructing formal specification, coding, generating test cases, executing the test, and debugging with the *GUI-aided approach* and the existing refinement approach respectively. Table 2 represents the result of each development time (hours). The

Table 1. Test result

	GUI-aided approach	Existing refinement
Functional test errors	3	11
Failure path test errors	6	3
Total number of errors	9	14

Table 2. The development hours

	GUI-aided approach	Existing refinement
Informal specification	1:54	3:06
GUI models & improve	8:33	-
Formal specification	11:52	9:48
Coding	33:08	42:02
Test & debug	1:05	3:11
Total time	56:32	58:07

constructing time for the formal specification by the *GUI-aided approach* is a seemingly inefficient, however it caused by difference of code lengths for the covered functions. The total codes have written 7.7 thousand of lines of code (KLOC), with the *GUI-aided approach*, 3.5 KLOC; with the existing refinement approach, 2.4 KLOC, and the other for the intersecting parts, 1.8 KLOC. Despite this, the coding was done in less time with the *GUI-aided approach* than the existing refinement approach. Especially, we should note that the executing test and debugging with the *GUI-aided approach* is cost-effective compared to the other.

5 Related Work

In human-computer interaction community, prototypes categorize low or high-fidelity [16]. Fidelity means a classification of similarity between a prototype and a final service or product. Low-fidelity prototypes are constructed quickly and low-cost, and provide restricted functions and little or no interactions to address screen layout issues for user-interface designs. In contrast, high-fidelity prototypes are fully interactive for describing the complete functionality of a product [17]. It supports usability testing for containing the similar behaviors of the final product. Consequently, a high-fidelity prototype is more expensive and more time-consuming than a low-fidelity prototype. It is a common limitation for model-based developments.

In spite of many tools' supports for low or high-fidelity prototypes, it is difficult for a designer to precisely express behaviors of a product to developers or other stakeholders. According to [18], 76% of the designers reported that communicating the design of behaviors to developers is more difficult than the appearance. After all, designers need to use any text document, such as annotation, with the pictures to serve as a specification for developers. For the purpose of covering this disadvantage, we believe formal methods strongly support to define functional behaviors in a formal specification.

6 Conclusions

This paper describes a comparative study between the *GUI-aided approach* and the ordinary refinement approach in SOFL for developing a travel reservation system. The *GUI-aided approach* aims to discover functional behaviors, including end-users' implicit requirements through GUI models. In our comparative study, we described a well-designed informal specification for the target system, developed GUI models for half of its functions defined in the informal specification using the *GUI-aided approach* and improved it. Next, a formal specification was defined by the *GUI-aided approach* and the existing refinement approach respectively. Then, the formal specification translated into a Java program. Finally, we generated test cases based on a specification based testing, executed the testing, and fixed errors.

The result of the comparative study shows the *GUI-aided approach* does not have a tremendous advantage in total development time, however, it is cost-effective compared to the other in test and debug. Also, we do not find any increase of the complexity of the system development. In fact, only one programmer carried out the implementation of the formal specification. This process is quite straightforward because the formal specification has clearly indicated the input variables, output variables, and the precise definition of the desired functions.

Meanwhile, we need more analyses for comparing the *GUI-aided approach* with other formal techniques in our future work. Since our experiment does not concentrate on the process of obtaining and applying the feedback from the end-user, nothing can be clearly reported in this paper. We will investigate what is added after getting the feedback, and how it alters the formal specification of the system behavior in more detail.

Acknowledgement. We would like to thank Daisuke Noguchi for developing GUI models, including writing the SOFL specifications and completing the implementation in Java. This work was supported by JSPS KAKENHI Grant Number 26240008.

References

1. Law, E.L.C., Roto, V., Hassenzahl, M., Vermeeren, A.P., Kort, J.: Understanding, scoping and defining user experience: a survey approach. In: Proceedings of the SIGCHI Conference on Human Factors in Computing Systems, CHI 2009, New York, NY, USA, ACM (2009) 719–728
2. Spohrer, J.C., Freund, L.E. (eds.): Advances in the Human Side of Service Engineering. Advances in Human Factors and Ergonomics Series. CRC Press, Boca Raton (2012)

3. Axure Software Solutions Inc.: Axure RP. https://www.axure.com/
4. Balsamiq Studios LLC.: Balsamiq mockups. https://balsamiq.com/
5. Liu, S.: A GUI-Aided approach to formal specification construction. In: Liu, S., Duan, Z. (eds.) SOFL+MSVL 2015. LNCS, vol. 9559, pp. 44–56. Springer, Cham (2016). doi:10.1007/978-3-319-31220-0_4
6. Nagoya, F., Liu, S.: A case study of a GUI-Aided approach to constructing formal specifications. In: Liu, S., Duan, Z., Tian, C., Nagoya, F. (eds.) SOFL+MSVL 2016. LNCS, vol. 10189, pp. 74–84. Springer, Cham (2017). doi:10.1007/978-3-319-57708-1_5
7. Gomaa, H.: The impact of rapid prototyping on specifying user requirements. SIGSOFT Softw. Eng. Notes **8**, 17–27 (1983)
8. Meghanathan, N., Chaki, N., Nagamalai, D. (eds.): CCSIT 2012. LNICSSITE, vol. 86. Springer, Heidelberg (2012)
9. Liu, S.: Formal Engineering for Industrial Software Development. Springer, Heidelberg (2004)
10. Miao, W., Liu, S.: A formal engineering framework for service-based software modeling. IEEE Trans. Serv. Comput. **6**, 536–550 (2013)
11. Li, M., Liu, S.: Integrating animation-based inspection into formal design specification construction for reliable software systems. IEEE Trans. Reliab. **65**, 88–106 (2016)
12. Liu, S., Nakajima, S.: A decompositional approach to automatic test case generation based on formal specifications. In: Proceedings of the 2010 Fourth International Conference on Secure Software Integration and Reliability Improvement, SSIRI 2010, IEEE Computer Society, Washington, DC, pp. 147–155 (2010)
13. Jones, C.B.: Systematic Software Development Using VDM. Prentice Hall International (UK) Ltd., Upper Saddle River (1986)
14. DeMarco, T.: Structured Analysis and System Specification. Prentice Hall PTR, Upper Saddle River (1979)
15. Reisig, W.: Petri Nets: An Introduction. Springer, New York (1985)
16. Rudd, J., Stern, K., Isensee, S.: Low vs. high-fidelity prototyping debate. Interactions **3**, 76–85 (1996)
17. Carter, A.S., Hundhausen, C.D.: How is user interface prototyping really done in practice? A survey of user interface designers. In: 2010 IEEE Symposium on Visual Languages and Human-Centric Computing, pp. 207–211 (2010)
18. Myers, B., Park, S.Y., Nakano, Y., Mueller, G., Ko, A.: How designers design and program interactive behaviors. In: Proceedings of the 2008 IEEE Symposium on Visual Languages and Human-Centric Computing, VLHCC 2008. IEEE Computer Society, Washington, DC pp. 177–184 (2008)

Missing Data Completion Using Diffusion Maps and Laplacian Pyramids

Neta Rabin and Dalia Fishelov(✉)

Department of Mathematics, Afeka - Tel Aviv Academic College of Engineering, Tel Aviv, Israel
{netar,daliaf}@afeka.ac.il

Abstract. A challenging problem in machine learning is handling missing data, also known as imputation. Simple imputation techniques complete the missing data by the mean or the median values. A more sophisticated approach is to use regression to predict the missing data from the complete input columns. In case the dimension of the input data is high, dimensionality reduction methods may be applied to compactly describe the complete input. Then, a regression from the low-dimensional space to the incomplete data column can be constructed from imputation. In this work, we propose a two-step algorithm for data completion. The first step utilizes a non-linear manifold learning technique, named diffusion maps, for reducing the dimension of the data. This method faithfully embeds complex data while preserving its geometric structure. The second step is the Laplacian pyramids multi-scale method, which is applied for regression. Laplacian pyramids construct kernels of decreasing scales to capture finer modes of the data. Experimental results demonstrate the efficiency of our approach on a publicly available dataset.

Keywords: Missing data · Dimensionality reduction · Diffusion maps · Laplacian pyramids

1 Introduction

A challenging problem in machine learning is preprocessing of the dataset that involves data normalization, detection of input outliers and handling missing data. This work focuses on completion of missing data, also known as imputation. There are several common ways to deal with this problem. Simple imputation techniques complete the missing data by replacing it with the mean or the median value of the column. Another simple imputation approach is to replace the missing values with a sample that is randomly selected from the same column [10,11]. A more sophisticated approach is to use a regression to predict the missing data in specific column from the rest of the columns. In case the dimension of the input data is high, it is reasonable to assume that the data columns are correlated, thus the input matrix **X** resides in a low-dimensional space. Therefore, a dimensionality reduction method can be applied to the set

O. Gervasi et al. (Eds.): ICCSA 2017, Part I, LNCS 10404, pp. 284–297, 2017.
DOI: 10.1007/978-3-319-62392-4_21

of full columns and the result may be used as the regression input to predict the missing values in the other columns. Such an approach was suggested in [1] for data imputation in road networks. UshaRani and Sammulal [17] apply dimensionality reduction and clustering for completion of missing medical data. Recently, Pierson and Yau [15] used a linear dimensionality reduction technique to fill in zero-values of single-cell gene expression data.

Dimensionality reduction methods can be separated into feature selection techniques, in which a subset of the columns is selected on the one hand, and feature extraction methods that construct latent combinations of all of the input columns on the other hand. Principal complainant analysis [14] is a common linear dimensionality reduction method that constructs linear combinations of the data columns while minimizing the reconstruction error. However, if the high-dimensional data contains non-linear relations, then linear methods fail to faithfully reduce the dimension of the dataset. Manifold learning methods that include Local Linear Embedding [20], Laplacian Eigenmaps [2,3] and Diffusion Maps [5], aim to reveal the intrinsic parameters that drive the data. These intrinsic modes parameterize the data in a low dimension space while preserving some properties of interest. In this work, diffusion maps are utilized for dimensionally reduction. There, the data is compactly re-organized data according to a diffusion distance metric that is defined by a random walk on the points in the ambient space. In the embedding space, the diffusion distance is the Euclidean distance between the embedded data points, thus the embedding is distance preserving.

Once the dimension of the data is reduced, several methods may be applied for imputations. Linear regression is a simple choice, but it relies on the assumption that there is a linear relationship between the function and the data. Regression using a k nearest neighbors is another simple regression approach, the drawback is that in case different columns need to be imputed, an optimal value for k should be set according to the smoothness of column's data. In this paper, we propose to use Laplcain pyramids for regression from the low-dimensional space to the column with the missing data. The Laplacian pyramids method was proposed in [19] and improved by adding an automatic stopping criteria in [7,9]. The method was applied in [6] for reconstruction data points in the ambient space, in [22] for analog forecasting and in [8] for meteorological data analysis. Here, we emphasize the advantage of this method to automatically stop at a suitable scale that fits the data, without manually tuning the scale parameters. This property is important when running many consecutive regressions for data completion.

2 Mathematical Background

This section describes the two central mathematical tools that are proposed for imputation. First, we review the diffusion maps framework, then the Lalpacian pyramids method is explained.

2.1 Diffusion Maps

Let $\mathbf{X} = \{x_1, \ldots, x_N\}$ be a set of data points in $\mathbb{R}^m$, hence x_i of size $1 \times m$ is the i-th row of $\mathbf{X}$. In order to construct a low-dimensional representation of $\mathbf{X}$, a graph $G = (\mathbf{X}, \mathbf{W})$ is built. The kernel $\mathbf{W}$, defined by $\mathbf{W} = (w(x_i, x_j))_{N \times N}$, contains the weights of the graph edges. We assume that the kernel is

- symmetric;
- positive preserving: $w(x_i, x_j) \geq 0$ for all $x_i, x_j \in \mathbf{X}$;
- positive semi-definite: for all real-valued bounded function f defined on $\mathbf{X}$, $\sum_i \sum_j w(x_i, x_j) f(x_i) f(x_j) \geq 0$.

Diffusion Kernels. Typically, kernels of the form $\mathbf{W}_\epsilon = \left(h\left(\frac{-\|x_i - x_j\|^2}{2\epsilon}\right)\right)$ are chosen since they are directionally independent. Here ϵ defines the width of the kernel. A common choice is the Gaussian kernel $\mathbf{W}_\epsilon = (w_\epsilon(x_i, x_j))$, where $w_\epsilon(x_i, x_j) = e^{\frac{-\|x_i - x_j\|^2}{2\epsilon}}$. The scale parameter ϵ can be defined (see [21]) by

$$\epsilon = median\{d_{ij}\}, \tag{1}$$

where $\mathbf{D} = (d_{i,j})_{N \times N}$ is the matrix of pairwise Euclidean distances of the set $\mathbf{X}$.

A general normalized form of the kernel, having a parameter α which controls the normalization type, was introduced in [5]. It is given by

$$w_\epsilon^{(\alpha)}(x_i, x_j) = \frac{w_\epsilon(x_i, x_j)}{q^\alpha(x_i) q^\alpha(x_j)}, \qquad q(x_i) = \sum_j w_\epsilon(x_i, x_j). \tag{2}$$

Then, a Markov transition matrix is defined by

$$\mathbf{P}^{(\alpha)} = \left(p^{(\alpha)}(x_i, x_j)\right), \qquad where \qquad p^{(\alpha)}(x_i, x_j) = \frac{w_\epsilon^{(\alpha)}(x_i, x_j)}{\sum_j w_\epsilon^{(\alpha)}(x_i, x_j)} \tag{3}$$

Three values of α that are commonly in use are $\alpha = 0, 1, 0.5$. When $\epsilon \to 0$, $\mathbf{P}^\alpha$ approximates the following operators:

1. $\alpha = 0$: the classical graph Laplacian [4];
2. $\alpha = 1$: the Laplace-Beltrami operator [5];
3. $\alpha = \frac{1}{2}$: the diffusion of the Foller-Planck equation [12].

In this paper α was set to be 1, denoting $\mathbf{P}^{(\alpha=1)}$ as $\mathbf{P}$.

Spectral Decomposition. The construction of the low-dimensional data representation involves the eigendecomposition of $\mathbf{P}$. This is computed by

$$p(x_i, x_j) = \sum_{k \geq 0} \lambda_k \psi_k(x_i) \phi_k(x_j). \tag{4}$$

Here $\{\lambda_k\}_{k=0}^{N-1}$ are the eigenvalues of $\mathbf{P}$ and $\{\phi_k\}_{k=0}^{N-1}$, $\{\psi_k\}_{k=0}^{N-1}$ are the corresponding left and right eigenvectors. We note that $\mathbf{P}$ is similar to a symmetric matrix $\mathbf{A}$, i.e., $\mathbf{A} = \mathbf{D}^{\frac{1}{2}}\mathbf{P}\mathbf{D}^{-\frac{1}{2}}$, where $\mathbf{D}$ is a diagonal matrix with values $\sum_j w_\epsilon^\alpha(x_i, x_j)$ on its diagonal. Thus, $\mathbf{P}$ and $\mathbf{A}$ share the same set of eigenvalues. The spectral decomposition of the matrix $\mathbf{A} = (a(x_i, x_j))$ is given by

$$a(x_i, x_j) = \sum_{k\geq 0} \lambda_k \mathbf{v}_k(x_i)\mathbf{v}_k(x_j). \tag{5}$$

Since $\mathbf{A}$ is symmetric the set eigenvalues $\{\lambda_k\}_{k=0}^{N-1}$ are real and the set of eigenvectors $\{\mathbf{v}_k\}_{k=0}^{N-1}$ are orthogonal. The left and the right eigenvectors of $\mathbf{P}$ are related to $\{\mathbf{v}_k\}_{k=0}^{N-1}$ by

$$\psi_k = \mathbf{D}^{-\frac{1}{2}}\mathbf{v}_k, \quad \phi_k = \mathbf{D}^{\frac{1}{2}}\mathbf{v}_k. \tag{6}$$

The orthonormality of $\{\mathbf{v}_k\}$ yields the biorthonormality of $\{\phi_k\}$ and $\{\psi_k\}$. This property is used for defining a metric on the data. In addition, since the eigenvalues decay fast to zero, the sum in Eq. (4) can be approximated by a small number of leading terms. These terms are used to define the diffusion maps embedding

$$\Psi(x_i) = (\lambda_1\psi_1(x_i), \lambda_2\psi_2(x_i), \lambda_3\psi_3(x_i), \cdots). \tag{7}$$

Diffusion Distances. This embedding (Eq. (7)) results in a compact representation of the data, in which the distances between the data points are determined by the geometric structure of the data. Following the definitions in [5,13], the diffusion distance between two data points x_i and x_j is the weighted L^2 distance

$$\mathbf{D}^2(x_i, x_j) = \sum_{x_l \in X} \frac{(p(x_i, x_l) - p(x_l, x_j))^2}{\phi_0(x_l)}, \tag{8}$$

where the value of $\frac{1}{\phi_0(x_i)}$ depends on the point's density. In this metric, two data points are close to each other if they are connected by many paths. Substituting Eq. (4) in Eq. (8) and using the biorthogonality properties, we obtain that the diffusion distance is expressed by

$$\mathbf{D}^2(x_i, x_j) = \sum_{k\geq 1} \lambda_k(\psi_k(x_i) - \psi_k(x_j))^2. \tag{9}$$

In these new set of diffusion maps coordinates, the Euclidean distance between two points in the embedded space represents the distances between the points as defined by a random walk.

2.2 The Laplacian Pyramid

The Laplacian pyramid is a multi-scale algorithm for approximating and extending an empirical function f, which is defined on a dataset $\mathbf{Z} = \{z_0, z_1, \ldots, z_n\}$, to new data points. In this algorithm, Gaussian kernels with descending widths are applied on the points in $\mathbf{Z}$ to construct a multi-resolution approximation of f. Then, this approximation can be extended to evaluate f for new points $\{\bar{z}\}$. An initial Gaussian kernel, having a relatively large scale σ_0, is defined on $\mathbf{Z}$ by $\mathbf{G_0} = (g_0(z_i, z_j))$ where

$$g_0(z_i, z_j) = e^{\frac{-\|z_i - z_j\|^2}{\sigma_0}}, \qquad z_i, z_j \in \mathbf{Z}. \tag{10}$$

Normalizing $\mathbf{G}_0$ results in a smoothing operator $\mathbf{K}_0 = (k_0(z_i, z_j))$, where

$$k_0(z_i, z_j) = q_0^{-1}(z_i) g_0(z_i, z_j), \quad \text{where} \quad q_0(z_i) = \sum_j g_0(z_i, z_j). \tag{11}$$

At a finer scale l, the Gaussian kernel $\mathbf{G}_l = (g_l(z_i, z_j))$ is defined by

$$g_l(z_i, z_j) = e^{-\|(z_i - z_j)\|^2 / (\frac{\sigma_0}{2^l})}, \qquad l = 1, 2, 3, \ldots \tag{12}$$

Normalization of $\mathbf{G}_l$ yields the smoothing operator $\mathbf{K}_l = (k_l(z_i, z_j))$, where

$$k_l(z_i, z_j) = q_l^{-1}(z_i) g_l(z_i, z_j), \qquad q_l(z_i) = \sum_j g_l(z_i, z_j), \qquad l = 1, 2, 3, \ldots \tag{13}$$

The Laplacian Pyramid representation of f is iteratively defined as follows. For the first level $l = 0$, a smooth approximation of f is

$$f_0(z_k) = \sum_{i=1}^{n} k_0(z_k, z_i) f(z_i), \qquad k = 1, \ldots, n, \qquad z_i, z_k \in Z. \tag{14}$$

Let

$$d_1(z_i) = f(z_i) - f_0(z_i), \qquad i = 1, 2, \ldots, n \qquad z_i \in \mathbf{Z},$$

then a finer representation of f is

$$f_1(z_k) = f_0(z_k) + \sum_{i=1}^{n} k_1(z_k, z_i) d_1(z_i), \qquad k = 1, \ldots, n.$$

In general, for $l = 1, 2, 3 \ldots,$

$$d_l(z_i) = f(z_i) - f_{l-1}(z_i), \qquad i = 1, \ldots, n, \tag{15}$$

$$f_l(z_k) = f_{l-1}(z_k) + \sum_{i=1}^{n} k_l(z_k, z_i) d_l(z_i), \qquad k = 1, \ldots, n, \tag{16}$$

where f_0 is defined in Eq. (14). Equation (16) approximates a given function f by the series of functions $\{f_0, f_1, f_2, \ldots\}$ in a multi-scale manner, going from a coarser to a finer representation. The functions $\{f_0, f_1, f_2, \ldots\}$ can be easily extended to a new point $\bar{z}$ in the following way.

$$f_0(\bar{z}) = \sum_{i=1}^{n} k_0(\bar{z}, z_i) f(z_i) \qquad \text{for} \quad l = 0 \tag{17}$$

$$f_l(\bar{z}) = f_{l-1}(\bar{z}) + \sum_{i=1}^{n} k_l(\bar{z}, z_i) d_l(z_i) \qquad \text{for } l = 1, 2, 3, \ldots, \tag{18}$$

where $d_l(z_i)$ is defined in Eq. (15).

The following example (taken from [6]) demonstrates the multi-scale approximation of the function

$$f(x) = \begin{cases} -0.02(x-4\pi)^2 + sin(x) & 0 \leq x \leq 4\pi \\ -0.02(x-4\pi)^2 + sin(x) + \frac{1}{2}sin(3x) & 4\pi < x \leq 7.5\pi \\ -0.02(x-4\pi)^2 + sin(x) + \frac{1}{2}sin(3x) + \frac{1}{4}sin(9x) & 7.5\pi < x \leq 10\pi \end{cases} \tag{19}$$

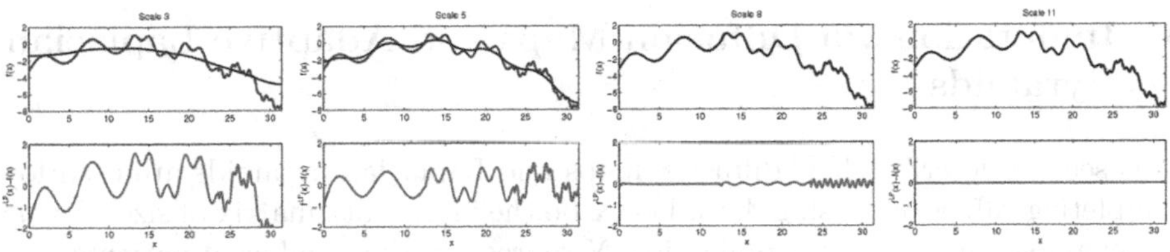

Fig. 1. Approximations of the function f that was defined in Eq. (19) for scales $l = 3, 5, 8, 11$ going from left to right. The function is plotted in blue in each of the top images, approximations f_l in black and the corresponding residuals d_l on the bottom row in red. (Color figure online)

Stopping Criteria and the Auto-adaptive Laplacian Pyramids. The Laplacian Pyramids iterations may be stopped by setting an admissible error to a small threshold *err*, for example by requiring $\|f - f_l\| < err$. When *err* is too large, then the iterations stop at a coarse scale, thus the approximation does not capture finer structures of the function f. If *err* is too small, then in finer scales a point may have few or no neighbors, thus over-fitting may occur. The auto-adaptive Laplacian Pyramids, which were introduced in [7,9], slightly modify the kernels constructed in Eqs. (10) and (12). This prevents over-fitting and provides a criteria for selecting a proper stopping scale l. The main modification is to replace the kernels $\mathbf{G}_l = (g_l(z_i, z_j))$ by $\tilde{\mathbf{G}}_l$, which are defined by

$$\tilde{\mathbf{G}}_l(z_i, z_j) = \begin{cases} \mathbf{G}_l(z_i, z_j) & i \neq j \\ 0 & i = j. \end{cases} \tag{20}$$

These yield the normalized operators $\tilde{k}_l(z_i, z_j) = \tilde{q}_l^{-1}(z_i)\tilde{g}_l(z_i, z_j)$, where $\tilde{q}_l(z_i) = \sum_j \tilde{g}_l(z_i, z_j)$ and the iterative construction

$$f_0(z_k) = \sum_{i=1}^{n} \tilde{k}_0(z_k, z_i) f(z_i) \qquad \text{for level} \quad l = 0 \tag{21}$$

$$f_l(z_k) = f_{l-1}(z_k) + \sum_{i=1}^{n} \tilde{k}_l(z_k, z_i) d_l(z_i) \qquad \text{for} \qquad l = 1, 2, \ldots. \tag{22}$$

Extension to new points is done in a similar manner, $\bar{z}$ replaces z_k in Eqs. (21) and (22).

By using the above modification, the pyramids are constructed using a Leave-one-out-cross-validation that is inherent in the algorithm, as each train point in $\mathbf{Z}$ is treated as test point. The approximation of f at z_i is built without using the value of the point itself, the contribution is only from $z_i's$ neighboring points. This modification makes the procedure more robust in the presence of noise. The stopping scale l is determined by computing the mean square error $err_l = \|f - f_l\|$ at each level and choosing the stopping scale l as the minimum value of the vector err_l. To conclude, this procedure is equivalent to running the Laplacian Pyramids algorithm in a Leave one out cross validation manner and choosing the scale where the error is minimal.

3 Imputation via Diffusion Maps and Adaptive Laplacian Pyramids

This section describes how diffusion maps and Laplacian pyramids are used for completing values of missing data. Let $\mathbf{X}$ be the input data matrix of size $N \times m$. For example, in a medical application N represents the number of patients and m is the number of medical results for each patient. For simplicity, we assume that missing data occurs in a single column $\mathbf{X}^{(j)}$ of the input matrix $\mathbf{X}$, and that the other columns $\{\mathbf{X}^{(k)}\}_{k \neq j}$ are complete. Thus, one can regress $\mathbf{X}^{(j)}$ using $\{\mathbf{X}^{(k)}\}_{k \neq j}$ or its corresponding low-dimensional representation.

Figure 2 provides an illustrative description of our approach. The input matrix $\{\mathbf{X}^{(k)}\}_{k \neq j}$ is given on the left of the figure in black. It is of dimension $N \times (m-1)$. The blue column next to it, $\mathbf{X}^{(j)}$, is of dimension $N \times 1$. Its red circles represent missing data while the blue line markers represent complete values. In accordance, the black lines in the matrix $\{\mathbf{X}^{(k)}\}_{k \neq j}$ indicate rows that have a value in $\mathbf{X}^{(j)}$, while the red lines correspond to rows with missing data in $\mathbf{X}^{(j)}$. In the center of the figure, the diffusion maps representation of the matrix $\{\mathbf{X}^{(k)}\}_{k \neq j}$ is plotted. Here, the black and red circles represent the embedding of black and red lines of the matrix $\{\mathbf{X}^{(k)}\}_{k \neq j}$, accordingly. On the right side of the figure, the column $\mathbf{X}^{(j)}$ is plotted as a function that is defined on the embedded points. Laplacian pyramids are constructed using the low-dimensional model on the right.

The proposed algorithm consist of the following three steps, a pre-processing step (Step 0) is also described.

Step 0: Preprocessing.

- For the simplicity of notations, sort the N points in $\mathbf{X}$ so that the first n, $n < N$ points are those that have a value in $\mathbf{X}^{(j)}$ and the following $N - n$ points have missing values in $\mathbf{X}^{(j)}$.
- In order to apply diffusion maps, the column of the input data matrix $\{\mathbf{X}^{(k)}\}_{k \neq j}$ should be of a comparable scale. Otherwise, if columns with large values dominate the pairwise distanced that are constructed in the kernel (see Sect. 2.1). A simple scale normalization can be performed by subtracting the

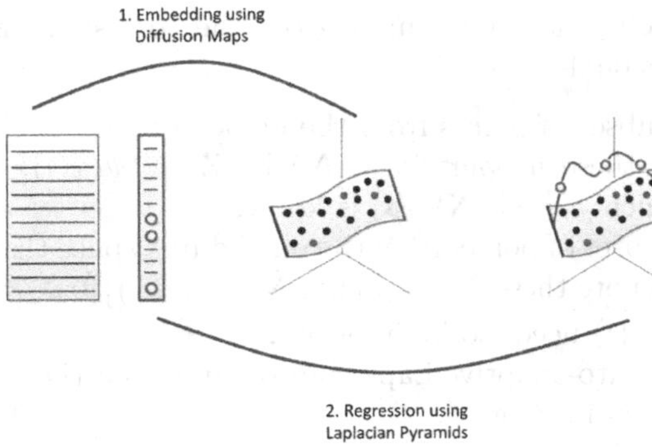

Fig. 2. Illustrative description of the proposed algorithm.

mean and dividing by the variance of each column. Another simple approach is to divide each column by it norm. A more sophisticated approach, which is based on diffusion kernels, was proposed in [18,19]. For the next step, we assume $\{\mathbf{X}^{(k)}\}_{k\neq j}$ holds columns of comparable scales.

Step 1: Reduce the dimension of the set $\{\mathbf{X}^{(k)}\}_{k\neq j}$ containing N rows and m columns via diffusion maps.

1. Construct a kernel

$$\mathbf{W}_\epsilon = w_\epsilon(x_i, x_j)) = e^{\frac{\|x_i - x_j\|^2}{2\epsilon}}.$$

The scale parameter ϵ can be computed as explained in Sect. 2.1. This results in an $N \times N$ symmetric matrix.
2. Use the Laplace-Bletrami normalization (see Sect. 2.1) and build

$$\mathbf{W}_\epsilon^{(1)} = w_\epsilon^{(1)}(x_i, x_j) = \frac{w_\epsilon(x_i, x_j)}{q(x_i)q(x_j)}, \qquad q(x_i) = \sum_j w_\epsilon(x_i, x_j).$$

3. Apply a row-normalization and obtain the Markov matrix

$$\mathbf{P} = p(x_i, x_j) = \frac{w_\epsilon^1(x_i, x_j)}{\sum_j w_\epsilon^1(x_i, x_j)}.$$

4. Compute the spectral decomposition of $\mathbf{P}$,

$$\mathbf{P} = p(x_i, x_j) = \sum_{k\geq 0} \lambda_k \psi_k(x_i)\phi_k(x_j).$$

5. Use the first $d << N$ leading diffusion coordinates

$$\Psi(x_i) = (\lambda_1\psi_1(x_i), \ldots, \lambda_d\psi_d(x_i)).$$

to embed $\{\mathbf{X}^{(k)}\}_{k\neq j}$ in a d-dimensional space.

Step 2: Set $f = \mathbf{X}^{(j)}$ and approximate from the corresponding points in the low-dimensional space.

1. Collect the subset of points from the embedding space that have a value in f. Denote these n point ($n < N$) by $\mathbf{Z} = \{\Psi(x_i)\}_{i=1}^{n}$, where x_i are points for which $f(x_i) = \mathbf{X}^{(j)}(i)$ is known.
2. Collect the subset of points from the embedding space that have a missing value in f. Denote these $N-n$ points $\bar{\mathbf{X}} = \{\Psi(x_i)\}_{i=N-n+1}^{N}$, x_i are points for which $\mathbf{X}^{(j)}(i)$ needs to be imputed.
3. Compute the auto-adaptive Laplacian pyramids for the function $f = \mathbf{X}^{(j)}$ using the points in Z by

$$f_0(z_k) = \sum_{i=1}^{n} \tilde{k}_0(z_k, z_i) f(z_i) \qquad \text{for} \quad l = 0$$

$$f_l(z_k) = f_{l-1}(z_k) + \sum_{i=1}^{n} \tilde{k}_l(z_k, z_i) d_l(z_i) \qquad \text{for} \quad l = 1, 2, \ldots$$

4. In each level store the error $err_l = \|f - f_l\|$.
5. Set the final (stopping) level $\tilde{l}$ to be $\tilde{l} = min(err_l)$.

Step 3: For each point $\bar{z} \in \bar{\mathbf{Z}}$, impute its value $f(\bar{z})$, where $f = \mathbf{X}^{(j)}$ by

$$f_0(\bar{z}) = \sum_{i=1}^{n} \tilde{k}_0(\bar{z}, z_i) f(z_i) \qquad \text{for} \quad l = 0$$

$$f_l(\bar{z}) = f_{l-1}(\bar{z}) + \sum_{i=1}^{n} \tilde{k}_l(\bar{z}, z_i) d_l(z_i) \qquad \text{for} \quad l = 1, 2, \ldots, \tilde{l}.$$

Set the missing values $f(\bar{z})$ to be $f_l(\bar{z})$.

4 Experimental Results

The application of the proposed algorithm is demonstrated on a mice protein expression dataset, taken from the UCI repository [16]. The data set consists of the expression levels of 77 proteins that produced detectable signals in the nuclear fraction of cortex. These types of biological datasets, including the above dataset, often have missing values. From this dataset, we select a subset of $N = 1000$ rows and $m = 66$ columns, which do not contain missing data. Assuming now that some data is missing in a particular column, then diffusion maps followed by Laplacian pyramids (see Sect. 3) is applied for imputation. We measure the efficiently of the proposed algorithm by calculating the mean and maximum errors. We compare our results to linear regression and K-nearest neighbors techniques. The evaluation is carried out using a 5-fold cross validation.

Let $\{\mathbf{X}^{(k)}\}_{k \neq 10}$ be the full dataset and $\mathbf{X}^{(10)}$ be a column with missing data. First, the data set $\{\mathbf{X}^{(k)}\}_{k \neq 10}$ is pre-processed according to Step 0 in Sect. 3 by substraction of the mean and division by the standard deviation of each

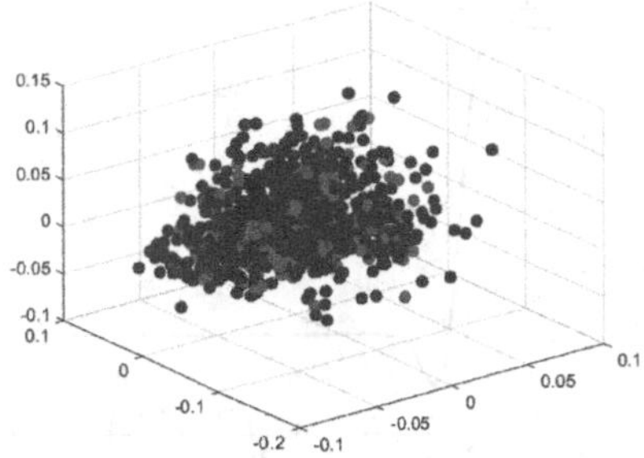

Fig. 3. Diffusion maps embedding of the set $\{\mathbf{X}^{(k)}\}_{k\neq 10}$. Points with known values in $\mathbf{X}^{(10)}$ are colored blue and points with missing values in $\mathbf{X}^{(10)}$ are colored red. (Color figure online)

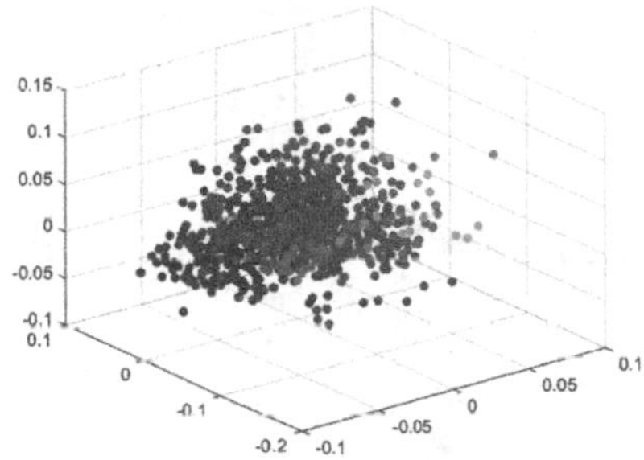

Fig. 4. Diffusion maps embedding of the set $\{\mathbf{X}^{(k)}\}_{k\neq 10}$ colored by the values of $f = \mathbf{X}^{(10)}$.

column. Next, diffusion maps is applied to the set $\{\mathbf{X}^{(k)}\}_{k\neq 10}$ and embedded in a 3-dimensional space. The values of $\mathbf{X}^{(10)}$ are known for $n = 800$ points and unknown for the remaining $N - n = 200$ points. Figure 3 presents the 3-dimensional embedding of $\{\mathbf{X}^{(k)}\}_{k\neq 10}$. The 800 points for which the values of $\mathbf{X}^{(10)}$ are known are colored in blue where as the 200 points that have missing values in $\mathbf{X}^{(10)}$ are colored in red. Coloring the embedding by the values of the function $f = \mathbf{X}^{(10)}$, we can induce from Fig. 4 that the function is smooth on the embedded data.

Next, the adaptive Laplacian pyramids are constructed with $n = 800$ known values. The stopping level $\tilde{l}$ for the Laplacian pyramids is set to be the iteration having the minimum error (as explained in Sect. 2.2). Figure 5 displays the values of err_l as computed for each iteration of the Laplacian pyramids; the minimum is reached for $l = 14$. This is the stopping scale used for extending the pyramids (see Step 3 in Sect. 3). Last, the $N - n = 200$ missing values are imputed with Laplacian pyramids and compared to linear regression and k-nearest neighbors. All three methods rely on the embedded data instead of the original high-dimensional space. Figure 6 displays the approximated values as imputed by the Laplacian pyramids (in blue), linear regression (in green) and k-nn with $k = 7$ (in yellow). The error bars as computed from a 5-fold cross validation are presented in Fig. 7. We plot the mean square error in the left

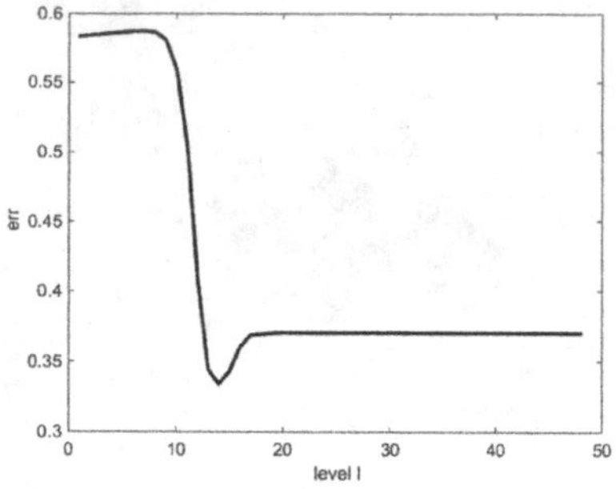

Fig. 5. Approximation errors for each scale l in the Laplacian pyramid construction. The stopping scale is set to be the level $\tilde{l}$ with the minimal error, here $\tilde{l} = 14$.

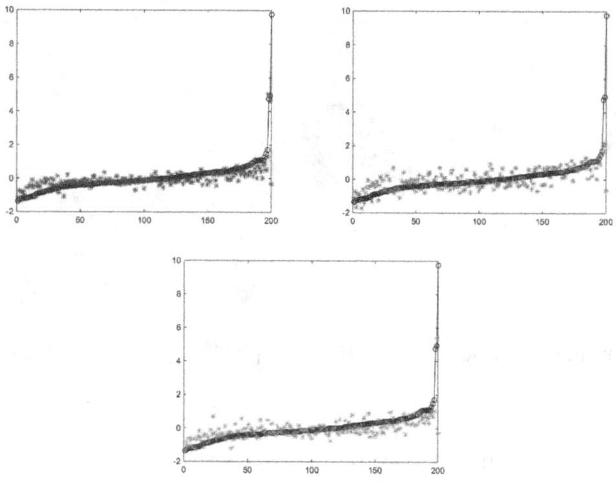

Fig. 6. Imputation of $f = \mathbf{X}^{(10)}$ (the missing values are sorted) using the adaptive Laplacian pyramids (blue), linear regression (green) and k-nn, $k = 7$ (yellow). The true values are in black. (Color figure online)

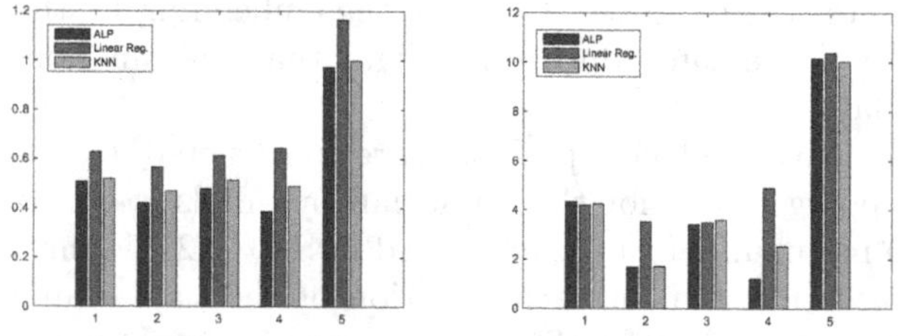

Fig. 7. Mean squared error (left) and maximum error (right) for the 5-fold cross validation imputation of $f = \mathbf{X}^{(10)}$.

chart and the maximum error in the right chart. It can be seen that the linear regression results with the largest error and that the pyramids based imputation yields the smallest errors.

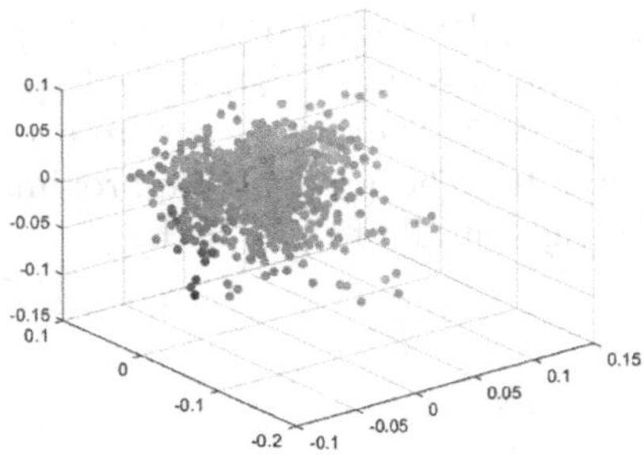

Fig. 8. Diffusion maps embedding of the set $\{\mathbf{X}^{(k)}\}_{k\neq 48}$ colored by the values of $f = \mathbf{X}^{(48)}$.

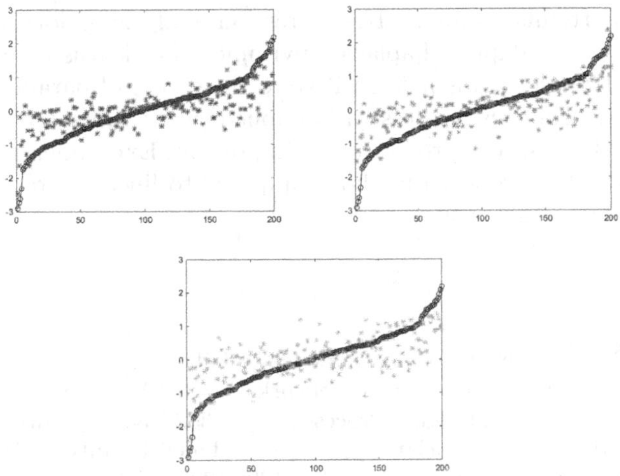

Fig. 9. Imputation of $f = \mathbf{X}^{(48)}$ (the missing values are sorted) using the adaptive Laplacian pyramids (blue), linear regression (green) and k-nn, $k = 7$ (yellow). The true values are in black. (Color figure online)

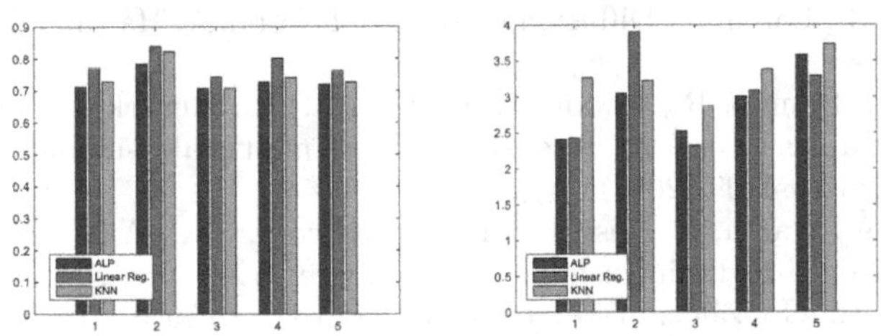

Fig. 10. Mean squared error (left) and maximum error (right) for the 5-fold cross validation imputation of $f = \mathbf{X}^{(48)}$.

We repeat the above example for another column $\mathbf{X}^{(48)}$. As before, the set $\{\mathbf{X}^{(k)}\}_{k\neq 48}$ is embedded into a 3-dimensional space. The function $f = \mathbf{X}^{(48)}$ is not as smooth as $\mathbf{X}^{(10)}$, this is displayed in Fig. 8, where the diffusion maps embedding is colored by $f = \mathbf{X}^{(48)}$. Figure 9 plots the imputation results for the missing values (sorted) as evaluated by the adaptive Laplacian pyramids, linear regression and k-nn with $k = 7$. The means square errors and the maximum errors are presented in Fig. 10. The pyramids approximations maintain small mean and max errors throughout all of the 5 folds.

5 Conclusions

In this paper, a two-step method was presented for data completion, which is suitable for high dimensional data. The proposed algorithm uses diffusion maps for reducing the dimension of the training samples with complete data in the first step. Next, a regression process is carried out in order to evaluate missing values from a particular column. The regression analysis is done with a multi-scale method named adaptive Laplacian pyramids that learns the suitable scale for regression. The method is not sensitive to the choice of parameters. Since in such imputation problems, regression is applied many times for filling in data from different columns, such properties are important. Experimental results show the advantages of the proposed method compared to linear regression and local k-nearest neighbors regression.

References

1. Asif, M.T., Mitrovic, N., Garg, L., Dauwels, J., Jaillet, P.: Low-dimensional models for missing data imputation in road networks. In: IEEE International Conference on Acoustics, Speech and Signal Processing, pp. 3527–3531 (2013)
2. Belkin, M., Niyogi, P.: Laplacian eigenmaps for dimensionality reduction and data representation. Neural Comput. **15**, 1373–1396 (2003)
3. Belkin, M., Niyogi, P.: Semi-supervised learning on Riemannian manifolds. Mach. Learn. **56**, 209–239 (2004)
4. Chung, F.R.K.: Spectral Graph Theory. AMS Regional Conference Series in Mathematics (1997)
5. Coifman, R.R., Lafon, S.: Diffusion maps. Appl. Comput. Harmon. Anal. **21**, 5–30 (2006)
6. Dsilva, C.J., Talmon, R., Rabin, N., Coifman, R.R., Kevrekidis, I.G.: Nonlinear intrinsic variables and state reconstruction in multiscale simulations. J. Chem. Phys. **139**(18), 184109 (2013)
7. Fernández, Á., Rabin, N., Fishelov, D., Dorronsoro, J.R.: Auto-adaptative laplacian pyramids for high-dimensional data analysis. arXiv preprint arXiv:1311.6594
8. Fernández, Á., González, A.M., Díaz, J., Dorronsoro, J.R.: Diffusion maps for dimensionality reduction and visualization of meteorological data. Neurocomputing **163**, 25–37 (2015)
9. Fernández, Á., Rabin, N., Fishelov, D., Dorronsoro, J.R.: Auto-adaptive Laplacian Pyramids. In: 24th European Symposium on Artificial Neural Networks. Computational Intelligence and Machine Learning, ESANN, pp. 59–64, Bruges, Belgium (2016)
10. Huisman, M.: Missing data in behavioral science research: investigation of a collection of data sets. Kwant. Methoden **57**, 69–93 (1998)

11. Little, J.A.R., Rubin, B.D.: Statistical Analysis with Missing Data, 2nd edn. Wiley, Hoboken (2002)
12. Nadler, B., Lafon, S., Coifman, R.R., Kevrekidis, I.G.: Diffusion maps, spectral clustering and eigenfunctions of Fokker-Planck operators. In: Neural Information Processing Systems (NIPS), vol. 18 (2005)
13. Nadler, B., Lafon, S., Coifman, R.R., Kevrekidis, I.G.: Diffusion maps, spectral clustering and reaction coordinate of dynamical systems. Appl. Comput. Harmon. Anal. **21**, 113–127 (2006)
14. Pearson, K.: On lines and planes of closest fit to systems of points in space. Philos. Mag. **2**(11), 559–572 (1901)
15. Pierson, E., Yau, C.: ZIFA: dimensionality reduction for zero-inflated single-cell gene expression analysis. Genome Biol. **16**, 241 (2015)
16. http://archive.ics.uci.edu/ml/datasets
17. UshaRani, Y., Sammulal, P.: An efficient disease prediction and classification using feature reduction based imputation technique. In: International Conference on Engineering & MIS (ICEMIS) (2016)
18. Rabin, N., Averbuch, A.: Detection of anomaly trends in dynamically evolving systems. In: 2010 AAAI Fall Symposium Series, pp. 44–49 (2010)
19. Rabin, N., Coifman, R.R.: Heterogeneous datasets representation and learning using diffusion maps and Laplacian pyramids. In: Proceedings of the 2012 SIAM International Conference on Data Mining, pp. 189–199 (2012)
20. Roweis, S.T., Saul, L.K.: Nonlinear dimensionality reduction by locally linear embedding. Science **290**, 2323–2326 (2000)
21. Schclar, A.: A diffusion framework for dimensionality reduction. In: Maimon, O., Rokach, L. (eds.) Soft Computing for Knowledge Discovery and Data Mining, pp. 315–325. Springer, Heidelberg (2008). doi:10.1007/978-0-387-69935-6_13
22. Zhao, Z., Giannakis, D.: Analog forecasting with dynamics-adapted kernels. Nonlinearity **29**, 2888 (2016)

Classification of Cocaine Dependents from fMRI Data Using Cluster-Based Stratification and Deep Learning

Jeferson S. Santos[1], Ricardo M. Savii[1], Jaime S. Ide[2], Chiang-Shan R. Li[2], Marcos G. Quiles[1(✉)], and Márcio P. Basgalupp[1]

[1] Instituto de Ciência e Tecnologia, Universidade Federal de São Paulo, São José dos Campos, SP, Brazil
{souza.jeferson,ricardo.manhaes,quiles,basgalupp}@unifesp.br
[2] School of Medicine, Yale University, New Haven, USA
{jaime.ide,chiang-shan.li}@yale.edu

Abstract. Cocaine dependence continues to devastate millions of human lives. According to the 2013 National Survey on Drug Use and Health, approximately 1.5 million Americans are currently addicted to cocaine. It is important to understand how cocaine addicts and non-addicted individuals differ in the functional organization of the brain. This work advances the identification of cocaine dependence based on fMRI classification and innovates by employing deep learning methods. Deep learning has proved its utility in machine learning community, mainly in computational vision and voice recognition. Recently, studies have successfully applied it to fMRI data for brain decoding and classification of pathologies, such as schizophrenia and Alzheimer's disease. These fMRI data were relatively large, and the use of deep learning in small data sets still remains a challenge. In this study, we fill this gap by (i) using *Deep Belief Networks* and *Deep Neural Network* to classify cocaine dependents from fMRI, and (ii) presenting a novel stratification method for robust training and evaluation of a relatively small data set. Our results show that deep learning outperforms traditional techniques in most cases, and present a great potential for improvement.

1 Introduction

According to the 2013 National Survey on Drug Use and Health, approximately 1.5 million Americans are currently addicted to cocaine. Despite treatment, individuals relapse to drug use at a very high rate, leading to prolonged, devastating impacts on human lives and the society. Understanding the neurobiology of cocaine dependence is therefore of extreme importance to public health. In particular, functional magnetic resonance imaging (fMRI) allows us to delineate brain dysfunctions in cocaine addiction.

Two main approaches have being proposed for the analysis of fMRI data: the standard mass-univariate approach known as general linear modeling (GLM) [5]

O. Gervasi et al. (Eds.): ICCSA 2017, Part I, LNCS 10404, pp. 298–313, 2017.
DOI: 10.1007/978-3-319-62392-4_22

and the multi-voxel pattern analysis (MVPA) using pattern recognition techniques [9]. GLM requires a priori task-design and is limited to observable or at least measurable task conditions [5]. On the other hand, MVPA has the ability to delineate complex associations between multiple voxel signals, stimuli, or mental states in a data-driven way. Unlike most previous work on brain decoding, our focus consists in predicting neuropathology, that is, identifying cocaine dependence and its associated neural features. This is a challenging task because of the commonly large inter-subject variability and small sample size of the data set [12].

In this context, the increasingly popular deep learning methods are potentially very promising tools for the classification task of addiction pathology using fMRI. However, the use of deep learning in fMRI with small data set is still an unexplored research area. This is critical because data from clinical populations are often times expensive to obtain and relatively small. In this work, we contribute by: (i) examining a small but rich fMRI data of 163 individuals (cocaine dependents and healthy individuals) performing a cognitive control task while in the MR scanner; (ii) delineating the optimal fMRI features for cocaine dependence classification. Traditional MVPA methods are based on high-dimensional voxel-wise data. Here, we deal with this challenge by using low-dimensional brain circuit-based features; (iii) investigating the performance of two deep learning models, and identifying the best architecture of these neural networks; and (iv) presenting a novel stratification strategy based on Growing Neural Gas (GNG) for robust learning.

This work is organized as follows. Section 2 presents a background about deep learning and fMRI. Section 3 describes cocaine dependence and cognitive control. Section 4 reports our experimental protocol, and Sect. 5 presents the experimental results and their analyses. Finally, Sect. 6 presents our conclusions and some future research directions.

2 Background

Bio-inspired techniques have diversified and been widely implemented throughout the scientific community, from parameters optimization with Genetic Algorithms [2] to classification and regression with Artificial Neural Networks (ANNs) [10]. Particularly the ANNs are a hot trend in machine learning community. Its evolution unveiled diverse applications in computer vision and voice recognition [16], among other fields [21,22].

An important limitation of ANN concerns the "Gradient vanish" in its learning algorithm. These limitations did not permit its use for high abstraction problems or deeper architecture [7]. In 2006, some of these problems were solved and Deep Learning emerged as a new field [11]. Following initial advancements, deep learning techniques were promoted by leading researchers and its effects began to reach out to the broader community [19].

Whilst deep learning's main improvements are in image understanding and voice recognition [16], its use is diversifying. Some researchers have applied deep models to fMRI data. For instance, Koyomada et al. [15] used deep learning techniques to identify human sentiments generated from visual stimuli. Plis et al. [23] used a deep learning neural network to classify healthy individuals and patients with schizophrenia. A common characteristic of previous image and voice processing work, as well as fMRI applications, is the data volume. Usually, deep networks are trained on data sets containing thousands or even millions of examples.

Large companies like Google, Facebook, Microsoft, Apple, and others focus on the use of deep learning techniques in big data [14]. However, in an open letter sent to one of the leading experts in deep learning Yann LeCun[1], Shalini Ananda pointed that most of the deep learning successes relies on big data scenarios, and drawn the attention to a gap in studies focused on smaller data sets. In the letter, Ananda observed that among the difficulties of dealing with small datasets are the extraction of information by the specialist and the configuration of suitable model parameters. To address these matters and to achieve success might be necessary an interdisciplinary approach. The interdisciplinarity requires expertise in both directions; for example, a deep learning engineer specialized in neuroscience and a neuroscientist with expertise in computational learning techniques.

3 Cognitive Control, Cocaine Dependence and Functional Connectivity

Cognitive control is a critical executive function, defined as the ability to withhold or modify actions in response to a dynamically changing environment. With a variety of laboratory paradigms, numerous studies have characterized deficits in cognitive control in chronic cocaine users. For example, in the stop signal task (SST) where ones respond quickly to an imperative "go" stimulus and must withhold response as instructed by an occasional "stop" signal, cocaine dependent individuals (CD) demonstrated diminished response inhibition and error processing, as compared to demographically matched healthy controls (HC) [18].

In a recent large scale study involving 461 subjects, authors demonstrated that individuals' lifestyle, demographic and psychometric measures are closely associated with pattern of brain connectivity [25]. Therefore, functional connectivity measures might be better predictors of cocaine dependence. In fact, previous work have shown disrupted/altered functional connectivity in cocaine abusers while performing a cognitive task [27] or even during resting state [26]. In Fig. 1, we depict relevant brain regions extracted from the current population in study and their associated functional connectivity.

4 Methodology

This section describes the methodology employed in this work. First, we present the dataset and its pre-processing. Then, we describe the deep learning models

[1] https://medium.com/@ShaliniAnanda1.

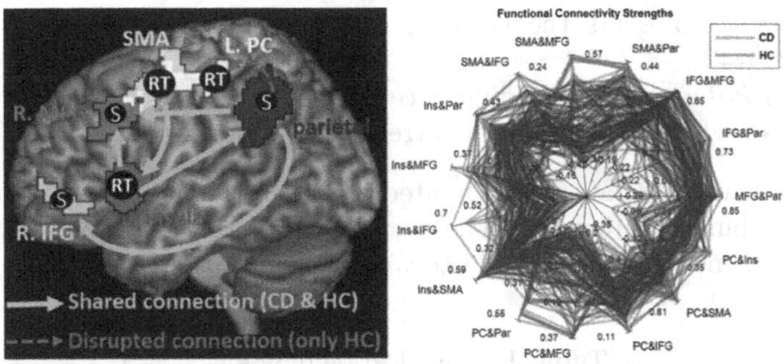

Fig. 1. Disrupted frontoparietal circuit in cocaine addicts. The frontoparietal circuit included six regions responding to Bayesian conflict anticipation [13] ("S") and regions of motor slowing ("RT"). (a) CD and HC shared connections (orange arrows); (b) We represent connectivity strengths between nodes for each individual subject in CD (red line) and HC (blue line) groups. (Color figure online)

used in this study. We also specify the baseline algorithms considered in the comparative analysis. Finally, we introduce the statistical tests undertaken in this investigation.

4.1 Data Acquisition and Pre-processing

One-hundred and sixty-eight subjects were recruited from the New Haven area, and MRI data were collected while subjects performed the stop signal task (SST), a common task used to study inhibition control [17]. The fMRI data were collected according to the protocol approved by the *Yale Human Investigation Committee*.

The data were preprocessed using Statistical Parametric Mapping 12 [4], following standard procedures. Further, fMRI time-series were extracted from particular regions of interest (ROIs) by averaging and linear detrending.

Unlikely most of the multivariate pattern analysis approaches [20], which use activation maps, we employed functional connectivity measures as features since based on our initial experiments, the later provided consistent and better results. Following previous studies [13], six ROIs were selected: bilateral parietal, left precentral cortex, supplementary motor area, inferior frontal *gyrus*, right medial frontal *gyrus* and bilateral *insula*. These regions were significantly activated in response to conflict expectation and are part of a frontoparietal circuit involved in attention and inhibitory control [13].

The time series of each ROI were concatenated through sessions and averaged across voxels. These 6 time-series were entered into a *multivariate Granger Casualty analysis* (mGCA), and cross-correlation computation. Finally, pairwise functional connectivity measures were generated as follows:

1. *F-Value* - represents the Granger causality strength estimated through standard F-statistics;

2. *F-Geweke* - represents the Granger causality strength computed through F-geweke [24];
3. *Degree of Influence* (DOI) - difference in F-geweke between two regions [24];
4. *Pearson's coefficients* - standard correlation coefficients;

Table 1 shows the 8 data sets generated from each original set of features (1–4) and their combinations (5–8), each data set containing 163 examples (each example represents one patient), with 75 cocaine dependent and 88 healthy controls.

Table 1. Classification data sets

Data-set	#Features	Acronym	Features
1	30	F-values	F-statistics
2	30	F-Geweke	Geweke coefficients
3	30	DOI	Degree of influence
4	15	Pearson	Pearson coefficients
5	45	***	F-values (30) + Pearson (15)
6	45	***	F-Geweke (30) + Pearson (15)
7	45	***	DOI (30) + Pearson (15)
8	105	***	F-values + F-Geweke + DOI + Pearson

4.2 Algorithms

Here, we took two deep learning models into account: the DBN (*Deep Belief Networks*) and the DNN (*Deep Neural Network*). (i) DBN is a model with unsupervised pre-training before supervised training [11]. The pre-training is done by applying a network called *Restricted Boltzmann Machine* (RBM), which captures the probability distribution of the set [11]. After the pre-training phase, the network is trained in a supervised fashion with the *Backpropagation* algorithm [10]. Following this step, networks are trained with more than two layers to avoid the *vanishing gradient* problem. With more layers, the abstraction capacity of the model is enhanced [16]. (ii) DNN is a supervised model which does not require pre-training. The DNN adopts a rectified linear unit activation function (ReLU) [7]. Once the ReLU is taken into account, the vanishing gradient problem is also avoided, which permits the setup of deeper topologies [7].

In this work, we adopted an unbiased exploratory strategy, and did not apply any particular method to optimize the parameters of the deep learning models. Instead, we empirically tested 229 different combinations by varying the following parameters: (i) number of layers (1–5); (ii) number of neurons per layer (100–600); (iii) learning rate (0.1, 0.01); and (iv) activation functions (sigmoid, ReLU and Softmax).

In order to make a comparative analysis, we used the following baseline algorithms available on WEKA [8] framework: J48, Random Forest, BayesNet,

Multilayer Perceptron (MLP), and SMO (SVM WEKA's implementation). All of the baseline classification algorithms were used with the default parameters values as adopted by WEKA, which typically represent robust values that work well across different datasets. None of the baseline classification algorithms had their parameter values optimized specifically for our fMRI data sets, since we consider the default values are already tuned for a high number of "generic" data sets. A more robust parameter optimization procedure, considering all methods, is open for future research.

4.3 Evaluation

A 10-fold cross-validation procedure was used to evaluate the seven algorithms (two deep learning methods plus five baselines algorithms). As the deep learning approaches are non-deterministic, results corresponded to an average over five executions. The main predictive measure used here was the accuracy. During the learning of the deep methods, for each fold the training data was divided into sub-training (80%) and validation (20%) sets, with the later being used to estimate the predictive performance as learning stop criterion.

In order to provide some reassurance about the validity and non-randomness of the obtained results, we present the results of statistical tests by following the approach proposed by Demšar [1]. In brief, this approach seeks to compare multiple algorithms on multiple data sets, and it is based on the use of the Friedman test with a corresponding post-hoc test. The Friedman test is a non-parametric counterpart of the well-known ANOVA, as follows. Let R_i^j be the rank of the jth of k algorithms on the ith of N data sets. The Friedman test compares the average ranks of algorithms, $R_j = \frac{1}{N}\sum_i R_i^j$. The Friedman statistic is given by Eqs. 1 and 2.

$$\chi_F^2 = \frac{12N}{k(k+1)}\left[\sum_j R_j^2 - \frac{k(k+1)^2}{4}\right] \quad (1)$$

is distributed according to χ_F^2 with $k-1$ degrees of freedom, when N and k are big enough.

$$F_f = \frac{(N-1)\times\chi_F^2}{N\times(k-1)-\chi_F^2} \quad (2)$$

which is distributed according to the F-distribution with $k-1$ and $(k-1)(N-1)$ degrees of freedom.

If the null hypothesis of similar performances is rejected, then we proceed with the Nemenyi post-hoc test for pairwise comparisons. The performance of two classifiers is significantly different if their corresponding average ranks differ by at least the critical difference

$$CD = q_\alpha\sqrt{\frac{k(k+1)}{6N}} \quad (3)$$

where critical values q_α are based on the Studentized range statistic divided by $\sqrt{2}$.

5 Experiments and Results

Our experiments and analyses are presented in two parts. The first part shows which datasets generated the best results with respect to both deep learning models (DBN and DNN), including a comparison with the baseline methods. The second part introduces a new stratification method for cross-validation, which considerably improved the results obtained by all methods.

5.1 Comparative Analysis

Figure 2 presents the results for all datasets. It is worth noting that the experiments with dataset number 4 achieved the best results. Moreover, the datasets 5–8, which represent a combination of dataset 4 with other features also deliver better results than those observed in datasets 1–3.

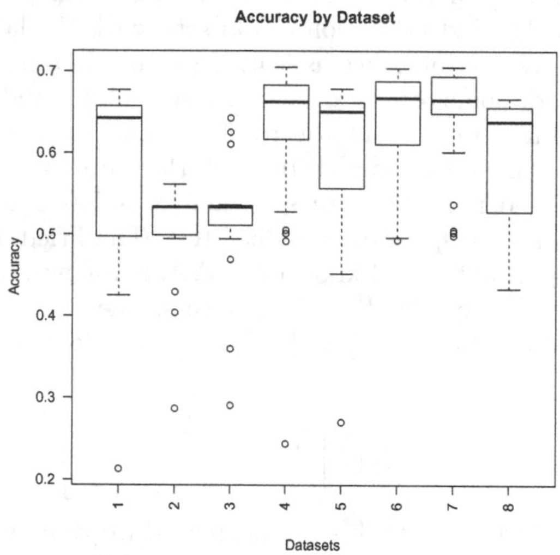

Fig. 2. Results for 8 data-sets

Table 2 depicts the comparison to the other classifiers. The columns Shallow and Deep refer to the Neural models considered in this work. The column Shallow contains the information about networks with a maximum of two layers; the Deep column shows the results with deeper architecture (more than two layers). All the results (except Shallow and Deep) were generated using software (workbench) that is a collection of machine learning algorithm called WEKA [8].

For the statistical analysis, the computed value of $F_f = 4.98$. Since $F_f > F_{0.05}(6, 42)$ $(4.98 > 2.32)$, the null-hypothesis is rejected, and thus we can argue that there are significant differences among the four methods regarding Accuracy. If we proceed with a post-hoc Nemenyi test, the critical difference is

Table 2. Average accuracy obtained by the deep approach and the baselines.

Data-sets	MLP	BayesNet	J48	RD	SVM	Shallow	Deep
1	0.620	0.533	0.573	0.653	0.627	0.651	0.677
2	0.633	0.553	0.627	0.687	0.640	0.561	0.533
3	0.593	0.560	0.593	0.607	0.627	0.643	0.535
4	0.687	0.693	0.633	0.673	0.700	0,705	0.691
5	0.687	0.673	0.680	0.713	0.727	0.659	0.679
6	0.680	0.673	0.593	0.700	0.700	0.704	0.704
7	0.667	0.673	0.600	0.687	0.707	0.705	0.700
8	0.673	0.647	0.620	0.713	0.680	0.665	0.659

$CD = 3.18$, and we can observe that SVM statistically outperforms both J48 and BayesNet, since the differences between SVM and J48 (3.625) and between SVM and BayesNet (3.438) are greater than CD (Fig. 3).

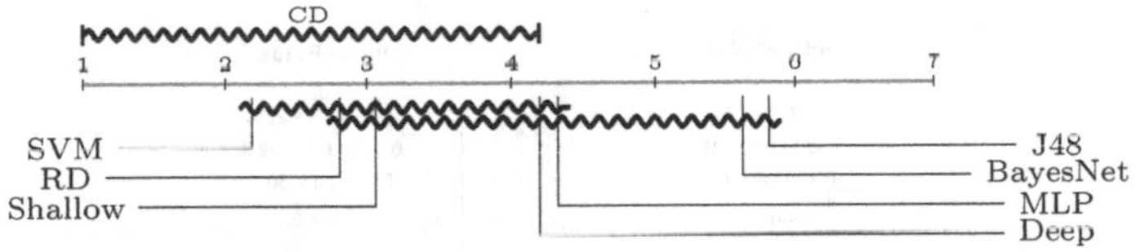

Fig. 3. Critical diagrams showing average ranks and Nemenyi's critical difference for accuracy.

Figure 3 shows the Nemenyi's critical diagram, as suggested by Demsăr [1]. In this diagram, a horizontal line represents the axis on which we plot the average rank values of the methods. The axis is turned so that the lowest (best) ranks are to the left since we perceive the methods on the left side as better. When comparing all the criteria against each other, we connect the groups of criteria that are not significantly different through a bold horizontal line. We also show the critical difference given by the Nemenyi's test above the graph. We can see that SVM is not connected to J48 and BayesNet (significant difference). However, all other methods are connected among them, which means that there is no statistically significant difference among these methods.

The low accuracy rate has motivated the investigation of the characteristics of the data sets, since we have noticed a large dispersion among folds. This evidence might be confirmed by analyzing the ranking presented in Fig. 4, which shows the top10 best results achieved by the DBN and the DNN models. Despite reaching an average accuracy of 70%, the dispersion is prominent with accuracy ranging from about 50% to 90%, i.e., a high standard deviation among the folds. Besides the ranking itself, this figure also shows which model (DBN or DNN), the number of layers, and the dataset considered in the experiments.

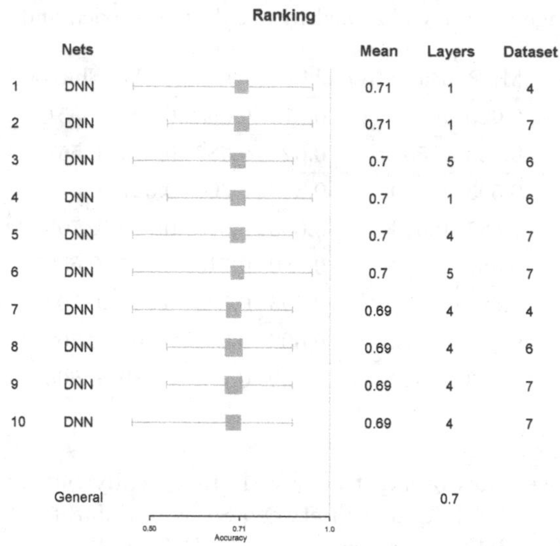

Fig. 4. General ranking - best performances.

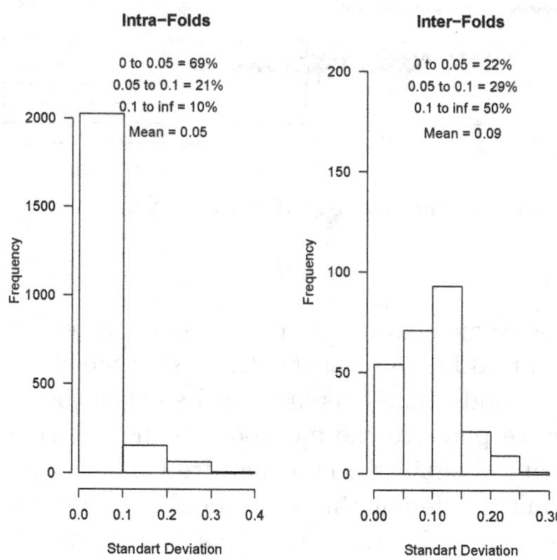

Fig. 5. Standard deviation levels intra-folds and inter-folds.

In Fig. 5 it is possible to observe the standard deviation levels intra-folds and inter-folds. By analyzing this figure, we can notice that the variance among (inter) folds generally does not repeat in the runs considering each fold (intra). This observation suggests that the high standard deviation values are a result of data split, possibly explaining the dispersion (between 50% and 90%). In the context of intra-folds, the majority of cases has a deviation lesser or equal to 0.05 in accuracy.

5.2 GNG-Based Stratification

To avoid the above-mentioned grouping problem, we propose a new method to stratify the examples during the cross-validation procedure. Instead of using the classes, we propose to use the "cluster information" to split the examples in different folds. To analyze this difference between folds, we take an unsupervised self-organizing neural network known as Growing Neural Gas (GNG) [6] into account.

The GNG has been widely applied to unsupervised problems for representing high-dimensional feature space into a low dimensional space. In particular, the GNG can learn the intrinsic topological relation between samples in an incremental fashion. During the learning phase, the GNG adjusts its topology by adding and removing neurons and links [6]. As a result, after the training phase, each GNG neuron might be considered as a prototype of a cluster it represents. Moreover, the GNG as a whole might be seen as a graph that links prototypes from each region of the feature space.

Thus, in contrast to splitting the folds based exclusively on the class information, which does not take the feature similarity of samples into account, the clusters might offer further information regarding representative samples that should be learned by the system.

In Fig. 6, we illustrate stratification problem. Consider we have a sample with two classes (circles and squares) and two groups (clusters) based on feature similarity (gray and black colors). When considering the samples in Fig. 6A and a traditional stratification, it is possible that all black squares might be assigned to the same fold (Fig. 6B). Thus, it would be impossible to classify correctly these samples since none of them is present in the training set (see Fig. 6C). Although these individuals (black squares) belong to class 1 (square), they are closer (in terms of similarity) to individuals from cluster 1, and also black but actually belong to class 2 (circle). This scenario is even more probable when dealing with small datasets, such as the one considered in this investigation.

Thus, the idea is not only split the data according to the class information but also based on the feature similarity by using the outcome of GNG network.

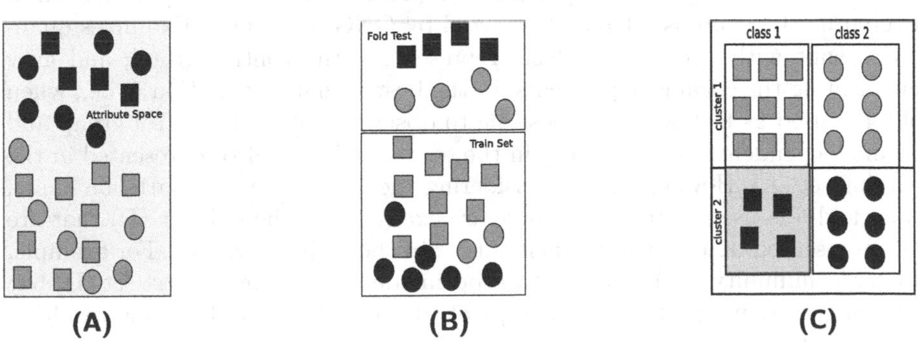

Fig. 6. Stratification problem.

Table 3. Neurons and community by experiment.

	Run 1	Run 2	Run 3	Run 4	Run 5
Neurons GNG	20	20	20	22	24
Community	4	4	3	2	3

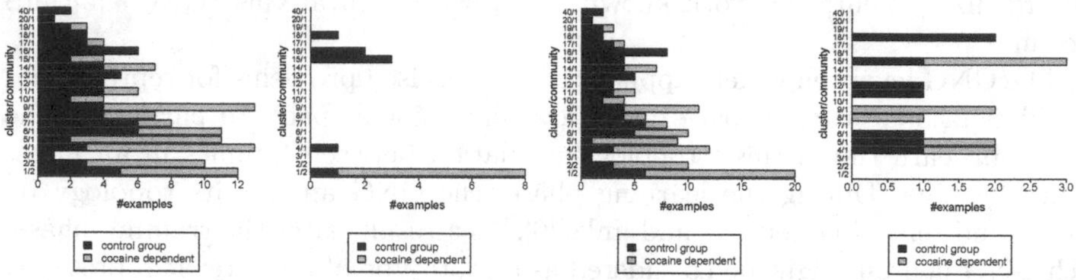

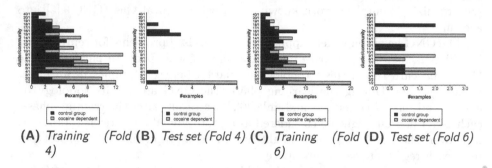

(A) *Training (Fold 4)* (B) *Test set (Fold 4)* (C) *Training (Fold 6)* (D) *Test set (Fold 6)*

Fig. 7. Distribution analysis by cluster/community from GNG/COM considering the original split (10 folds). (A) training set (all folds except 4); (B) test set (fold 4); (C) training set (all folds except 6); and (D) test set (fold 6).

Besides, assuming that the GNG is a graph and to reduce the number of clusters in our data, we run a community detection method, named Multilevel Community [3], to group the former GNG neuron clusters into a set of community of clusters. Here a community represents a set of densely connected (or densely link) GNG neurons. The communities might be considered as coarse-grained clusters of the feature space, which are the conglomerate of small clusters. From here, we will refer to these two stratified approaches as GNG and COM, respectively.

Table 3 shows the number of neurons generated and communities detected in five runs of these algorithms. All experiments from this section were executed for dataset number 4, which achieved the best results in Sect. 5.1.

Considering a specific run (4) from Table 3, Fig. 7 illustrates the distribution of the examples belonging to partitions 4 (best results) and 6 (worst results) covered by the neurons (clusters) detected by GNG and COM. Examples quantity is split with two colors: black, representing the control group; and gray representing the cocaine dependents. In the best scenario (partition 4, i.e., when fold 4 is used as test set), it is possible to observe that both control group and cocaine dependents, when present in the test set (B) are also represented in the training set (A). However, when considering the worst scenario (partition 6, i.e., when fold 6 is used as test set), there are examples in the test set (D) that are not represented in its correspondent cluster in the training set (C). For example, cluster/community 15/1 contains two cocaine dependents in the test set (D) but none in the training set (C). This type of situation hampers learning, resulting in low predictive accuracy.

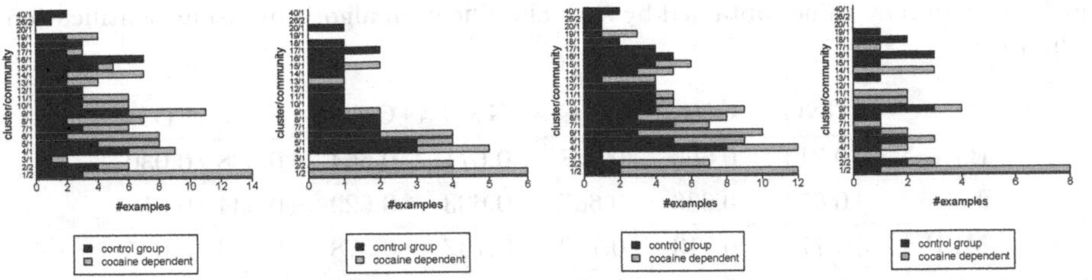

(A) *Training (Fold 1)* **(B)** *Test set (Fold 1)* **(C)** *Training (Fold 2)* **(D)** *Test set (Fold 2)*

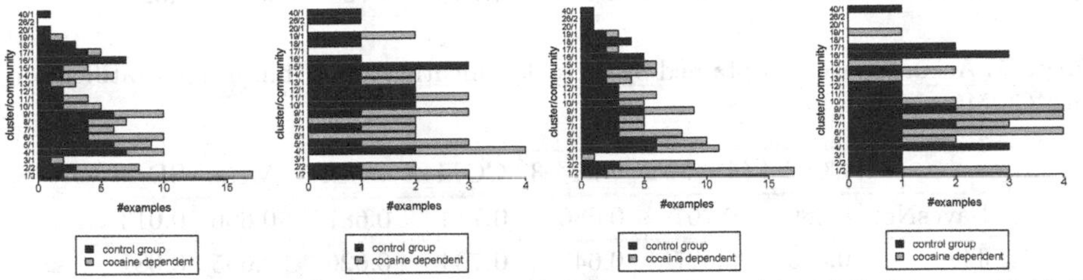

(E) *Training (Fold 3)* **(F)** *Test set (Fold 3)* **(G)** *Training (Fold 4)* **(H)** *Test set (Fold 4)*

Fig. 8. Distribution analysis by cluster/community obtained by GNG/COM with 4-fold cross-validation.

Since we are dealing with small datasets, some clusters contained a small number of examples after we applied GNG/COM stratification approaches. Thus, we used 4-folds cross-validation, instead of 10-folds, in order to increase the chances of having all clusters represented in each fold, and to produce reliable training. In contrast with the traditional stratification approach (Fig. 7), we can observe the data distribution considering GNG/COM approaches in Fig. 8. Now, in all partitions we can observe representativeness of both classes in almost all clusters, which, according to our hypothesis, can improve the learning process and consequently increase the prediction accuracy. Even thought some clusters/communities in the test sets still containing examples whose classes are not represented in the training set (for example, cluster/community 3/1 in Fig. 8D), the number of cases when it occurs is smaller, making possible (according to our hypothesis) to increase the accuracy rate.

In order to verify our hypothesis, we run the classification algorithms over the 4-fold cross-validation generated by GNG and COM. The results are presented in Tables 4 and 5, respectively. The improvement of the results obtained after applying both stratification approaches, GNG and COM, is notable when compared with the traditional stratification approach. This improvement is observed for all classification methods, but especially for DNN, which achieved an average accuracy of 0.765 (against 0.71 when using the traditional stratification). Moreover, it is possible to highlight that DNN obtained an accuracy of 0.855 in one execution (COM 4). This is an evidence that the model is able to achieve high

Table 4. Accuracy values obtained by each classification algorithm using stratification with GNG

	GNG 1	GNG 2	GNG 3	GNG 4	GNG 5	Mean	SD
BayesNet	0.714	0.665	0.728	0.671	0.664	0.688	0.030
J48	0.654	0.636	0.667	0.633	0.629	0.644	0.016
MLP	0.772	0.713	0.710	0.737	0.678	0.722	0.035
RD	0.712	0.646	0.679	0.683	0.664	0.677	0.024
SVM	0.701	0.702	0.665	0.683	0.722	0.695	0.022
DNN	0.776	0.722	0.741	0.730	0.728	0.739	0.022

Table 5. Accuracy values obtained by each classification algorithm using stratification with COM

	COM 1	COM 2	COM 3	COM 4	COM 5	Mean	SD
BayesNet	0.686	0.707	0.686	0.714	0.687	0.696	0.013
J48	0.583	0.627	0.643	0.794	0.626	0.655	0.081
MLP	0.784	0.713	0.735	0.849	0.699	0.756	0.061
RD	0.691	0.688	0.680	0.843	0.687	0.718	0.070
SVM	0.705	0.688	0.711	0.714	0.711	0.706	0.010
DNN	0.776	0.715	0.741	0.855	0.738	0.765	0.055

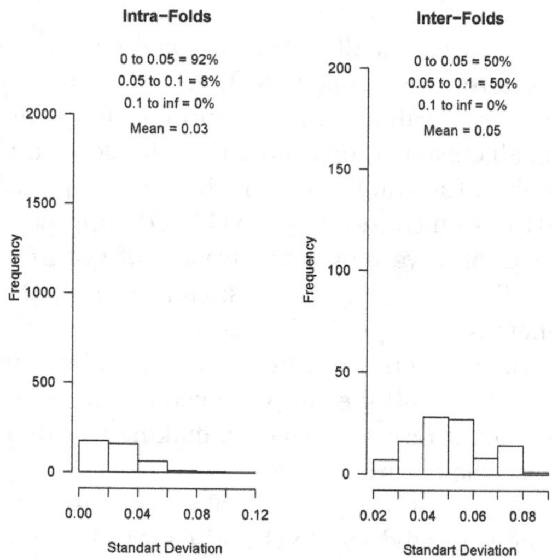

Fig. 9. Standard deviation levels intra-folds and inter-folds after applying the proposed stratification approach.

accuracy rate when using this stratification approach, and maybe even more so with a parameter optimization procedure, which is open for future research.

Figure 9 shows the standard deviation levels intra-folds and inter-folds considering the GNG stratification approach. Of note, the new standard deviation values are much lower than the ones presented in Fig. 5, when using the traditional stratification approach. This decrease is also notable when observing the inter-folds values, clearly suggesting that our hypothesis is correct.

6 Conclusion

fMRI is a widely used and accepted tool to scrutinize the organization of brain circuits and their association with several mental diseases. In this paper, we have investigated the use of deep learning models to classify fMRI signals. In particular, fMRI data of cocaine dependent and health subjects were used to train deep learning models, with different sets of features.

Our results have shown that the functional connectivity measures of the frontoparietal circuits are very informative features to classify cocaine dependence from fMRI data. This is in line with previous investigations highlighting the importance of those connection weights to identify brain pathology [25–27]. In particular, we observed that Pearson coefficients are the best set of features to describe neural circuit disruption in cocaine dependence. This conclusion is based on exhaustive experimental analyses performed. Moreover, instead of using traditional stratified cross-validation, here, we have introduced a cluster-based stratified cross-validation by applying a GNG neural network followed by a community detection algorithm. Experimental evaluation has demonstrated that this new approach improved the classification accuracy of the deep model. Even thought the problem became more difficult with 4-fold cross-validation, with less examples in the training sets, the final results significantly improved thanks to the new folds distribution. In fact, overall results with 4-folds cross-validation without GNG stratification (Table 6) are worst than the ones presented in Table 2 due to smaller training set. Our results also indicated that deep learning outperforms baseline methods even with a relatively small and heterogeneous dataset of cocaine dependents.

Table 6. Baselines results 4 folds

Dataset	BayesNet	J48	MLP	RD	SVM
1	0.500	0.600	0.569	0.562	0.519
2	0.519	0.600	0.588	0.606	0.606
3	0.550	0.588	0.469	0.562	0.556
4	0.662	0.612	0.712	0.650	0.662
5	0.669	0.612	0.656	0.700	0.662
6	0.612	0.612	0.619	0.688	0.675
7	0.612	0.581	0.600	0.662	0.694
8	0.606	0.612	0.594	0.631	0.619

Acnowledgement. The authors would like to thank CAPES, CNPq (458777/2014-5), FAPESP (2016/02870-0, 2016/16291-2) and NIH (grant R01DA023248) for funding this research.

References

1. Demšar, J.: Statistical comparisons of classifiers over multiple data sets. J. Mach. Learn. Res. **7**, 1–30 (2006)
2. Floreano, D., Mattiussi, C.: Bio-Inspired Artificial Intelligence: Theories, Methods, and Technologies. The MIT Press, Cambridge (2008)
3. Fortunato, S.: Community detection in graphs. Phys. Rep. **486**(3), 75–174 (2010)
4. Friston, K.J.: Statistical Parametric Mapping: The Analysis of Functional Brain Images. Elsevier/Academic Press, Amsterdam/Boston (2007)
5. Friston, K.J., Holmes, A.P., Worsley, K.J., Poline, J.P., Frith, C.D., Frackowiak, R.S.J.: Statistical parametric maps in functional imaging: a general linear approach. Hum. Brain Mapp. **2**(4), 189–210 (1995)
6. Fritzke, B., et al.: A growing neural gas network learns topologies. Adv. Neural Inf. Process. Syst. **7**, 625–632 (1995)
7. Glorot, X., Bengio, Y.: Understanding the difficulty of training deep feedforward neural networks. In: Aistats, vol. 9, pp. 249–256 (2010)
8. Hall, M., Frank, E., Holmes, G., Pfahringer, B., Reutemann, P., Witten, I.H.: The weka data mining software: an update. ACM SIGKDD Explor. Newslett. **11**(1), 10–18 (2009)
9. Haxby, J.V., Connolly, A.C., Guntupalli, J.S.: Decoding neural representational spaces using multivariate pattern analysis. Annu. Rev. Neurosci. **37**, 435–456 (2014)
10. Haykin, S.: Neural Networks: A Comprehensive Foundation, 2nd edn. Prentice Hall PTR, Upper Saddle River (1998)
11. Hinton, G.E., Salakhutdinov, R.R.: Reducing the dimensionality of data with neural networks. Science (New York, N.Y.) **313**(5786), 504–507 (2006)
12. Honorio, J.: Classification on brain functional magnetic resonance imaging: dimensionality, sample size, subject variability and noise. In: Chen, C. (ed.) Frontiers of Medical Imaging, pp. 266–290. World Scientific Publishing Company Pte Limited, Singapore (2014)
13. Ide, J., Shenoy, P., Yu, A., Li, C.: Bayesian prediction and evaluation in the anterior cingulate cortex. J. Neurosci. **33**(5), 2039–2047 (2013)
14. Jones, N.: Computer science: the learning machines. Nature **505**, 146–148 (2014)
15. Koyamada, S., Shikauchi, Y., Nakae, K., Koyama, M., Ishii, S.: Deep learning of fMRI big data: a novel approach to subject-transfer decoding. arXiv preprint arXiv:1502.00093 (2015)
16. LeCun, Y., Bengio, Y., Hinton, G.: Deep learning. Nature **521**(7553), 436–444 (2015)
17. Li, C.S.R., Huang, C., Constable, R.T., Sinha, R.: Imaging response inhibition in a stop-signal task: neural correlates independent of signal monitoring and post-response processing. J. Neurosci. **26**(1), 186–192 (2006)
18. Luo, X., Zhang, S., Hu, S., Bednarski, S.R., Erdman, E., Farr, O.M., Hong, K.I., Sinha, R., Mazure, C.M., shan, R., Li, C.: Error processing and gender-shared and -specific neural predictors of relapse in cocaine dependence. Brain **136**(4), 1231–1244 (2013)

19. Markoff, J.: Scientists See Promise in Deep-Learning Programs. New York Times, Manhattan (2012)
20. Norman, K.A., Polyn, S.M., Detre, G.J., Haxby, J.V.: Beyond mind-reading: multi-voxel pattern analysis of fMRI data. Trends Cogn. Sci. **10**(9), 424–430 (2017). http://dx.doi.org/10.1016/j.tics.2006.07.005
21. de Oliveira, F.A., Nobre, C.N., Zárate, L.E.: Applying artificial neural networks to prediction of stock price and improvement of the directional prediction index-case study of PETR4, Petrobras, Brazil. Expert Syst. Appl. **40**(18), 7596–7606 (2013)
22. Pahlavan, R., Omid, M., Akram, A.: Energy input-output analysis and application of artificial neural networks for predicting greenhouse basil production. Energy **37**(1), 171–176 (2012)
23. Plis, S.M., Hjelm, D.R., Salakhutdinov, R., Allen, E.A., Bockholt, H.J., Long, J.D., Johnson, H.J., Paulsen, J.S., Turner, J.A., Calhoun, V.D.: Deep learning for neuroimaging: a validation study. Front. Neurosci. **8**(August), 1–11 (2014)
24. Roebroeck, A., Formisano, E., Goebel, R.: Mapping directed influence over the brain using granger causality and fMRI. Neuroimage **25**(1), 230–242 (2005)
25. Smith, S.M., Nichols, T.E., Vidaurre, D., Winkler, A.M., Behrens, T.E.J., Glasser, M.F., Ugurbil, K., Barch, D.M., Van Essen, D.C., Miller, K.L.: A positive-negative mode of population covariation links brain connectivity, demographics and behavior. Nat. Neurosci. **18**(11), 1565–1567 (2015)
26. Tomasi, D., Volkow, N.D., Wang, R., Carrillo, J.H., Maloney, T., Alia-Klein, N., Woicik, P.A., Telang, F., Goldstein, R.Z.: Disrupted functional connectivity with dopaminergic midbrain in cocaine abusers. PLoS ONE **5**(5), 1–10 (2010)
27. Zhang, S., Hu, S., Bednarski, S.R., Erdman, E., Li, C.S.: Error-related functional connectivity of the thalamus in cocaine dependence. Neuroimage Clin. **4**, 585–592 (2014)

Linear Models for High-Complexity Sequences

Sara D. Cardell[1(✉)] and Amparo Fúster-Sabater[2]

[1] Instituto de Matemática, Estatística e Computação Científica, UNICAMP, Campinas, Brazil
sdcardell@ime.unicamp.br

[2] Instituto de Tecnologías Físicas y de la Información (CSIC), 144, Serrano, 28006 Madrid, Spain
amparo@iec.csic.es

Abstract. Different binary sequence generators produce sequences whose period is a power of 2. Although these sequences exhibit good cryptographic properties, in this work it is proved that such sequences can be obtained as output sequences from simple linear structures. More precisely, every one of these sequences is a particular solution of a linear difference equation with binary coefficients. This fact allows one to analyze the structural properties of the sequences with such a period from the point of view of the linear difference equations. In addition, a new application of the Pascal's triangle to the cryptographic sequences has been introduced. In fact, it is shown that all these binary sequences can be obtained by XORing a finite number of binomial sequences that correspond to the diagonals of the Pascal's triangle reduced modulo 2.

Keywords: Binary sequence · Period · Linear complexity · Difference linear equation · Binomial sequence · Pascal's triangle

1 Introduction

Stream ciphers are the fastest among the encryption procedures. Because stream ciphers tend to be small and fast, they are particularly adequate for applications with little computational resources, e.g. cell phones or other small embedded devices. GSM mobile phones [8] and RC4 stream cipher algorithm for encrypting Internet traffic [15] are some of the most relevant stream cipher applications in modern technologies.

Assuming that the information is given as binary sequences, stream ciphers encrypt bits individually [14]. The ciphertext is obtained by XORing the original message (*plaintext*) and a binary sequence (*keystream sequence*) with certain pseudorandomness characteristics. The decryption is performed in the same way, that is by XORing the ciphertext and the same keystream sequence. The main concern in stream ciphers is to generate from a short key a long keystream sequence that looks as random as possible.

In cryptographic terms, the keystream sequence is expected to exhibit some "good properties", such as long period and large linear complexity.

O. Gervasi et al. (Eds.): ICCSA 2017, Part I, LNCS 10404, pp. 314–324, 2017.
DOI: 10.1007/978-3-319-62392-4_23

The **linear complexity** of a sequence, denoted by LC, is defined as the length of the shortest Linear Feedback Shift Register (LFSR) [7] that can generate such a sequence. Indeed, linear complexity is related with the amount of sequence needed to reconstruct the whole sequence. Thus, in cryptographic terms linear complexity must be as large as possible. In fact, the Berlekamp-Massey algorithm [11] efficiently computes the length and characteristic polynomial of the shortest LFSR that generates a sequence given at least $2 \cdot LC$ of its bits. Due to their low linear complexity, LFSRs should never be used alone as keystream generators. Nevertheless, sequences generated from LFSRs have good statistical properties, such as a long period, good distribution of 0s and 1s or excellent autocorrelation [7], but their linearity has to be destroyed; that is, their linear complexity has to be increased before such sequences are used for cryptographic purposes.

In this work, we study the family of binary sequences with period 2^L, L being a positive integer. Among the keystream generators producing sequences with such a period, we can enumerate: the self-shrinking generator, the modified self-shrinking generator and the generalized self-shrinking generator (see [9,10,13]). Given a fixed value for L, it can be proven that the linear complexity of a sequence with period 2^L satisfies the inequality $2^{L-2} < LC \leq 2^{L-1}$ [9]. Although both properties (period and linear complexity) are adequate in cryptographic terms, sometimes these sequences present a weakness based on its linearity. For example, it has been proven that the self-shrinking generators can be modeled by linear structures based on cellular automata (see [1–4]). In this work, we prove that each sequence with period 2^L, can be obtained by XORing particular binary sequences known as binomial sequences. Such sequences are the binary diagonals of the Pascal's triangle [12] and their inherent linearity is a weakness that can be used to launch a cryptanalytic attack on the generators whose output sequences have period of value a power of 2. Besides, we propose an algorithm that computes the different combinations of binomial sequences that provide a given sequence of period 2^L. In particular, binomial sequences have also period 2^L and it can be seen that the structure of their set of combinations is the same as that one of the CA-image generated by rule 102 after having applied 15 iterations to the one-dimensional cellular automata (see [16]).

This paper is organized as follows: in Sect. 2, we introduce the basic concepts and definitions in order to understand the rest of the work. In Sect. 3, we prove that the sequences with period 2^L are solutions of a linear difference equation and that these sequences can be obtained by XORing binomial sequences. We also provide an algorithm that computes the different binomial sequences that characterize a given sequence depending on its starting point. Finally, in Sect. 4, some conclusions are given to close the paper.

2 Preliminaries and Basic Notation

Let $\mathbb{F}_2$ be the Galois field of two elements or the binary field. Let $\{a_i\} = \{a_0, a_1, a_2, \ldots\}$ be a sequence over $\mathbb{F}_2$ if its terms $a_i \in \mathbb{F}_2$, for $i = 0, 1, 2, \ldots$. The sequence $\{a_i\}$ is said to be periodic if and only if there exists an integer T, called **period**, such that $a_{i+T} = a_i$, for all $i \geq 0$.

Let L be a positive integer, and let $c_0, c_1, \ldots, c_{L-1}$ be given elements of $\mathbb{F}_2$. A sequence $\{a_i\}$ of elements in $\mathbb{F}_2$ satisfying the relation

$$a_{i+L} = c_0 a_i + c_1 a_{i+1} + \cdots + c_{L-2} a_{i+L-2} + c_{L-1} a_{i+L-1}, \quad i \geq 0, \tag{1}$$

is called a (L-th order) linear recurring sequence in $\mathbb{F}_2$. The first L terms $a_0, a_1, \ldots, a_{L-1}$ determine the rest of the sequence uniquely and are referred to as the **initial state**. A relation of the form given in expression (1) is called a (L-th order) linear recurrence relation.

The monic polynomial of degree L, given by $p(x) = c_0 + c_1 x + \cdots + c_{L-2} x^{L-2} + c_{L-1} x^{L-1} + x^L \in \mathbb{F}_2[x]$, is called the **characteristic polynomial** of the linear recurring sequence and this sequence $\{a_i\}$ is said to be generated by $p(x)$.

There exist several tools to generate linear recurring sequences. For example, the generation of this class of sequences can be implemented by **Linear Feedback Shift Registers** (LFSRs) [7]. Such devices are based on shifts and linear feedback, as its name indicates, and process information in the form of elements of $\mathbb{F}_2$. An LFSR has L memory cells (or stages), shift to the adjacent stage and linear feedback to the empty stage (see Fig. 1). If the characteristic polynomial of the linear recurrence sequence is primitive, then the LFSR is said to be maximal-length and its output sequence has period $2^L - 1$. This output sequence is called **PN-sequence** (pseudonoise sequence) or ***m*-sequence** (maximal sequence).

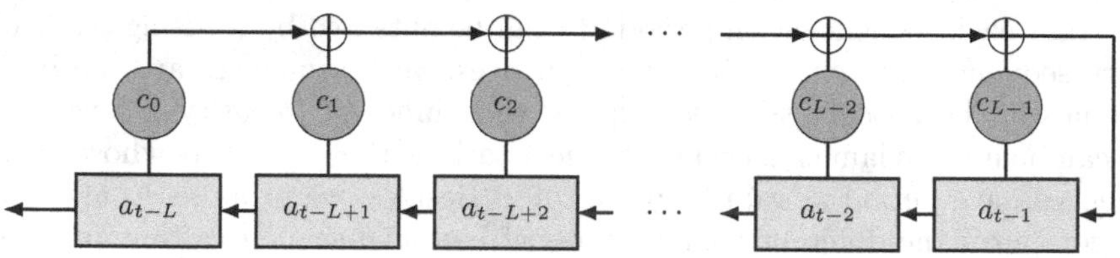

Fig. 1. An LFSR of length L

As we said before, the linear complexity of a sequence is related with the concept of LFSR. In fact, the linear complexity is the length of the shortest LFSR that generates such a sequence.

3 Linear Structures for the Sequences

Let E be the one-sided shift operator that acts on the sequence terms, that is $Ea_n = a_{n+1}$ and $E^k a_n = a_{n+k}$. Consider the equations of the form

$$(E^r + \sum_{j=1}^{r} a_j E^{r-j})^m z_n = 0, \quad \text{for } n \geq 0, \tag{2}$$

with m a positive integer and $p_m(x) = p(x)^m = (x^r + \sum_{j=1}^{r} a_j x^{r-j})^m$ its characteristic polynomial. If $p(x)$ is a primitive polynomial, then the roots of $p_m(x)$

are the same as those of $p(x)$, that is $\alpha, \alpha^2, \ldots, \alpha^{2^{r-1}}$, but with multiplicity m. Therefore, the solutions of Eq. (2) are given by

$$z_n = \sum_{i=0}^{m-1} \left[\binom{n}{i} \sum_{j=0}^{r-1} c_i^{2^j} \alpha^{2^j n} \right], \tag{3}$$

where $c_i \in \mathbb{F}_2$ and $\binom{n}{i}$ are binomial coefficients reduced modulo 2 (see [5]).

Now, a more simple type of difference equation is considered as well as the main result relating our sequences with linear difference equations is introduced.

Theorem 1. *A sequence $\{s_j\}$ with period $T = 2^L$ is a particular solution of the homogeneous linear difference equation*

$$(E+1)^{2^L} z_n = 0, \tag{4}$$

whose characteristic polynomial is $(1+x)^{2^L}$.

Proof. We know that the period T is a power of 2 and $(1+x)^T = 1 + x^T$ in $\mathbb{F}_2$. On the other hand, given $q(x)$ the characteristic polynomial of the shortest LFSR that generates $\{s_j\}$, then $q(x)|1+x^T$ what implies that the characteristic polynomial of the sequence has the form $q(x) = (1+x)^N$, where N is its linear complexity. It is a well known fact that the linear complexity of a periodic sequence is less or equal that its period [11]. Therefore, we know that $N \leq 2^L$ and $q(x)$ divides the polynomial $(1+x)^{2^L}$. Thus, $\{s_j\}$ satisfies Eq. (4) and is a particular solution of this homogeneous linear difference equation. □

According to (3), the solutions of Eq. (4) have the following form:

$$z_n = \binom{n}{0} c_0 + \binom{n}{1} c_1 + \cdots + \binom{n}{T-1} c_{T-1} \quad \text{for } n \geq 0,$$

where the coefficients $c_i \in \mathbb{F}_2$, $\alpha = 1$ is the unique root of the polynomial $(1+x)^T$, with multiplicity T and $\binom{n}{i}$ are binomial coefficients modulo 2. The sequence $\{z_n\}$ is just a bit-wise XOR of binary sequences weighted by coefficients c_i. Different choices of c_i will produce different sequences with different characteristics, but all of them have period 2^l, with $l \leq L$.

The binomial coefficient $\binom{n}{i}$ is the coefficient of the power x^i in the polynomial expansion of $(1+x)^n$. For every positive integer n, it is a well-known fact that $\binom{n}{0} = 1$ and $\binom{n}{i} = 0$ for $i > n$. When n takes successive values, each binomial coefficient modulo 2 defines a binary sequence with constant period T_i. We call these sequences (binary) binomial sequences. Table 1 shows the sequences and values of T_i for the first binomial coefficients $\binom{n}{i}$, $i = 0, 1, \ldots, 7$ [6].

It is worth noticing that if we arrange these binomial coefficients into rows for successive values of n, the generated structure is the Pascal's triangle (see Fig. 2). The most-left diagonal is the identically 1 sequence, the next diagonal is the sequence of natural numbers $\{1, 2, 3, \ldots\}$, the next one is the sequence of triangular numbers $\{1, 3, 6, 10, \ldots\}$, etc. Other fascinating sequences (tetrahedral

Table 1. Binomial coefficients reduced modulo 2, binary sequences and periods

Bino. coeff.	Binary sequences	T_i
$\binom{n}{0}$	11111111	$T_0 = 1$
$\binom{n}{1}$	01010101	$T_1 = 2$
$\binom{n}{2}$	00110011	$T_2 = 4$
$\binom{n}{3}$	00010001	$T_3 = 4$
$\binom{n}{4}$	00001111	$T_4 = 8$
$\binom{n}{5}$	00000101	$T_5 = 8$
$\binom{n}{6}$	00000011	$T_6 = 8$
$\binom{n}{7}$	00000001	$T_7 = 8$

numbers, pentatope numbers, hexagonal numbers, Fibonacci sequence, etc.) can be found in the diagonals of this triangle. On the other hand, if we color the odd numbers of the Pascal's triangle and shade the other ones, we can find the Sierpinski's triangle (see Fig. 3).

$$\binom{0}{0}$$
$$\binom{1}{0} \quad \binom{1}{1}$$
$$\binom{2}{0} \quad \binom{2}{1} \quad \binom{2}{2}$$
$$\binom{3}{0} \quad \binom{3}{1} \quad \binom{3}{2} \quad \binom{3}{3}$$
$$\binom{4}{0} \quad \binom{4}{1} \quad \binom{4}{2} \quad \binom{4}{3} \quad \binom{4}{4}$$
$$\binom{5}{0} \quad \binom{5}{1} \quad \binom{5}{2} \quad \binom{5}{3} \quad \binom{5}{4} \quad \binom{5}{5}$$

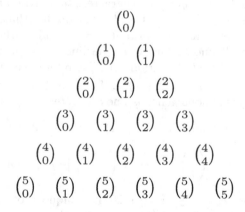

Fig. 2. Binomial coefficients arranged as the Pascal's triangle

In the following example, we apply the previous results to the sequences produced by a keystream generator known as modified self-shrinking generator [10].

Example 1. The **modified self-shrinking generator** (MSSG) was introduced by Kanso in 2010 for hardware implementations [10]. Given three consecutive bits $\{u_{2i}, u_{2i+1}, u_{2i+2}\}$, $i = 0, 1, 2, \ldots$ of a PN-sequence $\{u_i\}$, the output sequence $\{s_j\}$ is computed as

$$\begin{cases} \text{If } u_{2i} + u_{2i+1} = 1 \text{ then } s_j = u_{2i+2} \\ \text{If } u_{2i} + u_{2i+1} = 0 \text{ then } u_{2i+2} \text{ is discarded.} \end{cases}$$

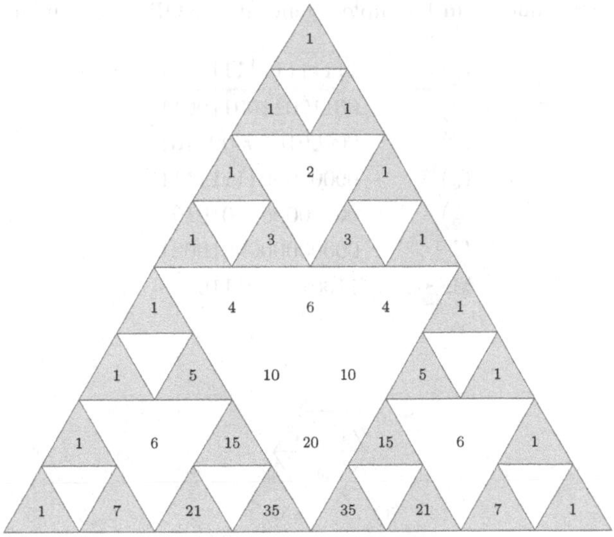

Fig. 3. Sierpinski's triangle (Color figure online)

We call the sequence $\{s_j\}$ as the **modified self-shrunken sequence** (MSS-sequence). If L (odd) is the length of the maximum-length LFSR that generates $\{u_i\}$, then the linear complexity LC of the corresponding modified self-shrunken sequence satisfies:

$$2^{\lfloor \frac{L}{3} \rfloor - 1} \leq LC \leq 2^{L-1} - (L-2),$$

and the period T of the sequence satisfies:

$$2^{\lfloor \frac{L}{3} \rfloor} \leq T \leq 2^{L-1},$$

as proved in [10], where L is the length of the LFSR that generates the PN-sequence $\{u_i\}$. As usual, the key of this generator is the initial state of the LFSR. Moreover, the characteristic polynomial of the register, $p(x)$, is also recommended to be part of the key.

Consider the LFSR generated by the primitive polynomial $p(x) = 1 + x^2 + x^5$ and consider the initial state 11111. The MSS-sequence generated with this polynomial is given by:

$$\{1\,1\,0\,0\,1\,0\,0\,1\,0\,1\,1\,1\,0\,0\,1\,0\}$$

This sequence can be obtained XOR-ing the successive binomial sequences: $\binom{n}{0}$, $\binom{n}{2}$, $\binom{n}{5}$, $\binom{n}{8}$, $\binom{n}{9}$, $\binom{n}{11}$ (see Table 2). As a conclusion, we can say that the MSS-sequence $\{s_j\}$ can be obtained XOR-ing binomial sequences corresponding to different diagonals in the (binary) Pascal's triangle.

It is worth noting that the binary sequences match exactly with the diagonals of the binary Sierpinski's triangle (see Fig. 4), but starting in a different bit.

Table 2. MSS-sequence in Example 1 generated XORing binomial sequences

$\binom{n}{0}$	1111111111111111
$\binom{n}{2}$	0011001100110011
$\binom{n}{5}$	0000010100000101
$\binom{n}{8}$	0000000011111111
$\binom{n}{9}$	0000000001010101
$\binom{n}{11}$	0000000000010001
MSS-seq.	1100100101110010

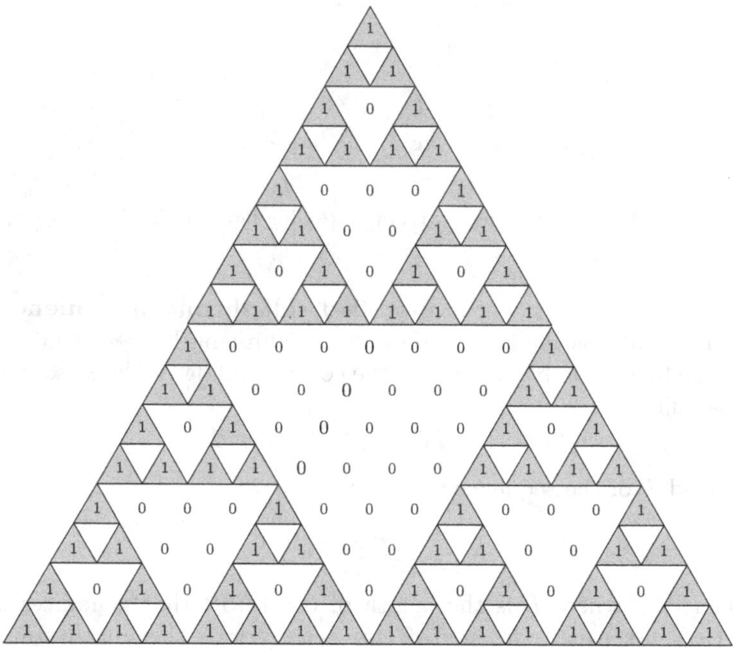

Fig. 4. Binary Sierpinski's triangle

For example, the sequence in red in Fig. 4 corresponds to the binomial sequence $\binom{n}{4}$. Therefore, every sequence of period 2^L can be obtained by XORing diagonals of the binary Sierpinski's triangle.

The set of binomial sequences necessary to obtain a sequence is called the **characterization** of such a sequence. We know that this characterization is different depending on the starting bit. Given a binomial sequence $\binom{n}{t}$, $t \geq 1$, if we shift such a sequence one bit, then we obtain $\binom{n}{t}+\binom{n}{t-1}$. If $t = 0$, the sequence remains the same (in this case the sequence is the identically 1 sequence). For example:

$$\binom{n}{2} : 0\,0\,1\,1\,0\,0\,1\,1\,\ldots$$

$$\binom{n}{2} + \binom{n}{1} : 0\,1\,1\,0\,0\,1\,1\,0\,\ldots$$

Therefore, given the characterization of a sequence $\sum_{i=0}^{LC-1} a_i \binom{n}{i}$, with $a_i \in \mathbb{F}_2$, if we shift one bit of the sequence we obtain the following characterization:

$$\sum_{i=1}^{LC-1} a_i \left[\binom{n}{i} + \binom{n}{i-1} \right] + a_0 \binom{n}{0}.$$

Since the period of the sequences we are dealing with is 2^L, we propose an algorithm that generates the 2^L characterizations of every sequence (given one characterization). The input of the algorithm will be one characterization of the sequence, represented by the binary coefficients $[a_0\ a_1\ \ldots a_{LC-1}]$ and the output will be a matrix where each row represents a different characterization (see Algorithm 1).

Algorithm 1. Characterizations of a sequence

Input: Characterization $\boldsymbol{a} = [a_0\ a_1\ \ldots a_{LC-1}]$

```
Store a in A: A = [a]
Store the length of a in aux: aux = length(a);
for j = 1 : 2^L − 1
   aux1 = [a(1), zeros(1, aux − 1)];
   aux2 = [zeros(1, aux)];
   for i = 2 : aux;
      if a(i) == 1;
         aux1(i) = 1;
         aux2(i − 1) = 1;
      endif
   endfor
   Store in a the new characterization: a = mod(aux1 + aux2, 2);
   Store a in A: A = [A; a];
endfor
```

Output:
Matrix A of characterizations

For example, in Table 3, we can find the 16 characterizations of $\binom{n}{15}$ provided by the algorithm.

Recall now, the definition of rule 102 according to Wolfram's notation [16] in theory of one-dimensional linear Cellular Automata (CA). For $k = 3$, k being the neighborhood factor, rule 102 is given by:

Table 3. Characterizations for $\binom{n}{15}$

$\binom{n}{0}$	$\binom{n}{1}$	$\binom{n}{2}$	$\binom{n}{3}$	$\binom{n}{4}$	$\binom{n}{5}$	$\binom{n}{6}$	$\binom{n}{7}$	$\binom{n}{8}$	$\binom{n}{9}$	$\binom{n}{10}$	$\binom{n}{11}$	$\binom{n}{12}$	$\binom{n}{13}$	$\binom{n}{14}$	$\binom{n}{15}$
0	0	0	0	0	0	0	0	0	0	0	0	0	0	0	1
0	0	0	0	0	0	0	0	0	0	0	0	0	0	1	1
0	0	0	0	0	0	0	0	0	0	0	0	0	1	0	1
0	0	0	0	0	0	0	0	0	0	0	0	1	1	1	1
0	0	0	0	0	0	0	0	0	0	0	1	0	0	0	1
0	0	0	0	0	0	0	0	0	0	1	1	0	0	1	1
0	0	0	0	0	0	0	0	0	1	0	1	0	1	0	1
0	0	0	0	0	0	0	0	1	1	1	1	1	1	1	1
0	0	0	0	0	0	0	1	0	0	0	0	0	0	0	1
0	0	0	0	0	0	1	1	0	0	0	0	0	0	1	1
0	0	0	0	0	1	0	1	0	0	0	0	0	1	0	1
0	0	0	0	1	1	1	1	0	0	0	0	1	1	1	1
0	0	0	1	0	0	0	1	0	0	0	1	0	0	0	1
0	0	1	1	0	0	1	1	0	0	1	1	0	0	1	1
0	1	0	1	0	1	0	1	0	1	0	1	0	1	0	1
1	1	1	1	1	1	1	1	1	1	1	1	1	1	1	1

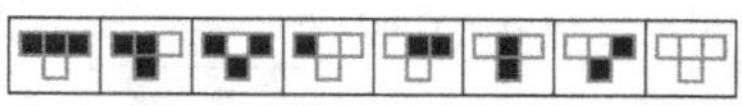

0 1 1 0 0 1 1 0

Fig. 5. Rule 102 depicted in terms of Wolfram's notation

Rule 102: $x_i^{t+1} = x_i^t + x_{i+1}^t$

111	110	101	100	011	010	001	000
0	1	1	0	0	1	1	0

The number 01100110 is the binary representation of the decimal number 102. In Fig. 5, it is possible to see this rule using the terminology introduced by Wolfram [16], where a white square represents the digit 0 and a black square represents the digit 1.

If we color the ones for every characterization of $\binom{n}{15}$ (Table 3), the general structure of the set of characterizations is the same as that one of the CA-image generated by rule 102 after having applied 15 iterations to the one-dimensional cellular automata (see Fig. 6).

On the other hand, we know that the representation of the binomial sequence $\binom{n}{t}$ showed in the binary Sierpinski's triangle starts in a different bit from the representation given in Table 1. In particular, the sequences start in the first non-zero bit, thus the characterization will be different. We can find the corresponding characterization in the $t+1$-th row of the matrix of characterizations provided by Algorithm 1. For example, for $\binom{n}{15}$, the characterization of the sequence in the binary Sierpinski's triangle will be the last one of Table 3: $\sum_{i=0}^{15}\binom{n}{i}$.

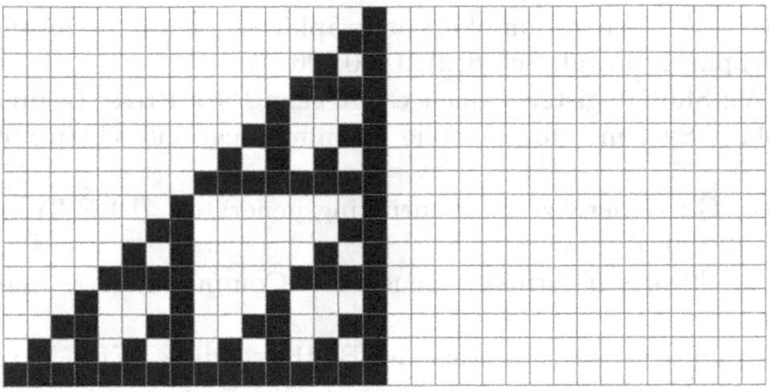

Fig. 6. CA-images generated with rule 102

4 Conclusions

The family of binary sequences considered in this work, sequences whose period is a power of 2, has good cryptographic properties such as long period and large linear complexity. However, we have seen that such sequences are simple solutions of linear difference equations with constant coefficients and can be obtained by XORing binomial binary sequences corresponding to diagonals of Pascal's triangle reduced modulo 2. Although different non-linear procedures, e.g. irregular decimation, are introduced to break the linearity of the LFSR-based sequence generators, this linearity is still visible in their output sequences. Consequently, such a linearity makes the generators that produce the previous sequences vulnerable against cryptanalysis and makes them not suitable as part of more complex cryptographic structures.

Acknowledgements. The first author was supported by FAPESP with number of process 2015/07246-0. The second author was supported by both Ministerio de Economía y Competitividad, Spain under grant TIN2014-55325-C2-1-R (ProCriCiS), and Comunidad de Madrid, Spain, under grant S2013/ICE-3095-CM (CIBERDINE). Both authors thank European FEDER funds.

References

1. Cardell, S.D., Fúster-Sabater, A.: Linear models for the self-shrinking generator based on CA. J. Cell. Autom. **11**(2–3), 195–211 (2016)
2. Cardell, S.D., Fúster-Sabater, A.: Recovering the MSS-sequence via CA. Procedia Comput. Sci. **80**, 599–606 (2016)
3. Cardell, S.D., Fúster-Sabater, A.: Modelling the shrinking generator in terms of linear CA. Adv. Math. Commun. **10**(4), 797–809 (2016)
4. Fúster-Sabter, A., Caballero-Gil, P.: Chaotic modelling of the generalized self-shrinking generator. Appl. Soft Comput. **11**(2), 1876–1880 (2011)
5. Fúster-Sabater, A., Caballero-Gil, P.: Linear cellular automata as discrete models for generating cryptographic sequences. J. Res. Pract. Inf. Technol. **40**(4), 47–52 (2008)

6. Fúster-Sabater, A.: Generation of cryptographic sequences by means of difference equations. Appl. Math. Inf. Sci. **8**(2), 1–10 (2014)
7. Golomb, S.W.: Shift Register-Sequences. Aegean Park Press, Laguna Hill (1982)
8. GSM, Global Systems for Mobile Communications. http://cryptome.org/gsm-a512.htm
9. Hu, Y., Xiao, G.: Generalized self-shrinking generator. IEEE Trans. Inf. Theor. **50**(4), 714–719 (2004)
10. Kanso, A.: Modified self-shrinking generator. Comput. Electr. Eng. **36**(1), 993–1001 (2010)
11. Massey, J.L.: Shift-register synthesis and BCH decoding. IEEE Trans. Inf. Theor. **15**(1), 122–127 (1969)
12. Mathematical Forum, Pascal's triangle. http://mathforum.org/dr.math/faq/faq.pascal.triangle.html
13. Meier, W., Staffelbach, O.: The self-shrinking generator. In: Santis, A. (ed.) EUROCRYPT 1994. LNCS, vol. 950, pp. 205–214. Springer, Heidelberg (1995). doi:10.1007/BFb0053436
14. Paar, C., Pelzl, J.: Understanding Cryptography. Springer, Berlin (2010)
15. Rivest, R.L.: The RC4 Encryption Algorithm. (RSA Data Sec. Inc., March 2002)
16. Wolfram, S.: Cellular automata as simple self-organizing system, Caltrech preprint CALT 68–938

PrescStream: A Framework for Streaming Soft Real-Time Predictive and Prescriptive Analytics

Marcos de Aguiar[1(✉)], Fabíola Greve[1], and Genaro Costa[2]

[1] Computer Science Department, Federal University of Bahia, Salvador, Brazil
marcos.deaguiar@gmail.com
[2] Humanities, Arts and Sciences Department,
Federal University of Bahia, Salvador, Brazil

Abstract. All the data volume generated by modern applications brings opportunities for knowledge extraction and value creation. In this sense, the integration of predictive and prescriptive analytics may help the industry and users to be more productive and successful. It means not only to estimate an outcome but also to act on it in the real world. Nonetheless, mastering these concepts and providing their integration is not an easy task. This work proposes PrescStream, a proof of concept framework that uses machine learning based prediction, and process this outcome result to do prescriptive analytics, allowing researchers to integrate predictive and prescriptive analytics into their experiments. It has a scalable, fault-tolerant microservices based architecture, making it ideal for cloud deployment and IoT (internet of things) applications. The paper describes the general architecture of the system, as well as a validation usage with result analysis.

Keywords: Predictive analytics · Prescriptive analytics · Microservices · Internet of things · Architecture frameworks

1 Introduction

Computing and storage costs are decreasing every year and on top of that computers are becoming more and more pervasive to our environment. Smartphones with internet access are a common view and more wearable devices, like smart watches and glasses, are being developed and pushed to the market; even traditional simple devices, such as camcorders, are becoming "smart" and saving extra data coming from sensors together with the video. The Internet of Things (IoT) seems to be the next wave of technology; some authors claims that more data will be generated by devices and objects than by humans [1].

This scenario, coupled with the traditional computer usage and the internet, is generating a massive amount of data as never seen in the history of mankind, in such a way that some are even considering the data as the 21st century Oil. The opportunity then, is not to only use all this data for operational and service purposes but also to be able to extract knowledge and value from it.

O. Gervasi et al. (Eds.): ICCSA 2017, Part I, LNCS 10404, pp. 325–341, 2017.
DOI: 10.1007/978-3-319-62392-4_24

The business industry is already benefiting from data analytics to extract knowledge from their data and harness new opportunities from it. Having specific knowledge can make a company focus on where it can gain the most. There are many simple examples, as of consumers tending to buy the same kind of products, which can be put together on the shelf, or to power a recommendation system on a web based business. These techniques are well known, and they get refined every day.

Much more challenging (and rewarding) is to use data for prediction. This kind of analytics may have many different applications: from companies, that predict a trend ahead of time and prepares to take advantage of it, to systems that can be used for detecting machine failures ahead of time. The list of real life projects that already employs these principles is vast. Some examples to illustrate are the following:

- Ford is developing a car that detects whether the driver is intoxicated and driving in harmful ways and fires an alarm [14];
- Target has developed a system to detect whether a client is pregnant, and sends a directed catalog to her;
- Many companies uses prediction to pre select customers more likely to respond to some advertisement or catalogs in order to save money on printing and mailing, as well as to increase the return of investment [14];
- Educational institutions are using predictive analytics to identify students more likely to drop out and intervene to prevent it [2].

Applications are endless, and every day people are finding new ways to get value out of predictive analytics.

On this paper, we focus on the design and implementation of a framework, namely PrescStream, that enable researchers to run experiments that will use predictive analytics coupled with prescriptive analytics. As far as we know, differently from most of the propositions in the literature, the system not only incorporates a predictive model, based on machine learning techniques, but also make prescriptive push like real-time actuation, based on processing coming from this model.

The PrescStream framework is generic enough to be coupled easily with most existing systems. Its microservice design allows users to plug to different backends and infrastructure to better suit their applications. It also provides ready to use components that plugs to known technologies, allowing users to use the system right away, just by extending a few classes. The result is a system exhibiting a flexible, scalable, fault-tolerant architecture, making it ideal for IoT and cloud computing scenarios. An implementation of the proposal and a test experiment are provided in order to validate the architecture.

The rest of the paper is organized as follows: Sect. 2 describes related works; Sect. 3 presents fundamental concepts; Sect. 4 describes PrescStream; Sect. 5 presents the experiment; Sect. 6 presents the conclusion and future works.

2 Related Work

Huang [8] uses prediction as a decision support tool for complex networks. It uses a trained system to plot graphs to help humans adjust power grids in order to have minimum impact in case of a partial failure. The paper presents a good case for predictive analytics, but the system is not real-time push enabled, as the one presented in this work.

Tönjes et al. [15] designs a generic framework to enable smart cities, in a way that it processes stream of data coming from Internet of Things (IoT) devices, does analytics about it, and provides models as API's (application programming interface) for services that need those information. The authors cite some ongoing projects that will use this tool, to enable better transportation, public safety, and open innovation for citizens interested in using these services and data to improve life quality and better city experience.

3 Model Construction, Validation and Concepts

The specification of the model is an important step to attain good results. The choices done in this phase must be carried out by someone with good knowledge in data sciences, able to make the right decisions about the method to use, and knowing how to validate it.

Firstly, it is necessary to decide what exactly the model will predict; this can be a generic business goal, such as, for example, which users will cancel a service, whose loans will default; what is the probability of a heart attack, given the actual conditions, etc. Depending on the question asked and the data that is available to answer, the model for prediction must be chosen.

The models are normally grouped into the following categories [17]:

- Classification: To assign labels to a result, one example would be to classify students by their probability of dropping out of a course; the classifications could be: Low Risk, High Risk. Another one, to classify animal pictures by species, like dog, cat, mice, etc.
- Scoring: Are tasks that have a continuous output. It could be, e.g., the prediction of a house price based on location, quantity of rooms and size; the algorithm learns the relationship between the input variables and predicts output for new instances of values. Normally, regression models are used for this task.
- Clustering: To group objects by their similarity and separate them from the ones that are different. It is very useful for exploratory experiments in order to find hidden patterns inside of the data.

The model defines what will be answered to the user; it could be a probability number, a score of some sort, or just the label of something, as in the case of classification. How these questions will be answered is the role of the algorithm. There is not a one to one algorithm for each model type; normally, the same problem can use a variety of algorithms, and the data scientist must test, choose and tune to decide which one is a better fit for the problem at hand.

Despite of the model, the method and the data, there is a common workflow that must be followed in every data science endeavors to ensure that the business goals will be met. Their phases are:

1. Acquire and clean data to be used for the model's training;
2. Acquire and clean data to be used in the model validation;
3. Train the model;
4. Validate the model and verify if it reached the desired precision, if not, repeat step 3;
5. Validate the tuned model;

The validation Phase 4 is very important and it is imperative to test the model with real world data, to make sure it will perform as expected when it goes to production. It would be a disaster to have a model that works perfectly with simulated data but performs poorly in the real world.

3.1 Training, Test and Validation Data

When building a model, the data is the most important ingredient. It has to be a good representation of the phenomenon that is being predicted. If the data is inaccurate and inconsistent, it will be very hard to build a model, and even if one is built, it will perform very poorly when in production. Once a dataset is acquired to be used to train the model, it should be split in two or three parts, depending on the case. They are:

- Training data: The dataset that will be used to train the model. It can be used over and over, until the model is well tuned and has a good precision on predicting it;
- Test data: It should be used only once, to test if the model is really working when the dataset changes. It serves as a "real world" test case for the model;
- Validation data (optional): It can be used as a sanity check, to double validate the model. Or, in some cases, whenever the model does not do well with the test data, it can act as a second test after the model is corrected due to failure in the test phase.

Normally, the option of having a validation dataset depends on how much data is available for creating the model. If there is very little data, the majority of it should be used to train the model, ensuring reasonably efficiency. The training, test and validation data must come from the same dataset. There are ways of randomly sampling it to get a more realistic partition for the different stages of the construction of the model. The rule of the thumb is:

- If you have a large dataset, use 60% for training, 20% for testing, 20% for validation;

- If you have a medium or small dataset, then use 60% for training and 40% for testing.

The reason to use the test and validation dataset only once is mainly to prevent the situation when a model did not perform well during the test and it was tuned again, taking the test results into consideration. This can lead to a situation where the test set served as a training dataset and when new, unseen, real world data is used, the model will perform poorly again.

3.2 Microservices Architecture

Microservices is a software architecture methodology that has as its corner stone the concept of breaking the application down into small independent services, instead of bigger monolithic applications [5]. On the conventional application architecture, one application has all the services within it, and when it needs to scale, all services are scaled together, having complete copies of the application, as show in Fig. 1.

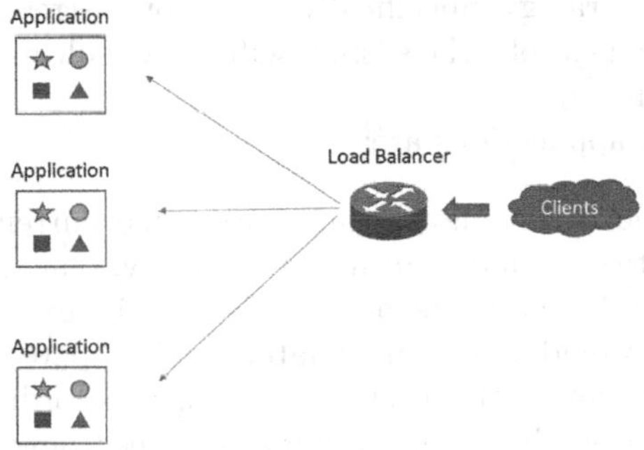

Fig. 1. Monolithic architecture.

In the case of microservices, parts of the application, the microservices, can be scaled independently, as show in Fig. 2. This allows a more efficient usage of resources, making it ideal for cloud computing scenarios.

Some other benefits of using this strategy are the following:

- Each microservice may have its own release cycle, so that it has not to wait for the whole testing of the rest of the application;
- The services can be written using different programming languages and be deployed on different platforms, providing extra flexibility to the developer that may choose the most appropriate tool for the situation;
- It is transparent to the final user;
- The different teams, responsible for their own services, can work independently and have more specialized skill sets.

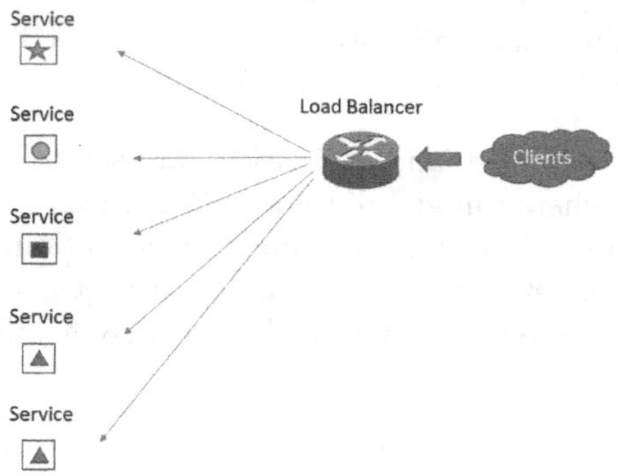

Fig. 2. Microservices architecture.

4 System Architecture

The model describes the predictive side of the whole solution. The other side is the prescriptive analytics strategy. In the case of this paper, it will be an actuation push based strategy. Specifically, based on a stream of incoming data, as near to real time as possible, the system will be consulting the model to guide a data driven decision.

Some examples of applications are:

- Real time health monitoring device that, based on occurrences of heart attacks during sport practices, could inform the athlete via an alarm that he/she is hitting a threshold, based on the age, body mass index and health history;
- Smart homes that would predict user habits and prepare the house for better comfort, as to determine the best temperature according to the weather, alarm the home owner when some activity deviates from the normal behavior of the people that lives in the house. Predict the arrival time in order to adjust the house temperature in a way that saves energy and the owner does not have to wait to the temperature to reach the ideal value;
- Trading systems that can analyze in real time the incoming data stream, and post orders that will maximize profit of the investors;
- Smart software that can be trained to identify malicious behavior of a system's intruder and alarm security personnel to take action, or try to fend off the attacker.

To be able to integrate with most systems and enable researchers to conduct experiments, the system must rely on open formats and protocols. In this case the API (Application Programming Interface) is accessible via REST (Representational State Transfer) [6], and the communication format is JSON (JavaScript Object Notation). For actuation or push based notification, an active REST request can be issued by the software, or an active connection can be maintained using Websockets protocol.

As shown in Fig. 3, the architecture has three layers:

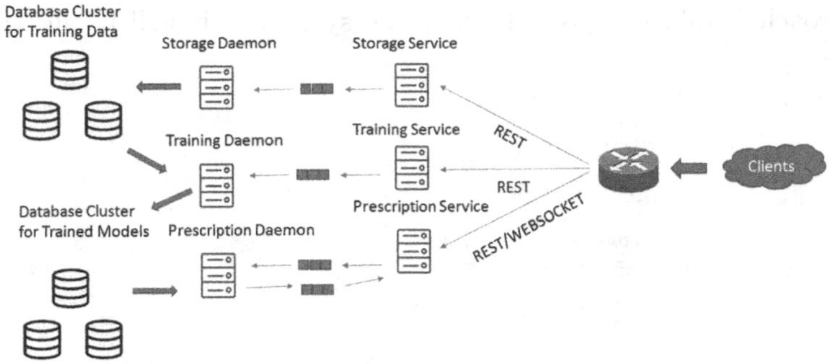

Fig. 3. System architecture.

- Front-end: REST and Websocket web servers that interact with the user, and also do the actuations via REST calls or Websocket push messages;
- Back-end: The Daemons where the developer extents the classes to configure its application and where all the processing happens;
- Storage System: The system assumes that there are two storage systems, one for the training data and another for persisting the trained model that will be used.

The front-end and back-end are connected trough message queues that have different communication patterns depending on the case. The PrescStream framework also uses the microservices approach, therefore each service can be scaled independently. For the client code, there are three major services: *Data Storage, Training Service* and *Prescription Service*.

All the three services can be independently horizontally scaled, and each of its instances can have a different back-end scaling as well. If the storage daemon, for instance, becomes a bottleneck, more back-end workers can be added to increase the throughput. As long as the web server is not the bottleneck, more workers can be added in the back-end up to when the web server becomes saturated. At this point, more web servers for that service can be instantiated with its own back-end workers. This approach fits very well for cloud deployment where failures are expected to happen, but it should not affect the clients [9].

Another positive aspect of this architecture is that it has the capacity for automatic failover, which is very important when deploying applications on the cloud. As shown in Fig. 4, if a back-end worker fails, the message is delivered again to another worker, that processes the message again, and only removes it from the queue when the task is accomplished. For the storage systems the PrescStream framework relies on the cluster solution for the storage technology being used. By default, the system comes with a Cassandra [4] back-end implemented for storing training data and a Redis [12] implementation for storing trained models.

It is worth mentioning that the flexible approach provides the system with default interfaces which allows for expansion and support for other technologies.

This approach is taken in every layer of the system as it will be shown in the next subsections.

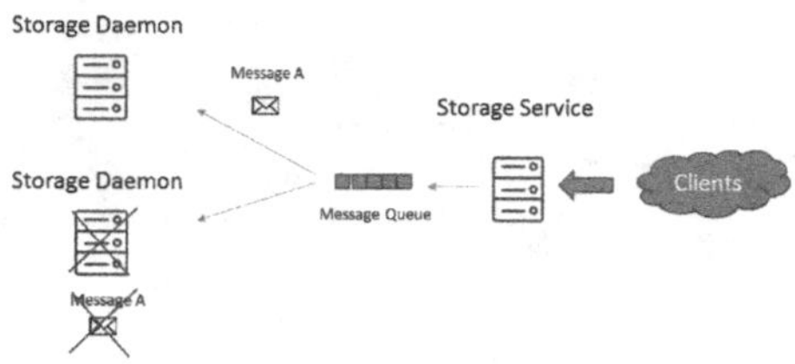

Fig. 4. Storage back-end daemon failover example.

4.1 Storage Service

The storage service is responsible for storing the training data for a given application. It is important to emphasize that the architecture on this module serves as the basis for the other services as well. As shown in Fig. 4, the web server communicates with the back-end via message queue. In our default implementation, RabbitMQ [11] is used, but as shown in the diagram in Fig. 5, this can be easily replaceable by other messaging solutions (i.e. Kafka, ZeroMQ, Storm).

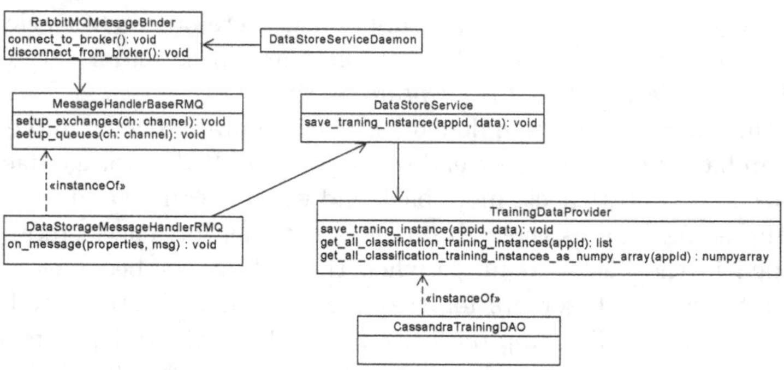

Fig. 5. Storage daemon internal design.

Every CSV (Comma Separated Values) instance is stored in a keyspace created by the user and in a table with the application name. For classification and regression every instance is unique for a given result; from a training perspective we should not have one instance with multiple results, this is enforced by the database constraint.

For the front-end REST API, the NodeJS [10] server was the choice because of its easy way to develop and deploy, and also due to the many packages available to integrate with other systems. That also does not mean that it cannot be replaced by another front-end, as long as the interaction with the message queue does not change. If higher throughput is desired, more data servers may be added to the cluster and more Storage Daemons may be added to the back-end. This is ideal for cloud scenarios and Big Data applications. The back-end daemons can also be seem as microservices.

For each application, this module supports two method calls done via REST by the client code:

- Store CSV instance: Adds an instance of CSV with its regression result or classification label, for a given application.
- Clear all data: Clear all training data for a given application, in case the user wants to start with a different dataset.

4.2 Training Service

This module is responsible for fetching training data, apply automatic tranformation to data, choose most relevant features, train with different algorithms, and choose the best trained model for the data. The model is then stored in the model storage cluster. The overall architecture is very similar to the Storage Service, which includes failover and back-end scaling (Fig. 6).

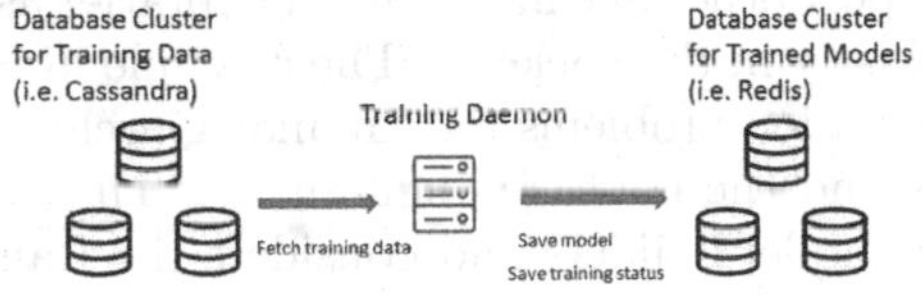

Fig. 6. Training services data access.

This service also supports two commands for the client:

- Train Model: Triggers the training for a given application and generates a model for the training data;
- Get Training Status: Gets the training status for a given application; these status can be: failed, success and training.

The training implementation has several steps that corresponds to the normal pipeline of model generation, which frees the researcher from a lot of the hard work. Using the framework philosophy enables the community to add more preprocessing logics and more algorithms to the system.

These are the steps that occur when the training is triggered:

- Verify columns that need transformation, such as ISO formatted dates (Fig. 7a), categorical variables and text instances, that need to be one hot encoded, as shown in Fig. 7b;
- Save the indexes of the data to be transformed and the encoders;

- Normalize the training data and save the normalizer encoder;
- Train different algorithms, with more than one parameter if necessary, and check for unnecessary features;
- Save the index of the features to be removed;
- Save the best scoring model to the model storage cluster.

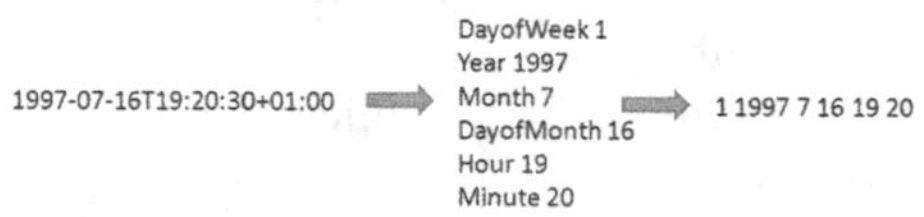

(a) ISO date extraction

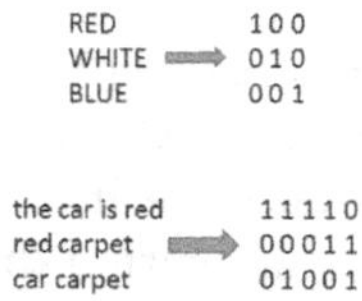

(b) One hot encoding

Fig. 7. Automatic data transformations.

In the default implementation, Redis [12] is used for the storage of the trained model; nonetheless, as seem in Fig. 8, the user can replace the storage backend. For model generation, the default implementation uses scikit-learn [13], but it can be replaced to support other machine learning libraries as well, even by using a high performance cluster in the backend. The example in Fig. 8 is showing the classification case; regression problems use an analog architecture, with changes in the model builders and the message queue name. The implementation done is currently using two model builders and considers the Random Forest [3] and the SVM [7] (Support Vector Machine) models.

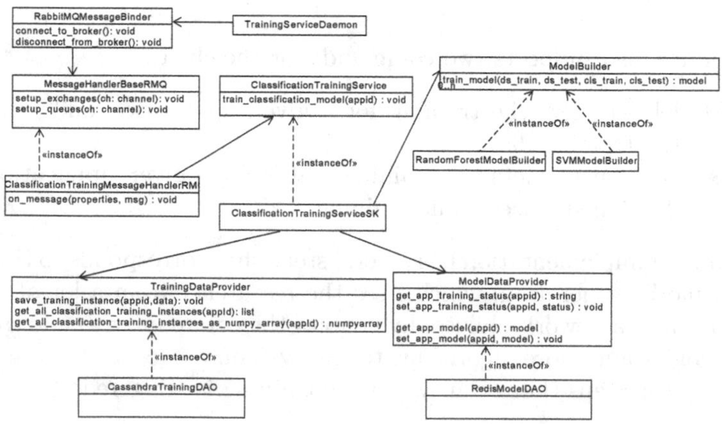

Fig. 8. Training daemon internal design.

4.3 Prescription Service

This service is responsible for receiving the stream of data and make prescriptions from it. As shown in Fig. 9, it has a slightly different architecture compared to the other two services. It has a return queue for RPC (Remote Procedure Call) that sends data back to the web server to be consumed by the clients.

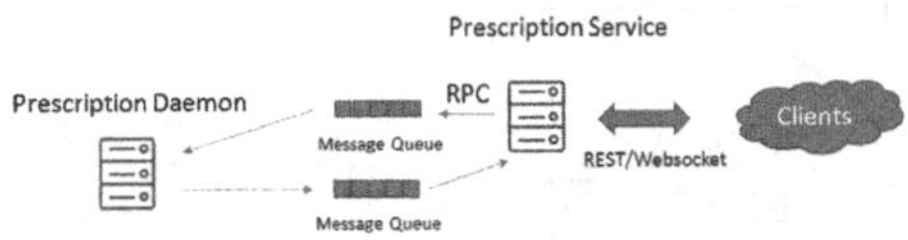

Fig. 9. Prescription services architecture.

Currently, it suports four types of communication:

- REST request reply communication: The client issues a REST call and receives the prescription as a response, as shown in Fig. 10a;
- Websocket request reply communication: It works analogous to the REST case, but it uses the Websocket protocol instead;
- REST actuation: The client sends the data either via REST or Websocket, and the server makes a REST call to another service provided by the user to actuate, as shown in Fig. 10b. The REST call is built by user defined logic and will be described in the next subsection;
- Websocket Publish-Subscribe actuation: In this pattern, when a user makes a request, either by REST or Websocket, the prescription goes to all Websocket subscribers for the application, as shown if Fig. 10c.

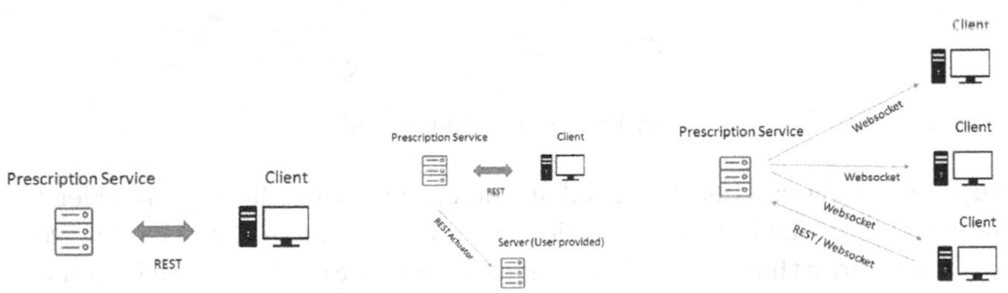

(a) *Request reply communication.* (b) *REST actuation.* (c) Publish subscribe actuation: *Websocket/REST.*

Fig. 10. Communication patterns.

Nothing prevents the client application to just get the prescription from the system and do the actuation itself. The idea behind this possibility is that the solution is generic enough to allow for all kinds of integration.

The internal design of the prescription service is shown in Fig. 11; the classes that the user extends are better explained in the next subsection.

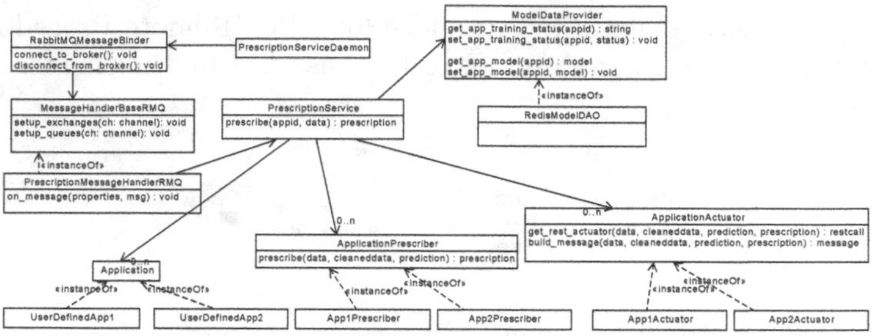

Fig. 11. Prescription daemon internal design.

4.4 Application Model

Figure 12 shows the major domain classes that are mostly important for the users of the PrescStream framework.

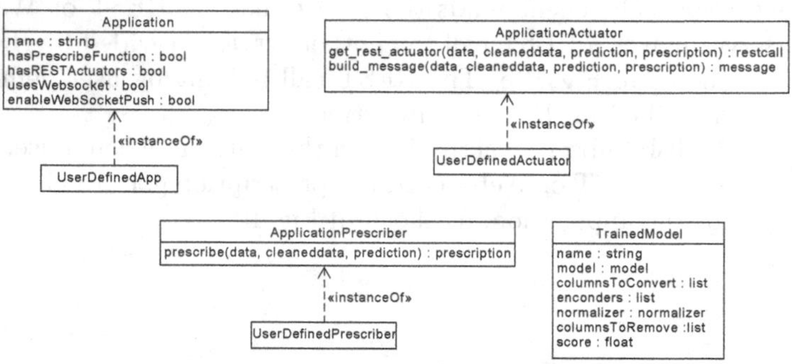

Fig. 12. Application model.

- Application class: This class needs to be extended by the user in order to set the configurations of the application. It is important to note that the servers may serve more than one application, as shown in Fig. 18, and each application may have distinct models and communication patterns. The properties are self explanatory, such as application name, whether it uses REST actuators or does publish subscribe communication;
- ApplicationPrescriber class: It needs to be extended by the user as well and it is where the optimization of prescriber model is plugged. Because this can be quite complex, there is not a default implementation for it and it is the user responsibility to implement it with the application relevant logic. The method prescribe receives the original data instance, the transformed data and the prediction, so that the user logic can make the prescription that will be send to the clients of the system;

- ApplicationActuator: Is the class that receives the outcome of the prescription and all previous data fed to it, afterwards it builds the REST actuator or the message to be pushed via Websocket.

Only by implementing these simple classes the framework user can build very powerful data driven actuating prescribing applications that are easy to integrate and scale on the cloud.

5 Experiment

In order to validate the architecture, an experiment was developed to show an example of how this structure can be used in a real life scenario. Data is the main fuel for any experiment involving analytics. Since the acquisition of such data would require setting up a lab, finding volunteers, using special devices and running several measurements, it would be too much for the author to go into all this length to test the architecture. The solution found was to use available public data, and simulate a real world scenario. All the calls to the system uses the same interface and protocols as a client application would. The same applies to all received push notifications; in this case, a software emulating a Websocket client was developed to verify that these notifications are coming as well.

Table 1. Experiment system configuration.

Processor:	Intel(R) Xeon(R) CPU E5-1607 v2 3.00 GHz
RAM:	16 Gb DDR3 1800 Mhz
Operating system:	CentOS 7.2
Python:	2.7.5
NodeJS:	4.4.3
RabbitMQ:	3.6.1
Cassandra:	3.0.5
Redis:	3.0.7
Scikit-learn:	0.17.0

5.1 The Data

Velloso et al. [16] conducted a series of experiments that use machine learning to quantify the quality of execution of a gym exercise. They used sensors in a belt, armband, glove and in the dumbbell, and had volunteers executing the exercise either correctly or in four other wrong postures. These data contains the sensor's pitch, roll, yaw, raw accelerometer, gyroscope and magnetometer [16]. All this information has been classified and made available to the public in CSV format.

5.2 The Test

The test emulates the idea that, in the near future, people will have such sensors embedded in their clothes which will communicate to a more complex device, such as a smart watch. This smart device will then be communicating to prescription services that incorporate models to prevent people from executing exercises the wrong way. The user would have a vibration feedback or an alarm beep when doing the exercise in an improper posture.

In order to emulate this behavior the dataset was filtered in order to remove features that are not necessary to build the model; empty and "NA" fields were removed as well. All the observations were then randomly split and a 30% of rows were separated to do the simulation. The training data was then fed to the system trough the REST API. To emulate such a client, a Python script has been developed to make the REST calls. After the loading of the training data, the model was generated triggered by the REST call. The best classification model was automatically chosen for the given data.

The prescription were set to get specific warning notifications whenever the instances would be classified as some wrong posture to the exercise being executed. These notifications would contain a prescription in what should be corrected to execute the movement in a correct fashion. The principle is to emulate what a smartwatch would give as advice for someone doing her workout. Table 2 shows the possible prescriptions.

Table 2. Experiment possible prescriptions.

Correct elbow position
Lift the dumbbell higher
Lower the dumbbell further
Correct hips position

A second Python script then made calls to get prescription using REST calls, passing the test instances separated in the data cleaning step. Another Python client was also developed to get these notifications via Websocket to make sure the push notification is also working. Latter on, it is measured how many prescriptions were expected, and how many actually came. This will be the number of the model precision in the end.

5.3 Results

From a black box perspective, considering that the application just uploaded the data, requested a model creation, set the necessary prescriptions, what should be analyzed is whether the prescriptions that arrived were correct or not. For that, the following three cases are considered:

- In a first case, it is measured how many correct prescriptions were expected, that is, whether the user was doing an exercise in an inadequate fashion and he should be notified to correct this execution.
- The second case considers the false positives and false negatives. A false positive's when the user is doing the activity correctly and it is prescribed to change what he is doing. A false negative's when the user is doing the exercise in a wrong manner and he is not notified to do it correctly.
- The last case considers the number of wrong prescriptions, when the user is executing the exercise incorrectly and receives a wrong prescription.

Since models are imperfect and probabilistic by nature, all the situations described are likely to occur. The Table 3 shows the results of the experiment.

Table 3. Experiment prescription results.

Total prescriptions	5886
Correct prescriptions	5749 (98%)
False positives	12
False negatives	26
Wrong prescriptions	99

Looking into a white box approach, it is interesting to observe that the logs on the system reported that first it tried to train a model using SVM (Support Vector Machine) algorithm [7], but it only yielded a 0.274 precision which would be an unusable model, then the system tested with the Random Forest algorithm [3] and got a 0.98 score, which is confirmed in the live tests. All this was transparent for the user of the system, that only had to upload the data and issue commands to train the model and configure the prescriptions.

There are other important aspects to observe in the results. The main one is related to the system's performance. Since this is a system designed to be able to handle IoT, and busy websites kinds of workload, the transactions per second must be high. Table 4 shows some timings. Considering the setup described in Table 1, the systems throughput is quite high.

Table 4. Experiment performance results.

CSV instance stores per second	1056
Prescriptions per sercond	82

6 Conclusion and Future Work

Data analytics is a preponderant tool in scenarios of generation of large volumes of data, like IoT. This paper presented PrescStream, a generic and flexible framework that allows developers to incorporate predictive and prescriptive analytics, while allowing timely action for the application user. The framework was developed with technologies that favor its modularity, integration, fault tolerance and performance. Its microservices architecture enables performance improvement by allowing the availability of computational power, precisely where bottlenecks happen, thus optimizing resource utilization.

Indeed, the system was designed to have a strong horizontal scalability, meaning that if the application's processing demand grows, it is possible to add more servers and the system can grow with it. There are many ways of scaling the system and one must identify where the bottleneck is, to add more resources.

A test experiment, based on simulations, was performed, demonstrating the feasibility of the framework considering scenarios with high data load. The troughput achieved was sufficient in order to have a heavy load application working with real time data streaming.

As future work, it is planned to implement the solution in real live settings with sensors streaming data, in addition to remote servers, in order to test the performance of the data flow under these conditions. Another interesting test would be the horizontal replication of the services, in order to verify the increase in transactions per second of the application, validating the architecture for Big Data and IoT scenarios. A number of future work may also involve the use of other model-building technologies such as deep learning models, image recognition applications and support for other storage systems.

References

1. Atzori, L., Iera, A., Morabito, G., Diee, A.: The internet of things: a survey. Comput. Netw. **54**(15), 2787–2805 (2010)
2. Barber, R., Sharkey, M.: Course correction: using analytics to predict course success. In: Proceedings of the 2nd International Conference on Learning Analytics and Knowledge, pp. 259–262. ACM (2012)
3. Breiman, L.: Random forests. Mach. Learn. **45**(1), 5–32 (2001)
4. Cassandra: Apache Cassandra. http://cassandra.apache.org/
5. Dragoni, N., Giallorenzo, S., Lafuente, A.L., Mazzara, M., Montesi, F., Mustafin, R., Safina, L.: Microservices: yesterday, today, and tomorrow (2016). arXiv preprint arXiv:1606.04036
6. Fielding, R.T., Taylor, R.N.: Principled design of the modern web architecture. ACM Trans. Internet Technol. (TOIT) **2**(2), 115–150 (2002)
7. Hearst, M.A., Dumais, S.T., Osman, E., Platt, J., Scholkopf, B.: Support vector machines. IEEE Intell. Syst. **13**, 18–28 (1998)
8. Huang, Z., Wong, P.C., Mackey, P., Chen, Y., Ma, J., Schneider, K., Greitzer, F.L.: Managing complex network operation with predictive analytics. In: AAAI Spring Symposium: Technosocial Predictive Analytics, pp. 59–65 (2009)

9. Jhawar, R., Piuri, V., Santambrogio, M.: Fault tolerance management in cloud computing: a system-level perspective. IEEE Syst. J. **7**(2), 288–297 (2013)
10. Node: NodeJS. https://nodejs.org/
11. Rabbit: Rabbitmq. https://www.rabbitmq.com/
12. Redis: Redis. http://redis.io/
13. Scikit: Scikit-learn. http://scikit-learn.org/
14. Siegel, E.: Predictive Analytics: The Power to Predict Who will Click, Buy, Lie, or Die. Wiley, Hoboken (2013)
15. Tönjes, R., Barnaghi, P., Ali, M., Mileo, A., Hauswirth, M., Ganz, F., Ganea, S., Kjærgaard, B., Kuemper, D., Nechifor, S., et al.: Real time iot stream processing and large-scale data analytics for smart city applications. In: Poster Session, European Conference on Networks and Communications (2014)
16. Velloso, E., Bulling, A., Gellersen, H., Ugulino, W., Fuks, H.: Qualitative activity recognition of weight lifting exercises. In: Proceedings of the 4th Augmented Human International Conference, pp. 116–123. ACM (2013)
17. Zumel, N., Mount, J., Porzak, J.: Practical Data Science with R. Manning, Greenwich (2014)

A Novel Approach for Supporting Approximate Representation of Linear Prediction Residuals in Pattern Recognition Tools

Alfredo Cuzzocrea[1(✉)] and Enzo Mumolo[2]

[1] DIA Department, University of Trieste and ICAR-CNR, Trieste, Italy
alfredo.cuzzocrea@dia.units.it
[2] DIA Department, University of Trieste, Trieste, Italy
mumolo@units.it

Abstract. The goal of this paper is to describe an optimization approach for selecting a reduced number of samples of the linear prediction residual. This can be extremely useful in pattern recognition tools. Sample determination is a combinatorial problem. Our approach addresses the combinatorial problem with simulated annealing based optimization. We show that better results than that obtained by a standard approximation approach, namely the multi-pulse algorithm, are obtained with our approach. Multi-pulse selects pulse locations by a sequential, suboptimal, algorithm and computes the pulses amplitudes according to an optimization criteria. Our approach finds the optimal residual samples by means of an optimization algorithm approach without amplitudes optimization. The compressed residual is fed to an all-pole model of speech obtaining better results than standard Multipulse modeling. We believe that this algorithm could be used as an alternative to other algorithms for medium-rate coding of speech in low complexity embedded devices. We also discuss performance and complexity issues of the described algorithm.

1 Introduction

Research on approximation of the linear prediction residue has been an active field for many years in the past. Important results have been obtained with the multi-pulse approach [1,2]. The multi-pulse approach has been extensively used to model the prediction residual in medium-rate speech coders. High-quality speech coders at bit-rates in the vicinity of 8 Kbits have been developed with the multi-pulse approach. However, it is very difficult to reduce the bit-rate with this approach, basically because of the quantization of the pulses amplitudes. CELP coders [3], which model the excitation signal with a binary sequence, can produce good-quality speech at bit rates as low as 4.8 Kbits, but they require a costly - from a computational point of view - code-book search. Multi-pulse algorithms, on the other hand, have to solve a nonlinear minimization problem for searching the pulse locations. Therefore, one problem of multi-pulse algorithms is the determination of optimal excitation sequences with a reasonable computational

O. Gervasi et al. (Eds.): ICCSA 2017, Part I, LNCS 10404, pp. 342–356, 2017.
DOI: 10.1007/978-3-319-62392-4_25

complexity. Such determination can be computationally simplified by using some approximations during the derivation, but this leads to suboptimal results. The classical multi-pulse algorithms described in [1,2] in fact use sequential, step-by-step procedures where the next pulse location is determined assuming that all the other locations remain constant. The high performance of the algorithm described in [1,2] is based on optimal adjustment of the pulse amplitudes during the location search.

CELP is based on the assumption that the short- and long-term LPC residual can be modeled with a zero-mean Gaussian process. A stochastic code-book filled with instances of a Gaussian process is then exhaustively searched for the minimum of the mean-squared error. However, standard CELP [3] is not without pitfalls. First, we should assume that the code-book is large enough to model any kind of voiced and unvoiced excitations. Therefore, the CELP performance depends on the size of the code-book and on the assumed statistical distribution. Improvements can be obtained using training algorithms for the design of the random code-books [4,5] and using adaptive codebooks [6]. Second, exhaustive search imposes high computational complexity, which can be reduced using suitable structures of the code-book [7,8]. Significant complexity reduction has been also described in [9–11], where a somehow different approach to low-bit rate coding is described.

A different approach to LP residual encoding is available from the year 2006, when Candès and Donoho proposed the Compressed Sensing or Compressed Sampling (CS) approach [12,13]. CS approach states the possibility to sample a signal well below the Nyquist rate without degradation of the recovered signal provided that two hypothesis concerning sparsity and incoherence are satisfied. The approach moves the complexity from the encoder to the decoder. Thus the receiver requires much more computational power than the transmitter, since signal recovery at the receiver is performed by solving an optimization problem. Compressed sampling of LP residual is appealing since the residual is a sparse signal. Giacobello *et al.* in [14,15] describe the computation of a sparse approximation of speech residual using compressed sensing.

In this paper we describe a different approach for the optimal determination of a sparse approximation of the residual sequence through a minimization carried out with simulated annealing. Standard multi-pulse finds pulse locations through a sub-optimal algorithm, trying to compensate the sub-optimality of pulse location with an optimal estimation of the pulses amplitudes. Our algorithm try to optimally estimate the sample locations while the amplitudes are the actual residual values at the found locations. This can be extremely useful in pattern recognition tools.

Quality achieved and complexity figures indicate that the algorithm described in this paper is an alternative to Multipulse and Celp approaches. Instead of searching a code-book like CELP, the proposed algorithm determines the optimum residual samples using heuristic optimization. The algorithm has many interesting features. First of all, besides the initial LPC analysis and the computation of auto- and cross-correlations, it requires only additions during the optimum samples determination. Secondly, the convergence of simulated annealing applied to this problem is easily reached with a simple exponential cooling schedule. The algorithm's objective performance (segmental SNR) have been compared with the standard multi-pulse performance and with an adaptive compressed sampling algorithm. Moreover, informal subjective tests conducted with the proposed algorithm compared with standard CELP show indistinguishable quality at a bit-rate of 5–6 Kbits. Finally, the algorithm can be implemented on a low-cost embedded system, as suggested by the simulation results.

The paper is structured as follows: in Sect. 2, the linear prediction of speech basis are briefly reviewed. In Sect. 3, we summarize the Compressed sensing algorithm to show its similarities with our approach. In Sect. 4, we briefly review the standard Multi-pulse approach which is compared with our approach. Section 5 describes our approach and Sect. 6 describes the optimization algorithm based on Simulated annealing. In Sect. 7 we report some experimental results about performance, and an analysis example of a speech frame. Finally, in Sect. 8, some concluding remarks and future work are discussed.

2 Preliminaries

According to the Linear Prediction (LP) of speech (e.g. [16]), in Z domain the LP residue is given by

$$E(z) = \left(1 - \sum_{i=1}^{p} a_i z^{-i}\right) X(z) = A(z)X(z) \tag{1}$$

where $E(z)$ is the LP residue, $X(z)$ is the speech signal, a_i are the linear prediction coefficients and p is the linear prediction order. Consequently, $X(z) = E(z)\frac{1}{A(z)}$. Turning to the discrete time domain, we have $x(n) = \sum_{i=0}^{p} h(i)e(n-i)$, where $x(n)$ is the speech signal, $h(i)$ is the impulse response of the all-pole system $\frac{1}{A(z)}$ and $e(n)$ is the LP residue. LP-based efficient coding of speech is based on finding an approximation $\tilde{e}(n)$ of the LP residual such that a high perceptual quality version $\tilde{x}(n)$ of the original speech signal is reconstructed at a lower bit rate using Eq. (2).

$$\tilde{x}(n) = \sum_{i=0}^{p} h(i)\tilde{e}(n-i) \tag{2}$$

An important residual approximation approach is multi-pulse [1], where the residual is approximated with a series of pulses whose amplitudes are optimized. Other popular approaches are based on the selection of the optimum residue from a code-book of residues randomly generated [3]. Recently, Giacobello et al. [14] use Compressed sampling for representing the LP residual in a compressed sampling framework. In this paper we propose to sample the residue by solving the related combinatorial problem with Simulated Annealing.

3 Compressed Sampling Principles

Candés *et al.* and Donoho in 2006 develop in [12,13] the Compressed Sampling (or sensing) theory (CS). By CS theory, under sparsity and incoherence hypothesis, a signal can be reconstructed from very few samples, well below the Nyquist-Shannon rate. Sparsity means that a signal frame $\boldsymbol{x} = [x(1), x(2), \ldots, x(N)]$ may be expanded onto a basis $\boldsymbol{\Psi} = [\psi_1, \psi_2, \ldots, \psi_N]$ so that $x(n)$ is represented by only K significant coefficients, $K \ll N$. The expansion is represented by Eq. (3) where only K coefficients in $\boldsymbol{c}$ are nonzero.

$$\boldsymbol{x} = \boldsymbol{\Psi c} \tag{3}$$

The signal is randomly sampled, so that $M < N$ random samples of x are taken, as described in Eq. (4).

$$\hat{\boldsymbol{x}} = \boldsymbol{\Phi x} = \boldsymbol{\Phi \Psi c} \tag{4}$$

The $M \times N$ measurement matrix $\boldsymbol{\Phi}$ is made by random ortho-basis vectors. By the CS theory, if $\boldsymbol{\Phi}$ and $\boldsymbol{\Psi}$ are incoherent, the original signal $\boldsymbol{x}$ can be reconstructed from $\hat{x}$ within the approximation error ϵ by solving the optimization problem described in Eq. (5).

$$\min_{c \in R^N} \sum_{n=1}^{N} |c(n)| \text{ such that } \sum_{i-1}^{N} [x(n) - \hat{x}(n)]^2 \leq \epsilon \tag{5}$$

Note that this optimization problem is solved at the receiver.

4 Review of LP Residue Approximation by Standard Multi-pulse Approach

In the standard multipulse algorithm, [1], the LP residual is approximated with an impulsive sequence $u(n)$. The standard multipulse algorithm uses a closed-loop procedure for computing $u(n)$, once the impulse response $h(n)$ of the all-pole filter $\frac{1}{A(z)}$ is determined. The reconstructed signal $\tilde{x}(n)$ is obtained by filtering the impulse excitation sequence $u(n)$ with the all-pole filter as reported in Eq. (6).

$$\tilde{x}(n) = \sum_{i=0}^{p} h(i)u(n-i) \tag{6}$$

The pulse sequence $u(n)$ is a model of the prediction residual signal and is given by Eq. (7) which is a linear combination of Kroneker delta functions.

$$u(n) = \sum_{i=0}^{M-1} \beta_i \delta(n - n_i) \tag{7}$$

In Eq. (7) β_i, n_i, and M are the pulse amplitudes, the pulse locations, and the number of pulses respectively. In [1] the error between the input signal and the reconstructed one, $r(n) = x(n) - \tilde{x}(n)$, is weighted with a *perceptual* filter derived from $A(z)$. The perceptual filter is intended to un-emphasize the error energy in the high-energy regions of the signal spectrum, according to auditory masking criteria [17]. The best values of pulse amplitude and locations β_i and n_i is attained by minimizing the energy of the weighted reconstruction error. Because the weighting filter is linear, the minimization problem is equivalent to minimizing the un-weighted mean squared error between the weighted speech signal $x_w(n)$ and the corresponding weighted synthetic signal $\hat{x}_w(n)$. The optimization problem of [1] is described by Eq. (8).

$$(n_i, \beta_i) = argmin \sum_{n=0}^{N-1} [x_w(n) - \hat{x}_w(n)]^2 \tag{8}$$

where N is the number of samples in the voice signal frame. The problem stated in Eq. (8) can be solved at different levels of optimality depending on the procedure for the pulse location search. The algorithm described in [1] uses a step-by-step procedure for finding the M locations of the best pulse sequence. On the other hand, the sequential approach to location space scanning does not guarantee the optimal solution to the minimization problem.

5 A Novel Approach for Sparse Approximation of the Speech LP Residue

Our Sparse Approximation of the residual is simply the selection of a reduced number of residual samples that is still able to give good signal reconstruction performance. It is an approach of compressed sampling of the residual. Compressed sampling on N points sequence may be viewed as a combinatorial problem, in the sense that a small number of K samples are selected out of N to represent the sequence according to an optimization criterion. The number of the possible sets of K samples Γ is given by Eq. (9).

$$\Gamma = \binom{N}{K} \tag{9}$$

In theory, the selection of the optimal set of samples would be performed as follows. All the possible sets of compressed samples are generated, and for each set an error measure is computed. The set corresponding to the minimum error is thus selected. However, let us consider a simple example. If we start from a sequence of 20 ms at a sampling rate of 11 K samples per second, we have a sequence of 220 points. If for example we want to select only 13 samples out of 220 points, the number of possible sets is about $3E^20$. Clearly, the computation of an error measure for each set is computationally impossible. For this reason we perform the compressed sampling with a Simulated Annealing optimization procedure [18].

The sampled residue $u(n)$ is given by Eq. (10), where $e(n)$ is the LP residue, n_k are the sampling locations and M is the number of samples.

$$u(n) = \sum_{k=0}^{M-1} e(n_k)\delta(n - n_k) \tag{10}$$

The sampled residue is fed to the all-pole filter whose impulse response is $h(n)$. Thus the reconstructed signal $\hat{x}(n)$ is described by Eq. (11).

$$\hat{x}(n) = \sum_{i=0}^{N} h(i)u(n-i) = \sum_{i=0}^{N} h(i) \sum_{k=0}^{M-1} e(n_k)\delta(n - i - n_k) = \sum_{k=0}^{M-1} e(n_k)h(n - n_k) \tag{11}$$

The reconstruction error E is reported in Eq. (12).

$$\begin{aligned}
E &= \sum_{n=0}^{N-1} [x(n) - \hat{x}(n)]^2 = \sum_{n=0}^{N-1} \left[x(n) - \sum_{k=0}^{M-1} e(n_k)h(n-n_k) \right]^2 \\
&= \sum_{n=0}^{N-1} x^2(n) + \sum_{i=0}^{N-1} \left[\sum_{i=0}^{M-1} e(n_i)h(n-n_i) \right] \left[\sum_{j=0}^{M-1} e(n_j)h(n-n_j) \right] \\
&\quad - 2 \sum_{n=0}^{N-1} x(n) \sum_{k=0}^{M-1} e(n_k)h(n-n_k) \\
&= \sum_{n=0}^{N-1} x^2(n) + \sum_{i=0}^{M-1} \sum_{j=0}^{M-1} e(n_i)e(n_j) \sum_{n=0}^{N-1} h(n-n_i)h(n-n_j) \\
&\quad - 2 \sum_{k=0}^{M-1} e(n_k) \sum_{n=0}^{N-1} x(n)h(n-n_k) \\
&= \sum_{n=0}^{N-1} x^2(n) + \sum_{i=0}^{M-1} \sum_{j=0}^{M-1} e(n_i)e(n_j)R_{hh}(n_i, n_j) - 2 \sum_{k=1}^{M-1} e(n_k)R_{xh}(n_k)
\end{aligned} \tag{12}$$

where $R_{hh}(n_i, n_j)$ is the auto-correlation of the impulse response at (n_i, n_j) and $R_{xh}(n_k)$ the cross-correlation at n_k.

5.1 The Simulated Annealing Algorithm

Simulated annealing (SA) is an heuristic optimization procedure proposed by Kirkpatrick in [19]. Initially applied to the solution of combinatorial problems. Other reported applications of SA, besides combinatorial problems, include pin assignment optimization [20], image segmentation [21] and image reconstruction in Electrical Impedance tomography [22]. The compressed samples locations determination problem is indeed a combinatorial one as M locations out of N (the number of samples in a frame) must be found according to a suitable MSE criterion. In [20–22] Simulated Annealing optimization is applied only once. In the present applications, however, SA shall be applied for each signal segment. Therefore, real-time computation issue is fundamental.

Roughly, the concept of Simulated Annealing comes from the physical process called annealing, where a solid is heated to melt. Subsequently the solid is slowly cooled to asses its reticule. In algorithmic terms, starting from an initial state, related to a starting temperature and associated to an energy definition, the state is randomly modified. If the energy associated to the new state is reduced with respect to the previous state, then the perturbed state is accepted and a new perturbation is performed. Otherwise, a new perturbation is performed. If the new state is in thermal equilibrium, as verified by the Boltzmann distribution, the perturbed state is accepted and the process continue. The sequence of random perturbations forms a Markov chain and it is generated until the thermal equilibrium is reached. Then, the temperature is reduced and another sequence of Markov chains is generated.

Thus, the algorithm consists in a sequence of Markov chains, each at a different decreasing temperature. The test for thermal equilibrium has been proposed by Metropolis *et al.* [23]. In summary, the Simulated Annealing algorithm is described in Algorithm 1.

Algorithm 1. Simulated annealing algorithm

(T_0, α)

procedure SA
    $T \leftarrow T_0$
    $E \leftarrow$ Energy definition
    Cooling Schedule $\leftarrow T = \alpha T$
    Current Model $\leftarrow$ random()
    while Not converged **do**
        New Model $\leftarrow$ *Randomly Perturbed Current Model*
        $\Delta_E \leftarrow$ *E(New Model) - E(Current Model)*
        if $\Delta_E \leq 0$ **then**
            Current Model $\leftarrow$ *New Model*
        else
            $r \leftarrow$ *randomnumber perturbation*
            if $(r \leq e^{-\frac{\Delta_E}{T}})$) **then**
                Current Model $\leftarrow$ *New Model*
            end if
        end if
    end while
    return *Current Model*
end procedure

6 Compressed Sampling of Linear Prediction Residue

The SA solution of the sample selection problem is described as follows: a starting set of locations is randomly chosen and a generation scheme is suitably defined so that, given a set of locations, another set of locations is obtained. For each new set of locations a cost function E is computed as the weighted mean squared error between the original and the reconstructed signal. The new set of locations is accepted if the Metropolis test is satisfied. This process continues until a minimum is obtained. The Metropolis test is based on the Boltzmann distribution and uses ΔE and T_k. These two parameters are, respectively, the variation of the cost between two iterations, and the temperature that controls the annealing process. The temperature T_k is updated at each iteration k.

6.1 Generation of Samples Locations

In theory, at each Simulated Annealing iteration the algorithm should manage a set of M different random locations. Recall that to each iteration corresponds a generation of a Markov chain where the M locations are randomly perturbed. Therefore, a multivariate random generator with a Gaussian distribution should be used. This approach would introduce this difficulty: each parameter might require a different annealing schedule for a good convergence and the computation of the cost function would be high. For these reasons, we developed a more robust samples locations generation scheme, where only one location at a time is modified according to a uni-variate Gaussian distribution. Therefore, the scheme we use to generate a new set of sample locations is the following:

1. Randomly select one pulse.
2. Perturb its location according to a uni-variate Gaussian distribution.

6.2 Algorithm for Determining Optimal Residual Samples

The initial value of the temperature, T_0, is set to the value of the standard deviation of the cost function computed during an initial free Markov chain. According to [18], the temperature update is realized as reported in Eq. 13.

$$T_{k+1} = \alpha_k T_k, \quad \alpha_k = e^{-0.4\frac{T_k}{\sigma^2}} \tag{13}$$

where σ^2 is the variance of the cost function during the k-th iteration.

The optimization algorithm for the determination of the optimal residual samples is described in the Algorithm 2.

The Metropolis test is based on the variation of the error function, as shown in Eq. (14).

$$\begin{aligned}\Delta_E =& E_{new} - E_{old} = \\ & \sum_{i=0}^{M-1}\sum_{j=0}^{M-1} e(n_i^{new})e(n_j^{new})R_{hh}(n_i^{new}, n_j^{new}) - 2\sum_{k=1}^{M-1} e(n_k^{new})R_{xh}(n_k^{new}) \\ & - \left[\sum_{i=0}^{M-1}\sum_{j=0}^{M-1} e(n_i^{old})e(n_j^{new})R_{hh}(n_i^{old}, n_j^{old}) - 2\sum_{k=1}^{M-1} e(n_k^{old})R_{xh}(n_k^{old})\right]\end{aligned} \tag{14}$$

Algorithm 2. Compressed sampling of the linear prediction residue $(T0, \alpha)$

procedure SA1
  $T \leftarrow T_0$
  Current Locations set $\leftarrow$ random() ▷ random samples initialization
  while Not converged **do**
    while Iterations not terminated **do**
      New Locations set $\leftarrow$ *Randomly Perturbed Current Locations set*
      $\Delta_E \leftarrow$ *E(New) - E(Current)*
      if $\Delta_E \leq 0$ **then**
        Current Samples Locations $\leftarrow$ *New Samples Locations*
      else
        $r \leftarrow$ *random()*
        if $(r < e^{-\frac{\Delta_E}{T}})$ **then** ▷ Metropolis test
          Current Locations set $\leftarrow$ *New Locations set*
        end if
      end if
    end while
    $T = \alpha T$ ▷ Temperature Annealing
  end while
  return *Current Samples Locations*
end procedure

However many of the terms before and after the minus sign in Eq. (14) are equal because of the samples locations generation scheme: only one location is perturbed at each iteration. Thus many terms are canceled and Eq. (14) reduces to Eq. (15)

$$\begin{aligned} \Delta_E = & E_{new} - E_{old} = R_{hh}(n_{new}, n_{new}) - R_{hh}(n_{old}, n_{old}) \\ & - 2e(n_{new}) \left[R_{xh}(n_{new}) - \sum_{k \neq old} e(n_k) R_{hh}(n_{new}, n_k) \right] \\ & + 2e(m_{old}) \left[R_{xh}(n_{old}) - \sum_{k \neq old} e(n_k) R_{hh}(n_{old}, n_k) \right] \end{aligned} \tag{15}$$

where E_{new} and E_{old} are the error functions related to the actual and the previous iterations.

6.3 Convergence Behavior

In Fig. 1 the convergence behavior of the algorithm is shown. This figure describes the segmental SNR versus Markov chain length for 13, 16, and 22 samples per frame. Recall that in our experiments we use 20ms long frames, which at 11025 sampling rate means 220 samples. Thus 22 samples per frame is a 90% residual compression. The correct convergence is ensured by the saturation in the SNR as the Markov chain length increases. The best length value has been taken in correspondence of the saturation edge, and it was set to 300. The number of temperature decrements was experimentally found to be about 20 in order to reach the minimum.

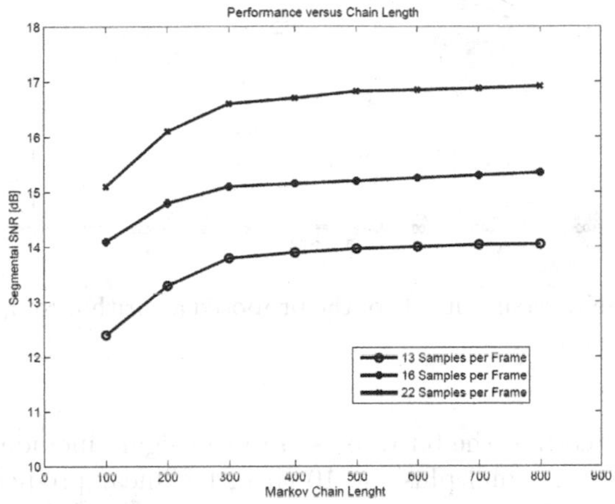

Fig. 1. Segmental SNR versus Markov chain length for various numbers of samples per frame.

All the products in Eq. (15) are computed prior to the Simulated Annealing iterations. Thus each Simulated Annealing iteration require only additions and the iterations are very fast.

7 Experimental Results and Analysis

Experimental results are obtained with sentences extracted from the Artic speech dataset from CMU [24]. One hundred sentences spoken by two US male speakers are extracted from the database and the results are averaged among the two speakers. The data is down-sampled at 11000 samples per second, and analyzed with 20 ms frames, namely 220 samples long. We used a 10th-order LPC analysis with a correlation approach. The segmental SNR is defined as shown in Eq. (16).

$$Segmental\ SNR = \frac{10}{M}\sum_{m=0}^{M-1} log\frac{\sum_{n=N\cdot n}^{N\cdot n+N-1} x^2(n)}{sum_{n=N\cdot n}^{N\cdot n+N-1}[x(n)-\hat{x}(n)]^2} \tag{16}$$

The performance results, averaged over all the utterances and speakers, are shown in Fig. 2. More precisely, Fig. 2 shows the objective performances of the proposed algorithm and of the multi-pulse algorithms described in [1] versus the bit rate.

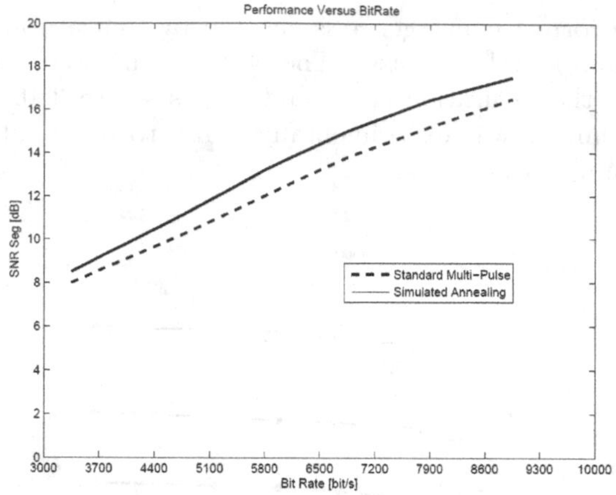

Fig. 2. Segmental SNR versus bit rate of the proposed algorithm compared to standard multipulse algorithm.

It should be noted that the bit rate used in this figure include the bits needed to code the excitation signal plus the 10-th order linear predictor, represented with Line Spectrum Pairs parameters and coded with 35 bits per frame. We should also note that no pitch prediction at all was used in obtaining all the data shown in Fig. 2.

We then implemented in Matlab the adaptive compressed algorithm described in [25] and executed on the same data. The SNR results are reported in Fig. 3, where the performance of our algorithm are included for comparison.

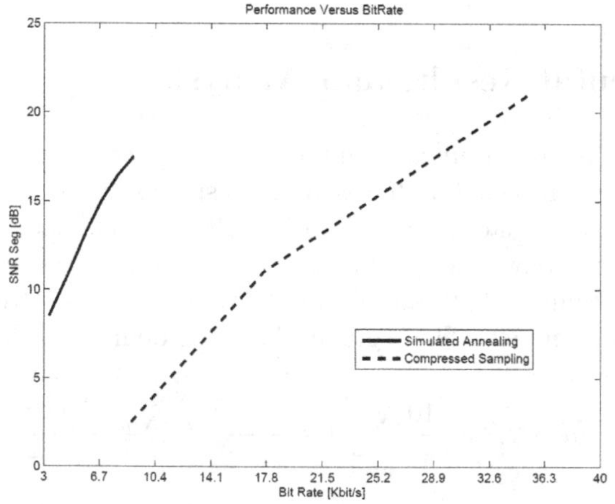

Fig. 3. Segmental SNR versus bit rate of the proposed algorithm compared to Compressed Sampling.

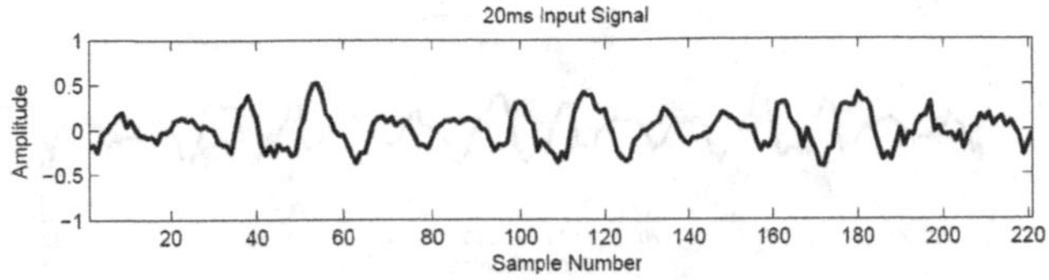

Fig. 4. Input 20 ms speech frame sampled at 11000 samples per second.

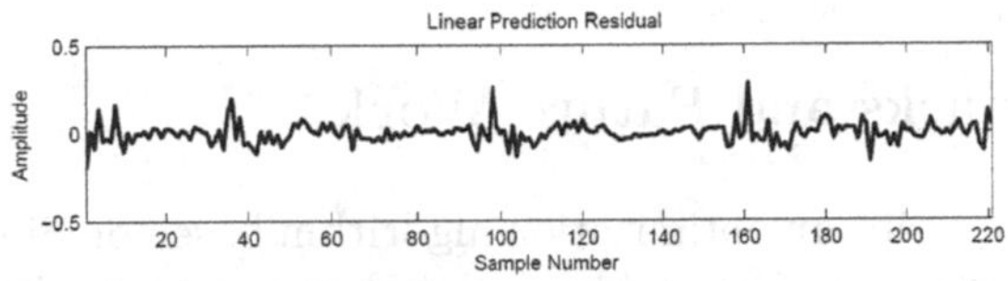

Fig. 5. 10-th order Linear Prediction residual.

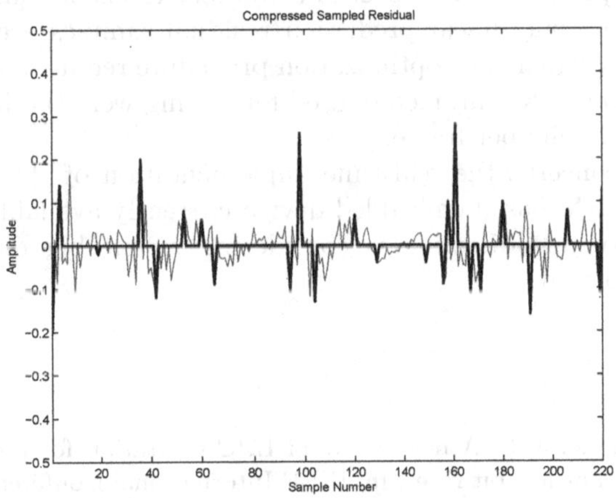

Fig. 6. Compressed sampling of the LP residual with 22 pulses per frame.

We finally report an example of an analysis performed with our algorithm on a frame of speech data. In Fig. 4 the input frame, 20 ms long, is reported.

The corresponding 10-th order linear prediction residue is shown in Fig. 5.

The Simulated Annealing based optimization algorithm selects the 22 samples per frame reported in Fig. 6, where the selected samples are overlapped with the residual signal. This compressed sampled residual is given in input to the $A(z)$ system to reconstruct the input signal.

The reconstructed signal is reported in Fig. 7 using thin line, overlapped with the original signal which is plotted using a strong line in order to see the differences.

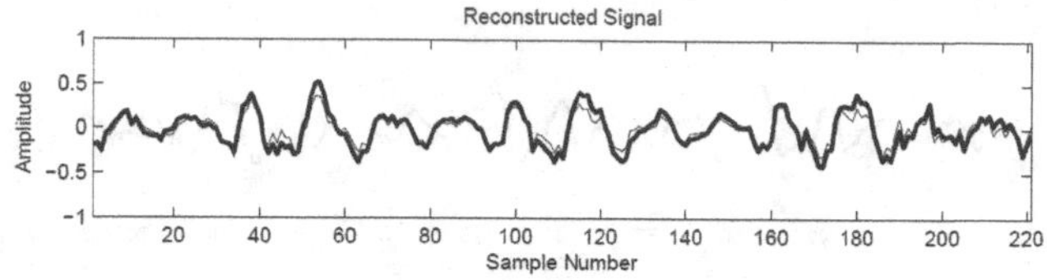

Fig. 7. Reconstructed signal with compressed LP residual (thin line) overlapped to the original signal (strong line).

8 Final Remarks and Future Work

In this paper, we describe an optimization algorithm based on Simulated Annealing for solving an optimization problem related to the approximate representation of the Linear prediction residual. We show that the approach turns out to be a form of compressed sampling of the residual signal. The algorithm described in this paper can produce coded speech at different levels of quality and bit rate depending on how many linear prediction residual samples are used to represent the excitation signal. The optimization procedure requires only additions in every iteration. The LPC parameters used for coding were the LSP, which have been coded with 35 bits per frame.

Future work concerns the real time implementation of the described algorithm on of the ARM-based embedded device currently available. On the other hand, another possible direction is considering recent big data issues (e.g., [26–28]) to be explored.

References

1. Atal, B.S., Remde, J.R.: A new model of LPC excitation for producing natural-sounding speech at low bit rates. In: IEEE International Conference on Acoustics, Speech, and Signal Processing, ICASSP 1982, Paris, France, 3–5 May, pp. 614–617 (1982)
2. Singhal, S., Atal, B.S.: Amplitude optimization and pitch prediction in multipulse coders. IEEE Trans. Acoustics Speech Sig. Process. **37**(3), 317–327 (1989)
3. Schroeder, M., Atal, B.S.: Code-excited linear prediction(CELP): high-quality speech at very low bit rates. In: Acoustics, Speech, and Signal Processing, IEEE International Conference on ICASSP 1985, vol. 10, pp. 937–940 (1985)
4. Sooraj, S., Anselam, A.S., Pillair, S.S.: Performance analysis of CELP codec for Gaussian and fixed codebooks. In: 2016 International Conference on Communication Systems and Networks (ComNet), pp. 211–215 (2016)
5. Manohar, K., Premanand, B.: Comparative study on vector quantization codebook generation algorithms for wideband speech coding. In: 2012 International Conference on Green Technologies (ICGT), pp. 082–088 (2012)
6. Stachurski, J.: Embedded CELP with adaptive codebooks in enhancement layers and multi-layer gain optimization. In: IEEE International Conference on Acoustics, Speech and Signal Processing, pp. 4133–4136 (2009)

7. Kiran, C.G., Rajeev, K.: A fast adaptive codebook search method for speech coding. In: TENCON 2008–2008 IEEE Region 10 Conference, pp. 1–4 (2008)
8. Romaniuk, P.D.R.: Sparse signal modeling in a scalable CELP coder. In: 21st European Signal Processing Conference (EUSIPCO 2013), pp. 1–5 (2013)
9. Sankar, M.S.A., Sathidevi, P.S.: Scalable low bit rate CELP coder based on compressive sensing and vector quantization. In: 2016 IEEE Annual India Conference, pp. 1–5 (2016)
10. Chen, F.K., Chen, G.M., Jou, Y.D.: Complexity scalability design in coding of the adaptive codebook for ITU-T G. 729 speech coder. In: 2011 8th International Conference on Information, Communications and Signal Processing, pp. 1–4 (2011)
11. Ha, N.K.: A fast search method of algebraic codebook by reordering search sequence. In: IEEE International Conference on Acoustics, Speech, and Signal Processing, pp. 21–24 (1999)
12. Candès, E.J., Tao, T.: Near-optimal signal recovery from random projections: Universal encoding strategies? IEEE Trans. Inf. Theory **52**(12), 5406–5425 (2006)
13. Donoho, D.L.: Compressed sensing. IEEE Trans. Inf. Theory **52**(4), 1289–1306 (2006)
14. Giacobello, D., Christensen, M.G., Murthi, M.N., Jensen, S.H., Moonen, M.: Retrieving sparse patterns using a compressed sensing framework: applications to speech coding based on sparse linear prediction. IEEE Sig. Process. Lett. **17**(1), 103–106 (2010)
15. Giacobello, D., Christensen, M.G., Murthi, M.N., Jensen, S.H., Moonen, M.: Sparse linear prediction and its applications to speech processing. IEEE Trans. Audio Speech Lang. Process. **20**(5), 1644–1657 (2012)
16. Atal, B.S., Hanauer, S.L.: Speech analysis and synthesis by linear prediction of the speech wave. J. Acoust. Soc. Am. **50**(2), 637–655 (1971)
17. Atal, B.S., Schroeder, M.R.: Predictive coding of speech signals and subjective error criteria. In: IEEE International Conference on Acoustics, Speech, and Signal Processing, ICASSP 1978, Tulsa, Oklahoma, USA, 10–12 April, pp. 573–576 (1978)
18. Hamacher, H.W.: Theoretical and computational aspects of simulated annealing (PJM van laarhoven EHL aarts) and simulated annealing: theory and applications (PJM van laarhoven and EHL aarts). SIAM Rev. **32**(3), 504–506 (1990)
19. Kirkpatrick, S., Gelatt, C.D., Vecchi, M.P.: Optimization by simulated annealing. Science **220**(4598), 671–680 (1983)
20. Li, Z.Y., Zhang, M.S., Long, Y.: Pin assignment optimization for large-scale high-pin-count BGA packages using simulated annealing. IEEE Trans. Compon. Packag. Manuf. Technol. **6**(10), 1465–1474 (2016)
21. Yang, Y., Wang, Y.: Simulated annealing spectral clustering algorithm for image segmentation. J. Syst. Eng. Electron. **25**(3), 514–522 (2014)
22. de Castro Martins, T., de Camargo, E.D.L.B., Lima, R.G., Amato, M.B.P., de Sales, M.: Image reconstruction using interval simulated annealing in electrical impedance tomography. IEEE Trans. Biomed. Eng. **59**(7), 1861–1870 (2012)
23. Metropolis, N., Rosenbluth, A.W., Rosenbluth, M.N., Teller, A.H., Teller, E.: Equation of state calculations by fast computing machines. J. Chem. Phy. **21**, 1087–1092 (1953)
24. CMU: CMU artic database. http://www.festvox.org/cmu_arctic/index.html/
25. Xu, Q., Yunyun, J.: Frame-based adaptive compressed sensing of speech signal. In: 7th International Conference on Wireless Communications, Networking and Mobile Computing, pp. 1–4 (2011)

26. Cuzzocrea, A.: Privacy and security of big data: current challenges and future research perspectives. In: Proceedings of the First International Workshop on Privacy and Secuirty of Big Data, PSBD@CIKM 2014, Shanghai, China, 7 November, pp. 45–47 (2014)
27. Cuzzocrea, A., Matrangolo, U.: Analytical synopses for approximate query answering in OLAP environments. In: Galindo, F., Takizawa, M., Traunmüller, R. (eds.) DEXA 2004. LNCS, vol. 3180, pp. 359–370. Springer, Heidelberg (2004). doi:10.1007/978-3-540-30075-5_35
28. Cuzzocrea, A., Fortino, G., Rana, O.F.: Managing data and processes in cloud-enabled large-scale sensor networks: state-of-the-art and future research directions. In: 13th IEEE/ACM International Symposium on Cluster, Cloud, and Grid Computing, CCGrid 2013, Delft, Netherlands, 13–16 May, pp. 583–588 (2013)

Genetic Estimation of Iterated Function Systems for Accurate Fractal Modeling in Pattern Recognition Tools

Alfredo Cuzzocrea[1(✉)], Enzo Mumolo[2], and Giorgio Mario Grasso[3]

[1] DIA Department, University of Trieste and ICAR-CNR, Trieste, Italy
alfredo.cuzzocrea@dia.units.it

[2] DIA Department, University of Trieste, Trieste, Italy
mumolo@units.it

[3] CSECS Department, University of Messina, Messina, Italy
gmgrasso@unime.it

Abstract. In this paper, we describe an algorithm to estimate the parameters of Iterated Function System (IFS) fractal models. We use IFS to model Speech and Electroencephalographic signals and compare the results. The IFS parameters estimation is performed by means of a genetic optimization approach. We show that the estimation algorithm has a very good convergence to the global minimum. This can be successfully exploited by pattern recognition tools. However, the set-up of the genetic algorithm should be properly tuned. In this paper, besides the optimal set-up description, we describe also the best tradeoff between performance and computational complexity. To simplify the optimization problem some constraints are introduced. A comparison with sub-optimal algorithms is reported. The performance of IFS modeling of the considered signals are in accordance with known measures of the fractal dimension.

Keywords: Iterated function systems · Fractal · Genetic optimization · Speech · EEG

1 Introduction

The interest in fractal models of signals arises from the observation that many signals, like images and speech, possess some degree of fractal properties [1]. A popular tool for describing and generating fractal objects is Iterated Function System (IFS) [2]. Very complex objects with fractal properties, such as self-affinity, are easily generated with IFS. However, the types of data an IFS can represent well are limited. Thus different IFS models, such as the piecewise self-affine [3] or the hidden-variable [4] models, has been developed in order to represent signals that are neither self-affine nor self-similar. In this paper, we use self-affine and piecewise self-affine IFS fractal models. The estimation of the parameters of an IFS model that reconstructs a given function, which

O. Gervasi et al. (Eds.): ICCSA 2017, Part I, LNCS 10404, pp. 357–371, 2017.
DOI: 10.1007/978-3-319-62392-4_26

is called inverse problem of the IFS, can be carried out in various ways. This can be successfully exploited by pattern recognition tools. In [5] an approach is proposed through wavelet decomposition and the moment method. Alternatively, the inverse problem can be formulated through least squares [6], and therefore a function minimization algorithm can be used. In [4], an optimization approach for the determination of the parameters of IFS models is used.

However, the optimization problem turns out to be quite complex, because the function to be minimized is nonlinear with many local minima and because many parameters are involved. In [7] a heuristic approach based on simulated annealing is used, but some drawbacks are pointed out, such as the very slow convergence to the global optimum and the critical tuning of the optimization algorithm. If constraints are imposed on the variables, however, efficient suboptimal solutions can be obtained and the definition of the best trade-off between complexity and performance becomes the major concern. In [8] a computationally efficient suboptimal inverse algorithm is described, but the search space is greatly limited. The main question would then be whether the imposition of different constraints could lead to better results. The answer to this question is not trivial because it requires extensive experimentation. The main goal of this paper is therefore to evaluate the trade-offs between performance and computational requirements related to different constraints in order to find the best one. The estimation is carried out with Genetic Algorithms (GAs) based optimization, which offers greater robustness with respect to other heuristic optimization algorithms, better convergence to the global minimum, and ease of programming, code maintenance, and updating. Moreover, GA's lead quite naturally to parallel implementations [9].

This paper is organized as follows. Section 2 reviews some concepts of fractal geometry and describes the self-affine and piecewise self-affine fractal models, useful to this work. Section 3 gives a brief description of GAs. Sections 4 and 5 describe, respectively, the tuning of the optimization algorithm based on GAs and the experimental results. Finally, Sect. 6 provides conclusions and future work.

2 Fractal Geometry, IFS, Linear and Piecewise Fractal Interpolation

Fractal geometry extends Euclidean geometry and describes objects characterized by a non-integer dimension. If we define $H(R^n)$ as the set of compact subsets of R^n, we can say that a fractal object is an element of $H(R^n)$. The usual distance between two compact subsets is the Hausdorff distance [10], which introduces a metric in $H(R^n)$. Let us consider the bi-dimensional case, that is, $n = 2$. A succession in the metric space $H(R^2)$ is obtained by an iterative application of affine transformations. An affine transformation is a deformation of elements of $H(R^2)$ realized by means of suitable maps, which can be represented in matrix form as shown in Eq. (1).

$$Y = \begin{bmatrix} a\ b \\ c\ d \end{bmatrix} X + \begin{bmatrix} e \\ f \end{bmatrix} \tag{1}$$

An important class of affine transformations is the class of contractive affine transformations [11]. An IFS is given by a complete metric space, in our case R^2, a distance measure (the Hausdorff distance), and a number N of affine contractive transformations on the metric space $w_n : R^2 \rightarrow R^2$, each with a contraction factor s_n. The contraction factor of the IFS is given by $s = \max\{s_n, n = 1, 2, \ldots, N\}$. Let us define a function $W : H(R^2) \rightarrow H(R^2)$ as reported in Eq. (2).

$$W(B) = \bigcup_{n=1}^{N} w_n(B) \tag{2}$$

Then W is itself a contractive transformation on the complete metric space $H(R^2)$. That is, calling $h(\ldots, \ldots)$ the Hausdorf distance, $\exists s,\ 0 < s < 1$, such that Eq. (3) holds, for each $B, C \in H(R^2)$.

$$h(W(B)) \leq s \cdot h(B, C) \tag{3}$$

Therefore the succession $A_n = W(A_{n-1})$ converges to a point of $H(R^2)$, called attractor of the IFS or fractal. The fractal $A = W(A)$ can be represented as shown in Eq. (4).

$$A = \bigcup_{n=1}^{N} w_n(A) \tag{4}$$

Two algorithms can be used for the generation of a fractal by means of IFS: the deterministic and the random iteration algorithm [3]. The deterministic algorithm is a direct application of the definition of attractor mentioned above. Assume that an initial set A_0 is given. Then the succession A_n, computed as reported in Eq. (5), converges to the attractor of the IFS.

$$A_n = \bigcup_{j=1}^{N} w_j(A_{n-1}) \quad n = 1, 2, \ldots \tag{5}$$

The random iteration algorithm is the following: let us take a given IFS, together with a set of probabilities p_i, one for each transformation. Starting from an initial point x_0, we then select one transformation w_i with probability p_i. A succession of points is then generated as reported in Eq. (6).

$$x_n = w_i(x_{n-1}) \tag{6}$$

The succession of points x_n converges to the attractor of the IFS. For the work described in this paper, a very important result is the so-called Collage Theorem [12,13], described in (7).

$$\text{given } L \in H(R^2),\ \epsilon \geq 0 \text{ and an IFS such that } h(L, W(L)) \leq \epsilon \rightarrow h(L, A) \leq \frac{\epsilon}{1s} \tag{7}$$

where A is the attractor of the IFS and s the contraction factor of the IFS. The importance of this theorem is that it provides a way to test an IFS without the need to compute the attractor. Because we want to represent a function, the affine transformations suitable for us are the so-called shear transformations, reported in Eq. (8).

$$w_i \begin{bmatrix} x \\ y \end{bmatrix} = \begin{bmatrix} a_i & 0 \\ c_i & d_i \end{bmatrix} \cdot \begin{bmatrix} x \\ y \end{bmatrix} + \begin{bmatrix} e_i \\ f_i \end{bmatrix} \tag{8}$$

A shear transformation maps vertical lines into vertical lines, and therefore it represents a single-valued function.

The fractal interpolation approach [3] is applicable to self-affine curves. Let us assume that a number of data points (u_n, v_n), $n = 0 \dots N$ and $u_n < u_{n+1}$ are given. A set of interpolation points (x_i, y_i), $i = 0 \dots M$ and $M < N$ are also given. The interpolation points are a sub-set of the data points with $(x_0, y_0) = (u_0, v_0), (x_M, y_M) = (u_N, v_N)$. Using fractal interpolation, an IFS can be built according to the constraints reported in Eq. (9).

$$w_i \begin{bmatrix} x_0 \\ y_0 \end{bmatrix} = \begin{bmatrix} x_{i-1} \\ y_{i-1} \end{bmatrix}; \; w_i \begin{bmatrix} x_N \\ y_N \end{bmatrix} = \begin{bmatrix} x_i \\ y_i \end{bmatrix}; \; i = 1, \dots M \tag{9}$$

For each interpolation point the system of four equations into five unknowns described in Eq. (10) can therefore be written.

$$\begin{array}{l|r} a_i \cdot x_0 + e_i = x_{i-1} & a_i \cdot x_m + e_i = x_i \\ c_i \cdot x_0 + d_i \cdot y_0 + f_i = y_{i-1} & c_i \cdot x_0 + d_i \cdot y_0 + f_i = y_{i-1} \end{array} \tag{10}$$

Equation (10) can be solved fixing one of the coefficients, usually the so-called contraction factors d_i. The are two ways of fixing these coefficients. The first one [14] consists in imposing a given fractal dimension to the final curve. In the second one [8], the contraction factor is computed through a least-squares minimization of the error given by the difference between the original data and the collage.

Once all the five parameters are determined, an IFS describing the data points is derived. We can state that this IFS is a fractal model of the signal (u_n, v_n), $n = 0 \dots N$. As stated in the introduction, for many signal a piecewise self-affine fractal model [8] is more appropriate. The piecewise self-affine model, which is merely an extension of the linear self-affine model, contains many degrees of freedom and is therefore extremely flexible. In the following, a brief description of the fractal piecewise model is given.

At each interpolation point $(x_i, y_i), i = 0 \dots M$ an affine map w_i and two points, called addresses in [8], described as $(\tilde{x}_i^1, \tilde{y}_i^1)$ and $(\tilde{x}_i^2, \tilde{y}_i^2)$ are associated. The affine maps w_i map the function between the addresses and the interpolation points. Therefore we can write the condition reported in Eq. (11).

$$w_i \begin{bmatrix} \tilde{x}_i^1 \\ \tilde{y}_i^1 \end{bmatrix} = \begin{bmatrix} x_{i-1} \\ y_{i-1} \end{bmatrix} \quad w_i \begin{bmatrix} \tilde{x}_i^2 \\ \tilde{y}_i^2 \end{bmatrix} = \begin{bmatrix} x_i \\ y_i \end{bmatrix} \quad i = 1, \dots M \tag{11}$$

Also in this case according to (7) a set of four equation into five unknown (the IFS's parameters) can be written [8], if the interpolation points and the addresses are known.

The generation of a fractal function by means of the piecewise model must be done only through the deterministic algorithm, because the random iteration algorithm does not guarantee that the points within a given interpolation section are necessarily contained within any address interval.

3 Genetic Algorithms

Genetic algorithms (GAs) are well-known population-based heuristic algorithms for the optimization of complex problems. In contrast with other optimization methods, GAs adopt a parallel approach in the search of the global optimum because during the search a "population" of candidates to the optimum is maintained. Each candidate solution is characterized by its fitness, which is a measure of the goodness of the solution. Some operators, which exploit an analogy with the processes of natural selection and sexual reproduction, are applied to the candidates during the process.

Each element of the population is given a "genetic code", usually a string built with characters taken from a given alphabet $V = a_0, a_l, \ldots, a_{n-1}$ called "chromosomes" (this is a first analogy with nature). During the elaboration the population is updated, old individuals being replaced with new ones. At each process iteration, called a "generation", the new solutions, forming the new population, are created applying suitable procedures (operators) to the chromosomes. These procedures are selection (the analogue of natural selection), cross-over (sexual reproduction) and mutation.

A basic concept of GAs is the "scheme", which describes the configurations evaluated during each iteration of the algorithm. Note that during the evaluation of a configuration's fitness, GAs also gain some information about all the hyperplanes to which that configuration belongs [15]. The techniques used in the choice of the parents of a new solution is of fundamental importance. The parents are chosen between the best scoring elements of the population, that is, the ones with the extreme values of the objective function. In the simplest procedure, the roulette wheel selection [15], parents are chosen according to a probability distribution based on the fitness of the population's elements. This technique, although very simple and immediate, is strongly inconvenient: it tends to lead to premature convergence owing to the lack of restorative pressure, which in terms of GAs means that the population loses its genotype diversity. In order to maintain good genotype diversity we have adopted a simple strategy, which we call in this paper *modified roulette wheel selection (MRWS)* and is described as follows. The elements of the population are chosen as in RWS, but the parents are selected only if the genotype difference is greater than a given threshold. The genotype difference between coded strings is computed with the Euclidean distance.

We have found that another procedure, the so-called local selection [16], gives better results. In this procedure the elements are arranged on a bi-dimensional grid and two parents are chosen through a random walk over the grid, starting from the element to be replaced and selecting the elements with the best fitness.

This kind of selection, while giving the chance of mixing the genetic patrimony, maintains a good genetic diversity, thus preserving the process from premature convergence.

In the context of GAs, the *cross-over operator* realizes the genetic recombination, which is fundamental because it allows for an efficient sampling in the space of the solutions. If the cross-over is too low, its effect is negligible and a premature convergence is eventually introduced. If it is too high, it introduces a schemes disruption that eventually rejects the high fitness schemes. The chromosome is generated concatenating the coded strings of the parameters. In the classical *1-point cross-over operator*, the cross-over can take place anywhere inside the genetic string, whereas in our *1-point modified cross-over* operator it can happen only at the boundary between groups of genes that represent an IFS transformation. This is like working with a high cardinality alphabet, since the disruptive effects are reduced and the convergence is speeded-up.

The continuous cross-over is an attempt to balance the two main phenomena of genetic computation, that is, exploration of the solution space and exploitation of solutions that have already been found. Another important GA operator is mutation, which is a change in an hereditary character. In this work, three approaches were considered: the soft, hard, and dynamic ones. *Soft mutation* is performed by incrementing (or decrementing) the integer value of one parameter, chosen with a given probability. *Hard mutation* was performed by changing the value of a parameter with a random one. Finally, *dynamic mutation* [17] is performed by modifying the value of a given parameter by a value that decreases, according to a given perturbation function, as the generations go on. The result of this approach is to make the exploration in the solutions space more and more local as the algorithm proceeds. The last issue concerning GAs is the way to handle constraints. In the case that a given solution does not satisfy a constraint, we added a penalty to the solution itself. Our coding performs a very complex mapping between genotypes and phenotypes, because some parameters are optimized in a genetic way and others are computed in closed form. Therefore, the constraints cannot be applied directly to the values of the genes.

4 Optimal Determination of the IFS Parameters

The parameters involved in linear self-affine models are interpolation points and contraction factors, whereas the piecewise self-affine models are controlled by interpolation points, contraction factors, and the addresses associated with each section of the discrete sequence. In order to reduce the number of variables, the contraction factors are computed in a closed form [8]. All the other parameters are estimated through the minimization of the fitness function reported in Eq. (12).

$$E = \sum_{n=1}^{N} [s(n) - c(n)]^2 \tag{12}$$

where $s(n)$ is the input signal and $c(n)$ the collage $W(L)$, defined in Eq. (12). Clearly, L is the input frame and N its dimension. It is worth noticing that, in virtue of the Collage theorem, described in (7), the Collage instead of the attractor is used.

As the parameters are not correlated to each other, a complete genetic optimization of all the parameters should be performed. However, this is the most critical situation for the GAs. In order to reduce the number of parameters to be optimized, constraints should be imposed on the parameters. The estimated solution is sub-optimal because the constraints limit the search space. In order to solve the minimization problem described in Eq. (12), algorithms have been developed [8] that perform an exhaustive search over the solution space, although they are subject to a number of constraints that keep the complexity low. For an exhaustive algorithm to be computationally affordable, the constraints on the search space must be quite heavy. For example, the algorithm described in [8] explores only 32 out of $1.32E^{3}6$ possible points (in the hypothesis of eight interpolation sections and four address intervals, with eight bit parameters). The introduction of GAs allows us to relax the constraints keeping the computational time sufficiently low.

In conclusion, depending on what parameters are included in the optimization process and on the type of constraints, a number of sub-optimal solutions can be obtained. Each solution is therefore a trade-off between required computational power and performances. For example, if the interpolation points are distributed in some known way and we want to determine the optimum d_i, and address positions $(\tilde{x}_i^1, \tilde{x}_i^2)$, the chromosome must be set as reported in Eq. (13).

$$\{d_0, \tilde{x}_0^1, \tilde{x}_0^2, \ldots, d_n, \tilde{x}_n^1, \tilde{x}_n^2\} \tag{13}$$

Then the genetic algorithm described in Sect. 3 is used to minimize Eq. (12).

In the following, we describe the tuning of the GAs used for the optimal determination of IFS parameters, that is, the optimal setup for coding, selection, mutation, and cross-over. Several experiments are performed to obtain the optimum set-up. Data for experiments is obtained for two class of signals, namely speech and electroencephalography signals. Speech data is extracted from the Artic speech dataset from CMU [18] and electroencephalography data is extracted from the KDD Archive from UC at Irvine, CA [19].

Speech data in Artic is sampled at 16 KHz. The signal is divided in 18.75 ms frames or 300 samples long. On the other hand, electroencephalography data is sampled at 256 Hz and the signal is divided in 400 ms frames or 102 samples.

One hundred of speech sentences from Artic and one hundred of electroencephalography measurements from the KDD Archive are selected and the results are averaged among them. In Fig. 1 the convergence behavior is shown. Three selection procedures "roulette wheel, modified roulette wheel, and local selection" are reported in this figure.

The best selection turns out to be the local selection, in terms of better convergence with smaller difference between minimum and average values. We have implemented and tested a number of variants of the classical genetic algorithm,

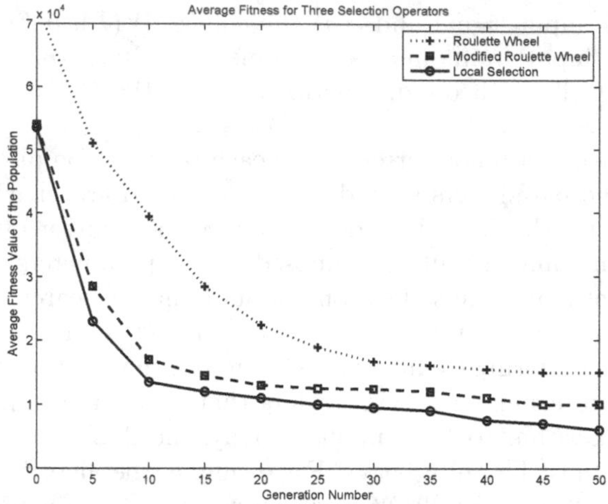

Fig. 1. Average value of the population for the three selection operators as in the figure.

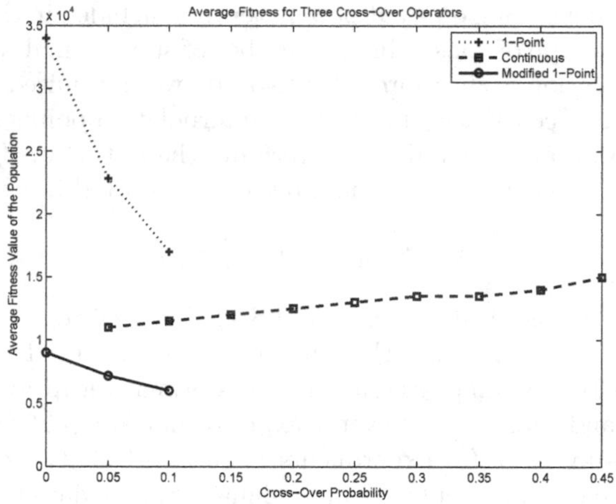

Fig. 2. Average value of the population for the three cross-over operators as in the figure.

including non binary coding and cross-over. Figure 2 reports the average value of the population versus cross-over probability, showing different kinds of cross-over procedures.

The procedure that achieves the lowest values is a variant of the classical 1-point cross-over procedure, namely the *1-point modified operator*, with a probability of 90%. Finally, we also tested several mutation operators. In Fig. 3 the average value of the population versus mutation rate for dynamic and hard mutation is shown.

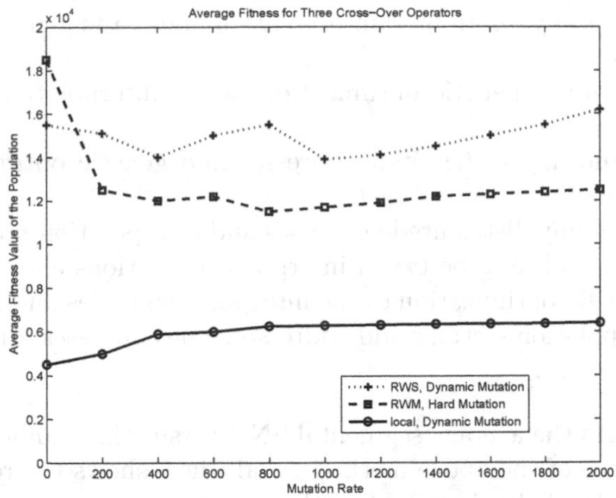

Fig. 3. Average value of the population for the three mutation operators as in the figure.

The best operator we found is a dynamical mutation at a 1/100 rate, where the amount by which a "gene" is changed is a function of time, starting with higher values at the beginning of the elaboration and decreasing towards the end.

5 Experimental Results and Analysis

Experimental results are obtained with the same data described before. The segmental SNR is defined as shown in Eq. (14).

$$Segmental\ SNR = \frac{10}{M}\sum_{m=0}^{M-1} log\frac{\sum_{n=N\cdot n}^{N\cdot n+N-1} x^2(n)}{sum_{n=N\cdot n}^{N\cdot n+N-1}[x(n)-\hat{x}(n))]^2} \tag{14}$$

where $x(n)$ is the original signals and $\hat{x}(n)$ is the attractor of the IFS estimated from the original signal by means of genetic optimization. In Sect. 4 the design of the GAs for the determination of the fractal models parameters (local selection, modified one-point cross-over with a 90% probability, dynamic mutation with 1/100 rate) has been made through experimental measurements. We first implemented the linear self-affine fractal interpolation with genetic optimization of the interpolation nodes (called method 1 in the following). For comparison purposes, moreover, the suboptimal procedure described in [4] and called the *Mazel/Hayes* method has been implemented. In summary, we implemented the following versions of the piecewise self-affine fractal interpolation:

- method 1: linear self-affine fractal interpolation with genetic optimization of the interpolation nodes

- method 2: uniformly distributed interpolation nodes and genetic optimization of the addresses
- method 3: complete genetic optimization of the interpolation nodes and of the addresses
- method 4: uniformly distributed addresses and genetic optimization of the interpolation nodes
- method 5:uniformly distributed addresses and interpolation nodes, with optimization of the indexing be-tween interpolation sections and addresses
- method 6: genetic optimization of the interpolation nodes and of the indexing between interpolation sections and addresses. The addresses correspond to the interpolation nodes.

Figure 4 reports the average segmental SNR versus the number of IFS transformations for each of the above methods, and Fig. 5 shows the relative computational complexity of the described methods.

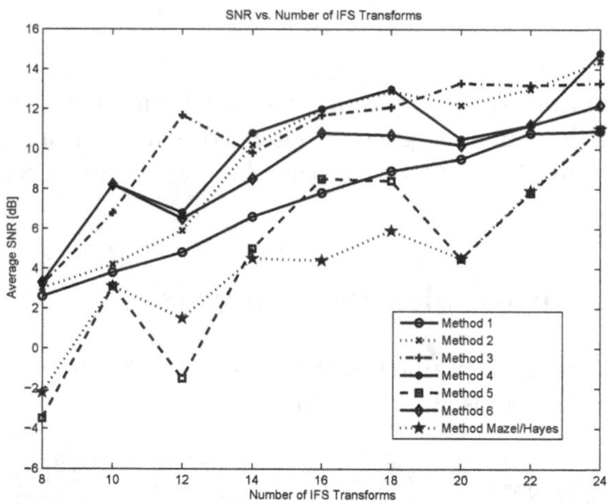

Fig. 4. Segmental SNR versus the number of IFS transformations for each method.

Let's make some considerations on Figs. 4 and 5. Methods 1 and 5 are worst in terms of SNR performance, as method 1 is not suitable for modeling sequences that are not self-similar, and method 5's constraints limit the search space too much. On the other hand, methods 2 and 4 behave quite similarly, suggesting that their setup is equivalent. The efficient *Hayes/Mazel* method, described in [8], though better than methods 1 and 5, gives worse performances than all the other methods. Moreover, method 3, which is a complete genetic optimization of the parameters, seldom gives very high results, supporting the conclusion that convergence, in this case, is quite difficult to reach. In any case, Fig. 5 shows that method 3 is the most expensive piecewise method in terms of required

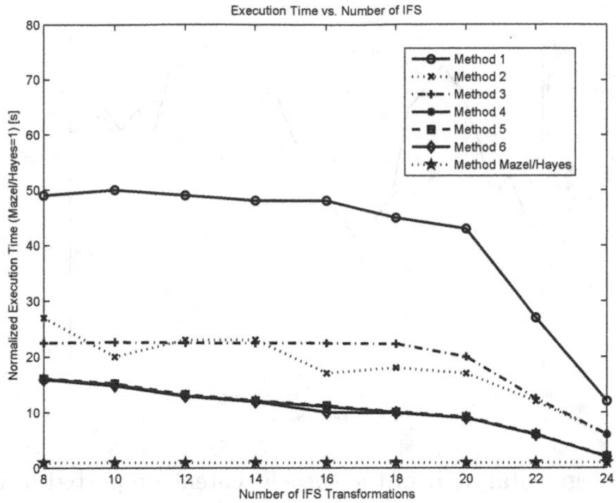

Fig. 5. Normalized execution time for each method.

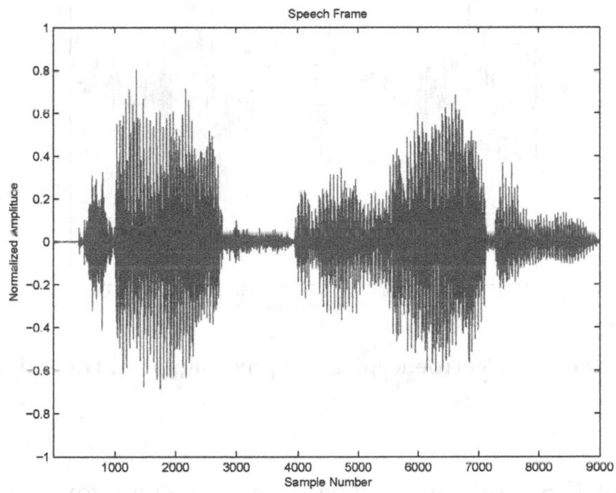

Fig. 6. Waveform of one speech sentence.

computational power. In summary, comparing the different methods, method 6 offers high performances without requiring too much computation, and thus it represents the best tradeoff between SNR, compression ratio, and computational complexity. We then used method 6 to model the data sequences.

The first results concern the modeling of speech. In Fig. 6 a speech sentence from the Artic database is reported. The signal is divided into frames which are separately-modeled with IFS.

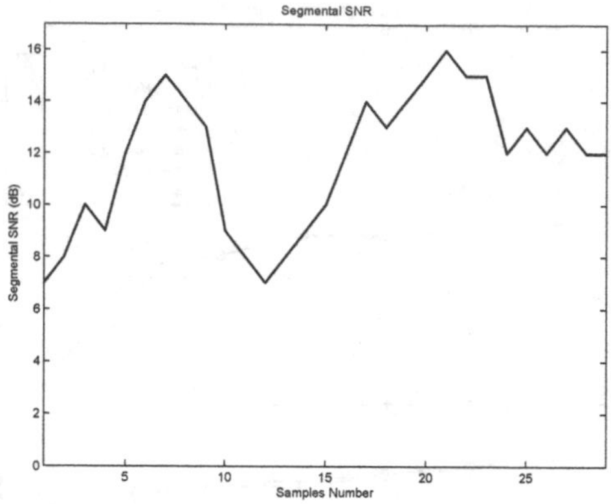

Fig. 7. Segmental SNR of the speech sentence reported in Fig. 6

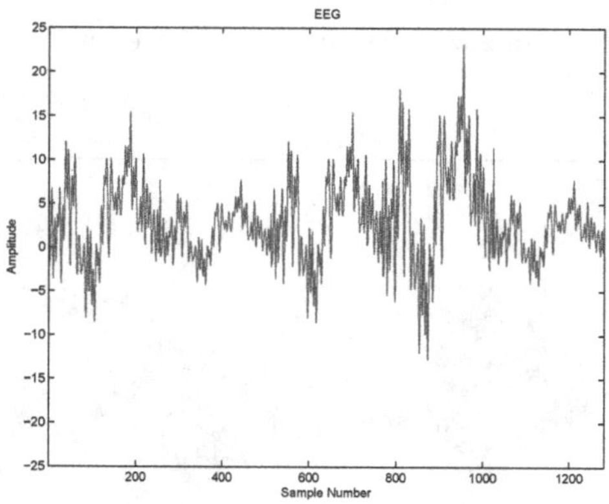

Fig. 8. Waveform of an Electroencephalographic signal extracted from the KDD dataset

The quality of IFS modeling is shown by reconstruction the original signal and comparing the two. In this example, 24 IFS maps for each frame were used. The segmental SNR (frame by frame) computed according to Eq. (14) is reported in Fig. 7.

Piecewise fractal model is then applied to a Electroencephalography (EEG) signals [20], which is a measure of the electrical activity of the brain. Using method 6 with 16 IFS transformations per 102 samples frame and 30 generations, we obtained an average SNR, among all the extracted recordings, of about 19 dB. The waveform of a section of an Electroencephalographic recording from one channel is reported in Fig. 8.

Figure 9 shows the frame-by-frame SNR results related to an analysis of an EEG signal. With the same parameters as before, that is, 16 IFS for each frame and 102 samples per frame, we obtained an average SNR of about 26 dB.

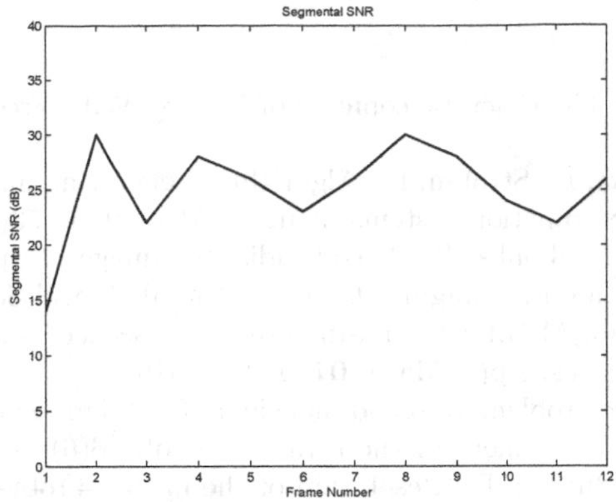

Fig. 9. Segmental SNR of the signal reported in Fig. 8.

6 Final Remarks and Conclusions

We have described an optimization approach to the determination of fractal models of speech and electroencephalography signals. Very good Segmental SNR results were obtained by estimating the IFS fractal models with genetic algorithms. It is worth noting that the SNR results depend on how much the signals possess fractal properties. This can be measured by computed the fractal dimension of the signals. There are several different definitions of fractal dimension [16,21]. In speech signals the fractal dimension depends on the phoneme type. The values of fractal dimension can range normally from 1.2 to 1.6 for vowels and from 1.6 to 1.8 for consonants [16]. For electroencephalography signals the values of fractal dimension can range normally from 1.7 to 1.9 depending on the pathological conditions of the subject [22]. Our electroencephalography data is related to alcoholic subjects and thus the fractal dimension is expected to be high. Turning to the Segmental SNR reported in Figs. 7 and 9, this is the reason why the values of SNR are higher for the electroencephalography data then speech data we considered.

A number of constraints on the search space were explored and the best tradeoff are experimentally defined. The proposed method gives much higher SNR performance than suboptimal techniques. On the other hand, even if suboptimal approaches are the most efficient, the proposed method requires the same order of computational effort as the number of IFS transformations increases. The problem of efficient coding of the estimated parameters for storing or transmission purposes are not considered in this paper. Future work may also comprise recent and emerging big data trends (e.g., [23–25]), which well cope with pattern recognition problems.

References

1. Mandelbrot, B.B.: The Fractal Geometry of Nature. W.H. Freeman and Company, New York (1977)
2. Jaros, P., Maslanka, L., Strobin, F.: Algorithms generating images of attractors of generalized iterated function systems. Numer. Algorithms **73**(2), 477–499 (2016)
3. Drakopoulos, V., Bouboulis, P., Theodoridis, S.: Image compression using affine fractal interpolation on rectangular lattices. Fractals World Sci. **14**, 1–11 (2006)
4. Akhtar, N., Prasad, M.G.P.: Graph-directed coalescence hidden variable fractal interpolation functions. Appl. Math. **07**, 1–11 (2016)
5. Abenda, S.: Inverse problem for one-dimensional fractal measures via iterated function systems and the moment method. Inverse Probl. **6**(6), 885 (1990)
6. Sarafopoulos, A., Buxton, B.: Resolution of the inverse problem for iterated function systems using evolutionary algorithms. In: IEEE International Conference on Evolutionary Computation, CEC 2006, Part of WCCI 2006, Vancouver, BC, Canada, 16–21 July 2006, pp. 1071–1078 (2006)
7. Dimri, V.P., Srivastava, R.P., Vedanti, N.: Fractal Models in Exploration Geophysics Applications to Hydrocarbon Reservoirs - Handbook of Geophysical Exploration Seismic Exploration, vol. 41, 1st edn. Elsevier Science, San Diego (2012)
8. Mazel, D.S., Hayes, M.H.: Using iterated function systems to model discrete sequences. IEEE Trans. Signal Process. **40**(7), 1724–1734 (1992)
9. López, N., Rabanal, P., Rodríguez, I., Rubio, F.: A formal method for parallel genetic algorithms[1]. In: Proceedings of the International Conference on Computational Science, ICCS 2015, Computational Science at the Gates of Nature, Reykjavík, Iceland, 1–3 June 2014, pp. 2698–2702 (2015)
10. Arutyunov, A.V., Vartapetov, S.A., Zhukovskiy, S.E.: Some properties and applications of the hausdorff distance. J. Optim. Theory Appl. **171**(2), 527–535 (2016)
11. Barnsley, M.F.: Fractals Everywhere. Academic Press, Cambridge (1988)
12. Honda, H., Haseyama, M., Kitajima, H., Matsumoto, S.: Extension of the collage theorem. In: Proceedings 1997 International Conference on Image Processing, ICIP 1997, Santa Barbara, California, USA, 26–29 October 1997, pp. 306–309 (1997)
13. Øien, G.E., Baharav, Z., Lepsøy, S., Karnin, E.D.: A new improved collage theorem with applications to multiresolution fractal image coding. In: Proceedings of ICASSP 1994: IEEE International Conference on Acoustics, Speech and Signal Processing, Adelaide, South Australia, Australia, 19–22 April 1994, pp. 565–568 (1994)
14. Wadströmer, N.: An automatization of Barnsley's algorithm for the inverse problem of iterated function systems. IEEE Trans. Image Process. **12**(11), 1388–1397 (2003)
15. Goldberg, D.E.: Genetic Algorithms in Search Optimization and Machine Learning. Addison-Wesley, Boston (1989)
16. Maragos, P., Potamianos, A.: Fractal dimensions of speech sounds: computation and application to automatic speech recognitiona. J. Acoust. Soc. Am. **105**(3), 1925–1932 (1999)
17. Chan, K.Y., Fogarty, T.C., Aydin, M.E., Ling, S., Iu, H.H.C.: Genetic algorithms with dynamic mutation rates and their industrial applications. Int. J. Comput. Intell. Appl. **7**(2), 103–128 (2008)
18. CMU: CMU artic database. http://www.festvox.org/cmu_arctic/index.html/
19. UCI: UCI KDD archive. http://kdd.ics.uci.edu/databases/eeg/eeg.html
20. Teplan, M.: Fundamentals of EEG measurement. Meas. Sci. Rev. **2**(12), 1–11 (2002)

21. Paramanathan, P., Uthayakumar, R.: An algorithm for computing the fractal dimension of waveforms. Appl. Math. Comput. **195**(2), 598–603 (2008)
22. Truong, Q.D.K., Ha, V.Q., Toi, V.V.: Higuchi fractal properties of onset epilepsy electroencephalogram. Comp. Math. Methods Med. **2012**, 461426:1–461426:6 (2012)
23. Cuzzocrea, A.: Privacy and security of big data: current challenges and future research perspectives. In: Proceedings of the First International Workshop on Privacy and Security of Big Data, PSBD@CIKM 2014, Shanghai, China, 7 November 2014, pp. 45–47 (2014)
24. Cuzzocrea, A., Matrangolo, U.: Analytical synopses for approximate query answering in OLAP environments. In: Galindo, F., Takizawa, M., Traunmüller, R. (eds.) DEXA 2004. LNCS, vol. 3180, pp. 359–370. Springer, Heidelberg (2004). doi:10.1007/978-3-540-30075-5_35
25. Cuzzocrea, A., Fortino, G., Rana, O.F.: Managing data and processes in cloud-enabled large-scale sensor networks: state-of-the-art and future research directions. In: 13th IEEE/ACM International Symposium on Cluster, Cloud, and Grid Computing, CCGrid 2013, Delft, Netherlands, 13–16 May 2013, pp. 583–588 (2013)

Evaluation of Frequent Pattern Growth Based Fuzzy Particle Swarm Optimization Approach for Web Document Clustering

Raja Varma Pamba[1(✉)], Elizabeth Sherly[2], and Kiran Mohan[3]

[1] Mahatma Gandhi University, Kottayam, India
pambaraj@gmail.com
[2] Indian Institute of Information Technology Management-Technopark, Thiruvananthapuram, Kerala, India
[3] Payszone LLC LTD, Dubai, UAE

Abstract. Soft and hard clustering efficiency evaluation of novel approach of frequent pattern growth based fuzzy particle swarm optimization for clustering web documents is studied and analyzed in this paper. The conventional approaches K-Means and Fuzzy c-means (FCM) fails with regard to random initialization and local minima hookups. To overcome this drawbacks, bio inspired mechanisms like genetic algorithm, ant colony optimization and particle swarm optimization (PSO) are used to optimize the K-means and FCM clustering. The major contribution of the novel method are three fold. Primarily in its ways to automatically find effective cluster numbers, cluster centroids and swarms for the bio inspired fuzzy particle swarm optimization. Second in yielding fuzzy overlapping clusters using the FCM objective function overcoming the drawbacks of the existing methods. Third, the methodology discusses in this paper prunes out the irrelevant elements from the search space and thereby retains all relationships with search query as semantic conditionally relatable sets. The evaluation results show that our proposed approach performs better for Adjusted Rand Index (ARI), Normalized Mutual Information (NMI) and Adjusted Concordance Index (ACI) against various distance based similarity measures and FCMPSO.

Keywords: Fuzzy · Information retrieval · Particle swarm optimization · Frequent pattern growth · Adjusted concordance index

1 Introduction

Clustering is the unsupervised strategy of partitioning data elements into disjoint sets by minimizing intra cluster similarity and maximizing inter cluster similarity. The traditional approaches in practice like K-Means, maximum margin clustering, mixture models mainly caters to hard clustering, which in real world applications fails for reasons of finding overlapping clusters as seen in web documents.

O. Gervasi et al. (Eds.): ICCSA 2017, Part I, LNCS 10404, pp. 372–384, 2017.
DOI: 10.1007/978-3-319-62392-4_27

Web Documents are dynamic, vague and unstructured. The huge influx of information makes the task of finding relevant information rather tenuous for the users. In the case of web documents the question is more complex as the same document may show its presence in diverse areas and domains. So the task of clustering becomes more complex when the demarcation is not possible. The option left to cluster web documents become more of soft clustering. Soft clustering in comparison to hard clustering allows a single document to belong to multiple clusters with fractional membership degrees. The end user can interpret the degrees by setting a threshold limits to turn a soft to hard clusters for further evaluation. Fuzzy clustering a soft partitional clustering technique is the widely used. The most popular of soft clustering algorithm in use is fuzzy c-means proposed by [1] and had been modified by [2] which raises the issue of local minima hookup with random centroid initialization by the user. In the fuzzy c-means (FCM) objective function, the difference between data objects and cluster centroids is the key deciding factor for final clustering of documents. The proposed methodology extends the Fuzzy particle swarm optimisation with FP Growth to finds the cluster numbers, centroids and swarm for fuzzy particle swarm optimization to cluster web documents. Particle swarm optimisation is a meta heuristic stochastic tool to overcome the drawbacks of FCM. Authors in [3] discussses on the combination of PSO and Fuzzy C-Means. There is also an integration of PSO and K-means algorithm (KPSO). While in [5] improves KPSO algorithm by proposing an enhanced cluster matching. Prior to the PSO updating process, the sequence of cluster centroids encoded in a particle is matched with the corresponding ones in the global best particle with the closest distance. A comprehensive use of PSO with FCM algorithm and their applications to clustering can be found in paper [4]. In [6] presented the concept of constant inertia weight.

Having discussed the issues of existing methodologies, the authors in their previous work in [10] proposes use of frequent pattern growth without candidate generation for improving fuzzy particle swarm optimization. The frequent pattern growth algorithm is used to prune the irrelevant with respect to the search query and thereby reducing the search space by retaining only items or terms and frequent item sets matching or in relevance with the search context. These frequent items sets can be treated as semantic conditionally relatable set giving all combinations of item sets matching to the search context. Once the system is learned to reduce the search space and retain items matching other user requirements we are left with highly related and relevant set of semantic related conditional terms matching user requirements.

These conditionally related sets then needs to be fuzzy clustered and imbibe in them a systematic herd behavior like honey bee swarming and fish schooling. This learned behavior can be inherited by new web documents as well.

The rest of the paper is organized as follows. Section 2 describes the paper objective. Section 3 discusses proposed methodology. The experimental results and discussions in Sect. 4. Conclusions are described in Sect. 5.

2 Paper Objective

The objective of the paper is to evaluate the proposed methodology that is modeled to cluster web documents with varying degrees of membership to all its respective clusters using frequent pattern growth approach based fuzzy particle swarm optimization. The method works by

1. Reducing search space by retaining the relationships with its search context.
2. Generate cluster centroids and particles automatically for fuzzy particle swarm optimization to begin.
3. Finally to evaluate and investigate into the effectiveness and efficiency of the designed model for NMI, ARI and ACI measures.

3 Proposed Methodology

Proposed methodology as discussed by the authors in their previous work in [10] uses frequent pattern approach and fuzzy particle swarm optimization for achieving the objectives mentioned as in Sect. 2. While each part mentioned in the objective is detailed in the following subsections.

3.1 Reducing Search Space Using Frequent Pattern Growth Algorithm

Frequent pattern growth approach commonly known as FP Growth algorithm is traditionally used for market basket analysis. The method works in two phases. First phase is the indexed compressed FP tree generation. In this the term document matrix (TDM) is considered as the transaction database used in FP growth. The transaction id and items in transaction database is rephrased to document id's and terms in term document matrix as given in Tables 1 and 2. The elements in each node of the FP tree are sorted and ordered based on the relevant threshold set to match with the support and confidence.

Table 1. Transaction database for FP growth

Transaction ID	Items in each transaction
TID1	A, B
TID2	B, C

Table 2. Term document matrix for FP growth

Document ID	Terms occurring in document
D1	Computer, server
D2	Usr, openwindows

This TDM will be used as input to the FP Growth algorithm to generate the indexed compressed data structure, frequent pattern tree (FP tree). From Table 2, each transaction say D1 is mapped to a path in the FP-Tree from the root (null) node. As different documents can have common terms, and if their path overlaps, the count of each such term is incremented by 1. Else a new node is created with a count of 1 for each new terms in the path. An example for FP tree generation as discussed in the paper [7] is given below in Fig. 1.

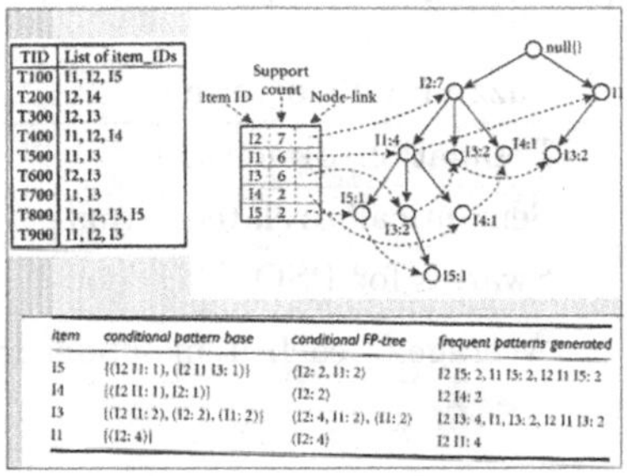

Fig. 1. Reference for FP tree generation

As a result of FP Tree generation a complete and compact set of frequently occurring terms relevant to the search context are retrieved and stored in FP Tree.

3.2 Semantically Related Conditional Pattern Base

For example few key terms chosen to illustrate the algorithm after pre-processing of text documents from 20Newsgroup dataset are sun, file, openwindows, xview, echo and usr. Denoted as A, B, C, D, E, F then after FP tree it shows the co-occurrence of each keyword and in every chain of keywords how many times the particular terms have been traversed. With FP growth, it finally generates all possible combinations like [sun, file], [sun, server], [usr, openwindows, echo], [sun, file, server, openwindows] etc. matching the threshold support. A template of the semantic relatable set retrieved from the FP growth algorithm is

Term	Conditional Pattern Base	Conditional FP Tree	Generate Frequent Patterns
A	{F,B,E}{F,B,E,D}{C,B,D},{C,F,B,D},{C,F,B,E,D}	{C:4},{F:5},{B:3},{E:3}	{A,C},{A,F},{A,B},{A,E},{A,C,F,B,E}
D	{F,B,E},{C,B},{C,F},{C,F,B},{C,F,B,E}	{C:4},{F:4},{B:4},{E:2}	{D,C},{D,F},{D,B},{D,C,F,B}
E	{F,B},{C,F,B}	{C:1},{F:2},{B:3}	*
B	{F},{C},{C,F}	{C:2},{F:2}	*
F	{C}	{C:1}	*
C	*	*	*

Fig. 2. Reference format for semantic conditional frequent patterns generation

as given below in Fig. 2. From Sects. 3.1 and 3.2, the proposed method generates all possible combinations of terms matching the user search context support and confidence threshold. This is considered as swarms or particles in the whole set of population for fuzzy particle swarm optimization (FPSO). Cluster centroids are calculated as average of all items under each frequent item sets. The transformation of outputs from frequent pattern growth to fuzzy particle swarm optimization are shown in Table 3.

Table 3. Translating frequent mining concepts to fuzzy particle swarm clustering.

FP growth concepts	Fuzzy PSO clustering
Items	Population/dimension
Itemsets	Semantically related terms
Frequent itemsets	Swarms for PSO
Cluster centroids	Average of each items under frequent itemsets

3.3 Fuzzy Particle Swarm Optimization

Fuzzy particle swarm optimization (FPSO) is a genre of particle swarm optimization (PSO) as discussed by authors in [9,12]. The FPSO as discussed by authors in [8,10,11] begins with the initialisation of population of particles and initial number of clusters whose positions represent the potential solutions for the studied problems and velocities are randomly initialized in the search space. The particle or swarms derived from Sect. 3.2 are assumed to be spread across different positions in the search space as potential solutions. At its current location each of the particle are evaluated for the fitness function. The fitness function chosen for the discussed approach is the inverse of objective function of fuzzy c-means algorithm as used by authors in [8]. Once the function is evaluated, the particle makes its next move in the search space comparing the current, individual best fitness so far obtained and social best position so far obtained by its neighbors in the swarm. This global and local search mechanism overcomes the issue of local minima hookups. Next iteration starts when all the particle have moved to its respective best positions which is a position near to the optimal cluster centroids computed as in Sect. 3.2. After every iteration the particles are nearing to its respective cluster centroids or attractive points. This is like birds searching for food or finding their nests. Here all documents take their transition from initial location in search for its respective domains which are the cluster centroids calculated as in Sect. 3.2.

FPSO being a part of fuzzy clustering, every particle is assigned a fuzzy membership degree to every other cluster centroids. The various parameters in use for the FPSO is as given below:

The fuzzy clustering of objects is described by a fuzzy matrix μ with n rows and c columns in which n is the number of data objects and c is the number of clusters. The element in the ith row and j^{th} column in μ_{ij}, indicates the degree of association or membership function of the ith object with the jth cluster. The characteristics of μ and J_m are as follows:

$$\mu_{ij} \in [0,1] \forall_i = 1,2\ldots n \forall_j = 1,2\ldots c \tag{1}$$

$$\sum_{j=1}^{c} \mu_{ij} = 1, \forall_i = 1,2,\ldots n \tag{2}$$

$$J_m = \sum_{j=1}^{c}\sum_{i=1}^{n} \mu_{ij}^{m} d_{ij} \text{ where } d_{ij} = \|o_i - z_j\| \tag{3}$$

$$z_j = \frac{\sum_{i=1}^{n} \mu_{ij}^{m} o_i}{\sum_{i=1}^{n} \mu_{ij}^{m}} \tag{4}$$

$$\mu_{ij} = \frac{1}{\sum_{k-1}^{c}} \frac{d_{ij}}{d_{ik}}^{\frac{2}{m-1}} \tag{5}$$

The position matrix X is redefined in this proposed algorithm, represent the fuzzy relation (membership function) between the frequent items sets (particles) in columns and cluster centers as rows. The position matrix is given below:

$$X = \begin{bmatrix} \mu_{11} & \cdots & \mu_{1c} \\ \vdots & \ddots & \vdots \\ \mu_{n1} & \cdots & \mu_{nc} \end{bmatrix} \tag{6}$$

In each of its iteration, the search for optimal solution is executed by updating the particle velocities and its position. The fitness value of each frequent item sets (particles or swarm) is determined using a fitness function as discussed in [10,11]

$$F = \frac{K}{J_m} \tag{7}$$

based on Euclidean distance measure and K is any constant.J_m is calculated as in Eq. 3. The velocity of each particle is updated using two best position, individual best position i_{best} and social best solution s_{best}. The individual best position i_{best} is the best position that particle has visited so far and s_{best} is the best position the swarm has visited so far. A particle velocity and position are updated as follows:

$$V(t+1) = w \otimes V(t) \oplus c_1 r_1 \otimes (i_{best}(t) \ominus X(t)) \oplus c_2 r_2 (s_{best}(t) \oplus X(t))) \tag{8}$$

$$K : 1,2,\ldots,P \tag{9}$$

$$X(t+1) = X(t) \oplus V(t+1) \tag{10}$$

where X and Y are position and velocity of particles respectively. w is inertia weight, c_1 and c_2 are constants, called acceleration coefficients which control the influence of i_{best} and s_{best} on the search process, P is the number of particles in the swarm derived from the frequent itemsets generated from FP Growth, r_1 and r_2 are random values in the range [0, 1].

3.4 Frequent Pattern Growth Based Fuzzy Particle Swarm Optimization

The authors in their previous work [10] details the hybrid method for web document clustering in two phases: first the frequent Pattern approach for swarm and cluster centroid generation as discussed in Sects. 3.1 and 3.2. Second stage is fuzzy particle swarm optimization running on the swarm received from first stage. Based upon the fitness function evaluated which is inversely proportional to the objective function J_m, the particle with least fitness function can be eliminated to help improve the search space.

A stage reaches where the position of the particle saturates with no changes in the position further. Here the algorithm converges to retrieve the final fuzzy cluster positions of individual documents to its respective clusters.

4 Experimental Result and Discussion

4.1 Description of Data Set

The proposed method for evaluation uses 20Newsgroup and Reuters-21578 from the site [22,23]. Table 4 gives the overview of the 20Newsgroup data set chosen for the study with the categories and number of instances taken respectively. After running frequent pattern based fuzzy particle swarm optimization, a small snippet of final fuzzy cluster positions of two categories computer and science (357 instances chosen with 3127 terms after pre-processing) of 20Newsgroup data set is as shown in Table 5. Set 1 and Set 2 of Table 5 represents the class science and computer in 20Newsgroup data set retrieved by the method automatically.

From Reuters-21578,300 documents from six classes each with 50 documents and 1232 terms (earn, money-supply sugar, trade, ship and gold) is selected.

Table 4. List of items and instances chosen from Newsgroup data set.

Group	Target.Group (#)	Category
2	Comp.graphics (54)	Comp
3	Comp.os.ms-windows.misc (52)	Comp
4	Comp.sys.ibm.pc.hardware (60)	Comp
5	Comp.sys.mac.hardware (63)	Comp
14	Sci.med (63)	Sci
15	Sci.space (65)	Sci

Translating the Table 5 to its respective terms and cluster combinations are given below in Table 6:

Table 5. Final document cluster membership positions

Message ID's	Set1	Set2
1	0.0508929	0.9491071
2	0.2453241	0.7546759
3	0.8683597	0.1316403
4	0.0000000	1.0000000
5	1.000000	0.0000000
6	0.4734385	0.5265615
7	0.0000000	1.0000000
8	0.0000000	1.0000000
9	0.000000	1.0000000
10	0.000000	1.0000000

Table 6. Final fuzzy cluster positions for the documents

Class set	Clusters	Message ID
Comp	{C1, C2, C3, C4}	{D1, D2, D4}, {D7, D10}, {D8}, {D9}
Sci	{C5, C6}	{D3, D5}, {D6}

4.2 Evaluation Measures

Finally to evaluate the performance of the frequent pattern based fuzzy particle swarm optimization we have chosen FCM, K-means [14], PSO [15] and FCMPSO [4]. From the studies conducted by [4] it can be inferred that FCMPSO method outperforms K-means and PSO. The proposed algorithm retrieves the membership degrees of each to its respective clusters with its respective positions in d dimensional space as shown in Table 5. These membership degrees needs to be converted to hard clusters by assigning the documents with highest membership degrees to one cluster. For example considering Table 5, message id 1 will be hard clustered to set 2 with a membership degree of 0.9491071. This assignment makes it possible to measure the hard clustering performance of these algorithms along with the soft clustering measures [13]. Hard clustering metrics used in the study are normalized mutual information (NMI) and adjusted rand index (ARI) [16,17]. As in [17,28] NMI is used to compare two hard partitions, A and B of a data set X with N objects. In our case A and B are set 1 and set 2 as shown in Table 5. Assume that A and B have m and n clusters, respectively. The probability P (i) that a randomly selected object from X falls into cluster A.

$$P(i) = \frac{\|A_i\|}{N} \tag{11}$$

The entropy, E(A) is then defined as

$$E(A) = \sum_{i=1}^{m} P(i) log P(i) \tag{12}$$

Let P(i, j) denote the probability that an object belongs to cluster A_i and B_i in class A and B respectively then

$$P(i,j) = \frac{\|A \cap B\|}{N} \tag{13}$$

The NMI [16,17] between the two hard partitions A and B can then be defined as

$$NMI(A,B) = \frac{\sum_{i=1}^{m} \sum_{j=1}^{n} P(i,j) log \dfrac{P(i,j)}{P(i)P(j)}}{\sqrt{(E(A)E(B)}} \tag{14}$$

For Rand Index the following cardinals need to be detailed with reference to the example in Tables 5 and 6:

a: number of data points belonging to the same class in A and to the same cluster in B.
b: number of data points belonging to the same class in A and to different clusters in B.
c: the number of data points belonging to different classes in A and to the same cluster in B.
d: the number of data points belonging to different classes in A and to different clusters in B. Rand Index (RI) is defined as:

$$RI = \frac{a+d}{a+b+c+d} \tag{15}$$

and the Adjusted Rand Index used in the study is given as:

$$ARI = \frac{a - \dfrac{(a+b)(a+c)}{a+b+c+d}}{\dfrac{(a+b)+(a+c)}{2} - \dfrac{(a+b)(a+c)}{a+b+c+d}} \tag{16}$$

Both NMI and ARI [24] gives the efficiency of the proposed algorithm when compared to the ground truth values of clustering.

While considering soft clustering measures as pointed out by [18,19], when soft clusterings is converted to hard clustering techniques by simple assignment it often fails to reflect the performance of soft clustering algorithms. For example, different fuzzy partitions may result in the same crisp partition for various cases as discussed in [19]. This loss of information due to the disposal of the fuzzy membership values makes the hard clustering measures intolerant to overlapping clusters.

To overcome these issues, [18] proposed a fuzzy extension of the Rand index, Adjusted Concordance Index (ACI) is used [26]. Whose parameters are obtained by rewriting the formulation of the adjusted rand index in Eq. 16 a fully equivalent form using following concepts discussed below. Given two fuzzy membership partition, A and B as shown in Table 5, the basic idea underneath the fuzzy extension of the Rand index, Adjusted Concordance Index (ACI) is to generalize the concept of concordance in the following way.

Considering a pair $(x, x^{'}) \in X$ as being concordant as A and B agree on its degree of equivalence, [20] define the degree of concordance as:

$$doc = 1 - \|EQ_A(x, x^{'}) - EQ_B(x, x^{'}) \quad \in [0, 1] \tag{17}$$

and the degree of discordance as:

$$dodc = \|EQ_A(x, x^{'}) - EQ_B(x, x^{'}) \quad \in [0, 1] \tag{18}$$

where EQ is defined as a fuzzy equivalence relation on X in terms of similarity measure as:

$$EQ = 1 - \|P(x) - P(x_{'}\| \tag{19}$$

EP is equal to 1 if and only if x and $x^{'}$ have the same membership pattern and is equal to 0 otherwise.

ACI quantities are defined as:

- a conc: objects x and x' are concordant because their degree of equivalence in A and in B is similar and their degree of equivalence in A and in B is high. $a = \top(1 - \|EQ(x, x') - EQ(x, x')\|, \top(\|EQ(x, x') - EQ(x, x')\|$
- b conc: the degree of equivalence of x and x' in A is larger than in B $b = max(\|EQ(x, x') - EQ(x, x')\|)$
- c conc: the degree of equivalence of x and x' in A is smaller than in B $b = max(\|EQ(x, x') - \|EQ(x, x'))$
- d conc: negation of a concordance $a = \top(1 - \|EQ(x, x') - EQ(x, x')\|, \perp(1 - \|EQ(x, x'), 1 - EQ(x, x')\|$

To measure the similarity between pair of documents we have used Cosine Similarity, Jaccard coefficient and Pearson correlation coefficient after normalizing the document vectors respectively. Finally all similarity functions are normalized for final evaluation results.

(1) **Cosine Similarity:** When documents are represented as term vectors, the similarity between them corresponds to the correlation between the two vectors. This is quantified as the cosine of the angle between vectors. Cosine similarity is one of the most popular similarity measure applied to text documents, such as in numerous information retrieval applications [21] and clustering too [27].
(2) **Jaccard Coefficient/Tanimoto Coefficient:** Compares the sum weight of shared terms to the sum weight of terms that are present in either of the

two document but are not the shared terms. The value of Jaccard coefficient ranges from 0 to 1, 1 when documents $\boldsymbol{t}_a = \boldsymbol{t}_b$ are equal and 0 when $\boldsymbol{t}_a \neq \boldsymbol{t}_b$.

$$F_{jc} = \frac{\boldsymbol{t}_a.\boldsymbol{t}_b}{|\boldsymbol{t}_a|^2 \times |\boldsymbol{t}_b|^2 - \boldsymbol{t}_a.\boldsymbol{t}_b} \tag{20}$$

(3) **Pearson Correlation Coefficient:**

$$F_{pc} = \frac{m \times (\boldsymbol{t}_a.\boldsymbol{t}_b) - TF_a \times TF_b}{\sqrt{AB}}, \tag{21}$$

where

$$TF_a = \sum_{t=1}^{m} tfidf(d_a, t), TF_b = \sum_{t=1}^{m} tfidf(d_b, t), \tag{22}$$

$$A = m\sum_{t=1}^{m} tfidf(d_a, t)^2 - TF_a^2, \quad B = m\sum_{t=1}^{m} tfidf(d_b, t)^2 - TF_b^2 \tag{23}$$

Table 7. Comparison of different similarity measure functions in terms of ARI

Dataset	F_{cs}	F_{jc}	F_{pc}	FCM	FCMPSO	FPFPSO
20Newsgroup	.152	.138	.148	.152	.150	.155
Reuters −21578	.293	.278	.283	.293	.301	.315

5 Conclusion

The proposed hybrid algorithm of extending the utility of frequent pattern growth algorithm to fuzzy particle swarm optimization is used for clustering text datasets to its respective clusters. The proposed algorithm is easy to implement and find ways by itself for initialization of cluster centers, reducing the redundancies and irrelevancies in the search space and get rid of local minima hookups. Obtained results are compared against conventional algorithms for similarity measures.

Figure 1 (taken from [7]) and Fig. 2 shows a frame of reference on how FP growth approach performs the search space reduction and how its retrieves semantically related conditional term base. These conditional term base with frequent item sets get morphed to swarms and cluster centroids as shown in Table 3. Final fuzzy clusters with its respective particle positions and varying membership degrees for different clusters are shown in Tables 5 and 6. This Table 5 forms the base for hard and soft clustering evaluation. For hard clustering highest membership degree is assigned to the data points. Evaluation results for ARI, NMI and soft clustering metric ACI and various similarity measures are shown in Tables 7, 8, and 9. The tables shows that the proposed algorithm is highly robust and offers a better efficiency in comparison to tested algorithms. The fact behind this success is attributed to FPSO for its global search mechanism and added to it is the use of FP Growth for retaining the relevant search space.

Table 8. Comparison of different similarity measure functions in terms of NMI

Dataset	F_{cs}	F_{jc}	F_{pc}	FCM	FCMPSO	FPFPSO
20Newsgroup	.401	.386	.402	.401	.408	.410
Reuters −21578	.426	.403	.421	.426	.437	.438

Table 9. Comparison of different similarity measure functions in terms of ACI

Dataset	F_{cs}	F_{jc}	F_{pc}	FCM	FCMPSO	FPFPSO
20Newsgroup	.282	.265	.280	282	.293	.301
Reuters −21578	.363	.331	.343	.363	.364	.368

References

1. Dunn, J.C.: A fuzzy relative of the ISODATA process and its use in detecting compact well-separated clusters. J. Cybern. **3**, 32–57 (1973)
2. Bezdek, J.C.: Pattern Recognition with Fuzzy Objective Function Algorithms. Kluwer Academic Publishers, Norwell (1981)
3. Liu, H., Pei, T., Zhou, T., Zhu, A.X.: Multi-temporal MODIS-data-based PSO-FCM clustering applied to wetland extraction in the Sanjiang Plain. In: International Conference on Earth Observation Data Processing and Analysis, Wuhan, China, vol. 7285 (2008)
4. Silva Filho, T.M., Pimentel, B.A., Souza, R.M.C.R., Oliveira, A.L.I.: Hybrid methods for fuzzy clustering based on fuzzy c-means and improved particle swarm optimization. Expert Syst. Appl. **42**(17–18), 6315–6328 (2015)
5. Lam, Y.-K., Tsang, P.W.M., Leung, C.-S.: PSO-based K-Means clustering with enhanced cluster matching for gene expression data. Neural Comput. Appl. **22**(7–8), 1349–1355 (2013)
6. Feng, Y., Teng, G.F., Wang, A.X., Yao, Y.M.: Chaotic inertia weight in particle swarm optimization. In: Second International Conference on Innovative Computing, Information and Control, pp. 475–501. IEEE (2008)
7. Han, J., Pei, J., Yin, Y., Mao, R.: Mining frequent patterns without candidate generation: a frequent-pattern tree approach. Data Min. Knowl. Discov. **8**, 53–87 (2004)
8. Izakian, H., Abraham, A.: Fuzzy C-means and fuzzy swarm for fuzzy clustering problem. Expert Syst. Appl. **38**(3), 1835–1838 (2011)
9. Kennedy, J.F., Eberhart, R.C., Shi, Y., NetLibrary, Inc.: Swarm Intelligence. Morgan Kaufmann Publishers, San Francisco (2001)
10. Pamba, R.V., Sherly, E., Mohan, K.: Automated information retrieval model using FP growth based fuzzy particle swarm optimization. Int. J. Comput. Sci. Inf. Technol. **9**(1) (2017)
11. Priyadharshini, S.P., Pujeri, R.V.: Performance analysis of fuzzy clustering. Int. J. Adv. Eng. Technol. (2014)
12. Zheng, Y., Qu, J., Zhou, Y.: An improved PSO clustering algorithm based on affinity propagation. WSEAS Trans. Syst. **12**(9), 447–456 (2013)
13. Huang, H.-C., Chuang, Y.-Y., Chen, C.-S.: Multiple kernel fuzzy clustering. IEEE Trans. Fuzzy Syst. **20**(1), 120–134 (2012)

14. Jain, A.K.: Data clustering: 50 years beyond K-means. Pattern Recogn. Lett. **31**(8), 651–666 (2010). Elsevier
15. Cui, X., Potok, T.E.: Document clustering analysis based on hybrid PSO+Kmeans algorithm. J. Comput. Sci. 27–33 (2005). Special Issue
16. Wu, J., Xiong, H., Chen, J.: Adapting the right measures for k-means clustering. In: Proceedings of the 15th ACM SIGKDD International Conference on Knowledge Discovery and Data mining, ser. KDD 2009, pp. 877–886 (2009)
17. Strehl, A., Ghosh, J.: Cluster ensembles - a knowledge reuse framework for combining multiple partitions. J. Mach. Learn. Res. **3**, 583–617 (2003)
18. Amodio, S., d'Ambrosio, A., Iorio, C., Siciliano, R.: Adjusted concordance index, an extension of the adjusted rand index to fuzzy partitions. STAD Research report 03 2015 (2016)
19. Campello, R.J.G.B.: A fuzzy extension of the rand index and other related indexes for clustering and classification assessment. Pattern Recogn. Lett. **28**(7), 833–841 (2007)
20. Hullermeier, E., Rifqi, M., Henzgen, S., Senge, R.: Comparing fuzzy partitions: a generalization of the rand index and related measures. IEEE Trans. Fuzzy Syst. **20**(3), 546–556 (2012)
21. Yates, R.B., Neto, B.R.: Modern Information Retrieval. Addison-Wesley, New York (1999)
22. http://qwone.com/jason/20Newsgroups/
23. Cardoso-Cachopo, A.: Datasets for single-label text categorization. http://web.ist.utl.pt/acardoso/
24. Labatut, V.: Generalized measures for the evaluation of community detection methods. https://arxiv.org/ftp/arxiv/papers/1303/1303.5441.pdf
25. https://arxiv.org/ftp/arxiv/papers/1303/1303.5441.pdf
26. https://arxiv.org/pdf/1509.00803.pdf
27. Larsen, B., Aone, C.,: Fast and effective text mining using linear-time document clustering. In: Proceedings of the Fifth ACM SIGKDD International Conference on Knowledge Discovery and Data Mining (1999)
28. Alok, A.K., Saha, S., Ekbal, A.: Development of an external cluster validity index using probabilistic approach and min-max distance. Int. J. Comput. Inf. Syst. Ind. Manag. Appl. **6**, 494–504 (2014)

A Differential Evolution Algorithm for Computing Caloric-Restricted Diets - Island-Based Model

João Gabriel Rocha Silva[1], Iago Augusto Carvalho[2], Leonardo Goliatt[1], Vinícus da Fonseca Vieira[3], and Carolina Ribeiro Xavier[3](✉)

[1] Universidade Federal de Juiz de Fora, Juiz de Fora, Brazil
[2] Universidade Federal de Minas Gerais, Belo Horizonte, Brazil
[3] Universidade Federal de São João del Rei, São João del Rei, Brazil
carolinaxavier@ufsj.edu.br

Abstract. A rich and balanced diet, combined with physical exercises, is the most common and efficient manner to achieve a healthy body. Since the classic Diet Problem proposed by Stigler, several works in the literature proposed to compute a diet that respects the nutritional needs of an individual. This work deals with a variation of the Diet Problem, called Caloric-Restricted Diet Problem (CRDP). The CRDP objective is to find a reduced caloric diet that also respects the nutritional needs of an individual, thus enabling weight loss in a healthy way. In this paper we propose an Island-based Differential Evolution algorithm, a distributed metaheuristic that evolves a set of populations semi-isolated from each other. Computational experiments showed that this island-based structure outperforms its non-distributed implementation, generating a greater variety of diets with small calorie count.

Keywords: Differential Evolution · Islands model · Diet problem · Calories · Obesity

1 Introduction

A high calorie diet and the lack of physical activity are the main causes of overweight and obesity [12]. An efficient treatment for obesity consists in the ingestion of a hypocaloric diet [1,6]. It is indicated that a hypocaloric diet with 1200 kilocalories (kcal) can be used by overweight woman to reduce their body fat [17]. Besides, a hypocaloric diet with 1000 to 1200 kcal can improve the life quality of obese individuals in less than 30 days [5]. However, these diets also need to provide all the necessary vitamins and nutrients, such as carbohydrates, iron, zinc and fibers, among others [20].

The Diet Problem (DP) [23], one of the most classical linear optimization problems, aims at computing a minimum cost diet that satisfies all nutritional needs of an individual. Despite its relevance for the operational research and linear programming communities, the results obtained in [23] can not be used in

O. Gervasi et al. (Eds.): ICCSA 2017, Part I, LNCS 10404, pp. 385–400, 2017.
DOI: 10.1007/978-3-319-62392-4_28

practice, as the generated diet lacks of food variety and palatability. Besides, it is only focused in healthy individuals, not being applicable to prevent or treat any disease.

The Caloric-Restricted Diet Problem (CRDP), a variation of DP that minimizes the calorie count of a diet, was introduced in [21]. Instead of the classic objective function that minimizes the financial cost, CRDP aims to build a diet that minimizes the number of ingested calories, while respecting the minimal amount of nutrients. Besides, it models a real life diet as a CRDP, representing a diet by six different meals that can be eaten in a daily basis.

CRDP was originally solved by a Differential Evolution algorithm (DE) [21]. In this work, we extend this algorithm by proposing an Island-Based Differential Evolution algorithm (IBDE). It distributes the DE population into several smaller subpopulations that evolves semi-isolated from each other. In such subdivision, a subpopulation communicates with another by means of a migration operator, exchanging solutions among them. We show that IBDE can generate diets with a smaller number of calories than DE with the same number of evaluations.

The remainder of this paper is organized as follow. Section 2 presents the studied problem. Section 3 presents the Island-Based Differential Evolution. Section 4 presents and discuss the computational experiments, comparing IBDE with its non-distributed version. Finally, the conclusions of this work are drawn in the last section.

2 The Caloric-Restricted Diet Problem

It is notable that a diet with a high amount of calories is correlated with weight gain [3, 16]. More than simple affecting the body weight, it can greatly affects the health of an individual, being responsible for a great number of chronic diseases, such as coronary heart diseases [7], diabetes [13] and kidney diseases [11].

It is difficult to develop a caloric-restricted diet that provides all necessary nutrients, such as proteins, zinc and iron, among others. The first attempt to develop an algorithm for computing a healthy diet was presented in [23]. However, it only seeks for a minimum cost diet, not being concerned with the application of such diet in a daily basis. Besides, the developed diet lacks of food variety and palatability, being hard to follow.

With the increasing number of overweight and obese individuals [26], the development of computational methods to generate diets that are healthy and tasty at the same time is needed. The work [21] presented the Caloric-Restricted Diet Problem (CRDP), an optimization problem that aims to compute caloric-restricted diets that are healthy and tasty at the same time. A diet was modeled as six different meals and each meal is represented by a set of foods.

Let T and N be respectively the set of available foods and the set of nutrients considered in the problem. The CRDP mathematical formulation is defined with binary decision variables y such that $y_i = 1$ if food $i \in T$ is included into the diet and $y_i = 0$ otherwise. Besides, auxiliary variables $p_i \in \mathbb{R}_+$ represent the amount of food $i \in T$ to be consumed. Also, let c_i represent the cost per portion of food $i \in T$, m_{ij} represent the amount of nutrient $j \in N$ contained in food $i \in T$, and b_j be the minimal requirement of the nutrient $j \in N$. Finally, let $kcal_i$ represents the calorie count of one portion of food $i \in T$. CRDP can be expressed by Eqs. 1–4. One can see that this is a Mixed-Integer Quadratic Programming problem. Therefore, it is a NP-Hard optimization problem [10].

$$\min \left| 1200 - \sum_i kcal_i \; p_i \; y_i \right|, \quad \forall i \in T \tag{1}$$

$$\sum_i \sum_j m_{ij} \; p_i \; y_i \geq b_j, \quad \forall i \in T, j \in N \tag{2}$$

$$y_i \in \{0, 1\}, \quad \forall i \in T \tag{3}$$

$$p_i \in [0.5, 3], \quad \forall i \in T \tag{4}$$

The objective function (1) aims at developing a diet with 1200 kcal. This number was fixed since a diet with 1200 kcal is the most indicated one to reduce weight [5,17], such that a higher amount of calories does not greatly contribute to this objective. On the other hand, a diet with a smaller amount of calories can represent a risk to the health of an individual. Inequalities (2) maintain all necessary nutrients at a required amount. Equations (3) and (4) define the domain of the variables y and p, respectively. This model considers that a portion contains 100 g of a food or 100 mL of a beverage. The value of p_i is limited within the interval [0.5, 3]. Therefore, neither a large quantity of some food is prescribed, thus dominating the diet, or a small portion of some food to be included. It prevents an impracticable real situation (imagine to cook only 10 g of fish at dinner, for instance).

Besides the number of calories of the diet, CRDP takes into account another nine nutrients, namely dietary fibers (Df), carbohydrates (C), proteins (Pt), calcium (Ca), manganese (Mn), iron (Fe), magnesium (Mg), phosphor (P), and zinc (Zn). Table 1 presents the daily intake recommendation of nutrients [24]. These minimum requirements are guaranteed by Inequalities 2.

Table 1. Nutrients daily intake recommendation [24]

Nutrient requirement (g)	Nutrient requirement (mg)
$Df \geq 25$	$Ca \geq 1000$
$C \geq 300$	$Mn \geq 2.3$
$Pt \geq 75$	$Fe \geq 14$
	$Mg \geq 260$
	$P \geq 700$
	$Zn \geq 7$

The nutrients of each food product used in this work were obtained from the Brazilian Table of Food Composition (TACO) (Tabela Brasileira de Composição de Alimentos, in Portuguese) [15]. TACO was elaborated by the University of Campinas (UNICAMP), Brazil. It contains data from a large number of foods and beverages, with a quantitative description about 25 of their nutrients. Each data represents a portion of 100 g (or 100 mL, when appropriate) of a product.

TACO displays information about foods in a variety of states. For instance, it presents uncooked meat, uncooked meat with salt, and cooked meat. Thus, only a subset of the foods from TACO (that can be served in a meal) are selected to be included in T. Moreover, this work only considers the most important nutrients, thus considering 9 from the 25 nutrients from TACO in the problem. The selected nutrients are displayed at the odd numbered columns of Table 1.

The selected subset of food products were classified into 9 categories, according to their characteristics. This procedure was carried out by a specialist in the area. These categories are displayed in Table 2. The first column displays the category name. The second column shows the symbol associated with each category. The third column indicates the number of foods in each category. The last column shows the interval of numbers associated with each category. This interval will be used to represent a solution for the Island-Based Differential Evolution (see Sect. 3.1).

Table 2. Different classification of products from TACO [15]

Product	Symbol	#	Interval
Beverages	B	21	1–21
Juices	J	11	22–32
Fruits	F	62	33–94
Lacteal	L	19	95–113
Carbohydrates 1	$C1$	21	114–134
Carbohydrates 2	$C2$	12	135–146
Grains	G	12	147–158
Vegetables	V	41	159–199
Proteins	P	95	200–294

Beverages, except natural juices and alcoholic beverages, are symbolized as B. Natural fruit juices are represented by the symbol J. Fruits, in general, are represented as F. Milk-derived products have the symbol L. Carbohydrates are separated into two different groups, symbolized by $C1$ and $C2$. $C1$ contains snacks, as bread, cookies and crackers, while $C2$ are the main meal carbohydrates, such as rice, potato and cassava. Grains and leguminous foods, such as lentils and beans, are represented as G and vegetables are symbolized as V. The last category contains high protein foods, such as meat, chicken and eggs. It is represented by the symbol P.

3 Island-Based Differential Evolution for CRDP

The Differential Evolution algorithm (DE) [18] is a classical evolutionary algorithm that originally deals with real-valued variables. It is a general algorithm that can be easily adapted to a wide range of problems. In the literature, it is possible to find DE applications in engineering [14], chemistry [2], biology [4], finances [19], and artificial intelligence [25], among others [8].

Despites other evolutionary algorithms, DE algorithm operators are applied in a different order, as shown in Algorithm 1. First, the parents are selected. Next, the mutation, recombination and evaluation phases are executed, in this order. Finally, the most fitted individuals are selected to go to the next generation.

Algorithm 1. DE pseudocode

```
Initialize the population
while stopping criterion is not met do
    Selection
    Mutation
    Recombination
    Selection
end while
return the best solution found
end
```

3.1 Representation of Solutions

This work uses the same representation of solution proposed in [21] for the CRDP. It proposes to represent a diet as a set of different meals. Thus, a solution is modeled as a combination of the most common types of products consumed in each meal. Since foods are divided in categories (see Table 2), it is possible to build a diet that has a great variety of products every day. Figure 1 displays such solution modeling.

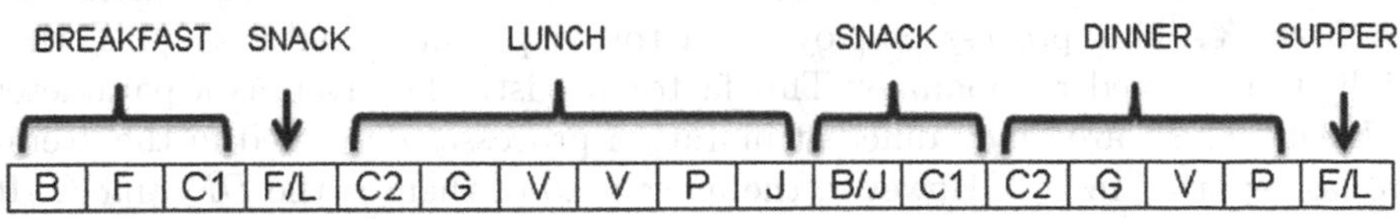

Fig. 1. The model of a solution. There are 6 different meals made up of 17 distinct food products.

The solution presented in Fig. 1 considers that an individual has 6 different meals in a given day, namely breakfast, two snacks, lunch, dinner, and supper. Breakfast is the first meal by the morning. Snacks are small meals that takes place between two major meals. Supper is the last meal of the day, after the dinner and before sleep, which is common to a great number of individuals.

0.6	2.1	2.9	1.2	1	0.8	2	2.7	1.9	1.1	0.6	1	1.6	1.1	1.7	2	0.5
13	52	16	17	19	7	33	12	84	10	18	6	16	11	31	9	41

Fig. 2. Representation of a IBDE solution for CRDP

From Fig. 1, it is possible to build an IBDE's solution representation as shown in Fig. 2. A solution consists in two vectors of 17 variables each, representing the 6 meals of Fig. 1. Each position of the vectors is said to be a gene. The first vector, called *portions vector*, represents the portions p_i of the selected food product. The second vector, denoted *ID vector*, contains an integer number in the range $[1, 294]$. A bijective function $f : \mathbb{N} \mapsto T$ that maps each number as a different food is applied to each gene of the *ID vector*. It is worth to mention that each gene of this vector is just allowed to assume the numbers that map products from it's own category, as shown in the fourth column of Table 2.

3.2 Population Initialization

IBDE initial population X is randomly generated with $|X| = pop$ individuals. For the *portions vector*, a random real number is generated in the interval $[0.5; 3]$ for each gene. Next, a random integer number is generated for each gene of the *ID vector*. This integer number is within the interval of its class, as mentioned above.

3.3 Mutation Operator

Mutation is the process that inserts random information in the population, thus expanding the search space. Mutation produces a new population of trial individuals $V = \{v_0, \ldots, v_{pop}\}$, such that $|V| = |X| = pop$. The IBDE for CRDP applies two different mutation operators, one for each vector of the solution.

The *portions vector* of the trial individual $v_s \in V$ is generated as the addition of an individual from X to the difference between two other individuals, also from X. This process employs a factor of perturbation $F \in [0; 2]$ that weighs the inserted randomness. This factor needs to be given as a parameter for IBDE. There are many different mutation processes described in the literature, as reported in [18]. However, the most used of them is the *DE/rand/1*. It is explained below.

Let X^g denote the IBDE population at generation g. Also, let x_α, x_β and x_γ be three distinct random individuals that belong to X^g and let x_j^k denote the k-th gene of individual $x_j \in X^g$. *DE/best/1* generates a trial individual v_s that belongs to the generation $g + 1$ as show in Eq. (5). This operator is applied to each gene separately. If v_s^k results in a value smaller than 0.5 or greater than 3, then it is rounded to the nearest acceptable value of the interval.

$$v_s^k = x_\alpha^k + F(x_\beta^k - x_\gamma^k) \tag{5}$$

The *ID vector* of the trial individual v_s is generated by a different operator. Each gene $k \in \{1,\ldots,17\}$ from v_s is generated by randomly selecting one of the k-th genes from solution x_α, x_β or x_γ with the same probability. Thus, each gene of the trial individual also respects its food category and does not compromise the solution's modeling.

3.4 Recombination Operator

The recombination operator intensifies the search into good solutions by reusing previously successful individuals. For each target individual $x_s \in X$ a trial individual $v_s \in V$ is generated from mutation. Then, an individual u_s, denominated offspring, is generated as

$$u_s^k = \begin{cases} v_s^k, & \text{if} \quad r \leq Rr \\ x_s^k, & \text{if} \quad r > Rr \end{cases}, \quad \forall i \in \{1,\ldots,17\}, r \in [0;1]$$

where u_s^k, v_s^k and x_s^k are the k-th gene of the individuals u_s, v_s, and x_s, respectively. These solutions are inserted into a new population $U = \{u_0,\ldots,u_{pop}\}$. The value $r \in [0;1]$ is randomly generated at each iteration of IBDE. *Rr* is a parameter required by IBDE that represents the *recombination ratio*, *i.e.* the probability that an offspring inherits the genes from the trial individual v_s. As r, *Rr* is in the interval $[0;1]$. When $Rr = 1$, the offspring will be equal to the trial individual v_s^k and IBDE does not intensify any solution. On the other hand, when $Rr = 0$, the offspring will be equal to the target vector x_s^k and IBDE does not insert any randomness in the population during its evolution process. If $0 < Rr < 1$, the offspring can receive genes from both v_s^k and x_s^k. One can see that this is a key parameter for IBDE, as it controls the exploration/intensification ratio.

3.5 Selection Operator

As the number of offsprings is the same of the number of individuals in the current generation, the selection operator compares one offspring with one individual from the current population. Then, the individual with the worst fitness is discarded and the other undergo to the next generation. This procedure ensures that generation $g+1$ always will be equal or better than the generation g.

3.6 Constraint Handling

When no constraint is violated, IBDE employs the objective function, described by Eq. (1), as the fitness function. However, when some constraint is violated, a mechanism to handle these violations must be adopted. Hence, the fitness function becomes the objective function plus a penalization term [9].

Let N' represents the set of violated constraints. Also let M be a big constant number, used as a penalty term for violated constraints. Besides, m_{ij}, p_i, y_i, and b_j are the same as denoted in the mathematical formulation described by Eqs. (1) and (4). The fitness function with penalization is stated as Eq. (6).

$$min \quad |Kcal - 1200| + \sum_j \left(\frac{|\sum_{i \in F}(m_{ij}\ p_i\ y_i) - b_j|}{b_j} \right) \cdot M, \quad \forall \in N' \quad (6)$$

One can see that there is no penalty for the value of y_i. The limits of y_i are handled by the mutation operator (see Sect. 3.3). Therefore, the value of y_i do not need to be penalized.

3.7 The Island-Based Implementation

Island models is a class of distributed evolutionary algorithms in which the population is split into multiple subpopulations, called islands. Separate EAs run independently on each island, but they interact by means of a migration operator [22]. Figure 3 graphically represents this model. The use of the island allows n populations in n islands, where each one can be designed with its own behavior or not.

This work implements an island model with 12 islands, where each island has a different behavior. Such behaviors are defined by a set of different parameters. Thus, some islands intensify their population, while other islands diversify the solution.

The set of parameters were selected according to scenarios applied by Silva *et al.* [21], in this way it is possible to compare the performance obtained in this work to the existing results. Table 3 shows the behavior of each island, *i.e.*, how each island is parametrized in the tests. In Table 3, F is the mutation factor and Fc is the crossover probability.

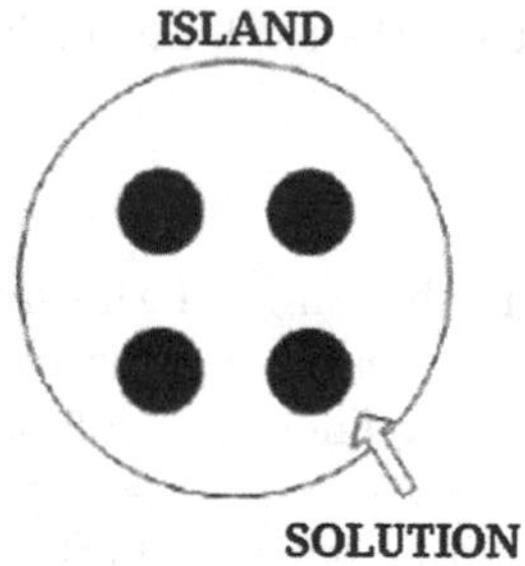

Fig. 3. Island representation.

Migration. The migration operator in islands model consists in sending information from a population x to another population y via the exchange of individuals. This procedure leads to a greater solution diversity, since an individual has characteristics of his own island. This operator can be beneficial or not, and this is what we want to check for this problem.

In literature we can find a migration rate of up to 25%. In this work, the migration rate is 10% of total individuals, which are randomly chosen. The

Table 3. Parameters by island.

Island	F	Fc
1	0.3	0.4
2	0.3	0.6
3	0.3	0.8
4	0.8	0.4
5	0.8	0.6
6	0.8	0.8
7	1.3	0.4
8	1.3	0.6
9	1.3	0.8
10	1.8	0.4
11	1.8	0.6
12	1.8	0.8

topology used in this work was a random ring topology. In this topology, islands are randomly disposed in a ring and the individuals migrate to the island in right hand side of their island, as illustrated by Fig. 4.

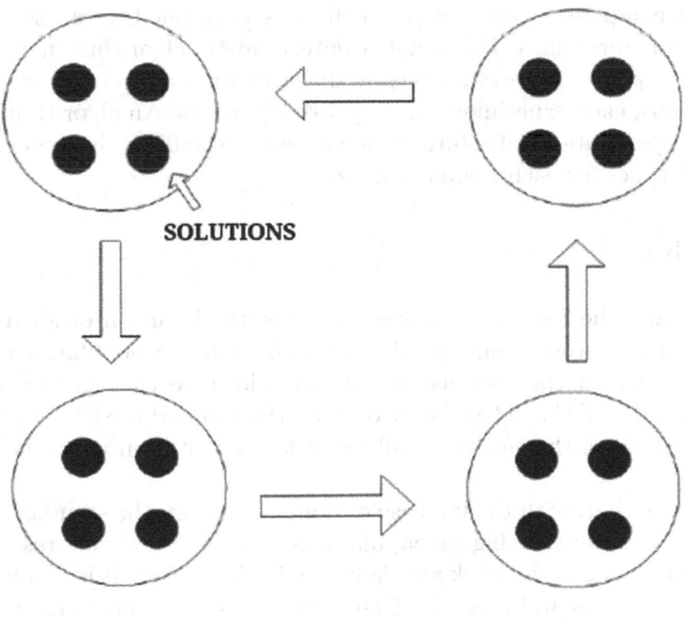

Fig. 4. Illustration of migration

In [21], the non-distributed version of this algorithm was executed for 100 generations and, for this reason, the experiments in this work also use 100 generations, with migration of individuals every 8 or 9 generations, until the number of migrations reach 12, the number of islands.

4 Computational Experiments

The computational experiments were performed an Intel Core i5 CPU 5200U with 2.2 GHz and 4 GB of RAM, running a Linux operating system. The ED algorithm was implemented in C and compiled with the GNU GCC 4.7.3 version.

4.1 Parameters

DE has three main parameters: the pertubation factor, the crossover probability, and the population size. The pertubation factor F varies in the range $[0; 2]$. The crossover probability Fc is a percentage value, and it varies in the range $[0; 1]$. The population size $|X|$ is an integer greater than 2. In order to adjust the algorithm, a complete experiment was developed with the three parameters in order to choose the best value for each.

The value of the perturbation factor F varies in the set $\{0.3, 0.8, 1.3, 1.8\}$. The crossover probability Fc varies in the set $\{0.4, 0.6, 0.8\}$. Finally, the population size $|X|$ is varied the set $\{10, 50, 100\}$.

Considering these values and assigning each combination of parameters to an island, a complete factorial experiment was performed in order to evaluate the DE parameters. Since DE is not a deterministic algorithm, it is necessary to run each experiment several times in order to mitigate its non-deterministic behavior. Thus, each experiment was repeated 20 times. An algorithm execution results in 12 populations of solutions one on each island, so the experiment was executed 20 times for each population size.

5 Results

Table 4 presents the results of 20 executions, with the mean of all results, the best result and the worst result by island, for the different population sizes. Note that the best mean of each population size (in bold) is reached by the first island, where the individuals have low factor of perturbation and medium-high crossing probability, showing that for this problem a lower mutation rate reaches better solutions.

One can see that, with the increase of population size, the standard deviation decreases, suggesting that bigger populations show more reliable results.

The main results of the work are shown in Table 5, where it is verified that for the same parameters and number of DE generations the introduction of island migration improves the average of the solutions.

The graphics in Fig. 5 present the evolution of best fitness by generation on island-based model for different population size, 10 (Fig. 5(a)), 50 (Fig. 5(b)) and

Table 4. Results obtained by the island-based algorithm.

Island	Poputation size	F	Fc	Mean	Best	Worst	Std deviation
1	10	0.3	0.4	1584.727	1526.087	1714.201	45.732
2	10	0.3	0.6	1576.349	1508.180	1641.985	34.808
3	**10**	**0.3**	**0.8**	**1566.870**	**1510.431**	**1680.196**	**51.341**
4	10	0.8	0.4	1579.300	1518.449	1667.008	42.668
5	10	0.8	0.6	1571.683	1497.961	1655.754	42.506
6	10	0.8	0.8	1574.803	1520.849	1670.674	33.787
7	10	1.3	0.4	1590.469	1515.916	1682.579	52.149
8	10	1.3	0.6	1583.563	1534.461	1677.312	35.338
9	10	1.3	0.8	1592.272	1526.582	1683.061	42.631
10	10	1.8	0.4	1589.054	1480.454	1659.647	50.069
11	10	1.8	0.6	1579.753	1451.473	1720.354	59.738
12	10	1.8	0.8	1583.554	1543.748	1637.938	27.478
1	50	0.3	0.4	1508.034	1479.616	1538.132	15.126
2	**50**	**0.3**	**0.6**	**1492.527**	**1451.569**	**1530.977**	**20.583**
3	50	0.3	0.8	1495.263	1457.470	1531.463	18.593
4	50	0.8	0.4	1514.669	1476.914	1549.676	20.433
5	50	0.8	0.6	1500.373	1470.604	1539.000	19.356
6	50	0.8	0.8	1511.098	1479.673	1550.212	20.334
7	50	1.3	0.4	1520.804	1487.961	1547.607	16.943
8	50	1.3	0.6	1520.687	1475.837	1565.738	25.748
9	50	1.3	0.8	1508.995	1459.175	1545.719	25.205
10	50	1.8	0.4	1514.497	1473.839	1542.139	20.410
11	50	1.8	0.6	1517.421	1484.751	1559.145	18.369
12	50	1.8	0.8	1514.563	1473.484	1542.221	14.471
1	100	0.3	0.4	1489.824	1453.768	1526.610	18.976
2	100	0.3	0.6	1473.319	1454.195	1499.285	15.598
3	**100**	**0.3**	**0.8**	**1462.901**	**1430.271**	**1502.436**	**17.696**
4	100	0.8	0.4	1494.771	1461.473	1539.541	20.197
5	100	0.8	0.6	1488.132	1457.017	1520.509	14.514
6	100	0.8	0.8	1484.361	1445.291	1519.070	20.279
7	100	1.3	0.4	1497.502	1468.686	1529.350	17.216
8	100	1.3	0.6	1486.994	1457.932	1522.724	16.831
9	100	1.3	0.8	1492.142	1469.386	1526.784	15.017
10	100	1.8	0.4	1499.688	1443.144	1529.011	18.304
11	100	1.8	0.6	1495.918	1468.918	1520.644	13.848
12	100	1.8	0.8	1495.496	1456.164	1539.773	22.698

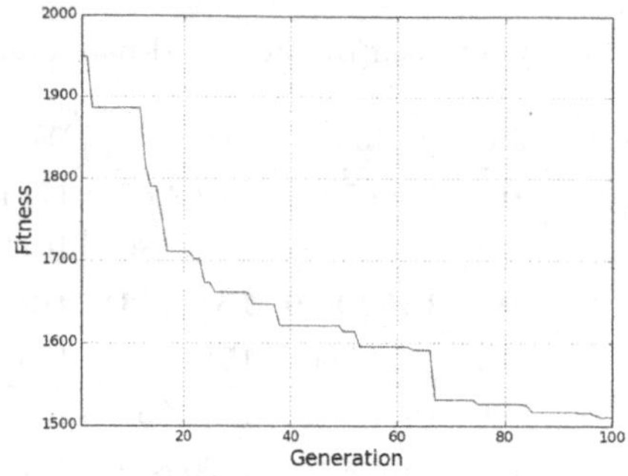

(a) Population Size 10

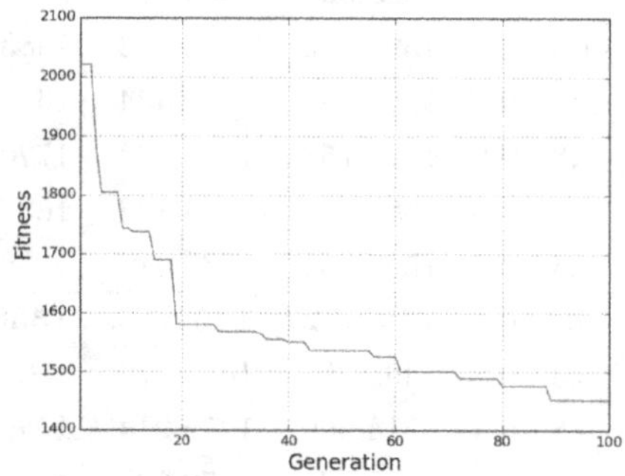

(b) Population Size 50

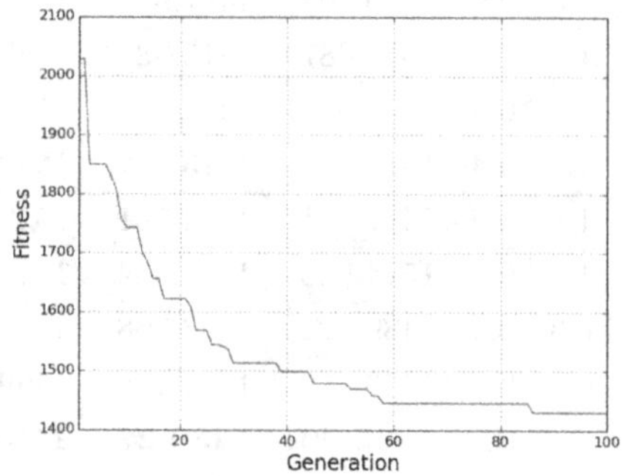

(c) Population Size 100

Fig. 5. Fitness by generation in island-based model

100 (Fig. 5(c)). We can see that the final results are reached early than generation number 100, being the main improvements between generation number 10 and generation number 60.

From Table 5, it is easy to see that the hypothesis of this work is verified and the proposed goals are achieved, since the proposed implementation using islands shows better results than the original model in 92% (33 of total of 36)

Table 5. Mean - island-based model × original model.

Population size	F	Fc	Mean island-based	Original	Difference
10	0.3	0.4	**1584.727**	1665.81	−81.083
10	0.3	0.6	**1576.349**	1761.16	−184.811
10	0.3	0.8	**1566.870**	1869.79	−302.95
10	0.8	0.4	**1579.300**	1631.13	−51.83
10	0.8	0.6	**1571.683**	1659.09	−87.407
10	0.8	0.8	**1574.803**	1675.09	−100.287
10	1.3	0.4	**1590.469**	1630.83	−40.361
10	1.3	0.6	**1583.563**	1633.73	−50.167
10	1.3	0.8	**1592.272**	1696.47	−104.198
10	1.8	0.4	**1589.054**	1650.69	−61.636
10	1.8	0.6	**1579.753**	1667.26	−87.507
10	1.8	0.8	**1583.554**	1749.45	−165.896
50	0.3	0.4	1508.034	**1507.89**	+0.144
50	0.3	0.6	**1492.527**	1508.93	−16.403
50	0.3	0.8	**1495.263**	1522.09	−26.827
50	0.8	0.4	**1514.669**	1519.45	−4.781
50	0.8	0.6	**1500.373**	1514.38	−14.007
50	0.8	0.8	**1511.098**	1522.015	−10.917
50	1.3	0.4	**1520.804**	1530.11	−9.306
50	1.3	0.6	1520.687	**1518.93**	−1.757
50	1.3	0.8	**1508.995**	1528.54	−19.545
50	1.8	0.4	**1514.497**	1517.64	−3.143
50	1.8	0.6	**1517.421**	1527.42	−9.999
50	1.8	0.8	**1514.563**	1531.6	−17.037
100	0.3	0.4	**1489.824**	1491.56	−1.736
100	0.3	0.6	**1473.319**	1479.08	−5.761
100	0.3	0.8	**1462.901**	1484.84	−21.939
100	0.8	0.4	**1494.771**	1496.23	−1.459
100	0.8	0.6	1488.132	**1487.97**	+0.162
100	0.8	0.8	**1484.361**	1493.18	−8.819
100	1.3	0.4	**1497.502**	1507.45	−9.948
100	1.3	0.6	**1486.994**	1499.52	−12.526
100	1.3	0.8	**1492.142**	1509.47	−17.328
100	1.8	0.4	**1499.688**	1511.4	−11.712
100	1.8	0.6	**1495.918**	1503.08	−7.162
100	1.8	0.8	**1495.496**	1507.61	−12.114

of all tests. Probably, the migration improves the diversity, which improves the results. Specially in tests with small population size, it helps the algorithm to scape from local solution. Tests with small population size considering the islands model achieve results similar to those obtained with larger population sizes with the original methodology.

Table 6 shows possibles diets generated by IBDE. We can see a large diversity of foods and drinks in all meals.

IBDE can be executed many times to solve CRDP in seconds, thus generating different diets on each execution. Therefore, a weekly diet can be generated in seconds. The great diversity of foods and drinks makes diets easy to be followed, so the taste will not be tired with repetitive meals, thus becoming easier to lose weight.

Table 6. Diet examples

Diet 1 1501.92 kcal		Diet 2 1507.72 kcal	
Breakfest			
Black tea	65 ml	Coconut water	175 ml
Gold banana (raw)	50 g	Pear (raw)	295 g
Corn cream	50 g	Bread with gluten	55 g
Snack			
Fermented milk	300 ml	Breadfruit	260 g
Lunch			
Cassava (cooked)	300 g	Sweet potatoes (boiled)	235 g
Peas in pod	50 g	Baked rosy beans	150 g
Catalan (raw)	300 g	Arugula (raw)	220 g
watercress (raw)	50 g	Lettuce	135 g
grilled sardines	50 g	Raw kibe	75 g
Lemon lime juice	260 g	Lemon lime juice	150 ml
Snack			
Orange juice	60 ml	Natural yogurt	50 ml
Whole bread	50 g	Corn cream	75 g
Dinner			
Potato (cooked)	50 g	Sweet potatoes (boiled)	80 g
Purple beans	250 g	Lentils (cooked)	250 g
Broccoli (cooked)	300 g	Alfavaca	150 g
Roasted kibe	50 g	Crab (cooked)	80 g
Supper			
Fuji apple	90 g	Acerola	115 g

6 Conclusions and Future Works

The key point to solve CRDP is to generate a great diversity of solutions. Thus, various diets can be easily generated by a DE. Deterministic methods, such as branch-and-bound approaches or mathematical programming methods, can not be adequately used to solve CRDP as it would always generate the same diet.

The use of the island-based model achieved great results, since it was able to reduce the amount of calories of the diets without violating the restrictions in majority of the scenarios. This work can be applied in real scenarios, as it can generate a lot of diets in short time.

For future works, we propose to analyze the migration rate from an island to another, test new topologies and scenarios to minimize the caloric values in diets.

Another idea for future direction is to develop a multi-objective version of CRDP, to minimize, in addition to the number of calories, the financial cost or the sodium consume, while maximizing the amount of fiber and protein, for example.

Acknowledgments. This work was partially supported by the Brazilian National Council for Scientific and Technological Development (CNPq), the Foundation for Support of Research of the State of Minas Gerais, Brazil (FAPEMIG), and Coordination for the Improvement of Higher Education Personnel, Brazil (CAPES).

References

1. Anderson, J.W., Konz, E.C., Frederich, R.C., Wood, C.L.: Long-term weight-loss maintenance: a meta-analysis of US studies. Am. J. Clin. Nutr. **74**(5), 579–584 (2001)
2. Babu, B., Angira, R.: Modified differential evolution (MDE) for optimization of non-linear chemical processes. Comput. Chem. Eng. **30**(6), 989–1002 (2006)
3. Bistrian, B.R.: Diet, lifestyle, and long-term weight gain. New Engl. J. Med. **365**(11), 1058–1059 (2011)
4. Chowdhury, A., Rakshit, P., Konar, A.: Protein-protein interaction network prediction using stochastic learning automata induced differential evolution. Appl. Soft Comput. **49**, 699–724 (2016)
5. Crampes, F., Marceron, M., Beauville, M., Riviere, D., Garrigues, M., Berlan, M., Lafontan, M.: Platelet alpha 2-adrenoceptors and adrenergic adipose tissue responsiveness after moderate hypocaloric diet in obese subjects. Int. J. Obes. **13**(1), 99–110 (1988)
6. Curioni, C., Lourenco, P.: Long-term weight loss after diet and exercise: a systematic review. Int. J. Obes. **29**(10), 1168–1174 (2005)
7. Dalen, J.E., Devries, S.: Diets to prevent coronary heart disease 1957–2013: what have we learned? Am. J. Med. **127**(5), 364–369 (2014)
8. Das, S., Mullick, S.S., Suganthan, P.N.: Recent advances in differential evolution-an updated survey. Swarm Evol. Comput. **27**, 1–30 (2016)
9. Datta, R., Deb, K.: Evolutionary Constrained Optimization. Infosys Science Foundation Series, vol. 1. Springer, Heidelberg (2015)

10. Garey, M.R., Johnson, D.S.: Computers and Intractability, vol. 29. W. H. Freeman, New York (2002)
11. Heerspink, H.J.L., Holtkamp, F.A., Parving, H.H., Navis, G.J., Lewis, J.B., Ritz, E., De Graeff, P.A., De Zeeuw, D.: Moderation of dietary sodium potentiates the renal and cardiovascular protective effects of angiotensin receptor blockers. Kidney Int. **82**(3), 330–337 (2012)
12. Ladabaum, U., Mannalithara, A., Myer, P.A., Singh, G.: Obesity, abdominal obesity, physical activity, and caloric intake in US adults: 1988 to 2010. Am. J. Med. **127**(8), 717–727 (2014)
13. Ley, S.H., Hamdy, O., Mohan, V., Hu, F.B.: Prevention and management of type 2 diabetes: dietary components and nutritional strategies. Lancet **383**(9933), 1999–2007 (2014)
14. Liao, T.W.: Two hybrid differential evolution algorithms for engineering design optimization. Appl. Soft Comput. **10**(4), 1188–1199 (2010)
15. Lima, D.M., Padovani, R.M., Rodriguez-Amaya, D.B., Farfán, J.A., Nonato, C.T., Lima, M.T.D., Salay, E., Colugnati, F.A.B., Galeazzi, M.A.M.: Tabela brasileira de composição de alimentos - TACO (2011). http://www.unicamp.br/nepa/taco/tabela.php?ativo=tabela. Accessed 16 Mar 2016
16. Mozaffarian, D., Hao, T., Rimm, E.B., Willett, W.C., Hu, F.B.: Changes in diet and lifestyle and long-term weight gain in women and men. New Engl. J. Med. **2011**(364), 2392–2404 (2011)
17. Pasquali, R., Gambineri, A., Biscotti, D., Vicennati, V., Gagliardi, L., Colitta, D., Fiorini, S., Cognigni, G.E., Filicori, M., Morselli-Labate, A.M.: Effect of long-term treatment with metformin added to hypocaloric diet on body composition, fat distribution, and androgen and insulin levels in abdominally obese women with and without the polycystic ovary syndrome. J. Clin. Endocrinol. Metab. **85**(8), 2767–2774 (2000)
18. Price, K., Storn, R.M., Lampinen, J.A.: Differential evolution: a practical approach to global optimization. Springer Science & Business Media, Heidelberg (2006)
19. Ravi, V., Kurniawan, H., Thai, P.N.K., Kumar, P.R.: Soft computing system for bank performance prediction. Appl. Soft Comput. **8**(1), 305–315 (2008)
20. Rolfes, S.R., Pinna, K., Whitney, E.: An overview of nutrition. In: Understanding Normal and Clinical Nutrition, 10th edn, pp. 3–34. Cengage Learning (2014)
21. Silva, J.G.R., Carvalho, I.A., Loureiro, M.M.S., Fonseca Vieira, V., Xavier, C.R.: Developing tasty calorie restricted diets using a differential evolution algorithm. In: Gervasi, O., et al. (eds.) ICCSA 2016. LNCS, vol. 9790, pp. 171–186. Springer, Cham (2016). doi:10.1007/978-3-319-42092-9_14
22. Skolicki, Z.M.: An analysis of island models in evolutionary computation. Ph.D. thesis, Fairfax, VA, USA (2007). aAI3289714
23. Stigler, G.J.: The cost of subsistence. J. Farm Econ. **27**(2), 303–314 (1945)
24. de Vigilância Sanitária, A.N.: Resolution RDC number 360. Diário Oficial da União (2003). https://goo.gl/wbZxgw. Accessed 16 Mar 2016
25. Wang, L., Zeng, Y., Chen, T.: Back propagation neural network with adaptive differential evolution algorithm for time series forecasting. Expert Syst. Appl. **42**(2), 855–863 (2015)
26. Yang, L., Colditz, G.A.: Prevalence of overweight and obesity in the United States, 2007–2012. JAMA Intern. Med. **175**(8), 1412–1413 (2015)

An Implementation of Parallel 1-D Real FFT on Intel Xeon Phi Processors

Daisuke Takahashi(✉)

Center for Computational Sciences, University of Tsukuba, 1-1-1 Tennodai, Tsukuba, Ibaraki 305-8577, Japan
daisuke@cs.tsukuba.ac.jp

Abstract. In this paper, we propose an implementation of a parallel one-dimensional real fast Fourier transform (FFT) on Intel Xeon Phi processors. The proposed implementation of the parallel one-dimensional real FFT is based on the conjugate symmetry property for the discrete Fourier transform (DFT) and the six-step FFT algorithm. We vectorized FFT kernels using the Intel Advanced Vector Extensions 512 (AVX-512) instructions, and parallelized the six-step FFT by using OpenMP. Performance results of one-dimensional FFTs on Intel Xeon Phi processors are reported. We successfully achieved a performance of over 91 GFlops on an Intel Xeon Phi 7250 (1.4 GHz, 68 cores) for a 2^{29}-point real FFT.

1 Introduction

The Intel Xeon Phi processor has emerged as an important computational accelerator in high-performance computing systems. Knights Landing processor [14] is the second-generation Intel Xeon Phi product. Cori, a system equipped with the Knights Landing processor, placed the 5th in the TOP500 list in November 2016 [1]. The fast Fourier transform (FFT) [6] is widely used in science and engineering. Parallel FFTs on many-core processors have been implemented [3,8]. Today, a number of processors have short vector SIMD instructions. These instructions provide substantial speedup for digital signal processing applications. Efficient FFT implementations with short vector SIMD instructions have also been investigated thoroughly [7,12,13,18]. Both vectorization and parallelization are of particular importance with respect to Intel Xeon Phi processors. We vectorized FFT kernels using the Intel Advanced Vector Extensions 512 (AVX-512) instructions. Furthermore, the proposed approach makes use of the parallelism of the Intel Xeon Phi processor by loop collapse.

In this paper, we propose an implementation of a parallel one-dimensional real fast Fourier transform (FFT) on Intel Xeon Phi processors. We implemented the parallel one-dimensional real FFT on an Intel Xeon Phi processor and herein report the resulting performance.

The remainder of this paper is organized as follows. Section 2 describes the vectorization of FFT kernels. Section 3 describes the six-step FFT algorithm and the real FFT algorithm. Section 4 describes the parallelization of the six-step FFT. Section 5 presents the performance results. In Sect. 6, we provide concluding remarks.

O. Gervasi et al. (Eds.): ICCSA 2017, Part I, LNCS 10404, pp. 401–410, 2017.
DOI: 10.1007/978-3-319-62392-4_29

2 Vectorization of FFT Kernels

The Intel AVX-512 [9] is a 512-bit single-instruction multiple data (SIMD) instruction sets on the Intel Xeon Phi processor. AVX-512 supports 512-bit wide SIMD registers (ZMM0–ZMM31).

The most direct way to use the AVX-512 is to insert the assembly language instructions inline into the source code. However, this can be time-consuming and tedious. Instead, Intel provides for easy implementation through the use of API extension sets referred to as intrinsics [10]. The latest version of the FFTW library (version 3.3.6-pl1) [7] uses the AVX-512 intrinsics to access SIMD hardware. The Intel C/C++ and Fortran compilers also use Intel AVX-512 to support automatic vectorization. In this paper, we use automatic vectorization.

An example of a vectorizable radix-2 FFT kernel is shown in Fig. 1. In this figure, arrays `A` and `B` are the input array and the output array, respectively. The twiddle factors [5] are stored in array `W`. The problem size n corresponds to $\mathtt{M} \times \mathtt{L} \times 2$. For the Intel Xeon Phi processor, memory movement is optimal when the data starting address lies on 64-byte boundaries. Thus, we specified the directive "`!DIR$ ATTRIBUTED ALIGN : 64`" for the arrays.

```
SUBROUTINE FFT(A,B,W,M,L)
COMPLEX*16 A(M,L,*),B(M,2,*),W(*)
COMPLEX*16 C0,C1
DO J=1,L
  DO I=1,M
    C0=A(I,J,1)
    C1=A(I,J,2)
    B(I,1,J)=C0+C1
    B(I,2,J)=W(J)*(C0-C1)
  END DO
END DO
RETURN
END
```

Fig. 1. Example of a vectorizable radix-2 FFT kernel [18]

In the radix-2 FFT kernel, the innermost loop lengths are varied from 1 to $n/2$ for n-point FFTs during $\log_2 n$ stages. For the first stage of the radix-2 FFT kernel in Fig. 1, the innermost loop length is 1. In this case, the double-nested loop can be collapsed into a single-nested loop to expand the innermost loop length, as shown in Fig. 2.

We use the radix-2, 4, 8, and 16 Stockham autosort FFT algorithm [15] for in-cache FFTs. Although the Stockham autosort FFT algorithm requires a scratch array of the same size as the input array, the algorithm does not include digit-reverse permutation. Table 1 shows the real inner-loop operations for radix-2, 4, 8, and 16 double-precision complex FFT kernels. In view of the Byte/Flop ratio,

```
SUBROUTINE FFT1ST(A,B,W,L)
COMPLEX*16 A(L,*),B(2,*),W(*)
COMPLEX*16 C0,C1
DO J=1,L
  C0=A(J,1)
  C1=A(J,2)
  B(1,J)=C0+C1
  B(2,J)=W(J)*(C0-C1)
END DO
RETURN
END
```

Fig. 2. First stage of a vectorizable radix-2 FFT kernel [18]

Table 1. Real inner-loop operations for radix-2, 4, 8, and 16 double-precision complex FFT kernels based on the Stockham FFT

	Radix-2	Radix-4	Radix-8	Radix-16
Loads	4	8	16	32
Stores	4	8	16	32
Multiplications	4	12	32	84
Additions	6	22	66	174
Byte/Flop ratio	6.400	3.765	2.612	1.984

the radix-16 FFT is preferable to the radix-2, 4, and 8 FFTs [17]. Although higher radix FFTs require more floating-point registers to hold intermediate results, the Intel Xeon Phi processor has 32 ZMM 512-bit registers. A power-of-two point FFT (more than or equal to 64-point FFT) can be performed by a combination of radix-8 and radix-16 steps containing at most three radix-8 steps. In other words, power-of-two FFTs can be performed as a length $n = 2^p = 8^q 16^r$ $(p \geq 6,\ 0 \leq q \leq 3,\ r \geq 0)$.

3 Six-Step FFT Algorithm and Real FFT Algorithm

The discrete Fourier transform (DFT) is given by

$$y_k = \sum_{j=0}^{n-1} x_j \omega_n^{jk}, \quad 0 \leq k \leq n-1, \tag{1}$$

where $\omega_n = e^{-2\pi i/n}$ and $i = \sqrt{-1}$.

If n has factors n_1 and n_2 $(n = n_1 \times n_2)$, then indices j and k can be expressed as

$$j = j_1 + j_2 n_1, \quad k = k_2 + k_1 n_2. \tag{2}$$

We define x and y as two-dimensional arrays (in Fortran notation):

$$x_j = x(j_1, j_2), \quad 0 \le j_1 \le n_1 - 1, \quad 0 \le j_2 \le n_2 - 1, \tag{3}$$

$$y_k = y(k_2, k_1), \quad 0 \le k_1 \le n_1 - 1, \quad 0 \le k_2 \le n_2 - 1. \tag{4}$$

Substituting the indices j and k in Eq. (1) with those in Eq. (2) and using the relation $n = n_1 \times n_2$, we derive the following equation:

$$y(k_2, k_1) = \sum_{j_1=0}^{n_1-1} \sum_{j_2=0}^{n_2-1} x(j_1, j_2) \omega_{n_2}^{j_2 k_2} \omega_{n_1 n_2}^{j_1 k_2} \omega_{n_1}^{j_1 k_1}. \tag{5}$$

This derivation yields the following six-step FFT algorithm [4,19]:

Step 1 : Transpose

$$x_1(j_2, j_1) = x(j_1, j_2).$$

Step 2 : n_1 individual n_2-point multicolumn FFTs

$$x_2(k_2, j_1) = \sum_{j_2=0}^{n_2-1} x_1(j_2, j_1) \omega_{n_2}^{j_2 k_2}.$$

Step 3 : Twiddle factor multiplication

$$x_3(k_2, j_1) = x_2(k_2, j_1) \omega_{n_1 n_2}^{j_1 k_2}.$$

Step 4 : Transpose

$$x_4(j_1, k_2) = x_3(k_2, j_1).$$

Step 5 : n_2 individual n_1-point multicolumn FFTs

$$x_5(k_1, k_2) = \sum_{j_1=0}^{n_1-1} x_4(j_1, k_2) \omega_{n_1}^{j_1 k_1}.$$

Step 6 : Transpose

$$y(k_2, k_1) = x_5(k_1, k_2).$$

In the six-step FFT algorithm, two multicolumn FFTs are performed in steps 2 and 5. The locality of the memory reference in the multicolumn FFT is high. Thus, the six-step FFT is suitable for cache-based processors because of the high hit rates in the cache memory. On the other hand, the three transpose steps (steps 1, 4, and 6) are typically the chief bottlenecks in cache-based processors. To reduce the number of cache misses in matrix transposition, we can use cache blocking. An example of matrix transposition with cache blocking is shown in Fig. 3. Parameter `NB` is the blocking parameter.

When the input data of the DFT are real, an n-point real DFT can be computed using an $n/2$-point complex DFT [5].

Let

$$x_j = a_{2j} + i a_{2j+1}, \quad 0 \le j \le n/2 - 1, \tag{6}$$

where $a_0, a_1, \cdots, a_{n-1}$ are n-point real input data.

We obtain the n-point real DFT as follows:

$$b_k = y_k - \frac{1}{2}(y_k - \overline{y}_{n/2-k})(1 + i\omega_n^k), \quad 1 \leq k \leq n/4 - 1. \tag{7}$$

$$\overline{b}_{n/2-k} = \overline{y}_{n/2-k} + \frac{1}{2}(y_k - \overline{y}_{n/2-k})(1 + i\omega_n^k), \quad 1 \leq k \leq n/4 - 1. \tag{8}$$

$$b_0 = \mathrm{Re}(y_0) + \mathrm{Im}(y_0). \tag{9}$$

$$b_{n/2} = \mathrm{Re}(y_0) - \mathrm{Im}(y_0). \tag{10}$$

$$b_{n/4} = y_{n/4}, \tag{11}$$

where $b_0, b_1, \cdots, b_{n/2}$ are $(n/2 + 1)$-point complex output data. Note that b_k $(n/2 < k \leq n - 1)$ can be reconstructed using the symmetry $b_k = \overline{b}_{n-k}$.

4 Parallelization of Six-Step FFT

When we parallelize the six-step FFT by using OpenMP, the outermost loop of each FFT step is distributed across the cores. In Fig. 3, the outermost loop length may not have sufficient parallelism for many-core processors. For an $n = 2^{18}$-point FFT, we assume `N1=N2=512` and `NB=8`. In this case, the outermost loop length is 64 (=512/8), which is less than the number of cores (68) on the Intel Xeon Phi 7250 processor. Thus, insufficient parallelism always leads to load imbalance.

A loop collapsing makes the length of a loop long by collapsing nested loops into a single-nested loop. When using the OpenMP collapse clause, which is supported from OpenMP 3.0 [2], the loops must be perfectly nested. Since the outermost nested loop in Fig. 3 is a perfectly nested loop, it can be collapsed into a single-nested loop. Figure 4 shows the parallelization of six-step FFT with loop collapse. Arrays `X` and `Y` are the input array and the output array, respectively. The twiddle factors ($\omega_{n_1 n_2}^{j_1 k_2}$) in Eq. (5) are stored in the array `W`. By using the OpenMP collapse clause, the parallelism of the outermost loop can be expanded from 64 to 4096 (= 64 × 64) for the $n = 2^{18}$-point FFT.

```
      COMPLEX*16 X(N1,N2),Y(N2,N1)
!$OMP PARALLEL DO PRIVATE(I,J,JJ)
      DO II=1,N1,NB
        DO JJ=1,N2,NB
          DO I=II,MIN(II+NB-1,N1)
            DO J=JJ,MIN(JJ+NB-1,N2)
              Y(J,I)=X(I,J)
            END DO
          END DO
        END DO
      END DO
```

Fig. 3. Example of matrix transposition with cache blocking

```
      COMPLEX*16 X(N1,N2),Y(N2,N1)
      COMPLEX*16 W(N1,N2)
!$OMP PARALLEL
! Step 1: Transpose N1*N2 to N2*N1
!$OMP DO COLLAPSE(2) PRIVATE(I,J,JJ)
      DO II=1,N1,NB
        DO JJ=1,N2,NB
          DO I=II,MIN(II+NB-1,N1)
            DO J=JJ,MIN(JJ+NB-1,N2)
              Y(J,I)=X(I,J)
            END DO
          END DO
        END DO
      END DO
! Step 2: N1 individual N2-point
!         multicolumn FFTs
!$OMP DO
      DO I=1,N1
        CALL IN_CACHE_FFT(Y(1,I),N2)
      END DO
! Steps 3-4: Twiddle factor
!            multiplication and
!            transpose N2*N1 to N1*N2
!$OMP DO COLLAPSE(2) PRIVATE(I,II,J)
      DO JJ=1,N2,NB
        DO II=1,N1,NB
          DO J=JJ,MIN(JJ+NB-1,N2)
            DO I=II,MIN(II+NB-1,N1)
              X(I,J)=Y(J,I)*W(I,J)
            END DO
          END DO
        END DO
      END DO
! Step 5: N2 individual N1-point
!         multicolumn FFTs
!$OMP DO
      DO J=1,N2
        CALL IN_CACHE_FFT(X(1,J),N1)
      END DO
! Step 6: Transpose N1*N2 to N2*N1
!$OMP DO COLLAPSE(2) PRIVATE(I,J,JJ)
      DO II=1,N1,NB
        DO JJ=1,N2,NB
          DO I=II,MIN(II+NB-1,N1)
            DO J=JJ,MIN(JJ+NB-1,N2)
              Y(J,I)=X(I,J)
            END DO
          END DO
        END DO
      END DO
!$OMP END PARALLEL
```

Fig. 4. Parallelization of six-step FFT with loop collapse

A block six-step FFT algorithm [16] improves performance by utilizing the cache memory more effectively. In the block six-step FFT, the multicolumn FFTs and the transpositions are combined to further reuse data in the cache memory by strip mining. Since the loops in the block six-step FFT are not perfectly nested, the loops cannot be collapsed. We emphasize parallelism rather than cache utilization and use the implementation of Fig. 4.

We combine the twiddle factor multiplication (Step 3) and the transposition (Step 4) to gain efficiency in utilizing the memory bandwidth. For one-dimensional real FFT, Eqs. (7) and (8) can be easily vectorized and parallelized.

5 Performance Results

To evaluate the implemented parallel one-dimensional real FFT, referred to as FFTE (version 6.2alpha), we compared its performance with that of the FFTW (version 3.3.6-pl1) [7] and the Intel Math Kernel Library (MKL, version 2017 Update 1) [3]. The FFTW and the MKL support the AVX-512 instructions.

The elapsed times obtained from 10 executions of real-to-complex FFTs were averaged. The input is in normal order, and the output is in a conjugate-even

Table 2. Specifications of the platform

Platform	Intel Xeon Phi processor
Number of cores	68
CPU type	Intel Xeon Phi 7250 Knights Landing 1.4 GHz
L1 Cache (per core)	I-Cache: 32 KB D-Cache: 32 KB
L2 Cache	1 MB (shared between two cores)
Main memory	MCDRAM 16 GB + DDR4-2400 96 GB
OS	Linux 3.10.0-327.22.2.el7.xppsl_1.4.1.3272.x86_64

order. The real FFTs were performed on double-precision real data, and the table for twiddle factors was prepared in advance. In the FFTW, the "measure" planner was used. The specifications of the platform used herein are shown in Table 2. We note that Hyper-Threading (HT) [11] was enabled on the Intel Xeon Phi 7250.

The compiler used was the Intel Fortran Compiler (`ifort`, version 17.0.1.132) for the FFTE and the MKL. The compiler options used were specified as `ifort -O3 -xMIC-AVX512 -qopenmp`. The compiler used was the Intel C Compiler (`icc`, version 17.0.1.132) for the FFTW, with the compiler options `icc -O3 -xMIC-AVX512 -qopenmp`. The compiler option `-O3` specifies to optimize for maximum speed and enable more aggressive optimizations. The compiler option `-xMIC-AVX512` specifies to generate AVX-512 Foundation instructions, AVX-512 Conflict Detection instructions, AVX-512 Exponential and Reciprocal instruc-

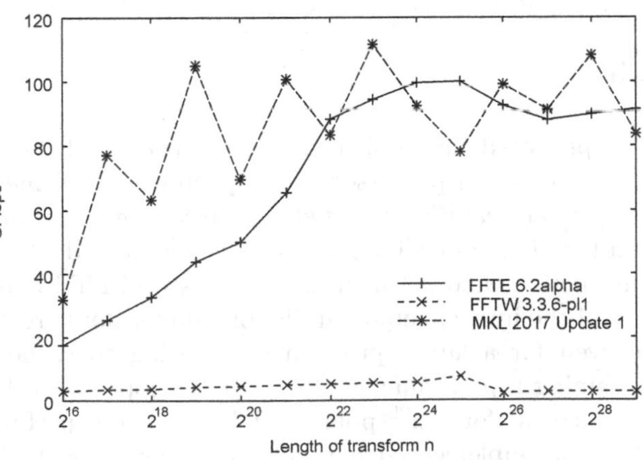

Fig. 5. Performance of one-dimensional real FFTs (Intel Xeon Phi 7250, 272 threads)

tions, and AVX-512 Prefetch instructions. The compiler option `-qopenmp` specifies to enable the compiler to generate multi-threaded code based on the OpenMP directives. The executions were performed in "flat mode" and "quadrant mode". The environment variable `KMP_AFFINITY=granularity=fine,balanced` was specified. All programs were run in MCDRAM.

Figures 5 and 6 compare GFlops of the FFTE, the FFTW, and the MKL. The GFlops values are each based on $5n \log_2 n$ for a transform of size $n = 2^m$. As shown in Figs. 5 and 6, the FFTE is faster than the FFTW. As shown in Fig. 5, the FFTE is faster than the MKL for the cases of $n = 2^{22}$, $2^{24} \leq n \leq 2^{25}$ and $n = 2^{29}$ on 272 threads. Figure 6 indicates that hyper-threading is effective for the FFTE.

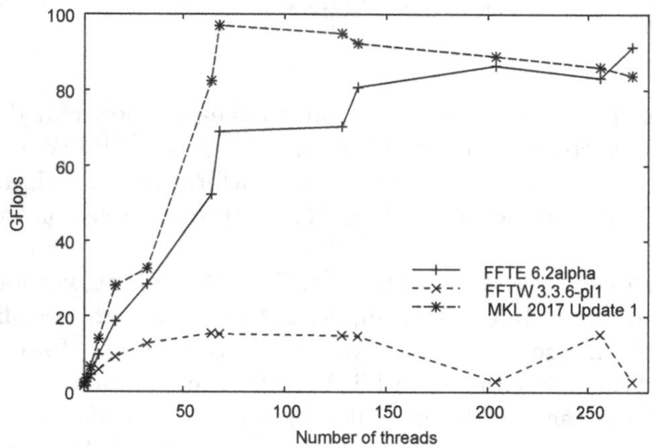

Fig. 6. Performance of one-dimensional real FFTs (Intel Xeon Phi 7250, $n = 2^{29}$)

6 Conclusion

In this paper, we proposed the implementation of the parallel one-dimensional real FFT on Intel Xeon Phi processors. The proposed implementation of the parallel one-dimensional real FFT is based on the conjugate symmetry property for the DFT and the six-step FFT algorithm. We vectorized FFT kernels using the AVX-512 instructions and parallelized the six-step FFT by using OpenMP. The performance of the implemented parallel one-dimensional real FFT remains at a high level even for a larger problem size, owing to cache blocking. We succeeded in obtaining a performance of over 91 GFlops on an Intel Xeon Phi 7250 (1.4 GHz, 68 cores) for a 2^{29}-point real FFT. These performance results demonstrate that the implemented parallel one-dimensional real FFT utilizes cache memory effectively and exploits the AVX-512 instructions.

Acknowledgments. This research was partially supported by Core Research for Evolutional Science and Technology (CREST), Japan Science and Technology Agency (JST).

References

1. http://www.top500.org/
2. OpenMP Application Program Interface. http://www.openmp.org/mp-documents/spec30.pdf
3. Intel Math Kernel Library Developer Reference (2017). https://software.intel.com/sites/default/files/managed/ff/c8/mkl-2017-developer-reference-c_0.pdf
4. Bailey, D.H.: FFTs in external or hierarchical memory. J. Supercomput. **4**, 23–35 (1990)
5. Brigham, E.O.: The Fast Fourier Transform and Its Applications. Prentice-Hall, Upper Saddle River (1988)
6. Cooley, J.W., Tukey, J.W.: An algorithm for the machine calculation of complex Fourier series. Math. Comput. **19**, 297–301 (1965)
7. Frigo, M., Johnson, S.G.: The design and implementation of FFTW3. Proc. IEEE **93**, 216–231 (2005)
8. Hascoet, J., Nezan, J.F., Ensor, A., de Dinechin, B.D.: Implementation of a fast Fourier transform algorithm onto a manycore processor. In: Proceedings of the 2015 Conference on Design and Architectures for Signal and Image Processing (DASIP 2015) (2015)
9. Intel Corporation: Intel architecture instruction set extensions programming reference (2016). https://software.intel.com/sites/default/files/managed/26/40/319433-026.pdf
10. Intel Corporation: Intel C++ compiler 17.0 developer guide and reference (2016). https://software.intel.com/en-us/intel-cplusplus-compiler-17.0-user-and-reference-guide-pdf
11. Marr, D.T., Binns, F., Hill, D.L., Hinton, G., Koufaty, D.A., Miller, J.A., Upton, M.: Hyper-threading technology architecture and microarchitecture. Intel Technol. J. **6**, 1–11 (2002)
12. McFarlin, D.S., Arbatov, V., Franchetti, F., Püschel, M.: Automatic SIMD vectorization of fast Fourier transforms for the Larrabee and AVX instruction sets. In: Proceedings of the 25th International Conference on Supercomputing (ICS 2011), pp. 265–274 (2011)
13. Püschel, M., Moura, J.M.F., Johnson, J.R., Padua, D., Veloso, M.M., Singer, B.W., Xiong, J., Franchetti, F., Gačić, A., Voronenko, Y., Chen, K., Johnson, R.W., Rizzolo, N.: SPIRAL: code generation for DSP transforms. Proc. IEEE **93**, 232–275 (2005)
14. Sodani, A., et al.: Knights Landing: second-generation Intel Xeon Phi product. IEEE Micro **36**, 34–46 (2016)
15. Swarztrauber, P.N.: FFT algorithms for vector computers. Parallel Comput. **1**, 45–63 (1984)
16. Takahashi, D.: A blocking algorithm for FFT on cache-based processors. In: Hertzberger, B., Hoekstra, A., Williams, R. (eds.) HPCN-Europe 2001. LNCS, vol. 2110, pp. 551–554. Springer, Heidelberg (2001). doi:10.1007/3-540-48228-8_58

17. Takahashi, D.: A radix-16 FFT algorithm suitable for multiply-add instruction based on Goedecker method. In: Proceedings of the 2003 IEEE International Conference on Acoustics, Speech, and Signal Processing (ICASSP 2003), vol. 2, pp. 665–668 (2003)
18. Takahashi, D.: An Implementation of parallel 2-D FFT using Intel AVX instructions on multi-core processors. In: Xiang, Y., Stojmenovic, I., Apduhan, B.O., Wang, G., Nakano, K., Zomaya, A. (eds.) ICA3PP 2012. LNCS, vol. 7440, pp. 197–205. Springer, Heidelberg (2012). doi:10.1007/978-3-642-33065-0_21
19. Van Loan, C.: Computational Frameworks for the Fast Fourier Transform. SIAM Press, Philadelphia (1992)

A Comprehensive Survey on Human Activity Prediction

Nghia Pham Trong[1(✉)], Hung Nguyen[1], Kotani Kazunori[1], and Bac Le Hoai[2]

[1] Japan Advanced Institute of Science and Technology, Nomi, Ishikawa, Japan
{phtrnghia,nvhung,ikko}@jaist.ac.jp
[2] University of Science, Vietnam National University Ho Chi Minh City, Ho Chi Minh City, Vietnam
lhbac@fit.hcmus.edu.vn
https://www.jaist.ac.jp/index.html

Abstract. Human activity recognition has been extensively studied and achieves promising results in Computer Vision community. Typical activity recognition methods require observe the whole process, then extract features and build a model to classify the activity. However, in many applications, the ability to early recognition or prediction a human activity before it completes is necessary. This task is challenging because of the lack of information when only a fraction of the activity is observed. To get an accurate prediction, the methods must have high discriminated power with just the beginning part of activity. While activity recognition is very popular and has a lot of surveys, activity prediction is still a new and relatively unexplored problem. To the best of our knowledge, there is no survey specifically focusing on human activity prediction. In this survey, we give a systematic review of current methods for activity prediction and how they overcome the above challenge. Moreover, this paper also compares performances of various techniques on the common dataset to show the current state of research.

Keywords: Activity prediction · Activity representation · Activity datasets · Review · Survey

1 Introduction

Human activity recognition has a lot of applications and attracts a significant amount of works. There are many methods which have been introduced to deal with various kind of video input and achieve good performance. The typical approach converts the input video data into a spatial-temporal representation, and then infer labels over these representations. These works use different types of information, such as human pose, interaction with objects, object shapes and appearance features.

However, these methods can only recognize an already completed activity and cannot be used to anticipate what can happen next. Recognition performance

O. Gervasi et al. (Eds.): ICCSA 2017, Part I, LNCS 10404, pp. 411–425, 2017.
DOI: 10.1007/978-3-319-62392-4_30

can be expected to be poor if we use these methods to classify partially observed activity video. In many real-world applications, the ability to predict an activity before it fully executed is important. It has a wide range of application such as surveillance, human-robot interaction (enable a robot to plan ahead to assist human) and prevention of dangerous event. Koppula and Saxena [1] made a real-life assistant robot application. When the robot sees a person walking toward a fridge and carrying an object, it will predict that she/he needs that object putting in the fridge and will make a response opening the door to help that person.

The overall goal of early activity recognition methods is illustrated in Fig. 1.

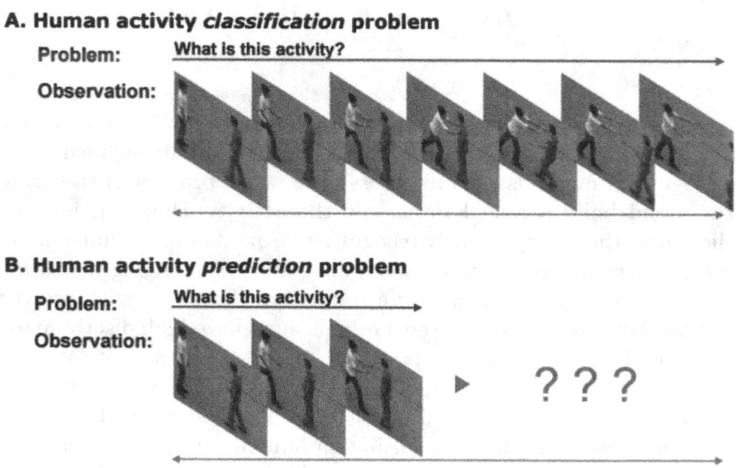

Fig. 1. The difference between activity classification and activity prediction [2]

Ryoo [2] gave a formal definition about the activity prediction as an inference of the ongoing activity given temporally incomplete observations. The system must detect an ongoing activity at its early stage after it starts but before it ends. The major difference between activity recognition and activity prediction is that the whole activity is observed in recognition, while only the beginning activity segment is provided in prediction. Therefore, to achieve a good accuracy, activity prediction system must have strong discriminative power with just the beginning part of the activity video input and it is also important to capture history activity information.

There is a huge amount of works about human activity recognition and also a lot of surveys on this topic [3–5]. However, methods specifically dealing with activity prediction are still limited and to the best of our knowledge, there is no review about activity prediction yet. Our main contribution in this paper is providing a comprehensive survey of human activity recognition methods about how they can capture activity information through feature extraction and modeling. Moreover, we introduce some activity datasets which is commonly used in this research field together with their strong and weak points. The benchmark of different activity prediction methods testing on these datasets also is provided to show the current state of the art.

The general idea of activity prediction is how to build a model to represent the activity. The decision at prediction phase is made based on the likelihood between partial observed video and activity model. General flowchart for human activity prediction consists of three steps: pre-processing, video representation, and activity modeling/classifier. In Fig. 2, we proposed a taxonomy for activity prediction methods.

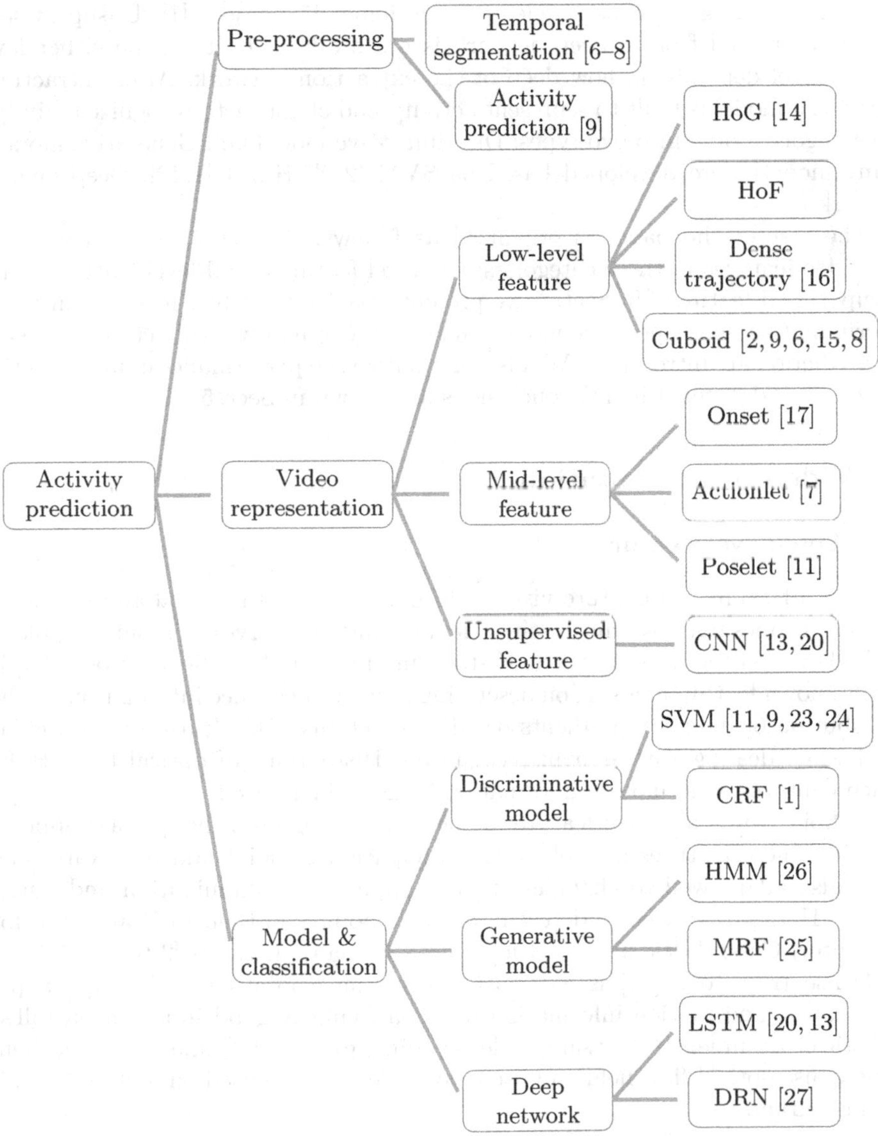

Fig. 2. Taxomony for human activity prediction methods

Many methods start with dividing video input into multi-temporal segments [6–8], each segment represents a sub-activity. Some methods require activity prediction - localize the activity of interest through bounding box or detector [9]. The purpose of video representation is to convert input video data into discriminative and robust features. There are various kinds of features extracted from a video which we categorize into three groups. Low level features capture visual information and temporal information such as Cuboid [10]. These features are usually clustered using a bag of visual word model to gain robust representation. Building on top of low-level feature is mid-level features which represent higher semantic meaning, such as poselet [11], actionlet [7], object [1]. Unsupervised feature extracted from a deep network is quite different. It can be either low or mid-level depends on how deep of the extraction network. After extracting features, a model is built to represent activity and classifier to recognize activity. We category model into two types: Discriminative model and Generative model. Many models were developed based on SVM [2,9], HMM [7,12], Deep neural network [13].

The rest of the paper is organized as follows: in Sect. 2, we review various video features in three categories: low-level features, mid-level features, and unsupervised features. In Sect. 3, we present activity modeling methods in both discriminative and generative model. In Sect. 4, commonly used activity dataset for prediction are introduced. We also summarize the performance of many methods on each dataset. Finally, conclusions are drawn in Sect. 5.

2 Video Representation

2.1 Low-Level Feature

This kind feature will capture visual information on both spatial and temporal domain. It does not have much semantic meaning, but serves as a building block for higher-level semantic features. Histogram of oriented gradients (HoG) [14] is a well-known feature focusing on describing static appearance information by the distribution of intensity gradients or edge directions. This feature is particular good when dealing with human recognition. Histograms of optical flow (HoF) which captures local motion information is also widely used.

Cuboid [10] is a sparse feature for behavior recognition in spatial-temporal domain. These features are robust in anticipation of noise and pose variation. They also adapt well to changing in size, appearance, illumination and image clutter. Harris3D is used to detect interesting point and then HoG and HoF for descriptor. Cuboid descriptor is widely used in many methods [2,6,8,9,15].

Dense trajectory [16] is a robust feature incorporating trajectory shape, appearance, and motion information. It is particularly good in an uncontrolled realistic environment. They sample dense points from each frame and track them using dense optical flow field and use a novel descriptor based on motion boundary histogram.

2.2 Mid-Level Feature

These kinds of features have some semantic meaning and are usually built from low-level feature extracted before. Onset signature [17] feature captures history activity information from its sub-actions. Onset activity is short and subtle human motion (waving and reaching) observable before the main activity (shaking a hand or throwing an object). Given a streaming video, they measure the similarity between each possible video segment and the onset activity, then record how this similarity changes over time. This onset feature is combined with typical visual features to predict activity.

Actionlet features capture sub-action information. In many methods, the first step is temporally segmented input video into semantic units. These segments are called actionlet which represents atomic actions (taking, cutting). Li and Fu [7] detected the boundary between actionlet by monitoring velocity changes of human action. They use Harris corner detector to extract interesting points, then generate a trajectory for these key points. Finally, they accumulate the trajectories to get velocity magnitude. Each hill in the graph represents a semantic atomic action. The velocity is lower at the start and end of each actionlet, and reach peak at the intermediate stage.

Poselet describes a particular part of the human pose under a given viewpoint [18]. Poselet is very helpful when dealing with clutter and occlusion. In activity prediction setting, poselet can capture partial pose which has strong discriminative information and provides a semantic, mid-level representation. In original work, Bourdev and Malik [18] used linear SVM to build a poselet classifier with HoG feature. However, HoG cannot capture fine local appearance details and motion information which are important for activity prediction. To overcome that drawback, [11] augment HoG-based poselets with BoW poselet which contains dense descriptors (SIFT, HoG, MBH). This poselet representation allows them to have a spatially localize representation and defines semantics poselet which are both generic (spanning multiple activities) and also action specific.

Context information such as objects (ball, milk, man), scene types (kitchen, field) is useful cues to help prediction activity. For example, when a person reaches a disk near the kitchen sink, we can tell that she/he prepares to wash. Li and Fu [7] applied human detection, object detection, and scene understanding to get these context information. Then they incorporated these clues in their context-aware prediction model by using sequential pattern mining.

2.3 Unsupervised Feature

Many activity prediction methods depend on hand-crafted features such as bag-of-words [2]. However, these kinds of features are not powerful enough to capture the salient motion information due to their loss of the global structure in the data. Unsupervised features have been shown that they perform better than the traditional hand craft-features in many computer vision problems [19], especially in the large scale dataset.

Ke et al. [13] used Convolution Neural Network (CNN) to capture both spatial and temporal information. They fed input frame images and a set of consecutive optical flow images to CNN to extract spatial and temporal feature respectively. They also divided an image into several local parts (left haft, right half, and the four corners), then fed its frame image and optical flow image into CNN to extract structural features. Ma et al. [20] also used CNN fed with each video frame to extract visual features.

3 Model and Classification

3.1 Discriminative Model

Discriminative models focus on modeling conditional probability distribution $P(y|x)$. It only models about target variables conditional on observed variables and less depends on distribution assumption of data.

Ryoo [2] proposed an integral bag-of-words model which can represent an activity as an integral histogram of spatio-temporal features and make a prediction based on the similarity between histogram representation of testing video and activity model. They also improved the above model by incorporating temporal relation among extracted features. This new model is called Dynamic Bag of words. However, both models are not really good when facing large intra-class variations and sensitive to outliers.

Cao et al. [6] used a spare coding (SC) method to learn feature base and build an activity model. They also extended SC methods by including more SC bases constructed from a mixture of segments with different lengths and locations (MSSC). This model overcomes the problem in Ryoo [2] work about outliers in training data and large intra-class variations. It also can handle the problem about limited training data.

Koppula [1] modeled the rich spatial-temporal structure of activities and object using CRF. Then they extended it to include future possible scenarios in a new model called anticipatory temporal conditional random field (ATCRF). This is CRF but augmented with trajectories, object affordances, and sub-activities.

Xu et al. [8] proposed a novel framework named combinatorial sparse representations based on 3D XYZ local interesting spatial-temporal descriptors. This method computes the likelihood of each segment labeled with a specific class based on combinatorial coefficients.

Wang et al. [15] considered human activity prediction as a Dynamic Time warping problem, a technique for measuring the similarity between two time-series which may vary in speed. They proposed a model called temporally-weight generally time warping (TGTW) extending of their previous work - generalized time warping (GTW) by adding temporal constraints over the warping path to encourage alignment in the early part of an activity sequence. The similarity derives from TGTW is put into the k-nearest neighbors algorithm to predict the activity class.

Hu et al. [21] focused on RGB-D video and showed that RGB-D yields better performance than only use RGB channel. They introduce a new RGB-D feature called local accumulative frame feature (LAFF) which can compute efficiently to make the whole process possible for the real-time applications. Their activity prediction system can identify ongoing activities under a regression framework with a soft label.

Support Vector Machine (SVM). SVM is a very strong classier and widely usage. A lot of approaches in activity prediction is developed based on SVM, especially Structure output SVM [22].

Typical methods for activity recognition are trained to detect a full-observed activity. When using these methods to recognize partial activity, it would lead to unreliable decisions because it does not consider the sequential nature of video data. Hoai and De la Torre [23] proposed Max-Margin Early Event Detector (MMED), a model based on Structured Output SVM, but extended to accommodate the nature of the sequential data. They stimulated the sequential data arrival for training time series and learn detector that correctly classifies partially observed sequences.

Kong et al. [33] proposed a multiple temporal scale support vector machine (MTSSVM) based on structure SVM for predicting action from a partially observed video. In their extend work [9], they introduced a novel model max-margin action prediction machine (MMAPM) which is formulated on structured SVM [22]. This model captures human activity at two different granularity. The fine granularity considers the sequential nature of human activity. The coarse granularity captures the history of activity information. This method allows them to capture both short-range and long-range dependency.

Raptis and Sigal [11] modeled activity as a sequence of keyframes, a representation of key states in the action sequences. They argue that to recognize human activity we do not need the whole video. A few important frames are enough discriminative power for activity recognition. This allows their model to focus on the most distinct parts of the action and remove frames that are not discriminative or relevant. This descriptor is not only spare and compact, but also rich and descriptive. Moreover, it increases robustness when action duration changes because these changes do not affect much when a video is represented by keyframes. To learn activity classifier, they use a max-margin discriminative framework where keyframes are treated as latent variables. A multi-class SVM is used to combine the scores obtaining from each activity model to compensate for the different bias of each model.

Lan et al. [24] propose Hierarchical Movemes (HM)- a new representation for predicting future action. The term moveme is used to capture the atomic component of human movements, such as reaching and grabbing. HM can describe human movements at multiple levels of granularities from coarse to fine. Given a hierarchy of moveme, they learn a classifier for each moveme. Feature descriptors from each level in the hierarchy are used to train SVM.

3.2 Generative Model

Generative model focuses on modeling conditional probability distribution $P(x, y)$. It tries to learn a full probabilistic model of all variables that generates the data by depending on an assumption about the distribution of data.

Chakraborty and Roy-Chowdhury [25] treated activity prediction problem as a graph inference problem on Markov Random Field (MRF) where each node is an individual activity. They model the simultaneous and sequential nature of human activity on the graph and combine with the interrelationship between scene cue and target trajectory known as activity and scene context.

Hidden Markov Model (HMM). HMM is good when dealing with sequential data which is the nature of human activity prediction. Therefore, many methods based on HMM have been proposed.

Li et al. [26] modeled a human activity by a complex temporal composition of constituent simple actions. They used a Probabilistic Suffix Tree (PST) to representing various order Markov Dependencies between action units, and a Predictive Accumulative Function (PAF) to predict each kind of activity.

Causality of action unit is an important clue to predict human activity [7]. To accurately predict activity, it needs to involve not only actions but also objects and their spatial-temporal arrangement with actions. For example, as long as we observe "a person grabbing a cup", we probably can tell that she/he is going to drink a beverage. They model this causality by using variable order Markov model (VMM). This model suits the activity prediction problem well because it can capture both large and small order Markov dependencies extracted from training data. Thus, it can encode richer and more flexible causal relationship.

Ding et al. [12] used a 3D skeletal joint location as input inferred from depth maps. A spatial-temporal pattern is learned by a Hierarchical Self-Organizing Map (HSOM), which consists of two self-organizing maps (i.e., action map and actionlet map) connected via associative links trained by Hebbian learning. Ongoing activities can be predicted by VMM.

3.3 Deep Neural Network

Modeling activity progression is important. The anticipation at each time not entirely depends on current observation at that time but also on the previous states which providing the temporal context for activity progression. Recurrent Neural network (RNN) is a good model to capture activity progression. However, RNN has a problem when dealing with long-term dependency. To overcome the above limitation, Long Short Term Memory (LSTM) was proposed based on RNN. The core of LSTM is a memory cell which can remove and add information over period of time. Therefore, it is able to capture the useful pattern of the previous observation and hidden points to provide long-range context.

Ma et al. [20] reasoned that the detection score should be monotonically non-decreasing when observing more activity. They proposed a novel ranking loss to penalize model when it violated the above idea. This ranking loss combines with classification loss in LSTM network to model activity progression. Ke et al. [13] used LSTM networks with a new ranking score fusion to combine the spatial, temporal and structural information.

Typically, an activity prediction method relies on extract feature, then use these features to build a supervised model to represent activity progression.

Finally, this model is used to predict future action. However, Vondrick et al. [27] proposed a different approach. At first, they predict future visual representation from video, and then apply recognition algorithms to their predicted representation to anticipate objects and actions. Deep regression network is used for this task. In training, they use a Euclidean loss to minimize the distance between network prediction and the representation of the future frame.

4 Related Datasets for Human Activity Prediction

This section introduces public datasets which are commonly used in human activity prediction researches. All these datasets are publicly available for research and useful for the comparison of various techniques.

Table 1 summarizes human activity prediction datasets in chronological order. We also represent the number of action classes, the number of videos (or length), setting, and the frame resolution.

Table 1. Human activity prediction datasets

Dataset	Year	#Action	#Video	Setting	Resolution
UT-Interaction dataset [28]	2010	6	20	Outdoor	720 * 480
CAD-120 dataset [29]	2012	10	120	Daily activities	240 * 320
BIT-Interaction dataset [30]	2012	8	400	Outdoor	320 * 240
JPL First-Person Interaction [31]	2013	7	57	Indoor	320 * 240
Activity Net [32]	2015	203	18994	Youtube	-various-

4.1 UT-Interaction Dataset [28]

This dataset contains videos of 6 classes of human-human interactions. Ground truth labels for these interactions are provided, including time intervals and bounding boxes. It is the most commonly used for activity prediction which contains high-level human activates of multiple actors and has sufficient temporal durations. However it has its own drawback: activities in this dataset executed in a random order without context (e.g., punching and then shaking hand). This is very unnatural and may affect methods relying on modeling causality of different actions.

Figure 3 provides a snapshot of this dataset and Table 2 shows performance comparison when observed half and full activity videos.

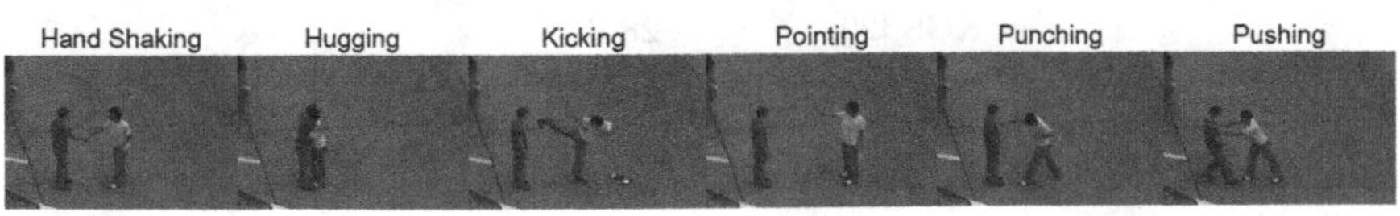

Fig. 3. Example of UT-Interaction dataset

Table 2. Performance comparison on UT-Interaction dataset

Method	Half observation (%)	Full observation (%)
Dynamic BoW [2]	70	85
Integral BoW [2]	65	81.7
Cuboid+SVMs	31.7	85
Kong [9]	78.33	95
Hierarchical movemes [24]	83.1	88.4
Ke [13]	83.33	0
Xu [8]	70	80
Poselet [11]	73.3	93.3

4.2 CAD-120 Dataset [29]

This dataset has 120 RGB-D videos of four different subjects performing 10 high-level activities. It has been annotated with object affordance, sub-activity, ground-truth object category, tracked object bounding box, and human skeleton. Figure 4 provides a snapshot of this dataset and Table 3 shows performance comparison when the system anticipates activity 3 s in the future based on micro P/R metrics [1].

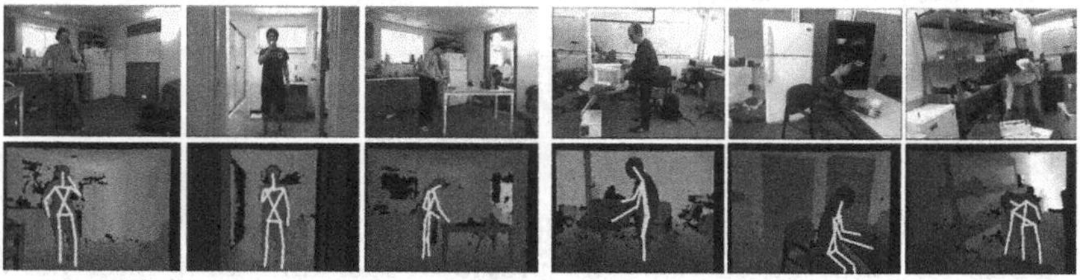

Fig. 4. Example of CAD-120 dataset

Table 3. Performance comparison on CAD-120 dataset

Method	Accuracy %
Random	10.0
Nearest neighbor	22.0
KGS [29]	28.6
ATCRF full [1]	49.6

4.3 BIT-Interaction Dataset [30]

This dataset consist of 8 activity types which are captured in a realistic environment with clutter background and a set of challenges including variation in subject appearance and behavior, scale, illumination condition and viewpoints. This dataset provides bounding box annotations about the human activity region. Figure 5 provides a snapshot of this dataset and Table 4 shows performance comparison when observed half and full activity video.

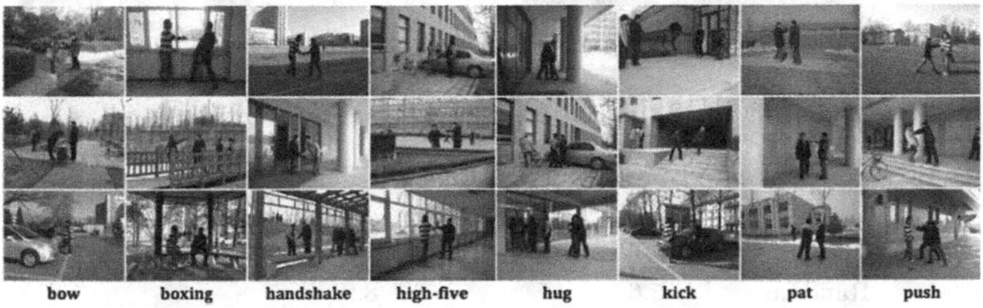

Fig. 5. Example of BIT-Interaction dataset

Table 4. Performance comparison on BIT-Interaction dataset

Method	Half observation (%)	Full observation (%)
Linear SVM	56.25	64.06
Dynamic BoW [2]	46.88	53.13
SC [6]	47.3	67.9
MSSC [6]	48.4	68
MTSSVM [33]	60	76.56
MMAPM [9]	67.96	79.69
LSTM Ranking loss fusion [13]	77.3	85.2

4.4 JPL First-Person Interaction Dataset [31]

This is the first-person dataset contains videos of interactions between humans and the observers. The observers, in this case, will directly involve in the interaction. Therefore, this kind of dataset is a good benchmark for methods related to robotics because visual input recording by a robot naturally is the first-person point of view. Annotations of this dataset consist of time intervals of activities and their corresponding activity class.

Figure 6 provides a snapshot of this dataset and Table 5 shows performance comparison when observed half and full activity video.

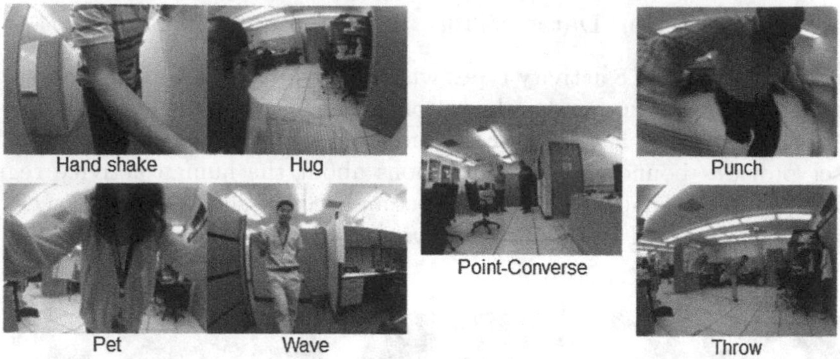

Fig. 6. Example of JPL First-Person Interaction dataset

Table 5. Performance comparison on JPL First-Person Interaction dataset

Method	Half observation (%)	Full observation (%)
Random	5.6	8.9
Bayersian	25.4	42.7
SVM	26.3	62.0
Integral BoW	28.0	41.8
Ryoo [31]	36.0	71.7
Ryoo [17]	47.3	72.9

4.5 ActivityNet [32]

This dataset contains 28k videos of 203 activity categories collected from YouTube. This is a very challenging dataset. A single video may contain multiple activities and periods with none of the annotated activities. This dataset is captured in uncontrolled environments, and variances within the same activity

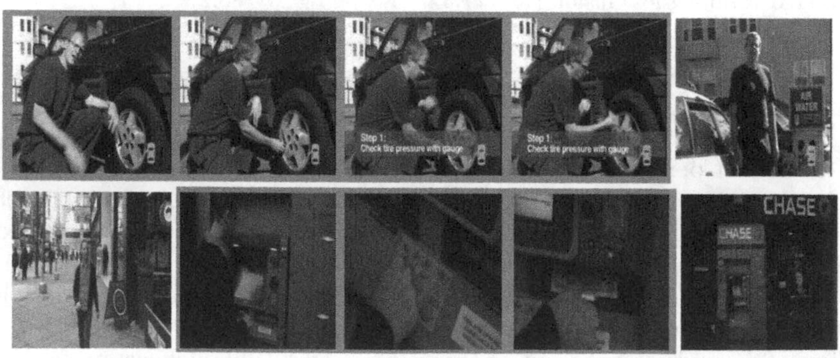

Fig. 7. Example of ActivityNet dataset (checking type and using ATM) [20]

Table 6. Performance comparison on ActivityNet dataset

Method	Accuracy %
CNN [34]	27.0
LSTM	49.5
LSTM with ranking [20]	55.1

category are often large. The viewpoint and foreground objects may change significantly within the same activity. Given these challenges, it is important that the model learns the progression of activities for accurate activity detection and early detection.

Figure 7 provides a snapshot of this dataset and Table 6 shows performance comparison when only 3/10 of each activity is observed.

5 Conclusion

While human activity recognition attracts a lot of researchers with many promising results, human activity prediction is still a new and unexplored area. In this paper, we review various kinds of activity prediction methods in both feature extraction and activity modeling. Common datasets for this topic and experimental results are also introduced to provide a general picture of this field.

Acknowledgments. This research is funded by Vietnam National Foundation for Science and Technology Development (NAFOSTED).

References

1. Koppula, H.S., Saxena, A.: Anticipating human activities using object affordances for reactive robotic response. IEEE Trans. Pattern Anal. Mach. Intell. **38**(1), 14–29 (2016)
2. Ryoo, M.S.: Human activity prediction: early recognition of ongoing activities from streaming videos. In: 2011 International Conference on Computer Vision, pp. 1036–1043, November 2011
3. Dawn, D.D., Shaikh, S.H.: A comprehensive survey of human action recognition with spatio-temporal interest point (STIP) detector. Vis. Comput. **32**(3), 289–306 (2016)
4. Poppe, R.: A survey on vision-based human action recognition. Image Vis. Comput. **28**(6), 976–990 (2010)
5. Vrigkas, M., Nikou, C., Kakadiaris, I.A.: A review of human activity recognition methods. Front. Robot. AI **2**, 28 (2015)
6. Cao, Y., Barrett, D., Barbu, A., Narayanaswamy, S., Yu, H., Michaux, A., Lin, Y., Dickinson, S., Siskind, J.M., Wang, S.: Recognize human activities from partially observed videos. In: The IEEE Conference on Computer Vision and Pattern Recognition (CVPR), June 2013

7. Li, K., Fu, Y.: Prediction of human activity by discovering temporal sequence patterns. IEEE Trans. Pattern Anal. Mach. Intell. **36**(8), 1644–1657 (2014)
8. Xu, K., Qin, Z., Wang, G.; Human activities prediction by learning combinatorial sparse representations. In: 2016 IEEE International Conference on Image Processing (ICIP), pp. 724–728, September 2016
9. Kong, Y., Fu, Y.: Max-margin action prediction machine. IEEE Trans. Pattern Anal. Mach. Intell. **38**(9), 1844–1858 (2016)
10. Dollar, P., Rabaud, V., Cottrell, G., Belongie, S.: Behavior recognition via sparse spatio-temporal features. In: 2005 IEEE International Workshop on Visual Surveillance and Performance Evaluation of Tracking and Surveillance, pp. 65–72, October 2005
11. Raptis, M., Sigal, L.: Poselet key-framing: a model for human activity recognition. In: The IEEE Conference on Computer Vision and Pattern Recognition (CVPR), June 2013
12. Ding, W., Liu, K., Cheng, F., Zhang, J.: Learning hierarchical spatio-temporal pattern for human activity prediction. J. Vis. Comun. Image Represent. **35**(C), 103–111 (2016)
13. Ke, Q., Bennamoun, M., An, S., Boussaid, F., Sohel, F.: Human interaction prediction using deep temporal features. In: Hua, G., Jégou, H. (eds.) ECCV 2016. LNCS, vol. 9914, pp. 403–414. Springer, Cham (2016). doi:10.1007/978-3-319-48881-3_28
14. Dalal, N., Triggs, B.: Histograms of oriented gradients for human detection. In: 2005 IEEE Computer Society Conference on Computer Vision and Pattern Recognition (CVPR 2005), vol. 1, pp. 886–893, June 2005
15. Wang, H., Yang, W., Yuan, C., Ling, H., Hu, W.: Human activity prediction using temporally-weighted generalized time warping. Neurocomputing **225**(C), 139–147 (2017)
16. Wang, H., Klser, A., Schmid, C., Liu, C.L.: Action recognition by dense trajectories. In: CVPR 2011, pp. 3169–3176 (2011)
17. Ryoo, M.S., Fuchs, T.J., Xia, L., Aggarwal, J.K., Matthies, L.: Robot-centric activity prediction from first-person videos: what will they do to me? In: Proceedings of the Tenth Annual ACM/IEEE International Conference on Human-Robot Interaction, HRI 2015, pp. 295–302. ACM, New York (2015)
18. Bourdev, L., Malik, J.: Poselets: body part detectors trained using 3D human pose annotations. In: 2009 IEEE 12th International Conference on Computer Vision, pp. 1365–1372, September 2009
19. Razavian, A.S., Azizpour, H., Sullivan, J., Carlsson, S.: CNN features off-the-shelf: an astounding baseline for recognition. In: The IEEE Conference on Computer Vision and Pattern Recognition (CVPR) Workshops, June 2014
20. Ma, S., Sigal, L., Sclaroff, S.: Learning activity progression in LSTMs for activity detection and early detection. In: 2016 IEEE Conference on Computer Vision and Pattern Recognition (CVPR), pp. 1942–1950, June 2016
21. Hu, J.-F., Zheng, W.-S., Ma, L., Wang, G., Lai, J.: Real-time RGB-D activity prediction by soft regression. In: Leibe, B., Matas, J., Sebe, N., Welling, M. (eds.) ECCV 2016. LNCS, vol. 9905, pp. 280–296. Springer, Cham (2016). doi:10.1007/978-3-319-46448-0_17
22. Joachims, T., Finley, T., Chun-Nam John, Y.: Cutting-plane training of structural svms. Mach. Learn. **77**(1), 27–59 (2009)
23. Hoai, M., De la Torre, F.: Max-margin early event detectors. Int. J. Comput. Vis. **107**(2), 191–202 (2014)

24. Lan, T., Chen, T.-C., Savarese, S.: A hierarchical representation for future action prediction. In: Fleet, D., Pajdla, T., Schiele, B., Tuytelaars, T. (eds.) ECCV 2014. LNCS, vol. 8691, pp. 689–704. Springer, Cham (2014). doi:10.1007/978-3-319-10578-9_45
25. Chakraborty, A., Roy-Chowdhury, A.K.: Context-aware activity forecasting. In: Cremers, D., Reid, I., Saito, H., Yang, M.-H. (eds.) ACCV 2014. LNCS, vol. 9007, pp. 21–36. Springer, Cham (2015). doi:10.1007/978-3-319-16814-2_2
26. Li, K., Hu, J., Fu, Y.: Modeling complex temporal composition of actionlets for activity prediction. In: Fitzgibbon, A., Lazebnik, S., Perona, P., Sato, Y., Schmid, C. (eds.) ECCV 2012. LNCS, vol. 7572, pp. 286–299. Springer, Heidelberg (2012). doi:10.1007/978-3-642-33718-5_21
27. Vondrick, C., Pirsiavash, H., Torralba, A.: Anticipating visual representations from unlabeled video. In: The IEEE Conference on Computer Vision and Pattern Recognition (CVPR), June 2016
28. Ryoo, M.S., Aggarwal, J.K.: UT-Interaction Dataset, ICPR contest on Semantic Description of Human Activities (SDHA) (2010). http://cvrc.ece.utexas.edu/SDHA2010/Human_Interaction.html
29. Koppula, H.S., Gupta, R. Saxena, A.: Learning human activities and object affordances from RGB-D videos. CoRR, abs/1210.1207 (2012)
30. Kong, Y., Jia, Y., Fu, Y.: Learning human interaction by interactive phrases. In: Fitzgibbon, A., Lazebnik, S., Perona, P., Sato, Y., Schmid, C. (eds.) ECCV 2012. LNCS, vol. 7572, pp. 300–313. Springer, Heidelberg (2012). doi:10.1007/978-3-642-33718-5_22
31. Ryoo, M.S., Matthies, L.: First-person activity recognition: what are they doing to me? In: IEEE Conference on Computer Vision and Pattern Recognition (CVPR), Portland, OR, June 2013
32. Ghanem, B., Heilbron, F.C., Escorcia, V., Niebles, J.C.: ActivityNet: a large-scale video benchmark for human activity understanding. In: Proceedings of the IEEE Conference on Computer Vision and Pattern Recognition, pp. 961–970 (2015)
33. Kong, Y., Kit, D., Fu, Y.: A discriminative model with multiple temporal scales for action prediction. In: Fleet, D., Pajdla, T., Schiele, B., Tuytelaars, T. (eds.) ECCV 2014. LNCS, vol. 8693, pp. 596–611. Springer, Cham (2014). doi:10.1007/978-3-319-10602-1_39
34. Simonyan,K., Zisserman, A.: Very deep convolutional networks for large-scale image recognition. CoRR, abs/1409.1556 (2014)

Using Ontology for Personalised Course Recommendation Applications

Mohammed Essmat Ibrahim[1(✉)], Yanyan Yang[1], and David Ndzi[2]

[1] School of Engineering, University of Portsmouth, Portsmouth, UK
{mohammd.ibrahim,Linda.Yang}@port.ac.uk
[2] School of Engineering and Computing, University of the West of Scotland, Paisley, Scotland
david.ndzi@uws.ac.uk

Abstract. The primary data source for universities and courses for students is increasingly becoming the web, and with a vast amount of information about thousands of courses on different websites, it is quite a task to find one that matches a student's needs. That is why we are proposing the "Course Recommendation System", a system that suggests the course best suited for prospective students. As there has been a huge increase in course content on the Internet, finding the course you really need has become time-consuming, so we are proposing to use an ontology-based approach to semantic content recommendation. The aim is to enhance the efficiency and effectiveness of providing students with suitable recommendations. The recommender takes into consideration knowledge about the user (the student's profile) and course content, as well as knowledge about the domain that is being learned. Ontology is used to both models and represent such forms of knowledge. There are four steps to this: extracting information from multiple sources, applying ontologies by using Protégé tools, semantic relevance calculation and refining the recommendation. A personalised, complete and augmented course is then suggested for the student, based on these steps.

Keywords: Recommendation systems · Semantic web · Ontology · Course selection · Semantic similarity

1 Introduction

When choosing a suitable university course, students need information from many external sources in order to improve their decision-making processes, including the web. The process of choosing a course can be extremely tedious and very complex. As students are required to choose from a wide range of courses, based on a series of decisions and recommendations [1], they frequently find it difficult to find a course that is suitable for them. It is possible to find courses that cover almost every domain of knowledge [2] and each university publishes information about this on their websites.

Such abundant information means that students need to find, organise and use resources that can match their individual goals and interests, as well as their current knowledge. This can be a slow task as it involves accessing each and every platform, searching for available courses, reading each of the course syllabuses carefully, and then choosing the appropriate one.

O. Gervasi et al. (Eds.): ICCSA 2017, Part I, LNCS 10404, pp. 426–438, 2017.
DOI: 10.1007/978-3-319-62392-4_31

There are many online systems that are currently available to find and search for courses [3]. These tools are based on either previous users' knowledge of courses or keyword-based queries. Just because more course information is now offered on university websites, it does not automatically mean that students possess the cognitive ability to evaluate them all. Instead, they are confronted with a problem that is generally termed as "information overloading" [4]. Studies also show that course choice decision is influenced by the student's background, as well as their personal or career interests [1].

By identifying the needs of the students and their areas of interest, it is possible to recommend an appropriate course. It is possible to help them to choose a course by developing methods that will both integrate data from multiple heterogeneous data sources and allow them to rapidly set valuable course-related information. This is based on their own preferences, such as electronic engineering [5].

In order to represent an area of knowledge, an ontology is used that formally describes a list of terms, each representing an important concept, such as classes of objects and the relationships that exist between them [6]. Ontologies provide formal semantics which can be used to process and integrate a range of information on the Internet. Ontology is described by Gruber [7] as an explicit specification of a conceptualisation.

Recommendation systems have recently offered personalised and more relevant recommendations. This is achieved by using information on the basis of situations, such as studying various objects, context and areas of interest, location and careers. For example, courses that are recommended to a student who wishes to work in IT, and is searching for "Computer Networking", will be different to those that are recommended to a student who aims to become an academic member of staff in the same area. This is because both their requirements and the level of education is different. These are treated as contextual data, which has been measured as a major source of the correctness of recommendations [8, 9].

This paper's proposed approach overcomes the overloading problem by using personalised search results. It extracts and integrates information about courses from many different sources, builds ontology mapping of the information and sorts it in the database. As designing ontology is the creation process of a lot of classes and relationships, the user will be able to gain clear knowledge about the course [3]. In this paper, we build a relationship between relevant information on the Internet, including course modules, job opportunities and users' interests. Ontology provides a vocabulary of classes and properties to describe a domain and emphasises the sharing of knowledge [6]. The use of semantic descriptions of the course and learner profiles allows for both qualitative and quantitative reasoning about the matching that is available, as well as the required courses and student interests that are needed to refine the process of deciding which course to take.

This present paper is structured as follows. In Sect. 2, we discuss the previous work that is relevant to this study. Section 3 presents the ontology model in order to express knowledge about the student profile, course content, job content and the domain that is being learned. Section 4 describes the ontology-based semantic recommendation in detail, and Sect. 5 describes the prototype implementation and preliminary results. Finally, Sect. 6 concludes the paper by pointing out the direction of future work.

2 Related Work

Recommendation systems are a promising way to effectively filter out an information overload. These are "software tools and techniques that provide the suggestions for items to be of use to a user" [11]. A variety of techniques have been used to perform a recommendation, such as content-based, collaborative, hybrid and other techniques [10–12]. The attention needed to develop the various recommender systems is still high because there is an abundance of practical applications that can help users to deal with the overload of information and provide a personalised service [13]. The objects that are influenced by recommender systems include a wide spectrum of artefacts, such as books, documents CDs, television programs and movies. Compared with these fields, and the emergence of the education field, course content recommendation is a new topic, which has only been investigated by several systems over the past few years. Many kinds of research into course recommendation systems that aim to help students to find courses that are suitable for them have been carried out [14–16]. Current course recommendation systems collect information from a single data source, including students, university databases, users' course ratings, course histories, past behaviour of students, historical enrolment data and previous students' work histories. The students, however, need to gain a clear knowledge of the relevant course that will meet both their personal needs and career interests.

Recently, a recommendation system and expert system was established that was generally based on domain knowledge and problem-solving methods, such as shared and reused knowledge. The recommendation system and expert system utilised an ontology in order to solve classification, annotation, rendering and to arrange different interpretations that make knowledge representation work efficiently.

We, therefore, proposed an approach that uses the knowledge-based semantic approach to making recommendations to students. We also support recommendation refining. We mainly consider the user's profile context for content recommendation, as we did in their learning goal and prior knowledge. This system extracts information from a number of sources about the content and then discovers semantic matching between the course information and user/student profile.

There are several techniques that have been employed to perform data matching in different applications. Two common measurements used to calculate the similarity of data records for matching are TF-IDF based methods [17] and String edit distance [18]. A support vector machine (SVM) [19] classifier, which has been trained with these similarity measures, is then used to identify instances that refer to the same real entities. This enables us to create semantic relations between different data sources.

3 Ontology Model

We use ontologies in the proposed approach to model knowledge about the course content (course profile), knowledge about the user (student profile) and the domain knowledge (the taxonomy of the domain being learned). Within the domain of knowledge representation, the term ontology refers to both the formal and explicit descriptions of domain concepts. These are frequently conceived as a set of entities, relations, functions, instances and axioms [7]. By enabling the users or contents to share a common understanding of the knowledge structure, ontologies give applications the ability to interpret the context of student profiles and course content features, based on their semantics. In addition, the hierarchical structure of the ontologies allows developers to reuse domain ontologies (for example, in computer science and programming language) to describe learning fields and build a practical model, without starting from scratch.

We constructed three ontologies in the proposed system. These are course ontology, student ontology and job ontology. To test our system, it was decided the domain ontology would be computer network courses. Knowledge, represented by ontologies, can be combined into one single ontology, as shown in this paper.

In addition, knowledge from different ontologies can be combined by merging ontologies. We have shown the merger of two ontologies in this paper: the first ontology is the course and the second is the student profile ontology. The latter contains details of the student developed in the educational domain. Protégé 5.1.0 tools were used to develop and merge by using the ontologies [20]. The course content ontology depicts various contexts about course information, including the course topic, type, duration, level and modules, as shown in Fig. 1.

The user/student ontology includes information about the student, such as personal information, academic information and general information. Refer to more details in [22], as shown in Fig. 2.

The job ontology includes information, such as the job topic, job requirements and location.

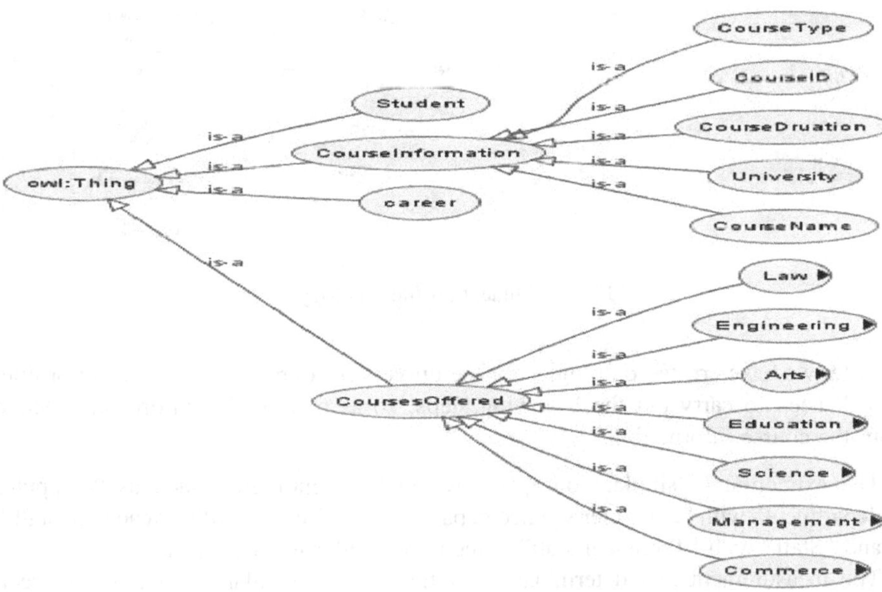

Fig. 1. Course ontology

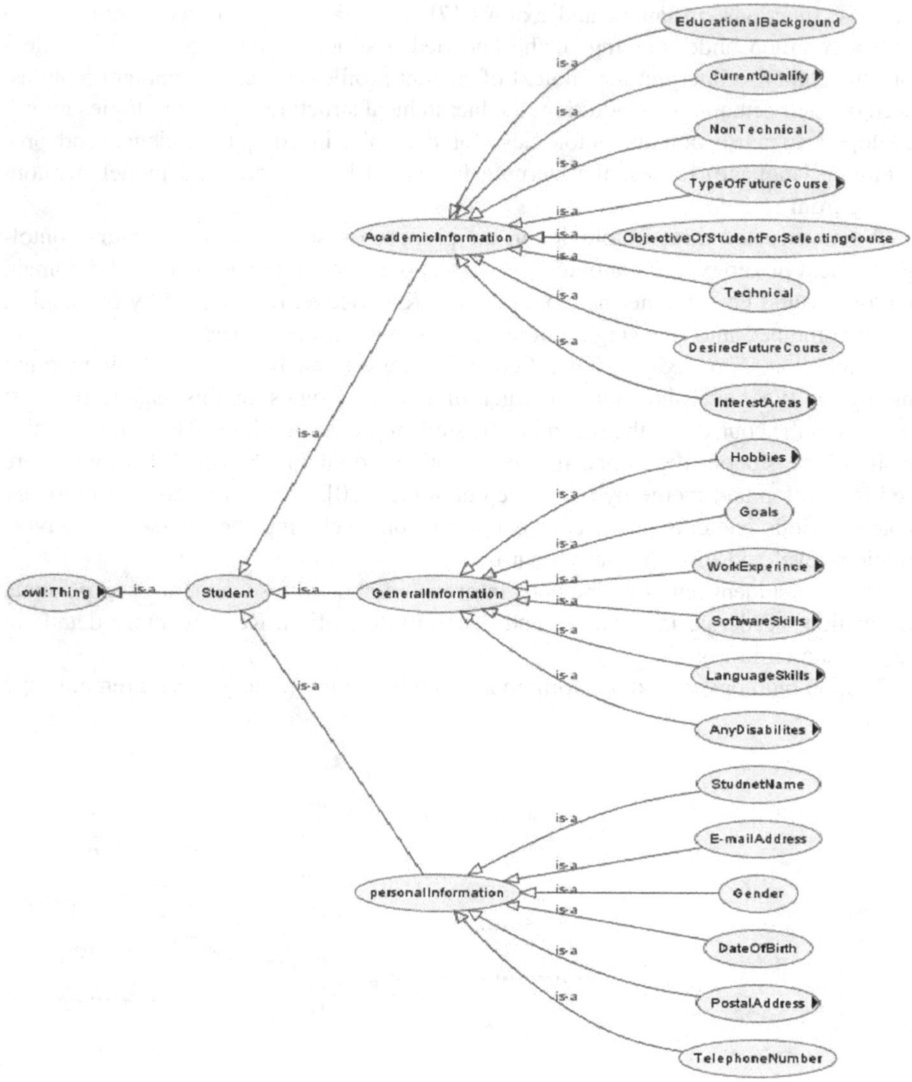

Fig. 2. Student profile ontology

After we have created ontologies for the university course and user/student profile, we will need to carry out the following steps, so as to provide uniform knowledge about the course information:

1. The extraction of similar concepts between two ontologies, such as "computer department" with "computer science department", "Faculty" with "Academic staff", and "Staff" with "Technical staff", which are similar to each other.
2. The measurement and determination of the type of similarity relations between terms. Each approach and algorithm could consider different types of similarity

relations between the terms (such as "Equivalent", "Less general", "More general" or "Overlapping". For example, "computer department" has an equivalent relationship with "computer science department" and "Master Program" has a less-general relationship with "Graduate Program"). This study considered using the String edit distance and TF-IDF methods to measure the similarity relation between the terms.
3. Representation of similar relations between terms. Similarities between terms are formalised in this step. For example, we should represent the similarity relation between "course" and "program" by one formal language. These formal descriptions are from similarity relationships and are called semantic mapping information.
4. Execution of semantic mapping between similar concepts. The concepts, which are similar to each other at this stage, are mapped together. For example, "computer department" is mapped to "computer science department".

Therefore, ontology semantic mapping is a difficult, complex process that requires the execution of an algorithm (for the detection and measurement of similarities), scripting language (for representing mapping information) and tools (for the execution of semantic mapping).

3.1 Ontology Mapping

The proposed system database consists of tables of course information, student profiles and job information. Each table consists of a set of attribute values. The attributes of tables are an RDF node. We define a semantics mapping as a process from a database to an RDF graph, in a final ontology. For example, let C1 be a course in the T1 and C1 be entity courseID, which is a primary key. All of the other attributes, such as course_modules, will be related to courseID if a student S1 has an attribute, such as Main_area_interest. The ontology mapping will be linked to the concepts in the course_title and course_modules, with student_main_area. The relationship between the concepts are based on the subjects' properties; for example, the domain of (has_select) property will be the person or student and the range will be the course. For the (leadTo) property, the domain will be the course and the range will be a job. The ontology representation of the database tables is shown in Fig. 3.

3.2 Ontology Features Matching

The use of ontology allows us to improve the methods that only compute string similarities between ontology instances.

Two ontology features are utilised in the proposed extended method. The first feature is the ontology hierarchy. With an ontology schema, we can compute the subsumption relations between concepts in the ontology schema by using a specific reasoner. A hierarchy of the concepts can then be constructed, which allows us to explore the "concept-level similarity" of instances. The quality and completeness of ontology data varies because different data sources contribute to it separately. There is no guarantee that the instances that refer to the same real-world entity are identified with exactly the same concept by different data sources.

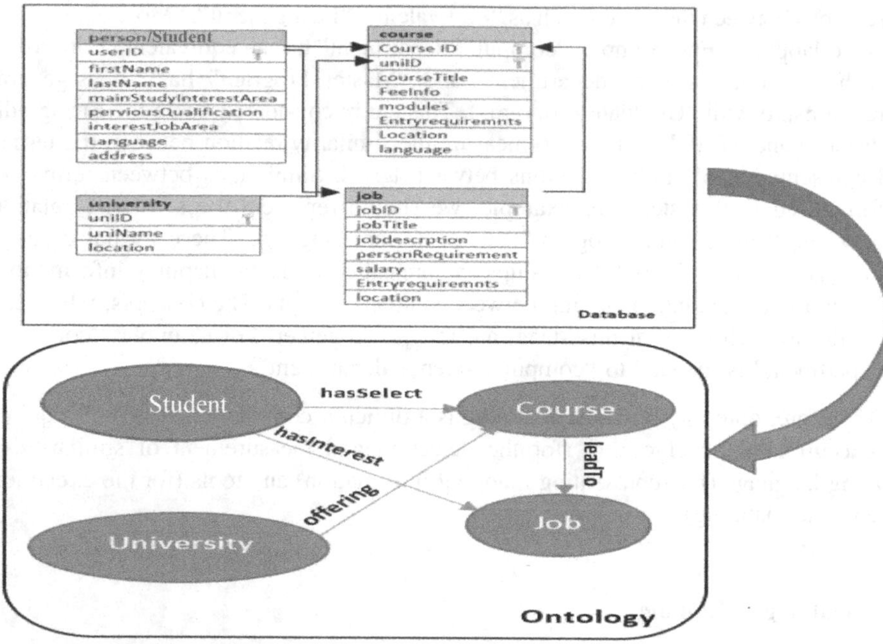

Fig. 3. Ontology represent of database tables

For example, any search for the course base on a "computer network", as a topic on the UCAS website [21], will give over 120 alternative courses that are similar in their topic concepts (including computer network security, Computer Network Technology, Computer Network Administration and Management, Network Computing, Network and Computer Systems Security, among offers). They provide similarity as an area of study, but when each topic was analysed, we found they had different modules or units that influenced the student's job fields. These modules or units were present as instances in ontology, as shown in Fig. 4. We need to share these concepts and define the relations for the modules with similar topics by the ontology.

We define "concept distance" in order to measure concept-level similarity. Suppose that two instances x, y are concept A and B, respectively. This can be referred to as A (x), B(y). The concept distance between x and y, referred to by ConceptDistance (x, y), is defined as follows:

$$ConceptDistance(x,y) = \begin{cases} 0 & A \equiv B \\ P(A,B) & A \sqsubseteq B \, or \, B \sqsubseteq A, \\ P(A,B)+k & A \not\sqsubseteq B, B \not\sqsubseteq A \\ \infty & A \sqcap B = \bot. \end{cases} \quad (1)$$

While P (A, B) means that the length of the path between concept A and B, according to the computed concept hierarchical tree, K is a penalty item and so is always given a positive number. If the concept distance of two instances is bigger, then naturally, the likelihood of it being the same would be less.

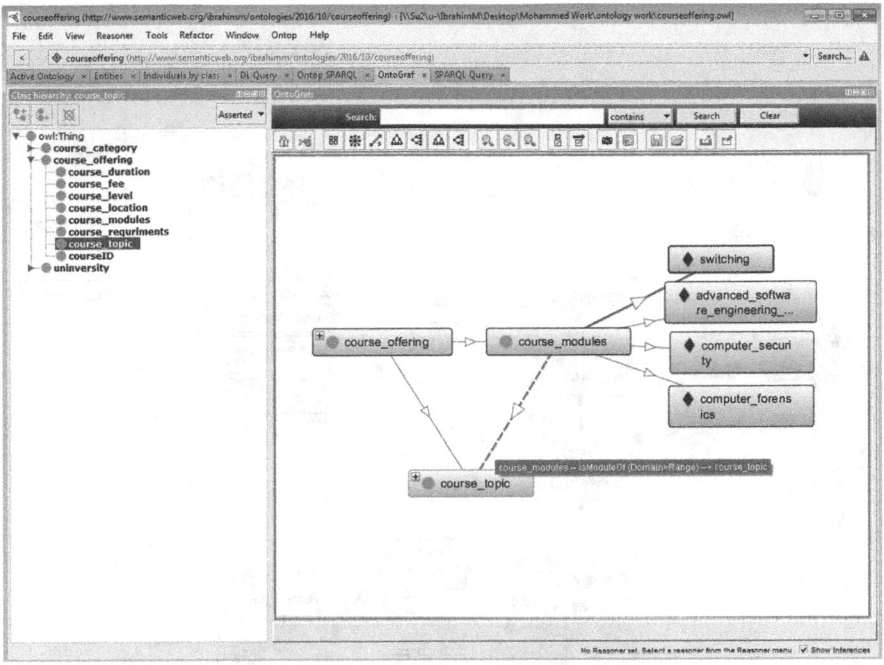

Fig. 4. Ontology represent of course topic and course modules

We also test the object properties of instances so as to compute "context similarity". Object properties enable users to create specific relations between instances. Before an object property is used to link instances with semantic relations, it generally has to be defined between concepts, with an option to specify its cardinality constraints. Moreover, an object property can be an inverse object property, as this allows the use of more flexible ways to describe ontology data. Inverse object properties are often very common among different data sources. For example, we tend to describe course modules that use a property as "has_modules", in order to relate them to their course instances. By reasoning on the inverse properties, and checking the cardinalities on them, we can compute the context similarity between instances.

4 Semantic Course Recommendation Design

In this section, we give an overview of the Course Recommendation System, which provides the user/student with relevant course recommendations. The proposed system contains two main parts – client side and server side, as shown in Fig. 5. The components will be handled sequentially, but iterations are planned, so each component can better accommodate the needs of the next.

The first part, which is the client side, will be implemented with a web interface module. This is responsible for taking users' queries, user information and user interactions (feedback) to the server side. A system interface will be available on a

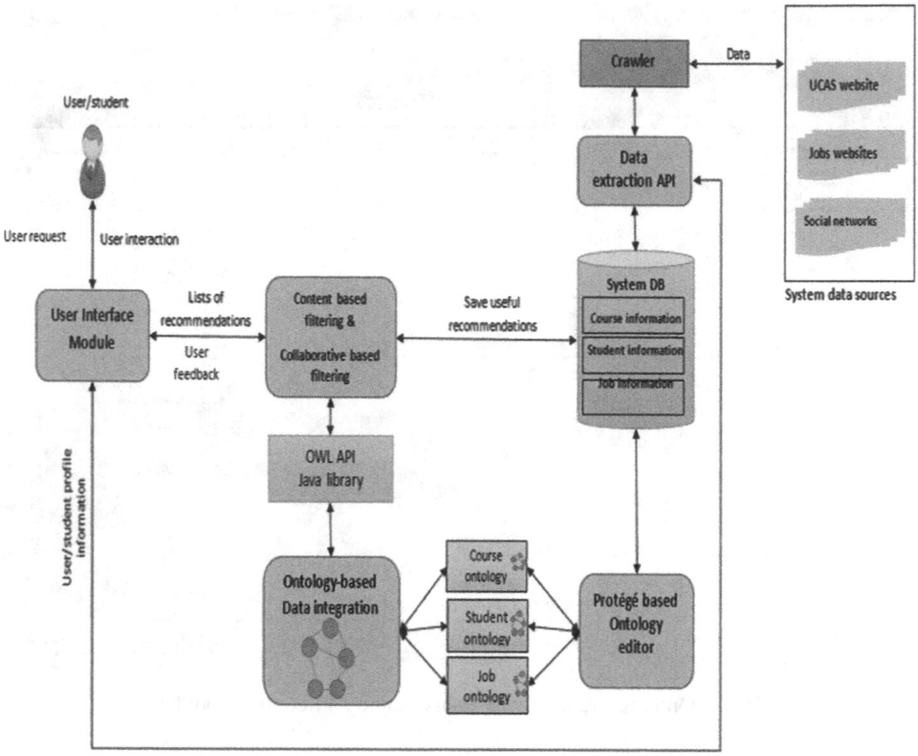

Fig. 5. Course recommendation system design

web-based user interface (WUI), which is transmitted via the Internet and viewed by a web browser program. The system data will take input from different information sources. Automated information extraction techniques will be applied and the results will consist of a list of features for each course, as well as relevant careers. All the information about the courses and users will be stored in a users' and course profiles' storage in the system database.

The second part of the system is the server side, which includes the following components:

- Data extraction API: we built this tool from scratch to extract specific information from UCAS website about postgraduate courses in the United Kingdom universities. At the other hand, this tool extract information about the jobs from job website in the United Kingdom.
- The ontology-based data integration component will gather course information, utilizing the web, through extraction of meta-information about the courses' attributes and will discover how these correlate to specific users' needs. In addition, it includes examining the ability of the new approach to data integration to translate the user' input to specific needs, and find the relationship between course information and different career goals. Along with this it exploit contextual and social data to create a meaningful profile. The data integration will be analyzed to create

(a)

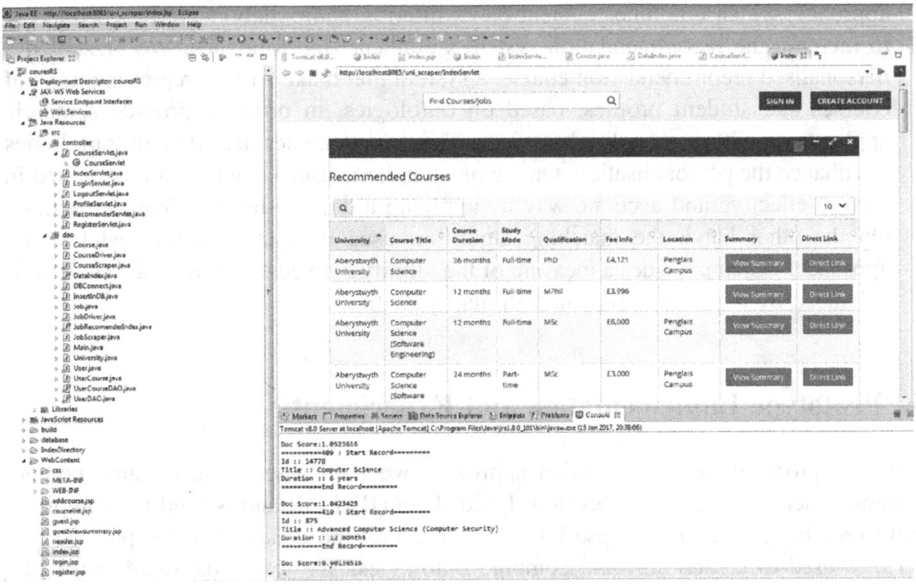

(b)

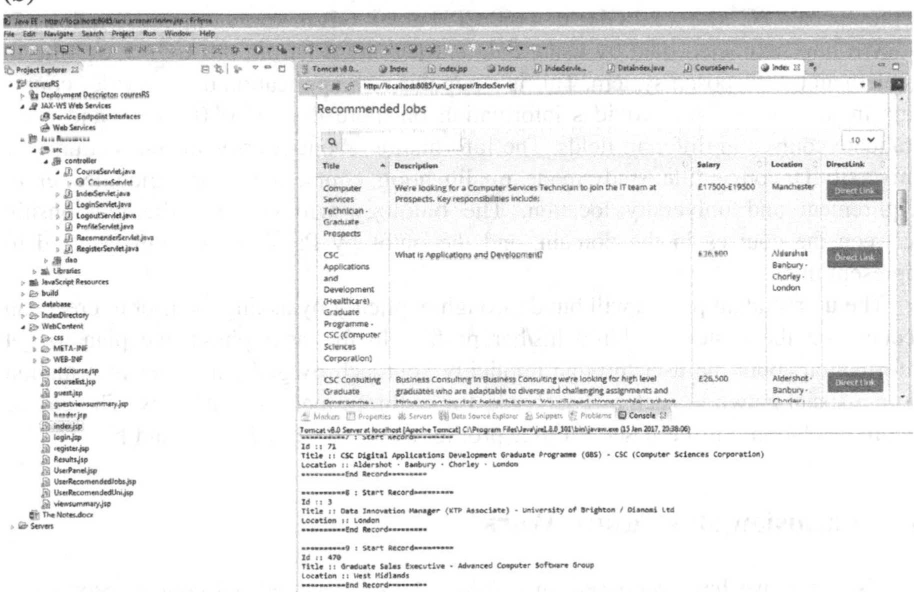

Fig. 6. (a) Initial results for course recommendation. (b) Initial results for relevant job for recommended course

more information about the whole course category, such as the discovery of the most important features, average or common feature values, and feature value to career relationship.

- Recommendation engine component. We developed a hybrid recommender filtering approach, which combines content bases filtering and collaborative based filtering to increase the recommendation system's efficiency performance. We presented a personalised recommendation course, a system that makes use of representations of courses and student profiles, based on ontologies, in order to provide semantic applications with personalised services. The recommender uses domain ontologies to enhance the personalisation. On the one hand, the user's interests are modelled in a more effective and accurate way by applying a domain-based inference method. On the other hand, the matching algorithm used by our content-based filtering approach, which provides a measure of the affinity between an item and a student, is enhanced by applying a semantic similarity method.

5 Prototype Implementation and Experiment

With the proposed recommendation approach, we built a semantic learning content recommender system. It was developed with Java (JDK1.8) and is used for compiling and executing java code. Eclipse IDE is used to edit the code, while the protégé tool 5.1.0 is used to create, edit and combine ontologies. We tested the overhead of the semantic content recommendation in terms of the response time. The experiment was deployed on a PC with 3.20 GHz i5-4460 CPU and 8 GB memory running Windows 7.

We have extracted data by using data extraction API from the UCAS website to implement the proposed system. This is a popular web application that provides course data in the UK. UCAS provides information on more than 78,000 under- and post-graduate courses in different fields. The information about courses includes courseID, universityID, course title, study mode, qualification, course fee, course modules, entry requirement and university location. The ontology will describe the relationship between the courses in the domain, and the ontology Protégé tool will be used to represent this.

The user/student profile will build through explicitly by asking the user to create an account on the system to build his/her profile. In the next phase, we plan to get information about the user through implicitly approach by gathering user information from a social network, such as Facebook, LinkedIn, to mention but a few. The course recommendation will be based on user profile, as depicted in Fig. 6a and b.

6 Conclusion and Future Work

In this paper, we have proposed an ontology-based personalised course recommendation system. The use of ontology can effectively improve the quality of service of a personalised recommendation, and we have also modelled a domain ontology to support semantic interoperation between the student's profile ontology and course ontology. Our experimentation has proved that this ontology-based recommendation approach can improve the recommendation's accuracy. This approach enables e-learning systems to easily reuse and share learning objects that have been published by various systems. It uses specific ontology to infer what course a student should study and what course content a system should look for automatically.

In future, we will enrich our repository by absorbing more course and user information and heterogeneous data sources. We will also further evaluate our approach and compare it with other related methods through simulation experiments that use more perspectives.

References

1. Le Roux, F., Ranjeet, E., Ghai, V., Gao, Y., Lu, J.: A course recommender system using multiple criteria decision making method. In: International Conference on Intelligent Systems and Knowledge Engineering. Atlantis Press, Chengdu (2007). doi:10.2991/iske.2007.238
2. Apaza, R.G., Cervantes, E.V., Quispe, L.C., Luna, J.O.: Online courses recommendation based on LDA. In: SIMBig, pp 42–48. Peru (2014)
3. Wang, S.C., Tanaka, Y.: Topic-oriented query expansion for web search. In: Proceedings of the 15th International Conference on World Wide Web, pp. 1029–1030. ACM, New York (2006). doi:10.1145/1135777.1135999
4. Zhang, Z.K., Zhou, T., Zhang, Y.C.: Personalized recommendation via integrated diffusion on user–item–tag tripartite graphs. J. Phys. A: Stat. Mech. Appl. **389**(1), 179–186 (2010)
5. Huang, C.Y., Chen, R.C., Chen, L.S.: Course-recommendation system based on ontology. In: International Conference Machine Learning and Cybernetics (ICMLC), pp. 1168–1173. IEEE, Tianjin (2013). doi:10.1109/ICMLC.2013.6890767
6. Yang, H., Cui, Z., O'Brien, P.: Extracting ontologies from legacy systems for understanding and re-engineering. In: Computer Software and Applications Conference, pp. 21–26. IEEE, Phoenix (1999). doi:10.1109/CMPSAC.1999.812512
7. Gruber, T.R.: A translation approach to portable ontology specifications. Technical report, Knowledge acquisition (1993)
8. Meehan, K., Lunney, T., Curran, K., McCaughey, A.: Context-aware intelligent recommendation system for tourism. In: Pervasive Computing and Communications IEEE International Conference, pp. 328–331. IEEE, San Diego (2013). doi:10.1109/PerComW.2013.6529508
9. Asabere, N.Y.: Towards a viewpoint of Context-Aware Recommender Systems (CARS) and services. Int. J. Comput. Sci. Telecommun. **4**, 10–29 (2013)
10. Burke, R.: Hybrid recommender systems: survey and experiments. J. User Model. User-Adap. Inter. **12**, 331–370 (2002). doi:10.1023/A:1021240730564. Springer
11. Ricci, F., Rokach, L., Shapira, B.: Introduction to Recommender Systems Handbook. Springer, New York (2011)
12. Balabanović, M., Shoham, Y.: Fab: content-based, collaborative recommendation. J Commun. ACM. **40**, 66–72 (1997). doi:10.1145/245108.245124. New York
13. Adomavicius, G., Tuzhilin, A.: Toward the next generation of recommender systems: a survey of the state-of-the-art and possible extensions. J. IEEE Trans. Knowl. Data Eng. **17**, 734–749 (2005). doi:10.1109/TKDE.2005.99. IEEE
14. Yu, Z., Nakamura, Y., Jang, S., Kajita, S., Mase, K.: Ontology-based semantic recommendation for context-aware e-learning. In: Indulska, J., Ma, J., Yang, Laurence T., Ungerer, T., Cao, J. (eds.) UIC 2007. LNCS, vol. 4611, pp. 898–907. Springer, Heidelberg (2007). doi:10.1007/978-3-540-73549-6_88

15. Tsai, K.H., Chiu, T.K., Lee, M.C., Wang, T.I.: A learning objects recommendation model based on the preference and ontological approaches. In: Advanced Learning Technologies Sixth International Conference, pp. 36–40. IEEE, Kerkrade (2006). doi:10.1109/ICALT.2006.1652359
16. Farzan, R., Brusilovsky, P.: Social navigation support in a course recommendation system. In: Wade, Vincent P., Ashman, H., Smyth, B. (eds.) AH 2006. LNCS, vol. 4018, pp. 91–100. Springer, Heidelberg (2006). doi:10.1007/11768012_11
17. Salton, G., Buckley, C.: Term-weighting approaches in automatic text retrieval. J. Inf. Process. Manag. **24**, 513–523 (1988). doi:10.1016/0306-4573(88)90021-0. Elsevier
18. Gusfield, D.: Algorithms on Strings, Trees and Sequences: Computer Science and Computational Biology. Cambridge University Press, Cambridge (1997)
19. Joachims, T.: Text categorization with support vector machines: learning with many relevant features. In: Nédellec, C., Rouveirol, C. (eds.) ECML 1998. LNCS, vol. 1398, pp. 137–142. Springer, Heidelberg (1998). doi:10.1007/BFb0026683
20. Cardoso, J.: The semantic web vision: where are we? J. IEEE Intell. Syst. (2007). doi:10.1109/MIS.2007.4338499
21. The Universities and Colleges Admissions Service in United Kingdom. https://www.ucas.com/
22. Ameen, A., Khan, K.U.R., Rani, B.P.: Ontological student profile. In: Proceedings of the Second International Conference on Computational Science, Engineering and Information Technology, pp. 466–471. ACM (2012). doi:10.1145/2393216.2393294

Accelerating Docking Simulation Using Multicore and GPU Systems

Everton Mendonça[1(✉)], Marcos Barreto[1], Vinícius Guimarães[2], Nelci Santos[2], Samuel Pita[2], and Murilo Boratto[3]

[1] Laboratório de Sistemas Distribuídos, Universidade Federal da Bahia, Salvador, Brazil
{evertonmj,marcosb}@ufba.br

[2] Laboratório de Bioinformática e Modelagem Molecular, Universidade Federal da Bahia, Salvador, Brazil
{vinicius,nelci,samuel.pita}@ufba.br

[3] Núcleo de Arquitetura de Computadores e Sistemas Operacionais, Universidade do Estado da Bahia, Salvador, Brazil
muriloboratto@uneb.br

Abstract. Virtual screening methodologies have been used to help drug researchers to discover new medicine. The main goal of these methodologies is to help in the docking phase, reducing the vast chemical space (usually referred to have 10^{60} molecules) to a small number that can be more easily processed and tested. The docking phase tests which molecules better interact with a drug target, such as an enzyme or protein receptor. This process is very time consuming, as we need to test all possible combinations. So, hybrid parallel architectures comprised by multicore processors and multi-GPUs can be a suitable approach to this problem, as they reduce the execution time whereas allow for the exploitation of huge libraries of candidate molecules. In this paper, we present a methodology to increase docking performance through the parallelization of the AutoDock tool over multiprocessor and GPU hardware. The results show our multicore implementation achieves a maximum speedup of 8 times, while our GPU implementation reaches a speedup of 35 times and the hybrid implementation provides a maximum speedup of 80 times.

Keywords: High-performance computing · Virtual screening · Bioinformatics · Autodock · CUDA

1 Introduction

Some recent advances in parallel hardware, specially hybrid architectures composed by multicore processors and multi-GPUs, have allowed their exploitation to execute virtual screening codes very efficiently. Virtual screening is a typical drug discovery application in which data sets (or libraries) of small molecules are compared in order to identify those structures which are most likely to couple to a given target, such as an enzyme or protein receptor [7,32].

O. Gervasi et al. (Eds.): ICCSA 2017, Part I, LNCS 10404, pp. 439–451, 2017.
DOI: 10.1007/978-3-319-62392-4_32

Prior the wide availability of hybrid architectures, traditional approaches to virtual screening were extensively based on sequential or small-scale MPI [6] codes. The major limitation of these approaches is the high computational costs to perform pairwise comparisons among candidate molecules, which imposes the use of a limited search space.

Today, it is usual to have computational systems formed by multicore processors and one or more graphics processing units (GPUs) for large-scale molecular dynamics simulations [17]. These systems are heterogeneous in terms of memory types and hierarchies, as well CPU and GPU processing speeds. This heterogeneity introduces new challenges to algorithm design and system software. From a programmer standpoint, the main challenge is to identify routines that can benefit from each type of parallel hardware and how to distribute tasks and data over this combined architecture. From a software perspective, we must count on specialized parallel libraries able to fully exploit the underlying hardware, giving the programmer the ability to build his code in a very abstract way.

In this paper, we propose a methodology for virtual screening representation on heterogeneous multicore and multi-GPU systems, which is based on OpenMP [27] and CUDA [26]. Such methodology provides very fast executions and impressive speedup over this kind of integrated architecture.

This paper is organized as follows: Sect. 2 briefly describes some related work and embeds our research question in the scientific literature. Section 3 explains the virtual screening application and some existing tools. Section 4 describes our parallel implementation. Section 5 presents our experimental results. Conclusions and future works are presented in Sect. 6.

2 Related Work

In this section, we present some related work mainly targeted to the parallelization of AutoDock over different parallel programming libraries. We also briefly discuss some proposals based on AutoDock Vina, which is an alternative to AutoDock targeted to explore multithreading and provide better accuracy.

2.1 Parallel AutoDock

In [15], the authors presented DOVIS, a parallelized pipeline cluster for virtual screening that uses AutoDock. Later [16], the same authors presented DOVIS 2.0, an updated version of their pipeline based on AutoDock 4.0, focusing on accuracy improvement, increased usability, through the compliance with some industry standards, and on better performance, reducing file system operations.

The work presented in [18] describes an approach to migrate AutoDock to NVIDIA GPUs using CUDA. In [29], a FPGA-based implementation to AutoDock [34] is proposed, which is later extended and compared with a GPU-based implementation [30].

VSDocker [31] is a parallel version of AutoDock based on MPI running over Microsoft Windows clusters. In [24], AutoDock is parallelized to OpenMP and MPI simultaneously aiming to achieve a better load distribution and performance. Other proposals based on MPI to improve AutoDock's performance and accuracy are presented in [1,19].

Recently, a parallel version of AutoDock to use CUDA and Intel Xeon Phi coprocessors was presented in [4], with a performance comparison between these two architectures.

2.2 Parallelization of AutoDock Vina

AutoDock Vina [33] was proposed as an alternative and more accurate version of AutoDock. The authors claim its efficacy due to some specialized binding mode predictors and the absence of restrictions related to the size of search space and the input data, as well the number of iterations to be performed during pairwise comparisons to match molecules to their drug targets.

A comparative study on the behavior of AutoDock Vina over different high-performance architectures, such as grids, small clusters, multicore processors, and Hadoop, is presented in [14]. The authors discuss several practical aspects a scientist needs to deal with when choosing an execution platform to run his virtual screening codes.

AutoDock Vina is compared with istar [20], a Web-based virtual screening tool that heavily relies on specialized techniques to predict the number of candidate molecules potentially able to link to a desired drug target. The authors claim that their proposal provides a set of functions commonly absent in other similar tools and is 8 times faster than AutoDock Vina for more than 35 different experiments.

As exposed, there are some approaches that explore parallel implementations of AutoDock using tools such as OpenMP and GPUs, but there are few comparisons involving these two approaches. The same occurs with solutions using OpenMP and CUDA simultaneously. Additionally, sometimes the environments used to execute such codes are not clearly explained, making difficult to reproduce and validate these studies.

Despite the existence of AutoDock Vina and its impressive performance and accuracy results, AutoDock is still considered a *de facto* standard within the Bioinformatics and Pharmaceutical Chemistry domains, with a large community of users and contributors.

Given this context, we decided to concentrate on AutoDock and provide a high accurate and faster code over hybrid parallel architectures. Our methods are able to simultaneously explore OpenMP and CUDA to provide the best data and task distribution.

3 Virtual Screening

Virtual screening can be defined as a methodology applied on pharmaceutical development that uses computational techniques to explore large molecule databases aiming to identify the best ones that bind to a specific target, usually an enzyme or receptor protein [2]. It reduces the number of molecules to an amount that can be manipulated, compared and tested within a laboratory, as the realizable exploratory space (i.e. the chemical universe) is considered very large.

As computational tools evolve, the amount of molecules that can be analyzed increases, allowing researchers to expand the number of iterations when executing screening routines. The greater the number of iterations, the greater the precision of final values. This fact directly affects the accuracy of final results, as well as influences the required time to execute the docking routines.

Considering this fact, the use of parallel approaches can be very helpful to reduce the execution time of virtual screening software. Some previous works, presented in Sect. 2, have implemented AutoDock and other similar software over different libraries (OpenMP, MPI, CUDA) and architectures (FPGA and Intel Xeon Phi), presenting interesting results in terms of performance and accuracy.

3.1 Virtual Screening Tools

There is a considerable number of different virtual screening tools, with many different approaches, focus and methods, such as BLAZE [5], DOCKTHOR [8], and GSA [21]. This paper prioritizes the study, adaptation and use of AutoDock, a free and open source software for virtual screening.

Our choice was based on the fact that AutoDock is a tool widely adopted by pharmaceutical researchers and accredited to provide accurate results. Besides that, it is free and open source, allowing researchers to adapt it according to their needs. Another important fact is that AutoDock is cited in various papers [13] and widely used for virtual screening research, having been adapted to various architectures and different environments.

3.2 AutoDock

AutoDock was developed by the Scripps Research Institute as a free software. It is a set of docking routines used in virtual screening process to predict bound conformations and bind energies between ligands and molecule targets. In this process, molecules presenting potential use as medicine for specific diseases can be selected for further investigation [23].

To allow searches within the massive conformational space of ligands around a specific protein, AutoDock uses a grid-based technique to allow fast analysis of binding energy of trial conformations. During the execution, the target protein is inserted into a grid. Then, a probe atom is sequentially placed in each grid point and the interaction (resulting energy) between this probe atom and the protein target is computed. The resulting value is hold in the grid and is available for search during the docking simulation [23].

The current version of AutoDock uses a semi-empirical free energy field to predict binding free energies of small molecules to macromolecular targets [10]. The force field relies on a comprehensive thermodynamic model that enables the incorporation of intra-molecular energies into the anticipated free energy of binding. This can be performed by evaluating energies for both states (bound and unbound). It additionally incorporates a replacement charge-based desolvation methodology that uses a typical set of atom types and charges.

4 Proposed Implementation

So far, most of the previous approaches to improve AutoDock performance are based on parallel libraries such as OpenMP [27] and MPI. In general, these adaptations are targeted to restricted and controlled computational architectures. For example, in [24], the authors present an adaptation of AutoDock for OpenMP and MPI running on IBM BlueGene/P [11] and IBM Power7 Server [12]. This kind of adaptation to a vendor-specific platform tends to limit the scope of use, posing some difficulties to pharmaceutical researchers and other users in their daily work.

Our goal is to provide an AutoDock version adapted to general-purpose programming libraries and hybrid architectures composed by multicore machines with multiple GPUs. This version should abstract from the users some tasks like data partitioning and data distribution during the execution of AutoDock over a hybrid parallel architecture. To achieve this, some auto-tuning routines are used to automatically calculate how to distribute tasks and data over the CPU and GPU cores available in the system.

Our approach is based on the adaption of AutoDock code to CUDA and OpenMP. Also, we ported AutoDock code to a hybrid architecture that simultaneously uses OpenMP and CUDA, aiming at improving the overall performance of AutoDock.

AutoDock allows users to define the number of iterations to be executed during the docking phase. In addition, AutoDock supports to different search algorithms, like Monte Carlo Simulated Annealing (SA), Genetic Algorithm (GA) and Lamarckian Genetic Algorithm (LGA). In general, LGA has a better performance in comparison with other algorithms and the work presented in [22] shows this superiority. Due to this, our choice was to use a LGA-based version of AutoDock.

The parallel models developed in this work are deterministic, partitioning the AutoDock workload between computational resources, like multicore processors and GPU cards. The AutoDock workload is mainly composed by the execution of LGA algorithm. It is important to emphasize that LGA is a stochastic and non-deterministic algorithm. Because of this, the greater the number of iterations, the greater the probability of finding a good result, although it is not guaranteed. Although unlikely, a good result can be found with few iterations.

4.1 Initial Tests

Before we start AutoDock adaptation, some executions were made with a test data set to investigate AutoDock's performance without modifications. To do this, the number of iterations was duplicated in each execution, ranging from 1 to 256, and the execution time was measured through the *times*() C function. This same approach was used in the other modified versions of AutoDock.

After that, *gprof* [9] was used to determine which parts of the code consume more resources during execution. These executions served as a starting point to guide AutoDock adaptation to parallel architectures. *gprof* determined that the most expensive function was *eintcal*(), that calculates the internal energy of a specific molecule, and *trilinterp*(), that performs a trilinear interpolation. Each function consumes, in average, 40% of the execution time, as depicted in Table 1. So, based on this analysis, we decided to parallelize these two functions in our implementations over CUDA and OpenMP.

Table 1. Most expensive functions of AutoDock.

Function	Execution time (sec.)	Time spent (%)
eintcal()	57.51	38.61
trilinterp()	46.88	31.47
torsion()	6.13	4.12
snorm()	4.95	3.32
RealVector :: *clone*()	4.41	2.96

4.2 OpenMP Implementation

After the initial tests, the application was adapted to use OpenMP and, following the same methodology, had its execution time measured. The tests showed the AutoDock adapted to OpenMP had a better performance.

According to [24], AutoDock uses a timestamp-based random number generator (RNG) that utilizes a deterministic algorithm (IGNLGI), which is part of a C library (RANLIB). OpenMP threads are created simultaneously with a non modified random number. Each thread will receive the same number and this will cause problems during the program execution. To solve this issue, a *thread_id* is included during the RNG step, allowing each thread to have an unique RNG.

After this, the program was parallelized. Moreover, it was necessary to add some critical points with specific OpenMP notation to avoid data interference between threads. The pseudo-code presented in Algorithm 1 shows how the implementation was coded.

Algorithm 1. Autodock – Multicore/OpenMP

```
Read configuration file
Prepares data before program execution
#pragma omp parallel for schedule(static) num_threads(n)
for interation total is not reached do
    Begin execution of docking with the Lamarckian Genetic Algorithm
    for individuals not yet analyzed do
        Begin Global Search
            Performs selection phase
            Performs crossover
            Calculate molecule energy and interpolation (calls to eintcal(), trilinterp() and operator() functions)
            Perform mutation
            Calculate molecule energy and interpolation (calls to eintcal(), trilinterp() and operator() functions)
            Select best individuals
        Begin Local Search
            Performs selection phase
            Performs crossover
            Calculate molecule energy and interpolation (calls to eintcal(), trilinterp() and operator() functions)
            Perform mutation
            Calculate molecule energy and interpolation (calls to eintcal(), trilinterp() and operator() functions)
            Select best individuals
        Analyze docking results
    end for
end for
Execution end and final reports
```

4.3 CUDA Implementation

In the CUDA adaptation of AutoDock, the most expensive functions (*eintcal*() and *trilinterp*()) were implemented as kernels. In this work, we have only one GPU to run the experiments.

Before execution, GPU memory must be allocated to store all data and kernels. Then, the program can proceed. The integer size of each thread block was fixed on 128. The grid size is determined dividing the block size (128) by the population size (defined in a configuration file), rounding up the final result. The pseudo-code presented in Algorithm 2 shows how the implementation was coded.

4.4 Hybrid Architecture Implementation

Finally, a hybrid OpenMP and CUDA version was developed. The intention with this hybrid version, named CUDAMP, was to take advantage of our two previous approaches and use an auto-tuning routine to define how data and tasks can be better distributed through a hybrid architecture aiming better results.

The CUDA version was utilized as a starting point and, following that, the previous OpenMP implementation was replicated in the application. The same parameters and models of the previous implementations were used in this new approach. The pseudo-code presented in Algorithm 3 shows how the implementation was coded.

Algorithm 2. Autodock – GPUs/CUDA

```
Read configuration file
Prepare data before the docking starts
GPU memory allocation
for iteration number is not reached do
    Begin execution of docking with the Lamarckian Genetic Algorithm
    for individuals not yet analyzed do
        Start Global Search
            Performs selection
            Performs crossover
            //calculation of molecule energy and interpolation implemented as CUDA kernels
            Calculate molecule energy and interpolation (calls to eintcal(), trilinterp() and operator() functions)
            Performs mutation
            Calculate molecule energy and interpolation (calls to eintcal(), trilinterp() and operator() functions)
            Select best individuals

        Start Local Search
            Perform selection
            Perform crossover
            //calculation of molecule energy and interpolation implemented as CUDA kernels
            Calculate molecule energy and interpolation (calls to eintcal(), trilinterp() and operator() functions)
            Perform mutation
            Calculate molecule energy and interpolation (calls to eintcal(), trilinterp() and operator() functions)
            Select best individuals

        Analyze docking results
    end for
end for
Execution end and final reports
```

5 Experimental Results

This section presents the experimental results obtained with the following environment:

> **[System** 1] Two Intel Xeon at 2.26 GHz and with 24 GB of DDR3 memory. Each one is a quadcore processor with 12 MB of cache memory. It contains two NVIDIA Tesla C2050 GPUs with 14 stream multiprocessors (SM) and 32 stream processors (SP) each, totalizing 448 cores.

In our experiments, several parameter values were used at installation time to estimate the best values for the algorithm. We do this through an auto-tuning routine [3] able to simulate different data and task distributions over the available number of CPU cores and GPU and find the best system and application parameters.

The available range of CPU cores varies from 1 to 16 in Systems 1, with Intel Hyper-Threading turned on. There are two important observations: (1) the number of CPU cores depends on the problem size, and (2) for each problem size and for various sizes of GPU blocks, we obtain a different optimum value to the execution environment. Being aware of this variability is essential to make good decisions during the selection of optimum parameters.

Table 2 and Fig. 1 show the execution with one thread (denoted by "Sequential"), while "CPU/cores" denotes the use of several CPU threads. The OMP

Algorithm 3. Hybrid Autodock – Multicore/OpenMP and GPU/CUDA

```
Read configuration file
Prepare data before the docking starts
GPU memory allocation
#pragma omp parallel for schedule(static) num_threads(n)
for iteration number is not reached do
    Begin execution of docking with the Lamarckian Genetic Algorithm
    for individuals not yet analyzed do
        Start Global Search
            Performs selection
            Performs crossover
            //calculation of molecule energy and interpolation implemented as CUDA kernels
            Calculate molecule energy and interpolation (calls to eintcal(), trilinterp() and operator() functions)
            Performs mutation
            Calculate molecule energy and interpolation (calls to eintcal(), trilinterp() and operator() functions)
            Select best individuals

        Start Local Search
            Perform selection
            Perform crossover
            //calculation of molecule energy and interpolation implemented as CUDA kernels
            Calculate molecule energy and interpolation (calls to eintcal(), trilinterp() and operator() functions)
            Perform mutation
            Calculate molecule energy and interpolation (calls to eintcal(), trilinterp() and operator() functions)
            Select best individuals

        Analyze docking results
    end for
end for
Execution end and final reports
```

Table 2. Execution time (sec) for different parameter values.

Iterations	Sequential	CPU/Cores	1GPU	CPU/Cores + 1GPU
1	15.97	**13.50**	37.26	42.12
2	30.33	**13.20**	37.72	35.26
4	60.25	**10.17**	38.40	33.79
8	119.66	**09.98**	39.70	34.27
16	237.61	**09.48**	43.01	41.94
32	473.04	76.79	**48.16**	**42.30**
64	958.89	295.24	**59.91**	**42.82**
128	1995.260	649.78	**82.85**	**43.42**
256	3807.390	1314.63	**129.04**	**44.56**

version distributes the calculation among threads and each thread runs exclusively on a CPU core. Version denoted by "1GPU" represents executions in one single GPU. The hybrid model ("CPU/Cores + 1GPU") uses all cores available in the heterogeneous system. In this model, the threads are executed by all machine elements, which correspond to the available number of CPU cores and one GPU.

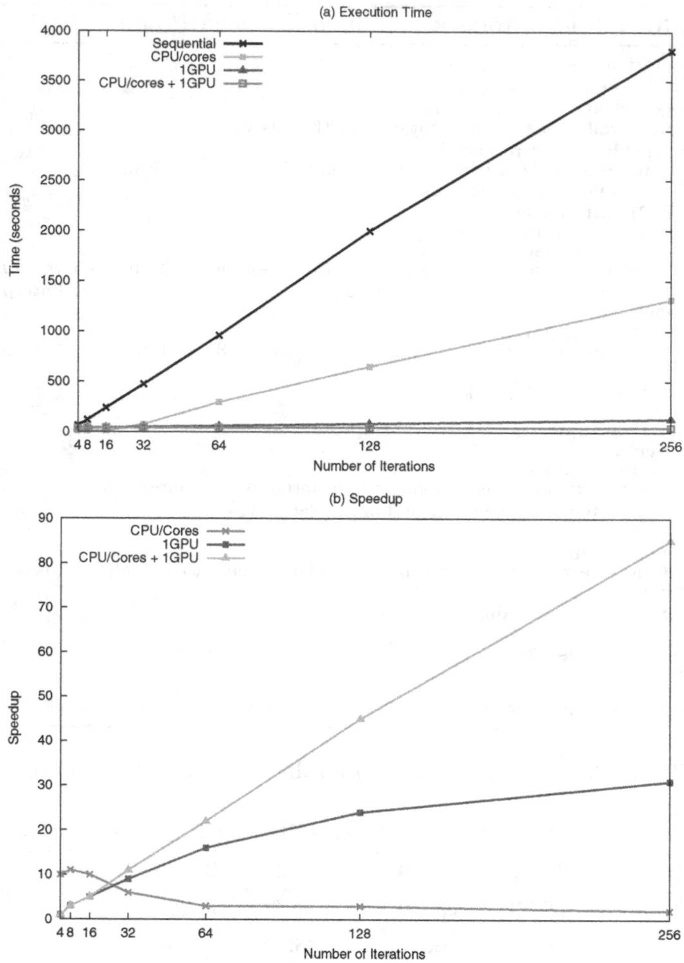

Fig. 1. Execution time (a) and speedup (b) of virtual screening on system 1.

We show in both plots of Fig. 1 the execution time and the speedup, respectively, for virtual screening with different number of iterations ranging from 4 to 256 in System 1. Executions were carried out independently for comparison purposes. Speedup was obtained considering the use of the sequential subsystem only.

The installation time spent on both platforms was the same (around 200 min). We did experiments with several combinations of parameters, considering small samples to obtain the model for a given platform.

Results show that the parallel CPU algorithm reduces the execution time significantly. Until 16 iterations, Sequential and CPU/Cores implementations outperform 1GPU and CPU/Cores + 1GPU. This fact occurs due to GPU memory allocation needed before the start of the docking process start. As we can observe in Fig. 1, the maximum speedup is around 8, matching the number of cores.

The performance of 1GPU is larger than the performance of "CPU/cores". After 32 iterations, 1GPU and CPU/Cores + 1GPU perform better than the other implementations. The best result was obtained with every resource available in the heterogeneous system. Speedup increases with the problem size reaching the theoretical maximum of 80, a number that has been obtained by comparing the computational power of one GPU against the CPU. The use of the GPU as a standalone tool provides benefits but does not allow to reach the potential performance that could be obtained by adding more GPUs and/or the CPU subsystem.

The results presented here in this section can be reproduced by adapting AutoDock, that is a open source software, following the models and algorithms presented in the Sect. 4, where the proposed implementation is presented. Besides this, the molecule datasets can be obtained in different sources like ZINC [35], NuBBE [25] and PDB [28]. It is important to emphasize that due to the stochastic nature of LGA algorithm, discussed in more detail in Sect. 4, the times obtained may have a small variation. However, this difference does not change the tendency of obtained times.

6 Conclusions and Future Directions

This paper presents a methodology to increase the performance of the docking phase of AutoDock, a tool widely used by pharmaceutical researchers to analyze energy bindings between molecules.

To achieve the desired goals, we ported AutoDock to multicore processors running OpenMP and to CUDA-based GPUs. We also developed a hybrid implementation combining these both versions. Prior to all these implementations, some tests with AutoDock were made to understand how docking occurs and to define the routines to be parallelized.

From our results, we can observe that the hybrid solution is 80 times faster than the sequential implementation indicating it is efficient and scalable.

As future directions, an hybrid implementation with more GPUs can be made to allow for even more scalability in terms of molecular data sets to be analyzed, thus increasing the chances to find better and more accurate results.

References

1. Atilgan, E., Hu, J.: Efficient protein-ligand docking using sustainable evolutionary algorithms. In: 2010 10th International Conference on Hybrid Intelligent Systems, HIS 2010, pp. 113–118 (2010)
2. Bohacek, R.S., McMartin, C., Guida, W.C.: The art and practice of structure-based drug design: a molecular modeling perspective. Med. Res. Rev. **16**(1), 3–50 (1996)
3. Boratto, M., Alonso, P., Giménez, D., Barreto, M.: Auto-tuning methodology to represent landform attributes on multicore and multi-GPU systems. In: Proceedings - 13th Symposium on Computing Systems, WSCAD-SSC 2012, pp. 9–16 (2012)

4. Cheng, Q., Peng, S., Lu, Y., Zhu, W., Xu, Z., Zhang, X.: mD3DOCKxb: a deep parallel optimized software for molecular docking with Intel Xeon Phi coprocessors. In: 2015 15th IEEE/ACM International Symposium on Cluster, Cloud and Grid Computing (Mic), pp. 725–728 (2015)
5. Cresset: Blaze: effective ligand-based virtual screening to dramatically increase your wet screening hit rate at a fraction of the cost (2015). http://www.cresset-group.com/products/blaze/
6. Message Passing Interface Forum: MPI: a message-passing interface standard, version 3. http://www.mpi-forum.org/
7. Galvez-Llompart, M., Zanni, R., Garcia-Domenech, R.: Modeling natural anti-inflammatory compounds by molecular topology. Int. J. Mol. Sci. **12**(12), 9481–9503 (2011)
8. GMMSB/LNCC: Dockthor: a receptor-ligand docking program (2015). http://dockthor.lncc.br/
9. GNU gprof: GNU gprof (2016). https://sourceware.org/binutils/docs/gprof/
10. Huey, R., Morris, G.M., Olson, A.J., Goodsell, D.S.: A semiempirical free energy force field with charge-based desolvation. J. Comput. Chem. **28**(6), 1145–1152 (2007). http://dx.doi.org/10.1002/jcc.20634
11. IBM: Ibm bluegene. http://www-03.ibm.com/ibm/history/ibm100/us/en/icons/bluegene/
12. IBM: Ibm power7. http://www-03.ibm.com/systems/power/hardware/775/index.html
13. The Scripps Research Institute: Autodock (2015). http://autodock.scripps.edu/
14. Jaghoori, M.M., Bleijlevens, B., Olabarriaga, S.D.: 1001 Ways to run AutoDock vina for virtual screening. J. Comput.-Aided Mol. Des. **30**, 237–249 (2016)
15. Jiang, X., Kumar, K., Wallqvist, A., Reifman, J.: DOVIS: A Tool for High-Throughput Virtual Screening. In: 2007 DoD High Performance Computing Modernization Program Users Group Conference 2007, pp. 421–424 (2007)
16. Jiang, X., Kumar, K., Hu, X., Wallqvist, A., Reifman, J.: DOVIS 2.0: an efficient and easy to use parallel virtual screening tool based on AutoDock 4.0. Chem. Cent. J. **2**, 18 (2008)
17. Jung, J., Naruse, A., Kobayashi, C., Sugita, Y.: GPU acceleration and parallelization of GENESIS for large-scale molecular dynamics simulations. J. Chem. Theory Comput. **12**, 101–108 (2016)
18. Kannan, S., Ganji, R.: Porting Autodock to CUDA. IEEE Congress on Evolutionary Computation, pp. 1–8, October 2010
19. Khodade, P., Prabhu, R., Chandra, N., Raha, S., Govindarajan, R.: Parallel implementation of AutoDock. J. Appl. Crystallogr. **40**(3), 598–599 (2007)
20. Li, H., Leung, K.S., Ballester, P.J., Wong, M.H.: istar: a web platform for large-scale protein-ligand docking. PLoS ONE **9**(1), 1–12 (2014)
21. LSMC: GSA: Stochastic dynamics through generalized simulated annealing (2015).http://www.cursosvirtuais.pro.br/gsa/
22. Morris, G., Goodsell, D., Halliday, R., Huey, R., Hart, W., Belew, R., Olson, A., Al, M.: Automated docking using a Lamarckian genetic algorithm and an empirical binding free energy function. J. Comput. Chem. **19**, 1639–1662 (1998)
23. Morris, G., Huey, R.: AutoDock4 and AutoDockTools4: automated docking with selective receptor flexibility. J. Comput. Chem. 30(16), 2785–2791 (2009)
24. Norgan, A.P., Coffman, P.K., Kocher, J.P.A., Katzmann, D.J., Sosa, C.P.: Multi-level parallelization of autodock 4.2. J. Cheminformatics **3**(1), 1–9 (2011)
25. NuBBE: Nubbe - núcleo de bioensaios, biosíntese e ecofisiologia de produtos naturais (2016). http://nubbe.iq.unesp.br/portal/index.html

26. NVIDIA: Parallel computation platform CUDA (2015). http://www.nvidia.com/object/cuda_home_new.html
27. OpenMP: The openmp api specification for parallel programming (2015). http://openmp.org/wp/
28. RCSB PDB - 1BZL: Crystal structure of trypanosoma cruzi (2016). http://www.rcsb.org/pdb/explore/explore.do?structureId=1BZL
29. Pechan, I., Feher, B., Berces, A.: FPGA-based acceleration of the AutoDock molecular docking software. 2010 Conference on Ph.D. Research in Microelectronics and Electronics (PRIME) (2010)
30. Pechan, I., Feher, B.: Molecular docking on FPGA and GPU platforms. In: 2011 21st International Conference on Field Programmable Logic and Applications, pp. 474–477 (2011)
31. Prakhov, N.D., Chernorudskiy, A.L., Gainullin, M.R.: VSDocker: a tool for parallel high-throughput virtual screening using AutoDock on Windows-based computer clusters. Bioinformatics **26**(10), 1374–1375 (2010)
32. Rester, U.: From virtuality to reality - virtual screening in lead discovery and lead optimization: a medicinal chemistry perspective. Curr. Opin. Drug Discov. Devel. **11**(4), 559–568 (2008)
33. Trott, O., Olson, A.J.: AutoDock Vina: improving the speed and accuracy of docking with a new scoring function, efficient optimization and multithreading. J. Comput. Chem. **31**(2), 455–461 (2010)
34. Xilinx: FPGA - field programmable gate array (2016). http://www.xilinx.com/training/fpga/fpga-field-programmable-gate-array.htm
35. ZINC: Zinc - a free database of commercially-available compounds for virtual screening (2015).http://zinc.docking.org/

Separation Strategies for Chvátal-Gomory Cuts in Resource Constrained Project Scheduling Problems: A Computational Study

Janniele Aparecida Soares Araujo[1,2(✉)] and Haroldo Gambini Santos[1]

[1] Computing Department, Federal University of Ouro Preto, Ouro Preto, Brazil
jannielе@decsi.ufop.br, haroldo@iceb.ufop.br
[2] Computing and Systems Department, Federal University of Ouro Preto, João Monlevade, Brazil

Abstract. The Resource Constrained Project Scheduling Problems is a well-known $\mathcal{NP}$-hard combinatorial optimization problem. A solution for RCPSP consists in allocating jobs by selecting execution modes and respecting precedence constraints and resource usage. One of main challenges that exact linear-based programming solution approaches currently face is that compact usually provide weak lower bounds. In this paper we propose use of general purpose Chvátal-Gomory cuts to strengthen the LP-based bounds. We observed that by using proper cut separation strategies, the produced bounds can compete with or improve bounds with those obtained with problem specific cuts.

Keywords: Chvátal-Gomory cuts · RCPSP · Project scheduling problem · Cutting plane algorithms

1 Introduction

This article discusses automatic Integer Programming reformulation strategies for non-preemptive Resource Constrained Project Scheduling Problems (RCPSPs) and its variants. From a theoretical point of view they are $\mathcal{NP}$-Hard [2] and are problems with significant academic and practical importance.

The most simplest RCPSP is the single-mode resource-constrained project scheduling problem (SMRCPSP), that involves only one processing mode for each job and a solution consists in allocating jobs respecting precedence and resource usage constraints. A generalization of SMRCPSP is the multi-mode resource-constrained project scheduling problem (MMRCPSP). In this version, it is possible to choose between different consumption-duration relationships to process the jobs. The most recent generalization for the MMRCPSP is the multi-mode resource-constrained multi-project scheduling problem (MMRCMPSP), which goes on to handle multiple projects and global resources. It was the subject of the Multidisciplinary International Scheduling Conference: Theory and Applications (MISTA) Challenge 2013.

O. Gervasi et al. (Eds.): ICCSA 2017, Part I, LNCS 10404, pp. 452–466, 2017.
DOI: 10.1007/978-3-319-62392-4_33

Figure 1 represents one instance of the most generalized version (MMR-CMPSP), in which there are two projects represented by each graph. The jobs of these projects are represented by the nodes, the execution modes are represented by the edges, and the direction of the edges represent the precedence relationships between activities. The weights represent the time-consumption, while the resource consumption is not included. In versions MMRCPSP and SMRCPSP, there is only one graph, while specifically for SMRCPSP, there is only a single edge representing a unique mode of execution.

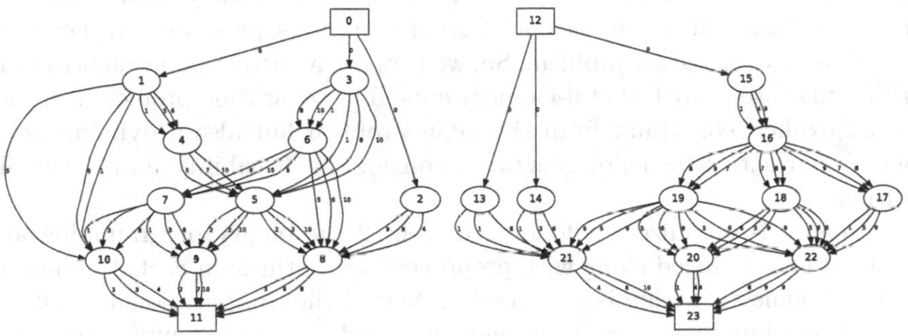

Fig. 1. Representation of one instance (MMRCMPSP)

Beside heuristics, integer programming methods have already been proposed for RCPSPs, but they are only able to solve restricted subsets of instances in feasible processing times. Artigues in [1], studied a set of six formulations for the RCPSP and concluded that the well-known time-indexed formulation pulse disaggregated discrete time (PDDT), which is based on a disaggregated way of modeling the precedence constraints, together with the step disaggregated discrete time (SDDT), where job i starts at time t or before, and the less well-known on/off time-indexed formulation (OODDT) are all equivalent in terms of LP-relaxation and belong to the family of strong time-indexed formulations. While the weak counterparts pulse discrete time (PDT), step discrete time(SDT) and on/off discrete time (OODT), based on an aggregated form of the precedence constraints, belong to a family of weak formulations and are also all equivalent in terms of LP relaxation. In this paper we consider the time-indexed formulation presented in [17] for the MMRCMPSP version, which is based upon the discrete time formulations proposed in [12,16]. Since this problem generalizes other RCPSP problems, computational experiments were also conducted for these more specific problems.

Some IP solution, like implicit enumeration, cuts and hybrids techniques, can be used to improve the limits of the exact models. Branch and bound (B&B) algorithms were presented for the RCPSPs in [3–5,8,14]. Zhu et al. [19] presented a Branch-&-Cut algorithm for the MMRCPSPs. The cuts used are stemmed from resource conflicts, where all resource constraints are in the form of constrained constraints with generalized upper bound (GUB). Cuts from precedence relationships, where for each precedence relationship of jobs (i, j), it can be said that if job i finishes after time t, then activity j can start no earlier than $t + d_{ij} + 1$, and considers d_{ij} the distance between these two jobs in a precedence graph. To make the problem resolution process faster, an adaptive branching scheme is created along with a bound adjustment scheme that is always executed interactively after branching. To optimize the solutions found in the first stage, they use a high-level neighborhood search strategy called Local Branching, these neighborhoods are defined by linear inequalities in the ILP model.

Rounding cuts, such as Chvátal-Gomory, has been widely used in different types of problems and have achieved good results as presented by Letchford et al. [15] for the stable set problem. So, we propose a cutting plane algorithm for RCPSPs that uses only Chvátal-Gomory cuts. The separation problem considers not only explicit constraint from the main problem but also a dynamic set of strong constraints which are generated considering feasibility and optimality conditions.

The paper is organized as follows: in Sect. 2 the binary programming approach used is introduced along with preprocessing routines; in Sect. 3.1 the overall cutting plane algorithm is presented; in Sect. 4 the computational results are presented and finally, in Sect. 5 we conclude and discuss the results of this work.

2 Binary Programming Approach

This section introduces the formulation used in the next sections to apply the cutting plane proposed herein.

The Total Project delay is defined as the difference between the Critical Path Duration (CPD), a theoretical lower bound on the earliest finish time of the project, and the actual project duration (makespan) minus the release date of this project. Wauters et al. [18] proposed the use of the sum of TPDs for all projects as the objective function for the generalized problem of multiple projects. Since for RCPSP variations for single projects (SMRCPSP or MMRCPSP), the TPD can be easily computed from the makespan, all costs presented in this paper will be computed as TPDs.

2.1 Input Data

This section shows the description of the input data used in the models:

- $\mathcal{P}$: projects set;
- $\mathcal{J}$: jobs set;
- $\mathcal{M}_j$: modes set for each job $j \in \mathcal{J}$;
- $\mathcal{J}_p$: jobs set on project p, such that $\mathcal{J}_p \subseteq \mathcal{J}\ \forall p \in \mathcal{P}$;
- $\mathcal{K}$: non-renewable resource set;
- $\mathcal{R}$: renewable resource set;
- $\mathcal{T}$: time horizon for all projects $p \in \mathcal{P}$;

$\mathcal{T}_{jm}$: time horizon for each job $j \in \mathcal{J}$ on mode $m \in \mathcal{M}_j$, defined after preprocessing;

S: immediate successors precedence relationship set between two jobs $j, l \in \mathcal{J}$;

d_{jm}: duration of the job $j \in \mathcal{J}$ on mode $m \in \mathcal{M}_j$;

q_{kjm}: required amount of the non-renewable resource $k \in \mathcal{K}$ to execute the job $j \in \mathcal{J}$ on mode $m \in \mathcal{M}_j$;

q_{rjm}: required amount of the renewable resource $r \in \mathcal{R}$ to execute the job $j \in \mathcal{J}$ on mode $m \in \mathcal{M}_j$;

$\bar{q}^k$: available amount of the non-renewable resource $k \in \mathcal{K}$;

$\bar{q}^r$: available amount of the renewable resource $r \in \mathcal{R}$;

u_p artificial job belonging to the project $p \in \mathcal{P}$, which represents the end of the project;

σ_p: the release date of project p;

2.2 Preprocessing

A critical point to reduce the search space is the definition of tight valid time windows to process jobs. A basic technique to define earliest starting times for jobs δ_j, $\forall j \in J$, consists of computing the Critical Path using the Critical Path Method (CPM) [11]. This computation also provides a lower bound for the completion time of each project. Consider for each project $p \in \mathcal{P}$, the release date σ_p, and the critical path or lower bound λ_p served as input data and the makespan value β_p was obtained from any feasible solution. Then, optimality conditions can be used to restrict the set of valid times where a job can be allocated. Initially consider the value α computed by Eq. 1, that represents an upper bound to the maximum total project delay:

$$\alpha = \sum_{p \in \mathcal{P}} (\beta_p - \sigma_p - \lambda_p) \tag{1}$$

Thus the maximum time window $\bar{t}$ to $\mathcal{T}$, can be obtained by Eq. 2

$$\begin{aligned} \bar{t} &= \max_{p \in P} (\sigma_p + \lambda_p + \alpha) \\ \mathcal{T} &= \{0, \ldots, \bar{t}\} \end{aligned} \tag{2}$$

Analogously, upper bounds can be computed for the processing times of jobs. These upper bounds can be strengthened if the selection of modes with different durations is also considered. A job j from a project p when processed at mode m will push forward all successor jobs in exactly d_{jm} time units. Thus, a set $\overline{S}_j$ containing the entire chain of successors of job j in the dependency path to the dummy completion u_p can be considered. Consider a lower bound $\mathcal{L}_{jm}$ indicating the total duration in this path, computed considering only the fastest processing modes for jobs in this chain. The maximum allocation time for a job j from a project p when processing in mode m $\bar{t}_{jm}$ to $\mathcal{T}_{jm}$ is stipulated by Eq. 3.

$$\begin{aligned} \bar{t}_{jm} &= \sigma_p + \lambda_p - \mathcal{L}_{jm} + \alpha \\ \mathcal{T}_{jm} &= \{0, \ldots, \bar{t}_{jm}\} \end{aligned} \tag{3}$$

Similar lower bounds can be stipulated for any two jobs in this path also considering the fastest processing modes for all jobs except the first one:

$\bar{d}_{jms}$: lower bound for distance between job j and direct or indirect successor job s considering the mode $m \in M_j$.
$\bar{d}^*_{js}$: lower bound for distance between the job j and direct or indirect successor job s considering j fastest mode.

2.3 Formulation

In this section, the Binary Programming (BP) formulation is presented. Binary decision variables are used to select the mode and starting time for jobs. They are defined as follows:

$$x_{jmt} = \begin{cases} 1 & \text{if the job } j \in \mathcal{J} \text{ is allocated on mode } m \in \mathcal{M}_j \text{ at the starting time } t \in \mathcal{T}_{jm} \\ 0 & \text{otherwise} \end{cases}$$

The objective function minimizes the total project delay over the project completion times for projects and its critical path.

Minimize:

$$\sum_{p \in \mathcal{P}} \sum_{m \in \mathcal{M}_{u_p}} \sum_{t \in \mathcal{T}_{u_p m}} [t - (\sigma_p + \lambda_p)] \cdot x_{u_p m t} \quad (4)$$

subject to:

$$\sum_{m \in \mathcal{M}_j} \sum_{t \in \mathcal{T}_{jm}} x_{jmt} = 1 \quad \forall j \in \mathcal{J} \quad (5)$$

$$\sum_{j \in \mathcal{J}} \sum_{m \in \mathcal{M}_j} \sum_{t \in \mathcal{T}_{jm}} q_{kjm} \cdot x_{jmt} \leq \bar{q}^k \quad \forall k \in \mathcal{K} \quad (6)$$

$$\sum_{j \in \mathcal{J}} \sum_{m \in \mathcal{M}_j} \sum_{q=(t-d_{jm}+1)}^{t} q_{rjm} \cdot x_{jmq} \leq \bar{q}_r \quad \forall r \in \mathcal{R}, \forall t \in \mathcal{T} \quad (7)$$

$$\sum_{m \in \mathcal{M}_j} \sum_{t \in \mathcal{T}_{jm}} (t + d_{jm}) \cdot x_{jmt} - \sum_{z \in \mathcal{M}_s} \sum_{u \in \mathcal{T}_{sz}} u \cdot x_{szu} \leq 0 \quad \forall j \in \mathcal{J}, \forall s \in \mathcal{S}_j \quad (8)$$

$$x_{jmt} \in \{0, 1\} \ \forall j \in \mathcal{J}, \ \forall m \in \mathcal{M}_j, \ \ \forall t \in \mathcal{T}_{jm} \quad (9)$$

In this formulation, constraints (5) ensure that each job is allocated for exactly one starting time and mode. Constraints (6) and (7) control the usage of nonrenewable and renewable resources, respectively. The constraints (8) force precedence relationships to be satisfied. Finally, (9) ensures that variables can only assume binary values.

3 The Chvátal Gomory Cut

Chvátal-Gomory cuts are well-known cutting planes for Integer Programming problems. The inclusion of these cuts allows to significantly reduce the integrality gap, even when only rank 1 cuts are employed, i.e. those obtained from original problem constraints [9].

Consider the integer linear programming (ILP) problem as $\min\{c^T x : Ax \leq b, x \geq 0 \text{ integer}\}$, where A is $m \times n$ matrix, $b \in \mathbb{R}^m$, and $c \in \mathbb{R}^n$, along with the two associated polyhedra: where $P := \{x \in \mathbb{R}^n_+ : Ax \leq b\}$ and $P_{\mathcal{I}} := conv\{x \in \mathbb{Z}^n_+ : Ax \leq b\} = conv(P \cap \mathbb{Z}^n)$. Where x is an integer variable, consider $\mathcal{I}$ and $\mathcal{H}$ the set of constraints and variables, respectively.

A Chvátal-Gomory cut [6] is defined as a valid inequality for $P_{\mathcal{I}}$: $\lfloor u^T A \rfloor x \leq \lfloor u^T b \rfloor$, where $u \in R^m_+$ is a CG multiplier vector and $\lfloor \cdot \rfloor$ is the lower integer part. An important factor is the choice of $u \in \mathcal{R}^+$ for deriving useful inequalities, and the strategies to choose them were detailed on Subsect. 3.2. Fischetti and Lodi [9] propose the MIP model for CG separation. The maximally violated inequality can be found by optimizing the following separation MIP model (10–15), where α_j, f_j and u_i are the variables.

Maximize:

$$\sum_{j \in \mathcal{H}(x*)} \alpha_j \cdot x*_j - \alpha_0 \tag{10}$$

subject to:

$$f_j = u^T \cdot A_j - \alpha_j, \ \forall j \in \mathcal{H}(x*) \tag{11}$$

$$f_0 = u^T \cdot b - \alpha_0 \tag{12}$$

$$0 \leq f_j \leq 1 - \delta \ \forall j \in \mathcal{H}(x*) \cup \{0\} \tag{13}$$

$$-1 + \delta \leq u_i \leq 1 - \delta \ \forall i = 1, \ldots, m \tag{14}$$

$$\alpha_j \ \mathbb{Z}, \ \forall j \in \mathcal{H}(x*) \tag{15}$$

where $\mathcal{H}(x*) := \{j \in 1, \ldots, n : x^*_j > 0\}$ and x^* are fractional values. To make the cut stronger, a penalty term $\sum_i w_i \cdot u_i$, where $w_i = 10^{-4}$ for all i, is applied on the objective function. To improve the numerical accuracy of the method, multipliers too close to 1 are forbidden ($u_i \leq 0.99, \forall i$).

3.1 Cutting Plane Algorithm

The cutting plane algorithm is based on the Chvátal-Gomory (CG) cuts proposed at [9], including several customizations to speedup the production of strong valid inequalities for RCPSPs. One important (and simple) problem structure that we explore is a time based decomposition: by defining a time window we can easily select a subset of constraints that share many variables due to the consumption of renewable resources. Thus, violated inequalities for fractional variables related to the processing of jobs in these intervals can be built by selecting the appropriate multipliers for these constraints.

The larger the size of these time windows, the more likely it is that violated inequalities will be found at the cost of a more expensive separation problem. To select time windows of restricted size and still produce violated inequalities, conflicts between different variables in this time window can be included in the separation problem. These conflicts, which can be computed considering feasibility or optimality conditions, are usually not included in the original problem due to the size of this conflict graph. In the separation problem, which is restricted to a time window, we can explicitly include all these conflicts.

The cutting plane algorithm has these main stages:

- identify the time windows with the most violated numbers of variables;
- create and optimize a separation problem that was built considering a subset of constraints and variables from the original model related to this time window, like constraints of modes, renewable and non-renewable resources, as well as strengthen the separation problem with conflict constraints; and,
- identify a violated CG cut.

In Algorithm 1 the overall cutting plane algorithm is described.

Algorithm 1. `cutting_plane`

Data: RCPSP BP Model $\mathcal{M}$, delay η, size interval λ, size jump Λ, percentage bigger to forward the interval σ, reduced cost θ, runtime
Result: object value *obj*, set of cuts $\mathcal{C}$, new model $\mathcal{M}'$, reduced cost θ

1 $\mathcal{I} \leftarrow \emptyset$;
2 **do**
3     $(x^*, o^*) \leftarrow$ `optimize_as_continuous`$(\mathcal{M})$;
4     $\mathcal{I} \leftarrow$ `rows`$(\mathcal{M})$;
5     $\mathcal{U} \leftarrow$ `find_variables_set`(x^*, θ, o^*);
6     $\mathcal{Z} \leftarrow$ `find_constraints`$(x^*, \mathcal{M}, \mathcal{I}, \mathcal{U}, \lambda, \Lambda, \sigma, \eta)$;
7     $\mathcal{C} \leftarrow$ `CG-SEP_cuts`$(x^*, \mathcal{Z}, \mathcal{M})$;
8     $\mathcal{M}' \leftarrow$ `add_cuts_to_model`$(\mathcal{C}, \mathcal{M})$;
9 **while** *(*$|C| > 0$ *or runtime is over)*;
10 *obj* $\leftarrow$ `obj_value`(x^*);
11 **return** *obj*, $\mathcal{M}'$;

At each iteration the linear programming relaxation of our BP formulation is solved and the fractional solution x^* along with reduced costs o^* are computed (line 3). To speed up the generation of violated valid inequalities, a subset $\mathcal{U}$ of variables is computed to be processed by the separation algorithms (line 5). This set contains variables which are non-zero in the current fractional solution and variables with the reduced cost less or equal to the reduced cost threshold θ.

One must observe that the inclusion of these latter variables will never contribute to the production of more violated inequalities. Nevertheless, stronger cuts can be generated when also considering these variables in the separation problem. Thus, the first step will be to find the row set from a set of constraints (line 6) and generate a Chvátal-Gomory cut. We optimize the separation problem until a time limit is reached or, in the case that violated inequalities are already available, a number of non-improvement nodes in the MIP search is processed[1]. This procedure is repeated until the timeout is reached (line 7). This is important because even though many violated inequalities can be discovered in the first nodes, the solver spends a much larger time trying to prove the optimality of the produced cuts. In this case, a better strategy is to include all violated cuts found in this restricted execution and re-optimize the linear program. Finally the cuts found in that separation are added to the model (line 8). These procedures are repeated until no more cuts are found or time is over, the stages are described in the following subsections.

3.2 Strategies to Define Time Window

The larger the set of non-redundant, tight constraints considered in the CG-SEP the more likely it is that violated inequalities will be found. On the other side, large separation problems can be hard to solve and the overall performance of the cutting plane method can degrade. Thus, we considered some strategies to find suitable sets of constraints. In our approach, we first define a time window used to filter the constraint and variable set.

To define the time window, we compute, for different time intervals, the summation of all infeasibilities for the integrality constraints for all variables of that interval. This value is composed by the sum of the nearest integer distance, to indicate how fractional the variables in that interval are. The following parameters were considered in this search: size of interval λ, size of jump Λ to go to the next interval and the percentage σ that allows to forward the interval. The interval only is forward if the violation was bigger than the value plus σ because it is better to find cuts first in the beginning of the window once the variables allocated in the first instants of time are responsible for advancing or delaying the allocations of the others in the preceding graph, then it is better to try to find integer values for the first ones. This procedure is summarized at line (2) of the Algorithm 2 Once the interval is found, it is time to find the important constraint to compose the set for the CG-SEP.

The renewable resources directly impact on the duration of the projects; so, the set is based mainly on this type of constraints (uRR lines 3–7). So in a first moment, all constraints and their variables that are in the interval were found. Also the following constraints from the original problem are included in the separation problem whenever they are related to the current time window: constraints which restrict the choice of only job (uMT lines 8–13) and constraints about capacity of non-renewable resources (uNR lines 14–19). To find

[1] 25 in our experiments.

Algorithm 2. `find_constraints`

Data: fractional solution x^*, RCPSP BP Model $\mathcal{M}$, rows set $\mathcal{I}$, size interval λ, size jump Λ, percentage bigger to forward the interval σ, delay η
Result: Set select rows $\mathcal{Z}$

$\mathcal{V} \leftarrow \emptyset$;
$(start, end) \leftarrow$ `identify_interval`$(x^*, \lambda, \Lambda, \sigma)$;
for $((r_1 \in \mathcal{I})$ & *is_in_interval*$(r_1, start, end))$ **do**
  $type \leftarrow$ `type_row`(C, r_1);
  if $(type = Ineq.\,7)$ **then**
    `add_row`$(\mathcal{Z}, r_1)$;
    $\mathcal{V} \leftarrow \mathcal{V} \cup$ `vars_row`(r_1);
for $(v \in \mathcal{V})$ **do**
  $O \leftarrow jobs(\mathcal{V})$;
  for $(j \in O)$ **do**
    $\mathcal{V}_j \leftarrow$ `vars_jobs`(V, j);
    $r_2 \leftarrow$ `create_constraints`$(\leq, 1)$;
    `add_set_var`$(r_2, \mathcal{V}_j)$; `add_row`$(\mathcal{Z}, r_2)$;
for $(v \in \mathcal{V})$ **do**
  $K \leftarrow$ `nonrenewable_resources`$(\mathcal{V})$;
  for $(k \in K)$ **do**
    $\mathcal{V}_j \leftarrow$ `vars_nonrenewable`(V, k);
    $r_2 \leftarrow$ `create_constraints`$(\leq,$ `capacity`$(k))$;
    `add_set_var`$(r_2, \mathcal{V}_j)$; `add_row`$(\mathcal{Z}, r_2)$;
for $(v_1 \in \mathcal{V})$ **do**
  for $(v_2 \in \mathcal{V})$ **do**
    $j_2 \leftarrow$ `job`(v_2); $j_1 \leftarrow$ `job`(v_1);
    $m_2 \leftarrow$ `mode`(v_2); $m_1 \leftarrow$ `mode`(v_1);
    $t_2 \leftarrow$ `time`(v_2); $t_1 \leftarrow$ `time`(v_1);
    if $(j_2 \in \bar{S}_{j_1})$ **then**
      $w \leftarrow \bar{d}^*_{j_1,j_2} -$ `job_min_dur`(j_1);
      if $(\,(end\_time(j_1) > t_2)$ *or* $(t_2 - end\_time(j_1) < w))$ **then**
        $r_2 \leftarrow$ `create_constraints`$(1, \leq)$;
        `add_var`(v_1, r_2); `add_var`(v_2, r_2); `add_row`$(\mathcal{Z}, r_2)$;
        continue;
    if $(proj(j_1) != proj(j_2)$ & $delay(j_1, m_1, t_1) + delay(j_2, m_2, t_2) > \eta)$ **then**
      $r_2 \leftarrow$ `create_constraints`$(1, \leq)$;
      `add_var`(v_1, r_2); `add_var`(v_2, r_2); `add_row`$(\mathcal{Z}, r_2)$;
return $\mathcal{Z}$;

additional constraints that represent conflicts between variables is a good strategy. A main conflict is analyzed, whereby the time window comprising variables that have precedence relationship and comprised two variables from different projects exceeding the delay are allowed to not lose the optimal solution (uTW and uP lines 20–23). The Algorithm 2 return a set of constraints to be used on CG-SEP (line 34).

4 Computational Results

This section presents the results obtained with the proposed cutting plane algorithm. An Intel ® Core i7-4790 processor with 3.6 GHz and 16 GB of RAM running SUSE Leap Linux was used during the experiments. All algorithms were coded in `ANSI C 99` and compiled with GCC 6.2.1 using flags *-Ofast* and *-flto*. The solver used is GUROBI 7.0.2 [10].

The experiments were executed over the most known and studied instance set, for the Project Scheduling Problem Library - PSPLIB [12] established in mid-1996. In 2013, a new instance set based on the instances of the PSPLIB emerged on MISTA Challenge [18]. The benchmark set for the experiments is composed of open instances for the RCPSPs. For the initial tests, we chose the sets of instances with parameter 13, since these instances are considered relatively difficult (see in [13]) due to their network complexity factor and resource capacity factor when being generated. For the SMRCPSP, these were `j6013, j9013, j2013`; for the MMRCPSP, it was `j3013`; and for the MMRCMPSP, it was `A` series, each set having ten instances.

Different values for the parameters $\lambda = \{10, 50, 100\}$ were tested, in order to identify the ideal size for the time window. Table 1 shows the average relative gaps closed[2] for different benchmark sets considering the best known solutions obtained with the original linear programming relaxations and the CG-SEP cuts.

Table 1. Results for Dual Limit CG-SEP Cuts, with $\lambda = \{10, 50, 100\}$, $\Lambda = 5$, $\sigma = 50\%$ and 3600 s.

Version	Benchmark set	ζ_{10}	ζ_{50}	ζ_{100}
SRCPSP	`j6013 (1:10)`	26,841	**27,269**	26,726
	`j9013 (1:10)`	22,596	**22,648**	22,513
	`j12013 (1:10)`	7,825	**8,307**	8,160
MMRCPSP	`j3013 (1:10)`	23,462	**24,897**	24,889
MMRCMPSP	`A (1:10)`	46,578	48,408	**48,444**

It can be noticed, in Table 1, that for the benchmark sets with instances that have a smaller number of variables, the best results were obtained with a medium size ($\lambda = 50$). However, the benchmark sets with instances that have a greater number of variables the best results were obtained with a larger window

[2] The relative gap closed is computed as follows: given an obtained dual bound $\underline{b}$ and the best known upper bound $\overline{b}$ the relative gap closed is $\underline{b}/\overline{b}$ if $\overline{b} \neq 0$ or zero otherwise.

size ($\lambda = 100$). These results show that our decomposition approach is valid for separating these cuts considering the complete formulation as in [7], but is not always the best option since larger time windows do not necessarily result in better cuts found in restricted times.

Table 2 shows the average relative gaps closed that were obtained with the original linear programming relaxations and with different combinations of cuts in a 3600 seconds time limit. The third column presents relaxation values without cuts, γ indicates results obtained with the traditional cover, Γ with precedence cuts, while Ω indicates results obtained with all the traditional cuts proposed by [19]. To represent the best results of the CG-SEP cuts is used ζ. It is possible to see that for the SRCPSP, the union of the standard cuts proposed by [19] was a little better, but if we compare them individually, CG was better than the traditional cover γ cuts. However, for instances with multi-modes and multi-projects the approaches with CG-SEP cuts performed significantly better than all approaches with traditional cuts. Different values for the parameters $\sigma = \{10\%, 20\%, 50\%\}$ were tested, but the results for the parameters printed in Table 2 were the best.

Table 2. Comparing results for the best Dual Limit CG Cuts with the traditional cuts.

Version	Benchmark set	$\emptyset$	γ	Γ	Ω	ζ
SRCPSP	`j6013 (1:10)`	26,234	26,240	27,898	**27,904**	27,269
	`j9013 (1:10)`	22,283	22,283	24,340	**24,343**	22,648
	`j12013 (1:10)`	7,792	7,792	**11,199**	**11,199**	8,307
MMRCPSP	`j3013 (1:10)`	19,726	19,98	21,080	21,302	**24,897**
MMRCMPSP	`A (1:10)`	42,603	42,665	43,397	43,457	**48,444**

To evaluate which multipliers u were the most useful ones to generate violated inequalities, we computed the number of times that multipliers related for each constraint type were activated to produce violated cuts in the search process see, on Fig. 2.

In Fig. 2, it is possible to see, for all of the benchmark set, the inclusion of the constraints related to distances in the precedence graph were especially useful, being activated quite often. It is important to remark that these constraints are implicit in the complete formulation. Another interesting observation about the u is that, for the constraints of nonrenewable resources, the variable u was activated just for the benchmark seta `A` and `j3013` because the other benchmarks do not use this type of resource. And the variable u was only activated for the constraints that work with the delay of projects for benchmark set A.

The Figs. 3 and 4 show the difference between the convergence on time for the cover cut γ, precedence cut Γ, both Ω and the CG-SEP ζ with $\lambda = \{10, 50, 100\}$, $\Lambda = 5$ and $\sigma = 50\%$, for the specific instances `j3013_9` and `A-4`.

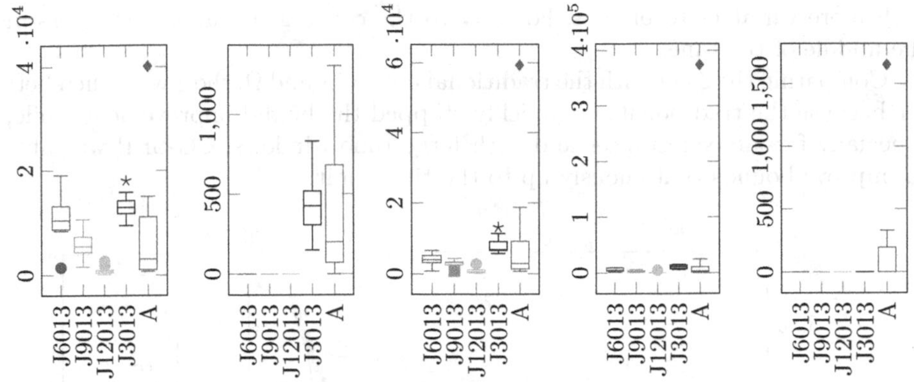

(a) uRR (b) uNR (c) uMT (d) uTW (e) uP

Fig. 2. Numbers of activated u for $\lambda = \{50\}$, $\Lambda = 5$, $\sigma = 50\%$ and 3600 s.

For the instance j3013_9 in Fig. 3 that has a smaller number of variables, it is possible to notice that both ζ with values 50 and 100 have the same convergence behavior.

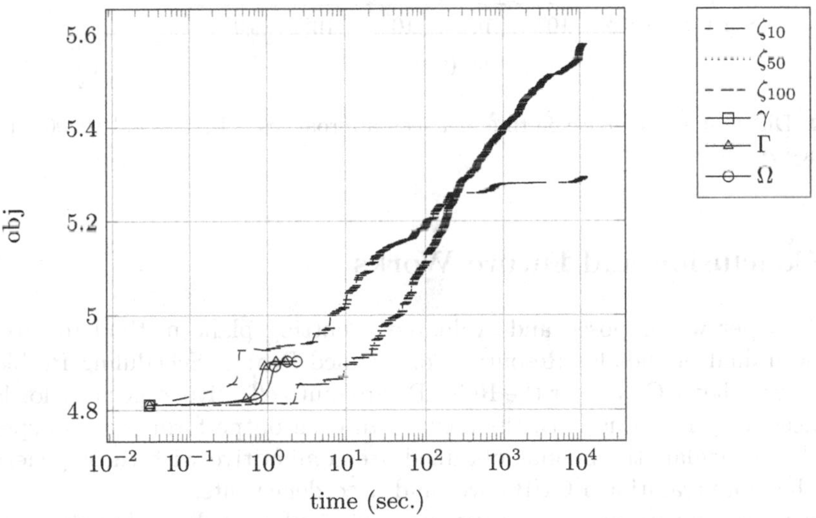

Fig. 3. Different convergence on time between approaches γ, Γ, Ω, ζ and 10000 s to the instance $j3013\_9.mm$

However, in the case of instance A-4 in Fig. 4, which has a greater number of variables, in the beginning $\zeta = 100$ converges a little more quickly, but at a given instant of time the $\zeta = 50$ can surpass the value. Perhaps with a longer time the $\zeta = 100$ can achieve better results, since as it comprises a larger time window with more variables to process. For $\zeta = 10$ the convergence almost staggers in the middle of the time.

Comparing the ζ cut with the traditional cuts λ, Λ and Ω, the ζ was much better because the traditional cuts quickly stopped the bound improvement while, especially for larger instances and with large time windows, CG-SEP was able to improve bounds continuously up to the time limit.

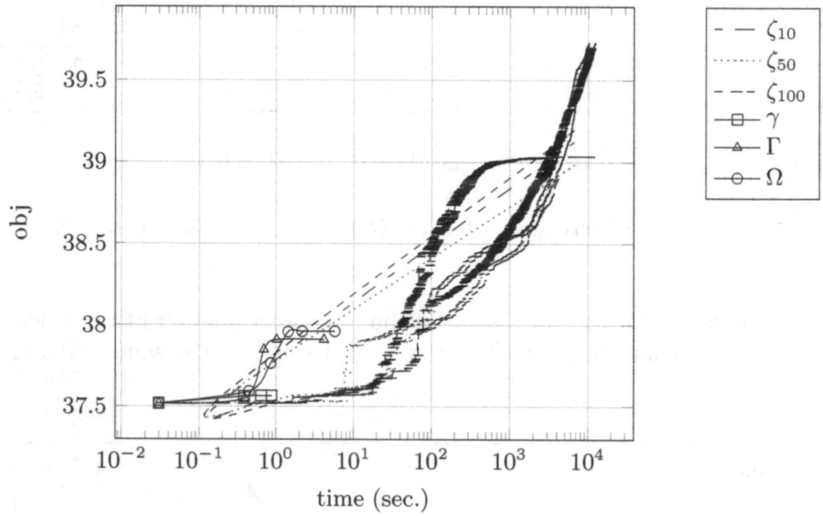

Fig. 4. Different convergence on time between approaches $\gamma, \Gamma, \Omega, \zeta$ and 10000 s. to the instance $A-4$

5 Conclusion and Future Works

In this paper we proposed and evaluated a cutting plane method to produce improved dual bounds for Resource Constrained Project Scheduling Problems. We showed that CG cuts for the RCPSPs are quite effective, especially for large instances. In some situations, these cuts can even outperform problem specific cuts. In particular, the bounds obtained are competitive with those generated with the more traditional GUB cover and precedence cuts.

Even though very positive results were obtained, we believe that our method still has significant room for improvement. Future works can evaluate the contribution and the feasibility of separating CG cuts with a rank greater than one for these problems. Another interesting future work is the inclusion of these cuts in a complete branch-and-cut environment.

Acknowledgements. The authors thank CNPq and FAPEMIG for supporting this research.

References

1. Artigues, C.: On the strength of time-indexed formulations for the resource-constrained project scheduling problem. Oper. Res. Lett. **45**, 154–159 (2017)
2. Blazewicz, J., Lenstra, J., Rinnooy Kan, A.: Scheduling subject to resource constraints: classification and complexity. Discrete Appl. Math **5**, 11–24 (1983)
3. Brucker, P., Knust, S., Schoo, A., Thiele, O.: A branch and bound algorithm for the resource-constrained project scheduling problem1. Eur. J. Oper. Res. **107**(2), 272–288 (1998)
4. Chakrabortty, R.K., Sarker, R.A., Essam, D.L.: Project scheduling with resource constraints: a branch and bound approach. In: International Conference on Computers and Industrial Engineering, vol. 45 (2015)
5. Christofides, N., Alvarez-Valdes, R., Tamarit, J.: Project scheduling with resource constraints: a branch and bound approach. Eur. J. Oper. Res. **29**(3), 262–273 (1987)
6. Chvátal, V.: Edmonds polytopes and a hierarchy of combinatorial problems. Discrete Math. **4**, 305–337 (1973)
7. Chvátal, V., Cook, W., Dantzig, G.B., Fulkerson, D.R., Johnson, S.M.: Solution of a Large-Scale Traveling-Salesman Problem. In: Jünger, M., Liebling, T.M., Naddef, D., Nemhauser, G.L., Pulleyblank, W.R., Reinelt, G., Rinaldi, G., Wolsey, L.A. (eds.) 50 Years of Integer Programming 1958-2008, pp. 7–28. Springer, Heidelberg (2010). doi:10.1007/978-3-540-68279-0_1
8. Demeulemeester, E.L., Herroelen, W.S.: Recent advances in branch-and-bound procedures for resource-constrained project scheduling problems. In: Summer School on Scheduling Theory and Its Applications (1992)
9. Fischetti, M., Lodi, A.: Optimizing over the first Chvátal closure. Math. Program. **110**(1), 3–20 (2007). http://dx.doi.org/10.1007/s10107-006-0054-8
10. Gurobi Optimization Inc.: Gurobi optimizer: Reference manual (2016). http://www.gurobi.com/documentation/7.0/refman.pdf
11. Kelley Jr., J.E., Walker, M.R.: Critical-path planning and scheduling. In: Papers Presented at the 1–3 December 1959, Eastern Joint IRE-AIEE-ACM Computer Conference, pp. 160–173, IRE-AIEE-ACM 1959 (Eastern), NY, USA. ACM, New York (1959)
12. Kolisch, R., Sprecher, A.: PSPLIB - a project scheduling problem library. Eur. J. Oper. Res. **96**, 205–216 (1996)
13. Kolisch, R., Sprecher, A., Drexl, A.: Characterization and generation of a general class of resource-constrained project scheduling problems. Manag. Sci. **41**(10), 1693–1703 (1995). http://www.jstor.org/stable/2632747
14. Land, A.H., Doig, A.G.: An automatic method for solving discrete programming problems. In: Jünger, M., Liebling, T.M., Naddef, D., Nemhauser, G.L., Pulleyblank, W.R., Reinelt, G., Rinaldi, G., Wolsey, L.A. (eds.) 50 Years of Integer Programming 1958–2008, pp. 105–132. Springer, Heidelberg (2010)
15. Letchford, A.N., Marzi, F., Rossi, F., Smriglio, S.: Strengthening Chvátal-Gomory cuts for the stable set problem. In: Cerulli, R., Fujishige, S., Mahjoub, A.R. (eds.) ISCO 2016. LNCS, vol. 9849, pp. 201–212. Springer, Cham (2016). doi:10.1007/978-3-319-45587-7_18
16. Pritsker, A.A.B., Watters, L.J., Wolfe, P.M.: Multi project scheduling with limited resources: a zero-one programming approach. Manag. Sci. **3416**, 93–108 (1969)
17. Toffolo, T.A.M., Santos, H.G., Carvalho, M.A.M., Soares, J.A.: An integer programming approach to the multimode resource-constrained multiproject scheduling problem. J. Sched. **19**(3), 295–307 (2016)

18. Wauters, T., Kinable, J., Smet, P., Vancroonenburg, W., Vanden Berghe, G., Verstichel, J.: The multi-mode resource-constrained multi-project scheduling problem. J. Sched. **19**(3), 271–283 (2016)
19. Zhu, G., Bard, J.F., Yu, G.: A branch-and-cut procedure for the multimode resource-constrained project-scheduling problem. INFORMS J. Comput. **18**, 377–390 (2006)

Data Mining Approach to Dual Response Optimization

Dong-Hee Lee(✉)

College of Interdisciplinary Industrial Studies, Hanyang University,
222 Wangsimniro, Seongdong-Gu, Seoul, Korea
dh@hanyang.ac.kr

Abstract. In manufacturing process optimization, analyzing a large volume of operational data is getting attention due to the development of data processing techniques. One of important issues in the process optimization is a simultaneous optimization of mean and variance of a response variable. It is called dual response optimization (DRO). Traditional DRO methods build statistical models for the mean and variance of the response variable by fitting the models to experimental data. Then, an optimal setting of input variables is obtained by analyzing the fitted models. This model based approach assumes that the statistical model is fitted well to the data. However, it is often difficult to satisfy this assumption when dealing with a large volume of operational data from manufacturing line. In such a case, data mining approach is an attractive alternative. We proposes a particular data mining method by modifying patient rule induction method for DRO. The proposed method obtains an optimal setting of the input variables directly from the operational data where mean and variance are optimized. We explain a detailed procedure of the proposed method with case examples.

Keywords: Dual response optimization · Design of experiments · Patient rule induction method

1 Introduction

The response surface methodology (RSM) consists of a group of techniques used in the empirical study between a response and a number of input variables. The researcher attempts to find the optimal setting for the input variables that either maximizes or minimizes the response (Myers et al. [1]). RSM assumes that that variance of the response is constant and focuses on the optimization of mean of the response. However, the constant variance assumption may not be valid in practice. In such cases, not only the mean response but also the standard deviation response should be considered in determining the optimum conditions for the input variables.

Dual response optimization (DRO) attempts to optimize both mean and variance of the response. Conventional approach to DRO is building statistical models for mean and standard deviation of the response by fitting the models to experimental data. Then, an optimal setting for the input variables is obtained by analysing the statistical models (Lee et al. [2], Lee and Kim [3]).

O. Gervasi et al. (Eds.): ICCSA 2017, Part I, LNCS 10404, pp. 467–477, 2017.
DOI: 10.1007/978-3-319-62392-4_34

This conventional approach may have two practical limitations. First, efforts for conducting experiments can be burden for process engineers. In order to build the statistical models, experiments are conducted and experimental data are gathered. In DRO, usually a large number of experiments are conducted because replicate experiments at each design point are needed to fit the statistical model for standard deviation of the response whereas replicate experiments are not mandatory in conventional RSM. Thus, efforts for conducting many experiments can be burden for process engineers. Second, mean and standard deviation of the response at the optimal setting for the input variable have uncertainty (Lee and Lee [4]). This is because the optimal setting is obtained from the prediction models. Care must be taken for the uncertainty issue when the prediction capability of the prediction models is poor. As the prediction model varies from the true model, the resulting optimal setting may be quite far from being optimal (Xu and Albin [5]). Also, it is reported that the prediction capability of prediction models is often poor when the models are fitted to operation data from manufacturing lines whose volume is typically large.

In many cases, it is common to have the operational data rather than experimental data. This phenomenon becomes very common these days due to the development of information technology such as IOT (internet of things) technology in manufacturing lines. In this environment, data from equipment in manufacturing lines are gathered automatically by sensors and stored. Analysing the operational data becomes more important in these days.

When a large volume of operational data such as data from manufacturing lines are available, data mining approach is a good alternative for resolving the aforementioned difficulties. Data mining approach does not use the prediction model, thus, it is free from the uncertainty issues. Also, it utilizes operational data, thus, no experiments are required. One of attractive data mining approach to process optimization is a patient rule induction method (PRIM) (Friedman and Fisher [6]). PRIM searches a set of sub regions of the input variable space within which the performance of the response is considerably better than that of the entire input domain (Chong et al. [7]).

PRIM has been adopted for process optimization in some literatures. Chong et al. [7] adopted PRIM for the process optimization. This work focuses on the optimization of mean of a single response variable assuming that variance of the response variable is stable. However, the stable variable assumption may not valid in practice. Lee and Kim [8] adopted PRIM to optimize multiple response variables. This work also neglects the optimization of the variability of the response variables.

Overall, no works have adopted PRIM for solving the dual response problem where mean and standard deviation of the response variable are optimized simultaneously. We believe that PRIM is an attractive alternative for solving the dual response problem because it can resolve practical difficulties resulted from building statistical models. In case of the operational data, building a statistical model for standard deviation of the response variable is often not possible because standard deviation can be only estimated when replicates data available. In PRIM, the replicates are not required because it does not build statistical models.

The purpose of the proposed method is to optimize the mean and standard deviation of the response variable simultaneously by adopting PRIM. The proposed method is "model-free approach," thus, it is free from the uncertainty issues resulted from the model building. Also, it utilizes the operation data, thus, no effort for conducting experimental data is needed.

The rest of the paper is organized as follows. In Sect. 2, the existing DRO methods and PRIM are reviewed. In Sect. 3, the proposed method is presented. In Sect. 4, the proposed method is illustrated by a case example. Finally, concluding remarks are made in Sect. 5.

2 Literature Review

2.1 Dual Response Optimization

DRO attempts to find a set of input variables that simultaneously optimize the mean and standard deviation of the response. In DRO, the mean and standard deviation of the response are modeled based on experimental data. They are often modeled as quadratic functions of the input variables $\mathbf{x}_i$ (i = 1, 2,…, p). The two models for the mean and standard deviations can be represented as follows:

$$\hat{\omega}_\mu(\mathbf{x}) = \hat{\beta}_0 + \sum_{i=1}^{p} \hat{\beta}_i x_i + \sum_{i=1}^{p} \hat{\beta}_{ii} x_i^2 + \sum\sum_{i<j}^{p} \hat{\beta}_{ij} x_i x_j,$$
$$\hat{\omega}_\sigma(\mathbf{x}) = \hat{\gamma}_0 + \sum_{i=1}^{p} \hat{\gamma}_i x_i + \sum_{i=1}^{p} \gamma_{ii} x_i^2 + \sum\sum_{i<j}^{p} \hat{\gamma}_{ij} x_i x_j,$$

where $\hat{\omega}_\mu(\mathbf{x})$ and $\hat{\omega}_\sigma(\mathbf{x})$ are the estimated mean and standard deviation of the response, respectively.

DRO was first presented by Myers and Carter [9] and popularized by Vining and Myers [10]. Vining and Myers [10] minimize the standard deviation response while keeping the mean response at the specified target. Lin and Tu [11] suggest a way of minimizing the mean squared error (MSE). The MSE consists of a squared bias component and a variance component:

$$MSE = (\hat{\omega}_\mu - T)^2 + \hat{\omega}_\sigma^2,$$

where T is the target of the mean response.

Most of the existing DRO methods can be considered as "model-based approach." The mean and standard deviation of the response are modelled in the form of "response surface" based on RSM framework. However, this model-based approach has the uncertainty issue described in Sect. 1. Furthermore, it is not appropriate for the large volume of operational data due to poor prediction capability. Thus, we propose a particular model-free approach for dual response optimization, called dual response PRIM (DR-PRIM).

2.2 Patient Rule Induction Method

PRIM searches a set of sub regions of the input variable space within which the performance of the response is considerably better than that of the entire input domain (Chong et al. [7]). Here, the response is considered a larger the better type of response.

Although PRIM originally allows the input variables to have both real and categorical values (Friedman and Fisher [6], Chong et al. [7]), only real valued input variables are considered here for the sake of simplicity in presentation.

For the p input variables $x_1, x_2, \ldots, x_p$, a p-dimensional box B is defined as the intersection.

$$B = s_1 \times s_2 \times \cdots \times s_p, \tag{1}$$

where s_j is a subrange of the jth input variable, x_j, denoted by $s_j = (l_j, u_j)$ where l_j and u_j are lower and upper limits, respectively.

Suppose that we have N observations denoted by $\{(y_i, \mathbf{x}_i), i = 1, 2, \ldots, N\}$ where y_i and $\mathbf{x}_i$ are values of the response variable and the input variables of the ith observation. $\mathbf{x}_i$ is a p-dimensional vector denoted by $\mathbf{x}_i = (x_{i1}, x_{i2}, \ldots, x_{ip})$. When $\mathbf{x}_i$ locates in the box B (i.e., $l_1 \le x_{i1} \le u_1, l_2 \le x_{i2} \le u_2, \ldots, l_p \le x_{ip} \le u_p$), it is denoted by $\mathrm{x}_i \in B$.

Given a box B and observations $\{(y_i, \mathbf{x}_i), i = 1, 2, \ldots, N\}$, there are two statistics that describe the properties of the box B. The first one is the support β_B which denotes the proportion of the observations located in the box B.

$$\beta_B = \frac{n_B}{N},$$

where n_B denotes the number of observations which are located inside the box B. The supports ranges zero to one. When all of observations locate in the box B, $\beta_B = 1$. The second one is the box objective Obj_B, the mean value of the response in the box B.

$$Obj_B = \bar{y}_B = \frac{1}{n_B} \sum_{\mathrm{x}_i \in B} y_i.$$

The main idea of PRIM is presented below. Given observations $\{(y_i, \mathbf{x}_i), i = 1, 2, \ldots, N\}$, an initial box B_0 is formed by defining $l_j = \min_{i=1,2,\ldots,N} x_{ij}$, $u_j = \max_{i=1,2,\ldots,N} x_{ij}$, $j = 1,2,\ldots, p$. From B_0, 2_p candidate boxes denoted by $\{C_{01-}, C_{01+}, C_{02-}, C_{02+}, \ldots, C_{0p-}, C_{0p+}\}$ are created. The candidate boxes C_{0j-} and C_{0j+}, $j = 1,2,\ldots, p$ are obtained by peeling $100\alpha\%$ of the observations in the box B_0. The parameter α determines peeling rate, typically its value is set to a small value between 0.05 and 0.1. Then, PRIM chooses the one having the largest box objective among the candidate boxes and lets this box be B_1. If the support of B_1 is greater than a stopping parameter, β_0, then PRIM continues to do more peelings to generate next box B_2. Otherwise, the algorithm ends. This iterative process continuous until the support of the box becomes less than the predetermined value, β_0. The value of β_0 determines stability of the box. When β_0 is too small, the box often becomes over fitted. According to Friedman and Fisher [6], it is recommended to set β_0 value between 0.05 and 0.1.

3 Proposed DR-PRIM

In this section, the proposed method, called dual response PRIM (DR-PRIM) is presented. DR-PRIM is a modified PRIM which attempts to optimize the mean and standard deviation of the response variable. As aforementioned in Sect. 1, the existing "model-based approaches" to dual response optimization have a limitation in that the prediction models are often difficult to explain the large number of operational data. The prediction capability of the models are often too low, thus, the obtained setting of the input variables may far from the optimal. In contrast, the proposed DR-PRIM is a "model-free approach" which is free from the low prediction capability issue.

Figure 1 shows the pseudo-code of DR-PRIM. It consists of 4 steps. Steps 2, 3 and 4 iterate until the stopping criterion is satisfied. Detail of each step is described below.

1: Set values of parameters	Step 1
2: **repeat**	
3: Generate candidate boxes to be peeled	Step 2
4: Select the candidate box having the smallest objective value as the box B_m	Step 3
5: **until** $\beta_m < \beta_0$	Step 4

Fig. 1. Algorithm of DR-PRIM

Step 1. Set Values of Parameters. In this step, the values for the two parameters α and β_0 are determined. Typically, α value is recommended to be set to a small value between 0.05 and 0.1 which is faithful to the original concept "patient" rule induction. Typically, β_0 is set to be 0.05. See Friedman and Fisher [6] for the discussion on the β_0 value. Next, the initial box B_0 is defined.

$$B_0 = s_{01} \times s_{02} \times \ldots \times s_{0p},$$

where $s_{0j} = [l_{0j}, u_{0j}]$ is a subrange of x_j, j = 1,2,…,p. Given observations $\{(y_i, \mathbf{x}_i), i = 1, 2, \ldots, N\}$, subranges are defined by $l_{0j} = \min_{i=1,2,\ldots,N} x_{ij}$, $u_{0j} = \max_{i=1,2,\ldots,N} x_{ij}$, j = 1,2,…, p. The box objective of the box B_0 is defined by

$$Obj_{B_0} = \left(\overline{y}_{B_0} - T\right)^2 + s_{B_0}^2, \quad (2)$$

where $\overline{Y}_{B_0}$ is the mean value of the response of the observations located in the box B_0. Because the box B_0 contains all of N observations in Step 1, $\overline{Y}_{B_0}$ is the mean of all y_i's. Here, we assume that the response is the nominal the best type of which target value is T. $s_{B_0}^2$ is the sample variance of the response of the observations located in the box B_0. $s_{B_0}^2$ is calculated by

$$s_{B_0}^2 = \frac{\sum_{i=1}^{n_{B_0}} \left(y_i - \overline{Y}_{B_0}\right)^2}{n_{B_0} - 1},$$

where n_{B_0} is the number of observations which are located in the box B_0. Similarly, $s^2_{B_0}$ is the sample variance calculated by all y_i's because the box B_0 contains all of N observations.

The box objective is sum of two terms one for bias and the other one for variance. It is called as mean squared error (MSE). By using MSE as the box objective, the variance is also considered in the optimization. This is one of the main differences from the original PRIM. MSE has been employed as an optimization criterion on various researches. In DRO, MSE was adopted as the optimization criterion in the researches of Lin and Tu [11], Ding et al. [12], Jeong and Kim [13], and Lee and Kim [3].

Step 2. Generate Candidate Boxes to be Peeled. For x_j, two candidate boxes denoted by C_{j-} and C_{j+} are generated.

$$C_{j-} = s_{01} \times \cdots \times s_{0j}^{(-)} \times \cdots \times s_{0p},$$

$$C_{j+} = s_{01} \times \cdots \times s_{0j}^{(+)} \times \cdots \times s_{0p}, j = 1, 2, \cdots, p,$$

where $s_{0j}^{(-)} = \left[x_{j(\alpha)}, u_{0j}\right]$ and $s_{0j}^{(+)} = \left[l_{0j}, x_{j(1-\alpha)}\right]$. $x_{j(\alpha)}$ is the α-percentile of x_j in the box B_0 (i.e., $\Pr\left(x_j < x_{j(\alpha)} | \mathrm{x} \in B_0\right) = \alpha$) (Lee and Kim [8]). A total of $2p$ candidate boxes are generated.

Step 3. Select the Candidate Box Having the Smallest Box Objective as the Box B_m. The box objectives of the candidate boxes are calculated by Eq. (2). A total of $2p$ box objectives are calculated. It should be noted that small value of MSE is preferred so the candidate box having the smallest box objective value is preferred. Therefore, the candidate box having the smallest box objective is selected as the box B_1.

Step 4. Check if the Stopping Criterion is Satisfied. If the support of the box B_1, β_{B_1}, is smaller than the predetermined threshold, β_0, the algorithm ends with the box B_1. Otherwise, the algorithm goes to Step 2 again and conducts the next peeling from Steps 2, 3, and 4 in order to generate the next box B_2 from B_1. This iterative procedure continues until the stopping criterion is satisfied.

4 Example: Concrete Compressive Strength Problem

In this section, the proposed DR-PRIM is illustrated via a case example, called concrete compressive strength problem (Yeh [14]). The response is concrete compressive strength (y) and it is affected by eight input variables (x_1, x_2,..., x_8) such as cement, blast furnace slag and so on. The purpose of this case study is to keep the mean of y at its target value, 60 (=T), while minimizing the standard deviation of y by controlling the input variables. We explain the optimization process by DR-PRIM and compare the optimization result with that of regression modelling approach.

4.1 Optimization of the Concrete Compressive Strength Problem

Step 1. Set Initial Parameters. Two parameters are set as $\alpha = 0.05, \beta_0 = 0.05$. The initial box B_0 is set as interaction of the sub ranges of the eight input variables (i.e., $B_0 = s_{01} \times s_{02} \times \ldots \times s_{08}$). The total number of observations, *N*, is 1030. The subrange of each input variable is given in Table 1. The mean and variance of response of observations are 35.82 and 279.08, respectively. Thus, the box objective of B_0 is calculated by $Obj_{B0} = (35.82 - T)^2 + 279.08 = 863.75$

Table 1. Subranges of input variables

	x_1	x_2	x_3	x_4	x_5	x_6	x_7	x_8
l_{0j}	102	0	0	121.8	0	801.0	594.0	1.0
u_{0j}	540	359.4	200.1	247.0	32.2	1145.0	992.6	365.0

Step 2. Generate Candidates to be Peeled. For each input variable, candidate box is generated. A total of 16 candidate boxes $C_{1-}, C_{1+}, C_{2-}, C_{2+}, \cdots, C_{8-}, C_{8+}$, are generated. For example, C_{3-}andC_{3+} are given below.

$$C_{3-} = s_{01} \times \cdots \times s_{03}^{(-)} \times \cdots \times s_{08},\ C_{3+} = s_{01} \times \cdots \times s_{03}^{(+)} \times \cdots \times s_{08},$$

where $s_{03}^{(-)} = \left[x_{3(0.05)}, u_{03}\right]$ and $s_{03}^{(+)} = \left[l_{03}, x_{3(0.95)}\right]$.

Step 3. Select the Candidate Box Having the Smallest Box Objective as the Box B_m. The box objectives of the 16 candidate boxes are calculated. They are $Obj_{C1-} = 831.30$, $Obj_{C2-} = 859.82$, …, $Obj_{C8-} = 700.02$, $Obj_{C1+} = 894.47$, $Obj_{C2+} = 863.69$,…, $Obj_{C8+} = 891.31$. The box having the smallest box objective is C_{8-}, thus, C_{8-} is defined as B_1. Its box objective is smaller than Obj_{B0} (i.e., 700.02 < 863.75). It should be noted that the support of B_1 is $\beta_{B_1} = 0.63$. Usually, β_{B_1} is 0.95 when the peeling rate is 0.05. However, in this particular example, 37% of observations were peeled off in the first iteration, thus, β_{B_1} becomes 0.63. This is because a lot of observations have same value of x_8, thus, they were excluded together by a single peeling process.

Step 4. Check if the Stopping Criterion is Satisfied. Because $\beta_{B_1} > \beta_0$ (i.e., 0.63 > 0.05) the algorithm goes back to Step 2. Then, peeling and candidate box selection are conducted to generate the next box B_2. This iterative process continues until the support becomes less than 0.1. As a result, a total of 24 iterations of Steps 2, 3, and 4 are conducted. The final box, B_{24}, is given in the last row in Table 2. The mean of B_{24} is 64.16 which quite close to its target value 60. The box objective of B_{24} is 113.4 which is quite smaller than that of B_0.

4.2 Comparisons of DR-PRIM with Original PRIM

As a further analysis, the proposed DR-PRIM is compared with the original PRIM proposed by Friedman and Fisher [6]. The original PRIM only considers the mean of the response variable. Table 2 compares the optimization results of PRIM and DR-PRIM. The mean of response by PRIM is 59.97 which is very close to the target value 60, however, the variance resulted by PRIM is 171.10 which is quite larger than that of DR-PRIM. As a result, the confidence interval (C.I) is wider than that of DR-PRIM. This means that the uncertainty of response value from PRIM is larger than that of DR-PRIM. Similarly, *MSE* resulted from PRIM is 171.1 and it is larger than that of DR-PRIM (i.e., 171.1 > 113.4).

Table 2. Comparison of DR-PRIM and modelling approach

Method	x*	$\bar{y}$	s_y^2	95% C.I
PRIM	$165 \leq x_1 \leq 475$, $11 \leq x_2 \leq 189$, $1.7 \leq x_5 \leq 22$, $852.1 \leq x_6 \leq 1081$, $56 \leq x_8 \leq 100$	59.97	171.10	[54.89, 65.05]
DR-PRIM	$165 \leq x_1 \leq 469$, $19 \leq x_2 \leq 262.2$, $x_3 \leq 159.9$, $126.6 \leq x_4 \leq 181.7$, $x_5 \geq 6.9$, $852.1 \leq x_6 \leq 1076$, $56 \leq x_8 \leq 100$	64.16	91.85	[61.21, 67.11]

We plotted peeling trajectory to see changes in mean and variance with respect to the support. Figures 2 and 3 show the peeling trajectories of PRIM and DR-PRIM, respectively. In case of mean (Fig. 2), the mean value of PRIM approaches the target value 60 rapidly as the support changes from 0.63 to 0.1. Then, it slowly approaches the target value as the support decreases from 0.1 and stops at 59.97 when the support reaches 0.05. In contrast, the mean value of DR-PRIM increases as the support decreases.

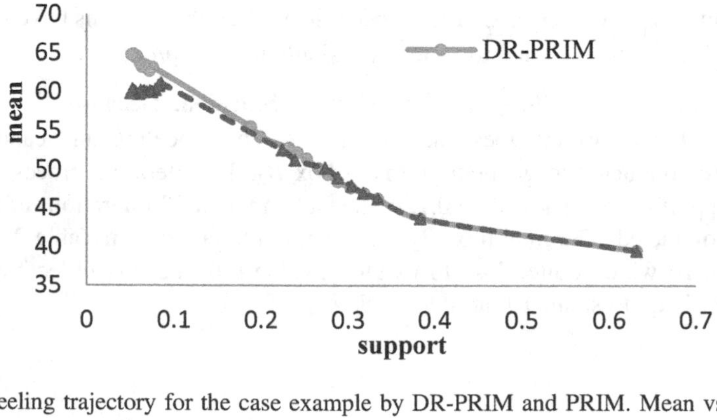

Fig. 2. Peeling trajectory for the case example by DR-PRIM and PRIM. Mean vs. support

In case of variance (Fig. 3), the variance value of PRIM decreases as the support decreases from 0.63 to 0.1. However, it increases rapidly as the support decreases from 0.1. This phenomenon implies that the boxes are unstable in the range where the support is less than 0.1. In contrast, the variance value of DR-PRIM continuously decreases as the support decreases which means that the boxes obtained by DR-PRIM are more stable.

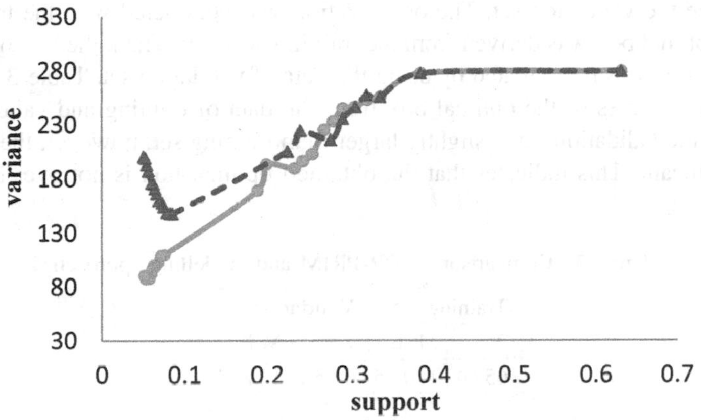

Fig. 3. Peeling trajectory for the case example by DR-PRIM and PRIM. Variance vs. support

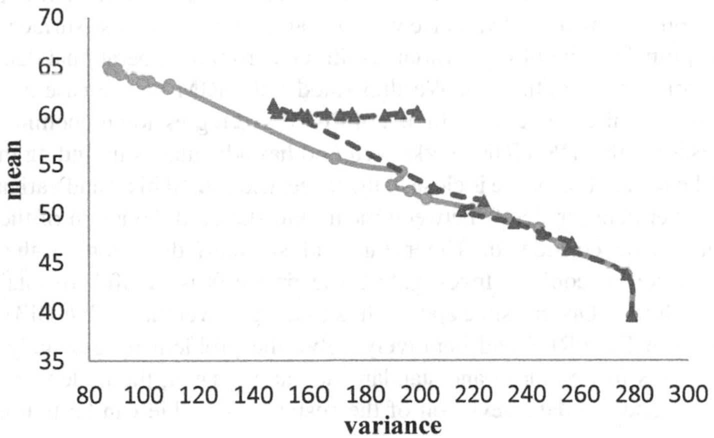

Fig. 4. Variance vs. mean for the case example boxes by DR-PRIM and PRIM.

Figure 4 shows the relationship between mean and variance of the response according to PRIM and DR-PRIM. In case of PRIM, the variance ranges roughly from 140 to 200 when the mean is near 60. This implies that PRIM is good at keeping mean at the target value 60, however, the variance is high and unstable. In case of DR-PRIM, the variance decrease to 91.85 and this value is quite less than that of PRIM.

4.3 Cross Validation of the Optimization

The obtained optimal box works well for the given data; however, it may not do for other data. When the obtained optimal box works well only for the given data, the obtained box can be considered to be over-fitted. In this regard, we conducted cross validation to check whether the obtained box is over-fitted or not.

First, we randomly split the existing entire data into the training set and the validation set. As a result, two-third of the entire data became the training set and the other data became the validation set. The optimization was conducted with the training set, thus, the optimal box was derived from the training set only. Then, the box objective of the optimal box was recalculated by using the data of validation set. Table 3 reports the box objective values of the optimal box from the data of training and validation sets. MSE from the validation set is slightly larger than training set, however, the difference is not significant. This indicates that the obtained optimal box is not over-fitted.

Table 3. Comparison of DR-PRIM and modelling approach

Training set		Validation set	
$\bar{y}$	MSE	$\bar{y}$	MSE
65.64	110.95	65.56	124.35

5 Conclusion

In this work, we proposed a modified version of PRIM, called DR-PRIM. DR-PRIM attempts to optimize mean and variance without building the response surface models. It obtains an optimal setting of input variables directly from the operational data in which mean and variance are optimized. We illustrated DR-PRIM with a case example and compared the optimization result with that of PRIM and regression modelling approach. It has been shown that DR-PRIM works well and has advantages in finding the optimal box where the mean of response is close to the target (i.e., small bias) and variance is low.

On the other hand, tradeoffs between mean and standard deviation of the response variable should be considered. The mean and standard deviation of the response variable are often in conflict. Investigating the tradeoffs is helpful to obtain a more satisfactory solution. One possible approach is to adopt a weighted *MSE* (*WMSE*) as the box objective in DR-PRIM and iteratively solve the problem by gradually changing weighting factors for the mean and standard deviation. Then, the tradeoff relationship between mean and standard deviation of the response variable can be figured out.

References

1. Myers, R.H., Montgomery, D.C., Anderson-Cook, C.M.: Response Surface Methodology. Wiley, Hoboken (2009)
2. Lee, D., Jeong, I., Kim, K.: A posterior preference articulation approach to dual-response surface optimization. IIE Trans. **42**(2), 161–171 (2010)

3. Lee, D., Kim, K.: Interactive weighting of bias and variance in dual response surface optimization. Expert Syst. Appl. **39**(5), 5900–5906 (2012)
4. Lee, H., Lee, D.: A solution selection approach to multiresponse surface optimization based on a clustering method. Qual. Eng. **28**(4), 388–401 (2016)
5. Xu, D., Albin, S.: Robust optimization of experimentally derived objective functions. IIE Trans. **35**, 793–802 (2003)
6. Friedman, J., Fisher, N.: Bump hunting in high-dimensional data. Stat. Comput. -LONDON- **9**(2), 123–142 (1999)
7. Chong, I., Albin, S., Jun, C.: A data mining approach to process optimization without an explicit quality function. IIE Trans. **39**, 795–804 (2007)
8. Lee, M., Kim, K.: MR-PRIM: patient rule induction method for multiresponse optimization. Qual. Eng. **20**(2), 232–242 (2008)
9. Myers, R., Carter, W.: Response surface methods for dual response systems. Technometrics **15**(2), 301–307 (1973)
10. Vining, G., Myers, R.: Combining Taguchi and response surface philosophies: a dual response approach. J. Qual. Technol. **22**(1), 38–45 (1990)
11. Lin, D., Tu, W.: Dual response surface optimization. J. Qual. Technol. **27**(1), 34–39 (1995)
12. Ding, R., Lin, D., Wei, D.: Dual-response surface optimization: a weighted MSE approach. Qual. Eng. **16**(3), 377–385 (2004)
13. Jeong, I., Kim, K., Chang, S.: Optimal weighting of bias and variance in dual response surface optimization. J. Qual. Technol. **37**(3), 236–247 (2005)
14. Yeh, I.: Modeling of strength of high performance concrete using artificial neural networks. Cem. Concr. Res. **28**(12), 1797–1808 (1998)

Integrated Evaluation and Multi-methodological Approaches for the Enhancement of the Cultural Landscape

Lucia Della Spina(✉)

Mediterranea University of Reggio Calabria, Reggio Calabria, Italy
lucia.dellaspina@unirc.it

Abstract. The paper presents an integrated assessment process for the identification of scenarios of cultural landscape sustainable valorization in a particularly significant area of southern Italy characterized by tangible and intangible resources. The decision-making process uses multi-methodological evaluations in order to support the development of scenarios and alternatives policy strategies, aimed at pre-order a territory development system subject of study. The methodological pathway is structured to allow the interaction among different techniques, which are selected in order to outline a decision support system, dynamic, flexible and adaptive, sensitive to the specificities of the context and oriented to the development of intervention strategies based on of experts and common knowledge, and on recognized and shared values. The selection of 'conscious actions' helps to reduce conflicts turning them in synergies, recognizing that the essential components of a landscape are multidimensional and complex and where interact different systems of values and relationships. Therefore, the strategies will be feasible or practicable in proportion to what projects will tend to achieve the idea of "scenario" for the site shared by social and institutional actors of the local system.

Keywords: Cultural landascape · Complex values · Decision support · Integrated evaluations · Multi-methodological evaluations · Multicriteria analysis

1 Introduction

Today the transformation of the territory represents complex decisional problems, and this highlights the need, more and more evident, to use appropriate evaluation tools for intervention projects. In fact, the plurality of possible solutions strongly require the question of the assessment as a fundamental tool for the comparison of different alternatives and the choice of best "compromise" solution, in order to ensure a progressive accumulation of knowledge. The evaluation of urban transformation alternative scenarios consequently needs to be placed in a correct evaluation framework, starting from the design and planning process, able to ensure the proper effects analysis of the strategic choices for the territory. Therefore, starting by an analysis of the

O. Gervasi et al. (Eds.): ICCSA 2017, Part I, LNCS 10404, pp. 478–493, 2017.
DOI: 10.1007/978-3-319-62392-4_35

applicable evaluation procedures, the paper highlightes how the evaluation may be seen as communication and knowledge production tool, capable of fully expressing its potential when organically integrated in the same project methodology [1]. The evaluation constitutes the basis of dialogue among knowledge and values, able to translate this dialogue in the selection of objectives and actions, in the identification of key values and associated meanings, exploring opportunities and building alternatives, analysing the possible impacts and effects and supporting the management of complex systems with multiple priorities. The integration of different values in the decision-making process helps to build greater acceptability and trust in public decisions [2], including the different perspectives and trying to reduce conflict [3].

In this paper, in order to identify a sustainable strategy for the valorisation of the cultural historical landscape of a particularly significant area in Southern Italy, characterized by tangible and intangible resources and values, a multi-methodologic evaluation process has been structured in order to support the development of scenarios and alternative intervention strategies. The study was conducted through the application of techniques and experimental approaches to a real case. On the operational side, the strategies will be feasible to the extent that the projects will tend to realize the idea of "scenario" of the site, shared by social and institutional actors of the local system. The activity of identifying the potentialities and criticality, or rather representing prefiguration "scenarios" of the immediate future, is finalized to pre-order a territory development system.

The purpose of this study is to provide an operational decisions support framework in order to support policy makers in their future strategic decisions by using a multi-methodological approach, allowing to justify the allocation of public resources with rational arguments able to better deal with critical steps and avoid preconceptions in decision-making. Multi-methodological approaches [4, 5] can be defined as a structured process designed to cope with multidimensional systems and complex problems using knowledges from different disciplines. Therefore, the multidisciplinary approaches deal with multidimensional systems, multi-stakeholder perspectives, using qualitative-quantitative approaches to better study alternative options. Although on the use of multi-methodological approaches there is a broad discussion in decision-making policies, best practices are still scarce [6]. In particular, this paper proposes a group-learning process as decision support methodological framework evolving through three main methods, Stakeholder Analysis, Cognitive Mapping and Multicriteria Analysis. Cognitive Mapping and Stakeholder Analysis have been used in the literature in combination with Multicriteria Analysis [7]. So far, there is no experiment in a real context of the joint use of the three proposed methods in this study, namely the Stakeholder Analysis [8]; Cognitive Mappings [9] and the specific Multicriteria of Regime [10–14]. The reasons that led to the choice of this specific method can be summarized as follows:

1. The Cognitive Mapping are considered one of the most promising tools for structuring problems before the application of Multicriteria Decision Aiding [7, 15].
2. The Stakeholder Analysis, in the form of interest matrix, is particularly suitable for completing Multicriteria technique, in a collaborative decision making process and in a context that does not effectively support reaching a consensus in the elicitation

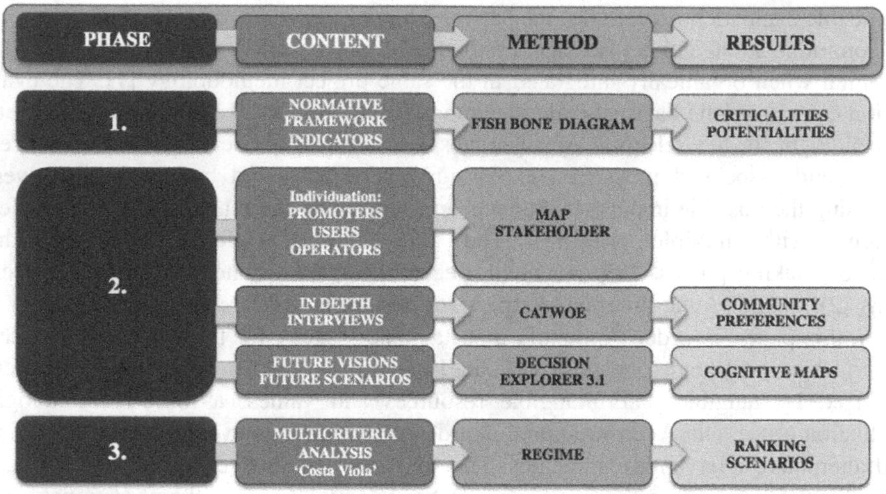

Fig. 1. Methodological framework of the multi-methodological evaluations

phase of preferences [7], referring the need to aggregate different points of view based on their different levels of importance. In addition, stakeholder analysis has shown to be a very important preliminary step in multi-dimensional decision making processes [8], as detailed in Sect. 2.

3. The Regime multicriteria method [10–14] is a method of supporting the decision that, in addition from the possibility of considering data of different nature (quantitative and qualitative) [16], offers the opportunity to assign different weights to the identified criteria, to manage the conflict between goals and to deduce the priorities among alternative options.

In particular, this paper presents the following structure: in Sect. 2 the results of a multi-methodological approach are presented (Fig. 1); the results have contributed to the development of five scenarios of intervention, alternatives with each other. For the evaluation purposes, Multi-criteria Analysis has been elaborated with reference to three objectives and five criteria, deducting a ranking of preferability among the suggested scenarios, applying the Regime multicriteria method (Sect. 2.1). In the conclusions (Sect. 3) the paper proposes final considerations on the opportunity to use a Multi-criteria valuation approach in the conservation and enhancement of the cultural landscape.

2 Multi-methodological Evaluations Applied to the Case Study

Starting from the definition of Historic Urban Landscape (HUL) [17, 18], this paper explores the theme of the historic urban landscape being understood as a 'common good', the result of a historical stratification of cultural values and natural, tangible and intangible components. This field of investigation and experimentation provides an innovative model of local development [19, 20].

Specifically, the landscape that is the subject of study, 'Costa Viola', consists of 15 municipalities situated in the province of Reggio Calabria. These are very heterogeneous, both in terms of location and territorial extension, and also in cultural and economic resources. The territory is characterized not only by small coastal municipalities, but also by more extended hilly municipalities, as well as municipalities that include both a coastal landscape and rural territory.

The fundamental characteristics to be considered in the pursuit of sustainable development of the landscape system are:

- Landscape heritage, characterized by high biodiversity and numerous cultural assets not networked among themselves
- A heterogeneous and fragile territorial system
- Hydrogeological risk
- Abandonment of agricultural terraces, which are a fundamental component of the cultural landscape system
- Agricultural terraces becoming fragmented
- Heterogeneity of the activities and of the local productions
- Seasonal tourism
- Many communities' funding often being spent without an integrated strategy

In the area being studied, in the years of past programming, a several negotiated programming tools were implemented using structural funds from the European Community. When studying the programming and planning tools currently in place, one finds a lack of overall unitary vision, in terms of the lack of coordination and integration of the projects with the different resources and landscape components, and also within the entire local system. However, this reconnaissance of the programming and measures in place allows us to compare the current situation with a more advantageous vision, and therefore to define the design scenarios towards which the future management of the site should be guided.

Therefore, activity preliminary to evaluation found it appropriate to engage a methodological path through a structured evaluation process, which combines different techniques for each of the phases of the decision-making process: this is coherent with applying the Systems Thinking Approach [21–24] to problem solving. The evolution of the ways of structuring decision-making processes has led to the combination of analysis techniques, evaluation and public involvement. Particular attention has been paid to building evaluation processes that can consider conflicts of interest, the plurality of viewpoints and different responsibilities; these processes are built through dialogue and discussion with to the entire community [25–29].

The multi-methodological evaluations selected are finalised to configure a decision support system oriented towards the elaboration of sustainable scenarios for transformation, enhancement and promotion. These must be able to reflect the interactive and dynamic dialogue between experts and communities' knowledge of values recognized and shared, and which can therefore manage the complexity of the interests and objectives involved.

As part of this strategic decision-making process, the strategic operational development plan for sustainable development appears to have value as a development tool, specifically for the launch of new and advantageous possibilities of land development on multiple levels and in further directions. As a strategic tool, the plan selects the short- and long-term objectives and how to achieve them, while the operative tool defines a system of actions to be implemented for sustainable local development [20, 30].

The methodological approach is structured in three phases, as identified in Fig. 1. The initial phase of analysis has resulted in the organization of the valid regulatory framework, the identification of the planning instruments currently in use in the territory, and the construction and selection of appropriate qualitative and quantitative indicators representative of the specific problems. The construction of the indicators was carried out with reference to specific thematic areas, such as population, the economy, tourism, transport, infrastructure (ground and underground), hydrosphere, landscape, cultural heritage, and services. On the basis of the selected indicators, the criticalities and potentialities of the territory have been defined in terms of the possible actions of the project, structured in accordance with the Fish Bone Diagram [31]. The diagram classifies the criticalities and the potentialities identified based on their importance, evaluated according to an appropriate rating scale (high, medium, or low importance).

In the central phase, Institutional Analysis has produced a map identifying the various stakeholders, dominant and important figures in the local culture, and characteristics of the places [29]. The stakeholders have been grouped into three prevalent groups: promoters, operators and users. The first group includes the institutions and experts, i.e. those who have a strong influence on making choices oriented towards the common good, due to their knowledge and skills, their strategic positioning, and their representativeness. The second group includes the operators of receptive and productive activity, but also the associations (operators in the dominant economic and social sectors). The third group, finally, comprises citizens and tourists who are also involved in policy making. To identify the views of stakeholders, in-depth interviews were carried out using the CATWOE approach [32], a useful tool for structuring the interviews and exploring the decision-making problem from multiple points of view. The interviews were structured on the basis of asking various questions considered significant for the sustainable enhancement programme of ‘Costa Viola’. This has highlighted the perception criticalities and potentialities, and identified future scenarios of transformation and their related implementation strategies. From the analysis of qualitative information (criticalities, potentialities, actions, future visions, obstacles, actors and environmental limits) contained in the interviews’ verbal protocol, it was possible to develop cognitive maps for the different categories of stakeholders (institutions, hotel managers, restaurateurs and traders, experts, associations, farmers, tourists and citizens). Each cognitive map was prepared using the Decision Explorer 3.1 software. Through the analysis and the comparison the results, it was possible to build the structure of explicit preferences of the various parties involved, and of the future scenarios. The revealed preferences made it possible to shape the future of the visions and of the enhancement scenarios for the historic urban landscape.

The final phase, that of interpreting the results of the two previous phases, has allowed the elaboration of five alternative scenarios (A, B, C, D, E) of possible intervention, constituted by a set of strategic actions integrated according to the interdependencies that characterize the spatial reference system. Specifically, the planned actions constitute specific interventions related to the local territorial system in its entirety; these can promote an actual integrated enhancement of the landscape [33].

In this context, five specific 'scenarios' have been imagined (see Sect. 2.1). These 'scenarios' have been constructed with reference to three different territorial dimensions:

- Cultural Heritage
- Natural Heritage
- Infrastructures of the territorial system

Each dimension has been defined by 'strategic actions' that respond to three strategic 'objectives', identified in relation to the same representative components of the landscape, as follows:

- Protecting and enhancing the Cultural Heritage
- Protecting and enhancing the Natural Heritage
- Improving and reinforcing the Infrastructure System

Finally, the Multicriteria Regime method has produced conclusive results that show the preferability of the future visions for the cultural historical landscape of 'Costa Viola': the overall assessment of the impacts for each scenario are defined with respect to each strategic action, obtaining a ranking of preferability among the scenarios by applying appropriate sensitivity analysis.

2.1 The Multicriteria Analysis

In the construction and evaluation of possible scenarios of development, multicriteria analysis [34] can play a fundamental role in structuring and supporting complex policy problems with multiple and often conflicting objectives. In this context, the 'Multi-criteria approach' can 'evaluate' future scenarios that are achieved through specific projects, and the related objectives of which are identified by the probable impacts on the local system. This enables the choice of strategic projects and their priorities, in order to achieve the goals set.

In addition, the multicriteria approach is able to consider the integration of the different dimensions that coexist in the local landscape, and moreover, can interpret current trends and the dialogue with the actors involved.

The multidimensional approach is necessary to represent the complexity of the landscape: in this way, the multiple dimensions of the landscape become the vital reference points for evaluating the conservation policies and redevelopment of the cultural and environmental heritage. In particular, it contributes to the definition of strategies, objectives and actions of the project, in order to overcome the conflict between environmental protection and development regarding the sustainability of territorial choices [35].

Table 1. The overall assessment of the impacts (Source: Elaborated by ICOMOS 2011)

Value of heritage asset	Intensity of change				
	No change	Negligible change	Minor change	Moderate change	Major change
Overall assessment of the impacts	Effect of change (positive or negative)				
	Neutral	Weak	Moderate	Strong	Very Strong
Very high	Neutral	Weak	Moderate/Strong	Strong/Very Strong	Very Strong
High	Neutral	Weak	Moderate/Weak	Moderate/Strong	Strong/Very Strong
Medium	Neutral	Neutral/Weak	Weak	Moderate	Moderate/Strong
Low	Neutral	Neutral/Weak	Neutral/Weak	Weak	Moderate/Weak
Negligible	Neutral	Neutral	Neutral/Weak	Neutral/Weak	Weak

The multicriteria evaluations allow the landscape to be 're-capitalized' as heritage, in order to build ethical development of the many tangible and intangible components of the place. Specifically, the inheritance of the past is enhanced to produce new wealth, which is not destructive of the consolidated values, but is able to determine 'territorial added value' [36].

The evaluation of the scenarios was conducted with reference to the guidelines for Heritage Impact Assessment (HIA) [17]. The impact evaluation on cultural landscape is a complex process involving several phases, which include the definition of the model and the evaluation of the impact, both direct and indirect. From an operational point of view, the overall evaluation can be achieved by combining the intensity of change with the effects, positive or negative; for this purpose, a five-point scale was used to evaluate the impact (from 'very strong' to 'negligible'), as reported in Table 1 [37].

According to the UNESCO guidelines for each scenario, an overall evaluation of the impacts has been produced, with reference to each strategic action and the following evaluation criteria

- Archaeological Heritage
- Built Heritage
- Historical Landscape
- Natural Heritage
- Infrastructural System
- Socio-Economic System

It should be noted that some strategic actions may be common to several scenarios (Tables 2, 3, 4); the empty cells indicate that the scenario is not affected by the strategic action. Since the assessment of the impacts is related to each strategic action, the resulting effect (expressed on the scale of 'very strong' to 'negligible') is the same for each scenario that contains that strategic action (Tables 5, 6, 7). The impacts are all positive and, for ease of reading, the empty cells show null impacts. To achieve an evaluation synthesis, a multicriteria approach has been adopted, which identifies a

Table 2. Strategic actions for the cultural heritage

Objective 1: to safeguard and to enhance the cultural heritage		Scenarios				
n.	Strategic actions	A	B	C	D	E
a1.1	Restoration of Agriculture Mosaics through the support of agricultural and non agricultural activities	O		O	O	
a1.2	Restoration of the system of terracements and its irrigation system	O		O	O	O
a1.3	Safeguard and recovery of the forest system, connected to the system of terraces and its supply chain	O		O	O	
a1.4	Protection of the distributed ancient settlements			O	O	
a1.5	Redevelopment of settlements and environment			O	O	
a1.6	Strengthening of the tourist accommodation and tourism services in the inner areas: identification of different well-equipped poles (central reception and information services; promotion and sale of local products; interchange station among the tourist buses, etc.)			O	O	O
a1.7	Integrated redevelopment of the main rural network of mule and trails (Rural Service) and complementary infrastructuring to that main rural tracks (hiking trails)			O	O	O
a1.8	Safeguard of the centralized ancient settlements				O	
a1.9	Consolidation and integration of territorial polarities consisting of historical and architectural interest assets			O	O	
a1.10	Integrated safeguard and enhancement of the historical architecture of civilian type and defensive military (such as watchtowers and defense along the coast)			O	O	
a1.11	Promotion of cultural network of the numerous historical and architectural heritage spread all over the territory, with the aim of a cultural tourist circuit, and scholastic circuit			O	O	
a1.12	Enhancement of the religious tourism circuit				O	
a1.13	Enhancement of the museum circuit				O	
a1.14	Enhancement of the archaeological tourist circuit			O	O	
a1.15	Enhancement of the early industrial architecture circuit	O		O	O	

preferability ranking among the different scenarios proposed. The evaluation was structured around the three key objectives (protect and enhance the cultural landscape; protect and enhance the natural resources; improve and strengthen the infrastructure system) and six criteria (archaeological heritage, built heritage, historic landscape, natural heritage, infrastructure system, socio-economic system) with respect to which the impacts were considered [38]. Specifically, the scenarios were compared by applying the Multicriteria Regime method [10–14]. In addition to the possibility of considering the different nature of data (quantitative and qualitative) [16], this method offers the opportunity to assign different weights to the identified criteria, in order to manage the conflict among the objectives, and to deduce priorities among alternative options.

Table 3. Strategic actions for the natural heritage

Objective 2: to safeguard and to enhance the natural heritage		Scenarios				
n.	Strategic actions	A	B	C	D	E
a2.1	Establishment of the ecological network in order to mitigate the effects of environmental fragmentation and to preserve biodiversity	O				
a2.2	Strengthening of prevention and mitigation of natural and anthropogenic risk factors related to landslides or flooding, as well as the pollution of water bodies (surface and groundwater) and marine waters	O				
a2.3	Mitigation of environmental risk (prevention and control of pollution of surface and groundwater bodies, monitoring and reduction of hydrogeological phenomena	O				
a2.4	Maintenance and reconstruction of the necessary hydraulic-forestry arrangement	O				
a2.5	Conservation and enhancement of the geological heritage	O				
a2.6	Networking of the various natural resources for nature tourism and scientific-educational or also for the recreation and free time	O		O	O	
a2.7	Safeguard of the landscape and environmental connotation of the landscape and environmental of the coastline through the safeguard and enhancement of the seabed	O		O	O	

The method of the regime was applied using the Definite 2.0 software (DEcision on a FINITE set of alternatives) [39]. In the first instance, the same weight was assigned to the three objectives (0.33 for each objective, with the sum of the weights equal to 1.00), and weights were assigned to the criteria by dividing the objective weight by the number of criteria (equal to 6): that is, assigning a weight of 0,055 to each evaluation criterion. The following lists of preferability were built: the first with equal weights for all objectives (Table 8a); the second obtains a set of rankings by assigning, to each goal in turn, more weight than the others, and equal weight to the two remaining objectives; this analyses the sensitivity of the rankings by varying the weights (Tables 8b, c and d). The rankings of preferability against objectives allow us to identify the following complete ranking of preferability among scenarios:

- I: Scenario D (score 1.00)
- II: Scenario E (score 0.75)
- III: Scenario C (score 0.50)
- IV: Scenario A (score 0.25)
- V: Scenario B (score 0.00)

It is specified, therefore, that the 'strength' of the results is not sensitive to the change in the distribution of weights assigned to the objectives, but to the characteristics of each scenario's performance with respect to the evaluation criteria.

Table 4. Strategic actions for the infrastructural system

Objective 1: to safeguard and to enhance the cultural heritage		Scenarios				
n.	Strategic actions	A	B	C	D	E
a3.1	Reorganization of "sea routes" through redevelopment, adaptation and the reinforcement of maritime infrastructure, like the promotion of "collective sea taxi" for the connection or excursions along the coast		O		O	O
a3.2	Specific development interventions on the integrated system of the not only touristic regional port infrastructure		O		O	O
a3.3	Redevelopment of the equipment for the coastal landing place services		O		O	O
a3.4	Strengthening of the equipment for the inter coastline link services		O		O	O
a3.5	Functional redevelopment of the landing place services for maritime links to the islands		O		O	O
a3.6	Restoring of the quay for docking of the hydrofoils and adjustment of the services for inter coastline link		O		O	O
a3.7	Strengthening of equipment for the interregional connecting services to public transport aims and tourist mobility		O		O	O
a3.8	Realization of an intermodal terminal (rail-road-sea-way), with realization of appropriate parking areas for private vehicles and tour buses				O	O
a3.9	Realization of an intermodal interchange station, equipped with reception and service infrastructure in order to dispose of the volume of vehicular traffic, optimize the connections and rationalise the flow of tourists in the area				O	O
a3.10	Integrated strengthening of the tourism services, redevelopment and rationalise of the touristic mobility, promotion of touristic services through urban and environmental redevelopment of poor quality existing settlements			O	O	O
a3.11	Facilitation of traffic flows identifying the entry points to the coast, of the necessary exchange areas and the related parking				O	O
a3.12	Improvement of road and rail infrastructure with relative interchange areas				O	O
a3.13	Strengthening of the existing road infrastructure system, through new parking spaces to service of the historical settlements, with interchange areas, pedestrian walkways, ecological bus, vectors, mechanical, etc.				O	O
a3.14	Realization of a mechanical vectors system for the connection among the coastal towns and the inner cores, like alternative and complementary mobility system				O	O
a3.15	Realization and adaptation of surface for air ambulance service and of civil protection for tourism				O	O

Table 5. Evaluation of the impacts for the cultural heritage

Strategic actions	Criteria						Scenarios				
	Arch. heritage	Built heritage	Historical heritage	Natural heritage	Infrastr. system	Soc. econsystem	A	B	C	D	E
a1.1			VS	S		S		O		O	O
a1.2			VS	S		M		O		O	O
a1.3			S	VS		M		O		O	O
a1.4		VS	S					O		O	O
a1.5		VS	VS	M				O		O	O
a1.6					S	VS		O		O	O
a1.7			M		M	W		O		O	O
a1.8		VS	S							O	O
a1.9	S	VS	M			VS				O	O
a1.10		VS	M			M			O	O	O
a1.11		S	M			M				O	O
a1.12		S	M			S				O	O
a1.13		S	M			VS				O	O
a1.14	VS					S				O	O
a1.15		S	M			M				O	O

VS – Very Strong; S – Strong; M – Moderate; W – Weak

Table 6. Evaluation of the impacts for the natural heritage

Strategic actions	Criteria						Scenarios				
	Arch. heritage	Built heritage	Historical heritage	Natural heritage	Infrastr. system	Soc. econ system	A	B	C	D	E
a2.1				VS				O		O	O
a2.2		M		VS	S			O		O	O
a2.3		M		VS	S			O		O	O
a2.4				S				O		O	O
a2.5				S				O		O	O
a2.6				S		M		O		O	O
a2.7				S		M		O		O	O

VS – Very Strong; S – Strong; M – Moderate; W – Weak

Table 7. Evaluation of the impacts for infrastructural system

Strategic actions	Criteria						Scenarios				
	Arch. heritage	Built heritage	Historical heritage	Natural heritage	Infrastr. system	Soc. econ system	A	B	C	D	E
a3.1					VS	M		O		O	O
a3.2					VS	M		O		O	O
a3.3					VS	M		O		O	O
a3.4					VS	M		O		O	O
a3.5					VS	M		O		O	O
a3.6					VS	M		O		O	O
a3.7					VS	M		O		O	O
a3.8					VS	M				O	O
a3.9					S	M				O	O
a3.10		M			S	S			O	O	O
a3.11					S	M				O	O
a3.12					S	M				O	O
a3.13					VS	M				O	O
a3.14					S	W				O	O
a3.15					M	W				O	O

VS – Very Strong; S – Strong; M – Moderate; W – Weak

Table 8. Evaluation multicriteria: ranking scenarios

(a) Equal weight to the three objectives		(b) Greater weight for objective: to protect and to enhance the cultural heritage	
	Regime		*Regime*
Scenario D	1.00	Scenario D	1.00
Scenario E	0.75	Scenario E	0.75
Scenario C	0.50	Scenario C	0.50
Scenario A	0.25	Scenario A	0.25
Scenario B	0.00	Scenario B	0.00
(c) Greater weight for objective: to improve and to reinforce the infrastructure system		**(d) Greater weight for objective: to protect and to enhance the natural heritage**	
	Regime		*Regime*
Scenario D	1.00	Scenario D	1.00
Scenario E	0.75	Scenario E	0.75
Scenario C	0.50	Scenario C	0.50
Scenario A	0.25	Scenario A	0.25
Scenario B	0.00	Scenario B	0.00

3 Conclusions and Discussion

The study explored the potential of an integrated approach in the elaboration of territorial development strategies, focusing on several specific values and on the complex resources that characterize the cultural historical landscape of 'Costa Viola'. The multi-methodological evaluation approach thus structured was an experimentation within a wider research course, aimed to delineate decision-making processes oriented towards the elaboration of shared design choices [1, 30, 40, 41]. The combined application of different methods and techniques originated from disciplines not necessarily those of the evaluation, and addressed a complex decision problem [20, 33], characterized by multiple variables and a high level of uncertainty, in an incremental and cyclical evaluation process that was characterized by constant feedback and by constant interactions. This method outlined a development plan conscious and shared for the transformation and enhancement, coherent with the principles underlying the approach HUL [17, 18]. This complex decision making requires an active collaboration between the various skills involved and the constant comparison among the territories and stakeholders. This implies that the length of the process depends on the time-period, and on the difficulties, obstacles and dynamics that arise in real contexts [19]. In an integrated decision-making approach, the need to examine 'complex values' [42] supports the structuring of a process of multicriteria evaluation, aimed towards the elaboration of strategic actions and objectives. This approach should be able to consider the material and immaterial values, objective and subjective, of use and non-use, as well as intrinsic values, and their synergistic and complementary relationships, in order to formulate actions. A methodological approach developed in accordance with the proposed model requires that the construction of the cognitive framework is developed over time and accompanies the development of the design choices, constantly drawing on new contributions. The methodological approach thus configured can provide a useful new stimulus to drive the following: the selection of information; the identification of values; the analysis of conflicts; and the construction of shared preferences oriented towards the development of transformation scenarios that respond to the needs of decision-making contexts that are characterized by complexity and uncertainty [30]. As part of the experiment applied to the case study, the use of an integrated and multi-methodological approach has considered the character of the Historical Urban Landscape with its different multi-dimensional components, the system of tangible and intangible relations, and its perception by stakeholders. This has allowed this research to identify the different priorities, and select those actions appropriate to the context, in order to reflect changes in an interactive and dynamic dialogue among communities, local expertise and experts [43].

This type of evaluation approach takes shape and is fed through the concept of HUL [17, 18], by the fielding of the tangible and intangible components and their mutual relations, and by developing in a dynamic and interactive process. The methodological approach proposed constitutes a possible means of constructing alternative intervention strategies in contexts that present the characteristics of the HUL. According to this concept, the system of tangible and intangible relations is an integral part of the local specific characteristics, and requires an integrated approach for its understanding, interpretation and evaluation. The approach proposed in this paper therefore has an innovative value that derives not only from experimenting with the mix of specific techniques to support decision making with a participatory approach, but also from testing these techniques in the context of public policy and cultural heritage management, where the combination of qualitative and quantitative methods seems to offer greater benefits [6]. Another interesting aspect of the work is linked to the use and demonstration of how prescriptive decision-analysis and participatory problem-structuring can generate new consensus alternatives in a real decision-making process. Therefore, the proposed integrated decision aid is thus expected to constitute a transferable framework to support policy makers in their strategic decisions.

In conclusion, a flexible and adaptive methodological course, combining complex evaluation techniques and stakeholder involvement techniques, can help build enhancement strategies and promote good governance [40]; these processes can improve the local deliberative democracy through effective collaboration among developers, operators and users. With the support of integrated evaluation approaches, it is possible to build shared actions in a long-term vision that is aimed at developing and making public decisions effectively.

Acknowledgements. This study has been supported by the project of the Mediterranea University of Reggio Calabria: Marine Energy Lab - PON03PE00012_1 - P.O.N. Ricerca e Competitività 2007-2013 Avviso n. 713/Ric. del 29/10/2010 - Titolo III - "Nuovi distretti e laboratori".

References

1. Della Spina, L., Scrivo, R., Ventura, C., Viglianisi, A.: Urban renewal: negotiation procedures and evaluation models. In: Gervasi, O., Murgante, B., Misra, S., Gavrilova, M.L., Rocha, A.M.A.C., Torre, C., Taniar, D., Apduhan, B.O. (eds.) ICCSA 2015. LNCS, vol. 9157, pp. 88–103. Springer, Cham (2015). doi:10.1007/978-3-319-21470-2_7
2. Stirling, A.: Analysis, participation and power: justification and closure in participatory multi-criteria analysis. Land Use Policy **23**, 95–107 (2006)
3. Liu, S., Sheppard, A., Kriticos, D., Cook, D.: Incorporating uncertainty and social values in managing invasive alien species: a deliberative multi-criteria evaluation approach. Biol. Invasions **13**, 2323–2337 (2011)
4. Creswell, J.W., Plano Clark, V.L.: Designing and Conducting Mixed Methods Research. Sage, Thousand Oaks (2011)
5. Morse, J., Niehaus, L.: Mixed Method Design: Principles and Procedures. Left Coast Press, Walnut Creek (2009)
6. Myllyviita, T., et al.: Mixing methods-assessment of potential benefits for natural resources planning. Scand. J. For. Res. **29**(1), 20–29 (2014)
7. Stewart, T.J., Joubert, A., Janssen, R.: MCDA framework for fishing rights allocation in South Africa. Group Decis. Negot. **19**, 247–265 (2010)
8. Dente, B.: Understanding policy decisions. PoliMI SpringerBriefs. Springer, Heidelberg (2014)
9. Eden, C.: Cognitive mapping: a review. Eur. J. Oper. Res. **36**, 1–13 (1988)

10. Hinloopen, E., Nijkamp, P., Rietveld, P.: The regime method: a new multicriteria technique. In: Hansen, P. (ed.) Essays and Surveys on Multiple Criteria Decision Making. LNE, vol. 209, pp. 146–155. Springer, Heidelberg (1983). doi:10.1007/978-3-642-46473-7_13
11. Hinloopen, E., Nijkamp, P., Rietveld, P.: Qualitative discrete multiple criteria choice models in regional planning. Reg. Sci. Urban Econ. **13**, 73–102 (1983)
12. Hinloopen, E.: De Regime Methode, MA thesis, Interfaculty Actuariat and Econometrics, Free University Amsterdam (1985)
13. Hinloopen, E., Smyth, A.W.: A description of the principles of a new multicriteria evaluation technique, The Regime Method. In: Proceedings Colloquium Vervoersplanologisch Speurwerk (1985)
14. Hinloopen, E., Nijkamp, P.: Quantitative multiple criteria choice analysis. Qual. Quant. **24**, 37–56 (1990)
15. Belton, V., Stewart, T.J.: Multiple Criteria Decision Analysis: An Integrated Approach. Kluwer Academic Publishers, Boston (2002)
16. Janssen, R., Nijkamp, P., Rietveld, P.: Qualitative multicriteria methods in the Netherlands. In: Bana e Costa, C.A. (ed.) Readings in Multiple Criteria Decision Aid, pp. 383–409. Springer, Heidelberg (1990). doi:10.1007/978-3-642-75935-2_17
17. UNESCO: Recommendation on the Historic Urban Landscape. UNESCO World Heritage Centre: Paris, France (2011)
18. Fusco, G.L.: Toward a smart sustainable development of port cities/areas: the role of the "Historic Urban Landscape" approach. Sustainability **5**, 4329–4348 (2013)
19. Calabrò, F., Cassalia, G., Tramontana, C.: The mediterranean diet as cultural landscape value: proposing a model towards the inner areas development process. In: Calabrò, F., Della Spina, L. (eds.) Book Series: Procedia - Social and Behavioral Sciences, vol. 223, 568–575 (2016)
20. Calabrò, F., Della Spina, L.: Innovative tools for the effectiveness and efficiency of administrative action of the metropolitan cities: the strategic operational programme. In: Bevilacqua, C., Calabrò, F., Della Spina, L. (eds.) Advanced Engineering Forum, vol. 11, pp. 3–10. Trans Tech Publications, Switzerland (2014)
21. Bánáthy, B.H.: Guided Evolution of Society: A Systems View (Contemporary Systems Thinking). Springer, Berlin (2000)
22. Jackson, M.: Systems Thinking: Creating Holisms for Managers. Wiley, Chichester (2003)
23. Checkland, P., Poulter, J.: Learning for Action. Wiley, Chichester (2006)
24. Ackoff, R.L.: Systems Thinking for Curious Managers. Triarchy Press, Gillingham (2010)
25. Medda, F., Nijkamp, P.: A combinatorial assessment methodology for complex transport policy analysis. Integr. Assess. **4**(3), 214–222 (2003)
26. Miller, D., Patassini, D. (eds.): Beyond Benefit Cost Analysis: Accounting for Non-market Values in Planning Evaluation. Ashgate, Aldershot (2005)
27. Fusco, G.L., Cerreta, M., De Toro, P.: Integrated planning and integrated evaluation. Theoretical references and methodological approaches. In: Miller, D., Patassini, D. (eds.) Beyond Benefit Cost Analysis: Accounting for Non-market Values in Planning Evaluation, pp. 175–205. Ashgate, Aldershot (2005)
28. Deakin, M., Mitchell, G., Nijkamp, P., Vreeker, R. (eds.): Sustainable Urban Development: The Environmental Assessment Methods, vol. 2. Routledge, Oxon (2007)
29. Munda, G.: Social Multi-criteria Evaluation for a Sustainable Economy. Springer, Heidelberg (2008)
30. Della, S.L., Calabrò, F.: Pianificazione strategica: valutare per programmare e governare lo sviluppo. LaborEst **11**, 3–4 (2015)
31. Gupta, K.: A Practical Guide to Needs Assessment. Wiley, Hoboken (2011)

32. Rosenhead, J., Mingers, J.: Rational Analysis for a Problematic World Revisited: Problem Structuring Methods for Complexity, Uncertainty and Conflict. Wiley, Chichester (2001)
33. Della Spina, L., Ventura, C., Viglianisi, A.: A multicriteria assessment model for selecting strategic projects in urban areas. In: Gervasi, O., et al. (eds.) ICCSA 2016. LNCS, vol. 9788, pp. 414–427. Springer, Cham (2016). doi:10.1007/978-3-319-42111-7_32
34. Figueira, J., Greco, S., Ehrgott, M. (eds.): Multiple Criteria Decision Analysis: State of the Art Survey. Springer, New York (2005)
35. Guarini, M.R., Battisti, F.: Application of a multi-criteria and participated evaluation procedure to select typology of intervention to redevelop degraded urban area. In: Calabrò, F., Della Spina, L. (eds.) Procedia Social and Behavioral Sciences, vol. 223, pp. 960–967 (2016)
36. Balletti, F., Soppa, S.: L'integrazione del paesaggio nelle politiche territoriali. Ricadute sugli strumenti di piano e sulle prassi concertative. S.I.U. - VIII Conferenza Nazionale, Mutamenti del territorio e innovazioni negli strumenti urbanistici, Firenze (2004)
37. ICOMOS: Guidance on Heritage Impact Assessments for Cultural World Heritage properties. Icomos, Paris, France (2011)
38. Rugolo, A., Calabrò, F., Scrivo, R.: The assessment tool for the design of the territories. The impacts from the road infrastructure in the Inland areas of the province of Reggio Calabria. In: Calabrò, F., Della Spina, L. (eds.) Procedia Social and Behavioral Sciences, vol. 223, pp. 534–541 (2016)
39. Janssen, R., van Herwijnen, M., Beinat, E.: DEFINITE (version 2) DEcision support system for a FINITE set of alternatives. Institute for Environmental Studies, Vrije Universiteit, Amsterdam (2001)
40. Calabrò, F., Della Spina, L., Sturiale, L.: Cultural planning: a model of governance of the landscape and cultural resources in development strategies in rural contexts. In: Proceedings of the XVII - IPSAPA Interdisciplinary Scientific Conference, vol. V, pp. 177–188. Rezekne Higher Educ Inst-Rezeknes Augstskola, Lettonia (2013)
41. Campanella, R.: Un progetto di territorio per il turismo sostenibile l'esperienza di ricerca applicata del PISL "Slow Life. Viaggio tra culture e natura nel Parco Nazionale d'Aspromonte, dal Tre Pizzi al Limina". LaborEst **10**, 17–22 (2015)
42. Fusco Girard, L., Nijkamp, P.: Le valutazioni per lo sviluppo sostenibile delle città e del territorio, Angeli, Milano (1997)
43. Mollica, E.: Le Aree interne della Calabria: una strategia e un piano quadro per la valorizzazione delle loro risorse endogene. Rubbettino, Soveria Mannelli (1996)

Study of Parameter Sensitivity on Bat Algorithm

Iago Augusto Carvalho[1], Daniel G. da Rocha[2], João Gabriel Rocha Silva[3], Viníčus da Fonseca Vieira[2], and Carolina Ribeiro Xavier[2(✉)]

[1] Department of Computer Science, Universidade Federal de Minas Gerais, Belo Horizonte, Brazil

[2] Department of Computer Science, Universidade Federal de São João del Rei, São João del Rei, Brazil
carolinaxavier@ufsj.edu.br

[3] Postgraduate Program in Computational Modeling, Universidade Federal de Juiz de Fora, Juiz de Fora, Brazil

Abstract. Heuristics and metaheuristics are known to be sensitive to input parameters. Bat algorithm (BA), a recent optimization metaheuristic, has a great number of input parameters that need to be adjusted in order to increase the quality of the results. Despites the crescent number of works with BA in literature, to the best of our knowledge, there is no work that aims the fine tuning of the parameters. In this work we use benchmark functions and more than 9 millions tests with BA in order to find the best set of parameters. Our experiments shown that we can have almost 14000% of difference in objective function value between the best and the worst set of parameters. Finally, this work shows how to choose input parameters in order to make Bat Algorithm to achieve better results.

Keywords: Bat algorithm · Sensitivity · Parameter analysis · Nonlinear optimization · Unconstrained optimization

1 Introduction

Bat Algorithm (BA) is a metaheuristic proposed in [1], based on bat echolocation behavior when hunting their prey. The metaheuristic is capable to converge to global optimum in many situations, not being trapped in local optima, proving to be an efficient optimization method. BA is being widely used in a great number of problems, like engineering problems [2], parameter estimation [3], vehicle routing problem [4], feature selection [5], numerical problems [6], inverse problems [7] and others. In addition to these applications, a greater number of variations of the BA are currently in development, like a Multi-Objective BA [8], a Binary BA [5] and a Fuzzy Logic BA [9].

Several works in literature presented studies regarding metaheuristic parameters sensitivity analysis [10–13]. Despite the great number of works with BA,

O. Gervasi et al. (Eds.): ICCSA 2017, Part I, LNCS 10404, pp. 494–508, 2017.
DOI: 10.1007/978-3-319-62392-4_36

to the best of our knowledge, the only work that analysis BA parameter's sensitivity is [14]. It shows that input parameters have a great influence in algorithm convergence, like in all other metaheuristics [15]. Variation of the BA's parameters, like loudness value, bat pulse rate, frequency and flying velocity, besides α and γ values has a significant influence on BA. The choice of appropriate parameters can accelerate convergence and makes easier for bats to escape from local optimum. On the other hand, inappropriate parameters can make the algorithm to stagnate, or harm the convergence.

In this paper, we extend the work of [14] performing an exhaustive number of tests with the variation of the BA parameters. We performed a combination of all parameters, running more than 9 million tests in order to measure the most appropriate set of parameters for BA in a set of mathematical benchmark functions.

This work is organized as follows. First, we describe the bat algorithm by idealizing the echolocation behavior of bats in Sect. 2. Next, we describe the test bed, including an overview of the benchmark mathematical functions used and the parameter variation scheme in Sect. 3. Finally, we present our results and discuss some implications and further studies in the last section.

2 Bat Algorithm

Bats, in especial microbats, use they echolocation to fly in the dark, avoiding obstacles and chasing a prey. They emit a very loud sound pulse and discover their surroundings by listening the echo that bounces back from surrounding objects. Such pulses typically travel no more than a few meters, depending on the actual frequency and can be as loud as 110 dB [16].

Bats sound pulse can be correlated to his hunting strategy: when they are just flying in the dark, each ultrasonic burst may last 5 ms to 20 ms, and microbats emit about 10 to 20 sound bursts every second. When a bat is hunting some prey, their can emit up to 200 pulses per second [17].

Bats have a incredible capacity to almost instantly process this great quantity of signals and discover exactly where are all objects in the ambient, building up a three dimensional scenario. In fact, microbats have the ability to avoid obstacles as small as human hairs. They can detect the distance and orientation of the target, and even the moving speed of the prey such as small insects using the time delay from emission and detection of the pulses, the time difference between their two ears and the loudness variation of the echoes. Some studies suggested that bats are able to discriminate targets by the variations of the Doppler effect [16,17].

It is possible to formulate bats echolocation making a association with an objective function to be optimized.

The original description of Bat Algorithm was presented by Yang [1]. In this section, we present an overview of the BA, explaining the basics of how it works, making a connection between BA and the bats echolocation behavior.

The objective function value corresponds to the distance that a virtual bat is from his prey. The heuristic consists in the chase of a swarm of virtual bats for a prey, localized in the global optimum. As closer virtual bats are from the prey, better is the objective function value.

Frequency f_{max} refers to the maximum bat pulse frequency, and is related to the pulse wavelengths size, being constant during bats flight. Loudness A_i refers to bats pulse loudness. Loudness varies from the loudest when a bat is searching for his prey to a quieter base when he closer to his prey. Pulse rate r_i refers to the rate that a bat emit his ultrasonic pulse, working together with loudness value. When bats are flying and searching for a prey, their have lower values pulse rate. When closer to his prey, then they need higher values of pulse rate in order to get a great accuracy about the prey position [17]. α and γ values adjust the pulse loudness and the pulse rate. The former makes the pulse loudness get lower when a bat is getting closer to they prey. The latter makes the pulse rate increases at the same time.

Velocity v_i is related to the velocity that each bat fly. It corresponds to the bat displacement. As higher is the bat velocity, faster he moves through the search space. Small velocities makes BA converge slower, but ensure that it will not be trapped in local optimum.

Idealizing some echolocation characteristics of bats, especially microbats, it is possible to develop various algorithms. The algorithm proposed by [1] uses three approximate rules:

1. All bats use echolocation to sense distance, and they also 'know' the difference between food/prey and background barriers in some magical way;
2. Bats fly randomly with velocity v_i at position x_i with a fixed frequency f_{min}, varying wavelength λ and loudness A_0 to search for prey. They can automatically adjust the wavelength (or frequency) of their emitted pulses and adjust the rate of pulse emission r in the range of $[0, 1]$, depending on the proximity of their target;
3. Although the loudness can vary in many ways, we assume that the loudness varies from a large positive A_0 to a minimum constant value A_{min}.

BA do not necessarily use the wavelengths themselves, instead. Instead, it can vary the frequency while fixing the wavelength λ. This is due λ and f relationship, as λf is constant. Therefore, just one parameter to be adjusted. As higher is the frequency, then quickly is the response from where is the prey.

We also assume that f has a value between $[0, f_{max}]$ and pulse rate in the interval $[0, 1]$, where 0 means no pulses at all, and 1 means the maximum rate of pulse emission. The closer the bats come from their prey, greater is the pulse rate value. They need the maximum emission pulse to be able to locate their prey with greater accuracy in small distances.

At last, the position x_i of a virtual bat corresponds to a value in the dimension i of the objective function used and x_* denotes the value of the best solution so far.

2.1 Movement of Virtual Bats

For each virtual bat, we need to define how it moves through the search space. We can define the rules in which their positions x_i and velocities v_i are updated in a d-dimensional search space, in a certain time step t, by the following equations:

$$f_i = f_{min} + (f_{max} - f_{min})\beta \tag{1}$$

$$v_i^t = v_i^{t-1} + (x_i^t - x_*)f_i \tag{2}$$

$$x_i^t = x_i^{t-1} + v_i^t \tag{3}$$

where $\beta \in [0, 1]$ is a random vector drawn from a uniform distribution. Here x_*^{t-1} is the last interaction global best location (solution) between all n bats.

In our implementation, we use $f_{min} = 0$ in all cases and f_{max} is a parameter that varies in the tests. Initially, a frequency value randomly drawn from the interval $[f_{min}, f_{max}]$ is to each bat. In the local search step, a new solution is generated locally for each bat using random walk, as follows.

$$x_i = x_i + \varepsilon A^t \tag{4}$$

where ε is a random number in the interval $[-1, 1]$, while $A^t =< A_i^t >$ is the average loudness of all bats at time step t.

2.2 Loudness and Pulse Emission

Loudness A_i values and the pulse emission rate r_i values have to be updated at each iteration of the algorithm. As the loudness usually decreases once a bat has found its prey, while the rate of pulse emission increases, the loudness can be chosen as any convenient value. For example, we can use $A_0 = 100$ and $A_{min} = 1$. When $A_{min} = 1$, a bat has found the prey and temporarily stop emitting any sound. Now we have:

$$A_i^t = \alpha A_i^{t-1} \tag{5}$$

$$r_i^t = r_i^{t-1}[1 - exp(-\gamma t)] \tag{6}$$

where α and γ are constants parameters that was varied in tests, always being $\alpha = \gamma$. These equations guarantees that BA reproduces the real bats behavior in nature.

Equation 5 enforces the loudness value to decrease as the algorithm converge. Equation 6 guarantees the pulse rate to increase as the algorithm converge, due the fact that $exp(-\gamma t)$ always result in a negative value. The number of generations also have influence at Eq. 6, making r_i to increase faster when BA is running for a long time.

3 Test Bed

This section present the test bed used for conducting the experiments in order to analyze the sensitivity of the parameter for Bat Algorithm. All tests were performed in a single core of PC with Intel® Xeon® CPU E5645 2.40 GHz and 32 GB memory.

3.1 Mathematical Benchmark Functions

Mathematical Benchmark Functions, or Artificial Landscapes, are useful to evaluate characteristics of optimization algorithms, such as velocity of convergence, precision, robustness and general performance.

We considered 15 benchmark unconstrained functions to evaluate the parameters variation for the BA. Benchmark functions were obtained from [12,18,19], which are presented in Table 1. Variable x corresponds to a d-dimensional decision vector, and the constants o are a d-dimensional vector that corresponds to the shifted global optimum.

The set of considered functions is presented in Table 1 covers a wide range of types of functions. Functions $f1$ to $f4$ are continuous, convex, bowl-shaped and unimodal functions. Function $f4$ have its coordinates rotated, making the original hyper-ellipsoid function harder to optimize. Functions $f5$ and $f6$ are continuous and unimodal, but not bowl-shaped. Function $f7$ to $f15$ are multi-modal functions, with a great number of local optimums. Function $f7$ and $f8$ are continuous, valley-shaped, and the global minimum lies in a very narrow valley. Function $f12$ and $f13$ are highly multi-modal, but locations of the minima are regularly distributed. Functions $f2, f6, f8, f11$ and $f13$ have their global optimums shifted. All functions are d-dimensional.

3.2 Parameters Variation

In order to test the sensitivity of BA, we vary the parameters of the algorithm in different ways, as shown in Table 2. Only one parameter on each run, resulting in 25200 sets of parameters for each function. Each set of parameters was executed 24 times in order to investigate the statistical relevance of the results.

According to [1], initial loudness A_i^0 can be typically $[1, 2]$, while the initial emission rate r_i^0 can be around 0. Furthermore, α and γ initial values are close to 1, acting similar to the cooling factor of a cooling schedule in the simulated annealing [20]. We fixed 40 individuals in population, as it demonstrated to archive good results in [21].

Each virtual bat have their initial parameters initialized independently. For pulse rate, frequency, loudness and α and γ, a value between 0 and the parameter value was randomly chosen for each virtual bat. For the loudness value, we randomly chose a value between 0 and the parameter value, then we added 1 to the result, in order to get a value between 1 and 2, like proposed in [1].

Table 1. Benchmark test functions used

	Name	Function	Domain
$f1$	Sphere	$\sum_{i=1}^{d} x_i^2$	±100
$f2$	Shifted sphere	$\sum_{i=1}^{d} (x-o)_i^2 + f\_bias$	±100
$f3$	Sum squares	$\sum_{i=1}^{d} ix_i^2$	±10
$f4$	Rotated hyper-ellipsoid	$\sum_{i=1}^{d}\sum_{j=1}^{i} x_j^2$	±5.12
$f5$	Schwefel problem	$max\ \|x_i\|, \forall i \in [1,\ldots,d]$	±100
$f6$	Shifted schwefel problem	$max\ \|x_i - o\| + f\_bias, \forall i \in [1,\ldots,d]$	±100
$f7$	Rosenbrock	$\sum_{i=1}^{d-1} [100(x_{i+1} - x_i^2)^2 + (x_i - 1)^2]$	±100
$f8$	Shifted rosenbrock	$\sum_{i=1}^{d-1} [100(x_{i+1} - x_i^2)^2 + (x_i - 1)^2] + f\_bias$	±100
$f9$	Schwefel ridge	$\sum_{i=1}^{d-1} \left(\sum_{j=1}^{i} x_j\right)^2$	±100
$f10$	Griewank	$\sum_{i=1}^{d} \frac{x_i^2}{4000} - \prod_{i=1}^{d} \cos\left(\frac{x_i}{\sqrt{i}}\right) + 1$	±600
$f11$	Shifted griewank	$\sum_{i=1}^{d} \frac{(x-o)_i^2}{4000} - \prod_{i=1}^{d} \cos\left(\frac{(x-o)_i}{\sqrt{i}}\right) + 1 + f\_bias$	±600
$f12$	Rastrigin	$10d + \sum_{i=1}^{d} [x_i^2 - 10\cos(2\pi x_i)]$	±5.12
$f13$	Shifted rastrigin	$10d + \sum_{i=1}^{d} [x_i^2 - 10\cos(2\pi x_i)] + f\_bias$	±5.12
$f14$	Salomon	$-\cos\left(2\pi\sqrt{\sum_{i=1}^{d} x_i^2}\right) + \sqrt{\sum_{i=1}^{d} x_i^2} + 1$	±100
$f15$	Levy #3	$\sin^2(\pi w_1) + \sum_{i=1}^{d} (w_i - 1)^2[1 + 10\sin^2(\pi w_i + 1)]$ $+(w_d - 1)^2[1 + \sin^2(2\pi w_d)],\ w_i = 1 + \frac{x_i - 1}{4}$	±10

Table 2. Parameters variation for BA

Parameter	Values						
Function dimension	2	5	15	30			
Frequency	1	2	5	10	15	20	30
Loudness	0.5	0.6	0.7	0.8	0.9	1.0	
Pulse rate	0.05	0.10	0.15	0.20	0.25	0.30	
Velocity	0.01	0.20	0.40	0.60	0.80		
α and γ	0.80	0.85	0.90	0.95	1.00		

4 Results and Discussion

In order to evaluate the best set of parameters to be used in BA, we made several tests with the algorithm. For these tests, we used the average results of each set of parameters after 24 executions. The stopping criterion used for each test was 100.000 function objective evaluations. The results are described in the following sections.

4.1 Statistical Analysis

Our results were subjected to statistical analysis, in order to evaluate some difference between parameters at the end of 100.000 evaluations of all functions.

Statistical analysis were performed in the R software. First, we made an ANOVA [22] analysis on our results. Then, a Tukey Honestly Significant Difference test (Tukey-HSD) [23] was developed with the ANOVA table data, in order to find out if any parameter value statistically differs from another. Tukey-HSD tests were performed with 95% confidence level. Results are shown on Figs. 1 and 2.

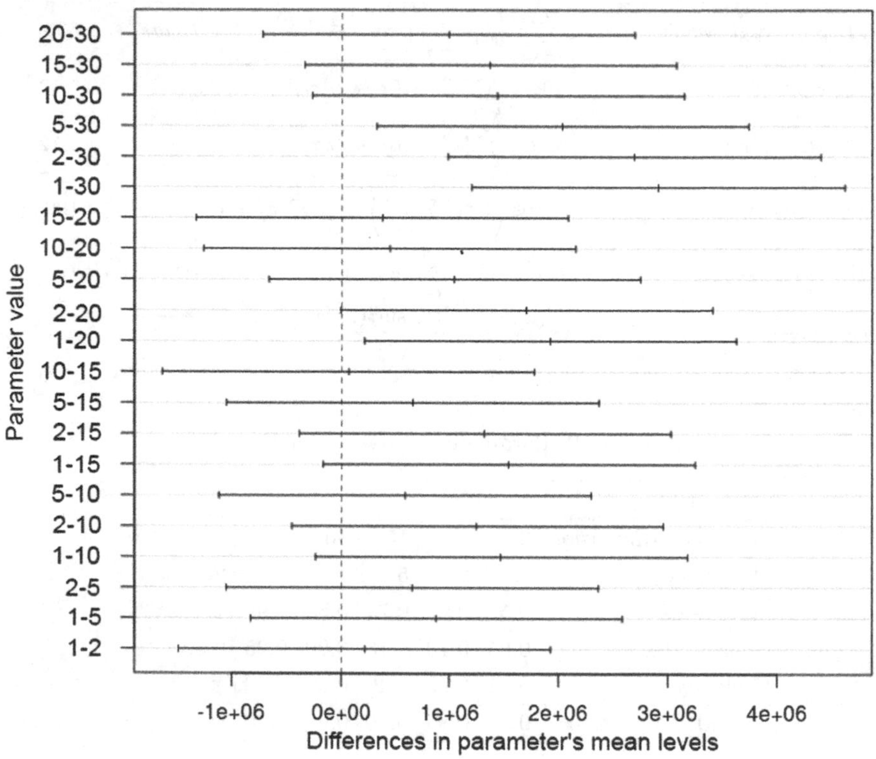

Fig. 1. Tukey-HSD test on BA Frequency. There is significant difference among parameter value, with 95% confidence level

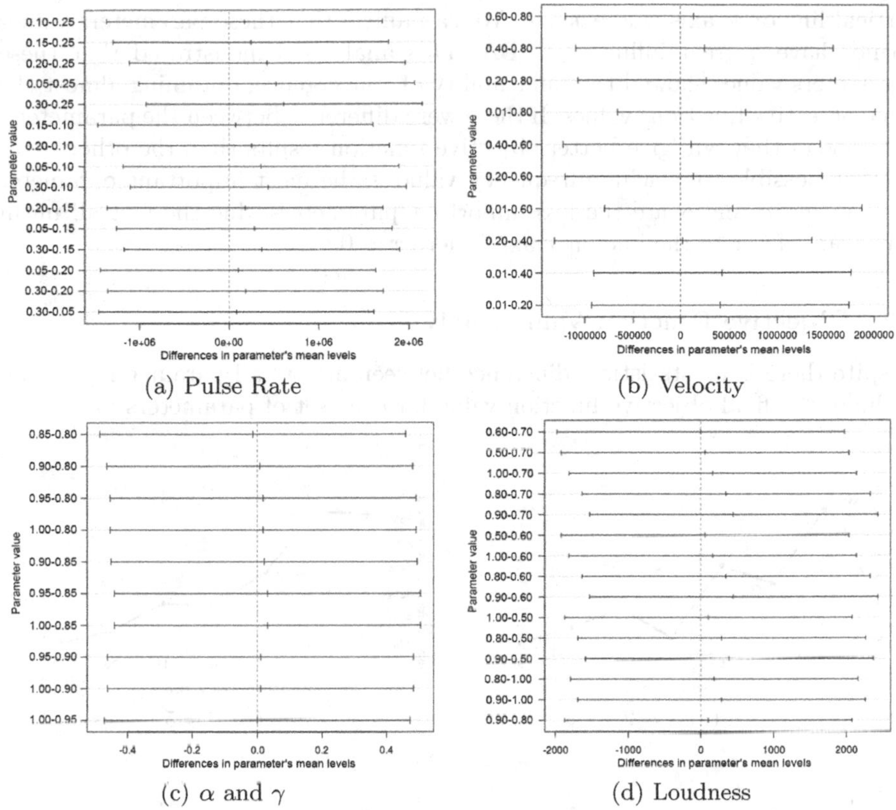

Fig. 2. Tukey-HSD test on some BA parameters. There were no significant difference among parameters values, with 95% confidence level

We analyzed the results of all fifteen functions together in order to verify statistical differences among the values of the parameters, in terms of objective function value. In each test, the different values of a single parameter were used as analysis factors for construction of ANOVA table and Tukey-HSD test. Figures 1 and 2 shows the Tukey-HSD tests for each parameter, where the x axis represents the test value and each value of y axis represents the comparison between two different parameter values.

Figure 1 shows that there is a statistical difference between pulse frequency values. The comparisons on Fig. 1 are not all over the vertical line on x axis where $x = 0$. This means that the choose of BA frequency value has a significant influence on the algorithm convergence. There are significant difference between the values 1–20, 1–30, 2–30 and 5–30. Furthermore, it is possible to see that exists a greater difference between the set 1–30, because they are more distant from the vertical line.

Figure 2 shows that there is no statistical difference between α and γ, pulse rate, pulse loudness and virtual bat's velocity, due all comparisons are over the vertical line on x axis where $x = 0$. We can affirm that these parameters values do not have a great influence in BA. This analysis demonstrated that these parameters values showed the same ability of convergence, obtaining almost the same objective functions values. If there were difference between the parameters, then one of they will give better objective function results than the other.

It is possible to conclude frequency value is the most important parameter on BA, and α and γ are the less important parameters, due the fact all mean values are closer to the line on x axis where $x = 0$.

4.2 Objective Function Value Analysis

Despite there is no statistical difference between almost all parameters, we can evaluate the final objective function value for each set of parameters.

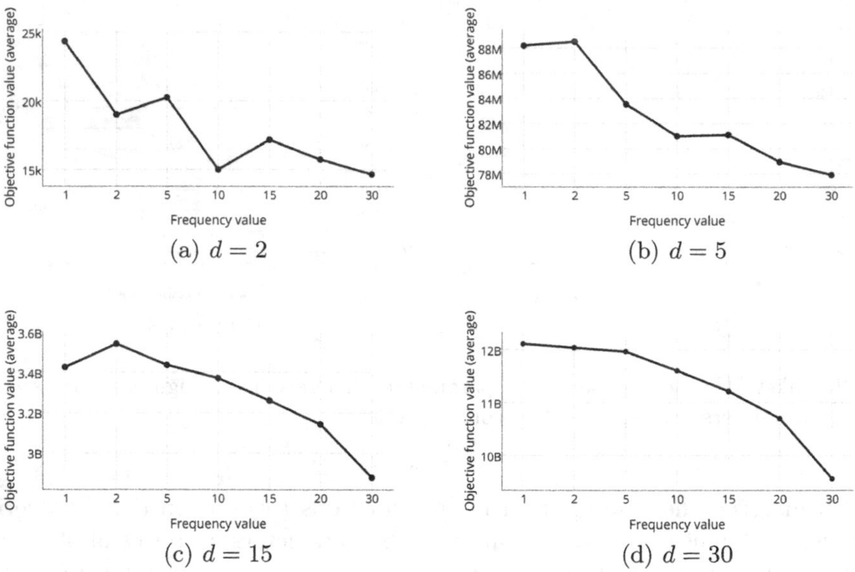

Fig. 3. Average objective function value for each value of frequency, for the sum of all functions of each dimensionality

Figure 3 shows the sum of average of the best objective function value after 100.000 evaluations, for each pulse frequency value. Each subfigure corresponds to a function dimensionality. There is significant difference between values of experiments with low pulse frequency and high pulse frequency. As higher is the pulse frequency, better is the objective function value. Clearly, high pulse frequencies make bats fly greater distances on each iteration, enabling a faster convergence. Higher values of pulse frequency makes BA more robust, being able to converge to better results with any evaluated benchmark function with any dimension.

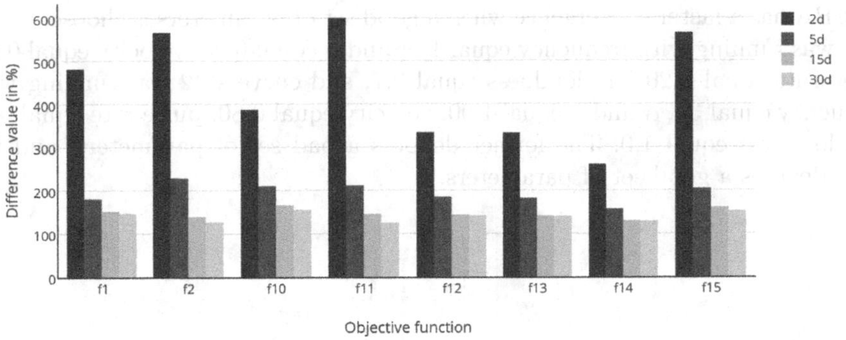

(a) Unimodal functions

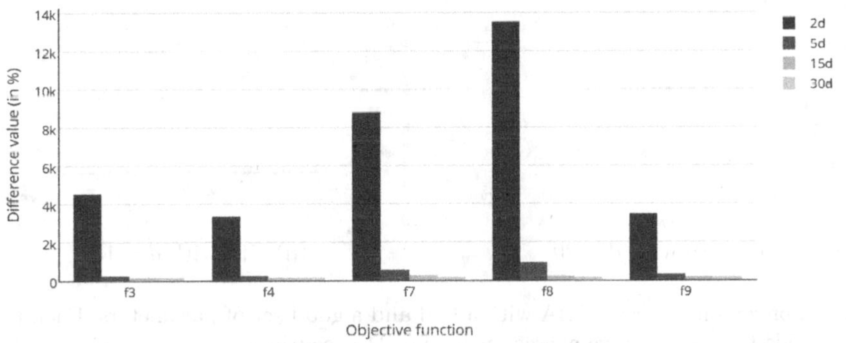

(b) Multimodal functions

Fig. 4. Difference between the best set of parameters and the worst set of parameters for each objective function, in %

Figure 4 illustrates the ratio between objective functions value after 100.000 evaluations, for each function of the test bed (presented as percentage). Figure 4 shows that, for simpler objective functions, the influence of the parameter set used is greater. Function $f8$ shows results obtained by the best set of parameters is almost 14.000% better than the results obtained by the worst set of parameters, with 2 dimensions, but only 169% better for 30 dimensions. Multimodal functions with 2 dimensions has a smaller difference than unimodal functions, but it is still show an improvement of more than 300% with the appropriate set of parameters. All high-dimensional functions have smaller differences than low-dimensional functions.

The difference in BA convergence speed with a good set of parameters and a bad set of parameters for some functions are shown at Fig. 5. Its makes clear that BA has a faster convergence when a good set of parameters is choose. Curve *set*1 was running with frequency equal 1, α and γ equal 0.95, velocity equal 0.01, pulse rate equal 0.20 and loudness equal 0.7, and curve *set*2 was running with frequency equal 30, α and γ equal 1.00, velocity equal 0.80, pulse rate equal 0.05 and loudness equal 1.0. The former denotes a bad set of parameters, and the later denotes a good set of parameters.

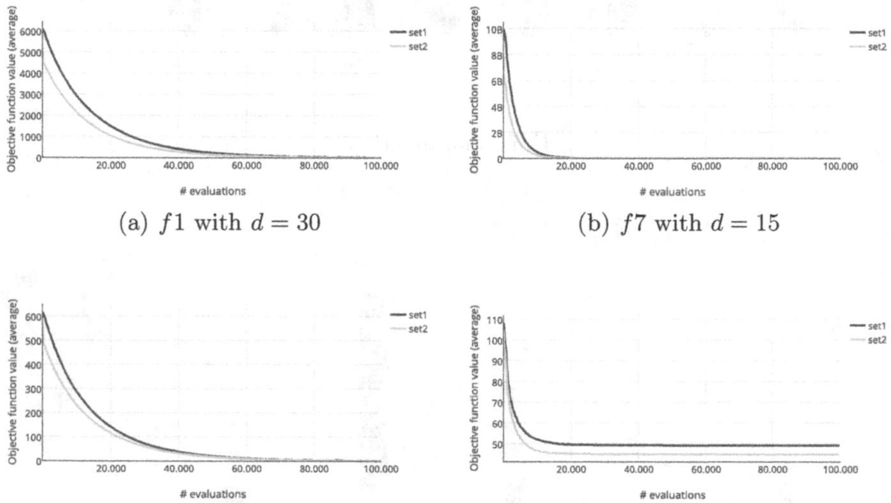

(a) $f1$ with $d = 30$

(b) $f7$ with $d = 15$

(c) $f10$ with $d = 30$

(d) $f15$ with $d = 15$

Fig. 5. Convergence curve of BA with a bad and a good set of parameters. Each point corresponds to the objective average value for 24 executions

In order to evaluate BA convergence, we can compare the metaheuristic with another well-established optimization method: the Differential Evolution algorithm (DE) [12]. DE is a direct search method that utilizes a population of d-dimensional parameter vectors as a population for each generation. Besides, it utilizes three basic operators, respectively called mutation, crossover and selection. A large number of works in the literature utilizes DE [24], that was initially developed for global optimization problems.

Figure 6 illustrates the convergence of BA and DE for some objective functions. DE parameters where choose according [25] and BA parameters set was the same *set*2 from Fig. 5.

Like shown in Fig. 6, BA convergence differs a lot from DE convergence. BA convergence is characterized by a curve that that decreases uniformly and slowly, while DE convergence curve has no defined behavior. In BA, each virtual bat "flies" just a small distance of the solution space to each generation when generating new solutions adjusting the virtual bat parameters. In DE, the mutation scheme allows great leaps in the search space between two consecutive generations.

Table 3. Frequency that parameters achieve the best and the worst values, for all objective functions

Frequency	Values	1	2	5	10	15	20	30
	Best	4.8%	5.2%	5.9%	7.8%	8.5%	12.2%	**55.3%**
	Worst	29.1%	28.8%	19.0%	11.1%	7.6%	5.1%	**2.9%**
α and γ	Values	0.80	0.85	0.90	0.95	1.00		
	Best	0%	0%	0%	0%	**99.9%**		
	Worst	27.7%	24.4%	24.7%	23.2%	**0%**		
Velocity	Values	0.01	0.20	0.40	0.60	0.80		
	Best	15.5%	17.3%	19.7%	21.5%	**27.7%**		
	Worst	30.5%	17.4%	**16.9%**	17.2%	17.7%		
Pulse rate	Values	0.05	0.10	0.15	0.20	0.25	0.30	
	Best	**19.4%**	17.8%	16.7%	15.1%	15.7%	15.0%	
	Worst	**14.6%**	15.4%	16.2%	16.9%	18.1%	18.6%	
Loudness	Values	0.5	0.6	0.7	0.8	0.9	1.0	
	Best	15.7%	17.1%	16.6%	16.6%	17.2%	**18.6%**	
	Worst	17.1%	16.6%	16.4%	**16.0%**	16.6%	17.5%	

Analyzing the average objective function value, for each function and dimensionality, DE achieves better results than BA. The former achieved better results in 36 sets of functions and dimensionality, and the latter gets better results in 24 sets. When analyzing different functions dimensionality, DE achieves better results in functions with $d = [2, 5]$, obtaining better results in 73% of objective functions sets, and BA achieves better results in high dimensional functions (i.e. functions with $d = [15, 30]$), obtaining better results in 53% of objective functions sets.

4.3 How to Choose a Good Set of Parameters

Input parameters have great influence on Bat Algorithm convergence curve. It is very important to choose the right set of parameters in order to obtain better results.

The most prominent parameter is the pulse frequency f_{max}. Greater values of pulse frequency can lead to better results, like shown in Fig. 3. Despite there is not significant difference in the others parameters, a wise choose of them can lead to better results.

By analyzing each BA input parameter separately, we infer the best set of parameters for the heuristic. We performed a comparison between each set of parameters, just modifying the value of one input. Results are shown in Table 3. Column "Best" indicates the percentage that this parameter value is best than all other parameter values, with all other parameters fixed. In the same way,

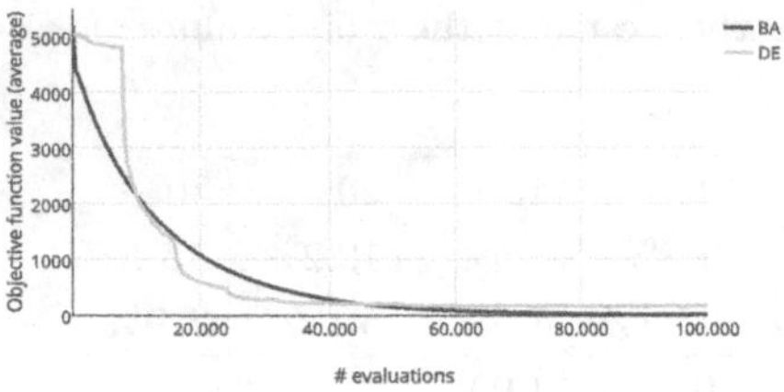

(a) $f1$ with $d = 30$

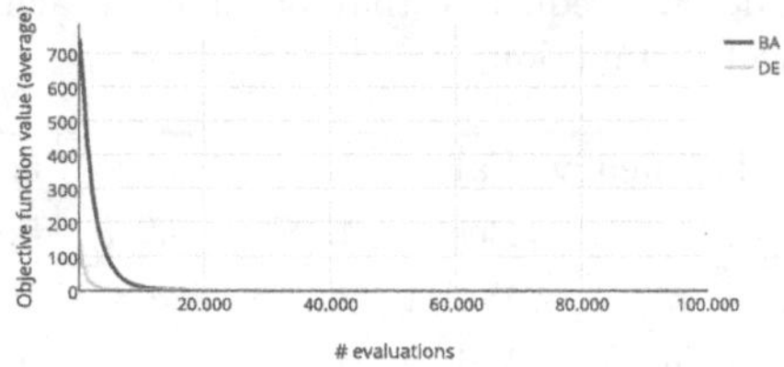

(b) $f4$ with $d = 15$

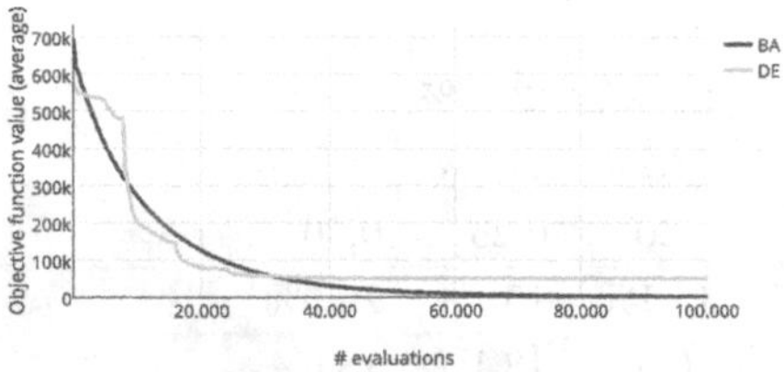

(c) $f4$ with $d = 30$

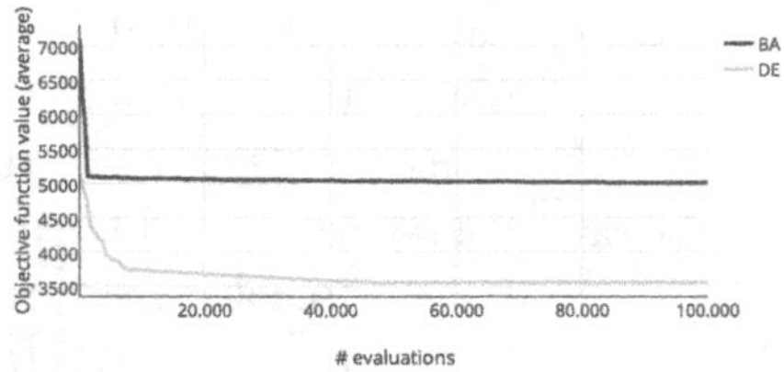

(d) $f8$ with $d = 2$

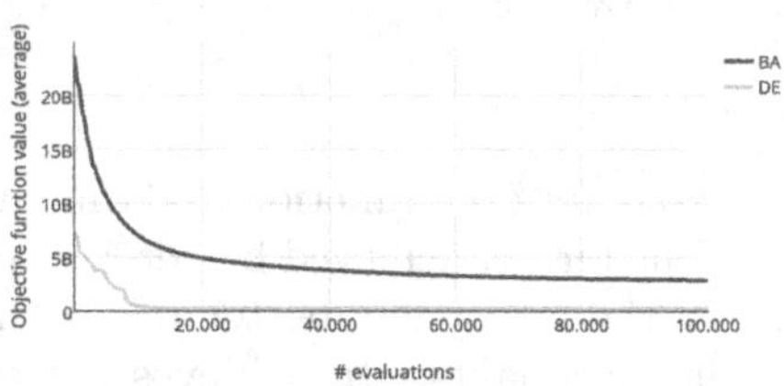

(e) $f8$ with $d = 15$

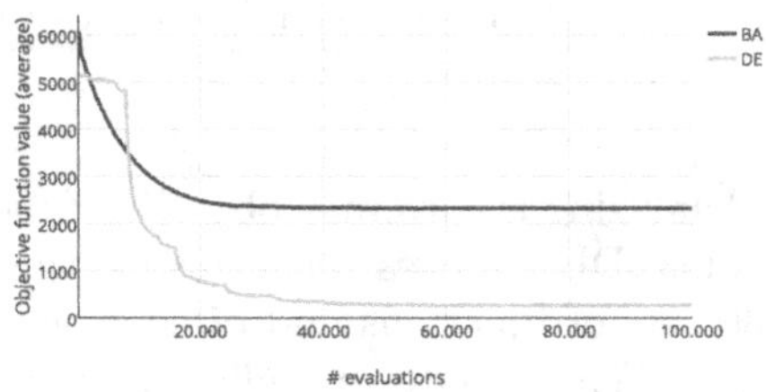

(f) $f13$ with $d = 30$

Fig. 6. Convergence curve of well tuned BA and DE. Each point corresponds to the objective average value for 24 executions

column "Worst" indicates the percentage that this parameter value is worst than all other parameter values, with all other parameters fixed.

In order to achieve the best results, according to Table 3, we can set high values of pulse frequency, alpha and gamma equal 1, small values of pulse rate and higher values of velocity. The pulse loudness value is insignificant: all loudness values tested obtain approximately almost the same results. A recommended set of parameters is frequency equal 30, α and γ equal 1.00, velocity equal 0.80, pulse rate equal 0.05, loudness equal 1.0 and population size = 40.

5 Conclusions

Bat Algorithm input parameters have great influence on the heuristic. We found results that the best set of parameters can be almost 14000 percent better than the worst set of parameters for an objective function.

The defined parameter set is robust enough to optimize all functions tested, enabling faster convergence and reaching better results when compared with another parameters sets. A well-tuned Bat Algorithm can optimize the evaluated functions as good as a well-tuned Differential Evolution algorithm, especially for high dimensional functions.

Further studies can be done at BA parameters, exploring an even larger amount of parameter sets and different applications. This is an important job, not only in BA, but for all others heuristics and metaheuristics. Therefore, one can extract the maximum potential of these algorithms. However, it is an exhaustive, costly and hard task. But the results worth all the hard work.

Acknowledgment. The authors thank CNPq, FAPEMIG and CAPES for the financial support.

References

1. Yang, X.-S.: A new metaheuristic bat-inspired algorithm. Stud. Comput. Intell. **284**, 65–74 (2010)
2. Yang, X.-S., Gandomi, A.H.: Bat algorithm: a novel approach for global engineering optimization. Eng. Comput. **29**(5), 464–483 (2012)
3. Lin, J.H., Chou, C.W., Yang, C.H., Tsai, H.L.: A chaotic levy flight bat algorithm for parameter estimation in nonlinear dynamic biological systems. J. Comput. Inf. Technol. **2**(2), 57–63 (2012)
4. Zhou, Y., Xie, J., Zheng, H.: A hybrid bat algorithm with path relinking for capacitated vehicle routing problem. Math. Probl. Eng. **2013**, 10 p. (2013). Article ID 392789. doi:10.1155/2013/392789
5. Nakamura, R.Y.M., Pereira, L.A.M., Costa, K.A., Rodrigues, D., Papa, J.P., Yang, X.-S.: BBA: a binary bat algorithm for feature selection. In: 2012 25th SIBGRAPI Conference on Graphics, Patterns and Images (SIBGRAPI), pp. 291–297. IEEE, Ouro Preto (2012)
6. Tsai, P.W., Pan, J.S., Liao, B.Y., Tsai, M., Istanda, V.: Bat algorithm inspired algorithm for solving numerical optimization problems. Appl. Mech. Mater. **148–149**, 134–137 (2011)
7. Yang, X.S., Karamanoglu, M., Fong, S.: Bat algorithm for topology optimization in microelectronic applications. In: 2012 International Conference on Future Generation Communication Technology (FGCT), pp. 150–155. IEEE (2012)
8. Yang, X.S.: Bat algorithm for multi-objective optimisation. Int. J. Bio-Inspired Comput. **3**(5), 267–274 (2011)
9. Khan, K., Nikov, A., Sahai, A.: A fuzzy bat clustering method for ergonomic ccreening of fofice workplaces. In: Dicheva, D., Markov, Z., Stefanova, E. (eds.) Third International Conference on Software, Services and Semantic Technologies S3T 2011. AINSC, vol. 101, pp. 59–66. Springer, Heidelberg (2011). doi:10.1007/978-3-642-23163-6_9
10. Akay, B., Karaboga, D.: Parameter tuning for the artificial bee colony algorithm. In: Nguyen, N.T., Kowalczyk, R., Chen, S.-M. (eds.) ICCCI 2009. LNCS, vol. 5796, pp. 608–619. Springer, Heidelberg (2009). doi:10.1007/978-3-642-04441-0_53
11. Lobo, F., Lima, C.F., Michalewicz, Z.: Parameter Setting in Evolutionary Algorithms, vol. 54. Springer Science & Business Media, Heidelberg (2007). doi:10.1007/978-3-540-69432-8

12. Price, K.V., Storn, R.M., Lampinen, J.A.: Differential Evolution: A Pratical Approach to Global Optimization. Natural Computing Series. Springer, Heidelberg (2006). doi:10.1007/3-540-31306-0
13. Coy, S.P., Golden, B.L., Runger, G.C., Wasil, E.A.: Using experimental design to find effective parameter settings for heuristics. J. Heuristics **7**(1), 77–97 (2001)
14. Cordeiro, J., Parpinelli, R.S., Lopes, H.S.: Análise de Sensibilidade dos Parâmetros do Bat Algorithm e Comparação de Desempenho. In: Encontro Nacional de Inteligência Artificial (ENIA), vol. 1, pp. 1–9 (2012)
15. Boussaïd, I., Lepagnot, J., Siarry, P.: A survey on optimization metaheuristics. Inf. Sci. **237**, 82–117 (2013)
16. Richardson, P.: Bats. Natural History Museum, London (2008)
17. Altringham, J.: Bats: From Evolution to Conservation. Oxford Biology. OUP, Oxford (2011)
18. Gavana, A.: Test functions index
19. Tang, K., Yao, X., Suganthan, P.N., MacNish, C., Chen, Y.-P., Chen, C.-M., Yang, Z.: Benchmark functions for the CEC'2008 special session and competition on large scale global optimization. Technical report, Nature Inspired Computation and Applications Laboratory, USTC, Hefei, China (2007)
20. Kirkpatrick, S., Gelatt, C.D., Vecchi, M.P.: Optimization by simmulated annealing. Science **220**(4598), 671–680 (1983)
21. Goel, N., Gupta, D., Goel, S.: Performance of firefly and bat algorithm for unconstrained optimization problems. Int. J. Adv. Res. Comput. Sci. Softw. Eng. **3**(5), 1405–1409 (2013)
22. Field, A.P.: Analysis of variance (ANOVA). In: Encyclopedia of Measurement and Statistics, 1st edn., pp. 33–36. SAGE Publications Inc. (2006)
23. Lane, D.M.: Tukey's honestly significant difference (HSD). In: Encyclopedia of Research Design, pp. 1566–1571. SAGE Publications Inc. (2010)
24. Das, S., Suganthan, P.N.: Differential evolution: a survey of the state-of-the-art. IEEE Trans. Evol. Comput. **15**(1), 4–31 (2011)
25. Ronkkonen, J., Kukkonen, S., Price, K.V.: Real-parameter optimization with differential evolution. In: Proceedings of the IEEE CEC, vol. 1, pp. 506–513 (2005)

A Control Mechanism for Live Migration with Data Regulations Preservation

Toshihiro Uchibayashi[1(✉)], Yuichi Hashi[2], Seira Hidano[3], Shinsaku Kiyomoto[3], Bernady Apduhan[4], Toru Abe[1], Takuo Suganuma[1], and Masahiro Hiji[1]

[1] Tohoku University, Sendai, Japan
uchibayashi@ci.cc, {beto, suganuma, hiji}@tohoku.ac.jp
[2] Hitachi Solutions East Japan Ltd., Kawasaki, Japan
yuichi.hashi.wg@hitachi-solutions.com
[3] KDDI Research Inc., Fujimino, Japan
{se-hidano, kiyomoto}@kddi-research.jp
[4] Kyushu Sangyo University, Fukuoka, Japan
bob@is.kyusan-u.ac.jp

Abstract. In this paper, we propose a data protection mechanism for live migration process. The proposed data protection mechanism verifies whether it is permitted to copy the data of from the migration source application to the migration destination based on the contents of the regulations concerning the use of the data issued by the organization and the country at the time of during the live migration procedures. This mechanism performs live migration only when copying is permitted. By applying this mechanism, it is possible to use appropriate data while protecting privacy during the live migration process. Detailed explanation and implementation evaluation of the data protection mechanism were carried out. As a result, we observed that the mechanism did not exhibit adverse effects on the live migration process.

Keywords: Cloud computing · Live migration · Audit system · Privacy policy · Data protection

1 Introduction

1.1 Background

Lately, the paradigm in which novel, useful knowledge is discovered through analysis of big data which is measured by many internet-connected devices is attracting wide attention. The widespread use of sensors and devices is particularly remarkable. Various sensors and devices for several purposes and with different shapes are now commercially available, including sensors for measuring environmental states and wearable devices for measuring individual health data. These sensors and devices are usually used in cooperation with cloud computing, i.e., huge amount of data are being stored in cloud storage every day. Such big data are analyzed by service providers to develop measures and information that are useful for the prediction of personal tastes and behaviors, and intervention in individual behaviors. Integrated analysis of data of

O. Gervasi et al. (Eds.): ICCSA 2017, Part I, LNCS 10404, pp. 509–522, 2017.
DOI: 10.1007/978-3-319-62392-4_37

different types which are stored on multiple cloud sites is expected to further enhance the usefulness of this data [1, 2]. Therefore, it is important to use data from various owners stored on each respective cloud site while properly complying the conditions permitted by the owners, the laws of the countries, and the organizations' rules.

Some studies have addressed the data security and privacy on cloud sites [3, 4]. However, these studies have been limited to organizing requirements and existing technologies for platforms from the viewpoint of general security and privacy and no implementation approaches on the appropriate use of data have been proposed.

1.2 Issues on iKaaS Platform

We have been conducting research and development of intelligent Knowledge as a Service (iKaaS), a platform for combining different types of data stored in multiple cloud sites for analysis [5, 6]. An iKaaS platform integrates and manages data on cloud sites to achieve transparent data access to applications. When data are accessed, the use of data with consideration on privacy is realized by providing data to applications according to the usage permission conditions for each dataset, the legal system of the country, and the organization's regulations [7]. Although iKaaS assures appropriate data use at data access, a new problem has arisen: the possibility of improper data movement during live migration of applications. In live migration, no limitation exists in the selection range of destination physical machines. Accordingly, live migration is performed not only between physical machines in the same rack or data center but between physical machines in different data centers and even physical machines across national borders. In a virtual machine, there can be some applications that can process and move data subject to the restrictions on data movement according to the laws and regulations of countries involved or the security policies of organizations. When a virtual machine with such an type of application is running and will be moved by live migration, it is necessary to verify whether the application have read the data stored in the physical memory in which data movement restrictions may apply, and to judge whether to proceed or not the live migration. However, in the present live migration mechanisms, there are no information and means to confirm the existence of such movement restrictions. Therefore, inappropriate data movement cannot be prevented at the request of live migration. For example, VMware vSphere can execute live migration of a virtual machine for long distances if the RTT delay between hosts does not exceed 150 ms [8]. There are applications that are allowed to process data whose movement is subject to restrictions based on the laws and regulations of the country and the security policy of organizations that are executed on a the virtual machine. In cases where a virtual machine that executes such an application will be moved by live migration, it is necessary to judge whether the application have read and store the data in the physical memory, and whether or not the live migration is possible.

However, conventional live migration mechanisms have no information or capability to impose such restrictions. Accordingly, improper data movement cannot be prevented when live migration is requested. A similar problem also occurs during cold migration. In the event of a disaster or a hazard, and to keep an application running, the operations manager of an organization must decide which host machines will migrate its running applications. However, it is difficult to determine which of the host machines which are all running applications can be migrated within a short time during an emergency. Some factors depends on the licenses used by the application, and the changing status of the place where you can migrate and the place where you cannot migrate. So, at this point, the administrator must properly decide a backup site. Therefore, it is necessary to have a mechanism which can decide where to move the data in compliance with the conditions set for each application data.

1.3 Our Contribution

To solve this problem, this paper checks whether data can be copied by comparing the data of a migration source application with the information of all host machines at the start of live migration, and duplicate the data when permitted. We propose a live migration mechanism with data protection mechanism to perform live migration only when it is permitted. By realizing this mechanism, it will be possible to migrate data between regions to promote global collaboration in knowledge discovery based on confidential data analysis. Migration secures security based on verification of trust chain between host machines. Trust chains are verified using attribute certificates issued by trusted third parties, i.e., Privacy Certification Authority (PCA). By conducting security analysis, it is possible to safely duplicate the data of the migration source application to the migration destination at the start of the live migration process. Accordingly, live migration is performed only when duplication is permitted.

The structure of this paper is as follows: Sect. 2 presents the general description of the iKaaS platform that realizes appropriate use of data; Sect. 3 introduce and discuss the implementation; Sect. 4 explains a the live migration mechanism realized on iKaaS platform that implements appropriate data use; Sect. 5 describes the implementation and evaluation of the data protection mechanism in the cloud environment, Sect. 6 concludes the paper.

2 iKaaS Platform

The iKaaS platform is privacy preserving and secure big data global cloud built atop multiple clouds (named local cloud) spanning across national borders. A local cloud administered by each cloud provider manages large scale data collected from various IoT devices and data sources on databases (local cloud DBs) and feeds them to applications under access control based on privacy and security policy settings. The database includes information of IoT devices: sensing data collected from the IoT devices, information of events on data collection, information of data owners, and the indicated subscription document of conditions during the collection of data. An application program runs on a global cloud that realizes transparent accesses to the local clouds and it acquires data from each local cloud via the global cloud. The iKaaS platform will be that it will consider the security, privacy and trust in a holistic way (named security gateway). The security gateway controls data flows between local cloud DBs and an application based on security and privacy settings provided by a policy database. Each application program running on a global cloud holds a privacy certificate issued by a privacy certification authority. The security gateway examines the privacy certificate in order to check the validity and data access rights of the application program. Three major components to ensure governance in terms of control and management of access to data are elaborated below.

2.1 The Privacy Certification Authority (PCA)

The PCA is establish in each country as an enforcement agency that takes charge of the regulations of various countries related to the handling of personal data. PCA issues privacy certificates based on various information related to regulations and applications (Table 1).

Table 1. Entries in a privacy certificate

CA country	Identifier of the home country of PCA
Application IP address	IP address of the host machine on which the concerned application is running
Application ID	Identifier of the application
LC countries	Identifier of countries where the application is permitted to access. Multiple entries are allowed
LC dataset IDs	List of dataset identifiers to which the application is permitted to access. The values are nested under each value of LC Countries. Multiple entries are allowed
Expires	Term of validity of the certificate
Application PK	Public key of the application
Signature	Signature generated using the secret key of PCA. The public key is distributed to a security gateway in advance

2.2 Security Gateway

Each local cloud has a single security gateway that receives a query from an application, sends it to a query function, and receives the retrieval result data obtained from the query function. To this, the application is provided only with data that are permitted to be duplicated based on the licensing contents and the list of identifiers of allowed countries and organizations which are attached to the retrieved data.

2.3 Query Function

The query function receives a query from the security gateway and transfers it to a local cloud DB. Then the query function extracts the database name out of the query, searches for a record having an identical database name in the data catalogue, and sets the local cloud DB designated by the IP Address of that record as the destination of the query.

When the retrieval result of a local cloud DB is received, the query function search the owner of the retrieval result data from the owner information database, obtains the contents of the user agreement and the list of identifiers of allowed countries and organizations, and then returns the retrieval result data to the security gateway.

3 Related Technologies and Works

Live migration on the cloud platform is performed under various conditions and policies. In this chapter, summarizes the existing methods of controlling live migration. There are four ways to control live migration. Role-based control [9–14], hardware requirement-based migration [15], network configuration-based migration [16–19], and policy-based live migration [20–22].

3.1 Role-Based Control

It is used to establish secure live migration. Execution of safe live migration can be controlled on the premise of the existence of a cloud environment with host machine, destination host machine, hypervisor, hardware, and secure communication channel. In this method, limited users defined by the policy of each virtual machine can perform live migration. When the user instructs the virtual machine to execute the live migration, it checks whether the user is permitted within the policy. This method prevents information leakage when an unauthorized user executes live migration. However, with this method, it is not possible to prevent the risk of executing live migration due to violation of the terms of consent of data usage and privacy laws and regulations by authorized users.

3.2 Hardware Requirements-Based Migration

It is a method used to perform virtual machine migration based on hardware requirements. Determines the destination host machine according to the policy definition condition of the hardware element. To make efficient use of physical hardware resources, cloud providers configure virtual machines. However, if an overload is detected, need to migrate the virtual machine to another host machine. Therefore, it can use this method to determine the destination host machine.

3.3 Network Configuration Based Migration

Determines the migration destination based on bandwidth, backbone network path, and the path between related virtual machines. Many cloud applications provide services on the network. The host machine's backbone network is an important element of virtual machines. During migration, the destination host machine is determined by the state of the backbone network.

3.4 Policy-Based Live Migration

Provide live migration method according to hypervisor type. Live migration can be categorized into three main methods: pre-copy, post-copy, and hybrid copy. In addition, the two major open source virtual machine monitors (VMM); Xen and KVM, are used by hypervisors to build virtual machines on the cloud platform. It can change the stop time and migration time during live migration by combining these live migration methods and hypervisor. Therefore, an optimal live migration method can be selected in consideration of the host machine having the virtual machine and the hypervisor of the destination host machine.

4 Live Migration Process with Data Protection Mechanism

Previous studies addresses migration control of a virtual machine, the backbone network surrounding the host machine, and live migration considering the state of the host machine. However, if migration is between regions, there is the risk that applications or data in the virtual machine will be transferred to unintended regions. Accordingly, it is necessary to execute migration considering the relationship between the destination host machine and the contents of the moving virtual machine. However, there is no mechanism to realize this yet.

Figure 1 depicts a live migration process with a data protection mechanism, and the newly appended Regulation attributes to the virtual machine (VM). When acquiring data by a query, the VM stores the list of identifiers of countries and organizations to which data movement is allowed based on the attached regulations attached to the data. The identifiers of countries and organizations are stored in the Regulation attribute in the form of

$$[\text{Country}_1, \text{Country}_2, \ldots, \text{Country}_n.\text{Organization}_1, \text{Organization}_2, \ldots, \text{Organization}_n]$$

The Country Code indicates the country to which data movement is permitted based on related regulations, and the Organization Code represents the organization that instantiates and uses the virtual machine. The list is attached to the data when acquired using a query.

In the context of the iKaaS platform explained in Sect. 2, the regulation attributes are set by the security gateway based on the security policy. In addition to the privacy certificate of the country where it is currently in operation, the virtual machine needs to acquire the PCA of the relocation destination. When data are acquired from the local cloud, the virtual machine transmits all the acquired privacy certificates to the security gateway. In cases where the use of data is permitted in countries designated by CA Country, the security gateway returns the values of the CA certificate and the value of Application ID written in the privacy certificate. Application ID is used instead of Organization Code. This process is executed for each privacy certificate, and all the sets are stored in the Regulation attributes.

The effectiveness of the Regulation attribute is guaranteed by the attribute certificate. The hash value of the program is also stored in the attribute certificate. Accordingly, the host machine can verify the effectiveness and regulation attributes of a program. The attribute certificate is issued by the PCA of a country where the virtual machine is currently located.

If the data can be moved to any country, the value of the Country Code is *All*, whereas if the data is available to any organization, the value of Organization Code is *All*. When data are newly acquired using a query, the logical sum (OR) of the list of identifiers of countries and organizations added to the newly acquired data and the list of the already stored identifiers is computed and stored.

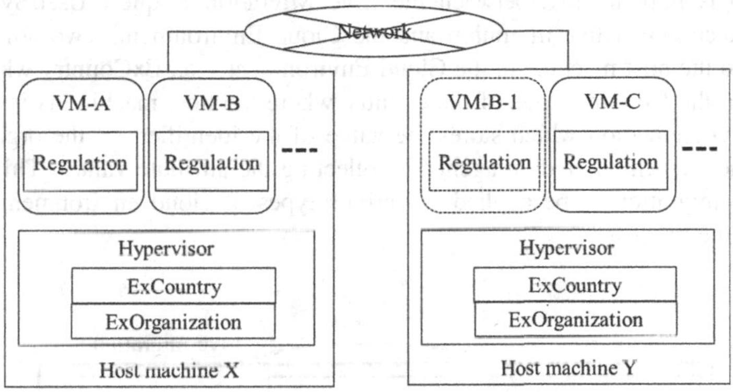

Fig. 1. Overview of live migration

The hypervisor has *ExCountry*, which stores the Country Code of the country where each host machine is in operation, and *ExOrganization*, which stores the identifiers of organizations that use each virtual machine. The data protection mechanism collects, stores in a database, and manages the values of *ExCountry* and *ExOrganization* of all the host machines that it administer.

In a live migration process, the hypervisor that manages the VM to be moved (VM-B in the figure) acquires the value of the Regulation attribute of the VM. Next, the hypervisor acquires the value of the Country Code of the ExCountry attribute of the destination hypervisor and verifies whether the acquired Country Code is included or not in the value of the Regulation attribute. If contained, then migration is executed; i.e., the data acquired by the moving VM can be moved to the country where the host machine of the migration destination is located. The execution of this live migration is similar to that of the conventional live migration process. Otherwise, if not contained, the data cannot be moved to the country where the destination host machine is located. In such a case, another host machine that allows movement should be sought.

When the destination host machine is unknown at the time of live migration execution, a record with the value of the acquired Regulation attribute is searched in the database. This search result provides candidate host machines to which the virtual machine can be moved. Finally, a destination host machine is selected from those qualified candidates, and then migration is executed.

5 Implementation and Discussion

This section addresses the implementation and evaluation of the data protection mechanism. We have been studying data protection based on the same ideas so far[]. It verified that the proposed method minimizes the impact on the conventional live migration process. The proposed method simply adds the verification of the destination address to the conventional live migration process that determines the destination host machine according to the conditions set by the host machine. By adopting this approach we were able to guarantee compliance with the regulations of the countries or organizations related to data usage with minimal impact.

Figure 2 depicts the model of the proposed data protection mechanism. This mechanism is implemented between the Live Migration Request User/System that requests execution of live migration and the Cloud Environment. Two attributes are appended to the host machine in the Cloud Environment; i.e., ExCountry which stores the value of the Country Code of the country where the host machine is in operation and the ExOrganization which stores the value of the identifiers of the organizations using the host machine, and an agent for collecting the attribute values. This addition allows live migration to be applied to various types of cloud environments without modifying the process.

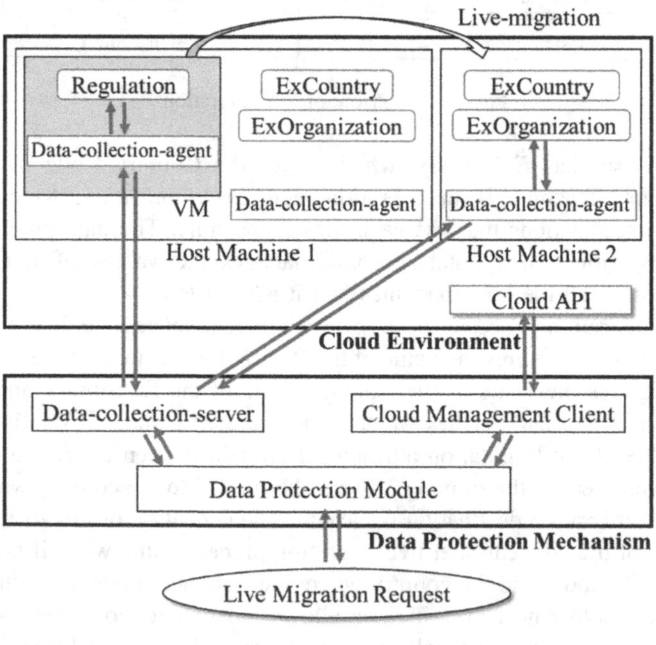

Fig. 2. Data protection mechanism model

The data protection mechanism is composed of three modules; namely, the *data collection*, *attribute database*, and *propriety decision*. The *data collection* module collects attribute values from agents that operate on each host machine and virtual machines and stores them in the *attribute database* which manages the collected attribute values. The *propriety decision* module determines whether live migration is permitted or not. The *data collection* module, the *attribute database*, and the agents are mounted using Zabbix. Figure 3 shows the prototype of the system environment. The OpenStack Infrastructure has one *Data Protection Server*, one *Management Node*, and ten host machines. The OpenStack Infrastructure is an existing cloud constructed by OpenStack Mitaka and is used to start virtual machines on the host machine.

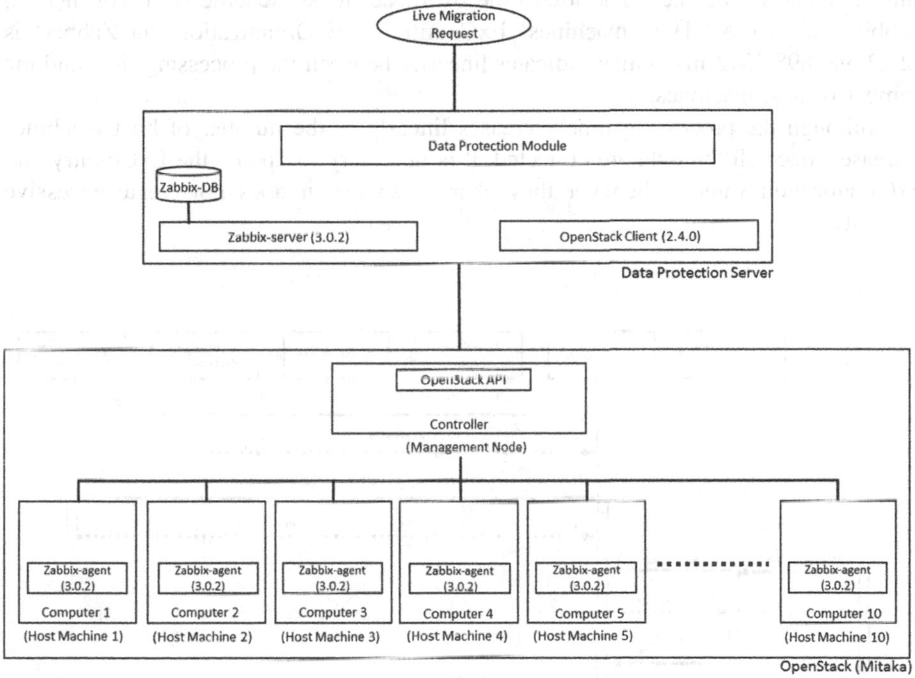

Fig. 3. System environment implementation

Zabbix 3.0.2 and OpenStack Client 2.4.0 are running in the Data Protection Server. Zabbix is a server monitor tool consisting of a Zabbix server and a Zabbix agent. The OpenStack Client is used to operate the OpenStack API. It evaluates the overhead cost in determining the destination host machine and executing migration in cases when the destination host machine of the virtual machine is undetermined during an emergency, such as a disaster.

5.1 Collection of Host Machine Information

The framework of collecting the values stored in ExCountry and ExOrganization on all host machines is explained next. Figure 4 illustrates the sequence diagram of the collection of host machine information. The Zabbix server requests acquisition of the values of ExCountry and ExOrganization to the Zabbix agent of each host machine. The Zabbix agent of a host machine obtains the values stored in ExCountry and ExOrganization and sends it to the Zabbix server. Table 2 shows the average of taking 10 measurements for each procedure. The number of host machines on the cloud is varied, i.e., #1, #5, and #10, and the effect of the data protection mechanism in the existing live migration is evaluated. The time required to Get Token from Zabbix and Get All Host machines' ExCountry & ExOrganization via Zabbix is linear to the number of host machines. The total time overhead of our scheme (Get Token from Zabbix and Get All Host machines' ExCountry & ExOrganization via Zabbix) is 123.3298–898.4542 ms, which indicates linearity between the processing time and the number of host machines.

Although the processing time increases linearly as the number of host machines increases, once all the data are collected, it is necessary to update the ExCountry and ExOrganization values whenever they change, so that it does not create excessive difficulty.

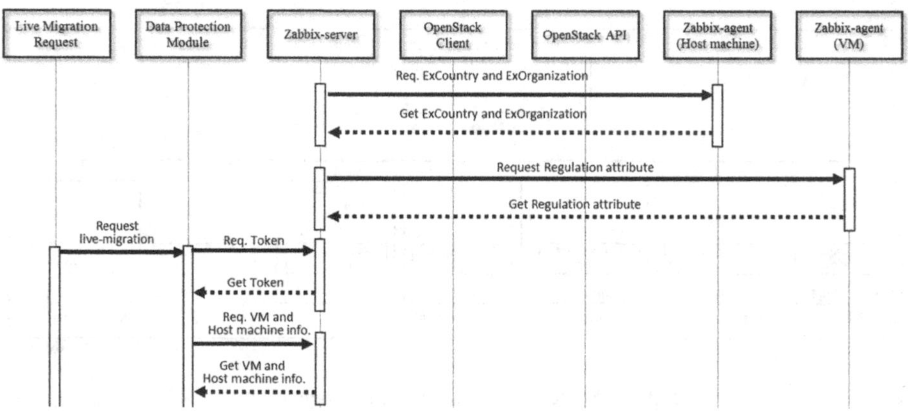

Fig. 4. Sequence diagram of host machine information collection

Table 2. Measurements for each processing

	#1	#5	#10
Get Token from Zabbix	35.5260	36.5231	35.3820
Get All Hostmachine's ExCountry & ExOrganization Data via Zabbix	87.8038	435.2219	863.0722

5.2 Determination of Destination Host Machine

Figure 5 illustrates the sequence diagram of the acquisition of the value of Regulation data from the virtual machine to be moved and the determination of the destination host machine. When the data protection module receives a request of live migration, it obtains a token for acquiring data from Zabbix, and then acquires Regulation from Zabbix. The required time to acquire a token from Zabbix is Get Token from Zabbix. The required time for acquisition of Regulation from Zabbix is Get All Hostmachine's ExCountry & ExOrganization Data via Zabbix. The data protection mechanism determines whether live migration is permitted by comparing the values of collected Regulation with the values of ExCountry and ExOrganization of all the host machines collected in Sect. 5.1. In cases where the permission is not granted, the data protection module returns a failure notification.

Otherwise, the data protection module sends a live migration command to the OpenStack Infrastructure via OpenStack Client and OpenStack API. The data protection module returns a success notification. Table 3 presents the average of 10 measurements for each procedure. The number of host machines on the cloud is varied and the effect of the protection mechanism is investigated. Get Token from Zabbix is almost constant irrespective on the number of host machines, although the required time for data comparison is linear to the number of host machines. The total overhead time and live migration time of our scheme (Get Token from Zabbix and Get All Host machines' ExCountry & ExOrganization via Zabbix) are 210.6121–217.9519 ms and 7063.5960–8948.7410 ms, respectively.

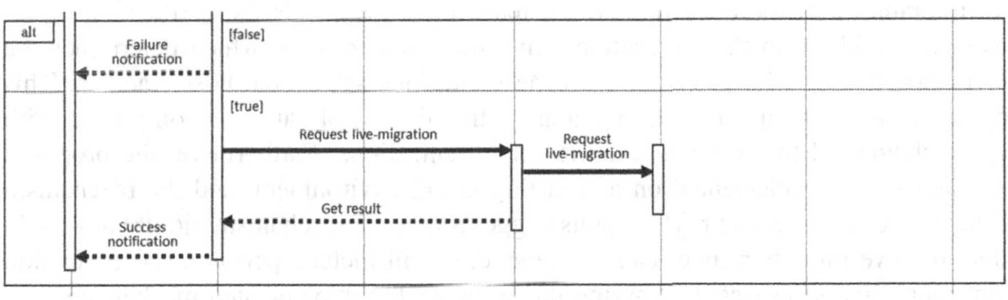

Fig. 5. Sequence diagram of acquisition in live migration

Table 3. Measurements for each processing

	#1	#5	#10
Get Regulation Data via Zabbix	175.0751	176.9121	182.5449
Compare	0.0110	0.0169	0.0250

Table 4 shows the time to collect the host machine information (Get Regulation via Zabbix), determination of the destination host machine (Compare Data), and live migration. The total overhead time and live migration time of our scheme are

298.4159–1081.0241 ms and 7063.5960–7960.9079 ms, respectively. The overhead part accounts for only about 7% of the live migration time. Accordingly, it is considered that appending the protection mechanism does not affect the live migration time.

Table 4. Total measurement time of protection mechanism and live migration

	#1	#5	#10
Collect Hostmachine Information	123.3298	471.7450	898.4542
Determination of Movable Hostmachine	175.0861	176.9290	182.5699
Live Migration	7907.4571	7960.9079	7063.5960

6 Conclusion and Future Work

The spread of IoT encourages activities to discover fresh knowledge by analyzing data from multiple owners. So that such data are used properly, it is fundamentally important to examine not only the owner's intentions, but also the regulations of relevant countries and organizations must be observed.

This paper demonstrates the possibility of infringement of regulations on data movement during live migration which is increasingly popular in cloud computing. Moreover, to prevent such regulation infringement, this paper presents a proposal of a live migration process with a data protection mechanism that regulates data movement before starting the live migration. In order to minimize the effects on the conventional live migration process, the proposed mechanism only appends the verification of the destination address to the conventional live migration process which determines the destination host machine according to the conditions set by the host machine. This approach has been implemented to guarantee that the use of data is in compliance with the regulations of the relevant countries or organizations. Furthermore, the proposed mechanism was implemented on an existing cloud environment, and the mechanism performance was evaluated. The results suggests that the mechanism did not adversely affect the live migration time. Our future studies will include performance evaluation from more diverse aspects by varying the status of the implemented mechanism.

Acknowledgment. The work is supported by the collaborative European Union and Ministry of Internal Affairs and Communication, Japan, Research and Innovation action: iKaaS. EU Grant number 643262.

References

1. EU FP7/ICT project 257115: OPTIMIS: Optimized Infrastructure Services, June 2010–May 2013
2. EU FP7/ICT project 287708: iCore: Internet Connected Objects for Reconfigurable Ecosystems, October 2011–September 2014

3. Meer, H., Pöhls, H.C., Posegga, J., Samelin, K.: On the relation between redactable and sanitizable signature schemes. In: Jürjens, J., Piessens, F., Bielova, N. (eds.) ESSoS 2014. LNCS, vol. 8364, pp. 113–130. Springer, Cham (2014). doi:10.1007/978-3-319-04897-0_8
4. Kiyomoto, S., Nakamura, T., Takasaki, H., Watanabe, R., Miyake, Y.: PPM: privacy policy manager for personalized services. In: Cuzzocrea, A., Kittl, C., Simos, D.E., Weippl, E., Xu, L. (eds.) CD-ARES 2013. LNCS, vol. 8128, pp. 377–392. Springer, Heidelberg (2013). doi:10.1007/978-3-642-40588-4_26
5. EU HORIZON 2020 project 643262: iKaaS: intelligent Knowledge as a Service (2014–2017)
6. Hidano, S., Kiyomoto, S., Murakami, Y., Vlacheas, P., Moessner, K.: Design of a security gateway for iKaaS platform. In: Zhang, Y., Peng, L., Youn, C.-H. (eds.) CloudComp 2015. LNICSSITE, vol. 167, pp. 323–333. Springer, Cham (2016). doi:10.1007/978-3-319-38904-2_34
7. http://kb.vmware.com/kb/2106949
8. Wang, W., Zhang, Y., Lin, B., Wu, X., Miao, K.: Secured and reliable VM migration in personal cloud. In: Computer Engineering and Technology (ICCET), pp. 705–709 (2010)
9. Aiash, M., Mapp, G., Gemikonakli, O.: Secure live virtual machines migration: issues and solutions. In: Advanced Information Networking and Applications Workshops (WAINA), pp. 160–165 (2014)
10. Rathod, N., Chauhan, S.: Survey: secure live VM migration in public cloud. Int. J. Sci. Res. Dev. (IJSRD) **2**(12), 271–274 (2015)
11. Shetty, J., Anala, M.R., Shobha, G.: A survey on techniques of secure live migration of virtual machine. Int. J. Comput. Appl. (IJCA) **39**(12), 34–39 (2012)
12. Gutierrez-Garcia, J.O., Ramirez-Nafarrate, A.: Policy-based agents for virtual machine migration in cloud data centers. In: Services Computing (SCC), pp. 603–610 (2013)
13. Upadhyay, A., Lakkadwala, P.: Secure live migration of VM's in cloud computing: a survey. In: Reliability, Infocom Technologies and Optimization (ICRITO), pp. 1–4 (2014)
14. Papadopoulos, A.V., Maggio, M.: Virtual machine migration in cloud infrastructures: problem formalization and policies proposal. In: Decision and Control (CDC), pp. 6698–6705 (2015)
15. Xianqin, C., Han, W., Sumei, W., Xiang, L.: Seamless virtual machine live migration on network security enhanced hypervisor. In: Broadband Network Multimedia Technology, pp. 847–853 (2009)
16. Shirazi, N., Simpson, S., Marnerides, A.K., Watson, M., Mauthe, A., Hutchison, D.: Assessing the impact of intra-cloud live migration on anomaly detection. In: Cloud Networking (CloudNet), pp. 52–57 (2014)
17. Kantarci, B., Mouftah, H.T.: Resilient design of a cloud system over an optical backbone. IEEE Netw. **29**(4), 80–87 (2015)
18. Cui, L., Tso, F.P., Pezaros, D.P., Jia, W.: PLAN: a policy-aware VM management scheme for cloud data centres. In: Utility and Cloud Computing (UCC), pp. 142–151 (2015)
19. Koto, A., Kono, K., Yamada, H.: A guideline for selecting live migration policies and implementations in clouds. In: Cloud Computing Technology and Science (CloudCom), pp. 226–233 (2014)
20. Majhi, S.K., Dhal, S.K.: A security context migration framework for Virtual Machine migration. In: 2015 International Conference on Computing and Network Communications (CoCoNet), pp. 452–456 (2015)
21. Yamada, H.: Survey on mechanisms for live virtual machine migration and its improvements. Inf. Media Technol. **11**, 101–115 (2016)

22. Vlacheas, P., Stavroulaki, V., Georgakopoulos, A., Kelaidonis, D., Biswas, A.R., Moessner, K., Miyake, Y., Kiyomoto, S., Yamada, K., Hashimoto, K.: An overview and main benefits of an intelligent knowledge-as-a-service platform. In: WWRF 34th Meeting (2015)
23. Hashi, Y., Uchibayashi, T., Hidano, S., Kiyomoto, S., Rahim, A., Suganuma, T., Hiji, M.: Data protection for cross-border live migration in multi-cloud environment. In: Proceedings of 4th International Symposium on Computing and Networking (CANDAR 2016), pp. 681–685 (2016)

Workshop on Challenges, Trends and Innovations in VGI (VGI 2017)

Determining the Potential Role of VGI in Improving Land Administration Systems in Iraq

Mustafa Hameed[1], David Fairbairn[1](✉), and Suzanne Speak[2]

[1] School of Civil Engineering and Geosciences, Newcastle University, Newcastle upon Tyne NE1 7RU, UK
david.fairbairn@newcastle.ac.uk
[2] School of Architecture, Planning and Landscape, Newcastle University, Newcastle upon Tyne NE1 7RU, UK

Abstract. Land is undoubtedly the most important natural resource and asset in any nation, contributing half to three quarters of the national wealth in most countries. Hence there is a need to manage it in a sustainable manner. The management of a land system is affected by a number of factors, including the political situation, the socio cultural environment, and economic aspects e.g. financial crises. In Iraq all these problems are present and further, the country has endured decades of internal and external conflict. The results of such factors include different waves of human population displacement, land ownership document forgery, seizure of public land by unauthorised groups, manipulation of urban planning by changing the use of the land, and subdivision of single parcels into multiple landholdings. By examining volunteered geographic information (VGI), we hypothesise that it has a role in enhancing the Iraqi land administration system. The paper presents a research design, including specific aims and objectives, for conducting such VGI application in Iraq and concludes with a framework for conducting challenging fieldwork in local communities which seek improvements in land administration.

Keywords: VGI · Land administration system · Fit-for-purpose

1 Introduction

This paper examines current issues and concerns related to land administration, volunteered geographic information (VGI) and the current state of socio-economic activity within the nation of Iraq. It begins by presenting a picture of current trends in land registration and cadastre development around the world, to justify the need for the development of varying methods. It then explores how different countries have tried to develop cadastres and the difficulties they have encountered. Finally, it focuses on the use of VGI and an understanding of how this can be conceptualized, and used, for land administration and cadastral purposes. With an international view in mind it identifies relevant issues for Iraq and aspects which might inform practical methodologies in data collection and management. The paper concludes with a definitive framework of research questions, tasks and applications in which the issues raised can be explored.

O. Gervasi et al. (Eds.): ICCSA 2017, Part I, LNCS 10404, pp. 525–540, 2017.
DOI: 10.1007/978-3-319-62392-4_38

2 General Definition and Components of Land Administration Systems (LAS)

Land administration is considered as the collective operations of determining, registering and propagating information about the value, tenure and use of land within a policy framework for implementing land management. It encompasses land registration, cadastral surveying and mapping, fiscal, maintaining legal and multi-purpose cadastres, and managing land information systems (Williamson 1985). As one important aspect of the land administration system, land registration is the procedure of recording all interests in land with regards to ownership, parcel size, location, land use, and value, and the preparation and structuring of such information in the public register (Ituen and Johnson 2014). Cadastral surveying is a specific term to describe the process of gathering the data, particularly measured survey data, about the land parcel and recording it. These data comprise the geometrical data, including the size, shape and location of the land parcel (Steudler et al. 2004). A further step of cadastral surveying is the production of cadastral maps. The main use for such a cadastral map is to produce a title registration system, the title being fundamental to simplify the process of land transaction. A cadastre is defined as the up-to-date information on the land parcel which contains records of the interests in land, such as rights, restrictions and responsibilities (Bennett et al. 2010). A fiscal cadastre is the cadastre that is organized mainly for valuation and fair taxation; a legal cadastre is established mainly for legal conveyancing, assisting in using and managing the land, facilitating environmental protection and sustainable development, and supporting the land market; a multipurpose cadastre meets multipurpose requirements and supports social, economic and environmental sustainability. The land information system is the digital data handling capability which contains information that is essential for decision making and managing of the land. Our study examines cadastral and registration systems which are responsible for demarcating plot boundaries and recording information of the land.

2.1 Current State of Land Registration and Cadastres

Recent technological developments in data handling and application of higher land surveying and data collection techniques in formal land administration systems in developed countries can ensure high security, easy access, and efficient services for citizens and their interaction with land. The majority of countries in the world, however, still do not have a functional land administration system and many people live under threat of the appropriation of their land and insecure tenure.

Only 25% (35–50 countries) have a complete land registration system, mostly in the industrial countries (McLaren 2013). The majority of the land occupiers in the remaining countries include the most vulnerable and poorest groups in society. The increasing proportion and numbers of urban population is likely to lead to serious problems in organising and recording land tenure in new densely populated areas. The abilities of the formal land administration systems to cover the excluded 75% of the world population are limited, so there is an urgent need for considering an alternative approach for land administration that is fast, cheap and supports community needs and services.

2.2 General Issues with Land Administration Systems in Developing Countries

The difficulties with formal land administration systems in developing countries which make them insufficient, ineffective, and unable to serve the need of the majority of the communities include the major reason of lack of funding (Adlington 2010). World Bank progress evaluation reports record a large number of land reform projects (about 30 countries) which had started directly after political change and economic transition from central to free market: 48% of these suffer from budget deficits and 17 projects are stalled due to resource issues. In Bulgaria, for example, the official cadastral system covers only 18% of the land parcels, and has been stopped due to the lack of funding (Basiouka and Potsiou 2012). A further factor leading to inefficiency is lack of trained staff: despite new technologies, the number of land professionals is insufficient (McLaren 2013). Enemark et al. (2014) exemplify this in Rwanda where there are very few qualified surveyors. In addition, Lemmen (2010) argued that an official tenure system may only consider the legal rights of ownership, while neglecting the millions of people whose tenures are predominantly *social* rather than *legal*. In sub-Saharan Africa, the minority of land (one-third of the total land) is considered under official systems, and knowledge from social tenure is ignored: neglect, during official cadastral field work, of opinions of land owners about the nature of their parcels and location of their boundaries is widespread. Further shortcomings in officially collected data are demonstrated by an official cadastral project in Tsoukalades, Greece ('Hellenic cadastre') which was repeated four times and did not succeed, due to fundamental errors in measured geometry (shape, location, parcel boundaries) and mis-registration (e.g. unknown owners being recorded for plots) (Basiouka and Potsiou 2012). Alemie et al. (2015) considered that no official system could efficiently cope with a change of regime or following a war, but even when a new country emerges following independence, or if formal regime-change results from a war, it is unwise to build a new official system from scratch, due to the time and cost needed. In such cases, it is appropriate to consider an interim land administration system that can serve the needs of the community, in an established 'Fit-For-Purpose' manner, until the official system develops.

Policy issues may promote land re-distribution: one example is in Zimbabwe, where in 1980 75% of the total land area was owned by 3% of the population, and re-allocation was promoted up to 1996 (Chitsike 2003). Further, disengagement often appears in developing countries when owners are unable to register their ownership or do formal transaction, due to the high taxes levied on such activities: in many circumstances people prefer not to register their ownership or may buy or sell their land without using the formal system. Fairbairn and Al-Bakri (2013) noted that there are shortcomings in official systems even where they do exist: these often relate to completeness of data, and lack of currency of information about numbers of plots, changes of use etc. Interaction with such agencies may be very slow and may not be appropriate for some communities (e.g. rural areas where land ownership practice may conflict with governmental views).

3 Experiences with the Iraqi Land Administration System

An example of an incomplete and inefficient land registration system can be seen in Iraq. Several problems are evident here, some internal to the system (notably forgery of title deed documents, and corruption within the agencies), and many external societal aspects (including extensive short and long range displacement of populations; unchallenged building expansion onto under-utilised public lands; expropriation of land by political regimes, including seizure of public property; military operations; sectarian violence; internal terrorism; economic crisis; and a general climate of fear throughout society. The pre-2003 Ba'ath regime, and the period of US-led Occupation after exemplify the issues.

3.1 Land-Registration Problems Before 2003

During the former regime, many people were forced out of Iraq completely or out of their original city as internally displaced people, and their land was taken by or given to other citizens illegally. For example, thousands of Tabaiya, people of Persian origin who had lived in Iraq for generations and speak Arabic, but accused as supporters of Iran during the Iran-Iraq war in 1980s, were evicted, their property sold cheaply mainly to the Ba'athists, who later resold it at a profit. These ethnic Persians were found mostly in the south near Karbala, Al Najaf, Hillah, and, to some extent, in what is now known as the Sadr City District in Baghdad. Since the occupation in 2003, many of those expelled have returned to reclaim their homes and land (USAID 2005).

Internal displacement was also the result of "Arabization" which forced out non-Arab inhabitants (mainly Assyrians, Kurds, Turkmen) from cities such as Kirkuk, and replaced them with Arabic people from different Iraqi cities. This policy aimed to reinforce the control of the Ba'athist regime over the oilfields and large tracts of fertile land in those areas (Isser and Van der Auweraert 2009).

The previous regime also distributed much land on an individual basis to enrich Ba'ath Party loyalists. These actions were hurried and caused stress in the land administration offices, preparing ownership documents on government land, and satisfying the need of party supporters, often without a proper survey of the land, which then caused many problems after the fall of the regime. It should be noted, however, that the land administration system was supported politically and working legally with little evidence of personal illegal behaviour.

3.2 Land-Registration Problems After 2003

After the fall of the Ba'athist regime, inherited problems became evident, and the weakness of the new government quickly became clear. *Post-war forced displacements of people* continued, but now caused by the victims of the previous regime, such as Kurds who expelled tens of thousands of long-standing Arab settlers, e.g. in Khanaqin, where 54,000 Arab settlers were expelled from 2003. The same happened in Kirkuk which caused such problems between old and new victims that the city continues to be socially and economically unstable till now. *Rigging of title deeds and changing the ownership of the land in the official records* happens due to the weakness of post-war regime: it is especially evident on land that belonged to Ba'athist party members who migrated after the fall of the regime, or those who fled after displacement. The national governmental land administration system has tried to establish a new mitigating policy to clamp down on such practices. *Seizing public buildings* is prevalent among those without property and displaced families, who occupy public buildings and change them into living spaces. For example, a notable area in central Baghdad (Salhiya) which used to house senior army officers from Saddam's Republican Guard, is currently occupied by homeless families and individuals (Isser and Van der Auweraert 2009). Some people have occupied public land and built their own house on it, violating any effective formal land registration system (USAID 2005). *Infringement on orchards and building on agricultural land within cities* is a phenomenon which increased rapidly after 2003 due to the weakness of municipal authorities and a clear advantage in the financial benefits derived from the exploitation of these areas for residential purposes, rather than agricultural purposes. This fragmentation has led to unplanned development of cities, decrease of the green zone, and inability of the government to provide services such as sanitation (Istabraq et al. 2016). *Subdivision of single plots into several sub-plots* is common as a result of the long-standing housing crisis in Iraq, with United Nations and World Bank estimates of 3 million housing units shortfall (Shaikley 2013). Such sub-division results in very small plots with complex ownership. After the February 2006 bombing of the Shia Al-Askari mosque (Saladin Governorate), *displacement from sectarian violence* accelerated significantly mostly in mixed Shia-Sunni areas close to Baghdad, affecting one and a half million Iraqis (Isser and Van der Auweraert 2009). More recently, *displacement resulting from the activities of Daesh (Isis)* has affected cities in the north and west of Iraq (Mosul, Ramadi, Tikrit and Fallujah). Continuing today, displacement both internal, to different Iraqi cities, and external, to different countries, causes further pressure on the land administration system, unable to monitor any occupancy in Isis-occupied cities, nor to prepare temporary places for migrant people to live.

The context is clearly, therefore, of a land administration system beset by problems of external societal issues, and internal resourcing and procedural issues affecting effectiveness, working methods, and results. It is proposed that engagement by the community and its participation in land administration systems can address these and other problems. Formal land tenure for the world's poorest people, including millions in Iraq, where recent wars have left the formal system broken and unable to secure land rights for the citizens, is a critical driver. Because cost and time are barriers to registering land, there is need for exploratory projects which seek to establish whether VGI can offer a low cost, inclusive approach to updating existing records, making the formal land administration system more efficient and meeting the needs of the community by creating more fit-for-purpose systems. A further impetus is the need to constantly maintain completeness of attribute data as well as spatial data.

4 Fit-For-Purpose (FFP) Land Administration

Systems which serve the need of the communities in countries where the official cadastral system is weak or does not exist at all, and where the cadastre is not multi-purpose (e.g. where it considers only legal ownership, neglecting other aspects) can be described as 'fit-for-purpose': these are not necessarily high geometric accuracy systems, but rather they can identify, rather than monument, land parcels, record ownership and provide security of tenure for underprivileged communities (Enemark et al. 2014).

These systems adopt principles such as 'general boundaries' rather than fixed boundaries, meaning the accuracy of the delineation process is not necessarily precisely determined, especially in the rural and semi-urban area, but rather concentrates on the information needed for day-to-day land administration purposes. A successful example that has adopted this approach is Land Tenure Regularisation in Rwanda in 2009 (Enemark et al. 2014). Data capture commonly uses satellite or aerial imagery rather than traditional surveys (with total station), three to five time cheaper overall yet yielding data suitable for most land administration purposes. The Tsoukalades, Greece project, already mentioned, offers printed 1:25,000 scale satellite images to non-professionals allowing annotation of land ownership details by pencil. In 'fit-for-purpose' systems, the required accuracy of information collected and maintained is aligned to its potential use, rather than some imposed high-order specification; further, the system is designed to optimise updating and embed flexibility to allow for improvements with time.

The 'fit-for-purpose' approach builds on several important elements (Enemark 2013): the *participatory nature of the system*, meaning geospatial data is collected and handled with the aid of the community; *inclusivity*, ensuring that all tenure types are considered without any exceptions, crucial to ensure coverage of non-traditional (and unrecorded in the majority of official land administration systems) tenure patterns; *attainability* ensuring realistic system-building depending on available resources and in a short timeframe; and *upgradability* meaning that the data can be readily updated and the system improved in response to developments in society and practical experiences.

The idea of 'fit-for-purpose' suggests that there is a role for less technically accurate data, but much research still needs to be done to assess its value for the many purposes of land administration, and to draw up the specifications of such systems. In order to focus on such developments, a spectrum of variables needs to be considered.

4.1 Variables Which Drive FFP Land Administration

The variables which drive fit-for-purpose land administration systems are themselves disparate and multi-faceted. Firstly, tenure may involve aspects such as occupancy, usufruct (official usage rights without ownership), informal rights, customary rights, indigenous and nomadic rights, and varying levels of security depending on their level and application. These levels differ from informal land rights to formal ones, but land tenure can migrate from one category to another, upgrading with time and changing towards better levels of security. For example, the informal settler may upgrade to an improved level of tenure if government begins to formally recognize certain group rights (Quan and Payne 2008). Secondly, the legal rights used in land administration may reflect a number of acceptable and practical systems such as tribal rights, religious rights embedded in the Islamic system, for example, and customary rights. A third variable is accuracy, which reflects different methods and techniques for collecting cadastral data by officials. This recognition of variability could be extended to public participation in collecting VGI, for example by field sketches, satellite imagery, or GPS observation. Volunteers may themselves exhibit variability - young amateurs may have technical experience whilst elders possess useful historical information; young people may prefer to use digital technologies (e.g. iPad or GPS) but older people may prefer to use sketch map for presenting their land information to a public participation system.

4.2 Public Participatory Work for Collecting Geographic Data

The nature of public participation in the collection of geographic data has been addressed by many researchers, and although the main perception for all is nearly the same, terminology can vary. Turner (2006) referred to 'neo-geography', meaning the use of geographical techniques and tools for personal and community activities, or by a non-expert group of users. Goodchild (2007) proposed the term 'citizen-as-sensor' for those numerous people who participate in collecting information on the weather and predicting catastrophe. A further term 'citizen science' describes the engagement of amateurs in any experimental and scientific observation (Bonney et al. 2009). In geographic terms, the use of the word 'crowdsourcing' can suggest the process of collecting information by the local citizen by means of mobile devices and the geoweb (Niederer and Van Dijck 2010), although it can also be used to refer to the validation of such data by multiple volunteers who can give a 'best estimate' on its trustworthiness and accuracy. For LAS, a relevant recent term is 'neo-cadastres' - citizen-built and legitimized land data handling systems maintained by volunteers rather than the government (de Vries et al. 2015). The work described in this project demonstrates the convergence of VGI with neo-cadastres: the focus will be mainly on the definition of VGI, its uses, issues related to quality and the methods for verifying VGI. A hypothesis and research questions relating to the role of VGI in LAS will be presented also.

5 Volunteer Geographic Information (VGI)

For centuries, the collection and use of spatial data has been reserved to official agencies (Goodchild 2008), but VGI offers the public the opportunity to be involved in data collection and use. This can be perceived as challenging to authorities and they may raise concerns about the data or the process. Overcoming such concerns, involves considering the motivation for adopting VGI in land administration projects, the quality of the data, and the ways of verifying the participatory data (Seeger 2008).

5.1 The Motivation for VGI

The Motivation for Using VGI Technologies in LAS. The ineffective nature of many LAS suggests that there is a need to build fast and inexpensive land administration systems that can activate stagnant land markets; and it may be that such initiatives are much more valuable than building highly accurate systems which may need time and money (Adlington 2010). Associated with this approach, it seems obvious that VGI can form a major part of such initiatives, and that contributions from people who live near to spatial phenomenon can release best knowledge on its characteristics (Bishr and Mantelas 2008). In terms of specific application, therefore, there is motivation to consider that official cadastral data provided by official expert structures can be completed and up-dated when combined with data that is provided by VGI. From a technological perspective, the expansion of internet coverage and its availability around the world, together with the developing of smart phones technology, allow for easy and quick techniques for collecting, picking, uploading, correcting and mapping of geospatial data by ordinary citizens without depending on GIS experts (Tulloch 2007).

Citizens' Motivation for Collecting and Editing Geospatial Data for LAS. The biggest motivation is that of volunteers without whom no VGI project can succeed. The urge of citizens to participate as volunteers for collecting geospatial data, over lengthy periods, not directly linked to financial gain, has been examined by Tulloch (2007) who identified "achieving a higher level of empowerment" as the main factor that motivates citizens to participate in the OpenStreetMap project (the highest profile VGI initiative). Goodchild (2007) has indicated that "self-promotion and personal satisfaction" are major factors also. Coleman (2010) suggests that positive factors might include altruism, professional and personal interest, intellectual stimulation, protection of personal investment, social reward, personal reputation, self-expression opportunity and (especially in the case of geographic information) pride in place. Negative factors include mischief, social, economic or political agenda, and malicious intent. Haklay and Budhathoki's study (2010) presents concepts such as 'fun', 'recognition', 'money', 'unique', 'ethos', 'reciprocity' and 'instrumentality' to characterise volunteer experience. Cotfas and Diosteanu (2010) saw volunteer participation as a recreational activity, the public not needing to be particularly aware or reward-driven for their participation.

However, Laarakker (2011) has suggested that the focussed recognition of a need for better public services and improved systems for land administration, might be a more positive driver for participation than such altruistic reasons. Basiouka and Potsiou (2012) who conducted the first practical cadastral mapping exercise using crowdsourcing techniques, have emphasised that the main reason for the motivation of their participants was a perceived need to overcome bureaucracy and to open the land market, which had been blocked for more than twelve years. The key opportunity for contribution might therefore be market-driven (professional or personal interest at an economic level); the enhancement to a personal job or project; being part of a large social network and being rewarded for having a strong personal presence online; the 'fun-factor' in working within a 'trendy' environment; and humanitarian and altruistic drivers etc. (Coleman et al. 2009; Winterbottom and North 2007; Genovese and Roche 2010).

The Authorities' Motivations. Although VGI projects are created and maintained by ordinary citizens, the role of authority in supporting and guiding VGI mapping projects can impact LAS. It is highly unlikely that a VGI-based cadastre can be successfully envisaged as a replacement for an official, formal system. McLaren (2013) has suggested that completion of unmapped areas or updating existing systems are the obvious incremental benefits of VGI in LAS, which can be recognised by government authorities.

Another scenario which might be considered is to apply VGI in a highly populated urban slum area with very small plots, where low land values might suggest that high accuracy for obtaining land boundaries is not so important for the authority. Such data can exemplify the ongoing flowline of land rights and boundary delineation information, which can be used for planning and updating social services and infrastructure (McLaren 2013). According to Enemark et al. (2014) crowdsourcing actually motivates land professionals by expanding their potential role and making them able to serve the whole population rather than focussing only on a small elite. They also pointed out that the role of the professional will be more managerial in relation to organising and using land related data, expanding their responsibility beyond just capturing it. It has also been suggested that VGI, as a new source of fresh data, will enhance the quality and quantity of information for professionals and decision makers (Seeger 2008).

Such considerations of motivation will directly help in the development of test data collection tools, and in the engagement during the fieldwork with both the authorities and communities. Once the test fieldwork data in this study project has been collected and analysed, it will be possible to liaise with the relevant authorities and explain the value of the VGI data, and approach, to their work.

5.2 Evaluating the Use of VGI in the Land Administration System

In recent years, the evaluation of the utility and quality of data that has been collected by participatory methods has been important. Its application has been considered as an opportunity for official land administration systems: de Vries et al. (2015) argue that the use of volunteer spatial data in cadastral systems is an opportunity for creating and maintaining geospatial data, because it changes the role of ownership from being passive to active, which can guarantee faster, cheaper and more fit-for-purpose techniques than traditional methods of registration. Bennett et al. (2010) also considered the importance of using VGI for cadastral purposes, notably as an interim solution that can serve the needs of the community by securing land rights, and/or transferring such rights to a different level of tenure security. Seeger (2008) refers to the opportunity of government cadastral authorities to obtain a new source of fresh data that may help them for planning and decision making. However, de Vries et al. (2015) point out that using VGI in cadastral systems may cause threats to official systems, because it may conflict with the rules and mechanism of official organization and the experts.

Goodchild and Li (2012) describe some quality issues with VGI even for multiply crowdsourced data: it may be difficult to control because of the fact that only a small number of people can verify the correctness of the information. In the case of a cadastre, those directly in touch with the land parcel can provide the correct boundary information.

Experience also counts: the majority of the VGI producers are amateurs who have little or no experience with the mapping process (Mummidi and Krumm 2008).

The usage of VGI for specific tasks in land administration has also been examined by Navratil and Frank (2013) who argued that although it is difficult to depend on VGI totally as an alternative option for an official cadastral system, as an important part of the LAS is land ownership which can only be verified by a limited number of people, it still has an important role to play in observing parameters which could be obtained by authorities only with considerable time and effort.

To conclude, de Vries et al. (2015) regard VGI as an opportunity and 'fit-for-purpose' as a reliable concept, ensuring servicing the needs of a community in case of under-development or absence of an official cadastral system. However, it clearly needs good methods for checking the quality of the VGI data.

5.3 VGI Quality Considerations

Land administration systems are only as good as the quality of the spatial, and other, data held therein. Because VGI is provided, in most cases, by people with little or no knowledge of the mapping process (Ciepłuch et al. 2010), it is necessary verify the quality of their data and balance the potential benefits. Data quality can be examined from a number of different perspectives.

- Positional accuracy, the Euclidean distance of coordinate values of a VGI feature (e.g. a point) to corresponding authoritative equivalent feature (Mullen et al. 2015).
- Thematic accuracy/Attribute accuracy, the reliable and reasonable correctness for attributes attached to points, lines and polygons features of the spatial database.
- Completeness, the comparison between two different data sets of the same area, to find which included and excluded features.
- Temporal accuracy, the agreement between encoded and 'actual' temporal coordinates (Veregin 1999).
- Logical consistency, the existence of logical contradictions within a dataset (Hashemi and Abbaspour 2015).

Goodchild and Hunter (1997) evaluated the accuracy of VGI by using traditional statistical methods (Root Mean Square Error (RMSE) and the standard error) to describe the spatial error of point features. More recently, Al-Bakri and Fairbairn (2012) reviewed the spatial match between VGI and government data, finding that RMSE of positions is consistently high for VGI. They attributed the errors to the prevalent low-precision devices, for example personal GPS units and commercial imagery services, which are commonly used in VGI collection.

Quality can vary depending on the location. For example, Zielstra and Zipf (2010) found that the completeness of VGI became worse the further it was collected from the urban core, when examining the differences between VGI and commercial data sources in Germany. Completeness is used to determine the difference between what is recorded (for example, number of houses or length of roads) and what is actually found in the real-world (Brassel et al. 1995). Haklay (2010) proposed a numerical assessment, by simply counting the total length of streets in OpenStreetMap (OSM) compared to official Ordnance Survey (OS) data sets for Great Britain. Jackson et al. (2013) undertook a study based on schools, first by simply counting them, which showed similarity across four data sets. However, when they repeated the study basing it on names (and in two cases on addresses) they noted that, although the numbers were similar, the schools themselves were not the same: simply basing the data collection on numbers was insufficient to assess completeness.

Using the same datasets of schools, Mullen et al. (2015) included demographic features in their specification (general population; economic status; educational attainment; and race/ethnicity). However, they failed to identify a clear association (or statistically significant correlation) between either positional accuracy or completeness, with any of the demographic properties.

It is clear that there is need for a better study of potential accuracy and quality of VGI in relation to LAS. It is interesting that there is agreement that completeness and currency are indeed measures of quality. Focussing on these, rather than simply on spatial accuracy, is important in assessing the approach to VGI in LAS.

Methods to Assess and Validate VGI When No Other Data Sets Are Available. It is important to also validate the process of data collection where there may be no formal or alternative data sets to assess the data against. In this situation, we must have confidence that the collection and collectors are reliable. Since the quality of VGI data depends on the reliability of the information that can be obtained from volunteers, it is crucial to find some ways for verifying the participatory information that aim to serve the need of the community and distinguish it from others who aim to defraud the system (Coleman et al. 2009). Methods of verification depend on criteria such as producer reputation, evaluating volunteer data by other users (Maué 2007). Another method assesses a producer's abilities in checking others' work (Bishr and Mantelas 2008). Haklay et al. (2010) show that eliminating the error in participatory mapping can be achieved by multiple volunteers – a principle of agreement. A social approach could be adopted, based on senior trusted users checking and correcting others.

Addressing concern about the validity of data and conflicts with existing formal systems are a focus of this study, which draws on these methods to analyze and validate VGI from field work. Some case study areas involve comparison with formally registered data but others rely on agreements within the community for verification.

5.4 Experiences of Using VGI for Land Mapping

Volunteer geographic information (VGI) techniques have been successfully used for many tasks around the world, and such studies can give insight into the mechanisms of VGI data collection and use. New cadastral maps have been created with the aid of the community in Juba, capital of newly independent and poorly mapped South Sudan (Haklay et al. 2014). Another example is a Canadian government project for correcting and updating topographic mapping using VGI technique (Bégin 2012), whilst cadastral data in New York City was enhanced with the aid of volunteers (Barth 2013). When government cannot offer an operational cadastral system or existing systems do not reflect the community understanding of their land boundaries and rights, a bottom-up approach to mapping can be initiated. The approach of the charity Shelter to slum mapping is described in Haklay et al. (2014). Extra-governmental participatory mapping practices and associated tenure allocation in communal areas in southern Ghana are reported in Olowu (2003) and Arko-Adjei (2011), and boundary mapping of indigenous land in Canada, reflecting the community's social, cultural and religious relation with the land is outlined by de Vries et al. (2015).

It is recognised that for some official systems, VGI is considered as threat depending as it does on a bottom-up approach with community roles and mechanisms of surveying and registration which are different from the authoritative approach. It is noteworthy, however, that assessment of such VGI has tended to evaluate the spatial and attribute data much more than consideration of the completeness or currency of data-sets. The project outlined here does address all of these concerns.

Further, the volunteers in this study of Al Hillah have been offered a range of data collection methods (GPS-enabled smartphone, iPad tablet, paper-printed satellite image) to assess motivation and accuracy, and understand ease of use, and impact of citizen characteristics (such as age, level of education, gender). The data collected is subject to a range of methods to test spatial accuracy and usability. This is important for a 'fit for purpose' system because it may be that very low-tech approaches, while not as precise or accurate as more complex methods (e.g. GPS) are valid in some situations where cost is an issue.

6 The Main Hypothesis and Research Questions

The main hypotheses being tested in this work are that:

- VGI can provide adequate, current and complete data to inform 'fit-for-purpose' land registration systems to secure land rights.
- Different individuals and geographic contexts require adoption of different methods of collecting and supplying VGI.
- Land professionals can be motivated to use VGI if they have more understanding of its use and potential.

Thus, the overarching research question has been:

> "What existing and potential roles do the professional and volunteer stakeholders have in collecting and handling geospatial data for land administration systems?"

This main research question is explored through a series of investigations:

Research tasks	Purpose	How answered
1. What is the range and scope of current land administration systems in developing countries?	To set the context and assess the potential need for VGI to support a 'fit-for-purpose' system	Through a literature review

(*continued*)

(*continued*)

Research tasks	Purpose	How answered
2. Do official land administration systems efficiently support customary and social tenure?	To identify if some issues might be better covered by VGI-based systems than current official LAS	Through a literature review
3. What difficulties do land administration professionals currently experience in the case study area?	To identify areas and tasks where VGI might be particularly valuable to professionals	From fieldwork, notably interviews with professionals
4. What is the current knowledge and perception of VGI and 'fit-for-purpose' land systems amongst land professionals in the case study area?	To identify barriers to professional acceptance	From interviews with professionals
5. What is the current knowledge of the local citizens about the importance of registering their land?	To assess motivation for engaging in VGI	From interviews with volunteers
6. What are the current technical and non-technical mechanisms being used by people participating in VGI?	To inform the methods to be used in the field work and develop data collection tools	From previous studies
7. Which methods and data collection tools work best in different contexts and for different individuals?	To analyze the usability and validity of different approaches	From practical VGI collection in the field
8. Which type of data can be provided by citizens for the system, and which types are not possible?	To acknowledge the limitations of VGI	From examination of the data flowline in the field and in the official office
9. How complete, current and accurate is VGI compared to more formally collected land data?	To acknowledge the limitations and strengths of VGI	From analysis of collected VGI and comparison with official data-sets
10. How can VGI to be incorporated into, or supplement, an official LAS?	To identify what changes formal systems might need to accommodate VGI	From final workshop with land professionals

7 Conclusion

This paper has reviewed issues of land administration systems in developing countries, focussing on an Iraqi case study. The country is suffering from population displacement, uncontrolled illegal development, seizure of public lands and buildings, fragmentation and parcel subdivision, forged ownership deeds, and other LAS issues. The motivation for, previous lessons from, and quality of VGI generally is considered, specifically its role in addressing such land administration issues. Finally, a detailed plan for undertaking fieldwork and data analysis in the challenging situations which pertain in Iraq is proposed, to determine the role of VGI in enhancing official systems.

References

Adlington, G., Tonchovska, R.: Good governance of tenure. FAO and World Bank support and future agendas. In: Proceedings of the 4th Regional Conference for Cadastre and Spatial Data Infrastructure, Bled, Slovenia (2010)

Al-Bakri, M., Fairbairn, D.: Assessing similarity matching for possible integration of feature classifications of geospatial data from official and informal sources. Int. J. Geogr. Inf. Sci. **26**(8), 1437–1456 (2012)

Alemie, B.K., Bennett, R.M., Zevenbergen, J.: Evolving urban cadastres in Ethiopia: the impacts on urban land governance. Land Use Policy **42**, 695–705 (2015)

Arko-Adjei, A.: Adapting Land Administration to the Institutional Framework of Customary Tenure: The Case of Peri-Urban Ghana (No. 184). IOS Press, Amsterdam (2011)

Barth, A.: New York City and OpenStreetMap Collaborating Through Open Data (2013). https://www.mapbox.com/blog/nyc-and-openstreetmap-cooperating-through-open-data/. Accessed 29 Mar 2017

Basiouka, S., Potsiou, C.: VGI in Cadastre: a Greek experiment to investigate the potential of crowd sourcing techniques in Cadastral Mapping. Surv. Rev. **44**(325), 153–161 (2012)

Bégin, D.: Towards integrating VGI and national mapping agency operations - a Canadian case study. In: The Role of Volunteer Geographic Information in Advancing Science: Quality and Credibility Workshop. GIScience Conference (2012)

Bennett, R., Rajabifard, A., Kalantari, M., Wallace, J., Williamson, I.: Cadastral futures: building a new vision for the nature and role of cadastres. In: FIG Congress, pp. 11–16 (2010)

Bishr, M., Mantelas, L.: A trust and reputation model for filtering and classifying knowledge about urban growth. GeoJournal **72**(3–4), 229–237 (2008)

Bonney, R., Cooper, C.B., Dickinson, J., Kelling, S., Phillips, T., Rosenberg, K.V., Shirk, J.: Citizen science: a developing tool for expanding science knowledge and scientific literacy. BioScience **59**(11), 977–984 (2009)

Brassel, K., Bucher, F., Stephan, E.M., Vckovski, A.: Completeness. In: Elements of Spatial Data Quality, pp. 81–108 (1995)

Chitsike, F.: A critical analysis of the land reform programme in Zimbabwe. In: 2nd FIG Regional Conference, pp. 2–5, December 2003

Ciepłuch, B., Jacob, R., Winstanley, A., Mooney, P.: Comparison of the accuracy of OpenStreetMap for Ireland with Google Maps and Bing Maps. In: Proceedings of the Ninth International Symposium on Spatial Accuracy Assessment in Natural Resources and Environmental Sciences, pp. 37–40, July 2010

Coleman, D.J.: Volunteered geographic information in spatial data infrastructure: an early look at opportunities and constraints. In: GSDI 12 World Conference, October 2010

Coleman, D.J., Georgiadou, Y., Labonte, J.: Volunteered geographic information: the nature and motivation of produsers. Int. J. Spat. Data Infrastruct. Res. **4**(1), 332–358 (2009)

Cotfas, L., Diosteanu, A.: Evaluating accessibility in crowdsourcing GIS. J. Appl. Collab. Syst. **2** (1) (2010)

De Vries, W.T., Bennett, R.M., Zevenbergen, J.A.: Neo-cadastres: innovative solution for land users without state based land rights, or just reflections of institutional isomorphism? Surv. Rev. **47**(342), 220–229 (2015)

Enemark, S.: Fit-for-purpose: building spatial frameworks for sustainable and transparent land governance. In: Annual World Bank Conference on Land and Poverty, January 2013

Enemark, S., Clifford Bell, K., Lemmen, C., McLaren, R.: Fit-for-purpose land administration. FIG Denmark (2014)

Fairbairn, D., Al-Bakri, M.: Using geometric properties to evaluate possible integration of authoritative and volunteered geographic information. ISPRS Int. J. Geo-Inf. **2**(2), 349–370 (2013)

Genovese, E., Roche, S.: Potential of VGI as a resource for SDIs in the North/South context. Geomatica **64**(4), 439–450 (2010)

Goodchild, M.F.: Citizens as sensors: the world of volunteered geography. GeoJournal **69**(4), 211–221 (2007)

Goodchild, M.F.: Commentary: whither VGI? GeoJournal **72**(3), 239–244 (2008)

Goodchild, M.F., Hunter, G.J.: A simple positional accuracy measure for linear features. Int. J. Geogr. Inf. Sci. **11**(3), 299–306 (1997)

Goodchild, M.F., Li, L.: Assuring the quality of volunteered geographic information. Spat. Stat. **1**, 110–120 (2012)

Haklay, M.: How good is volunteered geographical information? A comparative study of OpenStreetMap and Ordnance Survey datasets. Environ. Plan. B: Plann. Des. **37**(4), 682–703 (2010)

Haklay, M., Basiouka, S., Antoniou, V., Ather, A.: How many volunteers does it take to map an area well? The validity of Linus' law to volunteered geographic information. Cartogr. J. **47**(4), 315–322 (2010)

Haklay, M., Budhathoki, N.: OpenStreetMap–overview and motivational factors. Proceedings of Horizon Infrastructure Challenge Theme Day, University of Nottingham (2010)

Haklay, M.E., Antoniou, V., Basiouka, S., Soden, R., Mooney, P.: Crowdsourced Geographic Information Use in Government. Global Facility for Disaster Reduction and Recovery (GFDRR), World Bank, London (2014)

Hashemi, P., Abbaspour, R.A.: Assessment of logical consistency in openstreetmap based on the spatial similarity concept. In: Jokar Arsanjani, J., Zipf, A., Mooney, P., Helbich, M. (eds.) OpenStreetMap in GIScience. LNGC, pp. 19–36. Springer, Cham (2015). doi:10.1007/978-3-319-14280-7_2

Isser, D., Van der Auweraert, P.: Land, property, and the challenge of return for Iraq's displaced. United States Institute of Peace (USIP) (2009)

Istabraq, I., Mohammed, M., Hatem, S.: National Report of the Republic of Iraq for Habitat III. UN HABITAT, pp. 1–68 (2016). http://www.hlrn.org/img/documents/Iraq_National_Report.pdf

Ituen, U., Johnson, I.: Securing land title/ownership rights: a survey of the level of compliance with land registration in Akwa Ibom State, Nigeria. Development **4**(1) (2014)

Jackson, S.P., Mullen, W., Agouris, P., Crooks, A., Croitoru, A., Stefanidis, A.: Assessing completeness and spatial error of features in volunteered geographic information. ISPRS Int. J. Geo-Inf. **2**(2), 507–530 (2013)

Laarakker, P.: Person, Parcel, Power, Towards an Extended Model for Land Registration (2011)

Lemmen, C.: The Social Tenure Domain Model: A Pro-Poor Land Tool. International Federation of Surveyors (FIG) (2010)

Maué, P.: Reputation as tool to ensure validity of VGI. In: Workshop on Volunteered Geographic Information, December 2007

McLaren, R.: Engaging the land sector gatekeepers in crowdsourced land administration. In: Proceedings of World Bank Land and Poverty Conference, Washington, pp. 8–12 (2013)
Mullen, W.F., Jackson, S.P., Croitoru, A., Crooks, A., Stefanidis, A., Agouris, P.: Assessing the impact of demographic characteristics on spatial error in volunteered geographic information features. GeoJournal **80**(4), 587–605 (2015)
Mummidi, L.N., Krumm, J.: Discovering points of interest from users' map annotations. GeoJournal **72**(3–4), 215–227 (2008)
Navratil, G., Frank, A.U.: VGI for land administration-a quality perspective. ISPRS-Int. Arch. Photogramm. Remote Sens. Spat. Inf. Sci. **1**(1), 159–163 (2013)
Niederer, S., Van Dijck, J.: Wisdom of the crowd or technicity of content? Wikipedia as a sociotechnical system. New Media Soc. **12**(8), 1368–1387 (2010)
Olowu, D.: Local institutional and political structures and processes: recent experience in Africa. Public Adm. Dev. **23**(1), 41–52 (2003)
Quan, J., Payne, G.: Secure land rights for all. The United Nations Human Settlements Programme, UN-HABITAT (2008)
Seeger, C.J.: The role of facilitated volunteered geographic information in the landscape planning and site design process. GeoJournal **72**(3–4), 199–213 (2008)
Shaikley, L.K.: Iraq's housing crisis: upgrading settlements for IDPS (internally displaced persons). Doctoral dissertation, Massachusetts Institute of Technology (2013)
Steudler, D., Rajabifard, A., Williamson, I.P.: Evaluation of land administration systems. Land Use Policy **21**(4), 371–380 (2004)
Tulloch, D.L.: Many, many maps: empowerment and online participatory mapping. First Monday, **12**(2) (2007). http://dx.doi.org/10.5210/fm.v12i2.1620
Turner, A.: Introduction to Neogeography. O'Reilly Media, Inc., Sebastopol (2006)
USAID, Iraqi Local Governance Program: Land Registration and Property Rights in Iraq. RTI International, C. N: EDG-C-00-03-00010-00 Baghdad, Iraq (2005)
Veregin, H.: Data quality parameters. Geogr. Inf. Syst. **1**, 177–189 (1999)
Williamson, I.P.: Cadastres and land information systems in common law jurisdictions. Surv. Rev. **28**(217), 114–129 (1985)
Winterbottom, A., North, J.: Building an open access African studies repository using Web 2.0 principles. First Monday, **12**(4) (2007)
Zielstra, D., Zipf, A.: A comparative study of proprietary geodata and volunteered geographic information for Germany. In: 13th AGILE International Conference on Geographic Information Science, vol. 2010, May 2010

Workshop on Advances in Web Based Learning (AWBL 2017)

Unsupervised Learning of Question Difficulty Levels Using Assessment Responses

Sankaran Narayanan[1,2(✉)], Vamsi Sai Kommuri[1,2], N. Sethu Subramanian[1,2], Kamal Bijlani[1,2], and Nandu C. Nair[1,2]

[1] Amrita e-Learning Research Lab (AERL), Amrita School of Engineering, Amritapuri, India
{nsankaran,sethus,cnandu}@am.amrita.edu, vamsi5712@gmail.com, kamal@amrita.edu

[2] Amrita Vishwa Vidyapeetham, Amrita University, Coimbatore, India

Abstract. Question Difficulty Level is an important factor in determining assessment outcome. Accurate mapping of the difficulty levels in question banks offers a wide range of benefits apart from higher assessment quality: improved personalized learning, adaptive testing, automated question generation, and cheating detection. Adopting unsupervised machine learning techniques, we propose an efficient method derived from assessment responses to enhance consistency and accuracy in the assignment of question difficulty levels. We show effective feature extraction is achieved by partitioning test takers based on their test-scores. We validate our model using a large dataset collected from a two thousand student university-level proctored assessment. Preliminary results show our model is effective, achieving mean accuracy of 84% using instructor validation. We also show the model's effectiveness in flagging mis-calibrated questions. Our approach can easily be adapted for a wide range of applications in e-learning and e-assessments.

Keywords: e-assessments · Unsupervised learning · Personalized learning · Question bank · Difficulty levels

1 Introduction

Question Banks are a basic part of e-assessment systems. Questions in the bank span a wide-spectrum of subject matter. High quality questions can be effective in measuring the extent to which test takers have succeeded in meeting learning objectives [1]. The nature of questions directly influence assessment quality as well as the experience of test takers [2,10]. Precise estimation and dynamic calibration of question difficulty can be highly beneficial in personalized e-learning as well [3,4].

In e-assessments, examiners routinely define the proportion of questions to be chosen from each subject area, learning objective [1], difficulty level [2], and concept [13]. In randomized close-ended e-assessments, each test taker is given

O. Gervasi et al. (Eds.): ICCSA 2017, Part I, LNCS 10404, pp. 543–552, 2017.
DOI: 10.1007/978-3-319-62392-4_39

a random set of questions, selected according to these attributes. Quality of evaluation in randomized assessments thus depends on accuracy in mapping the attributes. Factors affecting the quality include large number of questions in the bank, author prejudice, and incoherent inputs.

We investigated the problem of improving accuracy in mapping of the question difficulty levels. Difficulty Level is a key parameter that determines assessment outcomes. In addition to higher assessment quality, accurate difficulty level mapping in question banks offers numerous benefits: better methods for personalized learning, adaptive testing, auto-generation of test questions and cheating detection. Existing work based on assessment responses falls in two categories: The first category leverages Natural Language Processing (NLP) to identify a set of patterns. A rules engine [5] or a supervised learning model [6,7] or both [8] help estimate the desired parameters based on the identified patterns. The second class of methods adopts statistical modeling techniques such as item response theory [2] to estimate learner skill levels [3,4], adaptive testing [9], and answer similarity detection [11].

We used test taker performance metrics derived from e-assessment responses to construct an unsupervised learning model [12] for precise estimation of question difficulty levels. We have shown that partitioning test takers into three groups based on assessment scores is an effective method for feature extraction. We tested our model on a real-world data set, generated from a university-level proctored e-assessment, covering two thousand students. Our model achieved 84% agreement with the instructor rating (external index), also validated using Cohen's Kappa Method [14]. Using the validation data set, we have shown our model's effectiveness in identifying mis-calibrated questions. Our technique can be easily adapted for personalized learning, adaptive testing, similarity detection, and automatic generation of questions.

Section 2 of the paper covers related work in this domain, followed by our proposed methodology in Sect. 3. Verification and validation of our approach is noted in Sect. 4, followed by conclusion in Sect. 5, and suggested follow-ups in Sect. 6.

2 Related Work

Question Difficulty Level affects both validity and reliability of assessments [2]. Validity measures the extent to which a test taker has succeeded in meeting learning objectives. Reliability indicates whether all test takers are evaluated consistently. Some attributes affect only validity (e.g., learning objective). However, incorrect difficulty levels impact both validity and reliability.

A set of prior methods were focused on leveraging Revised Bloom's Taxonomy (RBT) [1] and Natural Language Processing (NLP) techniques. Haris and Omar [5] used NLP to construct a set of rules to classify written questions and assess the cognitive level of test takers. Thomas et al. [8] calculated sentence similarity using structural and semantic analysis for automatic scoring of answers. Yahya et al. [6] constructed a supervised learning model using pre-selected questions and RBT-cognitive mappings. Sangodia et al. [7] have proposed integrating a similar supervised learning model into a question bank system.

A second class of methods have used statistical modeling [2] of assessment responses for learner level estimation [3,4], adaptive testing [9] and similarity detection [11]. Chen et al. [3] constructed a model based on study of material difficulty and learner level to facilitate personalized e-learning. Wang [4] used a pre-test approach to estimate learner level for adapting learning and assessment. Raykova et al. [9] capitalized learner response history for selection of adaptive testing questions. Wesolowsky [11] calculated measures of answer similarity in multiple choice exams for cheating detection.

Item Response Theory [2] is widely used for interpreting assessment responses. Difficulty Index (or p-value) [2,15] is the proportion of test takers answering a particular question correctly. Higher p-values indicate easier questions. Hence we will refer to the p-value as Easiness Index for the rest of this paper. ScorePak [15] classifies a question as 'Easy' if the easiness index is 0.85 or more, 'Moderate' if it's between 0.51 and 0.84, and 'Hard' otherwise. Discrimination Index [2] measures how well an assessment differentiates between top and bottom performers. These metrics need careful interpretation. For instance, if both easiness and discrimination indices are low, it could very well be that most test takers found the question Hard. We have generalized these metrics for use in e-assessment learning models.

3 Proposed Methodology

We have created a Question Classifier Engine that uses metrics derived from assessment data to improve mappings of difficulty levels of questions in the bank. We have also created a Verification Web Portal to validate and approve mapping changes suggested by the Question Classifier. Figure 1 shows a schematic representation of the system architecture. In the rest of this section, we define the features for our unsupervised learning model and discuss validation in the next section.

To estimate Question Difficulty Level based on unsupervised learning, we define it as an ordered attribute [12] with **l** levels: $D = \{d_1, d_2, \ldots, d_l\}$ on the question set: $Q = \{q_1, q_2, \ldots, q_q\}$ of **q** questions.

Based on the assessment responses, we can define three sets of metrics pertinent to test taker's experience: subject-level, question-level, and performance level. In our model, Subject metrics used were the maximum or highest score received, aggregate of assessment questions, aggregate of test takers for that subject. Question metrics were aggregate count of attempts made for that question, aggregate score for that question, and easiness index. Past work [2,15] has shown that the knowledge/cognitive level of test taker also plays a significant role in answering questions correctly. We use this observation to derive the Question-Level Performance metrics as will be shown below.

Let S be the subject for which assessment was conducted.
Set of test takers: $T = \{t_1, t_2, \ldots, t_t\}$
Set of questions: $Q = \{q_1, q_2, \ldots, q_q\}$

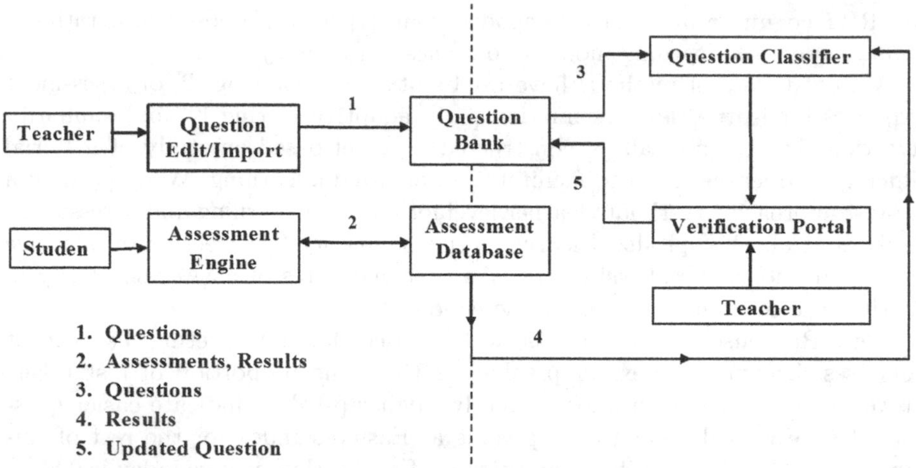

Fig. 1. System architecture

Total score of a test taker t_j: ts_j
Maximum score received in the subject: ms

Attempts for a given question q_i is defined as:

$$a(q_i, t_j) : \begin{cases} 1, & \text{if } t_j \textbf{ was given } q_i \\ 0 & otherwise \end{cases} \tag{1}$$

Score for a given question q_i is defined as:

$$s(q_i, t_j) : \begin{cases} 1, & \text{if } t_j \textbf{ answered } q_i \textbf{ correctly} \\ 0 & otherwise \end{cases} \tag{2}$$

Score Partitions for the subject S is composed of p partitions as defined by Eqs. 3 and 4:

$$SP = \cup_{k=0}^{p} sp_k \tag{3}$$

$$sp_k = \begin{cases} [\frac{\mathbf{ms}\times(k-1)}{p}, \frac{\mathbf{ms}*(k)}{p}], \text{ if } k \leq (p-1) \\ [\frac{\mathbf{ms}\times(k-1)}{p}, \frac{\mathbf{ms}*(k)}{p}] \ otherwise \end{cases} \tag{4}$$

Using Eqs. 1 and 4, we define the Attempts for q_i in Score Partition sp_k as:

$$a(q_i, t_{j,k}) : \begin{cases} 1, & \text{if } \ ts_j \in sp_k \\ 0 & otherwise \end{cases} \tag{5}$$

Similarly, using Eqs. 2 and 4 we define the Score for q_i in Score Partition sp_k as:

$$s(q_i, t_{j,k}) : \begin{cases} 1, & \text{if } ts_j \in sp_k \\ 0 & otherwise \end{cases} \tag{6}$$

As discussed before, easiness index is the proportion of test takers answering a question correctly. Using Eqs. 1 and 2, Easiness Index for a given question q_i is defined as:

$$E_0(q_i) = \frac{\sum_{j=1}^{t} s(q_i, t_j)}{\sum_{j=1}^{t} a(q_i, t_j)} \tag{7}$$

Similarly, the proportion of test takers answering a question correctly in any given score partition can now be defined using Eqs. 5 and 6. Easiness Index for q_i in Score Partition sp_k:

$$E_k(q_i) = \frac{\sum_{j=1}^{t} s(q_i, t_{j,k})}{\sum_{j=1}^{t} a(q_i, t_{j,k})} \tag{8}$$

To illustrate the suitability of partitioned Easiness Indices (Eq. 8) as features in a learning model, let us consider three partitions corresponding to Bottom, Medium and Top performing test taker groups. This partitioning is a widely accepted practice and hence serves as a good basis for validation. The example below shows the feature matrix formed by using E_1, E_2, E_3 as features:

$$\begin{bmatrix} e_{11} & e_{12} & e_{13} \\ e_{21} & e_{22} & e_{23} \\ \vdots & \vdots & \vdots \\ e_{q1} & e_{q2} & e_{q3} \end{bmatrix} \tag{9}$$

Row i of Eq. 9 is projection of question q_i as a point in $\mathbb{R}^3$. By clustering the projected points into l groups, we can create G as a proxy for D. This is achieved using the standard optimization function [12]:

$$\arg\min_{\mathbf{G}} \sum_{i=1}^{l} \sum_{\mathbf{e} \in G_i} \|\mathbf{e} - \boldsymbol{\mu}_i\|^2 \tag{10}$$

In this study, we have used the standard k-means algorithm to implement this optimization objective by creating three clusters for G. Since D is an ordered set, appropriate ordering of the cluster centroids of G provides us the mapping $Q \mapsto D$.

4 Validation Model

4.1 Data Set

The data used in this study was collected from a close-ended official e-assessment of first-year engineering students from a large university in India. Several thousand students were assessed at the same time in a proctored setting. Each student was evaluated on four subjects they had enrolled in the academic year. Fifteen multiple-choice questions were randomly selected per student from a confidential question bank that contained around hundred questions per subject. All questions had equal score of one mark. All questions had four choices. The test-scores contributed to the final grade of the students.

Question difficulty levels in the validation data set correspond to the first three levels of cognitive dimension in RBT – Remembering, Understanding, and Applying. Hence, we validated our model using a scheme with three difficulty levels: 'Easy', 'Moderate', and 'Hard'.

Table 1 shows the four subjects used for validation and the distribution of 'Easy', 'Moderate', and 'Hard' questions as provided by instructors. Table 2 shows the performance characteristics of the aggregate number of students assessed per subject. The aggregate attempts received for the entire subject is shown in column Agg. Attempts. The aggregate total score for the entire subject is shown in Agg. Score.

Table 1. Subjects question characteristics

Subject	Ques.	Easy	Moderate	Hard
Chemistry	99	19	64	16
Mechanics	97	28	41	28
Electronics	98	20	52	26
CS	99	28	54	17

Table 2. Student performance characteristics

Subject	Students	Agg. attempts	Agg. score
Chem	1893	28184	11528
Mech	2031	30185	12484
Elec	2374	35364	13855
CS	945	14139	7080

Figure 2 shows the easiness index distribution (Eq. 7) of the subjects. The mean Easiness Index is in the range 0.35 to 0.50 consistent with a norm-referenced test [2] designed to assess course material understanding and emphasize test taker differentiation.

Figure 3 shows a sample Question Classifier output for Mechanics. Points shown by 'Plus' represent 'Easy' questions, 'Circles' represent 'Hard' questions and the 'Triangles' represent the 'Moderate' questions. The misclassification errors and boundary cases are discussed subsequently.

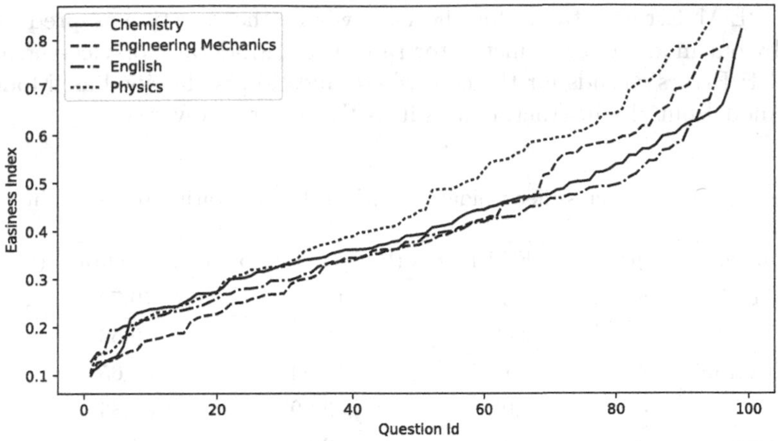

Fig. 2. Easiness index distribution

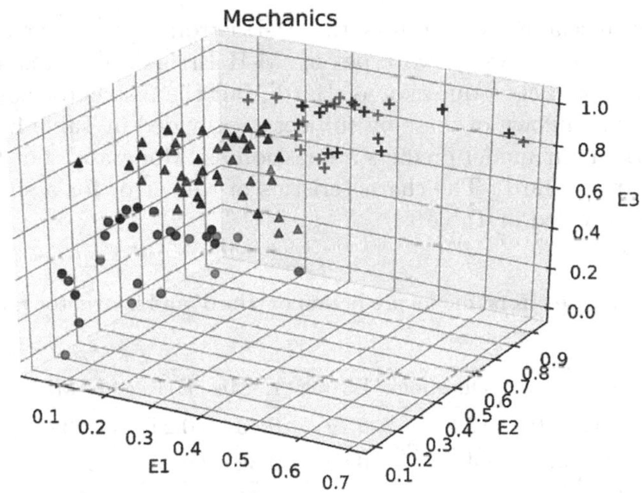

Fig. 3. Difficulty level classification of mechanics questions

4.2 Results

Table 3 shows a summary of the Question Classifier classification result for the four subjects obtained by partitioning test takers into three groups (ref Eqs. 3 and 8 in Sect. 3). The features that we can derive from a three-partition scheme are, E_1, E_2, E_3, the easiness indices corresponding to the bottom, middle and top performing groups of students respectively.

Column 'Accuracy %' shows the proportion of questions where there is agreement between the difficulty level predictions made by our model and the rating provided by the instructor (external index). Column 'E-M Errors' and 'M-H Errors' show the proportion of questions for which there is lack of agreement.

Column 'E-M Errors' stands for the case where the question is predicted as 'Easy' by our model but the instructor rates it as 'Moderate' or vice-versa. Column 'M-H Errors' stands for the case where question is classified as 'Moderate' by our model but the instructor rates it as 'Hard' or vice-versa.

Table 3. Question classifier validation with instructor rating (external index)

Subject	Questions	E-M Errors (%)	M-H Errors (%)	Accuracy (%)
Chemistry	99	7.07	13.13	79.79
Mechanics	97	2.06	8.2	89.69
Electronics	98	5.1	12.24	82.65
CS	99	7.07	9.09	83.83
Mean	98	5.3	10.6	84

Among the misclassification errors, the 'M-H Errors' account for 10.6% reduction in accuracy. There were two types of 'M-H Errors': The first type occurs when our model predicts a question as 'Hard', but the instructor has rated it as 'Moderate'. A breakdown of these by subject is provided in Table 4. The second type occurs when our model predicts a question as 'Moderate', but the instructor has rated it as 'Hard'. The characteristics of this error for a single subject, Mechanics, is provided in Table 5.

Table 4. Subject-wise errors: model prediction of 'Hard' and instructor rating of 'Moderate'

Subject	Questions	E3Reg	E3Mis	E1Reg	E1Mis
Chemistry	12	0.79	0.48	0.24	0.22
Mechanics	4	0.80	0.34	0.23	0.23
Electronics	9	0.80	0.51	0.26	0.18
CS	9	0.80	0.52	0.21	0.19

Table 4 shows the questions for which the prediction by our model was 'Hard' and the instructor supplied rating was 'Moderate'. The columns E3Reg and E1Reg show the mean easiness index of the top and bottom performing groups for correctly classified questions rated 'Moderate'. The columns E3Mis and E1Mis show the mean easiness index of the same set of groups for the misclassified questions. Using Chemistry as an example, we can see that the top performing students had a 79% mean success ratio for correctly classified questions. On the other hand, a similar performing group had 48% mean success ratio for the misclassified questions and hence the prediction outcome of 'Hard'. In this case, the performance properties for these misclassified questions are closer to those of

'Hard' questions. We can also see for the same Subject, the bottom performing students performed similar (24% vs 22%). This validates our proposed model.

Table 5. Mechanics: model prediction of 'Moderate', instructor rating of 'Hard'

S.no	Prediction	Instructor	E1	E2	E3
1	Moderate	Moderate	0.23	0.46	0.80
2	**Moderate**	**Hard**	**0.15**	**0.28**	**0.73**
3	Hard	Hard	0.11	0.22	0.39

Table 5 shows the mean easiness indices for the four questions in Mechanics for which the prediction by our model was 'Moderate' and the instructor rating was 'Hard'. Columns Prediction and Instructor show the difficulty levels as predicted by our Model and as rated by Instructor respectively. E1, E2, E3 show the mean easiness indices of bottom, middle, and top performing student groups respectively. Row 2 shows the misclassification, and we can observe that these questions have performance metrics higher than the typical 'Hard' question (Row 3) particularly for top performing students, but well below the typical 'Moderate' question (Row 1).

The first example shows the strength of our model in capturing test taker performance and translating them to accurate mapping of question difficulty levels. The second example demonstrates the boundary cases where we could further enhance the system using a rules-engine or a manual calibration workflow. The analysis for the 'E-M Errors' column in Table 3 is similar and hence not shown for brevity.

5 Conclusion

In this paper, we have proposed a flexible unsupervised learning approach to enhance consistency and accuracy in the assignment of question difficulty levels. We classify the test takers into three groups and derive performance metrics that can be used as features for unsupervised clustering. The predictions of our model achieve 84% agreement with instructor supplied difficulty levels, pursuing Cohen's Kappa Method. We also showed that the model is effective in identifying mis-calibrated questions. In closing, we would like to note that the proposed method can be an enabler for wider variety of e-learning and e-assessment scenarios requiring difficulty estimation such as personalized learning, adaptive testing, and anomaly detection.

6 Future Work

In this study we have used data obtained from a large university-level proctored e-assessment. We are actively evaluating model effectiveness in the presence of skews from test taker or question distributions. We have used three partitions of test taker performances; the efficacy of five partitions could be investigated. Questions could be automatically calibrated using a fine-grained difficulty level approach or by using a rules engine. Adapting this model for open-ended assessments is another interesting research problem.

References

1. Conklin, J., Anderson, L.W., Krathwohl, D., Airasian, P., Cruikshank, K.A., Mayer, R.E., Pintrich, P., Raths, J., Wittrock, M.C.: A Taxonomy for Learning, Teaching, and Assessing: A Revision of Bloom's Taxonomy of Educational Objectives Complete Edition. JSTOR (2005)
2. Reynolds, C.R., Livingston, R.B., Willson, V.L., Willson, V.: Measurement and Assessment in Education. Pearson Education International, Upper Saddle River (2010)
3. Chen, C.-M., Lee, H.-M., Chen, Y.-H.: Personalized e-learning system using item response theory. Comput. Educ. **44**(3), 237–255 (2005)
4. Wang, T.-H.: Developing an assessment-centered e-Learning system for improving student learning effectiveness. Comput. Educ. **73**, 189–203 (2014)
5. Haris, S.S., Omar, N.: A rule-based approach in Bloom's Taxonomy question classification through natural language processing. In: 2012 7th International Conference on Computing and Convergence Technology (ICCCT), pp. 410–414. IEEE (2012)
6. Yahya, A.A., Toukal, Z., Osman, A.: Bloom's taxonomy-based classification for item bank questions using support vector machines. In: Ding, W., Jiang, H., Ali, M., Li, M. (eds.) Modern Advances in Intelligent Systems and Tools. SCI, vol. 431. Springer, Heidelberg (2012). doi:10.1007/978-3-642-30732-4_17
7. Sangodiah, A., Ahmad, R., Ahmad, W.F.W.: Integration of machine learning approach in item bank test system. In: 2016 3rd International Conference on Computer and Information Sciences (ICCOINS), pp. 164–168. IEEE (2016)
8. Thomas, N.T., Kumar, A., Bijlani, K.: Automatic answer assessment in LMS using latent semantic analysis. Procedia Comput. Sci. **58**, 257–264 (2015). Elsevier
9. Raykova, M., Kostadinova, H., Totkov, G.: Adaptive test system based on revised Bloom's Taxonomy. In: Proceedings of the 12th International Conference on Computer Systems and Technologies, pp. 504–509. ACM (2011)
10. Haladyna, T.M., Downing, S.M., Rodriguez, M.C.: A review of multiple-choice item-writing guidelines for classroom assessment. Appl. Measur. Educ. **15**(3), 309–333 (2002)
11. Wesolowsky, G.O.: Detecting excessive similarity in answers on multiple choice exams. J. Appl. Stat. **27**(7), 909–921 (2000)
12. Friedman, J., Hastie, T., Tibshirani, R.: The Elements of Statistical Learning. Springer Series in Statistics, vol. 1. Springer, New York (2001)
13. Nair, N.C., Archana, J.S., Chatterjee, S., Bijlani, K.: Knowledge representation and assessment using concept based learning. In: 2015 International Conference on Advances in Computing, Communications and Informatics (ICACCI), pp. 848–854. IEEE (2015)
14. Stemler, S.: An overview of content analysis. Pract. Assess. Res. Eval. **7**(17), 137–146 (2001)
15. University of Washington: ScorePak: Office of Educational Assessment University of Washington, Seattle, USA

An Agents and Artifacts Metamodel Based E-Learning Model to Search Learning Resources

Birol Ciloglugil[1] and Mustafa Murat Inceoglu[2(✉)]

[1] Department of Computer Engineering, Ege University, 35100 Bornova, Izmir, Turkey
birol.ciloglugil@ege.edu.tr

[2] Department of Computer Education and Instructional Technology, Ege University, 35100 Bornova, Izmir, Turkey
mustafa.inceoglu@ege.edu.tr

Abstract. In this paper, an e-learning model based on Agents and Artifacts (A&A) Metamodel to search learning resources from multiple sources is proposed. Multi agent system (MAS) based e-learning models with the same functionality are available in the literature. However, they are mostly developed as standalone systems that contain a single agent responsible for searching and retrieving learning resources. With the highly distributed nature of learning resources over multiple repositories, giving this responsibility to only one agent decreases scalability. The proposed model exploits the A&A Metamodel to overcome this issue. A&A Metamodel focuses on environment modeling in MAS design and models entities in the environment as artifacts, that are first class entities like agents. From the perspective of MAS based e-learning systems, learning resources are the main components in the environment that agents interact with. Thus, an efficient solution can be achieved with an e-learning model that searches learning objects by using an e-learning environment model based on A&A Metamodel. The proposed e-learning system is developed with Jason and the e-learning environment model is implemented with CArtAgO framework. Finally, current limitations and future directions of the proposed approach are discussed.

Keywords: E-Learning · Learning resources · Multi agent systems · Agents and Artifacts Metamodel · Environment programming · CArtAgO

1 Introduction

Multi agent systems (MAS) technology have been applied to different application domains including e-learning. There are various efforts to provide MAS based e-learning systems [1–4]. These systems usually include specialized agents responsible for accessing learning resources, that are generally stored as learning objects in learning object repositories. The agents in these systems provide a mechanism to store, search and retrieve learning resources to other agents and

O. Gervasi et al. (Eds.): ICCSA 2017, Part I, LNCS 10404, pp. 553–565, 2017.
DOI: 10.1007/978-3-319-62392-4_40

learners. Therefore, these agents are generally referred to as repository agents, learning object agents or delivery agents [1,2]. Most of the MAS based e-learning systems have been developed as standalone applications, where agents access predetermined learning resources [1–4]. Thus, adaptability of MAS based e-learning systems to dynamically changing learning resources is an important issue to overcome.

Agents and Artifacts Metamodel is based on modeling of the components in the environment as first class entities like agents. There are many definitions of environment from different perspectives. One of the definitions focusing on the importance of its functionalities is as follows; "the environment is a first-class abstraction that provides the surrounding conditions for agents to exist and that mediates both the interaction among agents and the access to resources" [5]. Thus, environment modeling and the related Agents and Artifacts Metamodel is quite important for communication, interaction and coordination of agents situated in the same environment.

In this paper, first we introduce related standards and technologies and then, discuss how we can incorporate them to propose a new environment based e-learning model. The proposed system has two main components; an e-learning environment model based on A&A Metamodel [6] and a MAS based e-learning system that exploits it. The e-learning environment model is implemented as a prototype by using CArtAgO (Common Artifact infrastructure for Agent Open environment) framework [7]. The MAS based e-learning system is developed with Jason agent development framework and uses the e-learning environment model to search learning objects from different sources.

The rest of the paper is organized as follows; Sect. 2 introduces the e-learning standards as well as the agents and environment based technologies used in this study. Section 3 features details of the proposed e-learning model. First, the e-learning environment model is presented. Then, a MAS based e-learning system is examined by discussing how it exploits the e-learning environment model to search learning resources. Section 4 concludes the paper with a discussion and future work perspectives.

2 Related Standards and Technologies

Related standards and technologies will be discussed in the following three subsections. First, the e-learning standards related to learning resources will be introduced briefly. Then, Agents and Artifacts Metamodel and its implementation with CArtAgO framework will be examined, respectively.

2.1 E-Learning Standards

E-Learning standards can be categorized into two groups; metadata specification and content structure modeling. Metadata specification standards are involved with providing appropriate metadata to represent learning resources. IEEE LOM (Learning Object Metadata) [8] and Dublin Core [9] can be named as the two leading standards. The main content structure modeling standard is SCORM (Sharable Content Object Reference Model) [10]. SCORM standard is used to represent learning resources as compressed ".zip" file packages called learning objects to increase the reusability of learning resources. A SCORM package includes files as assets and/or SCOs (Sharable Content Objects) composed of assets; and a manifest file named "imsmanifest.xml". The "imsmanifest.xml" file contains metadata specification of the learning resources (assets and SCOs) and content organization of learning resources with the relationships between different learning resources in XML format.

SCORM uses IEEE LOM metadata standard for metadata specification of learning resources. Therefore, learning objects can be defined as learning resources packed according to SCORM standard that include appropriate metadata about the content of the learning resources with respect to IEEE LOM standard for reusability. IEEE LOM standard has nine categories (1. General, 2. Lifecycle, 3. Meta-Metadata, 4. Technical, 5. Educational, 6. Rights, 7. Relation, 8. Annotation and 9. Classification) and subcategories to model learning resources [8].

2.2 Agents and Artifacts Metamodel

In multi agent systems, the entities that are not modeled as agents can be considered as a part of the environment. These entities can be modeled as artifacts. From the agents' perspective, artifacts can be defined as the first class entities representing the resources and tools in the environment that can be dynamically created, used and managed by the agents to support their individual or collective activities [11]. Agents are autonomous and reactive, thus they perform operations related to the intelligent part of the system. Artifacts realize the computational entities that does not require intelligence and can be modeled as functions, i.e. database, web services and communication structures can be modeled as artifacts. As a summarization, it can be concluded that programming multi agent systems is a combination of agent and environment programming [12]:

"programming MAS = programming Agents + programming Environment"

Environment programming provides a "work environment" for agents to situate in and operate [13]. The notion of work environment is based on Activity Theory and Distributed Cognition [13–15]. Agents & Artifacts Metamodel is inspired by Activity Theory and Distributed Cognition to promote a methodology for modeling agents and artifacts as first class entities in a work environment.

Artifacts can have three different roles in the environment [5]; the first one is enabling agents to access the deployment context (hardware, software and external resources); the second one is providing a conceptual gap between the agent abstraction and the low level details of the deployment context; and the third one is supporting a regulation mechanism to mediate interaction between agents and access to shared resources. These roles enable artifacts to provide different levels of abstraction within the environment. Using artifacts in multi agent systems to provide abstraction has some benefits in terms of system design; firstly distinguishing the responsibilities of agents and artifacts helps supporting the separation of concerns in MAS. The second benefit is the decrease in system complexity and the increase in scalability as a result of the modeling of non-autonomous, computational entities as artifacts instead of agents.

2.3 CArtAgO Framework

CArtAgO (Common Artifact infrastructure for Agent Open environment) is the only available framework implementing Agents and Artifacts Metamodel [7]. CArtAgO provides an API to define artifacts; an API for agent programmers to interact with the artifacts and the work environment they belong to; and a run-time environment for dynamic management of the work environments. CArtAgO contains three basic building blocks; agent bodies, artifacts and workspaces. These building blocks and their relationship can be observed in Fig. 1; which demonstrates the abstract architecture of the CArtAgO framework as depicted in [12]. The first layer at the bottom contains the operating system and JVM. The right hand side of the second layer contains the CArtAgO framework that provides an execution platform for different MAS applications. Workspaces are the main CArtAgO building block that define the topology of the work environments and act as a logical container for artifacts. Artifacts are the basic blocks to structure the work environment. Agent bodies make it possible for agents to situate in work environments and create, use and manage artifacts provided by the workspace. The left hand side of the second layer features agent frameworks like Jadex and Jason to develop MAS applications as shown at the left hand side of the third layer. The right hand side of the third layer contains artifact based working environments developed with CArtAgO framework.

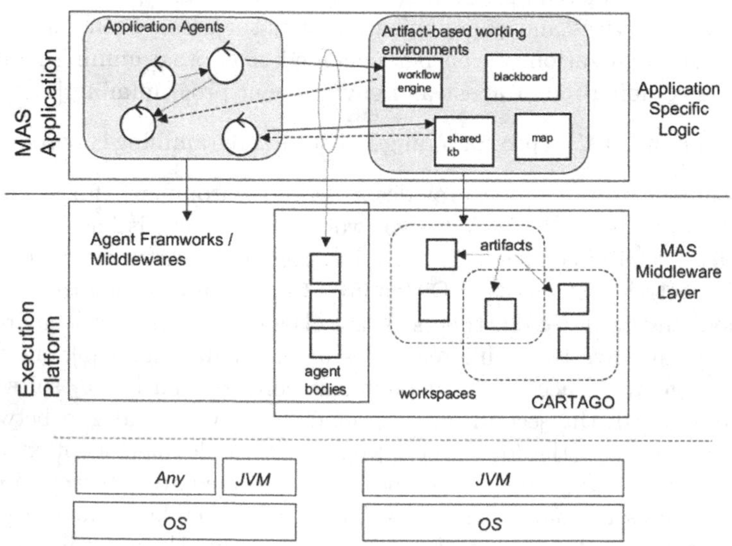

Fig. 1. The abstract architecture of the CArtAgO framework given in [12].

CArtAgO framework supports heterogeneous agent societies by making it possible for agents programmed with different agent development frameworks share the same work environment and interact with each other through artifacts. Agents are located in their own MAS systems while working on different work environments through the agent bodies. Agent bridge softwares are needed in order to provide a communication between agents and the agent bodies [12]. Jason [16] and Jadex [17] are BDI based agent development frameworks with bridge softwares ready to use. JADE [18] is a middleware-oriented agent platform that can be used to provide an infrastructure in CArtAgO, Jason and Jadex applications.

Agents and Artifacts Metamodel and CArtAgO framework are already applied in MAS contexts and real-world problems such as coordination, organisation oriented programming (OOP), system and technology integration and goal oriented use of artifacts [12]. Even though Agents and Artifacts Metamodel have already been used for system and technology integration, its usage in the e-learning domain for integration of different e-learning systems and technologies is relatively new [6].

3 The Proposed E-Learning Model Based on Agents and Artifacts Metamodel

Using agents in educational systems have advantages such as; the division of system's functionalities to various autonomous agents, the increase in system maintainability, the flexibility in different situations and the compatibility of different systems [1,2]. Hence, numerous MAS based e-learning systems have been proposed [1–4]. The architecture of the MAS based education system proposed by [2] is given in Fig. 2 as an example to illustrate the access mechanism to learning resources in typical MAS based e-learning systems. The system has five agent types; student agent, evaluation agent, record agent, modeling agent and learning object agent. The learning object agent manages the search and retrieval of learning objects to other agents and learners. It is the only agent responsible for this operation. This structure is similar in most of the MAS based e-learning systems that have been developed as standalone applications. This situation is one of the driving forces behind our environment based e-learning model proposal presented in [6]. By providing a common environment that combines various learning object resources and numerous agents from different MAS based e-learning systems, we can achieve a common infrastructure. This is important to increase interoperability of different e-learning systems and to have agents with the ability of adapting to dynamically changing learning resources.

The A&A Metamodel based e-learning model proposed in this paper has two main components; the A&A Metamodel based e-learning environment model [6] and the MAS based e-learning system exploiting it to search and retrieve learning resources. Each component is examined in a subsection, respectively.

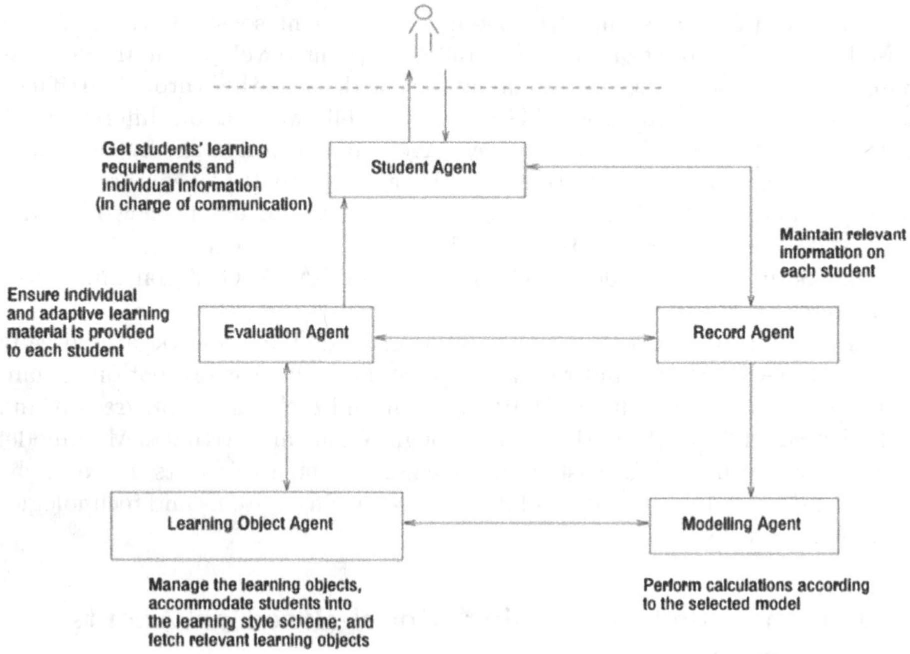

Fig. 2. The architecture of the MAS based education system proposed by [2].

3.1 The Agents and Artifacts Metamodel Based E-Learning Environment Model

The A&A Metamodel based e-learning environment model has been implemented with CArtAgO framework by using the SCORM and IEEE LOM e-learning standards [6]. The e-learning environment model supports searching learning resources stored as learning objects with three inputs. These are title, description and keywords, which correspond to the "1.2 Title", "1.4 Description" and "1.5 Keywords" metadata elements of the IEEE LOM standard, respectively.

The architecture of the Agents and Artifacts Metamodel based e-learning environment model presented in [6] is given in Fig. 3. Agent development frameworks and applications developed with these frameworks are at the left hand side of Fig. 3. The e-learning environment model is at the center containing a learning object repository management artifact and several artifacts of the two artifact types supported; local learning object repository and SCORM Cloud. This e-learning environment model runs at the right hand side of the third layer of the CArtAgO framework given in Fig. 1 as an artifact based working environment. The learning resources layer that contains two different types of repositories is shown at the right hand side of Fig. 3.

Learning resources can be stored in local drives or learning object repositories as learning objects. Agents and Artifacts Metamodel provides abstraction of various resources from different MAS applications by modeling them as artifacts.

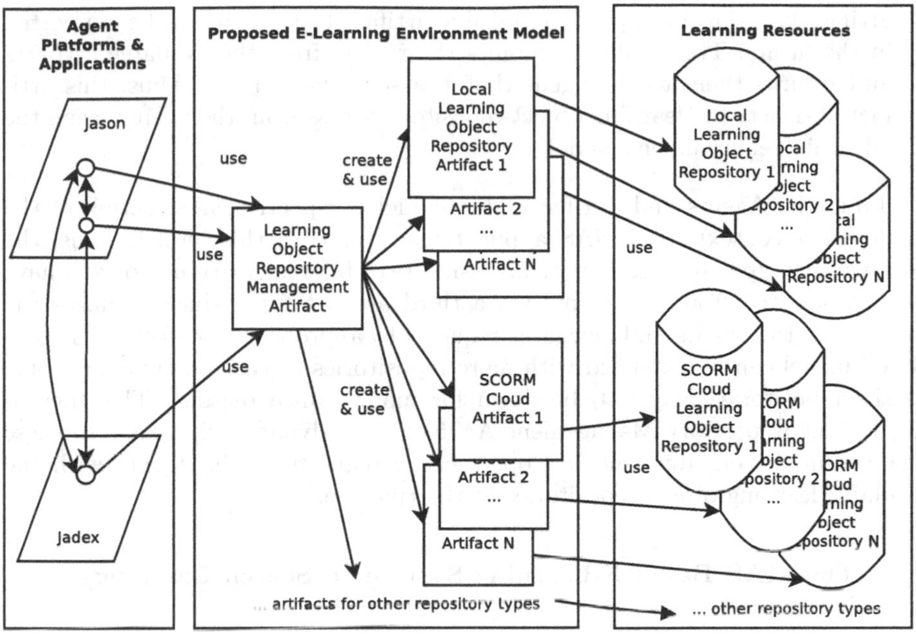

Fig. 3. The architecture of the Agents and Artifacts Metamodel based e-learning environment model presented in [6].

Retrieving learning resources from two different type of sources is sufficient to demonstrate the proposed model's support for multiple sources. Thus, two artifact types are illustrated in Fig. 3; one for accessing local files and one for accessing learning object repositories. SCORM Cloud API [19] is used as a learning object repository, because it provides a simple interface to store learning objects and a Web service based API to search and retrieve learning objects.

The e-learning environment model presented in [6] contains the following artifact types:

- "Local Learning Object Repository Artifact" is designed as a linkable artifact to provide search and retrieval of learning objects stored in local drives. This artifact parses the "imsmanifest.xml" file of each learning object to compare the supported metadata elements of each learning object with the search inputs and returns a list of the matching learning objects to "Learning Object Repository Management Artifact".
- "SCORM Cloud Artifact" is designed as a linkable artifact which is responsible for the search and retrieval of learning objects stored in the SCORM Cloud. "SCORM Cloud Artifact" uses the Web service API of the SCORM Cloud and returns the results it gets to "Learning Object Repository Management Artifact".
- "Learning Object Repository Management Artifact" is the linking artifact that links the linkable artifacts with the two previously defined linkable

artifact types or other possible linkable artifact types that can be supported in the future. This artifact combines the results from the available artifacts and returns them to the agent that has sent the request. Thus, this artifact abstracts the learning object repository access from the agents using the e-learning environment model.

From the Agents and Artifacts Metamodel perspective, abstraction of the deployment context from MAS applications is used in this model. Thus, the two artifact types responsible for accessing two different learning sources have been abstracted from the agents via a third artifact type which manages the requests of the agents and delegates requests to appropriate artifacts. The proposed model can be extended with more repositories at the run time and some of the repositories may also be down for maintenance reasons. The "Learning Object Repository Management Artifact" can dynamically adopt to these changes in the environment and operate the requests of the agents with the available learning object repositories at the runtime.

3.2 The MAS Based E-Learning System to Search Learning Resources

We proposed a MAS based e-learning system to search learning resources by using the A&A Metamodel based e-learning environment model introduced in Sect. 3.1. The proposed e-learning system has been developed with Jason. However, Jadex and Jason have bridge softwares available for CArtAgO, so other MAS based e-learning systems exploiting the e-learning environment model can be developed with both frameworks. In order to test the search functionality of the proposed e-learning system, 120 learning objects have been designed for the Logic Design course and stored in two repositories (one local and one on SCORM Cloud).

Figure 4 depicts the architecture of the proposed A&A Metamodel based e-learning model to search learning resources. Jason agent development framework; the LO search application developed with Jason as a MAS based e-learning system; and two example agents of the LO search application are shown at the left hand side of Fig. 4. The middle layer contains two environments. The A&A Metamodel based e-learning environment model containing three artifacts in total is at the upper part of the middle layer. Here, one artifact from each artifact type is shown for simplicity. LO search application's own environment reside at the bottom of the middle layer. The learning resources located in two different type of repositories are at the right hand side of Fig. 4.

In order to provide a GUI to learners to search learning resources and to represent the results, the proposed system has been developed to have its own CArtAgO environment with the following two types of artifacts:

- "LO Search Artifact" extends the "GUIArtifact" class of CArtAgO to provide a GUI for users to search learning objects with title, description and keywords. A screenshot of the search GUI of "LO Search Artifact" listing the results

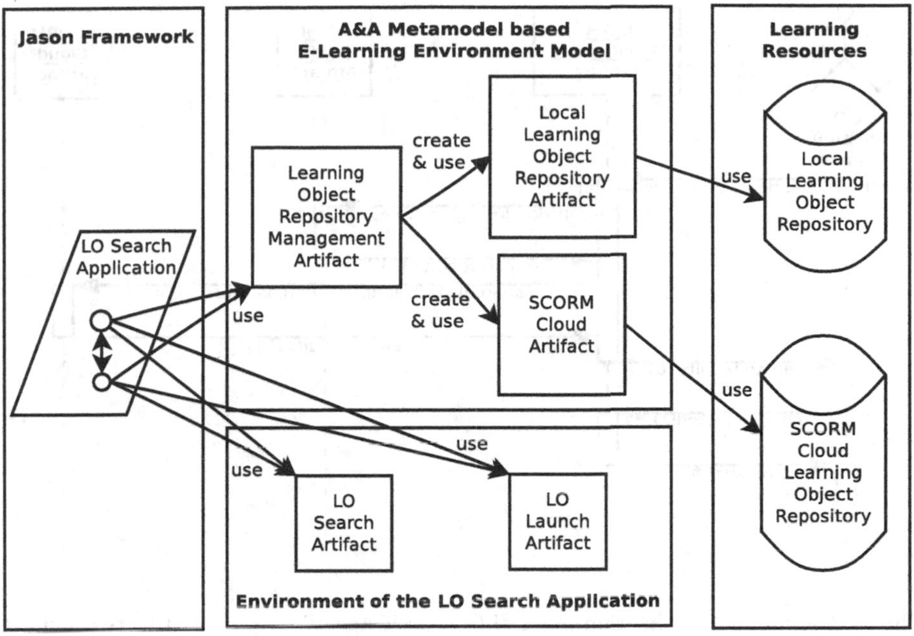

Fig. 4. The architecture of the proposed Agents and Artifacts Metamodel based e-learning model to search learning resources.

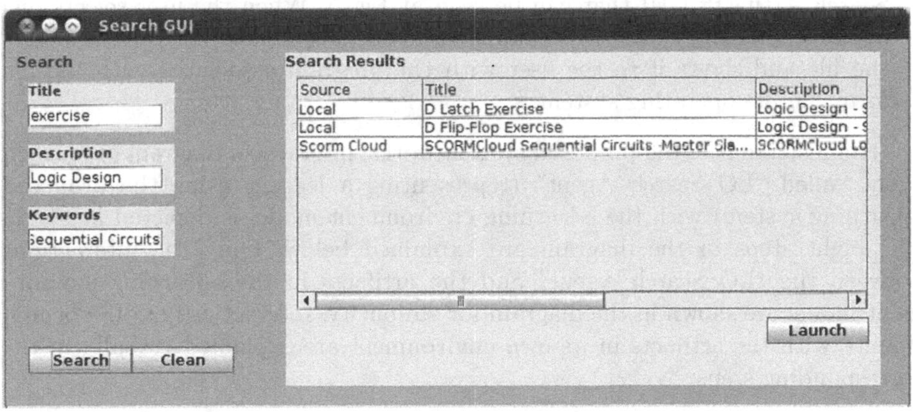

Fig. 5. A sample screenshot of the search GUI of "LO Search Artifact" listing the results for an example search scenario.

of an example search scenario is given in Fig. 5. In this scenario, learning objects with "exercise" as title, "Logic Design" as description and "Sequential Circuits" as keyword have been searched by a user. Three matching results, two from the local drive and one from the SCORM Cloud are also shown at Fig. 5.

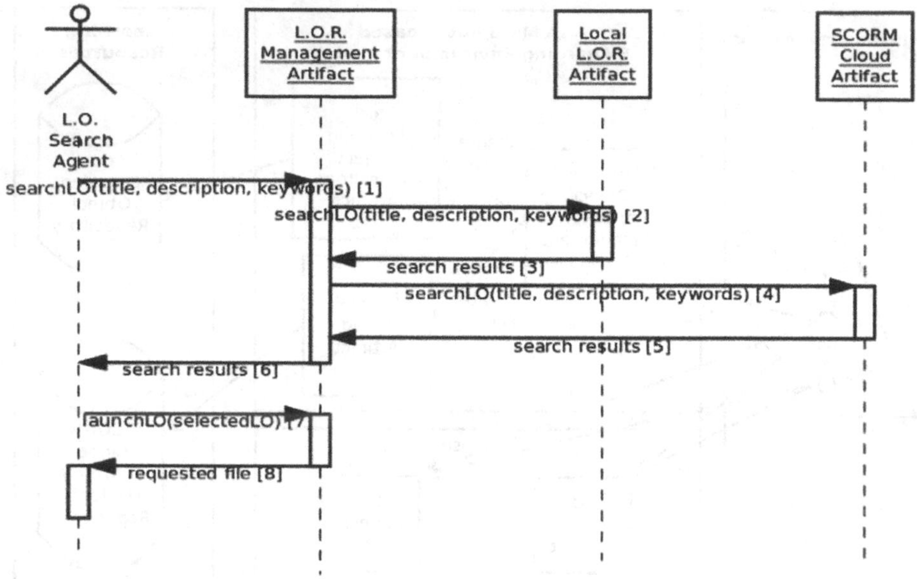

Fig. 6. The interaction diagram of an "LO Search Agent" with the artifacts provided by the e-learning environment model.

– "LO Launch Artifact" is associated with the "Launch" button of the "LO Search Artifact" GUI that can be seen at Fig. 5. When the user selects one of the results and clicks the "Launch" button, this artifact activates and gets the file and shows it to the user with the program associated with the file format in its operating system.

To illustrate how the proposed system works, interaction diagram of a Jason agent called "LO Search Agent" (representing a learner using the proposed e-learning system) with the e-learning environment model is depicted in Fig. 6. The eight steps in the diagram are explained below. Only the interactions between the "LO Search Agent" and the artifacts in the e-learning environment model are shown in the diagram for simplicity. Interactions of "LO Search Agent" with the artifacts in its own environment are explained verbally in the corresponding steps.

1. When "LO Search Agent" gets a search signal from the "LO Search Artifact" in its own environment, it runs a "joinWorkspace" command to join the workspace provided by the e-learning environment model to be able to use the available artifacts. Then, it calls the "searchLO" operation of the "LOR Management Artifact" with the requested parameters to search for learning objects.
2. "LOR Management Artifact" receives the search request and forwards it to all of the available artifacts connected to it with the link interface. In this scenario, "LOR Management Artifact" first sends the request to the "Local LOR Artifact" by calling its "searchLO" operation with the same parameters it has received.

3. "Local LOR Artifact" operates the request in the local repository and sends the matching results to "LOR Management Artifact".
4. "LOR Management Artifact" forwards the request to the second artifact connected to it with the link interface by calling the "searchLO" operation of the "SCORM Cloud Artifact" with the same parameters.
5. "SCORM Cloud Artifact" operates the request by using the Web service API of the SCORM Cloud and returns the matching results to the "LOR Management Artifact".
6. When "LOR Management Artifact" receives results from all of the artifacts it is connected with the link interface, it merges the results and sends them to the "LO Search Agent".
7. When "LO Search Agent" gets the search results from "LOR Management Artifact", it quits the e-learning environment by running the "quitWorkspace" command and gets back to its own environment. Then, "LO Search Agent" sends the search results to the "LO Search Artifact" to be presented to the user in its GUI. In this scenario, it is assumed that the user examines the results and decides to launch a learning object she wants to work with. Thus, she selects a learning object and presses the "Launch" button. In order to operate this request, "LO Search Agent" joins the e-learning environment workspace again and requests the selected file from "LOR Management Artifact".
8. "LOR Management Artifact" gets the file via the appropriate linking artifact and forwards it to "LO Search Agent". When "LO Search Agent" gets the file, it quits the e-learning environment workspace and gets back to its own environment to use the "LO Launch Artifact" for showing the requested file to the user.

4 Conclusion

In this paper, a multi agent e-learning system that searches learning objects by using an e-learning environment model based on Agents and Artifacts Metamodel has been presented. The e-learning environment model exploits Agents and Artifacts Metamodel to provide abstraction of learning resources. Thus, a new abstraction level between the agents and the e-learning resources in the proposed environment has been established with this model [6].

The proposed model provides a huge potential for extension from many aspects. Future studies for extension can be grouped in two categories. On one hand, the e-learning environment model can be expanded to provide more functionality to MAS based e-learning systems using it. On the other hand, new MAS based e-learning applications can be developed with different agent development frameworks.

The proposed multi agent e-learning system currently supports Jason agents capable of searching learning objects with title, description and keywords, because these are the only IEEE LOM metadata elements supported by the exploited e-learning environment model. In order to provide more search options in the proposed multi agent e-learning system, learning object search support of the e-learning environment model can be extended in the future with more IEEE LOM metadata elements. The proposed e-learning environment model can also be extended to support other metadata modeling standards and more learning object repositories with new learning object repository types. Furthermore, more artifacts with various functionalities can be added depending on the requirements of the MAS based e-learning applications using the e-learning environment model.

To spread the use of the e-learning environment model, new MAS based e-learning systems can be developed with different agent platforms. In this regard, a Jadex based MAS application can be developed and its agents can run in the same environment with the Jason agents already developed. Since the proposed e-learning environment model is based on Agents and Artifacts Metamodel and implemented with CArtAgO framework, integration of different MAS based e-learning systems developed with different agent development frameworks is supported.

References

1. Lin, F.O.: Designing Distributed Learning Environments with Intelligent Software Agents. Information Science Publishing, Hershey (2004)
2. Sun, S., Joy, M., Griffiths, N.: The use of learning objects and learning styles in a multi-agent education system. J. Interact. Learn. Res. **18**(3), 381–398 (2007)
3. Cruz, A.C., Valderrama, R.P.: Adaptive and intelligent agents applied in the taking of decisions inside of a web-based education system. In: Nguyen, N.T., Jain, L.C. (eds.) Intelligent Agents in the Evolution of Web and Applications. Studies in Computational Intelligence, vol. 167, pp. 87–112. Springer, Heidelberg (2009). doi:10.1007/978-3-540-88071-4_5
4. Ciloglugil, B., Inceoglu, M.M.: Developing adaptive and personalized distributed learning systems with semantic web supported multi agent technology. In: 10th IEEE International Conference on Advanced Learning Technologies, ICALT 2010, Sousse, Tunesia, 5–7 July 2010, pp. 699–700. IEEE Computer Society (2010)
5. Weyns, D., Omicini, A., Odell, J.: Environment as a first class abstraction in multiagent systems. Auton. Agents Multi-Agent Syst. **14**(1), 5–30 (2007)
6. Ciloglugil, B., Inceoglu, M.M.: Exploiting agents and artifacts metamodel to provide abstraction of e-learning resources. In: 17th IEEE International Conference on Advanced Learning Technologies, ICALT 2017, Timisoara, Romania, 3–7 July (2017, accepted)
7. Ricci, A., Viroli, M., Omicini, A.: CArtAgO: a framework for prototyping artifact-based environments in MAS. In: Weyns, D., Parunak, H.V.D., Michel, F. (eds.) E4MAS 2006. LNCS, vol. 4389, pp. 67–86. Springer, Heidelberg (2007). doi:10.1007/978-3-540-71103-2_4
8. IEEE-LOM, IEEE LOM 1484.12.1 v1 Standard for Learning Object Metadata - 2002 (2002). http://grouper.ieee.org/groups/ltsc/wg12/20020612-Final-LOM-Draft.html. Accessed 30 Mar 2017
9. DCMI: Dublin Core Metadata Specifications (1995). http://dublincore.org/specifications/. Accessed 30 Mar 2017

10. SCORM 2004, 4th edn. (2009). http://scorm.com/scorm-explained/technical-scorm/content-packaging/metadata-structure/. Accessed 30 Mar 2017
11. Omicini, A., Ricci, A., Viroli, M.: Artifacts in the A&A meta-model for multi-agent systems. Auton. Agents Multi-Agent Syst. **17**(3), 432–456 (2008)
12. Ricci, A., Piunti, M., Viroli, M.: Environment programming in multi-agent systems: an artifact-based perspective. Auton. Agents Multi-Agent Syst. **23**(2), 158–192 (2011)
13. Ricci, A., Omicini, A., Denti, E.: Activity theory as a framework for MAS coordination. In: Petta, P., Tolksdorf, R., Zambonelli, F. (eds.) ESAW 2002. LNCS, vol. 2577, pp. 96–110. Springer, Heidelberg (2003). doi:10.1007/3-540-39173-8_8
14. Kaptelinin, V.: Acting with Technology: Activity Theory and Interaction Design. MIT Press, Cambridge (2006)
15. Nardi, B.A.: Context and Consciousness: Activity Theory and Human-computer Interaction. MIT Press, Cambridge (1996)
16. Bordini, R.H., Hbner, J.F., Wooldridge, M.: Programming multi-agent systems in AgentSpeak using Jason, vol. 8. Wiley, Hoboken (2007)
17. Pokahr, A., Braubach, L., Lamersdorf, W.: Jadex: a BDI reasoning engine. In: Bordini, R.H., Dastani, M., Dix, J., Fallah-Seghrouchni, A.E. (eds.) Multi-Agent Programming. Multiagent Systems, Artificial Societies, and Simulated Organizations, vol. 15, pp. 149–174. Springer, New York (2005). doi:10.1007/0-387-26350-0_6
18. Bellifemine, F.L., Caire, G., Greenwood, D.: Developing multi-agent systems with JADE, vol. 7. Wiley, Hoboken (2007)
19. SCORM Cloud API (2002). http://cloud.scorm.com/doc/web-services/api.html. Accessed 30 Mar 2017

Workshop on Virtual Reality and Applications (VRA 2017)

Distributed, Immersive and Multi-platform Molecular Visualization for Chemistry Learning

Luiz Soares dos Santos Baglie[1(✉)], Mário Popolin Neto[2], Marcelo de Paiva Guimarães[3], and José Remo Ferreira Brega[1]

[1] Computing Department, School of Sciences, São Paulo State University (UNESP), Bauru, SP, Brazil
luizssb.biz@gmail.com, remo@fc.unesp.br

[2] Federal Institute of São Paulo (IFSP), Araraquara, SP, Brazil
mariopopolin@ifsp.edu.br

[3] Open University of Brazil-UNIFESP/Faccamp's Master Program, São Paulo, SP, Brazil
marcelodepaiva@gmail.com

Abstract. This paper presents Dimmol (acronym for Distributed Immersive Multi-platform Molecular visualization), a scientific visualization application based on UnityMol, developed with the Unity game engine, and that uses the Unity Cluster Package to enable distributed and immersive visualization of molecular structures across multiple device of different types, with support to Google VR, molecular trajectory files, and master-host-slave rendering. Its goal is to improve and facilitate the way educators and researchers visualize molecular structures with students and partners. In order to demonstrate a possible use scenario for Dimmol, better understand the contributions of each platform it can be executed in, and gather performance data, three molecular visualizations are loaded on it and distributed to a graphic cluster, a laptop, a tablet, and a smartphone. Other possible uses are also discussed.

Keywords: Virtual reality · Molecular visualization · Unity game engine · Distributed visualization

1 Introduction

The learning process is a considerably complex matter, requiring effort and dedication by the student. This ends up creating a need for educational resources and methods that, somehow, keep the student motivated along the process [29]. One particular factor that increases motivation (and thus the chance of assimilating the content studied) is the parity between the student's learning style and the way in which the information is received [23].

Among natural learning styles, the visual-spatial stands out the most [16]; in addition, understanding spatial relationships is a requirement in many areas of

O. Gervasi et al. (Eds.): ICCSA 2017, Part I, LNCS 10404, pp. 569–584, 2017.
DOI: 10.1007/978-3-319-62392-4_41

science, such as chemistry, where visualization is important to understand chemical processes [10] and the three-dimensional shape of molecules [22]. The lack of spatial instruction may make learning some concepts highly challenging [22].

It is expected, then, that educators look for new educational resources that satisfy these needs. One alternative can be found in educational software that use Virtual Environments (VEs) developed with Virtual Reality (VR) technology, which is highly motivating to students, grabbing and holding their attention, encouraging active participation instead of passivity, and is considered a natural evolution of computer-assisted instruction [25].

In fact, since the introduction of the capacity to create VEs using VR, some of their most important applications are focused on learning and training [6]; one reason for its use is the representation of impossible environments/situations [21]. An example is chemical modeling, where acquisition of knowledge about the models is facilitated by using 3D models and VR techniques [13].

In [18], VR is described in terms of software and hardware. As software, it can be described as a virtual environment that gives the user a sense of "being there". As hardware, it can be considered a natural evolution of 3D computer graphics, with advanced means and devices for input and output. Two advanced visual output media are [14]:

- Head Mounted Display (HMD), in which the user wears a special kind of spectacles with separate displays for each eye, in order to create the effect of stereoscopy. Current examples include Oculus Rift[1] and Google Cardboard[2]; and
- Projection technology systems, in which the images are produced by projectors on surfaces around the user(s). They allow multiple users within the running system. The main example is the CAVE Automatic Virtual Environment (CAVE) [9].

However, the use of VR in education is not trivial. On the financial side, there is price: even simple CAVE-like systems (e.g. [12,26]) may cost tens of thousands of dollars and powerful HMDs, although recently made easily available to the consumer market, still need reasonably capable PCs to be used; in the latter case, fortunately, there is Google Cardboard, a simple and affordable alternative that can even be assembled by the user [1]. On the practical side, it is still challenging to make VR technology easy to set up and intuitive to use [28].

Considering the use of VR in the scientific visualization of molecules and in education, this research project aims to improve the ways in which educators (and also possibly researchers) in chemistry-related areas visualize molecular structures and chemical reactions with their students and partners, and also improve the ways in which they coordinate these visualizations and collaborate in order to reduce or even eliminate barriers to understanding.

---

[1] Oculus Rift: https://www.oculus.com/rift/.

[2] Google Cardboard: https://vr.google.com/cardboard/.

A secondary objective is to provide a simpler and easier experience in configuring and preparing a scientific visualization distributed across multiple devices, including graphic clusters, which are the basis for CAVE-like systems [9,12,26].

More practically, we develop an application, Dimmol (based on the open-source software UnityMol [20] and using the Unity Cluster Package [24] components) which allows distributed visualization, coordination and synchronization of molecular structures and trajectories among a variety of devices, with support to VR and stereoscopic view (using the Google Cardboard HMD and CAVE-like systems).

The remainder of the paper is organized as follows. Section 2 discusses other studies related to the current research project. Section 3 presents the UnityMol application, which was used as the basis for Dimmol. Section 4 describes Dimmol, an application to facilitate chemistry teaching and discussion about molecular structures. Section 5 discusses how Dimmol can be used and the results of one particular configuration. In the end, Sect. 6 contains the conclusion of this study and future goals for Dimmol.

2 Related Work

Chastine et al. [7] developed AMMP-Vis, a system for immersive molecular modeling, supporting connectivity between local and remote devices and interactivity using HMDs and *data-gloves*. Collaboration was allowed synchronously or asynchronously. Although relevant to the current research project due to the distribution of molecular visualizations and use of HMDs, AMMP-Vis needed to connect to an AMMP server to run (not being able to function as a stand-alone application) and was not available for mobile devices.

Stone et al. [28] discuss the challenges of immersive visualization of molecules and the development of VMD (acronym for Visual Molecular Dynamics), a suitable application for this task. In considering future directions, it is acknowledged that a reduction in immersiveness and perceived complexity of VR components, coupled with the use of affordable equipment, would lead to accessible interactive molecular visualization simulations. The possibility of using smartphones as remote controllers for visualizations is brought to light because of their input capabilities (e.g. multi-touch screen, gyroscope and accelerometer), although no mention is made of using them for visualization like it is intended in Dimmol; there are no official builds of VMD available for mobile devices.

The studies of Liminiou, Roberts, and Papadopoulos [19] and Dias et al. [13] use a CAVE-like system to display molecules to students in a classroom, which is part of the motivation for the development of Dimmol. Both of them had positive results, reinforcing the viability of CAVE-like systems in molecular visualization, and, thus, showing a possible use scenario for Dimmol.

In [19], it is investigated how VR environments can increase interest and motivation in learning and promote chemistry learning by using chemical reactions animations and displaying them to a group of students using a 2D projector and then a CAVE. The results showed good acceptance of the CAVE by the students, as they felt they were inside the reactions, being able to observe them from different perspectives and points of view and better understand what was happening.

In [13], it was developed ChemCAVE3D, an application for molecular visualization in CAVE-like systems, in particular, the MiniCAVE [12], with interaction via mouse and Kinect[3]. In a pre-evaluation by educators, it was believed that it can be used as support for chemistry teaching. ChemCAVE3D does not seem to have been recently updated or made publicly available.

Similarly to these studies, but without the use of immersive visualization, Merchant et al. [22] explored how VEs built in Second Life[4] (SL) can teach chemistry and improve spatial skills and achievements in chemistry. The results showed higher effectiveness of the instruction passed in a 3D VE because students who had difficulties with 2D instruction became able to think in 3D and translate the information back to 2D; this highlights some of the benefits of offering ways to visualize chemistry-related content in three dimensions independently from the availability of stereoscopy or immersion, meaning that applications like Dimmol can still be beneficial even on common displays.

Anderson and Weng [4] developed VRDD (acronym for Applying Virtual Reality Visualization to Protein Docking and Design), a VR application for visualization of protein coupling and design, where the image is projected onto an ImmersaDesk, a semi-immersive display. The contribution was focused on the fact that, in regards to interaction, VR increases the engagement of the user's visual and motor senses; in addition, they also emphasized the importance of preventing the user from getting lost in the immersion provided by VR. These findings are important to the current study because its goals include this engagement increase and also because Dimmol's immersive visualization options may prove to be too much for some users.

Finally, Drouhard et al. [15] used an HMD for immersive and exploratory visualization of crystal structures and neutron scattering data, in order to decrease the entry barriers for exploratory analysis of complex scientific data. In the results, it is noted that collaborative visualizations with VR displays are particularly desired for one-of-a-kind installations and also that immersive VR holds promise for scientific visualization. This is important to the current study, as Dimmol was designed with the possibilities of collaborative and immersive visualizations in mind.

It is observed then that there are no means of molecular visualization that can be used in teaching and simultaneously attend to the factors of flexibility (allowing visualization in a single device or distributed among several devices of different types), multi-platform support, ease of use and maintenance, affordability and different modes of immersion. The current study addresses all these issues at once. In addition, it uses only software that are free of charge and (for most of them) open-source, the most important one being UnityMol.

[3] Kinect for Xbox 360: http://www.xbox.com/en-US/xbox-360/accessories/kinect.
[4] Second Life: http://secondlife.com.

3 UnityMol

UnityMol[5] is an open-source software[6] for visualizing molecular structures and biological networks, in a variety of representation styles (e.g. ball-and-stick, licorice, Van der Waals and HyperBall) along with many parameters [20,27]. It was developed using the popular Unity[7] game engine because of its multi-platform support and availability of easy-to-use interfaces for 3D application development [20]. Figure 1 shows four possible representations for a glycine molecule in UnityMol.

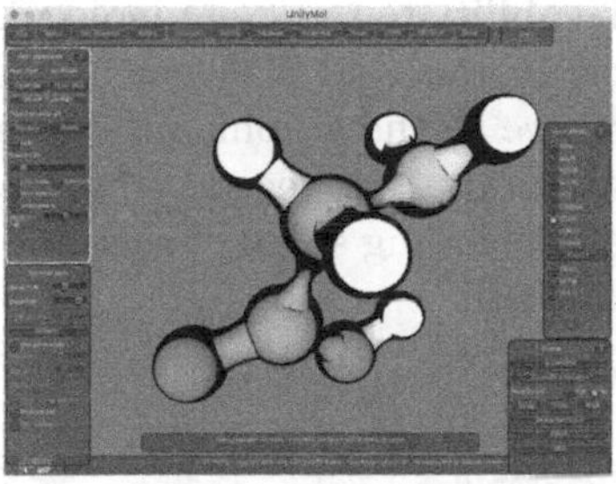

(a) HyperBalls

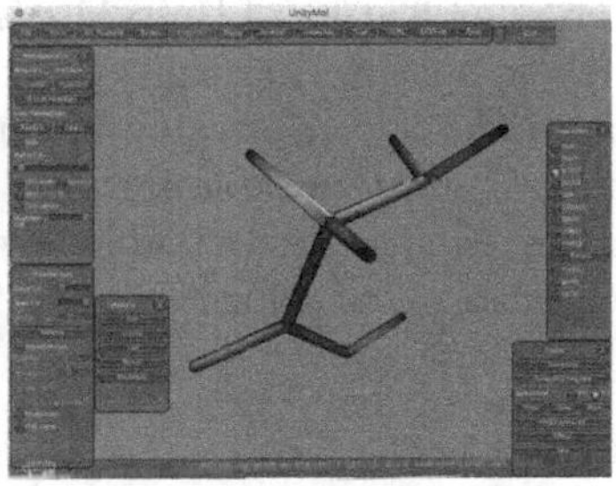

(b) Licorice

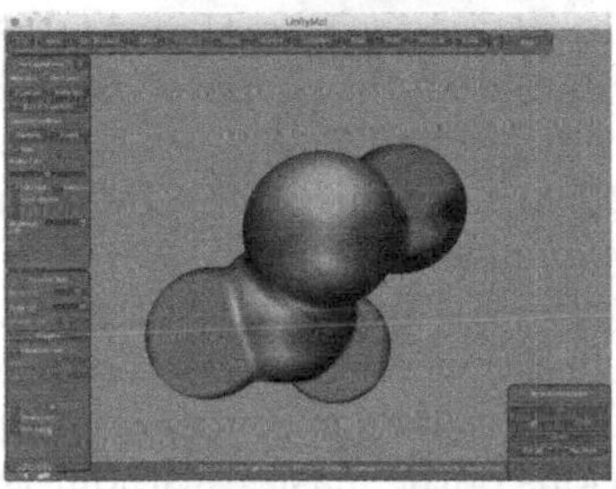

(c) Van der Waals hiding hydrogen atoms

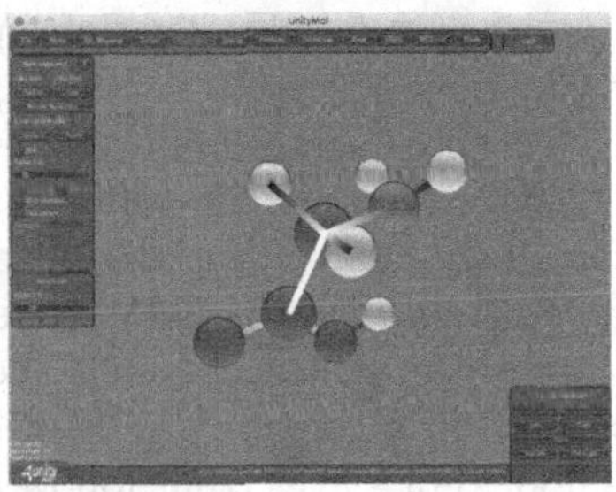

(d) Ball-and-stick using particles and lines

Fig. 1. Four possible visual configurations for a glycine molecule in UnityMol.

It is possible to load molecular structures by choosing a .PDB file or by searching the Worldwide Protein Data Bank[8] for a protein ID, which requires an internet connection. After the molecular structure is loaded, the user can manipulate the view (by spinning, moving and/or zooming it), change atoms' positions, and define aspects of their representation (metaphor, brightness, color, scale, texture, display/occlusion of hydrogen atoms, etc.). Other more advanced features such as surface processing and loading secondary structures are also offered.

[5] UnityMol: http://www.baaden.ibpc.fr/umol/.

[6] UnityMol official repository: https://github.com/bam93/UnityMol-Releases.

[7] Unity3D: https://unity3d.com.

[8] Worldwide Protein Data Bank: http://www.wwpdb.org.

By default, molecular structures are loaded using HyperBalls, a model proposed by Chavent et al. [8] where the representation of the structure is controlled by two parameters and allows the continuous change between several metaphors; this model also allows for a continuous aspect to the evolution of the bonds between atoms through the use of hyperboloids.

UnityMol was chosen as the basis for Dimmol because of two main reasons: it is a powerful visualization software for molecular structures and it was developed with the Unity game engine. The multi-platform support of this engine ensures that versions of the application for different platforms can share largely the same code base.

An interesting feature of UnityMol is the ability to represent molecular structures and their bonds using Unity's particle system and 2D elements. This representation model is considerably less intensive to the graphic processing of the application and most suitable for high frame rates per second (fps) in cases of large structures and/or less capable devices (including smartphones and tablets, two of the platforms at which Dimmol is aimed).

4 Dimmol

Dimmol[9] (acronym for Distributed Immersive Multi-platform Molecular visualization) is a scientific visualization application based on UnityMol that uses the Unity Cluster Package to enable distributed and immersive visualization of molecular structures across multiple device types.

Unity Cluster Package [24] (abbreviated in this paper as UCP) is a software package for Unity that allows the development of interactive VR applications for graphic clusters (such as CAVE-like systems). It uses the master-slave rendering architecture, which classifies the cluster's nodes as master (responsible for hosting and, in a general way, coordinating the execution of the application across the nodes) or slaves (that follow actions of the master, without triggering their own). With it, each node in the cluster executes at least one instance of the application and the UCP part of that instance takes care of connecting the slave nodes to the master node and synchronizing the view.

At first, UCP was integrated in UnityMol version 0.9.3 in order to allow the molecular structures to be displayed in CAVE-like systems. Several modifications have been made to UnityMol's source code so that it is possible to synchronize a molecular structure initially loaded on the master node, as well as changes applied to its view and representation, with the slave nodes. This synchronization was done using Unity's remote process call (RPC) capabilities in three main ways:

- Modifying key methods (static and in singletons) so that if they are invoked on the master node, they call themselves by RPC on the slave nodes;

[9] Dimmol repository: https://github.com/LuizSSB/dimmol.

- Changing key variables and properties (static and in singletons) so that in case their values are changed on the master node, special methods on the slave nodes are invoked by RPC to assign the new values to those variables/properties; and
- Creation and invocation by RPC of specific methods to synchronize certain changes, when the previous two cases are not applicable.

This approach involved identifying and changing methods, variables, properties, and key code fragments. However, it allowed the synchronization of data through nodes to be implemented with minimal interference to the actual logic of the application.

This feature was the basis upon which all other features were implemented.

4.1 Multi-platform Support

Using Unity's multi-platform support, versions of Dimmol for Windows, Mac OS X/macOS, Android 4.4+ and iOS 8.0+ have been created.

In general, all versions offer the same features and visual style, as they share the same code base. Visualizations can be distributed across devices of different platforms thanks to UCP, as it considers a graphic cluster any group of devices connected through the internet and executing the same application. In addition, all devices of all platforms may act as master or slave node.

Android and iOS versions, however, have one big difference: Google VR support, which allows for stereoscopic and immersive visualization using Google Cardboard [2] or another similar HMD. This feature is enabled/disabled individually on each device. When enabled, two virtual cameras are created side by side in the VE (their images displayed side by side in the device) what, together with the HMD, causes the stereoscopy effect.

Also, for the correct operation of GoogleVR, the user's head movement must be tracked. Because of this, integration with GoogleVR's software development kit (SDK) added to Dimmol the capability of manipulating the molecular structure view with 6 degrees-of-freedom (6DOF) using the gyroscope available in mobile devices, even when Google VR's stereoscopic visualization is not enabled.

4.2 UcpGui Scene

UCP makes use of a configuration file that describes how the node should behave (its type, view parameters, master node IP address, etc.) [24]. In order to make it easy to start an instance of Dimmol without having to manually edit this file, an initial scene (the equivalent to "screen", "page", or "level" in Unity) named UcpGui (acronym for Unity Cluster Package Graphical User Interface) was created.

It features a user interface mapping each configuration option of UCP to a visual control that can be edited by the device's user. It allows the user to properly start Dimmol using the values set in the controls and also to save them so that, later, changes made to the configuration are reloaded.

4.3 Master-Host-Slave Rendering Architecture

UCP has also been modified to support a rendering architecture called in this paper as master-host-slave.

In this architecture, the master node responsibilities are divided into two nodes, the host and the master. The host hosts the application and all other nodes connect to it, however, it has no other responsibility. The master node now connects to the host in the same way as a slave node, but it has the power to coordinate the application; all actions taken on the master are passed to the host, which then redistributes them to the other nodes. Lastly, the slave nodes work in the same way as before. Figure 2 shows a schematic of the architecture.

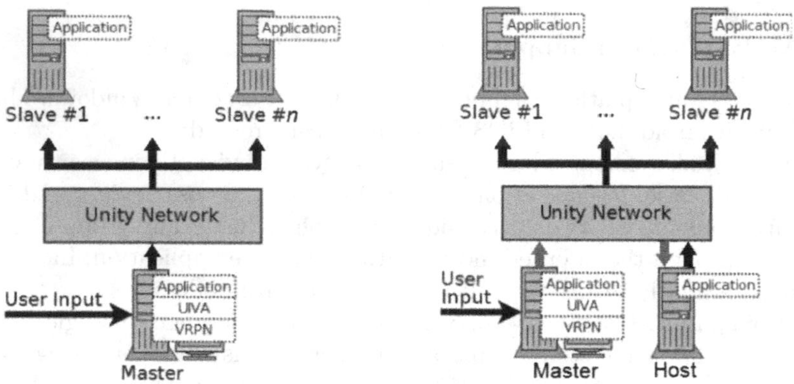

(a) Master-slave (b) Master-host-slave

Fig. 2. Comparison of rendering architectures. Adapted from [24].

With the master-host-slave architecture, application coordination can be unlinked from hosting, allowing less computationally powerful devices, such as smartphones and tablets, to coordinate the execution without the weight of serving it to all other nodes.

4.4 Free Visualization on Slave Devices

By default, when Dimmol is displaying a molecular structure, the master node forces its manipulation and representation of the view to the other nodes. In some cases, however, it is desired for slave nodes (or a subset of them) to have the power to manipulate their own view (bt spinning, moving and/or zooming it), without affecting the other nodes.

Therefore, this capability has been added to Dimmol. The master node is responsible for selecting the molecular structure and its representation, however, the slave nodes may have the power to manipulate their own view of the structure individually as its users desire.

While this feature may be problematic for visualization on CAVE-like systems, it can be used in cases where the nodes of the application are devices with independent displays (e.g. set of smartphones, each used by a different person). Because of this, the activation of this feature is done individually in each device.

As for the application controls, they are displayed and enabled only on the master node, in order to prevent other types of nodes from interfering with the execution of the application or leaving the correct synchronization state.

4.5 Trajectory Files

UnityMol version 0.9.3 supports loading molecular trajectories [20] (sequences of states of molecular systems) using the MDDriver [11] library, however, it is not possible to load trajectory files already on the disk. With this in mind, it was implemented in Dimmol the capability of loading trajectories files stored on disk or available at some web address.

Trajectories may be described using the XMOL format (which can be understood as a sequence of XYZ[10] descriptors) or contain the output of a geometry optimization operation calculated by GAMESS[11] [17].

Only one state of the trajectory is displayed at a time. It is possible to control which state is currently displayed or to animate the transition between states, so that it is possible to analyze the evolution of the system in a continuous way.

On the master node, after a trajectory file is selected or loaded from the web, Dimmol reads the entire file, identifying and extracting each state of the trajectory. This data is then synchronized with the other nodes. Depending on the number of atoms in the system, the number of states and the connection between the nodes, synchronization may take a while (in the users' perception); in spite of this negative, this approach later allows the transition between states to be more efficient, since, for that, only the index of the new state is synchronized, instead of all the information about the atoms' positions, leaving the raw work of loading and rendering the state for each node.

In any of the states of the trajectory, as well as during the evolution animation, total control is given to the user so that he/she can change the parameters of representation, point of view and other details, as he/she would with any other type of molecular structure loaded in UnityMol.

In the case of trajectories loaded from a GAMESS output, the file will also contain energy information for each state, so when loading these trajectories, Dimmol can display a vertical energy meter in which a pointer is placed in relation to the energy of the current state.

5 Results and Discussion

Because UCP considers a graphic cluster any set of devices connected to each other over the internet, there is a lot of flexibility in how Dimmol can be used to distribute a molecular visualization in a classroom or laboratory.

[10] XYZ (format): http://openbabel.org/wiki/XYZ.
[11] GAMESS: http://www.msg.chem.iastate.edu/index.html.

A visualization may be arranged taking into account the devices available to users, their processing power and the specific features on each of them. In this way, users can balance the distribution of the visualization between the processing power of PCs versus the stereoscopic visualization and 6DOF control via gyroscope of mobile devices.

The main usage proposal for Dimmol involves three elements: a control station, a CAVE-like system (or a single projector) and individual devices.

The control station may be a common PC controlled by the user in charge (e.g. teacher) and takes the role of the application's master node.

The CAVE-like system is located, preferably, near the control station and its nodes connect to the control station as slaves. The system's screens display the main continuous view of the molecular structure under study. In case there is no CAVE-like system available, a single projector may also be used; the projector may be connected to a separate PC or to the control station itself.

In front of the CAVE-like system/projector, there are the students or research partners. Each of them carries a device of their own (smartphone, tablet or laptop), which connects to the control station as a slave. These devices may have enabled free visualization on slave devices and/or immersive visualization through Google VR (if supported by the device).

While the purpose of the control station may be limited to control, the other elements have deeper roles. The CAVE-like system/projector provides a collective and generalized view of the molecular structure being studied, allowing users to view it from the same point of view and collaborate on what they see and what it means. In contrast, the use of individual devices allows users to explore the structure by following personal inquiries, without necessarily sharing it with colleagues if they do not think it is relevant yet. The possibility of stereoscopic visualization in both cases may also contribute to a better understanding of the structure as its three-dimensional properties are better exposed.

In order to validate this usage proposal, Dimmol was executed in the Laboratory of Interfaces and Visualization (LIV), part of the Computing Department (DCo) of the School of Sciences (FC) of São Paulo State University (UNESP), where the MiniCAVE is assembled [12]. In addition to the MiniCAVE cluster (six Windows 7 PCs, each with a Intel i7 processor and a NVIDIA FX 1800 758 mb video card [12]), the other devices involved were an Apple Macbook Pro (with Intel i7 2 GHz processor and Intel Iris Pro Graphics 1536 mb chip), an Apple iPad Mini 4 and an Asus Zenfone 2 Android phone. Figure 3 shows the MiniCAVE and how the devices are arranged.

Dimmol was installed and executed on all devices simultaneously, although, on the Zenfone 2 it was first configured for normal visualization (monoscopic), then later it was configured to use Google VR (stereoscopic). Three molecular structures were loaded during the application's execution:

- A .PDB file with the solution structure of Lactodifucotetraose (LDFT) beta anomer [3] (85 atoms and 86 bonds);
- A file with the output of a geometry optimization (calculated using GAMESS) of a glycine molecule (75 states with 10 atoms and 9 bonds per state); and

Fig. 3. Devices used with Dimmol and their arrangement. Each of MiniCAVE's three screens has area of $3.75\,m^2$ ($2.5\,m \times 1.5\,m$). Adapted from [24].

- A .XMOL file with states of molecular dynamics (calculated using Abalone[12]) between a chain of chitosan and a carbon nanotube (CNT) (66 states with 597 atoms and 789 bonds per state).

In the case of trajectories, the tests were performed while the transition between their states was animated. The last trajectory, in particular, was a simplified experiment inspired by the research by Aztatzi-Pluma et al. [5], in which they calculated the interactions between chitosan chains and CNTs, and which allows the user to see the chitosan chain wrapping the CNT.

Tables 1 and 2 present comparisons for fps rates of the devices in two possible representations of molecular structures. As can be observed, when visualizing structures with relatively few atoms and bonds rendered at the same time (most notably, the glycine optimization), there is no significant difference in fps rates between representations, however, as the number of atoms and bonds increases (most notably, on the trajectory with a chain of chitosan and a CNT), the difference in fps rates between the representations become noticeable, especially on mobile devices (iPad Mini 4 and Zenfone 2). This happens because the 3D objects of the HyperBall representation require more triangles to be drawn than the lines and particles, which can be drawn in squares with 2 triangles each, being more efficient for structures with large numbers of atoms and bonds [20]. For the Zenfone 2, it can also be observed that with a fairly low number of atoms and bonds there is not much difference in the fps rate in regards to the use of Google VR. With a considerably larger number of atoms and bonds, the difference is greater, and the monoscopic visualization performs better.

[12] Abalone: http://www.biomolecular-modeling.com/Abalone/index.html.

Table 1. Comparison of fps rates between devices executing Dimmol when using the HyperBall model/metaphor for atoms and bonds.

Device	LDFT beta anomer	Glycine optimization	Chitosan, CNT
MiniCAVE PC cluster	60 fps	60 fps	55 fps
Macbook Pro	60 fps	60 fps	53 fps
iPad Mini 4	59 fps	60 fps	18 fps
Zenfone 2	59 fps	60 fps	12 fps
Zenfone 2 (Google VR)	46 fps	56 fps	8 fps

Table 2. Comparison of fps rates between devices executing Dimmol when using the particle representation for atoms and line representation for bonds.

Device	LDFT beta anomer	Glycine optimization	Chitosan, CNT
MiniCAVE PC cluster	60 fps	60 fps	59 fps
Macbook Pro	60 fps	60 fps	59 fps
iPad Mini 4	60 fps	60 fps	31 fps
Zenfone 2	60 fps	60 fps	21 fps
Zenfone 2 (Google VR)	59 fps	60 fps	13 fps

It is important to notice, however, that there is no "right" way of using Dimmol. The usage proposal presented above is merely one way of executing Dimmol, using the resources available at the LIV.

Another possible configuration could be the use of only the control-station and CAVE-like system/projector, as seen in Fig. 4. This configuration could be useful to provide at least minimal immersion in schools where students don't have access to mobile devices powerful enough to execute Dimmol at satisfactory fps rates.

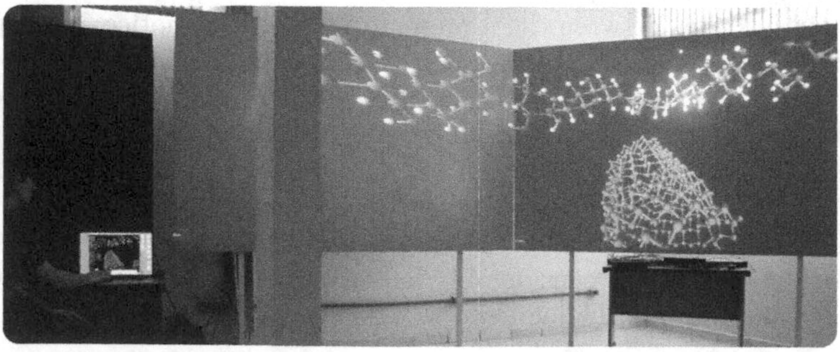

Fig. 4. Configuration using Macbook (master) and MiniCAVE cluster (slaves).

Yet another possibility could be the usage of a mobile device acting as master node with a CAVE-like system as slave, exemplified in Fig. 5. This allows the user to be inside in the visualization and "navigate" through it using 6DOF control offered by the mobile device's gyroscope.

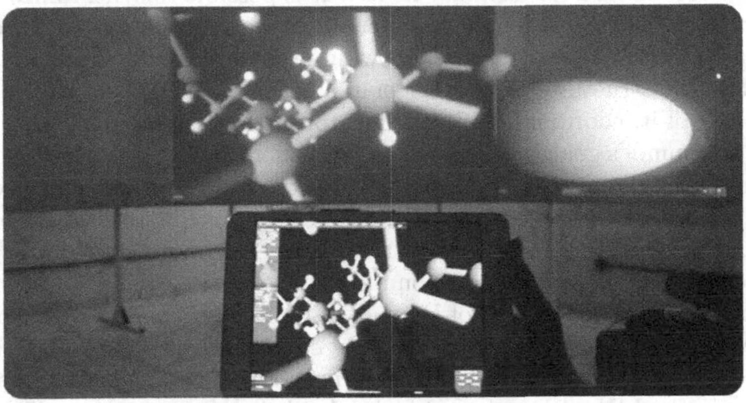

Fig. 5. Configuration using iPad (master) and MiniCAVE cluster (slaves).

A final possibility presented in this paper is to combine several different devices in order to obtain several different and simultaneous views of a molecular structure, as shown in Fig. 6. In it, the Macbook and the PC work together, one extending the screen of the other, while the iPad has its own view, allowing another point of view, and the Zenfone 2 provides stereoscopy, to better judge the three-dimensional details.

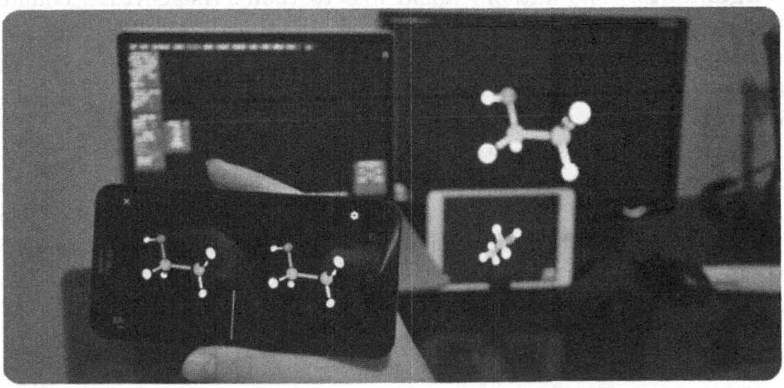

Fig. 6. Configuration using Macbook (master), Windows PC (slave), iPad (slave, free visualization), Android phone (slave, Google VR).

6 Conclusion and Future Work

The current study followed the development of Dimmol, an application for distributed and multi-platform visualization of molecular structures and their trajectories, with support to affordable and accessible immersion using Google VR and Google Cardboard (or a similar HMD). It is possible to synchronize a molecular structure between different devices and different platforms, with free view manipulation in some slave devices, allowing users of these devices to individually explore the structure using other points of view, while there is a collective visualization of it, coordinated by an educator or user in charge.

Two of Dimmol's goals were ease of use and affordability. A large part of them was achieved by developing Dimmol upon UnityMol, because it was developed using Unity, which allows compiling the application for different platforms. The creation of the UcpGui scene also contributes to ease of use, since users do not need to change the UCP configuration file manually.

The multi-platform support of Dimmol also allows users to choose how they visualize the molecular structure, leveraging each platform's particularities.

A pilot usability test for the main usage proposal of Dimmol is being planned to happen in the LIV with a class of high school chemistry students. This study will evaluate the perceived contribution of the application by students and teachers and provide feedback for future improvements or specifications.

Regarding new developments, the next step will be to update Dimmol to be based on version 0.9.6 of UnityMol, recently made available to the public in the official repository. It is also intended to make changes to the UcpGui scene in order to further improve its ease of use (some initial users reported difficulty understanding some of the technical terms used) and implementation of new controls for trajectory animation (e.g. speed, step size, direction), as well as other features from applications that display trajectories (e.g. VMD [28]).

Acknowledgements. The authors would like to thank the CAPES Foundation, a body of the Brazilian Ministry of Education, for financial support. Luiz Soares dos Santos Baglie was recipient of scholarship from CAPES.

The authors would also like to thank Paulo Noronha Lisboa Filho, Ph.D., and Francisco Carlos Lavarda, Ph.D., from the Physics Department of FC/UNESP, for their valuable feedback and suggestions during this study.

The copyright for UnityMol is held by the Centre National de la Recherche Scientifique (CNRS), France. UnityMol is developed by FvNano/LBT Team, and Marc Baaden, Ph.D. The source-code for version 0.9.3 of the software was downloaded from the official public repository.

References

1. Get Cardboard - Google VR. https://vr.google.com/cardboard/get-cardboard. Accessed Jan 2017
2. Google VR - Google Developers. https://developers.google.com/vr. Accessed Jan 2017

3. Solution structure of lactodifucotetraose (ldft) beta anomer (2014). http://www.rcsb.org/pdb/explore.do?structureId=2MK1
4. Anderson, A., Weng, Z.: VRDD: applying virtual reality visualization to protein docking and design. J. Mol. Graph. Model. **17**(3–4), 180–186 (1999)
5. Aztatzi-Pluma, D., Castrejón-González, E.O., Almendarez-Camarillo, A., Alvarado, J.F.J., Durán-Morales, Y.: Study of the molecular interactions between functionalized carbon nanotubes and chitosan. J. Phys. Chem. C **120**(4), 2371–2378 (2016)
6. Casu, A., Spano, L.D., Sorrentino, F., Scateni, R.: RiftArt: bringing masterpieces in the classroom through immersive virtual reality. In: Giachetti, A., Biasotti, S., Tarini, M. (eds.) Smart Tools and Apps for Graphics - Eurographics Italian Chapter Conference. The Eurographics Association (2015)
7. Chastine, J.W., Brooks, J.C., Zhu, Y., Owen, G.S., Harrison, R.W., Weber, I.T.: AMMP-Vis: a collaborative virtual environment for molecular modeling. In: Proceedings of the ACM Symposium on Virtual Reality Software and Technology, VRST 2005, pp. 8–15. ACM, New York (2005)
8. Chavent, M., Vanel, A., Tek, A., Levy, B., Robert, S., Raffin, B., Baaden, M.: GPU-accelerated atom and dynamic bond visualization using hyperballs: a unified algorithm for balls, sticks, and hyperboloids. J. Comput. Chem. **32**(13), 2924–2935 (2011)
9. Cruz-Neira, C., Sandin, D.J., DeFanti, T.A., Kenyon, R.V., Hart, J.C.: The CAVE: audio visual experience automatic virtual environment. Commun. ACM **35**(6), 64–72 (1992)
10. Davies, R.A., John, N.W., MacDonald, J.N., Hughes, K.H.: Visualization of molecular quantum dynamics: a molecular visualization tool with integrated Web3D and haptics. In: Proceedings of the Tenth International Conference on 3D Web Technology, Web3D 2005, pp. 143–150. ACM, New York (2005). http://doi.acm.org/10.1145/1050491.1050512
11. Delalande, O., Férey, N., Grasseau, G., Baaden, M.: Complex molecular assemblies at hand via interactive simulations. J. Comput. Chem. **30**(15), 2375–2387 (2009)
12. Dias, D.R.C., Neto, M.P., Brega, J.R.F., Gnecco, B.B., Trevelin, L.C., de Paiva Guimarães, M.: Design and evaluation of an advanced virtual reality system for visualization of dentistry structures. In: 2012 18th International Conference on Virtual Systems and Multimedia (VSMM), pp. 429–435, September 2012
13. Dias, D.R.C., Brega, J.R.F., Lamarca, A.F., Neto, M.P., Suguimoto, D.J., Agostinho, I., Gouveia, A.F.: Chemcave3d: sistema de visualizaçao imersivo e interativo de moléculas 3d. In: Workshop de Realidade Virtual e Aumentada, Uberaba-MG (2011)
14. Disz, T., Papka, M., Stevens, R., Pellegrino, M., Taylor, V.: Virtual reality visualization of parallel molecular dynamics simulation. In: Society for Computer Simulation, pp. 483–487 (1995)
15. Drouhard, M., Steed, C.A., Hahn, S., Proffen, T., Daniel, J., Matheson, M.: Immersive visualization for materials science data analysis using the oculus rift. In: 2015 IEEE International Conference on Big Data (Big Data), pp. 2453–2461, October 2015
16. Gardner, H.E.: Frames of Mind: The Theory of Multiple Intelligences. Basic Books, New York City (1983)
17. Gordon, M.S., Schmidt, M.W.: Advances in electronic structure theory: GAMESS a decade later, pp. 1167–1189. Elsevier, Amsterdam (2005)
18. Jayaram, S., Connacher, H.I., Lyons, K.W.: Virtual assembly using virtual reality techniques. Comput. Aided Des. **29**(8), 575–584 (1997)

19. Limniou, M., Roberts, D., Papadopoulos, N.: Full immersive virtual environment CAVETM in chemistry education. Comput. Educ. **51**(2), 584–593 (2008)
20. Lv, Z., Tek, A., Da Silva, F., Empereur-mot, C., Chavent, M., Baaden, M.: Game on, science - how video game technology may help biologists tackle visualization challenges. PLOS ONE **8**(3), 1–13 (2013)
21. Mantovani, F.: VR learning: potential and challenges for the use of 3D environments in education and training. In: Riva, G., Galimberti, C. (eds.) Towards CyberPsychology: Mind, Cognitions and Society in the Internet Age, pp. 207–225. IOS Press, Amsterdam (2003). Chap. 12
22. Merchant, Z., Goetz, E.T., Keeney-Kennicutt, W., Cifuentes, L., Kwok, O., Davis, T.J.: Exploring 3-D virtual reality technology for spatial ability and chemistry achievement. J. Comput. Assist. Learn. **29**(6), 579–590 (2013)
23. Mikropoulos, T.A., Natsis, A.: Educational virtual environments: a ten-year review of empirical research (1999–2009). Comput. Educ. **56**(3), 769–780 (2011)
24. Neto, M.P., Dias, D.R.C., Trevelin, L.C., Paiva Guimarães, M., Brega, J.R.F.: Unity cluster package – dragging and dropping components for multi-projection virtual reality applications based on PC clusters. In: Gervasi, O., Murgante, B., Misra, S., Gavrilova, M.L., Rocha, A.M.A.C., Torre, C., Taniar, D., Apduhan, B.O. (eds.) ICCSA 2015. LNCS, vol. 9159, pp. 261–272. Springer, Cham (2015). doi:10.1007/978-3-319-21413-9_19
25. Pantelidis, V.S.: Reasons to use virtual reality in education and training courses and a model to determine when to use virtual reality. Themes Sci. Technol. Educ. **2**(1–2), 59–70 (2009)
26. Pastorelli, E., Herrmann, H.: A small-scale, low-budget semi-immersive virtual environment for scientific visualization and research. Procedia Comput. Sci. **25**, 14–22 (2013)
27. Pérez, S., Tubiana, T., Imberty, A., Baaden, M.: Three-dimensional representations of complex carbohydrates and polysaccharides' sweetunitymol: A video game-based computer graphic software. Glycobiology **25**(5), 483–491 (2015)
28. Stone, J.E., Kohlmeyer, A., Vandivort, K.L., Schulten, K.: Immersive molecular visualization and interactive modeling with commodity hardware. In: Bebis, G., et al. (eds.) ISVC 2010. LNCS, vol. 6454, pp. 382–393. Springer, Heidelberg (2010). doi:10.1007/978-3-642-17274-8_38
29. Virvou, M., Katsionis, G., Manos, K.: Combining software games with education: evaluation of its educational effectiveness. Educ. Technol. Soc. **8**(2), 54–65 (2005)

Embedding Augmented Reality Applications into Learning Management Systems

Marcelo de Paiva Guimarães[1,2(✉)], Bruno Alves[2], Valéria Farinazzo Martins[3], Luiz Soares dos Santos Baglie[4], José Remo Brega[4], and Diego Colombo Dias[5]

[1] Open University of Brazil (UAB), UNIFESP, São Paulo, Brazil
marcelodepaiva@gmail.com
[2] Faccamp's Master Program, FACCAMP, Campo Limpo Paulista, Brazil
bruno_finus@hotmail.com
[3] Faculty of Computing and Informatics, Mackenzie Presbyterian University, São Paulo, Brazil
valfarinazzo@gmail.com
[4] Computer Science Department, São Paulo State University, UNESP, Bauru, SP, Brazil
luizssb.biz@gmail.com, remobrega@gmail.com
[5] Federal University of São João Del Rei, São João Del Rei, Brazil
diegocolombo.dias@gmail.com

Abstract. A tool is proposed to reduce the disparity between the state of the art of technologies and the time of maturity required for effective implementation, facilitating the insertion of augmented reality content into learning management systems. This tool uses didactic material based on augmented reality in the Sharable Content Object Reference Model (SCORM) that is a learning object standard. We tested this tool, generating a learning object based on augmented reality and sharing it to the Moodle platform. We also tested and shared this object to the repository SCORM Cloud.

Keywords: Augmented reality · Learning objects · SCORM · Moodle

1 Introduction

Recently, computational technologies have provided tools to modify traditional teaching-learning methods, resulting in effective, fun, and interactive learning experiences. In this new context, some technologies are not yet widely used, due to their peculiarities. In a general analysis, this lack of use is due to the disparity between the state of the art of these technologies and the time of maturity required for effective implementation. One possible technology that can be used as a learning tool technology is augmented reality (AR). AR technology works by superimposing virtual information on top of the real world, supplementing the user's reality instead of replacing it as other technologies (such as virtual reality) would [2]. AR applications can usually be visualized through a computer monitor, mobile device, or head-mounted display.

O. Gervasi et al. (Eds.): ICCSA 2017, Part I, LNCS 10404, pp. 585–594, 2017.
DOI: 10.1007/978-3-319-62392-4_42

Many studies have been made of AR as a learning tool (i.e., [5–7, 9, 15]), resulting in the conclusion that it has a positive impact on student motivation [7] and that it allows practical experimentation with a theoretical subject [5], possibly improving the application of learned knowledge [6].

Much of the investment made in the production of learning tools targets them to highly specific audiences [13]. However, it is desirable that learning materials (such as AR applications) be designed to be reusable and shareable by diverse educators and in different contexts. This approach is aligned with the aims of open educational resources [4], which can be understood as freely accessible and openly licensed documents, media, and applications that can be used and/or adapted by third parties for teaching, learning, and assessing. Open educational resource content can be distributed by packaging it as a learning object (LO), which is a set of didactic material with the following features: reusability (it can be adapted); interoperability (it can be supported by any hardware and software platform); accessibility (it can easily be stored and retrieved); and manageability (it can be updated over time) [13, 16]. LO repositories are databases that provide the means for educators to discover, exchange, and reuse the objects [13]. LOs are produced and distributed in standards defined by organizations [12], such as the Learning Technology Standards Committee (IEEE-LTSC)[1], the Alliance of Remote Instructional Authoring and Distribution Network for Europe (ARIADNE)[2], the IMS Global Learning Consortium[3], and the Advanced Distributed Learning[4] (ADL) initiative.

For educators to properly use LOs, they must use learning management systems (LMS) (e.g., Moodle, Blackboard, Edmodo, Skillsoft, Desire2Learn, and Schoology). According to Mahnegar [11], an LMS "is software used for delivering, tracking and managing training/education. LMSs range from systems for managing training/educational records to software for distributing courses over the Internet and offering features for online collaboration." With an LMS, the educator can track grades, attendance, and time spent by the students, and students can keep track of their grades and assignments and also submit homework and access course information [11].

This paper presents a tool that packages AR applications such as the Sharable Content Object Reference Model (SCORM), an LO standard, to be used within LMSs and/or be stored in repositories. We also demonstrate a case study that used our tool to embed an AR application to the SCORM standard and deploy it into Moodle. We also tested and shared this object to the repository SCORM Cloud. The AR application was developed using the authoring tool Flaras[5].

The major contributions presented in this paper are:

- a tool that packages AR applications into SCORM standard;
- the fact that a teacher can easily import LO based in AR to LMSs;
- the fact that AR content can easily be shared into LO repositories;

---

[1] IEEE-LTSC: http://www.ieeeltsc.org.

[2] Ariadne: http://www.ariadne-eu.org.

[3] IMS Global Learning Consortium: http://www.imsglobal.org.

[4] ADL: https://www.adlnet.gov.

[5] FLARAS: http://ckirner.com/faras2/.

- the fact that AR applications in LMSs provide a valuable didactic material;
- the fact that the packaging of LOs based on AR applications promotes the distribution and sharing of this content; and
- a case study illustrating an AR application developed with Flaras and running in Moodle;
- a case study illustrating an LO based on AR application stored into a repository.

This study is structured as follows. Section 2 discusses work related to this research. Section 3 presents the way to embed an LO based on AR into an LMS and also shows the developed packaging tool. Section 4 presents a case study of our packaging tool. Section 5 presents the conclusion and suggests future work.

2 Related Work

Baptista et al. [3] developed a tool that packages 3D content in the COLLADA[6] format to the SCORM package for use within Moodle. They used the Three.js[7] JavaScript library to embed the 3D content into an HTML document. Although their solution results in an LO based on 3D models, it is not AR content.

In [8], Gonen and Basaran dealt with the difficulty of teaching problem-solving skills in distance physics education. They created whiteboard math (WBM) movies that were packaged as SCORM packages and used within Moodle. A WBM movie is a screen recording of writing along with voice/text explaining a mathematical concept or solving a problem. In the end, they concluded, among other things, that the use of an LMS and SCORM improves content distribution capabilities and costs as well as the monitoring and evaluation of the students. However, it does not support AR content.

Liu et al. [10] developed the Handheld English Language Learning Organization (HELLO), a set of two systems designed to support English language learning. One of the systems is their own LMS that handles the learning content and offers a forum for discussion. The other is an AR application for PDAs that displays the content and allows students to perform the assignments; it also shows an AR virtual learning partner on top of the device's camera feed. The authors conducted a survey that indicated that AR is useful for providing context-aware experiences in learning activities. However, the insertion of the AR content they have created is compatible just with their own LMS.

Also related to the usage of AR with an LMS, de la Torre et al. [14] used the Easy Java Simulations[8] (EJS) tool to develop a virtual and/or remote laboratory (VRL) called Ball and Beam, which they later imported into Moodle using a packaging tool they also developed, called EJSApp. They also developed other tools to allow for more collaboration during experiments. The Ball and Beam VRL has an AR functionality that superimposes a graphical representation of the virtual system's behavior on top of a webcam feed of the real laboratory. They conducted an observational study that

[6] COLLADA: https://collada.org/.

[7] Three.js: http://threejs.org/.

[8] Easy Java Simulations: http://www.um.es/fem/EjsWiki/index.php/.

revealed a positive impact on students' learning when using the VRL and Moodle. However, although they inserted an AR application into Moodle, the solution does not promote content sharing.

3 Embedding LOs Based on RA into an LMS

The activity of developing teaching materials to support the teaching/learning process is a challenge for both educators and learners when it comes to novelties and techniques that result in a gain of learning. Included among the attempts to make this task easier is the LOs, which favor the development and reuse of didactic contents. However, some resources do not fully exercise the proposed role of teaching/learning support, due to difficulties such as the lack of a tool to help in the generation of LOs.

When the practice of e-learning on the web started to take over the use of educational compact discs, a set of standards (i.e., SCORM) was developed by the ADL initiative of the United States government with the purpose of addressing the challenges of the interoperability, reusability, and durability of educational content. At that time, each vendor/LMS used its own unique set of standards, which meant that organizations using these systems were forced to continue using them, or else they would have to produce all their content again from scratch.

SCORM allows the usage and distribution of educational content, according to the concept of "reusable, accessible, interoperable, durable" [3]:

- "reusable" implies the possibility of being used and modified by different tools;
- "accessible" implies the availability to meet the different needs of customers;
- "interoperable" implies support for different web browsers and operating systems; and
- "durable" implies not requiring changes in order to work with newer versions of software.

In order to insert didactic material based on AR into LMSs, a packaging tool was developed that packages an AR application into SCORM [1]. Figure 1 depicts the workflow proposed to create and deploy an LO based on AR into LMs or into a repository.

In the first phase, an AR authoring tool is used to create the AR application to be compatible with web technologies (i.e., HTML5, JavaScript, WebGL). This compatibility is necessary, because LMSs usually run on web browsers, avoiding portability problems. However, this is not a restriction, because advances in hardware and software technologies have enabled web browsers to run complex content, even 3D scenes in real time. Traditional web applications generally contain scripts, HTML, CSS files, and other media, such as video, images, and audio. However, if the application is also an AR application, it includes other files, such as 3D models and animations. Our solution also contains files referring to the configuration of the LO (i.e., XML, or eXtensible Markup Language, files). All files are packaged as an LO that is embedded into the LMS.

The second phase consists of packaging the AR application into an LO standard. As input, the application receives the application files and adds and updates files according to the LO standard adopted; it ends up generating an object that can be shared via

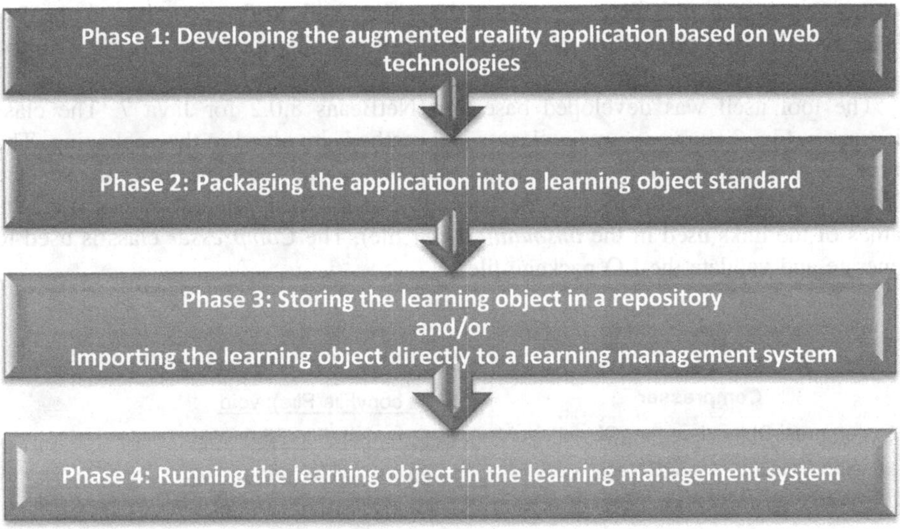

Fig. 1. Creation, packaging, and visualization of AR content.

repositories or be imported directly by LMSs. Our packaging tool was designed to accomplish this step. Figure 2 shows the user interface of our tool. The user specifies the path of the AR application (magnifying glass icon) and the destiny path to the LO (the text field below). The floppy disk icon creates the LO package in SCORM. The door icon closes the tool. This tool does not alter the application content; however, it adds some files, such the metadata file *imsmanifest.xml*, which describes the logical structure of the LO (i.e., it contains links to every piece of content inside the LO). This file also contains information about the cataloging and searching of the object, the tracking of ownership and attribution information, and the handling of rights management issues.

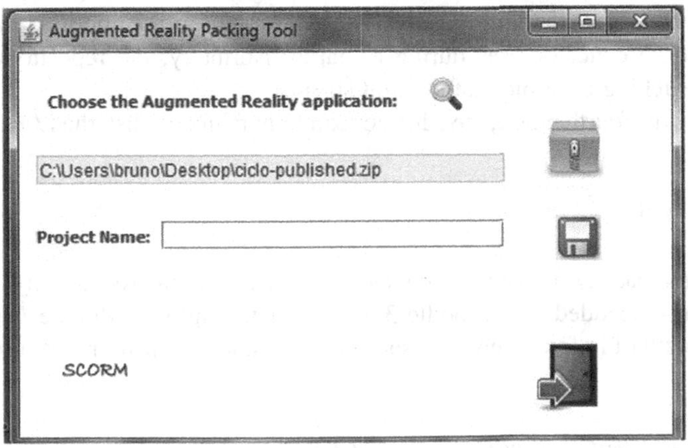

Fig. 2. User interface of the packaging tool.

So the *imsmanifest.xml* file ensures that the object will work according to the LO standards. The final product of the packaging tool is a single compacted file with the extensions "*.pkzip*" or "*.pif.*"

The tool itself was developed based on NetBeans 8.0.2 for Java 7. The class diagram in Fig. 3 shows the main classes and methods involved in the packaging. The *Main* class is initialized by the runtime and builds the user interface. It uses the *FileManager* class to add files to the package and set their internal values (i.e., the names of the links used in the *imsmanifest.xml* file). The *Compresser* class is used to generate and validate the LO package file.

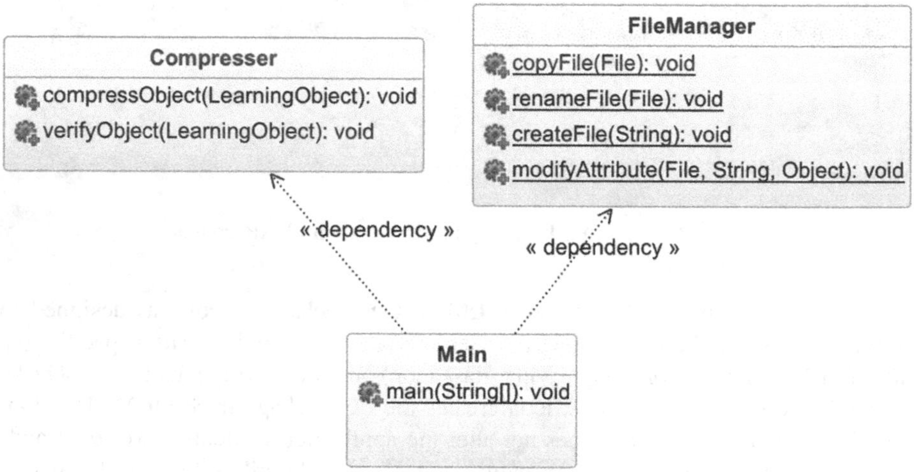

Fig. 3. Class diagram: packaging tool.

The third phase receives the LOs as input, and then the user can directly start step four or share the objects in repositories. In general, when the user is importing an LO file into a repository, it is necessary to describe their identifiers (i.e., audience, key words, hardware software requirements). These descriptions are important because they will be what the educator uses during a search. Normally, the repositories offer the means for searching by topic, author, and so on.

Finally, in the fourth phase, the educator can import and publish the LO into an LMS.

4 Case Study

To validate the packaging tool, we developed an educational AR activity, depicted in Fig. 4. It was embedded into Moodle 3.5.1. After the upload, Moodle was accessed, using the Mozilla Firefox[9] web browser and Chromium[10], in four operating systems:

[9] Mozilla Firefox: https://www.mozilla.org/en-US/_refox/new/.

[10] Chromium https://www.chromium.org/.

Windows 7, Windows 8, and Windows 10 (the 32-bit and 64-bit editions). The activity related to the LO was focused on aiding the process of teaching-learning with respect to the classification system used to identify animals for preschool education (mammals, birds, reptiles, and insects). The user's interaction with the LO is based on fiducial markers. The student points the camera at the underlying marker, and only after the marker is recognized (i.e., while acquiring camera data) is the three-dimensional bird shown by the application. Users can also use a mouse to interact with the RA content.

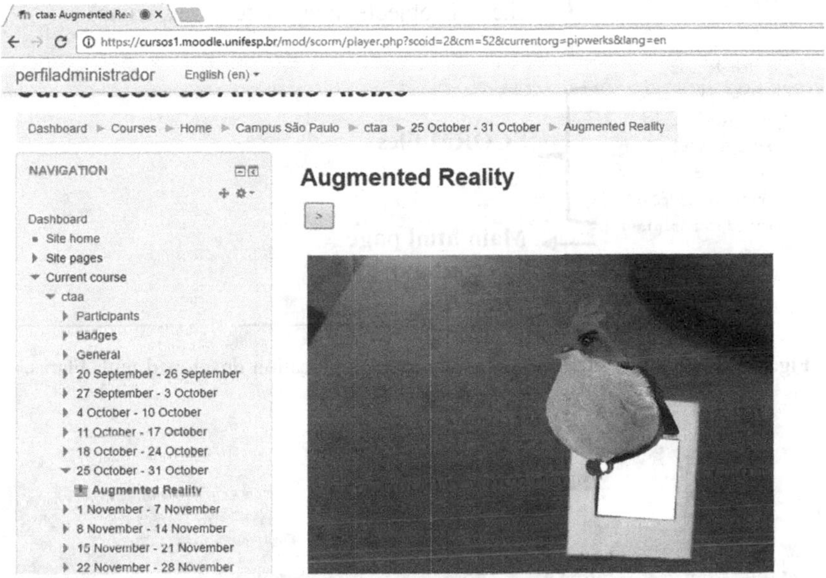

Fig. 4. Learning object based on augmented reality embedded into Moodle.

The AR application was developed using Flaras (the Flash Augmented Reality Authoring System), an authoring tool created using Adobe Air. This free open-source tool includes support for diverse media such as images, sounds, animations, and 3D objects. This tool was designed to assist nonexpert users with AR application development. It allows teachers to develop AR applications for the web with no programming knowledge. Thus, teachers can focus on content development rather than on the technology aspects. Flaras also includes tutorials based on texts and videos, FAQs, an e-book, and various examples accompanied by their respective projects. The AR application created with this tool can be run on users' web browsers through Adobe Flash[11]; it was designed to be simple enough so that lay people can use it by downloading 3D objects from 3D Warehouse[12] and importing them into their own applications.

[11] Adobe Flash: https://get.adobe.com/ashplayer/.

[12] 3D Warehouse: https://3dwarehouse.sketchup.com/index.html.

After the full activity was developed with Flaras, it was packaged as an LO, using our packaging tool. Figure 5 depicts its content. The root contains the files related to the SCORM standard, the main html page, and subfolders that contain Flaras files (i.e., sounds, images, textures, 3D files, and java scripts).

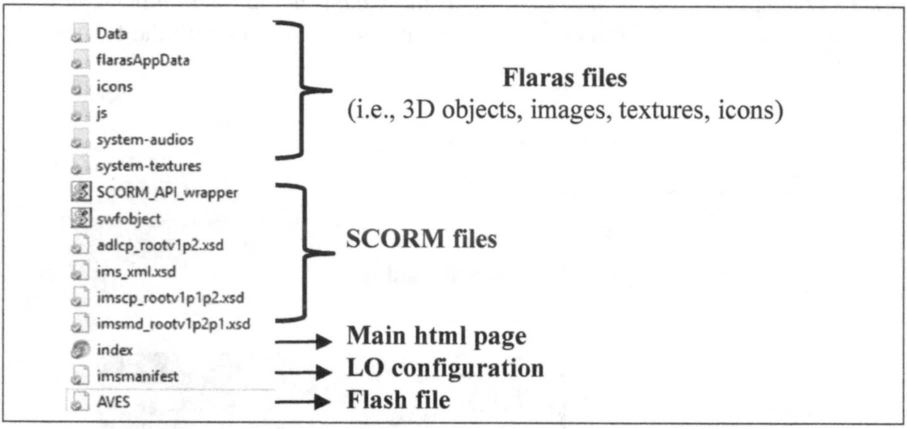

Fig. 5. Content of the SCORM based on AR application developed with Flaras.

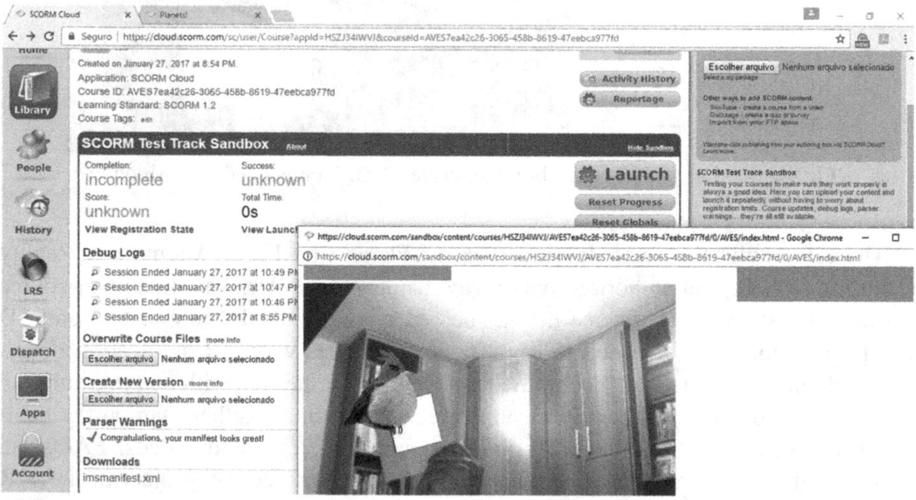

Fig. 6. The LO stored in the repository SCORM Cloud.

Validation and execution of the created LO were also published and tested in the repository SCORM Cloud (trial version), which is an online learning environment focused on storing and distributing e-learning content. It allowed the LO to be uploaded without any incident. Figure 6 depicts our LO deployed and running in the SCORM Cloud. If the LO is altered, all LMSs have it automatically updated; it does not require that[13] individual updates be sent to everyone who uses the content. This repository also tracks how much the LO content is being used, no matter which LMS the LO is stored in. SCORM Cloud is integrated with Moodle, so it is possible to use a LO based on AR within Moodle courses.

5 Conclusions

Until now significant effort has been made aiming to promote the reuse of didactic material. Even though there are several educational content repositories available, sharing material among educators still remains an open issue. Didactic material that follows an LO standard is a possible solution.

This paper presented a solution that allows the embedding of AR into LMSs. It can assist teachers using AR content to improve their classes, helping the students to understand the subjects. This proposal is based on LOs, which are didactic material projects designed to be reusable, interoperable, accessible, and manageable. We described a packaging tool based on SCORM, which is an LO standard.

We also introduced a case study that used our tool to create an LO based on AR and import it into Moodle. Using this tool, students can explore content in several different ways, allowing them to practice what they are studying in their own way. Students can use a marker and mouse to interact with the RA content. The LO developed was also deployed into and tested in the repository SCORM Cloud. We believe that the use of repositories can promote the shareability of educational content.

From the observed results, we can conclude that our tool has achieved its purpose. As future work, we intend to improve our packaging tool interface to support other LO standards (i.e., Tin Can API, AICC). We also aim to create AR applications with different authoring tools and package them using our solution.

References

1. Advanced Distributed Learning: SCORM ADL Net. https://www.adlnet.gov/adl-research/scorm. Accessed Jan 2017
2. Azuma, R.T.: A survey of augmented reality. Presence: Teleoperators Virtual Environ. **6**, 355–385 (1997)
3. Baptista, F.Q., Neto, M.P., Dias, D.R.C., de Paiva Guimarães, M., Brega, J.R.F.: 3D content generation to moodle platform to support anatomy teaching and learning. In: 13th ACS/IEEE International Conference on Computer Systems and Applications, pp. 1–6 (2016)
4. Belliston, C.J.: Open educational resources. Coll. Res. Libr. News **7**(5), 284–287 (2009). http://hdl.lib.byu.edu/1877/1437
5. Cardoso, A., Lamounier, E.: Aplicações na Educação e Treinamento. SBC, Brasil (2008). pp. 343–357
6. Dede, C.: Immersive interfaces for engagement and learning. Science **323**(5910), 66–69 (2009). http://www.sciencemag.org/cgi/content/abstract/323/5910/66

---

[13] SCORM Cloud: http://scorm.com/scorm-solved/scorm-cloud-features/.

7. Di Serio, Á., Ibáñez, M.B., Kloos, C.D.: Impact of an augmented reality system on students' motivation for a visual art course. Comput. Educ. **68**, 585–596 (2013)
8. Gonen, S., Basaran, B.: The new method of problem solving in physics education by using scorm-compliant content package. Turkish Online J. Distance Educ. **9**(3), 112–120 (2008)
9. Kaufmann, H.: Collaborative augmented reality in education. In: Keynote Speech at Imagina Conference, pp. 1–4 (2003). http://www.ita.mx/files/avisos-desplegados/ingles-tecnico/guias-estudio-abril-2012/articulo-informatica-1.pdf
10. Liu, T.Y., Tan, T.H., Chu, Y.L.: 2D barcode and augmented reality supported english learning system. In: 6th IEEE/ACIS International Conference on Computer and Information Science (ICIS 2007), pp. 5–10. IEEE (2007)
11. Mahnegar, F.: Learning management system. Int. J. Bus. Soc. Sci. **3**(12), 144–151 (2012)
12. Paulsson, F., Naeve, A.: Standardized content archive management-scam-storing and distributing learning resources. IEEE Learn. Technol. Newsl. **5**(1), 40–42 (2003)
13. Richards, G., Mcgreal, R., Hatala, M., Friesen, N.: The evolution of learning object repository technologies: portals for on-line objects for learning. Distance Educ. **17**(3), 67–79 (2002). http://www.ijede.ca/index.php/jde/article/view/297
14. de la Torre, L., Guinaldo, M., Heradio, R., Dormido, S.: The ball and beam system: a case study of virtual and remote lab enhancement with moodle. IEEE Trans. Ind. Inform. **PP**(99), 1 (2015). http://ieeexplore.ieee.org/lpdocs/epic03/wrapper.htm?arnumber=7120976
15. Wagner, D., Barakonyi, I.: Augmented reality kanji learning. In: Proceedings of the 2nd IEEE and ACM International Symposium on Mixed and Augmented Reality, ISMAR 2003, pp. 335–336, November 2003 (2003)
16. Wiley, D.A.: Connecting Learning Objects to Instructional Design Theory: A Definition, a Metaphor, and a Taxonomy. AECT, Bloomington (2000). pp. 343–357

Immersive Ground Control Station for Unmanned Aerial Vehicles

Glesio Garcia de Paiva[1(✉)], Diego Roberto Colombo Dias[2], Marcelo de Paiva Guimarães[3], and Luis Carlos Trevelin[1]

[1] Computer Science Department, Federal University of São Carlos, São Carlos, SP, Brazil
{glesio.paiva,trevelin}@dc.ufscar.br

[2] Computer Science Department, Federal University of São João del Rei, São João del Rei, MG, Brazil
diegodias@ufsj.edu.br

[3] Open University of Brazil/Post Graduate Program of Faccamp, Federal University of São Paulo – UNIFESP/Faccamp, São Paulo, SP, Brazil
marcelodepaiva@gmail.com

Abstract. Nowadays, the use of unmanned aerial vehicles, also known as drones, is growing in several areas, as military, civilian and entertainment. These vehicles can generate large volumes of data, such as 3D videos (three-dimensional) and telemetry data. A challenge is to visualize this data and take flight control decisions based on them. On the other hand, Virtual Reality provides immersion and interaction of users in simulated environments. Therefore, enriching the experience of drones' users with Virtual Reality becomes a promising possibility. This project aims at investigating how the features provided by virtual reality can be used in planning drone flight, allowing the flight plan creation - flight routes are determined by waypoints (georeferenced points), which contains altitude, latitude and longitude; and monitoring the entire path of mission - presenting information telemetry. To this end, a system of multiprojection is used, which inserts the user in an upcoming 3D environment videos captured by a Drone. The goal of this work is to develop an immersive and interactive control station for controlling, planing flights, and tracking.

Keywords: UAV · Virtual reality · Ground control station

1 Introduction

Virtual reality (VR) is a research area that offers numerous opportunities for scientific research and technological innovation. It enables the interaction and immersive involvement of users with simulations, be it, for example, using stereoscopic three-dimensional (3D) visualization devices, such as multiprojection environments, or employing touch devices and capturing user movements [8].

Unmanned aerial vehicles (UAVs), popularly called drones, have been widely used in military, civilian and entertainment applications [1]. The information

O. Gervasi et al. (Eds.): ICCSA 2017, Part I, LNCS 10404, pp. 595–604, 2017.
DOI: 10.1007/978-3-319-62392-4_43

generated by the drones is diverse, but the way they are presented is still a challenge. Information, such as 3D video generation, telemetry data, and route definition, can benefit from VR applications, since, depending on the visualization metaphors used, the user experience can be enriched. Flight routes, for example, are determined by waypoints, which contain altitude, latitude, and longitude. To track and define the flight route, we use systems software known as control stations, responsible for generating the flight plan and tracking the entire mission path, presenting telemetry information related to UAVs.

From the combination of UAVs with VR, there is the possibility of performing the flight planning through immersive environments, such as miniCAVE [4], so that the user, using a 3D virtual environment, can plan the whole flight route of the UAV using a highly immersive system for this purpose. Besides, videos captured by drone can be presented in 3D on miniCAVE, giving the user an immersive experience. Thus, the term immersive control station arises, an immersive platform for control, planning, and monitoring of flight mission for UAVs.

The need for collaborative immersive, interactive, and collaborative virtual environments is evident as it enables the use of VR-based software for employee training, product prototyping, plant visualization. In the VR area, several investments are being made by the industry in the production of hardware and software, which has generated and motivated its accelerated growth. In Precision Agriculture, for example, researchers and companies are looking for new technological resources for the improvement in the analysis and interpretation of images and decision making using UAVs in image acquisition. For increased user interaction through the images captured by UAVs, it is possible to make use of Augmented Reality (AR), that is, the insertion of virtual objects in a real environment.

Current applications make use of head mounted display (HMD) devices integration[1] but this can cause cybersickness (dizziness and nausea) to users [15]. Also, they have a limited area for viewing and presenting data and limit the option of interacting with other users because they are immersive in an environment that does not allow direct interaction with other users [11,14,16]. When using multiprojection systems for integration with UAVs, simulations are used for the environment where the aerial vehicle will fly [17]. These works do not allow waypoints (georeferenced points) flight planning to be done and tracked in an immersive multiprojection environment or HMD device.

This paper aims to investigate the use of an immersive control station to plan and monitor flights of UAVs. For this purpose, the objective of this paper is to propose an architecture for an immersive control station using the multi-projection environment called miniCave [4].

[1] A device coupled to the user's head, creating images for each eye and changing the view of the environment according to the position of the device.

2 Virtual Reality and Multiprojection System

In the literature, there are some definitions for VR. Hand [10] defines VR as a paradigm of computer interaction in a virtual environment that can be considered real at the moment of interaction with the user. For Hancock [9], VR is the most advanced form of the computer-generated three-dimensional user interface.

According to Kirner and Siscouto [12], VR is the most advanced way of user interaction with computer applications. However, some characteristics must be present, among them: immersion, interaction, and navigation with the elements of the virtual environment. For Leston [13], VR is a set of 3D graphical tools and techniques that allow users to interact with real-time computer generated environments, with some awareness that the interface is a virtual environment. Guimarães [8] explains that VR has demonstrated new ways to improve interface and interaction with users of computer systems, allowing immersion, interaction with computer-generated synthetic environments, exploring the senses of vision, hearing, touch and smell.

Multiprojection systems, usually composed of multiple screens, provide different points of view of the same environment to the user, allowing a highly immersive experience [4]. Multiprojection systems have been researched for more than a decade as a rendering solution for complex virtual environments [2,3,6,7].

Among the several existing multiprojection systems, we have the CAVE (CAVE Automatic Virtual Environment), which if implemented using clusters of computers can be deployed at low cost, depending on the devices used [5]. Graphic clusters are characterized by a set of nodes interconnected by a data network, giving the user the impression that the processing is performed by a single system [7]. The goal is to provide multiple views of the same data set, where each node process only the tasks associated with it, such as image generation from a given point of view and/or reception of user interactions.

The first fully immersive and interactive multiprojection system created in Latin America was called the Caverna Digital^{TM}, built by researchers at the Integrated Systems Laboratory of the Polytechnic School of the University of São Paulo. Caverna Digital^{TM} is composed of a graphic cluster responsible for managing and processing the data that is visualized in the multiprojection system. Behind each wall of the environment is arranged a mirror and projector so that the projection can be made. The user, when entering the multiprojection environment, needs 3D glasses so they can see the stereoscopic projections [8].

2.1 Graphic Cluster

Computer clusters are a group of computers with the ability to share resources among each other to achieve a common goal. They must have means for managing applications, having a proportional performance regarding the number of nodes used, thus, being able to perform computations with significant numerical, transactional and graphical complexities [5].

The centralized distribution architecture is based on the client–server logical model. Thus, all the current processing is performed on the server node.

Client nodes are responsible only for the presentation of graphical primitives to users. This type of architecture despite being a simple model has some problems, such as:

- **Scalability:** the fact that the server is the central point of synchronization and data distribution can cause problems with the number of client nodes supported by the environment. As the number of users increases, the server may not be able to handle the need for processing; and
- **Low Fault Tolerance:** the server is considered a single point of failure, so the entire system is stopped in the event of a failure. This type of failure can be circumvented by the use of replication techniques, however, harming the original simplicity provided by the architecture.

2.2 miniCAVE

To provide a low-cost multiprojection environment, Dias et al. [4] proposed an environment inspired by the CAVE [2]. However, having a different angle, aiming at immersing a larger number of users than the original CAVE, where few users can use it at the same time. This environment was called miniCAVE.

The environment is composed of a set of three screens with frontal projection. For imaging, conventional projectors are used. The polarization of the images is made through the use of polarizing lenses. The graphic cluster is composed of 4 machines, being 1 server and 3 clients. Figure 1 depicts the miniCAVE.

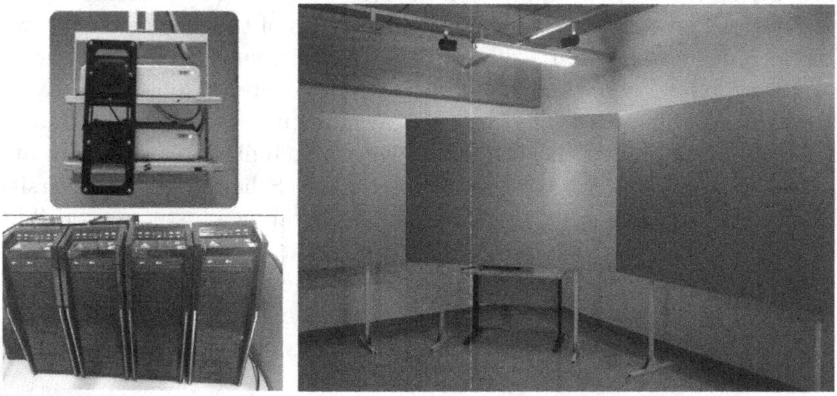

Fig. 1. miniCAVE – Graphic cluster, screens and multimedia projectors.

3 Related Work

In this section we present some works related to the use of UAVs with VR, making use of HMD devices and multiprojection environments.

Ikeuchi et al. [11] proposed and developed KinecDrone, an application to provide the sensation of flight to the user. The video captured by the Ar.Drone made by the company Parrot[2] is transmitted to the user using a Oculus Rift. Thus, even in a room, the user has the sensation of actually flying freely through the skies, being able to control the AR.Drone with natural gestures captured by a Kinect. The main purpose of this application is to increase the immersive user experience.

Pittman and La Viola [16] studied the capabilities of combined head tracking with an Oculus Rift as a mode of generating input stimuli to a robot. The authors used a Parrot AR.Drone to test techniques based on gestural metaphors such as Head Translation, Head Rotation, and Modified Flying Head. A case study was conducted with the purpose of observing the effectiveness of each of the techniques employed.

Mirk and Hlavacs [14] have used virtual tourism as an approach to the use of UAVs. The video obtained from the camera of a UAVs is presented to the user through an Oculus Rift. The user's head movement controls the orientation of the UAVs. In this way, the authors aim to provide the user with an immersive experience and without limitations to them. Experimental results were presented and analyzed about the compensation of delays generated by the data network used (Internet) in the experiment.

The papers presented make use of Oculus Rift, an HMD device that allows the user to immerse himself in an external environment giving the sensation of being in another place, as described by the researchers [11,14,16]. However, immersion by this devices limits the information capacity that can be displayed on display to the user and mission planning is performed in a non-immersive environment.

Walter et al. [17] propose a new design for an immersive control station using VR to simulate the environment where military-grade autonomous vehicles are controlled in a semi-autonomously way. To generate the VR environment, information already processed from the site, such as land surface information, is required. Through information gathered by sensors of the standalone vehicle, it is possible to add features of the environment. The autonomous vehicles used in this work are the military class, possessing an advanced technology on board.

4 The Immersive Ground Control Station – IGCS

The interaction between UAVs and a cluster of computers can be complex, so the specification of a conceptual architecture is a great help. This section presents an integration solution of graphic clusters and UAVs, resulting in the immersive control station that obtains images and videos, provided by a UAVs. Videos are rendered in real time, generating 3D renderings to be displayed on miniCAVE.

Figure 2 illustrates the proposed general architecture, which presents the components of the architecture: UAV, mobile device, graphic cluster (local and

[2] http://www.parrot.com.

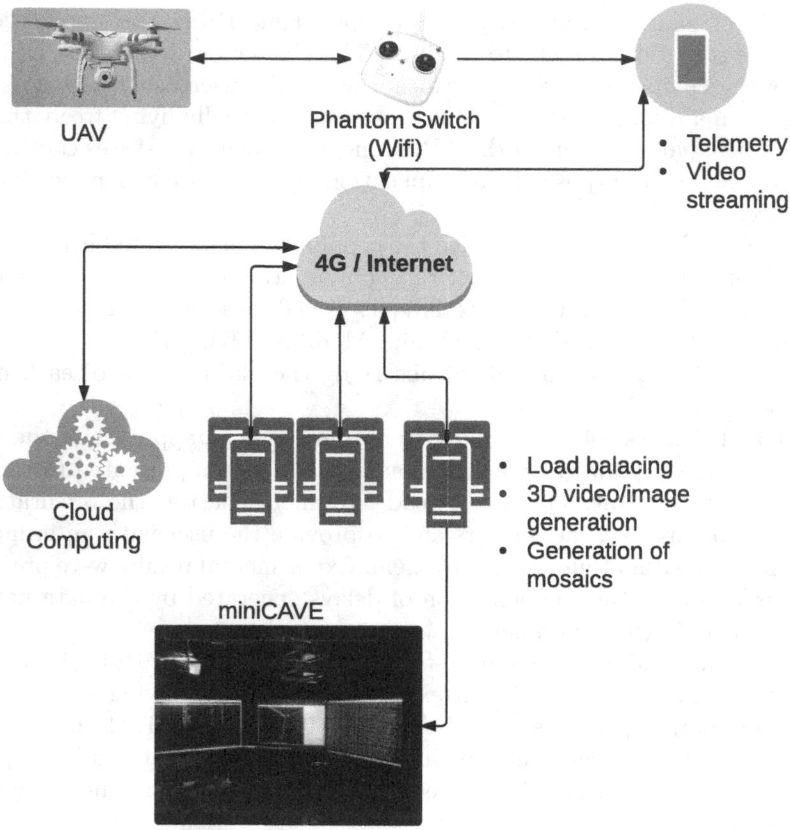

Fig. 2. Immersive ground control station.

distributed), and miniCAVE. The execution flow is described as follows: the UAV sends images/videos to the mobile device by means of a switch coupled to the control radio; the mobile device, besides allowing the user to visualize the image generated by the camera of the UAV, still carries out the sending to a remote location (graphic cluster), besides the telemetry data; if the video is sent to a local cluster, it is received by the cluster server, which is responsible for receiving the video and dividing it into several quadrants that are processed by the cluster nodes; soon after the division of quadrants, these are sent, selectively, to the nodes of the cluster. Thus, each node only processes what is its interest to optimize the processing of the images/videos.

Synchronization barriers are used to maintain image generation consistency. Massive data processing can also be done in a cloud solution, for example, Amazon EC2 (Elastic Compute Cloud). Each component is described in detail below.

- **UAV:** this component is responsible for capturing images/videos. In this project, a Phantom 2 Vision is used, since this is the model available for tests and simulations in BLIND;

- **Mobile devices:** this allows the user to view various information regarding the Phantom 2 Vision, even using its application provided by the manufacturer. However, using the SDK of the Phantom 2 Vision device, it is possible to develop applications that fit the requirements of the proposal, such as sending images/videos and telemetry data to different locations, in this case, through using a 4G network. Image and video compression algorithms were used to reduce the amount of data traffic across the network;
- **Graphic cluster:** this component will be responsible for the massive processing of data, as well as the presentation of information to users. The nodes of the cluster have graphics cards provided with a many-core architecture, which allows the data to be processed in parallel, to generate images/videos with just a little delay. The load balancing required for the graphic cluster was done using the libGlass [5]. Therefore, the solutions developed and proposed by Dias et al. [5] are used in the immersive control station. Aiming for a greater scalability and processing power, other distributed graphic clusters and cloud solutions should be used;
- **miniCAVE:** the immersive visualization is carried out through an environment composed of three screens, which gives users an immersive experience. The images generated by the graphic cluster are presented in miniCAVE. The interaction with the environment can take place through conventional devices such as mouse and keyboard or non-conventional such as Kinect and WiiRemote for example;
- **Web Service:** this component is responsible for receiving and transmitting data from a mobile device to the graphic cluster. Representational State Transfer (REST) was the chosen standard for building the web service; and
- **Control station:** this is responsible for planning the flight and monitoring the execution of the flight. For the deployment of this component was used the game engine Unity 3D[3].

Figure 3 demonstrates the execution and interaction flow of each component in the proposed architecture. The interaction between the UAV operator and the mobile device, UAV, and Radio Switch components can be performed in parallel to the execution of the Web Service components, Computer Cluster, and miniCAVE components.

The UAV Operator interacts with the mobile device to execute the mission tracking application and communication with the immersive solo control station via web service over a 4G data network. The cluster communicates with the mobile device via a web service. The control station operator plans the mission by interacting with the projection on the miniCAVE. Users track mission and telemetry and video data through the miniCAVE multiprojection environment.

The immersive control station presented in this paper has the objective of providing the user with a control interface that is not only remote but also immersive and interactive. Through this, it is possible to plan routes for UAVs

[3] https://unity3d.com.

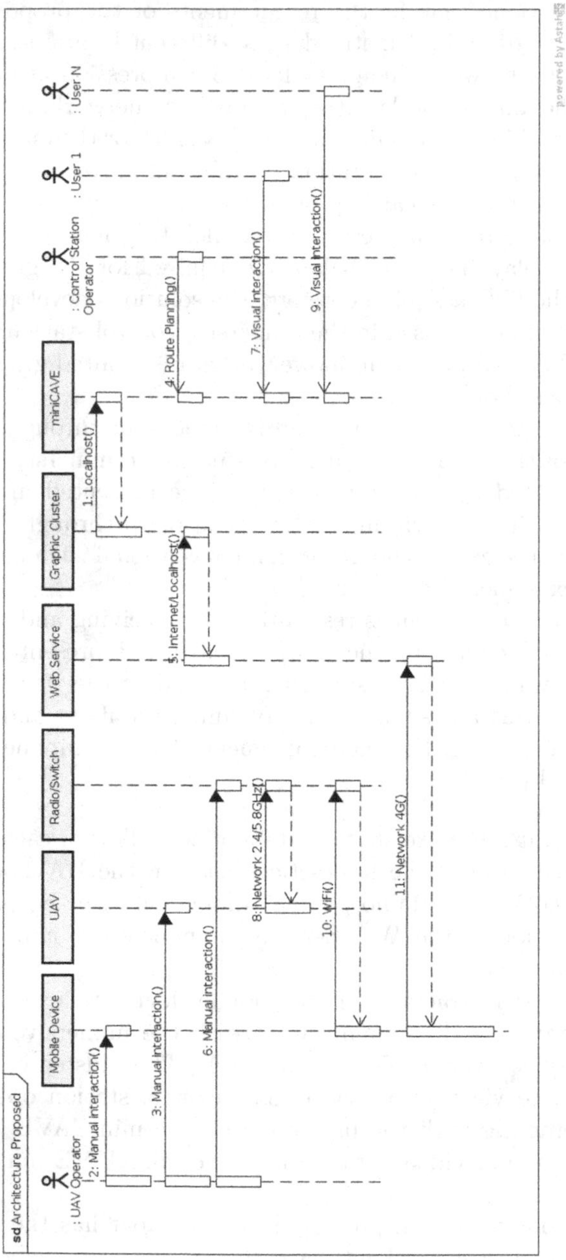

Fig. 3. Sequence diagram for proposed architecture.

and obtain information regarding the telemetry of the device. However, it is still possible to follow the route defined by waypoints in 3D using the miniCAVE or HMD based solutions.

This paper intends to implement the architecture for the immersive control station in multiprojection systems (miniCAVE) proposed. In this way, the implementation of this architecture allows the monitoring and planning of UAV missions in a remote and totally immersive environment. Thus, only the operator needs to be in the field, and the domain experts do not need to be in the field or have to wait for the mission to be fully completed so that the data collected can be analyzed. Among the areas in which this work can be applied are agriculture, monitoring of works, border patrol, search and rescue, inspection in transmission lines, and oil pipelines.

5 Concluding Remarks

This paper presented the proposal and development of an immersive control station for UAVs. The functionalities presented in the paper are innovative if compared to related papers.

VR has been employed in several areas. The authors presented a new possibility of using this type of interface, allowing UAVs, which are devices usually controlled and followed by a single user, to be monitored and accompanied by several people remotely. The possibility of remote flight planning is also something that the authors presented.

The immersive control station can be used in irrigation situations in areas of difficult access; monitoring works, power transmission lines, oil pipelines, events, workshops, tourism, and entertainment.

The initial experiments were performed using a private cluster. Since the required complexity is not high, this setup was sufficient for running the system. However, once new features and requirements arise, platforms like Amazon EC2 could be employed.

Acknowledgment. The authors gratefully acknowledge support from CAPES and CNPq.

References

1. Costa, F.G.: Integração entre veículos aéreos não tripulados e redes de sensores sem fio para aplicações agrícolas. Ph.D. thesis, Universidade de São Paulo (2013)
2. Cruz-Neira, C., Sandin, D.J., DeFanti, T.A., Kenyon, R.V., Hart, J.C.: The cave: audio visual experience automatic virtual environment. Commun. ACM **35**, 64–72 (1992)
3. DeFanti, T.A., Leigh, J., Renambot, L., Jeong, B., Verlo, A., Long, L., Brown, M., Sandin, D.J., Vishwanath, V., Liu, Q., et al.: The optiportal, a scalable visualization, storage, and computing interface device for the optiputer. Future Gener. Comput. Syst. **25**(2), 114–123 (2009)

4. Dias, D.C., La Marca, A., Moia Vieira, A., Neto, M., Brega, J., de Paiva Guimaraes, M., Lauris, J.: Dental arches multi-projection system with semantic descriptions. In: 2010 16th International Conference on Virtual Systems and Multimedia, Virtual Systems and Multimedia (VSMM), VSMM 2010, Seoul, pp. 314–317. IEEE Xplore, October 2010. ISBN 978-1-4244-9027-1
5. Dias, D.R.C., Guimarães, M.P., Kuhlen, T.W., Trevelin, L.C.: A dynamic-adaptive architecture for 3D collaborative virtual environments based on graphic clusters. In: Proceedings of the 30th Annual ACM Symposium on Applied Computing, SAC 2015, pp. 480–487. ACM, New York (2015)
6. Drolet, F., Mokhtari, M., Bernier, F., Laurendeau, D.: A software architecture for sharing distributed virtual worlds. In: Virtual Reality Conference, 2009, VR 2009, pp. 271–272. IEEE (2009)
7. Gnecco, B.B., de Paiva Guimaraes, M., Zuffo, M.K.: Um framework para computação distribuída. In: Simpósio Brasileiro de Realidade Virtual, Ribeirão Preto. SBC (2003)
8. Guimaraes, M.P.: Um ambiente para o desenvolvimento de aplicações de realidade virtual baseadas em aglomerados gráficos. Ph.D. thesis, Universidade de São Paulo (2004)
9. Hancock, D.: Viewpoint: virtual reality in search of middle ground. In: IEEE Spectrum (1995)
10. Hand, C.: Other faces of virtual reality. In: Brusilovsky, P., Kommers, P., Streitz, N. (eds.) MHVR 1996. LNCS, vol. 1077, pp. 107–116. Springer, Heidelberg (1996). doi:10.1007/3-540-61282-3_11
11. Ikeuchi, K., Otsuka, T., Yoshii, A., Sakamoto, M., Nakajima, T.: Kinecdrone: enhancing somatic sensation to fly in the sky with kinect and AR.Drone. In: Proceedings of the 5th Augmented Human International Conference, AH 2014, pp. 53:1–53:2. ACM, New York (2014)
12. Kirner, C., Siscoutto, R.: Realidade virtual e aumentada: conceitos, projeto e aplicações. In: Livro do IX Symposium on Virtual and Augmented Reality, Petrópolis (RJ), Porto Alegre: SBC (2007)
13. Leston, J.: Virtual reality: the it perspective. Comput. Bull. **38**(3), 12–13 (1996)
14. Mirk, D., Hlavacs, H.: Virtual tourism with drones: experiments and lag compensation. In: Proceedings of the First Workshop on Micro Aerial Vehicle Networks, Systems, and Applications for Civilian Use, DroNet 2015, pp. 45–50. ACM, New York (2015)
15. Neto, M.P., Agostinho, I.A., Dias, D.R.C., Rodello, I.A., Brega, J.R.F.: A realidade virtual e o motor de jogo unity. In: Tendências e Técnicas em Realidade Virtual e Aumentada, pp. 9–23 (2015)
16. Pittman, C., LaViola Jr., J.J.: Exploring head tracked head mounted displays for first person robot teleoperation. In: Proceedings of the 19th International Conference on Intelligent User Interfaces, IUI 2014, pp. 323–328. ACM, New York (2014)
17. Walter, B.E., Knutzon, J.S., Sannier, A.V., Oliver, J.H.: Virtual uav ground control station. In: AIAA 3rd Unmanned Unlimited Technical Conference, Workshop and Exhibit (2004)

SPackageCreator3D - A 3D Content Creator to the Moodle Platform to Support Human Anatomy Teaching and Learning

Fabrício Quintanilha Baptista[1], Mário Popolin Neto[2(✉)], Luiz Soares dos Santos Baglie[1], Silke Anna Theresa Weber[3], and José Remo Ferreira Brega[1]

[1] Computing Department, School of Sciences, São Paulo State University (UNESP), Bauru, SP, Brazil
fabricioqb@gmail.com, luizssb.biz@gmail.com, remo@fc.unesp.br
[2] Federal Institute of São Paulo (IFSP), Araraquara, SP, Brazil
mariopopolin@ifsp.edu.br
[3] Department of Ophthalmology, Otolaryngology and Head and Neck Surgery, Botucatu Medical School, São Paulo State University (UNESP), Botucatu, SP, Brazil
silke@fmb.unesp.br

Abstract. Understanding three-dimensional (3D) forms is very important for anatomy learning. Most methods of anatomy teaching are offered to students through two-dimensional (2D) resources, such as videos and images. In order to support anatomy teaching and learning, many software solutions have been developed to allow the interaction and visualization of 3D virtual models using several techniques and technologies. Even solutions that offer great visual resources lack common teaching and learning environments features. This article aims to present the SPackageCreator3D, a tool to create 3D SCORM packages to the Moodle platform, which is used by a large number of educational institutions, being one of the most popular teaching and learning management platforms, validated by educators and supported by pedagogical a methodology. In order to test the 3D SCORM packages created by the SPackageCreator3D, an evaluation with 32 students was performed, indicating potential uses and needed improvements.

Keywords: Virtual reality · Human anatomy · SCORM · Moodle

1 Introduction

Over the past years, anatomy teaching has been changing [6], many schools around the world have faced the challenge of improving the efficiency of human anatomy teaching by reducing the time taken to study it by up to 55% [13].

Some challenges can be found in the traditional methods, such as the number of synthetic anatomical models available to students, and religious, emotional,

O. Gervasi et al. (Eds.): ICCSA 2017, Part I, LNCS 10404, pp. 605–618, 2017.
DOI: 10.1007/978-3-319-62392-4_44

and ethical issues involving corpse manipulation [10]. Teaching methods may be inefficient in some cases due to lack of student experience and advanced skills in dissecting techniques and also regarding the size and complexity of the structures involved [1].

The understanding of anatomical science may be related to the acquisition, interpretation, and conceptualization of spatial information [30]. Positive and significant results concerning the usage of 3D virtual models to teach anatomy have been found in many studies [32]. The use of 3D virtual models makes it possible to visualize the spatial relationships of the human body organs from any desired angle, offering benefits over classical teaching methods, such as better understanding and information transmission. The possibility of accessing 3D virtual models anywhere and anytime is also an advantage, providing individual or group learning through an additional study to the traditional classes without the need to be in physical laboratories [17].

Nowadays students are known to be natives of the Technological Age, and in this case, more attention needs to be given to new teaching methodologies in order to take full advantage of the natural way in which this generation can adapt to technologies and modern resources [28]. New computer-supported and information-based teaching tools are increasingly used in the anatomy teaching environment, which attracts the attention of students that have electronic communication devices as first source of information [25].

The main contribution of this paper is to present the SPackageCreator3D, a prototype solution that allows integration of 3D virtual models and related content files into the Moodle platform through standardized content generation, aiming to reach an optimized environment [19] to support anatomy teaching and learning. The SPackageCreator3D combines these models and content files in an intuitive interface providing not only easy mouse interaction but also accessibility on the Internet in the Moodle, one of the most popular Virtual Learning Environment (VLE) platforms, including all its features such as interactive lessons, discussion forum, tasks, and structured exercises.

The remainder of the paper is organized as follows. Section 2 contains the related work to this research. Section 3 describes the Moodle and the 3D virtual models integration. Section 4 presents the tool for 3D SCORM packages generation. Section 5 presents the case study and evaluation methodology. Section 6 discusses the results. Section 7 presents conclusion and future works.

2 Related Work

In order to survey solutions for anatomy teaching and learning that makes use of 3D virtual models, a systematic literature review was conducted following Kitchenham's guidelines [18]. Although the results obtained in the systematic literature review did not point to a standardization or trend in the solutions found, with several programming languages and techniques being used, it is possible to classify most of the solutions in three different platforms: Mobile [16,23],

Web [5,8,15,20,21,29], or Desktop [2,3,7,9,11,14,24,26,31]. Other platforms can also be found, such as Virtual and Augmented Reality systems [12,22].

Some solutions have specific and inherent characteristics related to their target platform. As, for example, interactive touch-sensitive interface in the mobile solution presented by [23], which enables user interaction and access to information by taking advantage of tablets and smartphones screens that are touch-sensitive and offer the user an intuitive interaction.

Desktop solutions, on the other hand, can present better processing power and consequently better model rendering, generating more complex environments with higher visual quality, such as the virtual atlas presented by [14], although they may require specific hardware or training for their use, as well as software configuration.

Web solutions in turn allow on-line updating of 3D virtual models and content, and collaborative work with real-time interaction, such as the interactive web atlas presented by [8], besides being available at anytime and anywhere, these solutions do not require installation and software configuration or special hardware, being executed in web browsers.

Despite the specific characteristics of each platform, most of the solutions allow interaction with 3D virtual models and visualization at different angles and sizes, offering the possibility to visualize only specific parts, and presenting some information regarding the virtual model or the selected segment.

The SPackageCreator3D provides standardized content generation that organize 3D virtual models and related content files, which can be used as base for on-line human anatomy courses creation using Moodle, taking advantage of the web solutions features, such as being available anytime and anywhere, not needing previous software configuration, and exceptional hardware specifications. In addition to being integrated into Moodle, one of the most popular VLE platforms, that provides teaching and learning support, content packages generated by the SPackageCreator3D offers the main features presented by the other found solutions, such as interaction with the 3D virtual model, visualization of specific parts, and access to related content files. The Table 1 summarizes a comparison between solutions grouped by target platform.

Table 1. Comparison between solutions grouped by target platform.

Solutions	Attach content	High visual quality	Specific setup	On-line access	Integration with LMS	Structured activities
Desktop	No	Yes	Yes	No	No	No
Mobile	No	No	Yes	No	No	No
Web	No	No	No	Yes	No	No
Our solution	Yes	No	No	Yes	Yes	Yes

3 Moodle and 3D Virtual Models Integration

3.1 Moodle

The Modular Object Oriented Dynamic Learning Environment - Moodle is one of the most popular VLEs. It is open source under the GNU-GPL license and used in more than 70,000 schools, universities and corporations around the world, with more than 77 million users[1]. The Moodle project is supported by an active international community that aims to review and improve this popular platform.

The Moodle was developed based on the socio-constructivist pedagogical methodology for learning and teaching management, offering useful tools, resources and features to overcome the user's needs. Due to its scalability, it can be used in many types of environments, ranging from small groups to huge corporations. Moodle's main features are:

- Architecture: it provides services grouped into four layers. Hypertext Transfer Protocol – HTTP server, such as Apache[2]; Data Base Management System – DBMS, such as MySQL[3] or Oracle[4]; Files and directories server; and Operating Systems (OS), such as Linux, Solaris, Windows or Mac OS.
- Functionalities: it is divided into resources (Text Page, Web Page, Directory/Repository, Content Package, and Label) and activities (Portfolios, Interactive Lessons, Discussion Forum, Glossary, FAQs, Chat Channels, Collaborative Texts, Tasks and Exercises, Journals, and Evaluation Questionnaires).
- Users: it offers predefined user roles during platform set up (Designer, Guest, Authenticated User, Student, Professor, Administrator, and Course Editor) and also allows the creation of new types of user on demand.

3.2 3D Virtual Models Integration

There are basically two ways to integrate 3D virtual environments with the Moodle platform. In the first, the integration is performed through virtual worlds that access the platform's functionalities and students are usually represented by an avatar; And the second, 3D virtual models are inserted directly into the platform and allow the student to visualize, interact and manipulate these models [27].

The insertion of 3D virtual models into the Moodle pages proved to be more propitious [27], reaching an optimized learning environment [19], not requiring specific knowledge nor adaptation process for the usage of the interface. The 3D virtual models are inserted into the Moodle platform using the SCORM standard, organizing models in the Collada format and a WebGL-based Javascript API called Three.js, which has the ColladaLoader script, capable of loading Collada models into Web pages [27].

---

[1] https://moodle.org/.
[2] http://www.apache.org/.
[3] https://www.mysql.com/.
[4] http://www.oracle.com/.

4 SPackageCreator3D

The SPackageCreator3D was developed for the Microsoft Windows Operating System to create content packages in the SCORM standard, that organize 3D virtual models and related content files, in order to integrate these models and files into the Moodle platform to support anatomy teaching and learning.

The SPackageCreator3D interface is very simple and intuitive in a way that does not require advanced user training or specific knowledge. Figure 1 presents the interface, consisting of a text field for typing the package name and three buttons: "Load Collada Files" to choose the 3D virtual model in the Collada format, "Load Content Files" to choose the related content files to the 3D model, such as videos (e.g. surgery recordings) or PDF documents, and "Generate Scorm Package" to create and save the 3D SCORM package. All package creation process can be viewed at this video[5].

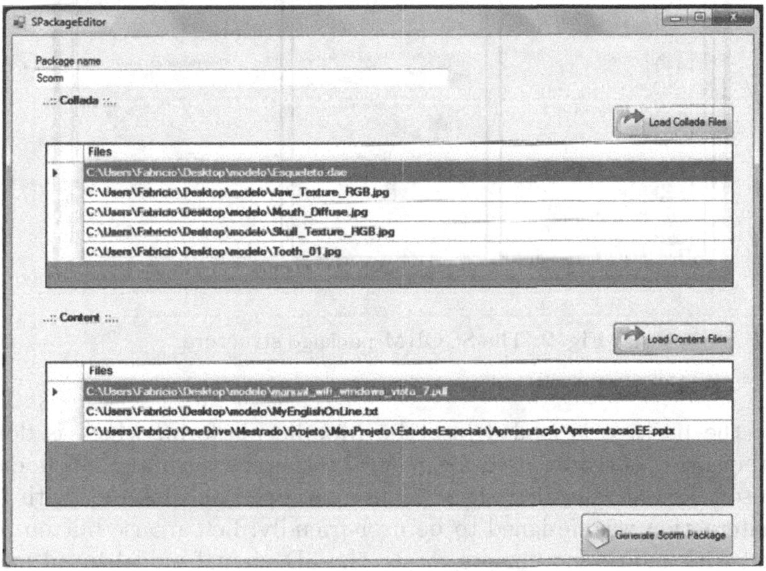

Fig. 1. SPackageCreator3D interface.

The root folder of a SCORM package created with SPackageCreator3D contains XSD (eXtensible Markup Language Schema Definition) files to ensure the validations of the standard, a subfolder for package resources, and the imsmanifest.xml file. The Fig. 2 shows the SCORM package structure. The SPackageCreator3D configures the imsmanifest.xml appropriately and organizes all files

[5] https://youtu.be/7Xd_1u7a1zc.

and folders [4]. After the package is created at the folder specified by the user, it is ready to be loaded into the Moodle. The SCORM package upload into the Moodle platform can be viewed at this video[6].

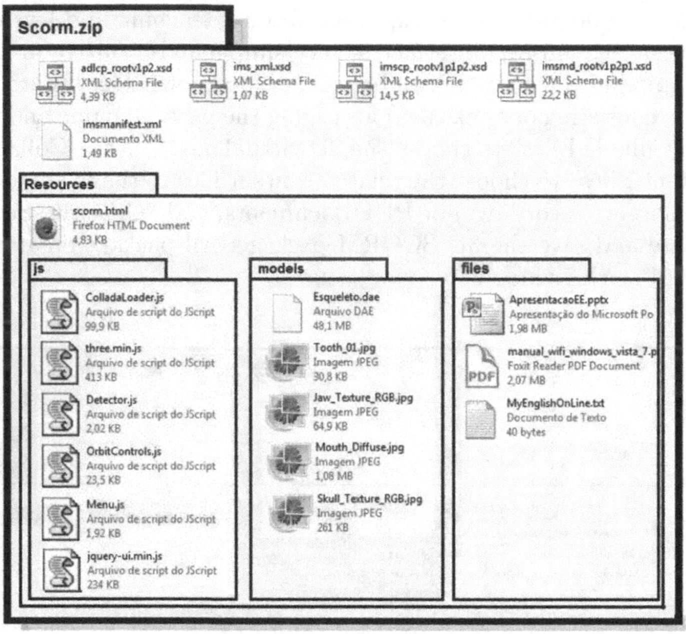

Fig. 2. The SCORM package structure.

Once the package is loaded into the Moodle platform, which is done following the same procedure used for general resources, such as PDF documents or Microsoft PowerPoint files, the students can open and interact with it. The mouse interaction was designed to be user-friendly: Left mouse button pressed combined with mouse movements rotate the 3D virtual model in all axes and all directions. Mouse scroll wheel allows to zoom in or out in the 3D virtual model. Right mouse button pressed combined with mouse movements move the 3D virtual model in the horizontal and vertical axes.

The SCORM package created by the SPackageCreator3D contains a menu interface where it is possible to access the related content files and set up the visible parts of the 3D virtual model. Figure 3 shows the three attached content files in a SCORM package loaded and opened into the Moodle and Fig. 4 shows the selective visualization provided by setting up a checkbox for the available geometries. The interaction procedures can be viewed at this video[7].

[6] https://youtu.be/VNhBJ9tCct4.
[7] https://youtu.be/WZSWhIuUzNA.

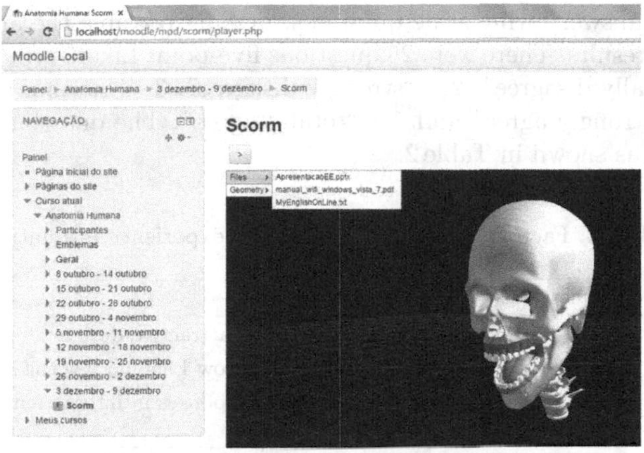

Fig. 3. Menu option to access the related content files.

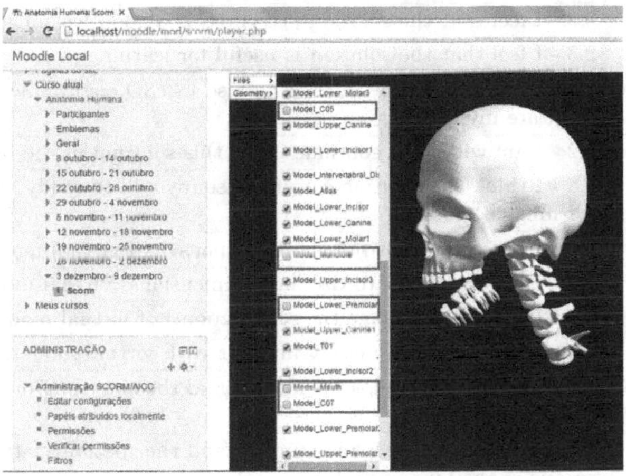

Fig. 4. Menu option for selective visualization.

5 Case Study

A case study was conducted to evaluate the user experience in the usage of a 3D SCORM package generated by the SPackageCreator3D. All participants were university students of the medical course.

University medical teachers used the SPackageCreator3D to create a class material using the 3D virtual model shown in Fig. 3, attaching content files related to this model and uploading the 3D SCORM package generated into the Moodle, so that students could interact with the 3D virtual model and access the additional content.

After using into the Moodle platform the 3D SCORM package created by the medical teachers using the SPackageCreator3D, students were asked to anonymously answer rating scale questions, making possible the analysis of the experimental results. There were 25 questions five-point Likert-type scale (range: 1–5): 1 – "totally disagree", 2 – "strongly disagree", 3 – "neither agree nor disagree", 4 – "strongly agree" and 5 – "totally agree". The questions are divided into 7 factors as shown in Table 2.

Table 2. Factors and questions for user experience evaluation.

Factor	Questions
1-Ease of use	1.1 - The class material features an attractive design
	1.2 - I feel it is easy to understand how I should use the solution
	1.3 - The solution provides quick responses in interactions with the environment (model and content)
	1.4 - I feel it's easy to learn how to interact with virtual models
2-Utility	2.1 - I feel that the solution helps in a better understanding of relationships and anatomical structures
	2.2 - I feel that the solution presented is a good learning system
	2.3 - I feel that the solution is useful for learning
3-Intention to use	3.1 - I would like other classes to also use 3D environment systems to facilitate my learning
	3.2 - I am willing to continue using this solution in the future
	3.3 - I think the solution can increase my will to study and help in learning the subject
4-Interaction	4.1 - I can easily move the three-dimensional virtual models
	4.2 - I can easily rotate the three-dimensional virtual models
	4.3 - I can easily change the scale (zoom) of virtual models
	4.4 - In general, I can easily interact with virtual models
5-Imagination	5.1 - The solution helps me understand the shapes of organs of the human body
	5.2 - The solution helps me understand the spatial relationship between organs of the human body
	5.3 - The solution helps me understand the position of organs in the human body
	5.4 - In general, the solution helps me better understand human anatomy
6-Immersion	6.1 - The solution has an immersive environment
	6.2 - The solution translates reality well in relation to the manipulation of three-dimensional models
	6.3 - The solution increases my attention at the time of study
	6.4 - In general, I feel involved in the learning environment
7-General	7.1 - I think it is important to access the three-dimensional virtual models in an unlimited way and without the need to be in a laboratory
	7.2 - I think it is important to integrate three-dimensional virtual models and content in the same environment
	7.3 - I think the use of LMS (Moodle) helps in the structuring of the activities, so that I do not feel lost when I'm studying or executing tasks

6 Results and Discussion

A total of 32 students, that have used into the Moodle platform the 3D SCORM package created for the case study, answered the 25 questions. Table 3 shows the frequency scores for each question and Table 4 shows the frequencies percentage,

Table 3. The frequency score of each question.

Factor	Question	1	2	3	4	5	Total
1-Ease of use	1.1	1	2	5	15	9	32
	1.2	1	4	6	14	7	32
	1.3	1	0	4	17	10	32
	1.4	0	4	8	9	11	32
	Global	3	10	23	55	37	128
2-Utility	2.1	0	1	3	15	13	32
	2.2	0	0	4	17	11	32
	2.3	0	0	6	12	14	32
	Global	0	1	13	44	38	96
3-Intention to use	3.1	0	0	5	13	14	32
	3.2	0	1	6	9	16	32
	3.3	0	1	7	12	12	32
	Global	0	2	18	34	42	96
4-Interaction	4.1	1	2	5	14	10	32
	4.2	0	1	6	14	11	32
	4.3	0	5	5	12	10	32
	4.4	0	1	4	19	8	32
	Global	1	9	20	59	39	128
5-Imagination	5.1	0	1	2	12	17	32
	5.2	0	1	2	13	16	32
	5.3	0	1	1	12	18	32
	5.4	0	0	3	14	15	32
	Global	0	3	8	51	66	128
6-Immersion	6.1	1	3	9	12	7	32
	6.2	1	2	4	17	8	32
	6.3	1	1	10	11	9	32
	6.4	0	1	9	14	8	32
	Global	3	7	32	54	32	128
7-General	7.1	1	1	3	9	18	32
	7.2	0	0	4	10	18	32
	7.3	0	1	6	16	9	32
	Global	1	2	13	35	45	96

Table 4. The frequency percentage of each question.

Factor	Question	1	2	3	4	5	Total	4 and 5
1-Ease of use	1.1	3.1	6.3	15.6	46.9	28.1	100	75
	1.2	3.1	12.5	18.8	43.7	21.9	100	65.6
	1.3	3.1	0	12.5	53.1	31.3	100	84.4
	1.4	0	12.5	25	28.1	34.4	100	62.5
	Global	2.34	7.81	17.97	42.97	28.91	100	–
2-Utility	2.1	0	3.1	9.4	46.9	40.6	100	87.5
	2.2	0	0	12.5	53.1	34.4	100	87.5
	2.3	0	0	18.8	37.5	43.7	100	81.2
	Global	0	1.04	13.54	45.83	39.58	100	–
3-Intention to use	3.1	0	0	15.6	40.6	43.8	100	84.4
	3.2	0	3.1	18.8	28.1	50	100	78.1
	3.3	0	3.1	21.9	37.5	37.5	100	75
	Global	0	2.08	18.75	35.42	43.75	100	–
4-Interaction	4.1	3.1	6.3	15.6	43.7	31.3	100	75
	4.2	0	3.1	18.8	43.7	34.4	100	78.1
	4.3	0	15.6	15.6	37.5	31.3	100	68.8
	4.4	0	3.1	12.5	59.4	25	100	84.4
	Global	0.78	7.03	15.63	46.09	30.47	100	–
5-Imagination	5.1	0	3.1	6.3	37.5	53.1	100	90.6
	5.2	0	3.1	6.3	40.6	50	100	90.6
	5.3	0	3.1	3.1	37.5	56.3	100	93.8
	5.4	0	0	9.4	43.7	46.9	100	90.6
	Global	0	2.34	6.25	39.84	51.56	100	–
6-Immersion	6.1	3.1	9.4	28.1	37.5	21.9	100	59.4
	6.2	3.1	6.3	12.5	53.1	25	100	78.1
	6.3	3.1	3.1	31.3	34.4	28.1	100	62.5
	6.4	0	3.1	28.1	43.8	25	100	68.8
	Global	2.34	5.47	25	42.19	25	100	–
7-General	7.1	3.1	3.1	9.4	28.1	56.3	100	84.4
	7.2	0	0	12.5	31.2	56.3	100	87.5
	7.3	0	3.1	18.8	50	28.1	100	78.1
	Global	1.04	2.08	13.54	36.46	46.88	100	–

including a special column that brings frequencies regarding satisfactory answers (4 and 5 scale points). The Fig. 5 shows the global percentages per factor.

According to Table 4, in general the results are satisfactory since the column "4 and 5" shows values greater than 70% for most of questions. The factors "1-Ease of use" and "4-Interaction" present good results, which indicates that the

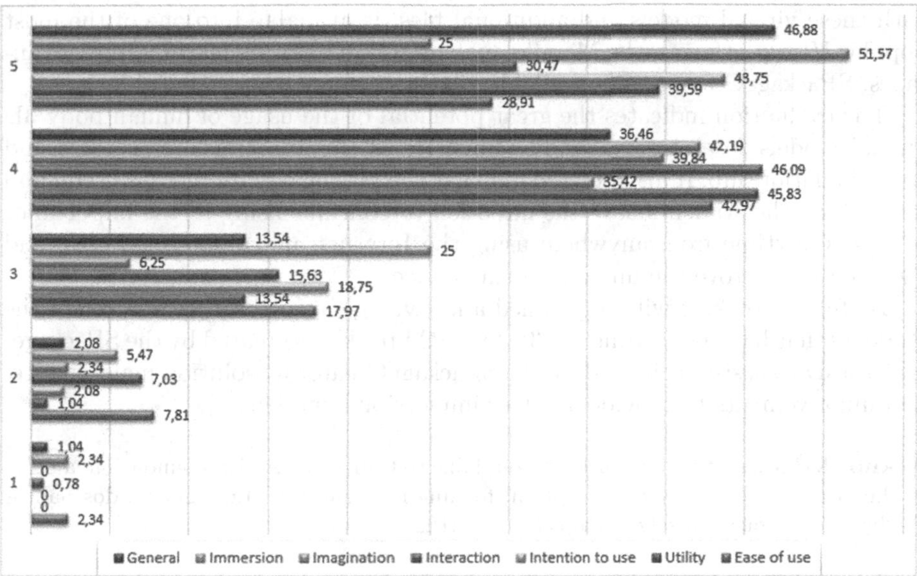

Fig. 5. Global percentages per factor.

virtual environment is easy to use and it provides interesting interactivity and fast feedback response concerning the 3D virtual model manipulation.

The factors "2-Utility", "3-Intention to use", and "7-General" achieved excellent results, which pointed out that the virtual environment is useful for learning and the students have clear intention to continue using it to assist their studies. In addition, it is observed as important features to access the 3D virtual model anytime and anywhere using the Internet, the related content files attached, and the integration into the Moodle (VLE platform) that organizes and facilitates the study.

The factor "5-Imagination" presented the best performance with all questions over 90% in the sum of scores 4 and 5, which indicates a great potential of the usage of 3D virtual models in order to help students to better understand the anatomical structures and spatial relationship. The results from the factor "6-Immersion" indicate that improvements must to be developed in the virtual environment in order to provide a better immersion experience to the user. The usage of common computers with non special displays may be related to the lack of immersion.

7 Conclusion and Future Works

Through the SPackageCreator3D it is possible to generate 3D SCORM packages arranging 3D virtual models and related content files, providing an improved teaching and learning environment that, besides visualization and interaction with these virtual models and additional files, is available into one of the most popular VLE platforms, the Moodle, making use of all of its features and advantages. SPackageCreator3D is available via Sourceforge project page[8].

The evaluation indicates the great potential of the usage of human body 3D virtual models to assist students to understand the anatomical structures and spatial relationship. It also pointed out that the virtual environment is useful and interactive, the students have the intention to continue using it, the importance of access anytime from anywhere using the Internet and Moodle features, and the need to improve the immersion experience.

As future work, it will be planned a knowledge acquisition evaluation of the students that have been using the 3D SCORM packages created by the SPackageCreator3D to assist their studies, the SPackageCreator3D solution maintenance, and improvements to provide a better immersion experience.

Acknowledgements. The authors would like to thank to CAPES Foundation, a body of the Brazilian Ministry of Education, for financial support. Luiz Soares dos Santos Baglie was recipient of scholarships from CAPES.

References

1. Adams, C.M., Wilson, T.D.: Virtual cerebral ventricular system: an MR-based three-dimensional computer model. Anat. Sci. Educ. **4**(6), 340–347 (2011)
2. Allen, L.K., Bhattacharyya, S., Wilson, T.D.: Development of an interactive anatomical three-dimensional eye model. Anat. Sci. Educ. **8**(3), 275–282 (2015)
3. Azkue, J.-J.: A digital tool for three-dimensional visualization and annotation in anatomy and embryology learning. Eur. J. Anat. **17**(3), 146–154 (2013)
4. Baptista, F.Q., Neto, M.P., Dias, D.R.C., de Paiva Guimaraes, M., Brega, J.R.F.: 3D content generation to moodle platform to support anatomy teaching and learning. In: 13th International Conference on Computer Systems and Applications (AICCSA) (2016)
5. Birr, S., Monch, J., Sommerfeld, D., Preim, U., Preim, B.: The liveranatomyexplorer: a webgl-based surgical teaching tool. Comput. Graph. Appl. IEEE **33**(5), 48–58 (2013)
6. Bleakley, A.: The curriculum is dead! Long live the curriculum! Designing an undergraduate medicine and surgery curriculum for the future. Med. Teach. **34**(7), 543–547 (2012)
7. Bochicchio, M.A., Longo, A.: Learning objects and online labs: the micronet experience. In: 2012 9th International Conference on Remote Engineering and Virtual Instrumentation (REV), pp. 1–7, July 2012
8. da Cruz, L.C., de Almeida Thomaz, V., de Oliveira, J.C.: Aicoh 3d: interactive atlas of human body. In: 2014 XVI Symposium on Virtual and Augmented Reality (SVR), pp. 24–27, May 2014
9. Codd, A.M., Choudhury, B.: Virtual reality anatomy: is it comparable with traditional methods in the teaching of human forearm musculoskeletal anatomy? Anat. Sci. Educ. **4**(3), 119–125 (2011)

[8] https://sourceforge.net/projects/spackagecreator3/.

10. de Azambuja Montes, M.A., de Souza, C.T.V.: Análise da taxa de reprovação na disciplina de anatomia humana em cursos da saúde (2007)
11. de Ribaupierre, S., Wilson, T.D.: Construction of a 3-d anatomical model for teaching temporal lobectomy. Comput. Biol. Med. **42**(6), 692–696 (2012)
12. Dias, D.R.C., Brega, J.R.F., Trevelin, L.C., Popolin Neto, M., Gnecco, B.B., de Paiva Guimaraes, M.: Design and evaluation of an advanced virtual reality system for visualization of dentistry structures. In: 2012 18th International Conference on Virtual Systems and Multimedia (VSMM), pp. 429–435, September 2012
13. Drake, R.L., McBride, J.M., Lachman, N., Pawlina, W.: Medical education in the anatomical sciences: the winds of change continue to blow. Anat. Sci. Educ. **2**(6), 253–259 (2009)
14. Hamrol, A., Górski, F., Grajewski, D., Zawadzki, P.: Virtual 3D atlas of a human body - development of an educational medical software application. Procedia Comput. Sci. **25**, 302–314 (2013). 2013 International Conference on Virtual and Augmented Reality in Education
15. Huang, H.-M.: A collaborative virtual learning system for medical education. In: 2011 3rd International Conference on Data Mining and Intelligent Information Technology Applications (ICMiA), pp. 127–130, October 2011
16. Huang, H.-M., Chen, Y.-L., Chen, K.-Y.: Investigation of three-dimensional human anatomy applied in mobile learning. In: 2010 International Computer Symposium (ICS), pp. 358–363, December 2010
17. Juanes, J.A., Ruisoto, P.: Technological advances and teaching innovation applied to health science education. In: Proceedings of the First International Conference on Technological Ecosystem for Enhancing Multiculturality, TEEM 2013, pp. 3–7. ACM, New York (2013)
18. Kitchenham, B., Brereton, O.P., Budgen, D., Turner, M., Bailey, J., Linkman, S.: Systematic literature reviews in software engineering - a systematic literature review. Inf. Softw. Technol. **51**(1), 7–15 (2009)
19. Kloos, C.D., Pardo, A., Organero, M.M., Ibáñez, M.B., Crespo, R., Merino, P.M., de la Fuente, L., Leony, D., Gutierrez, I.: Some research questions and results of UC3M in the eMadrid excellence network. In: 2010 IEEE Education Engineering (EDUCON), pp. 1101–1110, April 2010
20. Lu, J., Li, L., Sun, G.P.: A multimodal virtual anatomy e-learning tool for medical education. In: Zhang, X., Zhong, S., Pan, Z., Wong, K., Yun, R. (eds.) Edutainment 2010. LNCS, vol. 6249, pp. 278–287. Springer, Heidelberg (2010). doi:10.1007/978-3-642-14533-9_28
21. Melo, J.S.S., Brasil, L.M., Balaniuk, R., Ferneda, E., Santana, J.S.: Medical simulation platform (2011)
22. Meng, M., Fallavollita, P., Blum, T., Eck, U., Sandor, C., Weidert, S., Waschke, J., Navab, N.: Kinect for interactive AR anatomy learning. In: 2013 IEEE International Symposium on Mixed and Augmented Reality (ISMAR), pp. 277–278, October 2013
23. Noguera, J.M., Jiménez, J.J., Osuna-Pérez, M.C.: Development and evaluation of a 3D mobile application for learning manual therapy in the physiotherapy laboratory. Comput. Educ. **69**, 96–108 (2013)
24. Palomera, P.R., Méndez, J.A.J., Galino, A.P.: Enhancing neuroanatomy education using computer-based instructional material. Comput. Hum. Behav. **31**, 446–452 (2014)
25. Platt, A.: Teaching medicine to millennials. J. Phys. Assist. Educ. **21**(2), 42–44 (2010)

26. Neto, M.P., Agostinho, I.A., de Moraes, A.C., Brega, J.R.F., Dias, D.R.C., Weber, S.A.T.: Sistema de descrição semântica para visualização de modelos 3d da anatomia humana. In: Workshop de Realidade Virtual e Aumentada - WRVA 2013, Jataí, GO, Brazil (2013)
27. Neto, M.P., Sossai, I.A.B., Baptista, F., Santos, D., Braga, N.N., Weber, S., Brega, J.R.F.: Tecnologias na integração de ambientes virtuais tridimensionais e a plataforma de ensino e aprendizagem moodle. Revista do Simpósio Interdisciplinar de Tecnologias na Educação **1**, 148–156 (2015)
28. Prensky, M.: Digital natives, digital immigrants **9**(5) (2012)
29. Sander, B., Golas, M.M.: Histoviewer: an interactive e-learning platform facilitating group and peer group learning. Anat. Sci. Educ. **6**(3), 182–190 (2013)
30. Silén, C., Wirell, S., Kvist, J., Nylander, E., Smedby, O.: Advanced 3D visualization in student-centred medical education. Med. Teach. **30**(5), e115–e124 (2008)
31. Tworek, J.K., Jamniczky, H.A., Jacob, C., Hallgrímsson, B., Wright, B.: The lindsay virtual human project: an immersive approach to anatomy and physiology. Anat. Sci. Educ. **6**(1), 19–28 (2013)
32. Venail, F., Deveze, A., Lallemant, B., Guevara, N., Mondain, M.: Enhancement of temporal bone anatomy learning with computer 3D rendered imaging softwares. Med. Teach. **32**(7), e282–e288 (2010)

Workshop on Industrial Computational Applications (WIKA 2017)

Real-Time Vision Based System for Measurement of the Oxyacetylene Welding Parameters

Bogusław Cyganek[1(✉)] and Maciej Basiura[2]

[1] AGH University of Science and Technology,
Al. Mickiewicza 30, 30-059 Kraków, Poland
cyganek@agh.edu.pl

[2] Oil and Gas Institute - National Research Institute,
Ul. Lubicz 25A, 31-503 Kraków, Poland
basiura@inig.pl

Abstract. In the paper an original method of the oxyacetylene welding measurement and control is presented. The method is based on the computer processing of the flame images of the oxyacetylene torch. In this paper flame analysis is presented which is based on adaptive thresholding, statistical shape parameters computations, as well as color analysis of characteristic parts of a flame. The latter is done based on the proposed flame model. These parameters are then used as features in classification process. Thanks to this the proposed method is able to automatically determine in real-time parameters of a flame which can be used for automatic setup of the welding conditions.

Keywords: Computer vision · Oxyacetylene welding · Real-time image processing

1 Introduction

Welding is a process that joins metal materials by causing their fusion. This is achieved by melting of the base metal with simultaneous addition of the filler substance to the joint to form a pool of the molten material. After cooling, the parts are permanently bounded at the joint region. A necessity of melting of the metal parts makes welding different from the soldering or brazing processes [11, 12]. The common welding types are TIG (Tungsten Inert Gas), MIG (Metal Inert Gas), as well as oxyacetylene welding. The former two arc welding types became very popular due to many practical features. These became also highly automatized thanks to electronic arc control devices [19, 20]. Also, there are works on vision based pool shape control, as will be discussed. Nevertheless, there are applications in which oxyacetylene welding is still a necessity, such as oxy-acetylene cutting or metal depositing, bending or hardfacing, to name a few. However, the whole process, due to usage of the oxygen and acetylene gasses, produces high temperatures which are dangerous for an operating welder. Thus, there are many attempts to automatize this process as well, i.e. to design a system capable of oxyacetylene welding which does not require an operator at all or at least at the close proximity of the

O. Gervasi et al. (Eds.): ICCSA 2017, Part I, LNCS 10404, pp. 621–632, 2017.
DOI: 10.1007/978-3-319-62392-4_45

torch. One of the key building blocks of such a system is a control front-end equipped with a camera and a computer with proper software. In this paper we propose such a front-end, presenting details of the computer system for analysis of a type of a welding flame based on digital image analysis. More specifically, we propose an original method capable of the real-time analysis of a welding flame type in order to facilitate its control and an automatic settings. In further sections we discuss the three different flame types and present first architecture, then details of the image processing chain. In the rest of this section a short literature overview of the visual systems for control of flame, combustion and welding processes is presented.

Flame and combustion analysis with computer vision systems belongs to difficult tasks due to numerous factors. The first difficulty are dynamics of the combustion processes which requires special mathematical tools such as statistical methods and Markov chain modelling [6]. There are also environmental problems such as high temperature, sparks, to name a few. Fleury et al. [8] propose a computer vision system with Kalman filter to monitor nebulization quality of flames. Silva et al. analyze the problem of identification of the state-space dynamics of oil flames through computer vision and modal techniques [21]. Wang and Ren propose a vision based method of flame texture analysis for control of the rotary kiln combustion conditions [22]. Their method relies on computation of the grey-level co-occurrence matrix and application of the kernel PCA and neural network. On the other hand, the problem of pattern recognition methods in evaluation of the quality of gas fuel combustion is analyzed in the paper by Basiura [1]. Similarly, image flame analysis in order to make use of computer vision systems to evaluate and supervise the combustion process is presented in [2].

Visual computer systems are proposed to use in welding but mostly in the arc welding for monitoring of the pool geometry and melting. For instance, Gao and Wu present a method of experimental determination of weld pool geometry in gas tungsten arc welding based on a signal from the CCD camera equipped with a special filter [10]. After a simple edge position analysis, their system is able to compute the weld pool width, length and area. On the other hand, Kim and Park analyze an influence of welding parameters on weld bead in laser arc welding [16]. Their method is based on the coaxial monitoring system and image processing.

Lucas et al. present a computer vision technique to measure and control the upper surface of a weld pool size [18]. Their method is based on the image correlation technique and, as reported, can be used to operate with a number of different welding processes. For regulation a simple classical feedback control algorithm is proposed. Cyganek proposed methodology of using stereoscope system of cameras for weld pool analysis and welding control [7]. On the other hand, Jastrzebski et al. discuss theory of training of gas welders and solders as a key to copy welders motion into robots motion [15]. In their approach, a vision based goggle play important role in the training process.

Baskoro et al. propose a vision based computer system for welding speed control and monitoring [3]. Their method is suitable for the TIG process and allow good weld bead appearance, as reported.

2 Oxyacetylene Welding and Control of Its Parameters

The gas welding process is known and used for over a century. The apparatus for gas welding usually consists of an oxygen source and a fuel gas source, contained in the cylinders, as well as of two pressure regulators, two flexible hoses, and a torch [11, 12]. There are two basic types of a torch: a welding one and oxy-cutting one. Further on, a torch of appropriate size is chosen depending on a thickness of a material.

The combination of oxygen and acetylene results in a flame temperature over 3500 °C. This makes it appropriate for welding and cutting of different metals. In this research we concentrated exclusively on analysis of the oxygen-acetylene flames. Nevertheless, the proposed techniques can be used for other types of flames, used for welding, soldering or brazing, obtained with different combination of gases.

The flame region can be divided into three different color regions (see Fig. 1):

1. A long yellowish tip;
2. A blue middle section;
3. A whitish-blue intense section.

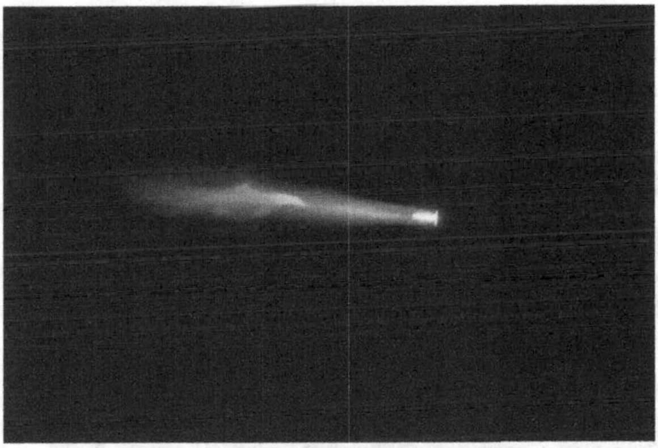

Fig. 1. An oxygen-acetylene flame with visible combustion regions of different shape and color. (Color figure online)

There are also three basic types of flames of a welding torch, as follows:

1. Oxidizing flame – in case of an excess of oxygen, the whitish-blue part of a flame becomes smaller than the blue part of a flame. This results in a higher temperature. A slightly oxidizing flame can be used in brazing, while a more strongly oxidizing flame in welding of certain brasses and bronzes.
2. Normal (neutral) flame – has no chemical effect upon the metal during welding. It is achieved by mixing equal parts oxygen and acetylene and is witnessed in the flame by adjusting the oxygen flow until the middle blue section and inner whitish-blue parts merge into a single region.

3. Reducing/carbonizing flame – a case with an excess of acetylene. The whitish-blue flame becomes larger than the blue part of a flame. The name of this type of flame comes from the fact that it contains white hot-carbon particles, which may be dissolved during welding. In effect, this type of flame removes oxygen from iron oxides in steel.

Figure 2 shows examples of the three basic types of the oxyacetylene flames.

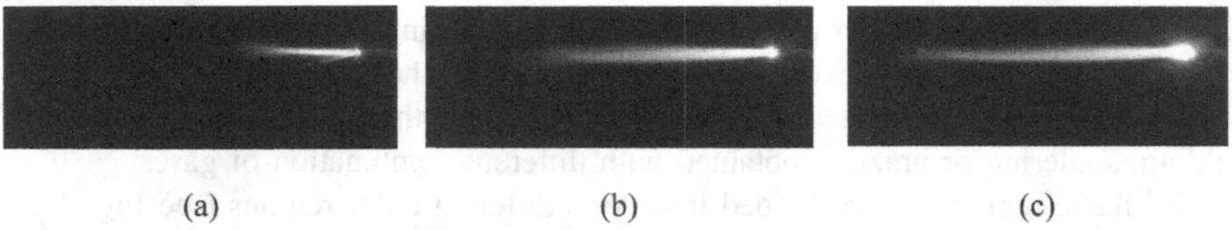

Fig. 2. Examples of the three basic types of flames of the oxyacetylene welding torch: oxidizing (a), normal (b), carbonating (c).

Last but not least, size of the flame naturally depends on a torch size and preset gas pressure. This is adjusted by a welder to fit thickness and type of the welded materials.

Figure 3 depicts our experimental setup. Visible is a welding torch with a flame, as well as the camera. For image registration a special markers were used – these are light points in the black background.

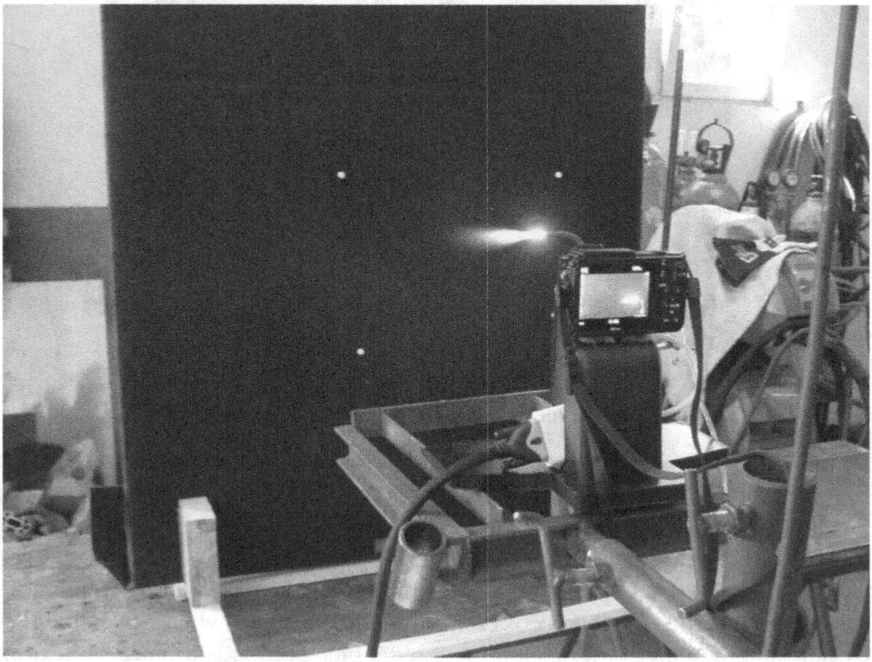

Fig. 3. Experimental setup. Visible a welding torch with a flame, as well as the camera. For image registration a special markers were used (light points in the black background).

In the experiments two types of cameras were used: Raspberry Pi camera with Sony IMX219 8-megapixel sensor, as well as Nikon COOLPIX AW130.

3 Computer Vision System for Welding Control

Figure 4 shows architecture of the proposed vision system for oxyacetylene flame control. Signal acquisition is done by a color camera module with preset gain and filter. In our experiments two different types of cameras were tested, as already described. The images were taken in a special setup equipped into the fiducial points which allow image registration. This facilitates detection of the torch and its end, as well as all parts of a flame. The characteristic feature of the flame are the combustion areas which are regions of a flame with different chemical characteristics and temperature. The latter usually rises, going from flame fringes (to the left) towards nose of a torch (right direction). This feature is used in our proposed method to build a so called intensity level-sets.

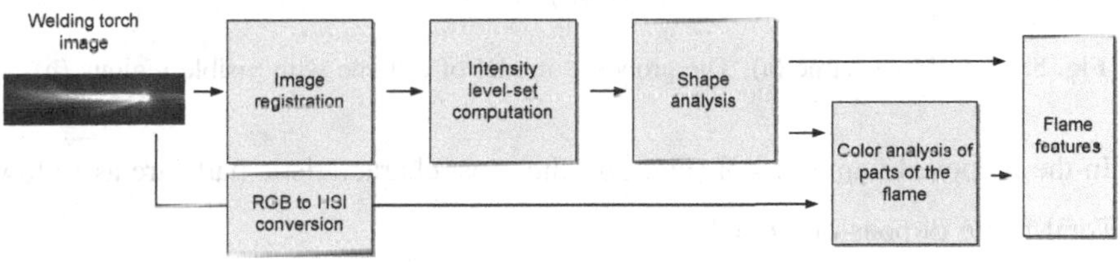

Fig. 4. Architecture of the proposed visual system for oxyacetylene flame analysis.

The aforementioned intensity level sets are of characteristic shape, which depends on a flame size and type. Thus, these are the features used to recognized flame type in our system. Shape analysis, as shown in Fig. 4, consists of measuring vertical and horizontal extensions of a shape, as well as computing its basic statistical shape features in a form of statistical moments, as will be described. Apart from the shape, also color in specific parts of a flame, plays an important role. In our system this is analyzed by computing two histograms of the hue (H) and saturation (S) channels of the HSI color space. All these constitute features which are used in flame measurement, as well as classification. For the latter the simple nearest-neighbor classifier is proposed.

3.1 Oxyacetylene Flame Model

As already mentioned previously, the oxyacetylene flames contain different combustion regions with temperature range 3000–3500 °C. There are characteristic flame regions from which usually three are well distinguished based on their shape and color distribution, as in the exemplary flame shown in Fig. 5a. These are: a long yellowish tip, a blue middle section, and a whitish-blue intense area (a kernel) close the torch nose. Based on this observation we propose a flame model shown in Fig. 5b.

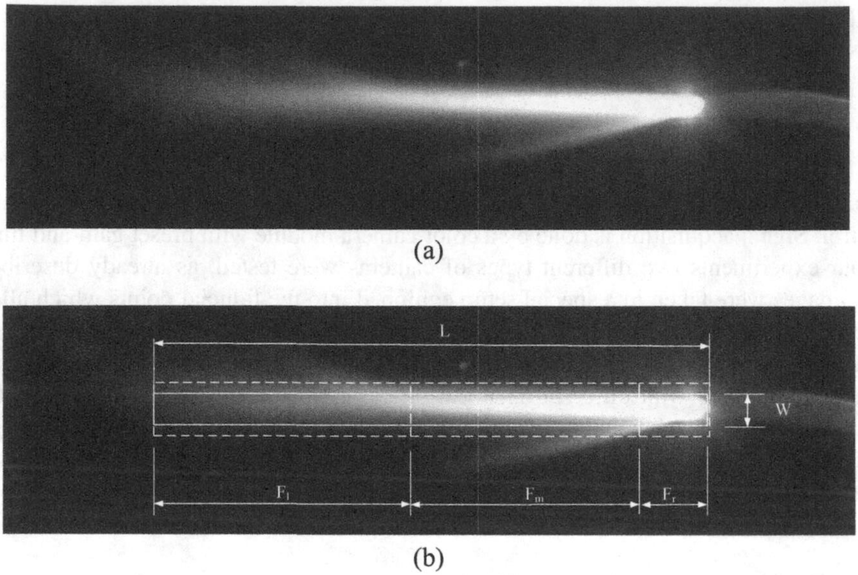

Fig. 5. A welding flame (a). The proposed model of a flame with visible regions (b).

In the proposed flame model (Fig. 5b.) the most characteristic parts are as follows:

1. Total flame dimensions $L \times W$;
2. Tip region F_l;
3. Middle region F_m;
4. Kernel region F_r.

Based on discussion with the expert welders we arbitrarily set F_l to $0.4\,L$, F_m to $0.4\,L$, and F_r to $0.2\,L$.

3.2 Intensity Level-Sets Computation and Statistical Moment Analysis

The key idea of the proposed visual method is to split a flame image into compact regions which correspond to areas of similar temperature due to similar chemical combustion processes. After converting a color image into its monochrome version, these are obtained by iterative thresholding in accordance with the following formula:

$$M_k(\mathbf{p}) = \begin{cases} 0 & if \quad I(\mathbf{p}) < \tau_k \\ 1 & if \quad I(\mathbf{p}) \geq \tau_k \end{cases}, \tag{1}$$

where $\mathbf{p} = (x, y)$ denotes coordinates of a pixel in the monochrome image I, and τ_k is a threshold at level k. However, although unusual, the above process can lead to many separate components. Therefore the map M_k is further processed and the largest connected component is further taken for further processing, as follows:

$$\hat{M}_k(\mathbf{p}) = \max_{\#M_k(\mathbf{p})} (M_k(\mathbf{p})), \tag{2}$$

where $\#M_k(\mathbf{p})$ denotes a number of connected points in the set M_k.

In our system k is in the range 7–15 and the threshold τ_k is changed automatically in the range 85–255 to form k equal steps. Certainly, these parameters depend on an expected type of flames, as well as camera exposition level. The latter should be chosen as to convey the maximal dynamic range of the observed scene, as will be further discussed. The intensity level-set method is depicted in Fig. 6. Their number and shape in high degree correspond to the size and type of a flame.

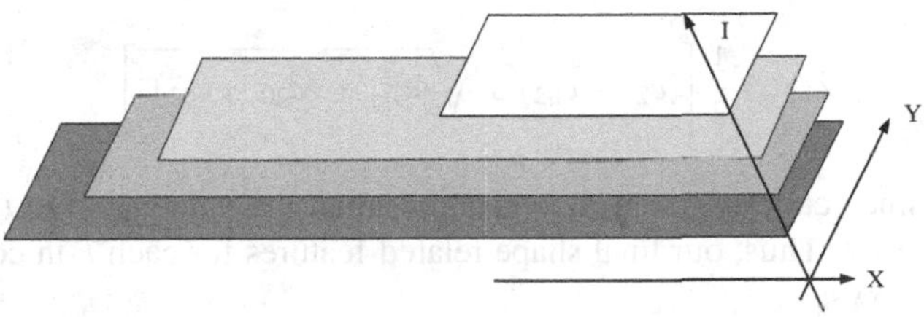

Fig. 6. Level sets of the intensity values extracted from the flame.

These way obtained shapes serve for computation of the statistical moments, which convey information characterizing the particular shape [6]. These constitute features which can be used during flame classification.

Computation of the statistical moment in a k-th connected component object obtained after (2), is defined as follows [6, 9, 17]:

$$m_{ab}^{(k)} = \sum_{x=1}^{R} \sum_{y=1}^{S} \hat{M}_k(\mathbf{p}) x^a y^b, \tag{3}$$

$$c_{ab}^{(k)} = \sum_{x=1}^{R} \sum_{y=1}^{S} \hat{M}_k(\mathbf{p}) (x - \bar{x})^a (y - \bar{y})^b, \tag{4}$$

where $\hat{M}_k$ is defined in (2) and R, S denote allowable coordinates of the connected component.

In the next step, a centroid point $(\bar{x}, \bar{y})$ is computed from the three moments defined in Eq. (3), as follows

$$\bar{x} = \frac{m_{10}}{m_{00}}, \quad \bar{y} = \frac{m_{01}}{m_{00}}, \quad \text{assuming } m_{00} \neq 0, \tag{5}$$

which is a center of gravity. Now, the so called inertia tensor can be computed as follows (here for clarity we skip the index "k") [6]

$$\mathbf{T} = \begin{bmatrix} c_{20} & -c_{11} \\ -c_{11} & c_{02} \end{bmatrix}. \tag{6}$$

A connected component can be further approximated by an ellipse with inertia tensor (6). Its parameters α, β, and φ are defined as follows

$$\alpha = 2\sqrt{\lambda_1},\ \beta = 2\sqrt{\lambda_2},\ \text{ and } \varphi = 0.5\, atan2(2c_{11}, c_{20} - c_{02}) \tag{7}$$

where function *atan*2 is an extended version of the arcus tangent function (see [6] for details), $\lambda_1 \geq \lambda_2$ are eigenvalues of the inertia tensor **T** which can be computed as follows

$$\lambda_{1,2} = \frac{1}{2}\left[(c_{20} + c_{02}) \pm \sqrt{4c_{11}^2 + (c_{20} - c_{02})^2}\right]. \tag{8}$$

However, since current system is registered, then the value of φ in (7) is skipped from the feature set. Thus, our final shape related features for each k-th component are proposed as follows:

$$\Theta_S^{(k)} = \left[m_{00}^{(k)}, \alpha^{(k)}, \beta^{(k)}\right]. \tag{9}$$

where their parameters are defined in (3) and (7), respectively.

3.3 Color Analysis of a Flame

Besides the shape, different regions of a flame are characterized by different color distributions. In the human visible spectrum, these are yellowish-red colors of the tip part F_l in our model, whereas the F_m is blueish. On the other hand F_r looks white-blueish. Thus, color plays important role, both for human welders, as well as for the automatic flame control system. In our method, after finding out the aforementioned three characteristic regions of a flame, color histograms are computed in the H and S channels independently but only for the region F_l obtained from the largest connected component object of a flame. A similar method was proposed to compute color distribution of the road signs [5]. Thus, color features are defined as follows

$$\Theta_C = [\{H\}, \{S\}]_B. \tag{10}$$

where $\{H\}$ and $\{S\}$ denote H and S histograms, respectively, computed for the chosen bin number B (in experiments we set $B = 64$). Finally, all features describing an oxyacetylene flame are proposed as follows:

$$\Theta = \left[\left\{\Theta_S^{(k)}\right\}_{1 \leq k \leq K}, \Theta_C\right]. \tag{11}$$

where K denotes a total number of obtained intensity level sets.

4 Experimental Results

The proposed visual front end was implemented in C++ in the Microsoft Visual 2015 IDE. The experiments were run on a laptop computer equipped with the Intel® Xeon® E-1545 CPU @2.9 GHz, 64 GB RAM, and OS 64-bit Windows 10. The *DeRecLib* library was used to provide the basic image processing procedures in C++ [14].

Our database was obtained in the welding workbench shown in Fig. 3. Six different torch settings were tested in these experiments: oxidizing, normal, carbonating for two different flame sizes. Each of these sets was limited to a 100 of color images. Thus, we tested 600 images. For comparison we made our database accessible from the Internet [13]. We conducted two types of experiments which can be characterized as follows:

1. Accuracy of flame dimensions measurements – results obtained by the system are compared to the manually measured ones; The purpose of this experiment is to answer if the system is able to automatically measure size of the flame;
2. Accuracy of flame recognition – in this case the *k*-nearest-neighbor simple classifier was tested (*k* was set to 1) with the Euclidean distance; That is, for each test images its *k* nearest neighbors were found. Then the winning class was reported; The purpose of this experiment is to check discriminating capabilities of the proposed features of the flames;

The employed *k*-NN classifier, although simple, in practice shows many useful features, such as simple implementation, fast operation on small and medium size datasets, and good accuracy, as shown for example in the sign recognition system [4].

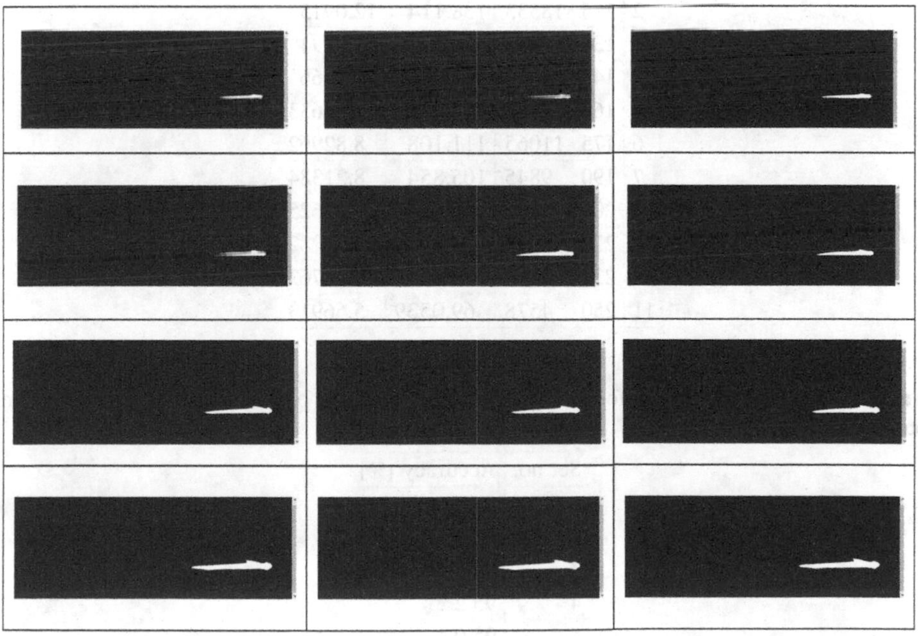

Fig. 7. Connected components from the intensity level-sets computed for the flame in Fig. 2b.

Figure 7 shows exemplary connected components from the intensity level-sets computed for the flame in Fig. 2b. On the other hand, Fig. 8 shows exemplary histograms of the tip region F_l from the same image. Finally, values of the statistical parameters are presented in Table 1. As expected, for the higher values of the threshold τ_k of the intensity level-set in (1), the flame region becomes gradually smaller.

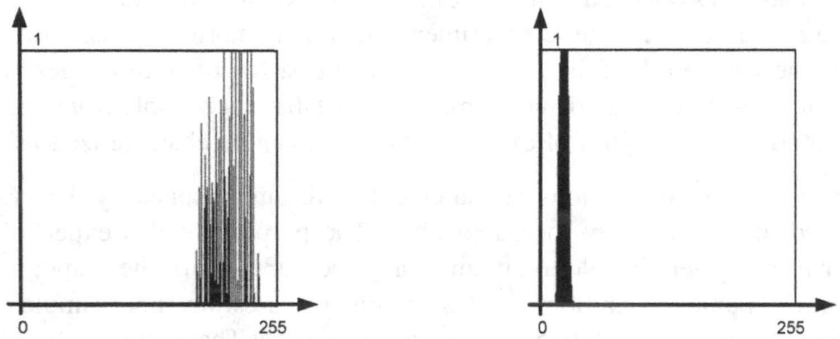

Fig. 8. H and S histograms of the tip region F_l of the flame from Fig. 2b.

Table 1. Exemplary statistical shape features for the flame from Fig. 2b.

k	τ_k	$m^{(k)}00$	$\alpha^{(k)}$	$\beta^{(k)}$
0	85	26193	157.902	15.0251
1	100	21735	148.047	13.2706
2	115	18533	138.114	12.0915
3	130	16017	129.293	11.0994
4	145	14023	122.354	10.2066
5	160	12460	116.907	9.47633
6	175	11065	111.108	8.82962
7	190	9845	105.854	8.21384
8	205	8706	100.934	7.58625
9	220	7538	94.3914	7.02282
10	235	6178	83.2511	6.58762
11	250	4578	69.0539	5.66913

Table 2. Classification results for all 6 groups of different type of the oxyacetylene flames from our database.

Set no.	Accuracy [%]
1	94.5
2	90.3
3	89.5
4	95.2
5	95.0
6	90.3

The measurement results show ±5 mm error which is acceptable in our system. Results of the second type of experiments are presented in Table 2.

It can be observed that the accuracy almost in all cases is above 90%. Some errors are due to similarity of some images among different classes of flames. However, these results show that the proposed system is capable of successful recognition of an oxyacetylene flames.

5 Conclusions

In the paper, a visual system for oxyacetylene flame measurement and classification is proposed. The system is aimed at automatic control of the oxyacetylene welding in order to release human welders from this dangerous process. The proposed system relies on real-time acquisition of the color images of the flames. The processing method is based on two paths. The first one is responsible for flame shape extraction and computation of its parameters. For this purpose a novel intensity level-set construction of the flame is proposed which relies on incremental properties of intensity of the oxyacetylene flame. Also useful is our proposed model of a flame which divides a flame into three characteristic regions. The obtained level-set shapes are then used to compute their characteristic statistical moments which constitute descriptive features. A second path assumes measurement of the color distribution of the tip region of the oxyacetylene flame. For this purpose two histograms of the H and S channels are computed. All these constitute the proposed flame features. The presented experiments show that this method offers discriminative features which can be used for flame classification. They can be also used for measurement of the flame dimensions. All these can be further used for automatic control of the flame and welding process. Further research will be concentrated on acquisition and processing of high-dynamic range images of the flames, as well as on development of further descriptive features.

Acknowledgement. This work was supported by the National Science Centre, Poland, under the grant no. 2014/15/B/ST6/00609.

References

1. Basiura, M.: Attempt to use pattern recognition methods to evaluate the quality of gas fuel combustion. NAFTA-GAZ, Rok LXXI, No. 5, pp. 314–319 (2015). (in Polish)
2. Basiura, M.: Image flame analysis in order to make use of computer vision systems to evaluate and supervise the combustion process. NAFTA-GAZ, Rok LXXII, No. 12, pp. 1119–1123 (2016). (in Polish)
3. Baskoro, A.S., Rahman, A.Z., Haikal: Automatic welding speed control by monitoring image of weld pool using vision sensor. ARPN J. Eng. Appl. Sci. **12**(4), 1052–1056 (2017)
4. Cyganek, B.: Recognition of road signs with mixture of neural networks and arbitration modules. In: Wang, J., Yi, Z., Zurada, Jacek M., Lu, B.-L., Yin, H. (eds.) ISNN 2006. LNCS, vol. 3973, pp. 52–57. Springer, Heidelberg (2006). doi:10.1007/11760191_8

5. Cyganek, B.: Soft system for road sign detection. In: Melin, P., Castillo, O., Ramírez, E.G., Kacprzyk, J., Pedrycz, W. (eds.) Analysis and Design of Intelligent Systems Using Soft Computing Techniques. Advances in Soft Computing, vol. 41, pp. 316–326. Springer, Heidelberg (2007). doi:10.1007/978-3-540-72432-2_32
6. Cyganek, B.: Object Detection and Recognition in Digital Images: Theory and Practice. Wiley, Hoboken (2013)
7. Cyganek, B.: An introduction to scene depth measurement with stereoscopic set of cameras. In: Przegląd Spawalnictwa, pp. 38–43 (2013)
8. Fleury, A.T., Trigo, F.C., Martins, F.P.R.: A new approach based on computer vision and non-linear Kalman filtering to monitor the nebulization quality of oil flames. Expert Syst. Appl. **40**(12), 4760–4769 (2013)
9. Flusser, J., Suk, T., Zitová, B.: Moments and Moment Invariants in Pattern Recognition. Wiley, Hoboken (2009)
10. Gao, J., Wu, C.: Experimental determination of weld pool geometry in gas tungsten arc welding. Sci. Technol. Weld. Joining **6**(5), 288–292 (2001)
11. https://en.wikipedia.org/wiki/Oxy-fuel_welding_and_cutting
12. https://en.wikipedia.org/wiki/Welding
13. home.agh.edu.pl/ ~ cyganek/OxAcDataBase_2017.zip
14. http://home.agh.edu.pl/ ~ cyganek/Projects.zip
15. Jastrzębski, R., Cyganek, B., Przytuła, J., Jastrzębska, I., Szczyrbak, K.: Theory of training of gas welders and solders as a key to copy welders motion into robots motion. Dozor Techniczny **4–5**, 74–82 (2014). ISSN 0209-1763. (in Polish)
16. Kim, T.W., Park, Y.W.: Influence of welding parameters on weld bead in laser arc hybrid welding process using coaxial monitoring system and image processing. Mater. Res. Innov. **18**(2), 898–901 (2014)
17. Klette, R., Rosenfeld, A.: Digital Geometry. Morgan-Kaufman, Burlington (2004)
18. Lucas, W., Bertaso, D., Melton, G., Smith, J., Balfour, C.: Real-time vision-based control of weld pool size. Weld. Int. **26**(4), 243–250 (2012). http://www.tandfonline.com/doi/abs/10.1080/09507116.2011.581336
19. Melton, G., Schuler, C., Houghton, M.: Laser diode based vision system for viewing arc welding, In: Eurojoin 7, Venice Lido, Italy (2009)
20. Mulligan, S.J.: Evaluation of a low-cost camera for monitoring arc welding processes. TWI Members Report 873 (2007)
21. Silva, R.P., Fleury, A.T., Martins, F.P.R., Ponge-Ferreira, W.J.A., Trigo, F.C.: Identification of the state-space dynamics of oil flames through computer vision and modal techniques. Expert Syst. Appl. **42**, 2421–2428 (2015)
22. Wang, J.-S., Ren, X.-D.: GLCM based extraction of flame image texture features and KPCA-GLVQ recognition method for Rotary Kiln combustion working conditions. Int. J. Autom. Comput. **11**(1), 72–77 (2014)

A Data-Mining Technology for Tuning of Rolling Prediction Models: Theory and Application

Francesco A. Cuzzola(✉) and Claudio Aurora

Danieli Automation SpA, Via Bonaldo Stringher 4, 33042 Buttrio, UD, Italy

Abstract. The realization of physical modeling of the rolling process is proposed as a material hardness virtual sensor and represents a valid tool for data exploration. The use of unsupervised clustering technology is here proposed and explored so as ease the material grouping process that might be strictly required for technological maintenance purposes.

Keywords: Data-exploration · Unsupervised clustering · Mathematical models calibration

1 Introduction

The mathematical approach applied to rolling mills for flat metal production can be considered a reliable tool for many purposes. However, the comparison between the predictions and the field feedback is an activity to be carried out with attention in order to guarantee the validity of the predictions in a wide range of working points: in [10], for instance, the authors propose a technology to verify by data-regressions the existence of a correlation between a whatever process variable and the prediction error and possibly automatically compensate it. In other words, the approach described in [10] aims at annihilating any systematic prediction error in front of any existing correlation *w.r.t.* any measured process variable: this approach is extremely efficient when the considered process variable has a *continuous* nature. Some of the inputs of the model have not a continuous nature but instead represent *discrete* inputs to the rolling model. The most important are:

1. the *Material Grade*, representing the quality of the rolled material;
2. the *Lubrication Type*, representing the quality of the lubricant used in the roll bite to guarantee the stability of the rolling process and influencing the roll bite friction μ.

Of course, this list of possible model discrete inputs is not necessarily complete (faults, work roll changes, wear, etc. could be included), but this issue is out of the scope of this paper. For instance, it is worth mentioning that it is a common situation that the same kind of material, if provided by different *suppliers*, may exhibit different hardening properties. For this reason it often happen that the material supplier is to be considered a non negligible discrete input of the rolling model.

The high cardinality of the realm of some discrete inputs is an important complexity factor to be considered in advance: for instance, in *steel* production the number of

O. Gervasi et al. (Eds.): ICCSA 2017, Part I, LNCS 10404, pp. 633–647, 2017.
DOI: 10.1007/978-3-319-62392-4_46

material grades is extremely large (near to 5000 for common steels only, see *e.g.* [2]); in the *aluminum* production the number of alloys is about 8000 (see *e.g.* [3]) and new alloys are continuously experimented and introduced in the market. This observation leads to the conclusion that when the *cardinality* of the material grades to be treated in the same plant is very large, then the tuning task of a mathematical model can turn out an extremely complex and time consuming activity for the human supervision.

Grouping of Materials. On one side, it is typical that a very large variety of material grades are to be processed in the same plant, on the other, many materials could turn out to be similar for the rolling process. This is due to the fact that *Material Grades* are identified through their commercial classification (grades are defined by international standards), but their hardness (*i.e.* their Yield Stress) depends on many factors [11–15] and consequently the rolling practice shows that different material grades may offer the same mechanical resistance to deformation despite the different material naming.

For this reason, the strategy of grouping affine materials into families or groups of materials aims at reducing the mentioned complexity. Indeed, first of all, it speeds up the model commissioning phase, by allowing for a more organic analysis of feedback data. Secondly, it allows for more efficient plant management, because any rolling practice is defined for few material groups rather than for all possible material grades. Unlucky, grouping of materials is not a trivial task. In particular, manual classification of grades is a time consuming and error prone activity. Of course, a wrong grouping might have a negative impact on both technological and production issues.

The paper is organized as follows. After introducing a self-tuning of the material Yield Stress model (Sect. 2), some basic concepts about the PDDP approach are recalled in Sect. 3 whereas the corresponding application technology to the material classification problem is the subject of Sect. 4. Before the conclusions, a real application is presented together with a methodology for performance evaluation, which is described in Sect. 5.

2 Self-tuning of Flow-Stress Models

In order to reduce the complexity of technological maintenance tasks for a rolling mill, the use of a mathematical modeling approach of the process is commonly proposed not only for setup and control optimization but also as a virtual sensor for evaluating the material hardness, see [10]. More, precisely once a consistent feedback from the rolling mill is collected the inversion of a rolling mill can be implemented so as to estimate the corresponding material *YS* (Yield Stress):

Algorithm 1:

$$[YS] = FindYSColdRolling(F, W, R, H_{en}, H_{ex}, S_{en}, S_{ex}, V, \mu)$$

where F is the Rolling Force, W the material Width, R the Work Roll Radius, H_{en}/H_{ex} the entry/exit material thickness, S_{en}/S_{ex} the entry/exit material Stresses, V the Rolling Speed and μ the Roll Bite Friction Coefficient.

For this reason, a database approach that allows recursively estimating the constitutive characteristics of a material can be proposed, see Fig. 1.

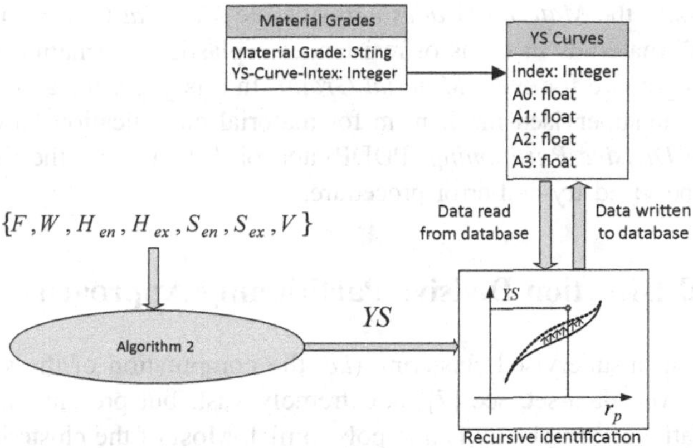

Fig. 1. Recursive identification of YS hardening curves by means of Algorithm 1.

In particular, as for the cold rolling case, in [10] the authors proposed to recursively estimate the Yield Stress hardening curve with an algorithm that has the following template:

Algorithm 2:
$[A, S] = TuningColdRolling(Material\ Grade, H_{ann}, F, W, R, H_{en}, H_{ex}, S_{en}, S_{ex}, V, \mu)$
Implementation:

1. Load from Database the coefficients $A = [A_0 \ \ \ldots \ \ A_3]$ corresponding to the *Material Grade* and the associated covariance matrix *S*.
2. Execute Algorithm 1, $[YS] :=$ $FindYSColdRolling(F, W, R, H_{en}, H_{ex}, S_{en}, S_{ex}, V, \mu)$.
3. Compute the progressive reduction: $r_p := 1 - H_{ex}/H_{ann}$.
4. Update the coefficients of Eq. (1) through a RLS regression:

$$[A, S] := RLS\left(A, S, r_p, YS\right).$$

5. Save $[A, S]$ into the database, correspondingly to the considered *Material Grade*, for the next use

   where the coefficients $A = [A_0 \ \ \ldots \ \ A_3]$ correspond to a polynomial representation of an hardening curve of this type:

$$YS = A_0 + A_1 r_p + A_2 r_p^2 + A_3 r_p^3 \tag{1}$$

   and $r_p = 1 - H_{ex}/H_{ann}$ is the *progressive reduction* and H_{ann} is the original thickness of the material after the last annealing. Of course, the same technology can be extended to other rolling cases.

The Algorithm 2, if used to estimate a *YS* hardening curve for every distinct *Material Grade*, can be produce conservative results and can be inefficient when the *cardinality* of the materials to be managed is very large. For this reason in the authors propose to classify the *Material Grades* into suitable *Material Groups*, i.e. *families* of "homogeneous" materials in terms of resistance to plastic deformation and then estimate a hardening curve for every *Material Group*. In this paper the authors would like to introduce an unsupervised mechanism for material classification based on a *Principal Direction Divisive Partitioning* (PDDP) approach that avoid the risks connected to a human supervised try-and-error procedure.

3 Principal Direction Divisive Partitioning Approach

The literature on unsupervised clustering (*i.e.* the computation of the splitting of the partition of a given data-set, see [7]) is extremely vast, but presents also algorithms whose computational complexity is non-polynomial. Most of the clustering algorithms available in the literature require the computation of the eigenvalues and eigenvectors of a covariance matrix. When the considered data-set has a large cardinality then this represents a computationally expensive activity that is substituted in PDDP (see e.g. [1, 4, 5]) by the execution of *Singular Value Decomposition* (SVD) of the same matrix.

Consequently, PDDP is here proposed as a suitable technology for automation system where real-time constraints need to be satisfied and therefore, in view of this, particularly efficient for the material classification problem: the data-set subject to clustering is the output of Algorithm 2 that is maintained in a database whose cardinality can be of considerable dimensions, for instance but not limited to, in case the catalog of material grades to be processed is huge. It is again remarked that, between all the plant-types involved in flat metal transformation, the *steel cold rolling* case (see [6, 16] and references therein) is adopted in the following as a reference case whereas the same approach can be easily extended to other cases.

3.1 The PDDP Technology and Its Application to Metal Hardness Testing During Rolling

The PDDP algorithm is applied to a data-set represented by a $n \times a$ matrix D whose rows d_i (with $i = 1 \ldots n$) are the data-samples. The result of the algorithm is a partition of matrix D represented by a number of matrices ("leafs") D_k with $k = 1 \ldots c$ $(c \leq n)$. The most common version of the PDDP algorithm has the following form:

Algorithm 3:

$$[c, D_k] = PDDP(D, c_{\max}).$$

It is worthwhile pointing out that, contrarily to many other applications, in the material classification problem the cardinality of the optimal partition (*i.e.* c) is hardly a-priori known since it varies from plant to plant. On the contrary, it is expected that a fair value for the input parameter $c_{\max}$ depends only on the metal type to be treated. Specifically, in the steel case, even if several thousands of *Steel Grades* are defined by the international standards, it is reasonable to distinguish no more than $c_{\max} = 20$ *Material Groups*, i.e. no more than 20 *YS* characteristic hardening curves.

4 Data Modeling and PDDP Application to Material Classification

4.1 Notation

- Given a generic vector v of dimension n then $v(i)$ represents the i-th element. Moreover, $M_{\pi}[v]$ denotes the following $n \times \pi$ matrix:

$$M_{\pi}[v] := \begin{bmatrix} 1 & v(1) & v(1)^2 & \dots & v(1)^{\pi} \\ \dots & \dots & \dots & \dots & \dots \\ 1 & v(n) & v(n)^2 & \dots & v(n)^{\pi} \end{bmatrix}$$

- The symbols $^+$ and ^{T} applied to a matrix will represent the More-Penrose pseudo-inverse and the transposed matrix, respectively.
- The operator $\lceil \cdot \rceil$ applied to a vector v represents the number of elements of the vector.
- Given a Boolean expression b then

$$\langle b \rangle := \begin{cases} 0 \text{ if } b \text{ is false} \\ 1 \text{ if } b \text{ is true} \end{cases}$$

4.2 The Steel Cold Rolling Case

Cold rolling implies several reduction steps to obtain the target thickness. In tandem mills, the total reduction is achieved by processing the strip in many (usually 5) consecutive stands (see *e.g.* [8]). In reversing mills, the target thickness is achieved by means of several rolling passes (usually from 1 to 9) through the same stand(s). So, independently of the plant installation, for every rolled coil the data-samples size is given by the number of reduction steps n_s. More precisely, after rolling each coil, the feedback is represented by the n_s-sized arrays of progressive reductions r_p and the corresponding YS estimations, see [9]:

$$YS_{\sigma} \text{ and } r_{p\sigma} \tag{2}$$

where $\sigma = 1 \dots n_c$ is the generic coil index, and n_c is the number of rolled coils (see Eq. (1)). The application of whatever PDDP technique on the raw coil feedback data (2) generally does not lead to a meaningful grouping since, for instance, it would not be possible to distinguish between materials that share a part of the hardening curve. Indeed, Fig. 2(A) shows two hardening curves having an intersection, but the achieved classification is technologically meaningless. Figure 2(B) represents a realistic situation where human intuition is immediately able to produce the correct classification into two material groups: the two characteristics are well defined, but a straightforward application of PDDP or whatever alternative technology known to the authors would be unsuccessful.

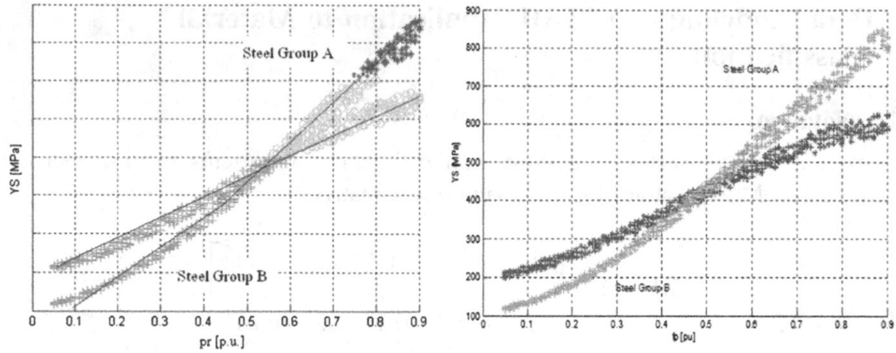

(A) (B)

Fig. 2. Benchmark example. The example refers to two materials whose hardening curves have an intersection. (A) is the result of the PDDP applied on raw data - the identified clusters are represented with different colors and symbols - (B) is the correct result obtained with the algorithm presented in the following sections. (Color figure online)

So as to achieve an effective PDDP-based technique, it is necessary to resort to a preliminary elaboration step. Since the association of each coil to a nominal *Steel Grade* is known then it is possible to build the following function

$$SG(\sigma) := \text{ Steel Grade Index for coil } \sigma$$

and, consequently, the raw data of Eq. (2) can be organized in the following vectors:

$$YS_{SG} \text{ and } r_{pSG} \tag{3}$$

where $SG = 1\ldots m$ is the generic steel grade index, m is the number of rolled steel grades and n_{RAW} is the length of these vectors. The organization of the data as in Eq. (3) can be exploited in order to derive a polynomial representation in the form of Eq. (1) by *Least Mean Squares*. More precisely, let define the vector A_{SG}^{π} of dimension $\pi + 1$ as follows:

$$A_{SG}^{\pi} := M_{\pi}\left[r_{pSG}\right]^{+} YS_{SG} \tag{4}$$

where $\pi = 1\ldots\pi_{\max}$ and $\pi_{\max} = 3$ like in Eq. (1). After data pre-elaboration, the PDDP Algorithm 3 can be executed by imposing as input the matrix $D := D_{\pi}$ where D_{π} is an $m \times (\pi + 1)$ where the generic row is given by the vector $A_{SG}^{\pi T}$. In the following, the results of each PDDP execution on matrix D_{π} will be denoted by:

$$\left[c_{\pi}, D_{\pi}^{g}\right] \tag{5}$$

where the integer c_{π} is the number of identified *Material Groups* (partitions), $g = 1\ldots c_{\pi}$ is the steel group index, D_{π}^{g} is a "leaf" matrix of dimension $m_g \times (\pi + 1)$, being m_g the number of *Steel Grades* associated to the group g, so that $\sum_{g=1}^{c_{\pi}} m_g = m$. The clustering results represented by (5) can be used to define the so-called *Material Groups* and the corresponding identifiers:

$$MatGroup_{\pi}(SG) := \{ j \text{ s.t. a row of matrix } D_{\pi}^{j} \text{ is } A_{SG}^{\pi} \} \in [1 \ldots c_{\pi}] \tag{6}$$

More precisely, the symbol $MatGroup_{\pi}(SG)$ is used in order to denote the *Material Group* index that contains the generic *Steel Grade SG*.

Note 1: Possible uncertainties associated to the choice of the polynomial order approximation π.

The application of the PDDP algorithm described in the previous section can lead to different results (see Eqs. (5)–(6)) according to the chosen value for π. In most cases an effective choice is represented by $\pi = 3$ but the PDDP algorithm can be executed with several values for π with a negligible elaboration-time. Then, if the results obtained from multiple runs with different choices of π are incongruent among them, a voting-based or similarity-based policy can be applied to the obtained results to automatically select the most reliable grouping. For the sake of simplicity, in the following it will be assumed that the index π is fixed.

5 Field Results and Performance Evaluation

In order to evaluate the fairness of the obtained material classification in *Material Groups*, it is necessary to re-organize the original data (3) used to perform the clustering. Therefore, it is necessary to build a number of vectors $r_{p\pi}^{g}$ and YS_{π}^{g} with $g = 1 \ldots c_{p}$:

$$\begin{cases} r_{p\pi}^{g} & := \bigcup_{SG=1}^{m} r_{pSG} \langle MatGroup_{\pi}(SG) = g \rangle \\ YS_{\pi}^{g} & := \bigcup_{SG=1}^{m} YS_{SG} \langle MatGroup_{\pi}(i) = g \rangle \end{cases} \tag{7}$$

The obtained vectors contain all the raw data-points corresponding to material grades that belong to the same group. The classification is to be considered fair if the reorganized data (7) do not present dispersion, with respect to their centroid, significantly higher than the dispersion associated to the original data (3). In order to evaluate the dispersion of a set of data with respect a centroid curve, the following performance index is proposed:

$$J\left(r_{p}, YS\right) = Mean\left(\left|YS - M_{3}\left[r_{p}\right]\hat{A}\right| ./YS\right) \text{ where } \hat{A} := M_{3}\left[r_{p}\right]^{+} YS \tag{8}$$

where the symbol ./ represents the element-wise division between vectors.

The coefficients $\hat{A}$ of Eq. (8) represent the centroid curve of the analyzed data $\left(r_{p}, YS\right)$. The cost function $J\left(r_{p}, YS\right)$ is nothing but a normalized average distance of the selected data with respect to the centroid curve obtained with a 3*rd* order polynomial regression (the 3*rd* order is considered a fairly good approximation of an hardening curve with the Roberts's method of Eq. (1), see [6]). So as to have an evaluation of the fairness of the obtained classification, the following criterion is desired:

$$\forall g = 1\ldots c_p, \frac{J\left(r^g_{p\pi}, YS^g_\pi\right)}{\underset{SGroup_\pi(SG)=g}{Max} J\left(r_{pSG}, YS_{SG}\right)} \leq (1+\varepsilon) \tag{9}$$

where ε is an arbitrarily small positive coefficient representing the tolerated deterioration in terms of data-dispersion when the grouping is executed. Notice in passing that, if $x\%$ is the desired accuracy of the rolling model to be tuned (*i.e.* the error between the predicted and the real rolling force is desired to be lower-equal than $x\%$), then ε can be set to $0.0x$. In other words, acceptance criterion expressed by Eq. (9) states that the data-dispersion obtained after grouping $J\left(r^g_{p\pi}, YS^g_\pi\right)$ cannot be excessively bigger than the data dispersion without any grouping.

5.1 Field Results

In this section, the results obtained with field data collected from a *Double Stand Cold Reversing Mill* (2SCRM, see e.g. [8]), are presented. Such a plant type is equipped with 3 XRay gauge-meters to measure the material thickness (installed at the entry/exit side of each stand). The 2SCRM is a reversing mill: it executes 1 or 2 (very seldom 3 or 4) rolling passes for each coil to reach the final target thickness (its minimal value is around 0.15 mm for steel production). This implies that for every coil the vectors of Eq. (2) have a dimension varying from 2 to 8, corresponding to the reduction steps executed (*i.e.* 2 reduction steps for each rolling pass).

Figure 3 presents the results obtained from Algorithm 2 with a data-log (see Eq. (2)) collected during 3 months of production: in the considered case the plant

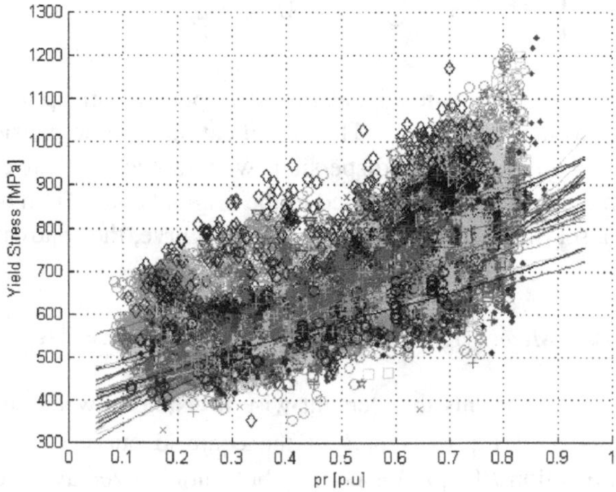

Fig. 3. Presentation of YS estimations obtained from Algorithm 2 with a 3-months data log. (Color figure online)

treated about 30 different steel grades that in some cases can be assimilated to the same material qualities and therefore it correspond to a case for which a material grouping activity is meaningful. On the other hand, the hardest material reaches a *YS* value of 1200 MPa that is about twice the *YS* value reached at the same reduction by the softest material. Therefore the classification cannot be trivial and, due to the data-set dimension, difficult to be executed through human supervision only.

It is worth pointing out that the considered plant actually manages a significantly higher amount of material grades in a larger time-span: for the sake of clarity the analysis here proposed is focused on the 3 months data-log presented in Fig. 3 *i.e.* on a

Table 1. Presentation of the coefficients A_{SG}^{3} derived from Eq. (4).

Material database index	A(0) [MPa]	A(1) [MPa]	A(2) [MPa]	A(3) [MPa]
57	182.36	190123.22	−34903201.03	2593205318.92
59	518.79	−23543.54	9075914.73	−171177639.78
76	469.33	21959.23	−2626995.87	701889827.95
23	62.95	318287.01	−67550608.79	5262688942.77
67	325.24	90707.93	−14382357.28	1347931768.69
28	340.62	111796.80	−22887196.10	2156921953.68
38	675.34	−85836.90	22245403.76	−994504873.30
52	744.27	−166303.32	43829938.82	−2768913892.34
74	329.15	103320.07	−17053896.80	1506859429.75
71	491.88	−30111.05	9813177.82	−165125836.18
61	438.10	41097.66	−5523060.84	881868015.21
22	437.35	37440.36	−7171447.03	1094421667.56
72	478.50	−52288.26	19118040.85	−1035821497.22
68	404.07	56591.01	−11939330.33	1410991629.43
69	400.64	36638.49	−1640931.14	343881758.64
56	−1774.78	1560109.19	−330492112.60	22699448416.22
65	501.69	−23586.29	9644312.57	−257898092.75
26	1427.76	−666021.68	155003568.74	−10633894568.52
7	386.30	25733.94	2215781.20	−134370445.61
75	426.21	28063.03	1334418.46	46344618.91
51	60.35	332880.41	−76552937.73	6231281832.84
55	535.96	−121896.35	50628085.89	−4466804360.50
58	561.49	−92319.82	30733044.17	−2165495108.04
66	626.44	−130753.13	38164613.79	−2392844524.83
27	−75.06	355045.44	−66345526.68	4550950428.30
29	542.51	−55979.83	25952468.64	−2068155904.99
63	886.47	−329337.24	82823255.25	−5845046669.56
64	263.70	139498.29	−23123231.67	1590343626.20
30	893.26	−316396.76	91336426.39	−6984388133.41
24	695.00	−219399.63	71311511.59	−5929988277.34
73	183.23	322163.28	−74162233.95	6106756374.06

case with a reduced production variety. In Table 1 the list of materials inside the considered production data-log is presented together with the coefficients A^3_{SG} obtained from the regression of Eq. (4).

The PDDP algorithm described in the previous sections can be fed with the coefficients A^{π}_{SG} with π that can assume a value from 1 to 3 and a material classification is obtained. The classification result is graphically depicted in Fig. 4 for the cases $\pi = 1$ and $\pi = 2$. Every group is represented with a corresponding symbol and color. In all the executed PDDP runs, the unsupervised clustering leads to the definition of 8 main *Material Groups* over 30 *Steel Grades* contained in the production data-log. In order to evaluate the classification obtained the dispersion performance coefficients $J\left(r^g_{p\pi}, YS^g_{\pi}\right)$ and $J\left(r_{pSG}, YS_{SG}\right)$ are computed for all the considered *Material Groups* and *Steel Grades* (see Table 2). As it can be seen, the acceptance criterion (9) is satisfied with $\varepsilon = 0.08$ that is fairly acceptable for most purposes.

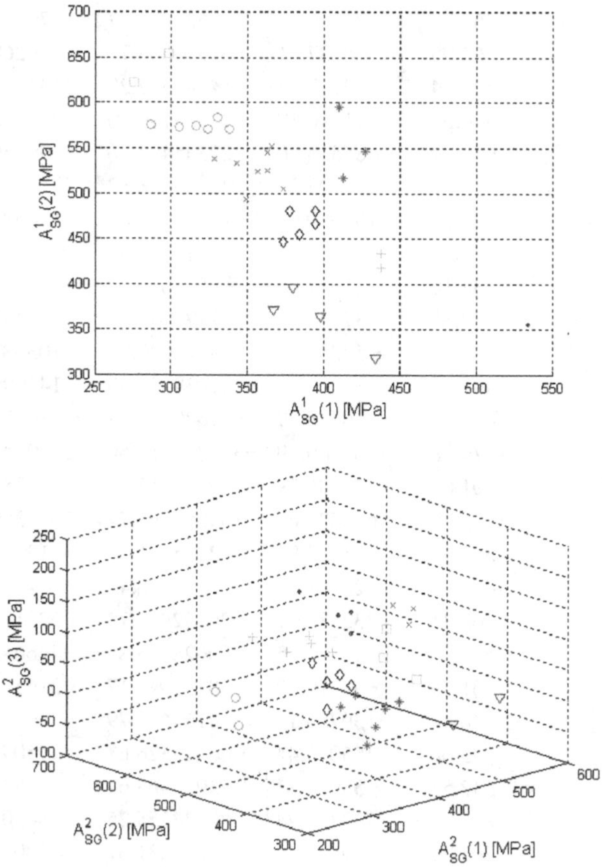

Fig. 4. Presentation of the clustering algorithm result with 30 different steel grades (for $\pi = 1$ and $\pi = 2$) – every group is associated to a specific symbol (e.g. 'Δ', '□', 'O', '•' and '*') and color. (Color figure online)

Table 2. Verification of classification validity through evaluation of data-dispersion.

Material group Id (# of steel grades)	$J\left(r^g_{p\pi}, YS^g_\pi\right)$	List of contained steel grades	$J\left(r_{pSG}, YS_{SG}\right)$ for each steel grade
1 (2)	0.0598	26	0.0615
		29	0.0562
2 (1)	0.0436	73	0.0436
3 (4)	0.0629	56	0.0621
		7	0.0628
		63	0.0513
		64	0.0589
4 (3)	0.0482	38	0.0460
		52	0.0342
		30	0.0431
5 (2)	0.0621	28	0.0488
		71	0.0612
6 (6)	0.0644	57	0.0485
		76	0.0541
		61	0.0602
		65	0.0532
		55	0.0526
		66	0.0528
7 (6)	0.0495	67	0.0424
		74	0.0431
		22	0.0431
		72	0.0273
		68	0.0467
		27	0.0445
8 (7)	0.0609	59	0.0434
		23	0.0512
		69	0.0449
		75	0.0437
		51	0.0333
		58	0.0479
		24	0.0560

5.2 Computational Complexity

As mentioned previously, one of the most important parameter to be evaluated for the selection of an unsupervised clustering technique is represented by the computational complexity. It is quite important to make available to the process supervisor a result in a real-time way so as to leave him the possibility to explore effectively several grouping scenarios.

Table 3 presents the computational time required by a MATLAB implementation based on the "`svd`" command. As it can be seen the computational time is almost linearly growing with the amount of the considered *Steel Grades*. This represents a promising result because it goes in the direction of getting a result in a predictable amount of time.

Table 3. Computational time with a MATLAB implementation (running on a PC with an Intel Core Duo CPU @3.16 GHz).

Number of steel-grades	Computational time for $\pi = 1$ [secs]	Computational time for $\pi = 2$ [secs]
30	2.3886	4.7492
60	4.3722	8.0049
100	5.9641	12.3903
200	12.9490	28.0428

5.3 The Implementation

The proposed unsupervised clustering technology leads to a result that can be validated by means of a visual application like to the one presented in Fig. 6.

The process of grades classification is initialized by the user with the button *"Grades Analysis"* that corresponds to loading of all the available feedback data and the execution of calculation $\hat{A} := M_3 \left[r^g_{p\pi} \right]^+ YS^g_\pi$ (identification of Yield Stress Curves, see Eqs. (7) and (8)) for all the Material Grades. The page allows the user to visually inspect the identified Yield Stress Curves corresponding to all the Material Grades and all the Grade Groups, together with the feedback points of Eq. (3). This preliminary analysis phase does not perform yet any re-classification of the materials but it just aims at evaluating the current classification.

The reclassification through PDDP algorithm is started by pressing the button *"Automatic Classification":* the operator decides the maximum number of Grade Groups. The algorithm produces a set of "suggested" Grade Groups. In other words, the result of the algorithm consists of a set of suggestions for the users: a list of already existing groups to be maintained or updated, a set of new groups to be created with respect to the current classification and, finally, a set of groups which according to the new classification will not be anymore used and that can be removed from the Production Database (Fig. 5).

The graphic tool allows the user to fully inspect the results of the automatic classification and to make manual adjustments before validating the classification. In particular, the visual tool shows: (1) for each Material Grade the current assigned Grade Group and the target one (*i.e.* the one suggested by the algorithm), (2) for each Grade Group the current Yield Stress Curve and the new one that, in turn, depends on the list of materials associated to the group. At the end of the validation process, the operator can press the button *"Apply Changes"* which saves the classification into the technological database.

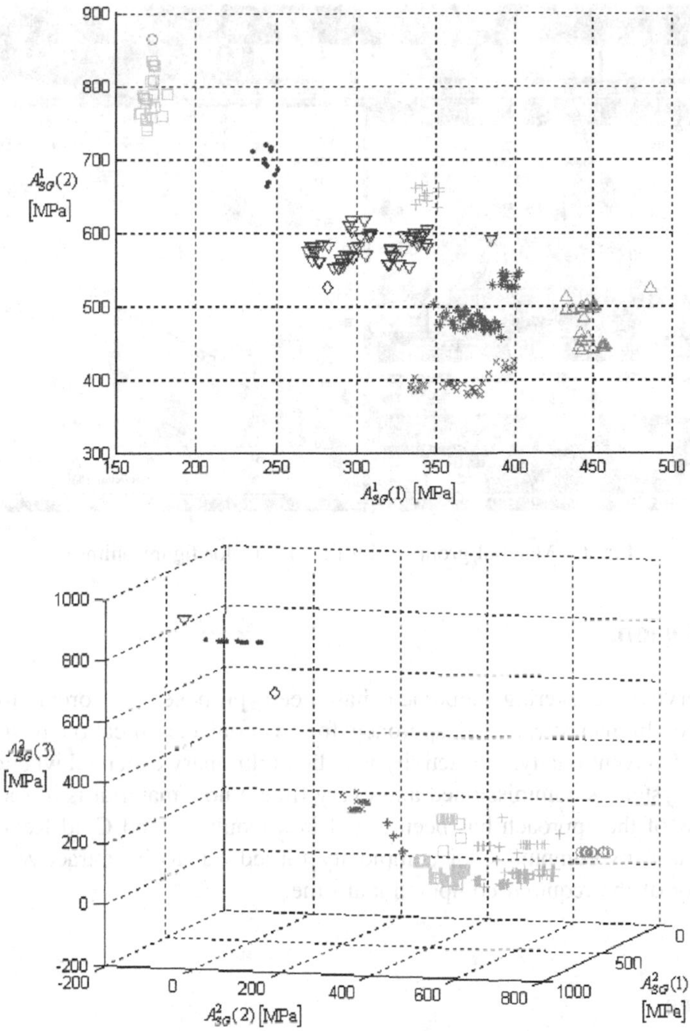

Fig. 5. Presentation of the clustering algorithm result with 200 different steel grades (for $\pi = 1$ and $\pi = 2$) – every group is associated to a specific symbol (e.g. 'Δ', '□', 'O', '•' and '*') and color. (Color figure online)

Some color–codes are introduced to highlight the suggested changes or the need of a manual interventions by the user: for instance, materials with no feedback data available are shown in grey color; materials and groups which are planned to be re-classified are shown in yellow; material grades for which the automatic classification algorithm cannot suggest a target grade group (*e.g.* because the available feedback data points are not enough) are presented in red.

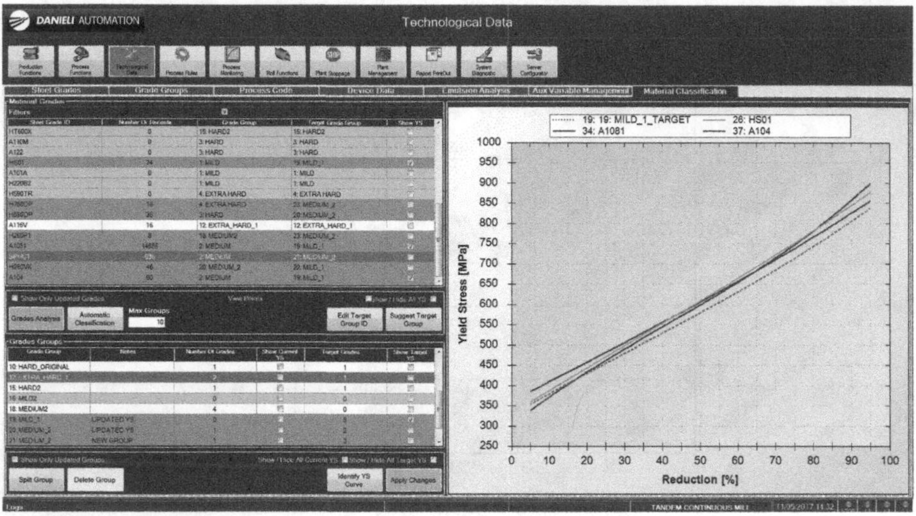

Fig. 6. Material grouping by PDDP. (Color figure online)

6 Conclusions

An unsupervised clustering approach has been proposed in order to avoid a trial-and-error human supervised approach for material classification in a flat metal rolling mill. Conventionally, this activity is to be preliminary executed whenever a new automation system is commissioned and every time a new material is introduced. The effectiveness of the approach has been tested in a Double Stand Cold Reversing Mill installation and its computational complexity turned out to be attractive due to the predictability of the required computational time.

References

1. Boley, D.: Principal direction divisive partitioning. Data Min. Knowl. Discov. **2**(4), 325–344 (1998)
2. http://en.wikipedia.org/wiki/Steel_grades
3. http://en.wikipedia.org/wiki/Aluminium_alloy
4. Tasoulis, S.K., Tasoulis, D.K.: Improving principal direction divisive clustering. In: 14th ACM SIGKDD International Conference on Knowledge Discovery and Data Mining, Workshop on Data Mining using Matrices and Tensors, Las Vegas, USA (2008)
5. Teukolsky, S.A., Vetterling, W.T., Flannery, B.P.: Support Vector Machines, Numerical Recipes: The Art of Scientific Computing, 3rd edn. Cambridge University Press, New York (2007)
6. Roberts, W.L.: Cold Rolling of Steel. Marcel Dekker, New York (1978)
7. Loiacono, D., Marelli, A., Lanzi, P.L.: Support vector regression for classifier prediction. In: Genetic and Evolutionary Computation, London (2007)

8. Cuzzola, F.A., Parisini, T.: Automation and control solutions for flat strip metal processing. In: Levine, W. (ed.) The Control Handbook. CRC Press, Boca Raton (2010)
9. Cuzzola, F.A., Aurora, C., Sclauzero, D.: An unsupervised clustering approach for yield stress prediction during flat rolling. In: IFAC Workshop on Automation in the Mining, Mineral and Metal Industries, Gifu, Japan (2012)
10. Aurora, C., Cuzzola, F.A.: Use of offline computational tools for plant data analysis and setup model calibration: a perspective in the industry of flat metal production. In: 39th International Convention on Information and Communication Technology, Electronics and Microelectronics (MIPRO) (2016)
11. Lee, Y.: Constitutive relation of alloy steels at high temperatures. Int. J. Precis. Eng. Manuf. **6**(4), 55–59 (2005)
12. Misaka, Y., Yoshimoto, T.: Formulation of mean resistance of deformation of plain carbon steel at elevated temperature. J. Jpn. Soc. Technol. Plast. **8**, 414–422 (1967)
13. Andorfer, J., Auzinger, D., Hriberning, G., Hubmer, G., Luger, A., Schwab, P.: Full metallurgical control of the mechanical properties of hot-rolled strip - a summary of more than two years operational experience. In: Palmiere, E., et al. (ed.) International Conference on Thermomechanical Processing: Mechanics, Microstructure and Control, pp. 164–168. Sheffield (2002)
14. Lissel, L., Engberg, G.: Prediction of the microstructural evolution during hot strip rolling of Nb microalloyed steels. In: Proceedings of the Third International Conference on Recrystallization and Grain Growth, Jeju Island, Korea, pp. 1127–1132 (2007)
15. Herrmann, K., Irle, M.: IMPOC: an online material properties measurement system. In: Ginzburg, V.B. (ed.) Flat-Rolled Steel Processes Advanced Technologies, pp. 265–270. CRC Press, Boca Raton (2009)
16. Garber, E., Traino, A., Kozhevnikova, I.: Novel mathematical models for cold-rolling process. In: Ginzburg, V.B. (ed.) Flat-Rolled Steel Processes Advanced Technologies, pp. 179–190. CRC Press, Boca Raton (2009)

Workshop on Web-Based Collective Evolutionary Systems: Models, Measures, Applications (IWCES 2017)

Structural and Semantic Proximity in Information Networks

Valentina Franzoni[1,2(✉)] and Alfredo Milani[1,3]

[1] Department of Mathematics and Computer Science, University of Perugia, Perugia, Italy
{valentina.franzoni,milani}@dmi.unipg.it

[2] Department of Computer, Control and Management Engineering, University of Rome La Sapienza, Rome, Italy

[3] Department of Computer Science, Hong Kong Baptist University, Kowloon Tong, Hong Kong

Abstract. This research includes the investigation, design and experimentation of models and measures of semantic and structural proximity for knowledge extraction and link prediction. The aim is to measure, predict and elicit, in particular, data from social or collaborative sources of heterogeneous information. The general idea is to use the information about entities (i.e. users) and relationships in collaborative or social repositories as an information source to infer the semantic context, and relations among the heterogeneous multimedia objects of any kind to extract the relevant structural knowledge. Contexts can then be used to narrow the domains and improve the performances of tasks such as disambiguation of entities, query expansion, emotion recognition and multimedia retrieval, just to mention a few.

There is thus the need for techniques able to produce results, even approximated, with respect to a given query, for ranking a set of promising candidates. Tools to reach the rich information already exist: web search engines, which results can be calculated with web-based proximity measures. Semantic proximity is used to compute attributes e.g. textual information. On the other hand, non-textual (i.e. structural, topological) information in collaborative or social repositories is used in contexts where the object is located. Both web-based and structural similarity measures can make profit from suboptimal results of computations. Which measure to use, and how to optimize the extraction and the utility of the extracted information, are the open issues that we address in our work.

Keywords: Group similarity · Semantic distance · Data mining · Collective knowledge · Knowledge discovery

1 Introduction: Problem Description and Potential Impact

With the pervasive diffusion of social digital interactions, network analysis became a critical task, as an increasing amount of economical, personal and physical transactions is digitally mediated, recorded and monitored, and can be used to predict and influence the way people or entities interact, work and, in general, behave. A huge quantity of

O. Gervasi et al. (Eds.): ICCSA 2017, Part I, LNCS 10404, pp. 651–666, 2017.
DOI: 10.1007/978-3-319-62392-4_47

information is generated daily, using heterogeneous multimedia data, by the spontaneous collaborative activity of whom we will just call users. For instance, images, videos or tweets are put by users in the Cloud, through social (e.g. Facebook) or collaborative (e.g. Wikipedia) platforms on the Internet, eventually using devices that add automatic metadata, e.g. date or geolocation. Often, data are stored in a semi-structured form following, or not, standards or recommended practices. Typical activities performed by users on such data, e.g. sharing, selecting, commenting, tagging and linking, reflect dynamic social relationships, from which to extract latent knowledge, valuable for several applications, e.g. advanced recommendation, filtering, or retrieval systems. Many systems in nature can be modeled as complex networks i.e., structures consisting of nodes and vertices connected by link or edges. Complex network science is applied in various disciplines, offering a common language to let them interact, e.g. computer science, sociology, biology, co-authorship networks. What distinguish network science from graph theory is the empirical nature, focused on data: each tool developing abstract mathematical theories is tested on real data, where the value is in the system's structure or evolution (Chiancone 2016).

Social network analysis, as it is in Artificial Intelligence, is a novel field, but its impact is all around us: starting from the economic impact, e.g. in web search, which relies on the network characteristics of the Internet, or in recommendation advertising, going on to the management impact, which is based on the structure of organizations and has the aim of increasing productivity, to health or security management, where it is possible to predict and augment information through diffusion schemes, and apply prevention strategies. Our research includes the investigation, design and experimentation of models and measures of semantic and structural proximity for knowledge extraction and link prediction. The aim of this research is to measure, predict and elicit, in particular, data from social or collaborative sources of heterogeneous information. The general idea is to use collaborative/social repositories with the information about entities/users and relationships as an information source to infer semantic knowledge and relations among heterogeneous/multimedia objects of any kind (Liu et al. 2012), with the aim of extracting the relevant semantic context. (Dey 2001) Contexts can then be used to narrow the domains and improve the performances of tasks such as disambiguation of entities, query expansion, (Franzoni et al. 2013b; Natsev et al. 2007) emotion recognition and multimedia *retrieval*, just to mention a few. There is thus the need for techniques able to produce results - even approximated - with respect to a given query, for ranking a set of promising candidates. (Lieberman 1995) Tools to reach the rich information already exist: web search engines, which results can be calculated with web-based proximity measures (Bollegala et al. 2011; Cilibrasi and Vitanyi 2004; Turney 2001), that compute especially textual information, and collaborative/social repositories which store relationships among objects/concepts, social media, or users (Dey and Abowd 1999), which non-textual (i.e. structural, topological) information can be used in contexts where the object is located (Milne and Witten 2008). Both web-based and structural similarity measures make profit from suboptimal results of computations, where approximations to evaluate frequencies and probabilities can be used to calculate semantic proximity or distance. Which measure to use, and how to optimize the extraction and the utility of the extracted information, are open issues. The ability of understanding the relevant semantic contexts underlying a situation is a major issue for machine intelligence applications. Many proposed techniques of context generation for web-applications (Ceri et al. 2007; Schilit et al. 1994; Dey 2001) rely on analysis and observation of explicit user actions, interactions and queries. However, only few seed concepts are usually available from direct user observations. Let an example explain that, in a search engine scenario where few search terms, say `cat` and `mouse`, are initially known: the goal being to generate the underlying contextually implied concepts, say e.g. `predator`, `catch`, `eating`. In order to define a framework for semantic context generation, a suitable knowledge source is needed, providing semantic relationships among concepts, and a suitable technique for extracting news concepts related, relevant and consistent with respect to the initial seed context. In text analysis, starting from the assumption that concepts with high co-occurrence in similar contexts are in relation, algebraic models are applied, in order to represent textual documents as vector of indexed terms (e.g. VSM, LSA) (Turney 2001) and weighting functions (e.g. tf-idf). A promising approach proposes to define web-based semantic proximity measures (e.g. NGD, PMI) (Cilibrasi and Vitanyi 2004), which use the estimation of probability distributions, obtained by queries on a search engine, without the need of an *in-deep* analysis of computationally heavy corpora. Based on a classical content-based approach, similarity is evaluated through the extraction of vectors of low-abstraction-level features, numeric or qualitative, extracted from the multimedia object (e.g. colour distribution, sound peaks), to which proximity measures are then applied. This approach presents computational limits (e.g. in case of large repositories) and difficulties in the extraction of high-level semantic features. In order to overcome the limits of this perceptive approach, state of the art annotation techniques focus on the optimization of the analysis of the (textual or non-textual) semantic context in which the object is located, measuring proximity of similar entities. These approaches usually have the limitation of not comparing objects of different abstraction level and size, focusing on a word-to-word or sentence-to-sentence measurement (Navigli et al. 2016). This paper capitalizes all the works of the authors on the topic of last three years, refines the unifying vision and details definitions.

Among the contributions of this work there are:

(a) the extension to Context-based Image Semantic Similarity of the comparison of semantic proximity measures for semantic annotation and the development of a new group distance, to express the semantic distance among objects of any size. A new class of similarity measures is thus defined, varying with respect to the different combination scheme and the different relatedness measure used for the elementary concepts, where an extended quantitative evaluation by experiments compared human similarity evaluation to similarity evaluation obtained with the proposed Context Similarity;

(b) the study of distances based on random walk, which are computationally efficient and robust for the identification of the semantic context of an object/entity in large and unknown graphs; the improvement of the Heuristic Semantic Walk (HSW) for context extraction from multipath chaining, in particular defining the path filter for the path aggregation, experimenting on collaborative/social networks. The application of the method was experimented both on textual and to non-textual domains and the evaluation was done in terms of convergence, path length, minimal solution and ranking.

(c) the investigation of models of semantic proximity which combines the local analysis of the network topology and the information on the nodes, i.e. integrates structure-based similarity with web-based similarity; web-based measures are integrated with the Quasi-Common Neighbours (QCN) general scheme, which have shown promising performance.

(d) the techniques and models descripted above are experimented on standard accepted benchmark data sets in order to assess their performances, including bibliographic data networks, recommender systems, social networks and semantics for the association to terms/sentences of context or emotional tags.

2 Results Achieved and Proposed Techniques

2.1 Previous Research

In previous research, we focused on the study and comparison of simple web-based proximity measures, and of their capability to take advantage of results of web-based search engines. In a preliminary study, we focused on performing an in-deep analysis and comparison to the already existent web-proximity measures and we also extended the research to other proximity measures (such as confidence, χ-square and others), that are suitable to be used as web-based measures. In particular, we designed a new proximity measure, the *PMING Distance* (Franzoni et al. 2012; Franzoni 2017), a web-based proximity measure between two terms. Integrating the measuring power of *Pointwise Mutual Information* (PMI) (Turney 2001) and *Normalized Google Distance* (NGD) (Cilibrasi and Vitanyi 2004), the PMING Distance improves sensibly their measuring performance with respect to clustering, local contexts and human evaluation (Milani 2016). PMING Distance has been applied, with success, to the evaluation of proximity of pairs of words that are attached to tagged images (Li 2014; Li 2017). The performance of PMING has been compared to web-based proximity measures (e.g. Confidence, Chi-square) and its application to image web-based similarity has been investigated. A proximity model, the *Heuristic Semantic Walk* (HSW) (Franzoni et al. 2014a; Franzoni et al. 2014b) has been then designed, using *random tournaments* (Gori and Pucci 2006) for ranking candidate nodes in the online navigation of a graph. The web-based proximity measures have been used as heuristics to guide the navigation, where the random tournament was run several times in the experiments, and then a function of aggregation was used to return the candidate node rank (Franzoni et al. 2014c).

2.2 Research Question

This paper mainly focuses on three tasks, related to proximity ranking and information networks:

- designing set proximity measures, in order to evaluate the distance among sets of terms, using web-based basic proximity measures and on the measures of centrality
- investigating algorithms for navigating the network graph and ranking the neighbour nodes, eventually using an online/randomized heuristic search
- evaluating the common/similar neighbourhood/attributes of a pair of nodes in a network, assessing the likelihood of a new link, with the goal of designing new models and measures for similarity evaluation.

Among the several possibilities, the applications on which we focus are context extraction from collaborative networks (i.e. Wikipedia) (Franzoni et al. 2015b), bacterial diffusion (Franzoni et al. 2016b; Franzoni et al. 2017), and emotion recognition (Franzoni et al. 2013; Franzoni et al. 2016f; Biondi 2016), but the models are aim to be applicable to any semantic data or social/collaborative network.

2.3 Random Walk Traces: Context Extraction

The automated generation of meaningful semantic contexts in a given application domain is an open problem for many research areas related to Artificial Intelligence.

In this line of research, the semantic context underlying a set of given input concepts is defined by the relevant multiple explanation paths connecting the input concepts in a collaborative network. The huge dimension and the online characteristic of the network may not allow using classical path-finding algorithms, which require preloading the graph in the local memory (Kurant et al. 2010). Moreover, an intractable number of multiple paths of different length may exist between a pair of seed concepts.

Our method is based on *random walk traces* (Franzoni et al. 2016; Milani 2013), a pheromone-like model (Baioletti et al. 2009) for the detection and the extraction of paths of explanation between seed concepts. The exploration of the online collaborative network of explanation uses a heuristic-driven random walk (Newell et al. 1957) based on semantic proximity measures. Random walks deposit pheromone on the traversed arcs. Pheromone distribution is then used to evaluate the relevance of concepts in the explanatory paths. The traces of most traversed paths during multiple runs of the heuristic weighted random walk are used to determine the semantic context underlying the source and target concepts. The relevance ranking is then used to filter and extract a subnetwork, which represents the actual context. The proposed methodology applies to collaborative semantic networks techniques inspired by the randomized evolutionary computation methods, used in Ant Colony Optimization (ACO): heuristic-driven search based on random tournament for generating explanatory path between concepts, and a pheromone-like model which is used to analyze and extract the relevance of context members. The main contributions of this work to the issue of context extraction consist on:

1. applying the Online Randomized Heuristic Semantic Walk, to determine the relevance of the intermediate nodes (Franzoni et al. 2014b);
2. defining a context as a subnetwork extracted from multiple random runs, instead of focusing on the notion of a single best path (Franzoni et al. 2016e);

3. assessing the relevance of the nodes belonging to the context by a pheromone-based technique, which considers centrality measures (Franzoni et al. 2016a).

The relevance of intermediate nodes is influenced by different factors as: distance from the seeds, path length, frequency in different paths.

2.3.1 Definitions

The following definitions are given to formally characterize the scheme for the semantic path finding and context extraction problem.

Definition: a *semantic network* or *semantic graph* $G = (V, E)$, is defined by a pair where N is a set of nodes/concepts and $E \subseteq (V \times V)$ is a set of oriented edges, representing the links between concepts in the network.

Definition: the semantic path finding problem or `Path(s = start, t = target)`, given a semantic network $G = (V, E)$ and two nodes $s, t \in V$, consists in finding if a sequence of concepts/nodes $(v_0, v_1, \ldots, v_n)$ exists, such that $v_0 = s, v_n = t$ and for each $i \in [0, n-1]$ *the edge* $(v_i, v_{i+1}) \in E$.

Definition: A *semantic proximity measure* is any function $\eta : V \times V \rightarrow \Re$ will be based on statistics drawn from search engines or repositories which, for our purposes, are formally defined below.

Definition: a search engine/repository S is a function $S : \wp(V) \rightarrow \mathcal{N}$, where *S(Q)* returns the number of occurrences of the concepts in the query Q according the engine/repository i.e., the statistics will reflect the frequency contextual use of terms in Q, according to the community related to the domain that feeds objects to S.

The requirements on $\wp(V)$ imply that an engine/repository S should be able to return statistics for co/occurrences of any subset of objects. Most engines/repositories are in fact able to return statistics for queries like $Q = (a \wedge b \vee \neg c)$. Since we are only interested in the number of occurrences returned by S, the actual returned objects are not relevant.

Definition. A *Heuristic Semantic Walk*, or HSW, is an algorithm for `Path` problems characterized by the triple (G, S, η). Where $G = (V, E)$ is a *semantic graph*, S a search engine, which can return statistics about the occurrence of elements in $\wp(V), \eta$ a proximity measure defined in terms of those statistics.

Definition. A *Context* is a subgraph of G which nodes pass semantic filtering conditions.

In the following *HSW(S, η, G)* denotes an instance of HSW for a semantic graph G where the randomized online search is based on the heuristics h_η computed from search engine source S. For instance *HSW(Bing, NGD, Wikipedia)* denotes a randomized search algorithm for Wikipedia that uses a heuristics derived from the Normalized Google Distance (NGD) measure, computed by using statistics from the search engine Bing.

A node *n* traversed by HSW during a run with seeds {s, t} is a candidate to context extraction if it appears in at least one explanatory path from *s* to *t*. The relevance of a candidate node *n* is estimated by different factors: the *number of traversing paths* from *s* to *t* passing through *n*, i.e. the more the explanation chains contain *n* the more chance *n* has to appear in the context of {s, t}; the *path length* of the node from/to the start/target nodes, i.e. a smaller distance is the clue of a closer causal/explanatory relationship; the *semantic proximity* measure between *n* and the start/end nodes. In order to evaluate these factors, the context extraction algorithm accumulates and updates on the nodes at each run different types of pheromone information, which represents *random traces*:

1. C(n, s, t), number of semantic chains from start node s to goal node t, in which node n appeared.
2. M(n, s, t), minimum distance either from start or to goal nodes over all the generated chains which traversed n.
3. A(n, s, t), average position in the chains, it represents the traversing velocity toward the goal node.
4. d(s, n) and d(t, n) are the bi-directional averaged proximity measures of the node n from the start node s and from the target node t, the bi-directional average is used for normalizing asymmetric measures while it is irrelevant for symmetric proximity measures.

These random traces can be related to formal notions such as centrality/connectivity degree (1), minimum (2) and average path distance of the candidate node from the start/goal pair (3), of which they represent an approximation.

2.3.2 Multipath Extraction

The semantic context extraction technique is based on the analysis of the multiple paths pointed out by random traces (Franzoni et al. 2016e).

A first filtering phase is based on pheromone information then a second filtering phase takes place, which simply consists on a rank based selection.

In the first phase, two different *path filters* can be applied:

Definition. Semantic Chain filter (pathFilter$_1$). A node *n* is filtered out from context $C_{s,t}$ if $C(n, s, t) < C(s, s, t)$or equivalently $C(n, s, t) < C(t, s, t)$.

Definition. Minimum Path filter (pathFilter$_2$). A node *n* is filtered out from context $C_{s,t}$by if $M(n, s, t) < avg(d(t, s), d(s, t))$and $d(t, n) < avg(d(t, s), d(s, t))$.

The semantic chain and minimum path filters aim at discarding those outlier nodes which are not *semantically within s* and *t.* pathFilter$_1$ aims at considering nodes which appear more often in a semantic chain, i.e. common concepts, while pathFilter$_2$ prefers nodes which are closer either to the start or target concept.

Definition rankFilter. A node *n* is filtered out from context $C_{s,t}$ if $Rank(n, s, t) < \theta$, where θ is a threshold.

The rank phase assigns a relevance coefficient to the nodes, which combines both their distance and their frequency in the paths between *source* and *target*, i.e.

approximated centrality. Different relevance coefficients for node ranking have been considered:

$$rankFilter_1(n,s,t) = C(n,s,t)/M(n,s,t)$$
$$rankFilter_2(n,s,t) = C(n,s,t)/A(n,s,t)$$

Note that $RankFilter_1$ emphasizes the occurrence of n in shortest paths, while $RankFilter_2$ gives more importance to bi-directional proximity to source and target.

It is also worth noting that $PathFilter_1$ and $PathFilter_2$ are parameters free, while $RankFilter_1$ and $RankFilter_2$ uses a simple threshold cutoff.

2.4 Set Similarity and Image Similarity

Set distance has been applied to image retrieval (Franzoni et al. 2015c); other promising application domains are in the field of expertise finding and attribute inference. The main idea behind set-based similarity is to leverage on the context in which an object is located, for instance the set of associated tags in a collaborative social media repository. We argue that the context carries meaningful semantics elements which can contribute to assess the similarity. The approach has been experimented on two levels of object size: an object-to-object level, with image similarity, and an object-to-concept level with a model for emotion recognition (Franzoni et al. 2013a), but more generally applies to objects or any type of web entity, and to the rich contexts provided by social networks. On the other hand, in traditional content-based image retrieval, similarity measurements of low-level image features drive the retrieval process. The aim of those approaches is to entail the similarity from the user perceptual point of view. However, deep semantic relationships among images cannot be entailed, because they require integrating content-based similarity with the deeper notion of concept-based similarity, which is certainly more effective for retrieval purposes. In this line of research, set proximity measures, the goal is to define a suitable semantic distance among objects basing on the set of terms associated to them (e.g. metadata, label, tags or keywords in a social network). (Strube et al. 2006; Volkel et al. 2006; Li 2014). The starting point is the observation that textual information related to the context of images, such as captions, labels, tags and comments, convey more meaningful concept related information than image low-level feature analysis. (Wu et al. 2008) We decided then to leverage on the textual context of an image (Franzoni et al. 2016d), to extract a group of text-based concepts, which semantically characterize it. The aim of a set similarity measure is then to compare groups of concepts related to the contexts of different images. Similarity measurement for textual documents based on text analysis and statistics has been extensively studied in Information Retrieval, while the advent of search engines, continually exploring the Web, represents a natural source of information on which to base a modern approach to semantic annotation. In Content Based Image Retrieval (CBIR), similarity measurements of low-level features are used to derive the retrieval process (e.g. image pixels or point clouds). The main objective of the CBIR approach is to entail the similarity from the usual perceptual point of view.

However, deep semantic relationships among images cannot be explored by CBIR since they require to integrate content-based similarity with the deeper notion of concept-based Image similarity, which represents high-level concepts with semantic or metadata terms, trying to extract the cognitive concept of a human ontology and maps to it the low-level image features, to fill the semantic gap. The main idea behind our Context-Based Image Semantic Similarity (CISS) is to leverage on the knowledge carried by the context in which a concept related to an image is located, e.g. the associated captions, labels, tags and comments in a collaborative social media repository; we argue that the context carries meaningful semantics elements which can contribute to similarity. such as captions, labels, tags and comments, convey more meaningful concept related information than image low-level feature analysis. (Wu et al. 2008) In order to assess set similarity, elementary proximity measures among terms have shown to be necessary (Li 2014) in order to overcome the limitations of the classical proximity measures among sets (e.g. Jaccard coefficient, edit-like distances etc.). A typical drawback with classical set proximity measures, like Jaccard similarity index (JS) between two sets S_1 and S_2, is that they are strongly oriented to weight common elements, i.e. elements in $S_1 \cap S_1$, and they do not weight any semantic aspect of the elements. If for instance S_1 = {"cat", "cheese"} S_2 = {"mouse", "cheese"} S_3 = {"rat", "milk"} then JS(S_1, S_2) = 1/3 since S_1 and S_2 have one common element over a total of 3, while JS(S_2, S_3) = 0 and JS(S_1, S_3) = 0 since they have no common elements. In other words the Jaccard index is not able to capture the intuitive similarity among the pair *rat-mouse* and *milk-cheese* (similar drawbacks can be observed in the phenomenon of zero-tail, see next paragraph on *link prediction*). On the other hand, we observe that web-based measures provide perfect tools for evaluating such semantic similarity. *Statistic-based measures* (e.g. PMI, NGD, mutual confidence) (Cilibrasi and Vitanyi 2004; Turney 2001) can be easily calculated from search engine statistics while ontology based measures require to navigate static ontologies (e.g. WordNet) or dynamic ontologies (e.g. Wikipedia). (Strube et al. 2006; Volkel et al. 2006).

2.4.1 Definitions

In the *context-based image similarity* we have introduced a general *set similarity scheme* in order to compare the set of concepts associated to an image I_i to the set associated to another image I_j. The scheme composes the elementary web-based similarities η of the elements in the first and second group of concepts. The set similarity scheme is general and can be applied to objects connected to set of elements for which a proximity measure η exists, moreover it is parametric with respect to the elementary similarity η and to the aggregation function SEL where $\eta \in \{PMI, PMING, CHI^2, Conf\}$. Formally the proposed scheme defines a class of Hausdorff distances.

Definition. Set similarity scheme. Let suppose image I_i and I_j a pair of images to be compared. $T_i = \{t_{i1}, t_{i2}, \ldots, t_{im}\}$ is the set of terms associated to the image I_i, while $T_j = \{t_{j1}, t_{j2}, \ldots, t_{jn}\}$ defines the set of terms associated to image v.

Then $D(T_i, T_j)$, the Context-based Image Semantic Similarity (CISS), which measures the mutual similarity of image I_i and image I_j, is defined by the parametric scheme:

$$D(T_i, T_j) = AVG2\{AVG1[\forall t_i \in T_i\ SEL(\forall t_j \in T_i\ \eta(t_i, t_j))], AVG1[\forall t_j \in T_j\ SEL(\forall t_i \in T_i\ \eta(t_j, t_i))]\}$$

where *SEL* defines the used aggregation function (e.g. *MAX*, *AVG* or *MIN*). η is the similarity calculated by any elementary web-based semantic similarity (e.g. *PMING*, *NGD*, and *mutual confidence*).

It is worth noticing that the set similarity scheme define a class of semantic similarity functions, where a different elementary distance and a different composition operator generate a single semantic similarity function of the class. In the experiments, we focused on the most significant elementary proximity measures which have been previously proved in their effectiveness in application domains. On the other hand, we decided to not generalize the scheme on the component AVG1 and AVG2, which use standard average function, and to limit the meta-operator SEL only to the proposed operators MIN, MAX, AVG. It is easy to prove that, with the given limitations, for any elementary similarity η and any composition operator SEL, the following properties hold for CISS: (1) if $0 \leq (x, y) \leq 1$ then $0 \leq D(T_i, T_j) \leq 1$; (2) for similarities (equiv. for distances) if $\forall x.d(x, x) = 1$ (equiv. d(x, y) = 0) then $\forall I.D(I, I) = 1$ (equiv. D(I, I) = 0) with max similarity normalized to 1; (3) D is symmetric $\forall SEL, \forall d, \forall Ii, \forall I_j : D(I_i, I_j) = DC(I_j, I_i)$, the property holds also if d is asymmetric; (4) D is a similarity(dissimilarity measure) when d is a similarity(dissimilarity).

2.5 Set Distance and Emotion Recognition

For the application of web-based proximity measures to emotion ranking for emotion recognition tasks, the goal is to evaluate the emotions contained in an emotionally rich textual object (Franzoni et al. 2016f) (e.g. a tweet or a news title, e.g. not an ingredients list). The approach we investigated is based the use of a web-based proximity measure η (e.g. confidence, PMI and PMING (Franzoni et al. 2012)) to evaluate the similarity of each word of the object and each of the emotional words of a psychology emotion model $\mathcal{E}$ (e.g. Ekman or Plutchick) (Ekman 1993; Plutchick 1991), where the set S of terms to consider is extracted from the object eventually with a pre-processing phase (tokenization, stop words, filtering) (Franzoni et al. 2013a). An automated search is then performed on a search engine, and the frequencies returned are scraped from the results page. The similarity values of terms in S are then calculated and aggregated, to rank the emotions for the entire sentence. Different from sentiment analysis, our approach works at a deeper level of abstraction, aiming to recognize specific emotions and not only the positive/negative sentiment, in order to predict emotions as semantic data. Such affective information can be used in various personalized systems like behaviour models, recommender systems, human-machine interfaces, and in decision-support systems or social robots.

2.5.1 Definitions

An emotional model is characterized by a set of m elementary emotions or emotional dimensions, say $\mathcal{E} = \{e_1, e_2, \ldots, e_m\}$ (Ekman 1993; Plutchick 1991). The emotional content of a term t_i can be encoded in the vector space model by a vector $v\_t_i = [\eta(t_i, e_1), \eta(t_i, e_2), \ldots, \eta(t_i, e_m)]$ of the distances from the basic emotions. The emotional content of a set of terms $\mathrm{S} = \{t_1, \ldots, t_n\}$ is represented by a vector v_S in the same space. The vector corresponding to the set S is obtained by aggregating separately the values in each single dimension by an operators SEL.

$v\_S = [\mathrm{v\_S}_1, \mathrm{v\_S}_2, \ldots, \mathrm{v\_S}_\mathrm{m}]$ where for each dimension $i, \mathrm{v\_S}_\mathrm{i} = \mathrm{SEL}\, j \in 1, n\{v_{i1}, \ldots, v_{in}\}$

more explicitely,

$$v\_S = [\mathrm{SEL}\, j \in 1, n\{\eta(\mathrm{t}_1, \mathrm{e}_1), \ldots, \eta(\mathrm{t}_\mathrm{n}, \mathrm{e}_1)\}, \ldots, \mathrm{SEL}\, j \in 1, n\{\eta(\mathrm{t}_1, \mathrm{e}_\mathrm{m}), \ldots, \eta(\mathrm{t}_\mathrm{n}, \mathrm{e}_\mathrm{m})\}].$$

The prevalent emotion or main emotion of a set S is obtained by selecting the elementary emotion $e_{max} \in \mathcal{E}$ which is maximal in the vector v_S, its value is:

$$v\_e_{max} = \max\{\mathrm{SEL}\, j \in 1, n\{\eta(\mathrm{t}_1, \mathrm{e}_1), \ldots, \eta(\mathrm{t}_\mathrm{n}, \mathrm{e}_1)\}, \ldots, \mathrm{SEL}\, j \in 1, n\{\eta(\mathrm{t}_1, \mathrm{e}_\mathrm{m}), \ldots, \eta(\mathrm{t}_\mathrm{n}, \mathrm{e}_\mathrm{m})\}\}.$$

Note that v_e_{max}, with SEL = *max*, is equivalent to compute the set distance $D_{\eta,\max}(\mathrm{S}, \mathcal{E})$, previously introduced, between the two sets, set $\mathrm{S} = \{t_1, \ldots, t_n\}$ representing the object, and set $\mathcal{E}$,i.e. representing the emotional model, where AVG1 and AVG2 are replaced in the scheme D by the *max* aggregator function. The emotion recognition model is then fully characterized by the triple ($\mathcal{E}$, η, SEL).

2.6 Topological Similarity for Link Prediction

Through the state of the art of the topological similarity measures applied to the *link prediction* problem for social networks, (Liben and Kleinberg 2003) we analysed link prediction and *attribute inference*, looking at feasibility and importance of the topic in the scientific engineering point of view. Among topics that recently gained visibility, where link prediction can be applied, are trust management, terrorism prevention, and disambiguation in co-authorship networks, besides viral diffusion of bacteria and information (Franzoni et al. 2016b; Franzoni et al. 2017).

2.6.1 Topological Similarity for Link Prediction: Quasi Common Neighbours

Common Neighbours-based (CN) rankings (e.g. Jaccard, Adamic-Adar etc.) (Leskovec et al. 2010; Lada et al. 2003) represent a class of measures for link prediction, which efficiently assess the likelihood of a new link based on the neighbours frontier of the already existing nodes. Although the CN measures in link prediction are more performant than others (e.g. preferential attachment, path-based distances) (Newman 2001; Katz 1953), they present the drawback of returning a large zero-tail. In fact, a zero-rank value is given to links of nodes, which have no common neighbours. On the other hand, those links may be potentially good ones for link prediction. A general technique can be proposed to evaluate the likelihood of a linkage, iteratively applying a given ranking measure to the *Quasi-Common* Neighbours of the node pair (Franzoni et al. 2015a), i.e. iteratively considering paths between nodes, which include more than one traversing step. The practical effect is that many significant links are lift up by the zero-tail (Franzoni et al. 2016c).

Let a *training* network $G = (V, E)$ at time *t*, with nodes N and undirected arcs $E \subseteq V \times V$, and let a given *test* network $G' = (V, E')$, where G' is an extension of G at time t + k with $E' \supseteq E$. The basic link prediction problem asks to find a rank function r (G), producing a ranking of the E_{pot} set, i.e. the set of the links of the complete graph, which could appear in G', such that the new links of G' are ranked first. $E_{pot} = V \times V \backslash E$ is thus the set of potentially new arcs, present in the complete graph but not in G, and $E_{new} = E' \backslash E$ is the set of *new arcs* of test network G'. A perfect rank function (Kendall 1938) would return all the E_{new} links in the first $|E_{new}|$ ranking positions. the ranking is then defined as:

$$KtA = \frac{1}{|A_{pot}| - 1 - |A_{new}|} * \sum_{1=|A_{new}|}^{|A_{pot}|-1} (r_1 - |A_{new}| + 1)$$

where the summation is done for each positive link l such that $|A_{new}| \leq r_l < |A_{pot}|$.

The main idea is to shorten the *zero tail*, which is typical of common-neighbour-based ranking approaches, e.g. Adamic-Adar (AA) (Lada et al. 2003), Jaccard index (JA), resource allocation (RA), preferential attachment, Katz distance (Katz 1953). In particular, the rationale behind Adamic-Adar (AA) is that the informative role of neighbours of a node depends on the node's degree, i.e. popularity: having in common a rare node represents a stronger bound between two nodes, while having in common a node that has a lot of neighbours, i.e. a very common rapport, is less informative (Watts and Strogatz 1998). Path distances focus on information of the overall network topology instead of only the local context; in some network domains the information about the connectivity of *A* and *B* can be useful, such as variations of *random walks* (Barabasi and Albert 1999; Erdos and Rényi 1959) and *Katz Path distance* (Katz 1953). While CN distances are very accurate where they actually have neighbours to evaluate, *path distances* are able to elicit an informative ranking on link *(A, B)* even when $\Gamma(A) \cap \Gamma(B)$ is empty. In the ranking induced by CN, JA, AA a large number of links lie in the zero tail: these ranking functions are not able to distinguish among those links (Chiancone 2016).

2.6.2 Definitions

Given a common neighbours-based ranking measure $\varphi : V \times V \times V \rightarrow \Re$ then a *Quasi-Common-Neighbours extension measure* $QCN\varphi : V \times V \times V \rightarrow \Re$ can be iteratively defined as:

$$\mathrm{QCN}_{\varphi} = \sum_{i=0}^{\max} \left[\frac{1}{(|\Gamma(A)| + |\Gamma(B)|)^{i}} * \varphi(A, B, \pi(A, B, i)) \right]$$

where $\pi(A, B, i)$ is the set of quasi-common-neighbours of A and B of level i. $\pi(A, B, i)$ consists of the nodes which are in a path of length $i + 2$ or more between A and B, but do not appear on paths shorter than $i + 2$.

2.7 Disambiguation of Bibliographic Networks Through Link Prediction

Disambiguation and link prediction in bibliographic data is the task of identifying entities referring to the same object in a set of candidates (Strotmann and Zhao 2012a; Ferreira et al. 2012). The problem of identifying and linking/grouping different manifestations of the same real world object is important for several applications, starting from simple deduplication tasks, to the extraction of expertise of people or institutions/labs. Entity resolution in bibliographic data is thus a promising application of link prediction techniques and of similarity measures. State of the art approaches to entity resolution include a wide range of areas, where explicit information and implicit evidence is computed, with supervised and unsupervised techniques (Morillo et al. 2013). Among others, pairwise comparativeness, set similarity functions (e.g. Jaccard, cosine, edit distance) and graph-based distances can be leveraged to exploit both semantic and structure-based evaluations. In this point of view, our current work aims at converging and synthetizing the different results and techniques developed in a single hybrid model for network similarity, which combines, in the same framework, structure-based proximity and web-based proximity. Among the possible application domains for a hybrid network similarity model, we are focusing on bibliographic data networks for the elicitation of expertise. This domain provides interesting challenges i.e., expert finding, team formation, name/affiliation disambiguation. Some key points, we are currently exploring for the hybrid network similarity, include to provide a clear formal and computationally sound definition of the mechanisms for mutual reinforcement of similarities, i.e. to define the synergy between semantic (i.e. node feature) and topological (i.e. network structure) aspects, in order to improve the precision of typical network tasks, such as node recognition, disambiguation and link prediction. Another aspect we are currently exploring is to apply the method of randomize heuristic navigation to bibliographic networks using skills heuristics, under the assumption that similar skills are topologically close in the network. A preliminary form of the hybrid similarity model can be found in the research on Heuristic Random Walk for context extraction (Franzoni et al. 2014b). At same extent, the Heuristic Random Walk approach already combines structural aspects (random tournament on the node's neighbours) and semantic aspects (evaluation of the distance of the candidate nodes from the target term). The HSW currently evaluates only the proximity between a single feature for each. In the case of bibliographic network the evaluation of multiple features sets between nodes is desirable.

Acknowledgements. Authors thank Proff. Daniele Nardi, Marco Schaerf and Roberto Basili for the useful feedback given during the PhD school presentations, and Dr. Yuanxi Li, Andrea Chiancone and Giulio Biondi for the fruitful discussions on previous works.

References

Baioletti, M., Milani, A., Poggioni, V., Rossi, F.: Optimal planning with ACO. In: Serra, R., Cucchiara, R. (eds.) AI*IA 2009. LNCS, vol. 5883, pp. 212–221. Springer, Heidelberg (2009). doi:10.1007/978-3-642-10291-2_22

Barabasi, A.L., Albert, R.: Emergence of scaling in random networks. Science **286**(509), 509–512 (1999)

Bollegala, D., Matsuo, Y., Ishizukain, M.: A web search engine-based approach to measure semantic similarity between words. IEEE Trans. Knowl. Data Eng. **23**, 977–990 (2011)

Biondi, G., et al.: Web-based Semantic Similarity for Emotion Recognition in Web Objects. CoRR abs/1612.05734 (2016)

Ceri, S., Daniel, F., et al.: Model-driven development of context-aware web applications. ACM TOIT **7**, 2 (2007)

Cilibrasi, R; Vitanyi, P.: The Google Similarity Distance. ArXiv.org (2004)

Chiancone, A.: Ph.D. thesis; Link Evolution in Social Networks using Topological Features, University of Perugia (2016)

Abowd, G.D., Dey, A.K., Brown, P.J., Davies, N., Smith, M., Steggles, P.: Towards a better understanding of context and context-awareness. In: Gellersen, H.-W. (ed.) HUC 1999. LNCS, vol. 1707, pp. 304–307. Springer, Heidelberg (1999). doi:10.1007/3-540-48157-5_29

Dey, A.K.: Understanding and using context. Pers. Ubiquit. Comput. **5**, 4–7 (2001)

Ekman, P.: Facial expression and emotion. Am. Psychol. **48**(4), 384–392 (1993)

Erdos, P., Rényi, P.: On random graphs. Publicationes Mathematicae Debrencen **6**, 290–297 (1959)

Ferreira, A.A., Goncalves, M.A., Laender, A.H.: A brief survey of automatic methods for author name disambiguation. SIGMOD Rec. **41**(2), 15–26 (2012)

Franzoni, V.: Just an Update on PMING Distance for Web-based Semantic Similarity in Artificial Intelligence and Data Mining. CoRR abs/1701.02163 (2017)

Franzoni, V., Milani, A.: PMING distance: a collaborative semantic proximity measure. In: 2012 IEEE/WIC/ACM International Conference on Web Intelligence and Intelligent Agent Technology, WI-IAT, vol. 2, pp. 442–449 (2012)

Franzoni, V., Poggioni, V., Zollo, F.: Automated Book Classification According to the Emotional Tags of the Social Network Zazie. ESSEM, AI*IA, CEUR Workshops, vol. 1096, pp. 83–94, CEUR-WS (2013)

Leung, C.H.C., Li, Y., Milani, A., Franzoni, V.: Collective evolutionary concept distance based query expansion for effective web document retrieval. In: Murgante, B., Misra, S., Carlini, M., Torre, C.M., Nguyen, H.-Q., Taniar, D., Apduhan, B.O., Gervasi, O., et al. (eds.) ICCSA 2013. LNCS, vol. 7974, pp. 657–672. Springer, Heidelberg (2013). doi:10.1007/978-3-642-39649-6_47

Franzoni, V., Mencacci, M., Mengoni, P., Milani, A.: Heuristics for semantic path search in wikipedia. In: Murgante, B., et al. (eds.) ICCSA 2014. LNCS, vol. 8584, pp. 327–340. Springer, Cham (2014). doi:10.1007/978-3-319-09153-2_25

Franzoni, V., Milani, A.: Heuristic semantic walk for concept chaining in collaborative networks. Int. J. Web Inf. Syst. **10**(1), 85–103 (2014)

Franzoni, V., Mencacci, M., Mengoni, P., Milani, A.: Semantic heuristic search in collaborative networks: measures and contexts. In: Proceedings - 2014 IEEE/WIC/ACM International Joint Conference on Web Intelligence and Intelligent Agent Technology - Workshops, WI-IAT 2014, vol. 1, pp. 187–217 (2014)

Franzoni, V., Niyogi, R., Milani, A.: Improving link ranking quality by quasi-common neighbourhood. In: Proceedings - 15th International Conference on Computational Science and Its Applications, ICCSA 2015, pp. 21–26, IEEE Press (2015)

Franzoni, V., Milani, A.: Semantic context extraction from collaborative networks. In: Proceedings of the 2015 IEEE 19th International Conference on Computer Supported Cooperative Work in Design, CSCWD 2015, pp. 131–136. IEEE Press (2015)

Franzoni, V., Leung, C.H.C., Li, Y., Mengoni, P., Milani, A.: Set similarity measures for images based on collective knowledge. In: Gervasi, O., Murgante, B., Misra, S., Gavrilova, Marina L., Rocha, A.M.A.C., Torre, C., Taniar, D., Apduhan, Bernady O. (eds.) ICCSA 2015. LNCS, vol. 9155, pp. 408–417. Springer, Cham (2015a). doi:10.1007/978-3-319-21404-7_30

Franzoni, V., Milani, A.: A pheromone-like model for semantic context extraction from collaborative networks. In: Proceedings - 2015 IEEE/WIC/ACM International Conference on Web Intelligence and Intelligent Agent Technology, WI-IAT 2015, 2016-January, pp. 540–547. IEEE Press (2016)

Franzoni, V., Chiancone, A., Milani, A., Poggioni, V., Pallottelli, S., Madotto, A.: A multistrain bacterial model for link prediction. In: Proceedings - International Conference on Natural Computation, 2016-January pp. 1075–1079. IEEE Press (2016). doi:10.1109/ICNC.2015.7378141

Chiancone, A., Franzoni, V., Li, Y., Markov, K., Milani, A.: Leveraging zero tail in neighbourhood for link prediction. In: Proceedings - 2015 IEEE/WIC/ACM International Joint Conference on Web Intelligence and Intelligent Agent Technology, WI-IAT 2015, pp. 135–139. IEEE Press (2016)

Franzoni, V., Milani, A., Pallottelli, S., Leung, C.H.C., Li, Y.: Context-based image semantic similarity. In: 2015 12th International Conference on Fuzzy Systems and Knowledge Discovery, FSKD 2015, pp. 1280–1284. IEEE Press (2016) doi:10.1109/FSKD.2015.7382127

Franzoni, V., Milani, A., Pallottelli, S.: Multi-path traces in semantic graphs for latent knowledge elicitation. In: Proceedings - International Conference on Natural Computation, 2016-January, pp. 281–288. IEEE Press (2016). doi:10.1109/ICNC.2015.7378004

Biondi, G., Franzoni, V., Li, Y., Milani, A.: Web-based similarity for emotion recognition in web objects. In: Proceedings - 9th IEEE/ACM International Conference on Utility and Cloud Computing, UCC 2016, pp. 327–332 (2016)

Franzoni, V., Chiancone, A., Milani, A.: A multistrain bacterial diffusion model for link prediction. Int. J. Pattern Recognit. Artif. Intell. IJPRAI **31**(11), 123–136 (2017). doi:10.1142/S0218001417590248. (World Scientific)

Gori, M., Pucci, A.: A random-walk based scoring algorithm with application to recommender systems for large-scale e-commerce. In: 12th ACM SIGKDD International Conference on Knowledge Discovery and Data Mining (2006)

Katz, L.: A new status index derived from sociometric analysis. Psychometrika **18**(1), 39–43 (1953)

Kendall, M.: A new measure of rank correlation. Biometrika **30**, 81–89 (1938)

Leskovec, J., Huttenlocher, D., Kleinberg, J.: Predicting positive and negative links in online social networks. In: Proceedings WWW 2010, ACM Press (2010)

Kurant, M., Markopoulou, A., Thiran, P:. On the bias of BSF. ITC (2010)

Adamic, L.A., Adar, E.: Friends and neighbours on the web. Soc. Netw. **25**(3), 211–230 (2003)

Lieberman, H.: An agent that assists web browsing, IJCAI95 (1995)
Liu, J., et al.: Intelligent social media indexing and sharing using an adaptive indexing search engine. ACM Trans. Intell. Syst. Technol. **3**(3), 221–238 (2012). ACM Press
Liben-Nowell, D., Kleinberg, J.: The link prediction problem for social networks. In: Proceedings of the Twelfth International Conference on Information and Knowledge Management, CIKM 2003, pp. 556–559 (2003)
Li, Y.X.: Semantic Image Similarity Based on Deep Knowledge for Effective Image Retrieval. Ph.D. thesis of Hong Kong Baptist University (2014)
Li, Y., et al.: Semantic Evolutionary Concept Distances for Effective Information Retrieval in Query Expansion. CoRR abs/1701.05311 (2017)
Franzoni, V., Milani, A.: Heuristic semantic walk. In: Murgante, B., Misra, S., Carlini, M., Torre, C.M., Nguyen, H.-Q., Taniar, D., Apduhan, B.O., Gervasi, O. (eds.) ICCSA 2013. LNCS, vol. 7974, pp. 643–656. Springer, Heidelberg (2013). doi:10.1007/978-3-642-39649-6_46
Franzoni, V., Milani, A.: A semantic comparison of clustering algorithms for the evaluation of web-based similarity measures. In: Gervasi, O., et al. (eds.) ICCSA 2016. LNCS, vol. 9790, pp. 438–452. Springer, Cham (2016). doi:10.1007/978-3-319-42092-9_34
Milne, D., Witten, I.H.: An effective, low-cost measure of semantic relatedness obtained from Wikipedia links. WIKIAI (2008)
Morillo, F., Aparicio, J., Gonzalez-Albo, B., Moreno, L.: Towards the automation of address identification. Sci-entometrics **94**(1), 207–224 (2013)
Natsev, A., Haubold, A., et al.: Semantic concept-based query expansion and re-ranking for multimedia retrieval. In: 15th ACM Conference on Multimedia, pp. 991–1000, USA (2007)
Jurgens, D., Pilehvar, M.T., Navigli, R.: Cross level semantic similarity: an evaluation framework for universal measures of similarity. Lang. Resour. Eval. **50**(1), 5–33 (2016)
Newell, A., et al.: Empirical explorations of the logic theory machine: a case study in heuristic. In: Proceedings of IRE-AIEE-ACM 1957, pp. 218–230. ACM, New York, USA (1957)
Newman, M.E.J.: Clustering and preferential attachment in growing networks. Phys. Rev. Lett. E, 1–8 (2001)
Plutchick, R.: The Emotions. University Press of America, Inc., Lanham (1991)
Schilit, B.N., Adams, N., Want, R.: Context-aware computing applications. In: Workshop on Mobile Computing Systems and Applications (1994)
Strapparava, C., Mihalcea, R.: SemEval-2007 task 14: affective text. In: Proceedings of the 4th International Workshop on Semantic Evaluations (SemEval 2007). Association for Computational Linguistics, USA, pp. 70–74 (2007)
Strotmann, A., Zhao, D.: Author name disambiguation: what difference does it make in author-based citation analysis? JASIST **63**(9), 1820–1833 (2012)
Strube, M., Ponzetto, S.P.: WikiRelate! computing semantic relatedness using wikipedia. In: Proceedings of the Twenty-First National Conference on Artificial Intelligence. AAAI Press, July 2006
Turney, P.D.: Mining the web for synonyms: PMI-IR versus LSA on TOEFL. In: Raedt, L., Flach, P. (eds.) ECML 2001. LNCS, vol. 2167, pp. 491–502. Springer, Heidelberg (2001). doi:10.1007/3-540-44795-4_42
Volkel, M., Krotzsch, M., Vrandecic, D., Haller, H., Studer, R.: Semantic wikipedia. In: WWW 2006: Proceedings of the 15th International Conference on World Wide Web, pp. 585–594, New York, NY, USA, 2006. ACM
Watts, D., Strogatz, S.: Collective dynamics of small-world networks. Nature **393**, 440–442 (1998)
Wu, L., Hua, X.-S., Yu, N., Ma, W.-Y., Li, S.: Flickr distance. In: MM 2008: Proceedings of the 16th ACM International Conference on Multimedia, New York, NY, USA, pp. 31–40. ACM Press (2008)

Data Protection Risk Modeling into Business Process Analysis

António Gonçalves[1(✉)], Anacleto Correia[2], and Luis Cavique[3]

[1] INESC-ID, UL-IST, Lisboa, 1048-001 Lisbon, Portugal
antonio.goncalves@inesc-id.pt
[2] CINAV, Alfeite, 2810-001 Almada, Portugal
cortez.correia@marinha.pt
[3] Univ. Aberta and MAS-BioISI, FCUL, Campo Grande, Lisbon, Portugal
luis.cavique@uab.pt

Abstract. We present a novel way to link business process model with data protection risk management. We use established body of knowledge regarding risk manager concepts and business process towards data protections. We try to contribute to the problems that today organizations should find a suitable data protection model that could be used in as a risk framework. The purpose of this document is to define a model to describe data protection in the context of risk. Our approach including the identification of the main concepts of data protection according to the scope of the with EU directive data protection regulation. We outline data protection model as a continuous way of protection valued organization information regarding personal identifiable information. Data protection encompass the preservation of personal data information from unauthorized access, use, modification, recording or destruction. Since this kind of service is offered in a continuous way, it is important to stablish a way to measure the effectiveness of awareness of data subject discloses regrading personal identifiable information.

1 Introduction

With the General Data Protection Regulation (GPDR), applicable from beginning 2018 on Europe [1], organizations must have a privacy configuration for their services by including in their operation the data protection capabilities necessary for regulatory compliance and attainment user belief, and therefore maintain organizations competitiveness.

The key to organizations adapt to the new GDRPR requirements is the ability to change the way it interacts with suppliers, partners, competitors, and customers to achieve with the new data and security protection requirements [1]. Improvement on the way organization operates should be done to confirm the new organization objectives imposed by regulators. The European Data Protection Legislation is a complex issue, whose techno-regulation transfers a bureaucratic overhead to system developers. However, GDPR is more linked with the data privacy requirements.

Business Processes [2] defined as a set of inter-related events, activities and decision points that involve several actors and objects which collectively pursue a

O. Gervasi et al. (Eds.): ICCSA 2017, Part I, LNCS 10404, pp. 667–676, 2017.
DOI: 10.1007/978-3-319-62392-4_48

business objective and policy goal, are a suitable way to capture the organization reality [3] and using a Business Processes model is possible to establish a Process discovery, i.e., gathering information about an existing process and organizing it in terms of an 'as-is' process model risk management to analyze security and data protection concerts of an organization [2].

In this paper, we discuss the foundations of quality assessment of data protection in business process models. Our main contribution is an approach considering human behavior aspects observed in process models to calculate a degree of possibility of data protection concerns. We validate the approach using some business process model. The outcomes highpoint which benefits organizations can have from artifacts used for data protection violation detection.

The remainder of the paper is organized as follows. Section 2 presents Event driven Process Chains (EPCs) [4], a modeling language to specify the temporal and logical relationships between activities of a business process that we use to exemplify our model. Section 3 presents our data protection concepts. Section 4 addresses the problem of data protection risk management. In Sect. 5, the previous techniques are combined resulting in the proposed risk data protection model. Finally, in Sect. 6, we draw some conclusions.

2 Business Process Modeling

We define a business process as a collection of inter-related events, functions, decision points, business objects and IT entities that involve several actors and that collectively lead to an outcome that is of value to at least one customer [2, 5].

A function corresponds to a task which needs to be executed. Events define the state before and after a function is executed. Connectors can be used to connect functions and events. There are three types of connectors: and, exclusive or and or. Business objects can be input data serving as the basis for a function, or output data produced by a function and finally, IT Entities are used to describe IT input elements which are needed to perform the process.

Figure 1 depicts the ingredients of this definition and their relations.

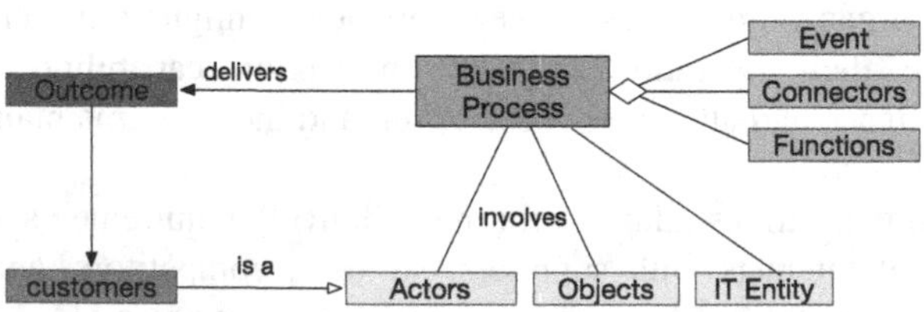

Fig. 1. Elements of a business process.

To model a business process, we need some modelling language. The Event Driven Process Chains (EPC) is a business process modeling language that was presented in [6] and are used by us to describe organization process models [7].

The use of EPC as one its main purposes, to provide a notation understandable by different kinds of process modelers and users: (1) business analysts that sketch the initial documentation of business processes; (2) process developers which are responsible for implementing business processes; (3) business users which are accountable for business processes' instantiation and monitoring.

EPCs have some similarities with flowcharts but they differ from flowcharts in that they treat events as first-class citizens. EPCs specify the temporal and logical relationships between activities of a business process throw control flow [2].

EPCs offer offers the following element types: function type (i.e. activity that is executed in a process), event type (represent pre and post-conditions of functions) and connector type. All elements are linked via control flow arcs. In EPCs there are three distinctive kinds of connectors: AND, XOR, and OR. They may be used as either join connectors. Connectors have either multiple incoming and one outgoing arc (join connectors) or one incoming and multiple outgoing arcs (split connectors).

The semantics of an EPC connectors can be described as follows [6]. The AND-split activates all subsequent branches in a concurrent manner. The XOR-split represents a choice between one of several alternative branches based on conditions. The OR-split triggers one, two or up to all multiple branches based on conditions. For XOR-splits and OR-splits, the activation conditions are given in events after the connector. The AND-join waits for all incoming branches to complete, then it propagates control to the subsequent EPC element. The XOR-join merges alternative branches. The OR- join synchronizes all active incoming branches. The next description enacts EPC adapted from [5].

Definition 1 (EPC). An Event-driven Process Chain is a seven-tuple (*E, F, C, O,T, l, A*) such that:
-***E*** is a finite, non-empty, set ($E \neq \emptyset$) of event;
-***F*** is a finite, non-empty, set ($F \neq \emptyset$) of functions;
-***C*** is finite set of connectors;
-O is finite set of business objects;
-***T*** is finite set of IT entities;
-$l \in C \rightarrow \{\wedge, XOR, \vee\}$ is a function which maps each connector onto a connector type;
-$A \subseteq (E \times F) \cup (F \times E) \cup (E \times C) \cup (C \times E) \cup (F \times C) \cup (C \times F) \cup (C \times C) \cup (F \times O) \cup (O \times F) \cup (T \times F)$ is a set of arcs.

Definition 1 shows that arcs of an EPC cannot connect two events or two functions directly, a well-formed EPC should satisfy other additional requirements.

Those expressions define that each event is at highest preceded by one input node and at highest succeeded by one output node. Every function has just one input and one output node. EPC needs at least one start event that is not preceded by any other node and one end event that is not succeeded by any other node. Connectors must have at least one input and one output node, but they can have numerous input nodes or numerous output nodes, with some restriction. In a well-formed EPC, there should be no paths connecting two events or two functions only via connector nodes in between.

Describe a Business process takes a significant part in an organization. It helps specify standard (as-is) and improved process of organization (to-be). Capture elements of business processes such us participants, their communications, resources contribute to organizational competiveness and could be a way to capture the data protection requirements. Thus, understanding and modelling of data protection becomes an important activity during a business process modeling.

3 Data Protection

The concept of data protection differs among different communities. We adopt the concept of data protection related with general legislation (EU directive 95/46/EC [8]) and privacy principles described by ISO/IEC 29100:2011 [9], which is mostly related with protecting personal data (i.e., Personal Identifiable Information or PII). The key concern of data protection is link to a person and related the protection of data that can be connected to an individual (i.e., data subject) [10].

Data can take on many forms. It can be printed or written on paper, stored electronically, transmitted by post or electronic means, shown on films, conveyed in conversation, etc. Normally organization try to use anonymized to avoid the require privacy protection. However, total anonymity is difficult, sometimes impossible [11].

The aim of data protection program at organization is to guarantee business continuity and minimize business damage by controlling the impact of data protection security incidents by implementing a framework according with a defined organization policy and consent compliance properties, according, for example, with EU directive [8].

Guarda and Zannone [12] describes the European Data Protection Directive in the following main principles: Fair and Lawful Processing; Data can only be collected and processed only if the data subject has given his explicit consent; lawful and legitimate use of personal data, data minimum necessary for achieving the specific purpose, Information Quality, Data Subject Control and Information Security.

A policy states general rules, determined by the stakeholders of the system, with respect to data protection. The policy and consent compliance property guarantees that the organization policy and the user consent are implemented and prescribed.

Several researcher emphasizes that data protection is not only a technology solution, but always encompass a process [11, 13]. For example, if we look at data security as a strictly technical issue, we also must take care of the process of securing these technical issues. Hence, it is necessary to evolve to extend beyond only the technical.

Data protection introduces an important set of definitions in terms of the properties that information manipulation should be concerned regarding privacy. These include personal identifiable information, PII (information which can be linked back to an individual), item of interest, IOI (information related to an individual) data subject (individual that is linked to the PII), unlinkability (not being able to distinguish whether two IOI are related), anonymity (not being able to identify the subject within a set of subjects), plausible deniability (being able to repudiate having performed an action), undetectability (not being able to distinguish whether an IOI exists), unobservability (undetectability against all subjects involved), confidentiality (authorized restrictions on information access and disclosure), awareness (being conscious about consequences of sharing PI information) and Compliance (following regulations and internal business policies) [14].

Regarding business process and data protection we can define, adapting from requirements engineering [15], data protection of business process (DPBP) as the elicitation, evaluation, specification, analysis and evolution of privacy objectives and constraints to be achieved by a business. A main concern of DPBP is to identifies potential threats and determine which threats are in fact applicable. To achieve that we should implement a Data Protection Risk Management to direct and control the risk.

4 Data Protection and Risk Management

The notion of risk is related to uncertainty from an expected organization objective and can be quantified as a positive or negative deviation. There are several ways to outline a risk. One of the possible ways is by linking it to the events that may happen, their consequences and the likelihood of the occurrence. The lack of information regarding the event occurrence, its consequence, or likelihood, is what drives to the state of uncertainty that underlies risk [16, 17].

Data Protection Risk management is an artefact that includes a set of coordinated activities performed to direct and control the risk of threads regarding properties that information manipulation should be concerned described in Sect. 3. It includes a set of plans, relationships, accountabilities, resources, processes that provide the policy and objectives to manage data privacy risk. The risk management policy addresses the aims and strategy of the organization regarding risk management [16, 17] (Fig. 2).

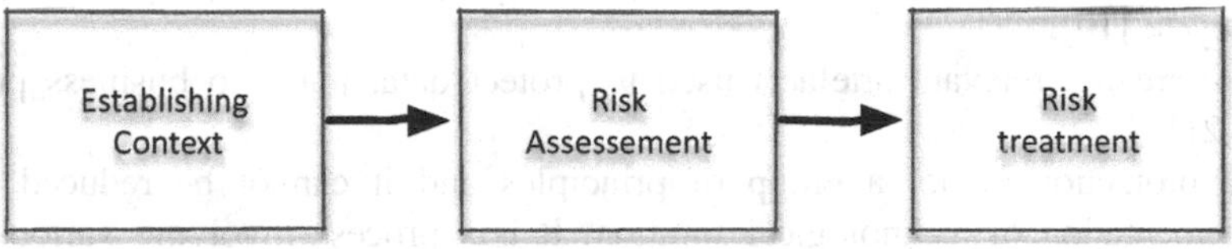

Fig. 2. Risk management

A risk management framework can be understood as a system whose purpose is to ensure the fulfilment of the goal of risk management. It should also include a risk management process, and the resources and principles used in its implementation, as represented on Fig. 1. These features can vary, however, the most important one in practice are grouped into three main stages (Fig. 1) which can result in multiple solutions depending on the technical and technological support available to the risk management. The key concept of data protection risk, capture from ISO guide 73 [17], is present in Table 1.

Table 1. Risk management concepts

Concept	Description
Risk	Effect of uncertainty on objectives
Risk management	Coordinated activities to direct and control an organization about risk
Risk management process	Systematic application of management policies, procedures and practices to the activities of communicating, consulting, establishing the context, and identifying, analyzing, evaluating, treating, monitoring and reviewing risk
Risk management framework	Set of components that provide the foundations and organizational arrangements for designing, implementing, monitoring, reviewing and continually improving risk management throughout the organization

5 Proposed Model for Data Protection Risk

In Sect. 3, we analyze the privacy principles described by EU data protection regulation and ISO/IEC 29100:2011 and in Sect. 4 we analyze the data protection risk management principles from the viewpoint of ISO guide 73 [17]. Both concepts are going to be used to arise the data protection model. The foremost goals with this approach is to contribute to minimize the following problems:

1. Despite the importance data protection for organizations, and therefore it enclosure on business processes, it is possible to find a set of common problems inside enterprises regarding protection, since security and data protection are integrate into organization in an ad-hoc way, often during the implementation process phase [18], or during the system administration phase of a business process management life-cycle [19];
2. There are not relevant artefacts used to protect data, made to business process. [19–21];
3. Data protection is not a group of principles and it cannot be reduced to the implementation of technological solution. It is a process involving various technological and organisational components, which implement privacy and data protection principles [22].

It is necessary to develop data security models that considers several organization security perspectives, such as static, about the processed information, functional, from the viewpoint of the system processes, dynamic, about the data security requirements from the life cycle of the objects involved in the business process, organizational, used to relate responsibilities to acting parties within the business process and the business processes perspective, that provides us with an integrated view of all perspectives with a high degree of abstraction.

Our approach is a model-based technique. That is relies on a representation of business process (BP). Since BP can be constructed to describe to viewpoint 'as-is' and 'to-be', our approach can be used to analysis the current data protection model or the future data protection model of a business and inforce the concept of privacy.

However, the definition of privacy varies depending on context, stakeholder interests. General privacy meanings comprise the right to informational self-determination and allowing individuals to control, edit, manage, and delete information about themselves and decide when, how and to what extent that information is communicated to others.

Our model, in Fig. 3, integrated the main concepts related with information security applied to data protection and should be a baseline to implement and verify the stage of privacy.

A threat represents a possible violation of the security of a business asset with some negative impact [23] while vulnerability is a real security flaw which makes, an organization open to an attack. So, an attack is a use of a vulnerability to realize a threat. We can this combination an event. Threat modelling for data protection can support classify the threat, their attack surface and the entry or access points on business assets. A business asset is an element of business process that has value to the organization in terms of its business model and is necessary for achieving its objectives (e.g., function, business object, IT Entity). Data protection property on business assets characterizing their data protection requirements. Data protection property describe as a meter to measure the significance of risk. We adopt the taxonomy of privacy proposed by Pfitzmann [24] and from LINDDUN methodology [25], adapted as data protection property: (i) Unlinkability means that all data processing is operated in such a way that the privacy-relevant data are unlikable to any other set of privacy-relevant data outside of the domain; (ii) Transparency means that all, privacy relevant, data processing, can be understood and reconstructed at any time. The information should be available before, during, and after the processing takes place; This allows that the data subject could have access to information requested from an organization; (iii) Intervenability ensures intervention is possible concerning all ongoing or planned privacy-relevant data processing, by those persons whose data are processed. It allows the possibility of a data subject to request to rectification and erasure of data; (iv) Anonymity refers to hiding the link between an identity and an action or a piece of information and (v) Confidentiality refers to hiding the data content or controlled release of data content.

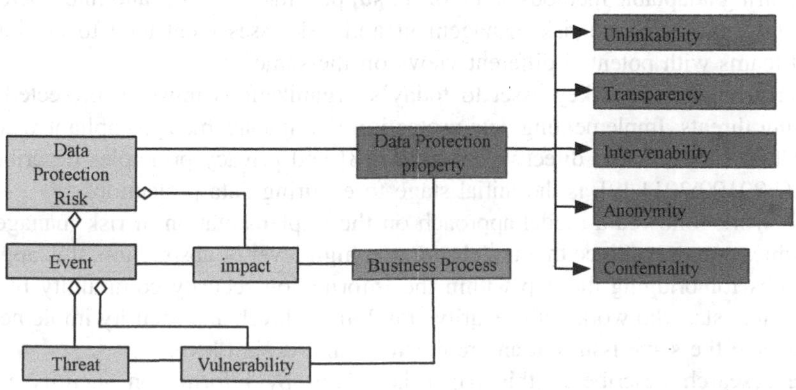

Fig. 3. Data protection risk model

We believe that from the business process perspective business analysts can integrate their view about business data security. Concerning the data protection requirements that can be modeled in business processes, it is necessary to consider that data protection requirements in any application at the highest level of abstraction will tend to have the same basic kinds of valuable and potentially vulnerable assets [29].

We can use risk data protection model in several stages of a risk management. risk data protection model centers on the potential weaknesses that might allow to someone cause the damage of a business asset. We can apply at diverse levels of abstraction, depending on the business assets considered. In other words, assets can be more abstract, such as in information objects our assets are more tangible, like IT components.

Besides, the development of a risk data protection model it is necessary to contemplate that apprehending the data protection of a Business process is a difficult work. One the benefices of using associated with a business process, is that it offers a structure view that can be used as a basis for the specification of data protection requirements. Business process model may present different levels of abstraction. Consequently, we believe that business analysts can integrate their view about business security into the business process perspective and in addition security requirements, since any application at the highest level of abstraction will tend to have the same basic kinds.

6 Conclusion and Future Work

Privacy can be defined as "the right of the individual to decide what information about himself should be communicated to others and under what circumstances" [26]. This description relates privacy to the right to control the information that is revealed to others. This is an issue that organization has concern with the new EU data protection regulation regarding data subject regrading personal identifiable information.

Data protection risks exist universally and can have costs every day, whether it is recognized by the organization affected by them. One of the main challenges that the organization must address is on the modelling of risk data protection using their context, i.e. business process model. Data protection modelling involves a highly heterogeneous set of business assets: events, methods, stakeholders and responsibilities, requiring adaptable methods and tools to support the exchange and interoperability of risk information, since risk management and risk assessment tend to be done by distinct teams with potential different views on the same risks.

Data protection as a key asset to today's organizations must be protected from increasing threats. Implementing data protection risk management compliant with ISO 31000:2009 [16, 17], EU directive 95/46/EC [8] and privacy principles described by ISO/IEC 29100:2011 [9], is the initial stage to ensuring data protection.

This work followed a model approach on the implementation of risk management and business process. Since the models are at a high level of abstraction, this approach contributes for bridging the gap within the information security community between domain analysts, who work with security at a domain level, and security implementers, who analyze the same issues at an architectural and design levels.

The research describe in this paper is driven by information security in any potential scenario that deals with information. A common way to model and address complex business systems is the use of model. Thus, we intend to bring the concepts and strategies of modelling into this subject. Modelling information risk is a complex task, especially in scenarios where protection of information is not a unique concern of the organization. To cope with information security risks, this there are need to have a cooperation between risk management and information security department, making it possible to align information security concerns with other, potentially relate, organizational concerns.

Future work intends to extend risk management and data protection models to include the dynamic perspective of information security, and conduct empirical studies for assessing the usability and efficacy of our approach in the risk management and information security domains.

References

1. The European Parliament: The European Council: General Data Protection Regulation (2016)
2. Dumas, M., La Rosa, M., Mendling, J., Reijers, H.A.: Fundamentals of Business Process Management. Springer, Heidelberg (2013)
3. Becker, J., Kahn, D.: The process in focus. In: Becker, J., Kugeler, M., Rosemann, M. (eds.) Process Management, pp. 1–12. Springer, Heidelberg (2003). doi:10.1007/978-3-540-24798-2_1
4. Scheer, A.-W., Thomas, O., Adam, O.: Process modeling using event-driven process chains. Process. Inf. Syst. 119–146 (2005)
5. Van Dongen, B., Dijkman, R., Mendling, J.: Measuring similarity between business process models. In: Bellahsène, Z., Léonard, M. (eds.) CAiSE 2008. LNCS, vol. 5074, pp. 450–464. Springer, Heidelberg (2008). doi:10.1007/978-3-540-69534-9_34
6. Curran, T., Keller, G., Ladd, A.: SAP R/3 Business Blueprint: Understanding the Business Process Reference Model. Prentice Hall PTR, Upper Saddle River (1998)
7. Becker, J., Kugeler, M., Rosemann, M.: Process Management: A Guide for the Design of Business Processes. Springer Science & Business Media, Heidelberg (2013)
8. EU Directive: 95/46/EC of the European Parliament and of the Council of 24 October 1995 on the protection of individuals with regard to the processing of personal data and on the free movement of such data. Off. J. EC. 23 (1995)
9. Drozd, O.: Privacy pattern catalogue: a tool for integrating privacy principles of ISO/IEC 29100 into the software development process. In: Aspinall, D., Camenisch, J., Hansen, M., Fischer-Hübner, S., Raab, C. (eds.) Privacy and Identity 2015. IAICT, vol. 476, pp. 129–140. Springer, Cham (2016). doi:10.1007/978-3-319-41763-9_9
10. Parlamento Europeu, Conselho da União Europeia: GDPR - EUR-Lex - 32016R0679 - EN. J. Of. da União Eur. 59 (2016)
11. Tucker, P.: Has big data made anonymity impossible? MIT Rev. **116**, 64–67 (2013)
12. Guarda, P., Zannone, N.: Towards the development of privacy-aware systems. Inf. Softw. Technol. **51**, 337–350 (2009)
13. Laudon, K.C., Laudon, J.P.: Management Information Systems 13e (2013)
14. Pfitzmann, A., Hansen, M.: A terminology for talking about privacy by data minimization: anonymity, unlinkability, undetectability, unobservability, pseudonymity, and identity management (2010)

15. Sommerville, I., Kotonya, G.: Requirements Engineering: Processes and Techniques. Wiley, Hoboken (1998)
16. ISO: 31000: 2009 risk management–principles and guidelines. Int. Organ. Stand. Geneva, Switz. (2009)
17. ISO Guide: 73: 2009. Risk Manag. (2009)
18. Backes, M., Pfitzmann, B., Waidner, M.: Security in business process engineering. In: Aalst, W.M.P., Weske, M. (eds.) BPM 2003. LNCS, vol. 2678, pp. 168–183. Springer, Heidelberg (2003). doi:10.1007/3-540-44895-0_12
19. El-Attar, M., Luqman, H., Karpati, P., Sindre, G., Opdahl, A.L.: Extending the UML statecharts notation to model security aspects. IEEE Trans. Softw. Eng. **41**, 661–690 (2015)
20. Nunes, F.J.B., Belchior, A.D., Albuquerque, A.B.: Security engineering approach to support software security. In: 6th World Congress Services, pp. 48–55 (2010)
21. Abie, H., Aredo, D.B., Kristoffersen, T., Mazaher, S., Raguin, T.: Integrating a security requirement language with UML. In: Baar, T., Strohmeier, A., Moreira, A., Mellor, Stephen J. (eds.) UML 2004. LNCS, vol. 3273, pp. 350–364. Springer, Heidelberg (2004). doi:10.1007/978-3-540-30187-5_25
22. Danezis, G., Domingo-Ferrer, J., Hansen, M., Hoepman, J.-H., Metayer, D.L., Tirtea, R., Schiffner, S.: Privacy and data protection by design-from policy to engineering. arXiv Preprint arXiv:1501.03726 (2015)
23. Oladimeji, E.A., Supakkul, S., Chung, L.: Security threat modeling and analysis: a goal-oriented approach. In: Proceedings of the 10th IASTED International Conference on Software Engineering and Applications, SEA 2006, pp. 13–15 (2006)
24. Pfitzmann, A., Kiel, U.L.D.: Pseudonymity, and identity management – a consolidated proposal for terminology. Management 1–83 (2008)
25. Wuyts, K., Scandariato, R., Joosen, W.: Empirical evaluation of a privacy-focused threat modeling methodology. J. Syst. Softw. **96**, 122–138 (2014)

Suitability of BPMN Correct Usage by Users with Different Profiles: An Empirical Study

Anacleto Correia[1(✉)] and António Gonçalves[2]

[1] CINAV, Alfeite, 2810-001 Almada, Portugal
cortez.correia@marinha.pt

[2] INESC-ID, UL-IST, Avenida Rovisco Pais, 1, 1048-001 Lisbon, Portugal
antonio.goncalves.pt@gmail.com

Abstract. A declared purpose of the BPMN standard was to provide a business process modeling language, amenable of being used for modelers regardless of their technical background. This aim was intended to be achieved by extensive documentation of the syntax rules of the notation, as well as by proposed best practices for process modeling from practitioners. The wide acceptance of BPMN standard seems to accomplished the mentioned purpose, namely when considering its usage in business oriented process documentation and improvement scenarios, as well as in IT implementation of process diagrams supported by software tools. However, a relevant question can be raised regarding the correctness of business process diagrams produced by modelers with different profiles. This issue is important since the conformance of produced process diagrams to the syntax rules of the language determines the quality of the modeling process whatever its purpose is. Therefore, the main aim of this work was to gather statistical evidence that could validate the assertion that, BPMN diagrams, they have the same level of correctness, irrespective of the technical profile of people involved in modeling tasks. This paper reports a between-groups empirical study with business-oriented and IT-oriented profiles modelers.

Keywords: Business process modeling · BPMN · Empirical study

1 Introduction

The Business Process Model and Notation (BPMN) standard [1], published by the Object Management Group (OMG), has as one of its main purposes, to ensure a visual language that could be understandable by all types of modelers and users, namely: (1) business analysts that sketch the initial documentation of business processes; (2) process developers which play an essential role in actual implementation of business processes; (3) business users which are responsible for the instantiation and monitoring of ongoing business processes.

The requirement for accurate specification of process diagrams is relevant for several operations in the organization, that is when delivering specifications of processes to meet legal and regulatory conditions (such as BASEL III and Sarbanes-Oxley Act), as well as for analysis, design and development of information systems

O. Gervasi et al. (Eds.): ICCSA 2017, Part I, LNCS 10404, pp. 677–692, 2017.
DOI: 10.1007/978-3-319-62392-4_49

supporting business process [2], based on service-oriented architectures [3], and implemented by web services or alike technologies [4].

Although BPMN notation can be used for the mentioned purposes, the produced diagrams are quite different regarding their business or technical nature. In general business diagrams construction has the main focus on documentation, and on a better understanding of the regular process flow. Hence, the emphasis on the so-called *happy path*, i.e. the common flow that main activities take [5]. Abnormal situations are often bypassed and exception handling constructs avoided. On other hand, business processes' execution capabilities, available in the BPMN language, are relevant when IT-oriented diagrams are built. Process developers focus on BPMN diagrams that can be translated into machine-readable languages. For those professionals, the target is the enactment of process diagrams in distributed environment (e.g. integrating BPEL and web services standards), shared across multiple domains, and supported by multiple technologies. Though, the process modeling language BPMN supports different uses. These utilizations rely on different kinds of elements either for a graphical depiction of diagrams, as for configuration of processes' simulation and execution. Nevertheless, both kinds of BPMN diagram must be syntactically correct and well-formed. To validate the suitability of BPMN to be used by distinct profiles of modelers, as stated by OMG documentation, empirical evidence should be collected.

In this paper, the authors reported an experiment conducted with two kinds of modelers (business-oriented and IT-oriented profiles), in order to assess the adequacy of BPMN as modeling language irrespective of the technical background of the modeler. To gather statistical evidence regarding this issue, a between-groups empirical study was conducted with modelers with the two different profiles.

This work is structured as follow. In Sect. 2 the applied research method is detailed. The empirical study is described in Sect. 3, by going through the research definition (Sect. 3.1), planning (Sect. 3.2), execution (Sect. 3.3), data analysis (Sect. 3.4), and analysis of results (Sect. 3.5). In Sect. 4 previous related work on process modeling is outlined. Finally, Sect. 5 concludes the paper and suggest future work.

2 The Research Method

The process of making science requires that other researchers being able of replicate, under identical conditions, previous research. Therefore, the selection of the research method to be followed is an important requirement for attaining grounded conclusions of the phenomenon under study. The framework of the research method provides guidelines for the planning of activities to be developed, resources to be assigned, and also allows envisioning the factors that influence the phenomenon under analysis.

In this work, we chose a research method based on the scientific method, customized according the frameworks used in the Experimental Software Engineering field [6, 7]. With this assumption, the following main activities were proposed to be developed in the context of the current work:

- *Research Definition* – this set of activities addresses the research problem by formulating the research question concerning expected results, the objective of the experiment, as well as the characterization of the context of the experiment to be carried out (Sect. 3.1);
- *Research Planning* – these activities details how the experiment will be performed. The formulation of the hypothesis under study is the basis for the experiment's design in order to get the answer regarding the research question. Also necessary is the selection of the set of independent/dependent variables to be used in the statistical tests, as well as the selection of participants in the study. The experiment should be designed to filter out external factors and to avoid possible bias regarding the results. The instrumentation of the experiment and the preliminary evaluation of validity close this phase (Sect. 3.2);
- *Research Execution* – in this phase, the previously established plan, constrained by the specific circumstances for the actual experiment, is instantiated. Concomitantly, data is collected in order to gather empirical evidence (Sect. 3.3);
- *Research Data Analysis* – the activities in this phase consist of the description and reduction of the data set, as well as the test of the hypothesis formulated during the experiment's planning (Sect. 3.4);
- *Results Analysis* – this set of activities consists in packaging the results so that they can be used by the community. The phase aims to ensure the comparability of the current research with results achieved by future researchers that would eventually repeat the experiment. This consists in documenting the whole experimental process, discuss the experiment's results validated by the statistical analysis, address the results' interpretation, discuss the limitations of the study, the adequacy of the inferencing process to be generalized to the population of BPMN modelers, and highlight of lessons learned for future experiments (Sect. 3.5).

The following sections detail the actual empirical study took place that instantiated the above-mentioned activities.

3 Empirical Study on BPMN Correct Usage

3.1 Definition of the Empirical Study

Following the guidelines of the framework previously proposed, this section addresses the activities that constitute the experiments' definition. Those activities are:

- specification of the *research problem* and justification of the relevance of the empirical study;
- formulation of the *research objective*, highlighting the aim of the study; and
- *definition of the experiment's context*, by characterizing the environment that constrains the research problem and the objective.

Research Problem

The research problem under investigation is concerned with the assessment of the effectiveness of different kinds of modelers being able of delivering BPMN process diagrams with the same quality (i.e., process diagrams with the same degree of correctness). Therefore, the sample to be tested should come from BPMN diagrams produced by modelers with different technical backgrounds. The degree of correctness of process diagrams produced by a modeler with specific profile should be measured and compared with the level of correctness of diagrams built by modelers with different background.

From the attained results it is expected to confirm (or not) whether the process modeling language BPMN, can be used to produce process diagrams with the same degree of correctness, irrespective of modelers technical background.

Research Objective
For addressing in a rigorous and systematic way the above-mentioned problem, and to delimit the empirical study boundaries, the *research question* was formulated as follow:

What is the likelihood of having BPMN diagrams with the same level of correctness, produced by modelers with different technical profiles?

The activities described next were developed in order to get the answer to the stated research question. First, the research question was refined into a research goal, which in turn, drove to the formulation of a research hypothesis. To provide a grounded answer to the research question, one had to analyze a sample of process diagrams built by modelers with distinct technical background.

The setup of the experiment was based on the Goal-Question-Metric (GQM) framework [8]. The GQM model has a tree structure, with a top-level goal as the root of the tree. Starting from this goal several questions are derived by breaking down a problem into its major parts. For each question, a metric is derived. The same metric can be used to answer different questions regarding the same goal. The GQM framework is operationalized through a *template* with five topics, regarding the research *object of study, purpose, quality focus, viewpoint, and environment*. The template mark out the experiment's boundaries by focusing on the relevant goal, specifying the elements to measure, the dependent and the independent variables, and the hypothesis to be raised.

For the current empirical study, the instantiation of the framework was made by matching the research objective of the study with the GQM's top-level goal and considering this goal in the terms outlined by the above-mentioned template. So, the GQM for the empirical study was formulated as:

Analyze BPMN diagrams (*object of study*),
for the purpose of checking compliance of BPMN syntax by business analysts and developers (*purpose*),
with respect to the assessment of diagrams' degree of correctness (*quality focus*),
from the point of view of the OMG standard document (*viewpoint*),
in the context of an experimental study **constrained by** using surrogates of process modelers (*environment*).

The last item of the GQM instantiation, the context of the experiment, raises the question whether the experimental results are generalizable to a broader context. It was considered relevant to underline the context for the sake of comparability of the results since each experiment has its distinct context.

Definition of the Context Study

The study presented herein was developed by collecting a sample of BPMN diagrams produced in the context of a case study, in an experiment performed with students attending a course on process modeling taught at an academic institution.

Although the conclusions of this empirical study can be generalizable to other contexts of process modeling, caution must be taken, and further research is required to confirm the attained results.

3.2 Empirical Study Planning

This set of activities aims to the developing the topics regarding the definition of the research goal, formulation of hypothesis under study (together with the dependent and independent variables to be used for hypothesis' testing), the definition criteria for selection of participants in the experiment, the design and instrumentation of the experiment, as well as the assessment of the experiment's validity.

Research Goal

The current empirical study has only one research objective and therefore no sub-goal.

Hypothesis and Variables

The hypothesis presented here is intended to verify the existence of empirical evidence that could corroborate the claim that BPMN is a general purpose process modeling language, amenable of be used by any kind of process modeler. By not rejecting this hypothesis it was expected to contribute to back up the assertion of BPMN being suitable both for business analysts and process developers. Thus, the goal previously settled drove to the null hypothesis formulation in the following terms:

- H_0: the quality of BPMN diagrams, measured in terms of the degree of correctness, is not significantly different for modelers (business analysts or developers) with distinct technical background.

The variables (independent and dependent) used for the H_0 hypothesis are listed in Table 1. The dependent variable is related to the number of errors detected in the BPMN diagrams, as result of violations of the language syntax defined by the BPMN standard. The description of each variable in Table 1, refers the attribute to be measured, the applied computation rule, and the assigned measurement unit.

Table 1. Independent and dependent variables for H_0.

Variable role	Variable name	Description
Independent	Modeler type	Identifies the technical background of the modeler (business analyst or process developers) in the sample. Nominal variable
Dependent	Total of errors	The number of BPMN syntax rules violated in the model. An absolute scale variable

Selection of Participants in the Study

A blended strategy for sampling was used characterized by a *simple organization*, with all participants treated equally in the sample, and a *convenience sampling*, with participants chosen due to their easy availability.

The experiment was conducted at an academic institution with participants, although attending two different degrees, having a course with the same content. The subjects were undergraduate students at the first cycle, in the 2nd semester of the academic year. The features of the two degree programs were the following:

- *Technology and Industrial Management* - a degree with courses including subjects covering information systems (with BPMN as process modeling language) and industrial processes. Since the topics taught were more business and industry oriented, the students attending the 2nd year of the degree were chosen as surrogates of business analysts;
- *Informatics Engineering* - this degree includes computer science subjects, covering procedural, object-oriented and functional programming languages, systems analysis and design (supported by UML) and process modeling (supported by BPMN). Since there were more topics related to software development in this degree, students from the 3rd year of this degree were chosen as surrogates of process developers.

Students of both degrees did attend a simultaneous class on BPMN process modeling, so this was a harnessed opportunity to them together in the same experiment.

The course on process modeling was taught on a blended-learning context with students making their first initiation to the BPMN language, with regular classes supported by an e-learning platform. Course materials (course's notes and tools' tutorials) were provided to the students regarding BPMN diagramming, as well as the course's assignments. All the assignments contributed to the final score of the course with a certain weight. One of the mandatory assignments the students had to accomplish was the BPMN case study that was used as part of the current experiment.

The total number of participants in the experiment was fifty-one (with eighteen acting as developers and thirty-three as business analysts). The experiment was developed in two phases:

- *Initial Training* - this phase was designed to be a period for students get acquainted with the BPMN language. So, the outputs generated in this phase were discarded from data analysis. Detailed specifications about the financial services supported by Automated Teller Machines (ATMs), were provided to the students. A deadline of one week was given to the students for accomplishing the first assignment, consisting of drawing a first draft of an ATM system using BPMN.

The intent of this phase was mainly that students could: (1) learn with the syntax of BPMN constructs and the subtleties of their usage; (2) practice process modeling concepts taught in regular classes when eliciting the activities of the modeled system; (3) get familiarized with a CASE tool provided, amenable of basic syntactical checking of BPMN diagrams.

- *Instantiation of the Experiment* – for the actual tasks of process modeling, required by the experiment, a laboratory class, with a time frame of two hours, was scheduled. At class's beginning, the instructor explained to the students what they were expected to do for accomplishment of the assignment, namely: (1) a carefully read of the business process case study description (an extract of the overall case previously made available); (2) complete the BPMN diagram and storing it within the database of the BPMN tool. As part of this database, a baseline of the solution was provided to the participants in the experiment, in order to narrow the set of diagrams delivered by students, to a specific part of the overall problem.

As previously mentioned, the primary goal of the experiment was to evaluate of BPMN diagrams' correctness assuming that they were built by modelers with two distinct modeling profiles (business analysts and IT developers).

Experimental Design

In this section the design of the experiment is described, as well as the role of students, playing as surrogates of professional modelers, on artifacts production.

For the non-experimental study conducted to test the H_0 hypothesis, a quasi-experimental design was chosen. The quasi-experimental experiment design considers the selection of groups related to the variable under test, irrespective of any random pre-selection phase. So, in quasi-experimental studies, even the researcher considers the study as an experiment, although, strictly speaking, it is not. Because the treatment and control groups cannot be randomized or matched, there is no control group or the independent variable cannot be manipulated, so the researcher has limitations on drawing conclusions [9].

In empirical studies carried out in academic contexts, it is common match groups with classes and therefore mounting quasi-experiments with non-equivalent groups. So, for a case study such the one described here, the time and resources made available for the experiment, as well as the constraints of academic procedures, turns out unfeasible to constitute a control group. On the other hand, without students' randomization, the time and resources assigned to the experimentation could be reduced. Given the mentioned constraints, the quasi-experimental design seemed so, the best option to use. After all, the gathered figures allowed some sort of statistical analysis.

Therefore, the participants in the experiment were assigned to non-equivalent groups and submitted to an equal treatment. The splitting of participants between groups was *convenient* in order to reduce the disruption in classes to the minimum. Although the probabilistic equivalence of groups was lost, they could still be compared. To compose the non-equivalent groups, the students were assigned based on the academic year they belonged. The two groups, each one with modelers with different profiles, delivered BPMN diagrams during the experiment. The testing factor they use was the same: the assigned case study. One group, with business analysts' profile, was more knowledgeable regarding the domain and had more skills related to organizational processes. The other group, with developers' surrogates, had previous modeling knowledge and proficiency on the UML graphical notation, as well as programming skills. The expected results were that any divergence would be due to the fact that, unlike advocated by the BPMN standard [1], the notation was not equally suitable for business analysts and process developers.

For assessment of the effectiveness of modelers' interventions a post-test was used. The two groups received the treatment, for the same period of time, and got exactly the same support from the instructor in charge of monitoring the experiment. We relied on statistical analysis to measure the extent by which the profile of the groups had a significant relevance on the attained results.

The applied post-test quasi-experiment with non-equivalent groups and between-subjects design can be represented as:

$\mathbf{N_1\ X\ O}$
$\mathbf{N_2\ X\ O}$

This notation means that each non-equivalent group (N_1 and N_2) was subject to a treatment (X) and the observation (O) of the responses.

Collection Procedure

The data collected from the sample was generated by students, acting as surrogates of modelers, in a time frame of two hours. Their assignment was to complete a baseline, provided in a repository with part of the case study solution, by drawing the absent BPMN diagrams. The new diagrams had to be delivered in the same repository, a file format readable by the CASE tool used for depicting the BPMN model.

After collecting all the files with the solutions delivered by the subjects participating in the experiment, the files were loaded into a models' checker tool, in order to verify the BPMN diagrams for syntax violations. The models' verification was automatically made, through the BPMN diagram checker tool described in [10]. This tool allows for the verification of BPMN diagrams against the BPMN metamodel, enriched with well-formedness rules and best practices suggested by practitioners, implemented as OCL invariants. The BPMN syntax violations detected in diagrams were recorded in a text file. Next, those files were processed and transformed into a data file readable by the SPSS statistical tool. Eventually, the statistical treatment required for testing the raised hypothesis could be run.

Instrumentation

For instrumentation of the empirical study there were some off-the-shelf tools required. Each one of the tools was used independently of the other, following the pipes and filters software architectural style [11]. The role of each tool is detailed next:

- *CASE Tool (Enterprise Architect)* - an editor of BPMN diagrams used by modelers to build the solution for the case study;
- *Eclipse* - the IDE for building the BPMN2USE Java application used for querying and extracting from the Enterprise Architect repository, the solution built by each process modeler;
- *SPSS* - this tool is a statistical package used for processing and analysis of the collected sample data, as well as to perform the hypothesis test;
- *UML based Specification Environment* (USE) - a tool that allows the specification of models and metamodels using the UML class diagram supplemented by OCL constraints and invariants that enforce models' integrity [12].

- *BPMN2USE* - a transformation tool that takes as input a BPMN diagram produced with the CASE tool and allows the instantiation of the BPMN metamodel through the USE specific syntax [13].

Furthermore, as part of the instrumentation process, the participants in the experiment received support material concerned to the BPMN language, as well as a baseline with the template of the ATM case study solution.

Threats to Validity
One of the kinds of threats that is inherent to empirical studies is an inconsistent administration of the treatment for different groups. These variations could happen if different people apply the treatment to the group, as it was the case, since different instructors taught the classes that were part of the study.

3.3 Empirical Study Execution

After the planning phase, the experimental work was carried out on a laboratory class of the course. As previously mentioned, the case study assignment was the basis for data collection regarding the experiment together with the pedagogical objective of practicing the BPMN concepts acquired on the course. The students that participated in the experiment concluded it, so there were no subjects' mortality regarding the experiment. Nondisclosure of individual responses was ensured to all participants.

Preparation
Before starting the experiment, at the beginning of the class lab, the students received the case study's description, as well as the repository of the CASE tool with the baseline of the assignment's solution. Students were also briefed about the data that would be collected during the class. However, they were not informed about the details of the research, since this could threaten the results' validity. Since the case study contributed to students' final score in the course, there was a motivation for performing well the process modeling assigned tasks.

Data Collection
The results' verification took place after the students' participation in the experiment, as part of the lab class. The verification of the BPMN diagrams was off-line, after the class, so it did not interfere with the solutions' delivery by students. A panel of instructors carried out the verification process, using mainly the BPMN2USE and USE tools. After finishing the verification of the experiment's outputs, information regarding the number of errors found in the diagrams was made available for analysis. For the present experiment was not considered relevant the severity of the errors but how much of them were found in a process model.

3.4 Empirical Study Data Analysis

In this phase collected data regarding design errors, incurred by modelers with different backgrounds, were analyzed. The set of those activities is detailed in the next sections by the description of data sets, as well as the test of hypothesis raised in the experiment's planning.

Data Description

In this section, data exploration begins by describing the variables in the data set. In Table 1 we already identify the dependent variable *Total of Errors* (total of found errors found in diagrams) and the independent variable *Modeler Type* (possible modelers' background). The positive skewness and kurtosis (Table 2) of the dependent variable indicates an asymmetric (*skewness* = 0.792 > 0) and leptokurtic distribution (*kurtosis* = 1.19 > 0), with higher frequency of lower values (see Table 2). This behavior of global errors distribution (irrespective of process modeler background) is similar to the errors distribution considering each particular type of modeler.

Table 2. Descriptive statistics.

Statistic	Total of errors
N	51
Mean	8.12
Median	8
Mode	8
Std. deviation	3.09
Skewness	0.792
Kurtosis	1.19
Minimum	3
Sum	414

Table 3 summarizes the results of the normality test. The null hypothesis states that the sample comes from a Gaussian (normal) distribution. Conversely, the alternative hypothesis assumes that the sample comes from a non-normal distribution. The non-normality of the variable is confirmed, since the p-value (sig) < 0.05 for both tests regarding the *Total of Errors* variable. Therefore non-parametric tests must be used to assess H_0 hypothesis.

Table 3. Normality test.

	Kolmogorov-Smirnov[a]			Shapiro-Wilk		
Variable	Statistic	df	Sig.	Statistic	df	Sig.
Total of errors	0.143	51	0.011	0.952	51	0.038

[a]Lilliefors Significance Correction

For verification of the H_0 hypothesis, the collected data must be submitted to a statistical test. The result of the test must show whether there is enough statistical evidence for not rejecting the claim that the average number of errors in BPMN diagrams is not significantly different for the two considered types of modelers (business analysts or process developers).

The *Mann-Whitney U* (M-W U) test for two unpaired samples was the non-parametric statistical test chosen for verifying whether the means of two samples differ significantly. The considered samples were the two distinct sets of BPMN diagrams built by surrogates of business analysts and process developers. The aim was to test whether the means of *Total of Errors* for both sets of BPMN diagrams differ significantly. By running this hypothesis test, it was gathered statistical evidence regarding the quality (correctness) of BPMN diagrams, namely whether they diverge significantly considering the technical background of modelers.

Hypothesis Testing

As previously mentioned, the hypothesis H_0 aims to verify the relevance of the discrepancy between on the number of syntax errors found in process diagrams built by modelers with distinct profiles. To conclude about the hypothesis it was necessary to compare the means of the two unpaired groups of BPMN diagrams and test whether the two statistics differ significantly. The independent variable considered was *modeler type* variable. This variable was used to separate the BPMN diagrams in two groups: one regarding diagrams built by business analysts and the other considering diagrams delivered by process developers.

The Mann-Whitney U test as a non-parametric test aims to assess whether the two samples came from the same population. In Table 4 is summarized the information concerning the computed ranks for the errors in the BPMN diagrams' sample. The computation of the test started by ranking all the errors observed in models, with values sorted in descending order, disregarding the modeler type of the diagram. The columns of Table 4 present the statistics attained by modeler type (1- Process Developer, 2- Business Analyst), the corresponding counting of BPMN diagrams (*N*) delivered by each group, the errors mean (*Mean Rank*) and the total sum of errors found (*Sum Ranks*) in the diagrams produced by each group.

Table 4. Ranks for H_0.

Variable	Modeler type	N	Mean rank	Sum of ranks
Total of errors	1- Process Developer	18	27.81	500.5
	2-Business Analyst	33	25.02	825.5
	Total	51		

Eighteen of the BPMN diagrams analyzed were producede by developers, while the remaining thirty-three came from business analysts. The M-W U test is summarized in Table 5, which lists: the *Mann-Whitney U* statistic, the *Wilcoxon W* statistic, the test's *Z score*, and the 2-tailed asymptotic significance (*Asymp.Sig (2tail)*). This test allows the conclusion for non-rejection of the null hypothesis at 5% of level of significance (since *Asymp. Sig.*, i.e. p-value greater than 0.05).

The Two-Sample Kolmogorov-Smirnov test confirmed the results from the Mann-Whitney test. The Two-Sample Kolmogorov-Smirnov test is also a non-parametric test that measures the difference in the shapes of the distributions of errors for the two groups of BPMN diagrams. This test also uses the rank classification mentioned regarding Table 4.

In Table 6 is presented the summary of the test's results. One can see that the short values of the most extreme difference for absolute (*Absolute*) and positive (*Positive*) values (0.177), as well for the most extreme difference negative (*Negative*) of -0.061. A *Kolmogorov-Smirnov Z* score of 0.603 and a 2-tailed asymptotic significance *Asymp. Sig. (2 tail)*, i.e. p-value greater than 0.05 corroborate the results presented for the Mann-Whitney U test.

The achieved results indicate that the errors in BPMN diagrams produced by business analysts do not significantly different from the syntax rule violations found in BPMN diagrams produced by process developers.

Table 5. Mann-Whitney U test for the *Total of Errors* variable (Modeler type is the grouping variable).

Statistic	Value
Mann-Whitney U	264.5
Wilcoxon W	825.5
Z	−0.645
Asymp. Sig. (2-tailed)	0.519

Table 6. Two-sample Kolmogorov-Smirnov test for the *Total of Errors* variable (Modeler type is the grouping variable).

Most extreme differences	Absolute	0.177
	Positive	0.177
	Negative	−0.061
Kolmogorov-Smirnov Z		0.603
Asymp. Sig. (2-tailed)		0.86

3.5 Empirical Study Results

The final discussion refers to the interpretation of the results of this empirical study, the extent of which the results found can be considered representative of the population of process modelers, as well as with the identification of the learned lessons.

Interpretation

The outcome of the hypothesis test was that the empirical evidence did not allow to reject the following hypothesis: *diagrams built by business analysts and process developers using the BPMN language have the same degree of correctness.*

This evidence is in line with the claimed suitability of BPMN [1], as process modeling language, be usable by modelers from both business and IT. Indeed, data collected corroborate that business analysts are able to cope with BPMN constructs and rules as the process developers. However, since both incur in faults, during process modeling, one can also conclude that automatic BPMN model checker would be beneficial for both modelers giving hints and alerting for syntactical errors occurring throughout the modeling process.

Inferences

The evidence collected throughout the experiment suggests that the results achieved by participants, which revealed similar BPMN diagramming skills regardless the modelers' technical profile, can be extrapolated for undergraduate students from the two degrees of the academic institution they came from. Assuming that these students have basically a similar academic profile of those of other undergraduate institutions with equivalent degrees, the results may also be generalized to the students of those institutions. However, this assumption should be tested by replicating this experiment in those institutions.

One could expect that results found could hold as well for the entire population of modelers irrespective of their profiles. However, evidence is needed to be collected that could confirm that results attained by students, with a similar profile of the ones of the current experiment, are equivalent to those achieved by novice professionals. This inference should be supported through the replication of the conducted empirical study in a professional environment. Extrapolating the observed behavior for seasoned experimenters require also that other studies must be conducted.

Lessons Learned

A great deal of time and effort was required for experiments' preparation and execution. After collecting BPMN diagrams a pipeline of applications was feed in to prepare data for analysis. Data converters and transformers were also used for automating repetitive tasks.

While conducting the empirical study, we realize that some steps on the experimental process could be improved. Also, different approaches could be used, in future replications of this experiment, to face some of the challenges found, namely:

- Narrow the scope of the case study, in order to include in the analysis not only the counting of the number of errors in diagrams but also the degree of coverage of presented solutions, of functional requirements specified by the case study;
- Consider a new version of the case study, with a treatment consisting in using an automatic model checker by participants. Modelers should also be able to decide which rules they want to enforce upon BPMN diagrams (e.g. control-flow vs. data flow, or sets of best practices or standard rules);
- Measure the effects on students' learning curve regarding BPMN, when using an automatic model checker for verifying in real time the diagram being modeled.

The details of the experimental process, pursued in the empirical study, were recorded. Also registered was the feedback provided by students while learning and applying BPMN to the case study. Those data were particularly valuable for packaging the experiment and writing this paper, as well as for a future replication of the empirical study.

4 Related Work

Empirical studies regarding BPMN characteristics for different users' profiles are not found. However, we could found some empirical studies analyzing BPMN characteristics, as well as comparing BPMN with other modeling languages. Those studies are presented next highlighting the main differences between those works and the one presented here.

In [14] a set of measures is presented to evaluate the structural complexity of business process diagrams at a conceptual level. However, conversely to the work presented herein, there is only an experimental plan and not the actual results. The plan refers the development of a family of experiments to be applied by experts in business analysis and software engineering. The intention is validating some proposed metrics and evaluating, at a conceptual level, quality aspects of the business process diagrams.

The previous study was further developed in [15], through the empirical validation of the measures with a linear regression analysis aimed at estimating process model quality in terms of modifiability and understandability. The study was applied to a homogeneous sample of participants (students of computer science and information systems), which differs from the heterogeneous groups of students used in our study, with surrogates of process developers and business analysts. As result of carrying out a correlation and a multiple linear regression analysis from the data collected, it was identified a reduced group of measures useful in predicting several aspects when evaluating the understandability and modifiability of business process diagrams expressed with BPMN. From the analyzed measures, after carrying out a correlation and a principal components analysis of the variables, the authors concluded that 12 of the measures are useful for predicting aspects of understandability of a business process model. With regard to modifiability, other measures were identified as good predicting variables. By crossing the data obtained, the authors also conclude that some measures can be considered good predictors for both dependent variables. According to the authors, the regression diagrams obtained represent a guideline for defining understandable and modifiable processes or for predicting such characteristics in those which already exist. The measures seem to be also useful for guiding process improvement initiatives.

In [16] was presented the results of an empirical study that examines the BPMN and the UML Activity Diagram (UML-AD) modeling by business users during a model creation task. This study, differently of the empirical study presented here, compare BPMN with another modeling language, and the results indicate that the (UML-AD) is at least as usable as BPMN since neither the characteristics of user effectiveness, efficiency, nor satisfaction differ significantly.

In a study presented in [17] BPMN is compared against Event-Driven Process Chains (EPC). The study measured the comprehension and problem-solving capabilities of students as surrogates of business users. No significant differences were identified between the process modeling languages, and the study recognized to have a different focus: examine teaching effects and therefore compare the performance of trained participants in a EPC group versus untrained participants in a BPMN group.

5 Conclusions

This work intended to validate, through an empirical study, whether the BPMN is a process language suitable for modelers with different technical skills. This was done by assessing whether diagrams produced by people with different technical profiles have the same degree of compliance with BPMN syntax rules.

As a research method, a customized version of the scientific method was used, based on previous frameworks applied to experimental software engineering.

The collected empirical evidence did not allow to reject the hypothesis that BPMN is also suitable for business analysts and process developers since the diagrams built by both profiles have the same degree of correctness.

Future works will be focused on verifying the results presented here, by replacing surrogates for actual professionals working as business analysts and process developers in the industry.

Acknowledgment. This work was supported by Portuguese funds through the Center of Naval Research (CINAV), Portuguese Naval Academy, Portugal.

References

1. OMG: Business Process Model and Notation (BPMN). dtc/2010-05-04 (2011)
2. Dumas, M., et al. (eds.): Process-Aware Information Systems. Wiley, Hoboken (2005)
3. Erl, T.: SOA: Principles of Service Design. Prentice Hall, Upper Saddle River (2007)
4. Indulska, M., et al.: Measuring method complexity: the case of the business process modeling notation. BPM Center Report (2009)
5. Silver, B.: BPMN Method and Style. Cody-Cassidy Press, New York (2009)
6. Jedlitschka, A., Pfahl, D.: Reporting guidelines for controlled experiments in software engineering. In: Proceedings of the 4th International Symposium on Empirical Software Engineering (ISESE 2005), pp. 95–104. IEEE Computer Society, (2005)
7. Kitchenham, B.A., et al.: Preliminary guidelines for empirical research in software engineering. IEEE Trans. Softw. Eng. **28**(8), 721–734 (2002)
8. Basili, V.R., et al.: The goal question metric approach. Encycl. Softw. Eng. **2**, 528–532 (1994)
9. AcademyHealth: Research Methods and Techniques. http://www.hsrmethods.org/glossary.aspx. Accessed May 2014
10. Correia, A.: Quality of process modeling using BPMN: a model-driven approach. Ph.D. thesis, DI, UNL-FCT, Lisboa (2014)

11. Bass, L., et al.: Software Architecture in Practice. Addison-Wesley/Pearson Education, Boston/Upper Saddle River (2003)
12. Richters, M., Gogolla, M.: Validating UML models and OCL constraints. In: Evans, A., Kent, S., Selic, B. (eds.) UML 2000. LNCS, vol. 1939, pp. 265–277. Springer, Heidelberg (2000). doi:10.1007/3-540-40011-7_19
13. Gogolla, M., Bohling, J., Richters, M.: Validation of UML and OCL models by automatic snapshot generation. In: Stevens, P., Whittle, J., Booch, G. (eds.) UML 2003. LNCS, vol. 2863, pp. 265–279. Springer, Heidelberg (2003). doi:10.1007/978-3-540-45221-8_23
14. Aguilar, E.R., et al.: Evaluation measures for business process diagrams. In: Proceedings of the 2006 ACM Symposium on Applied Computing, pp. 1567–1568. ACM (2006)
15. Rolon, E., et al.: Prediction diagrams for BPMN usability and maintainability. In: IEEE Conference on Proceedings of the Commerce and Enterprise Computing, CEC 2009, pp. 383–390. IEEE (2009)
16. Birkmeier, D., Overhage, S.: Is BPMN really first choice in joint architecture development? An empirical study on the usability of BPMN and UML activity diagrams for business users. In: Heineman, G.T., Kofron, J., Plasil, F. (eds.) QoSA 2010. LNCS, vol. 6093, pp. 119–134. Springer, Heidelberg (2010). doi:10.1007/978-3-642-13821-8_10
17. Recker, J.C., Dreiling, A.: Does it matter which process modelling language we teach or use? An experimental study on understanding process modelling languages without formal education (2007)

Analysis of Tweets to Find the Basis of Popularity

Rajat Kumar Mudgal and Rajdeep Niyogi(✉)

Department of Computer Science and Engineering,
Indian Institute of Technology Roorkee, Roorkee 247667, India
rajatmudgal17@gmail.com, rajdpfec@iitr.ac.in

Abstract. Smart and intelligent recommendation systems can be designed based on analyzing the tweets. Our work is aimed at analyzing the tweets to find the basis of popularity of a person. Although there are some works that have analyzed tweets to detect popular events, not much emphasis has been given to find out the reason behind the popularity of a person based on tweets. In this paper, we suggest an algorithm to find out the reason behind the popularity of a person. We have implemented our algorithm using 2,18,490 tweets of 5 different countries. The results are quite encouraging.

Keywords: Popular person · Twitter user · Popularity · Event detection

1 Introduction

The advent of online social media and its continuous growing popularity has provided a new channel and arena for exchange and/or sharing of information [1, 2]. People on online social media have got an open platform to share opinions, viewpoints, and information on any topic. Over the last few years, Twitter, a micro-blogging service, has gained popularity as one among the most prominent information dissemination and news source agent. Exchange of messages on social media [3] increases considerably with the occurrence of an event that may be related to, for example, social cause, disaster, politics, or a particular person. Users sign in to their Twitter or other social media accounts, to either spread the information or to get updates about the information. Twitter can thus be used to analyze the ongoing situation since it is being used by public and thus it has the potential to provide real-time information. Content on Twitter supplies rich information related to the occurred activity. However, such abundant information is often not trustworthy since it may also contain fake information.

There has been a lot of interest to analyze twitter content [4] which includes, for instance, work in the field of event detection, user selection, and classification of tweets. Besides knowing information about popular people on twitter, it may be useful to know what event has caused the popularity. Such information can let the users know about the arena of popular person and other attributes of the person which can enhance the knowledge of users about famous personalities [5–7]. Moreover, a user may be

O. Gervasi et al. (Eds.): ICCSA 2017, Part I, LNCS 10404, pp. 693–704, 2017.
DOI: 10.1007/978-3-319-62392-4_50

interested to get suggestion about the people she wishes to follow on the basis of the area. The users would like to be kept updated with the currently ongoing events that may lead to the rise or fall of a known figure.

There are multiple sources of information like television, newspapers, social network sites or mouth to mouth words from friends and family [8]. A user may be interested to know about the person who is on everybody's mind in recent times and also wishes to know the reason behind it. To achieve this, initially the current popular persons are obtained from data. In order to find the reasons of popularity, categorization of tweets is carried out since a person may be popular because of more than one reason at a time but there would be one prime reason. By applying all these techniques we can provide better information to the users.

The aim of this paper is to design a method for detecting the popularity of a person and the reason causing the popularity. We use the tweets of different users related to a particular person. We used Twitter4j api in Java to collect the tweets, initially for user selection, and then later to get data about that user. This approach uses nouns in the tweets as their keyword and combines tweets together into a single reason when their match score is above some threshold. Classification of tweets to which category (like business, politics, technology etc.) is realized by categorizing keywords used in each tweet.

The paper is organized as follows. Section 2 describes the related work. Section 3 describes our method for popularity detection. Section 4 describes the implementation details and results obtained by our method. Conclusion and future works are given in Sect. 5.

2 Related Work

A considerable amount of work has been done in classification of tweets, sentiment analysis, and detection of events from tweets. Different approaches have been proposed for sentiment analysis, finding sentiments in words, sentences, topics. Some approaches use natural language processing, some uses pattern based approach and some takes into account machine learning.

In [9], a technique for constructing a Key Graph is suggested using the keywords in the tweets to detect events. This approach is dependent on the interdependency between the keywords. The Key Graph is comprised of nodes and edges where nodes correspond to keywords and the occurrence or the existence of two keywords simultaneously in a tweet is represented by an edge between the nodes. Clusters are created from the Key Graph by clustering different nodes together using a community detection algorithm. In [10], the authors suggest an algorithm called NED (new event detection) to detect events. It consists of two subtasks that are online and retrospective; online NED detects new events in the stream of text while in retrospective NED, unidentified events are detected.

Wavelet transformation is used for event detection in [11]. The problem of identifying events and their user contributed social media documents as a clustering task, where documents have multiple features, associated with domain-specific similarity metrics [12] and pheromone based techniques [13–15]. A general online clustering framework, suitable for the social media domain is proposed in [16]. Several techniques for learning a combination of the feature-specific similarity metrics are given in [16] that are used to indicate social media document similarity in a general clustering framework. In [16] a clustering framework is proposed and the similarity metric learning technique is evaluated on two real-world datasets of social media event content.

Location is considered in [17] with every event as incident location and event are strongly connected. The approach in [17] consists of the following steps. First, preprocessing is performed to remove stop words and irrelevant words. Second, clustering is done to automatically group the messages in the event. Finally, a hotspot detection method is performed.

TwitInfo is a platformfor exploring Tweets regarding to a particular topicis presented in [18]. The user had to enter the keyword for an event and TwitInfo has provided the message frequency, tweet map, related tweets, popular links [19, 20] and the overall sentiment of the event. The*TwitInfo*user interface contained following thing: the user defined name of the event with keywords in the tweet, timeline interface with y axis containing the volume of the tweet, Geo location along with that event is displayed on the map, Current tweets of selected event are colored red if the sentiment of the tweet is negative or blue if the sentiment of the tweet is positive and Aggregate sentiment of currently selected event using pie charts.

TwitterMonitor system is presented in [21] that detect the real time events in defined time window. This is done in three steps. In first step bursty keywords are identified, i.e. keywords that are occurring at a very high rate as compared to others. In second step grouping of bursty keyword is done based on their occurrences. In third and last step additional information about the event is collected.

A news processing system for twitter called as *TwitterStand* is presented in [22]. For users, 2000 handpicked seeders are used for collecting tweets. Seeders are mainly newspaper and television stations because they are supposed to publish news. After that junk is separated from news using the naïve Bayes classifier. Online clustering algorithm called leader-follower clustering to cluster the tweets to form events. A statistical method *MABED* (mention-anomaly-based event detection) is proposed in [23]. The whole process of event detection is divided in three steps. In first step detected the events based on mention anomaly. Second, words are selected that best describes each event. After deleted all the duplicated events or merged the duplicate events. Lastly, a list of top k events is generated.

In [24] a co-relation between clustering and event detection is shown. An aggregate trend change is similar to event detection. To find the popular event, authors of [24] have used algorithms based on community detection. In [26] to find the clusters the authors have suggested a hierarchical clustering of tweets along with the dynamic cutting and rating of resultant clusters is used, a similar technique has been applied in systematic search of maximal length codes [27]. In [28] a technique for finding bursty words is used for detecting events and location recognition using modules.

In [25] it has been stated that an event is associated with the message context but also with the location information, since location is also an important factor of an event. Localized events like any emergency event or any public event, emergency would be more accurately messaged or tweeted by the users closer to the event location in comparison to other users. Hence such users can play the role of sensors – human sensors for briefing an event.

A considerable amount of work has also been carried in the field of sentiment analysis that stresses on finding the sentiments in topics, sentences and the words. Various approaches have been suggested to carry out the sentiment analysis, these approaches either make use of natural language or pattern based processing or machine learning.

In [29] for sentiment analysis authors have suggested a sentiment tree bank approach that is based on a recursive neural network. It calculates in a bottom up manner the parent node vectors and takes advantage of a composition function and also the node vector that features for that node. In [30] an approach has been suggested for finding the sentiment score of informal, short text and also the sentences that consists of phrases within themselves.

Two methods for classification of the Twitter trending topics are proposed in [31] first, based on textual information and the other based on the network structure. In text based model all the hyperlinks are removed from the tweet and then a tokenizer removes stop words and delimited character. Since there is a limitation of 140 characters in a tweet, people use acronyms for words and so a vocabulary is used that has the full form of these words (e.g., BR is used to represent best regard). The network based approach uses a similarity model to find out the trending topic say X. It searches for five topics that are similar to the topic X and finds out the similarity index [5].

Most of the above works are related to sentiments, recommendation systems, trending topic and considered temporal context of messages and classification of tweets. However, these works do not discuss about the rising or decreasing popularity of a person and the reasons behind it. Our approach is different from others as we first look for the popular person and also let the users know the reason behind the popularity.

3 Proposed Methodology

An approach to extract a popular person from tweets is to find a person's name and storing tweet counts corresponding to the person. In order to find the reasons behind the popularity of a person we are using keywords of tweets corresponding to the person.

3.1 Architecture

Figure 1 shows the basic flow diagram of our method. First, we download tweets of different users from different countries and then we look for the person that has been most talked about among those tweets. Then we fetch tweets of that specific person from our database. To detect the reason of popularity we divide all the tweets related to that person into keywords and separate hashtags. Keywords in a tweet are names of things (e.g., name of a person, name of a city). Hashtag is represented using the symbol # followed by some meaningful word like 'Olympics2016'. If two tweets have the same hashtag, it means that these tweets are related and the tweets can be merged into one single tweet.

First we will check hashtag of tweet with events which are already found. Then we pass keywords of that tweet with keywords of events, which are already found into a function called similarity. Similarity we are finding as number of common keywords divided by number of total different keywords. And for every found event with which event, similarity is maximum and greater than threshold then we add tweet into that event. Like this for all tweets algorithm is performed. In the end we find out main reasons behind popularity of person. Then we classify tweets of that person for showing the interest of users towards that popular person means what general users think about that person. Here user is the twitter user, whose tweets are downloaded from twitter.

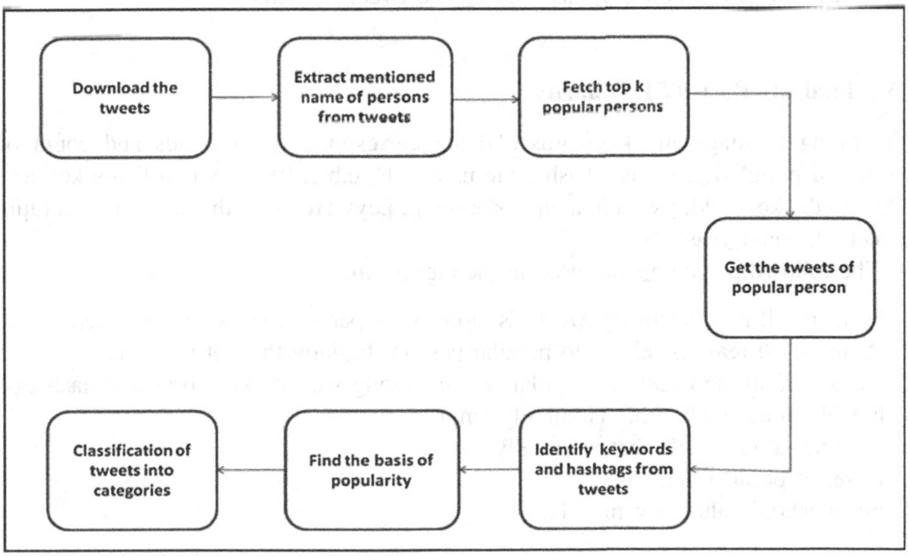

Fig. 1. Overview of the proposed method

3.2 Data Collection

We collected 2,18,490 tweets of 5 different countries from September, 2016 to November, 2016 using Twitter4j API [33]. Tweets were downloaded by taking latitude and longitude values of countries. We took news channels (CanadaNews, bbcnews) into consideration because news channels are reliable sources of data; news channels produce more data than simple twitter users.

3.3 Extraction of Names of Persons from Tweet

We used Stanford Named Entity Recognition (NER) tagger [32] for extracting the names of persons from tweets. NER labels sequence of words into a text which contains names of things, such as name of person, name of company, and name of place. Every tweet is passed through the NER tagger and it returns names of things for every tweet. We store only names of persons, and for this we used a hash function.

3.4 Fetching Top k Popular Persons

For storing name of a person and the number of occurrences of the names, we use a hash table named h_table, that has two fields: key and value. In the key field, we store person name; in the value field, a tuple <tweet_id, count_name>. If a name of a person does not exist, the count of the person is set to 1 and add the corresponding tweet id. Otherwise, increment count by one and update tweet id field.

3.5 Find the Basis of Popularity

For storing hashtags and keywords and the corresponding tweet ids and count of reasons of popularity, we use hash table named H_table, that has two fields key and value. In the key field, we store a tuple <hashtag, keywords>; in the value field, a tuple <tweet_id, count_reason>.

The following symbols are used in the algorithm.

S: set of all tweets storing tweet ids along with person mentioned in tweet.
R: set of all reasons related to popular person. Initially this set is empty.
PT: set of all the tweets of popular persons along with its keywords and hashtags.
h_table: a hash table that is initially empty.
H_table: a hash table that is initially empty.
P: set of popular persons.
m: threshold value, $0 < m < 1$.

```
Algorithm for discovering reason of popularity
Input: S        output: R
Step 1: Get the tweet count for each person
    for each tweet t ∈ S do
        for each name ∈ t do
            if name exists in h_table then
                    add tweet_id in h_table[name][0];
                    increment the count field of h_table[name][1] by1;
            else    create new entry h_table[name] ;
                    add tweet_id in h_table[name][0];
                    set h_table[name][1] to 1;
        end for
    end for
Step 2: Sort h_table on the basis of count in descending order. Find the top k
        popular persons and store them into P.
Step 3: Get all the tweets of P from S and break down these tweets into keywords
        and hashtags and store them into PT.
Step 4: Get all the reasons related to first k popular persons.
    for each person p ∈ P
            for each tweet t of p in PT
                tag := hashtag of t; kw := keyword of t;
                flag := false;
                for each key k of H_table
                        if tag = k[0] then
                                flag := true;
                                k[1] := k[1] ∪ kw;
                                increment H_table[k][1] by 1;
                        else if similarity (kw, k[1]) > m then
                                flag := true;
                                k[1] := k[1] ∪ kw;
                                increment H_table[k][1] by 1;
                end for
                if flag = false then
                        add tweet id into H_table[(tag, kw)][0]
                        set H_table[(tag, kw)][1] to 1;
            end for
    end for
Step 5: sort H_table according to field of tweet count in descending order and store
into R. Find the top popular reasons r from the set R having maximum tweet count.
```

4 Implementation and Results

To implement the algorithm, we collected 2,18,490 tweets of 5 different countries, using Twitter API. First, a user provides the value of n i.e., top n popular persons according to the downloaded tweets. Table 1 shows the output when a user provides the value of n = 4.

Table 1. Top n (n = 4) popular persons and their tweet count

Sl. no.	Person_name	Tweet_count
1.	Donald Trump	13117
2.	Hillary Clinton	9934
3.	Justin Trudeau	5432
4.	Malcolm Turnbull	5048

Once the user gets the top n popular persons, she can select any one person from the results to get more details of the selected person. In this interface, on selecting one person it will show all the tweets of that person. User can get more information about the person using these tweets. Figure 2 shows the output of selecting one person.

Person_Name	tweet_id	tweet
Donald Trump	788376631698153472	What If Donald Trump Won't Concede? 'Rigged' Elect...
Donald Trump	788434479341772800	Obama tells Donald Trump to 'stop whining' about e...
Donald Trump	788576404476665856	Donald Trump bringing Barack Obama's brother to 3r...
Donald Trump	788581947060748290	Michael Moore announces release of surprise Donald...
Donald Trump	788604340625936386	6 witnesses corroborate Canadian writer's account ...
Donald Trump	788697204789768193	Donald Trump, Hillary Clinton supporters get to ch...
Donald Trump	788730166528925696	As final debate looms, Hillary Clinton opens wides...
Donald Trump	788740736644775936	Donald Trump's Childhood Home In NYC Heads To Auct...
Donald Trump	788755058208931845	Donald Trump and Hillary Clinton ready for final d...
Donald Trump	788757621272637441	The Only Good Thing To Come From Donald Trump http...

Fig. 2. Tweets corresponding to the selected popular person

For the selected person, the reasons of popularity are given in Table 2. The table lists person name, all the popularity reasons, and the corresponding tweet counts.

Table 2. Reasons of popularity of the selected person

Person_name	Popularity_reason	Tweet_count
Donald Trump	Election2016	1387
Donald Trump	Campaign	948

The pie chart in Fig. 3 shows users' interest towards the selected popular person (Donald Trump). Since a large percentage of tweets are related to politics, this indicates that users are showing interest in political aspects of the person.

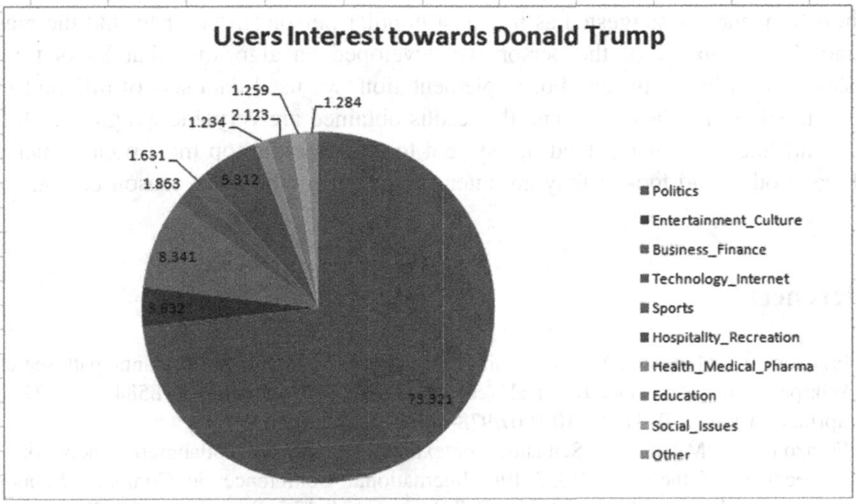

Fig. 3. Pie chart representing classification of tweets according to users' interest toward popular person Donald Trump

We can compare users' views for two different popular persons. Figure 4 shows users' views for Donald Trump and Malcolm Turnbull. From this Figure we can conclude that in politics, users are more interested toward Trump than Turnbull.

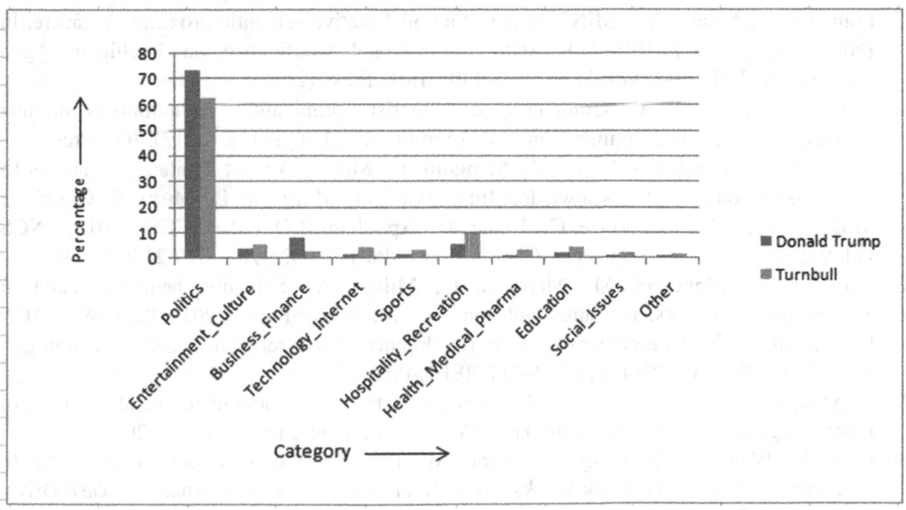

Fig. 4. Comparison between the tweets related to Trump and Turnbull

5 Conclusion and Future Work

In this paper, we suggested an approach to get the popular person from the gathered tweets and obtained the reason behind the popularity of that person. In our approach, we first look for the names mentioned in the tweets and the name that occurs with highest frequency is suggested as the most popular person. In order to find the reason behind the popularity of the person we developed an algorithm that looks for the possible events in the tweets. For implementation, we used data sets of different time frames to showcase the output and the results obtained are very encouraging. In future we would like to further extend our system to compare the top most popular persons with each other and look if they are inter connected by the same reason or not.

References

1. Franzoni, V., Mencacci, M., Mengoni, P., Milani, A.: Heuristics for semantic path search in Wikipedia. In: Murgante, B., et al. (eds.) ICCSA 2014. LNCS, vol. 8584, pp. 327–340. Springer, Cham (2014). doi:10.1007/978-3-319-09153-2_25
2. Franzoni, V., Milani, A.: Semantic context extraction from collaborative networks. In: Proceedings of the 2015 IEEE 19th International Conference on Computer Supported Cooperative Work in Design, CSCWD 2015, pp. 131–136. IEEE Press (2015)
3. Leung, C.H.C., Chan, A.W.S., Milani, A., Liu, J., Li, Y.: Intelligent social media indexing and sharing using an adaptive indexing search engine. ACM Trans. Intell. Syst. Technol. **3**(3), 221–238 (2012). ACM Press
4. Leung, C.H.C., Li, Y., Milani, A., Franzoni, V.: Collective evolutionary concept distance based query expansion for effective web document retrieval. In: Murgante, B., Misra, S., Carlini, M., Torre, C.M., Nguyen, H.-Q., Taniar, D., Apduhan, B.O., Gervasi, O. (eds.) ICCSA 2013. LNCS, vol. 7974, pp. 657–672. Springer, Heidelberg (2013). doi:10.1007/978-3-642-39649-6_47
5. Franzoni, V., Milani, A.: PMING distance: a collaborative semantic proximity measure. In: Proceedings - 2012 IEEE/WIC/ACM International Conference on Intelligent Agent Technology, IAT 2012, vol. 2, pp. 442–449. IEEE Press (2012)
6. Milani, A., Santucci, V.: Community of scientist optimization: an autonomy oriented approach to distributed optimization. AI Commun. **25**(2), 157–172 (2012). IOS Press
7. Franzoni, V., Leung, C.H.C., Li, Y., Mengoni, P., Milani, A.: Set similarity measures for images based on collective knowledge. In: Gervasi, O., Murgante, B., Misra, S., Gavrilova, M.L., Rocha, A.M.A.C., Torre, C., Taniar, D., Apduhan, B.O. (eds.) ICCSA 2015. LNCS, vol. 9155, pp. 408–417. Springer, Cham (2015). doi:10.1007/978-3-319-21404-7_30
8. Franzoni, V., Mencacci, M., Mengoni, P., Milani, A.: Semantic heuristic search in collaborative networks: measures and contexts. In: Proceedings - 2014 IEEE/WIC/ACM International Joint Conference on Web Intelligence and Intelligent Agent Technology - Workshops, WI-IAT 2014, pp. 187–217. IEEE Press (2014)
9. Sayyadi, H., Hurst, M., Maykov, A.: Event detection and tracking in social streams. In: Proceedings of the Third International ICWSM Conference, pp. 311–314 (2009)
10. Dou, W., Wang, X., Ribarsky, W., Zhou, M.: Event detection in social media data. In: Proceedings of IEEE VisWeek Workshop on Interactive Visual Text Analytics-Task Driven Analytics of Social Media Content, pp. 971–980 (2012)

11. Weng, J., Yao, Y., Leonardi, E., Lee, F.: Event detection in Twitter. In: Proceedings of the 5th International AAAI Conference on Weblogs and Social Media, pp. 401–408 (2011)
12. Becker, H., Naaman, M., Gravano, L.: Learning similarity metrics for event identification in social media. In: Proceedings of the Third ACM International Conference on Web Search and Data Mining, pp. 291–300 (2010)
13. Baioletti, M., Milani, A., Poggioni, V., Rossi, F.: Experimental evaluation of pheromone models in ACOPlan. Ann. Math. Artif. Intell. **62**(3–4), 187–217 (2011). Springer
14. Ukey, N., Niyogi, R., Singh, K., Milani, A., Poggioni, V.: A bidirectional heuristic search for web service composition with costs. Int. J. Web Grid Serv. **6**(2), 160–175 (2010). Inderscience
15. Milani, A., Poggioni, V.: Planning in reactive environments. Comput. Intell. **23**(4), 439–463 (2007). Wiley
16. Becker, H., Naaman, M., Gravano, L.: Beyond trending topics: real-world event identification on Twitter. In: Proceeding of Fifth International AAAI Conference on Weblogs and Social Media (2011)
17. Unankard, S., Li, X., Sharaf, M.A.: Emerging event detection in social networks with location sensitivity. In: Proceedings of World Wide Web, pp. 1–25 (2014)
18. Marcus, A., Bernstein, M.S., Badar, O., Karger, D.R., Madden, S., Miller, R.C.: TwitInfo: aggregating and visualizing microblogs for event exploration. In: Proceedings of the SIGCHI Conference on Human Factors in Computing Systems, pp. 227–236 (2011)
19. Chiancone, A., Franzoni, V., Niyogi, R., Milani, A.: Improving link ranking quality by quasi-common neighbourhood. In: Proceedings - 15th International Conference on Computational Science and Its Applications, ICCSA 2015, pp. 21–26. IEEE Press (2015)
20. Franzoni, V., Milani, A.: Heuristic semantic walk. In: Murgante, B., Misra, S., Carlini, M., Torre, C.M., Nguyen, H.-Q., Taniar, D., Apduhan, B.O., Gervasi, O. (eds.) ICCSA 2013. LNCS, vol. 7974, pp. 643–656. Springer, Heidelberg (2013). doi:10.1007/978-3-642-39649-6_46
21. Mathioudakis, M., Loudas, N.: TwitterMonitor: trend detection over the twitter stream. In: Proceedings of the 2010 ACM SIGMOD International Conference on Management of data, pp. 1155–1158 (2010)
22. Sankaranarayanan, J., Samet, H., Teitler, B.E., Lieberman, M.D., Sperling, J.: TwitterStand: news in Tweets. In: Proceedings of the 17th ACM SIGSPATIAL International Conference on Advances in Geographic Information Systems, pp. 42–51 (2009)
23. Guille, A., Favre, C.: Event detection, tracking, and visualization in twitter: a mention anomaly based-approach. Proc. Soc. Netw. Anal. Min. **5**(1), 1–18 (2015)
24. Aggarwal, C.C., Subbian, K.: Event detection in social streams. In: Proceeding of SDM, vol. 12, pp. 624–635 (2012)
25. Abdelhaq, H., Sengstock, C., Gertxz, M.: EvenTweet: online localized event detection from twitter. Proc. VLDB Endow. **6**(12), 1326–1329 (2013)
26. Ifrim, G., Shi, B., Brigadir, I.: Event detection in Twitter using aggressive filtering and hierarchical Tweet clustering. In: Proceedings of SNOW WWW Workshop (2014)
27. Marcugini, S., Milani, A., Pambianco, F.: NMDS codes of maximal length over Fq, $8 \leq q \leq 11$. IEEE Trans. Inf. Theory **48**(4), 963–966 (2002). IEEE Press
28. Wang, X., Zhu, F., Jiang, J., Li, S.: Real time event detection in Twitter. In: Wang, J., Xiong, H., Ishikawa, Y., Xu, J., Zhou, J. (eds.) WAIM 2013. LNCS, vol. 7923, pp. 502–513. Springer, Heidelberg (2013). doi:10.1007/978-3-642-38562-9_51
29. Socher, R., Perelygin, A., Wu, J.Y., Chuang, J., Manning, C.D., Ng, A.Y., Potts, C.: Recursive deep models for semantic compositionality over a sentiment treebank. In: Proceedings of the Conference on Empirical Methods in Natural Language Processing, pp. 1631–1642. Citeseer (2013)

30. Kiritchenko, S., Zhu, X., Mohammad, S.M.: Sentiment analysis of short informal texts. J. Artif. Intell. Res. **50**(1), 723–762 (2014)
31. Lee, K., Palsetia, D., Narayanan, R., Patwary, M.M.A., Agrawal, A., Choudhary, A.: Twitter trending topic classification. In: 11th IEEE International Conference on Data Mining Workshops, pp. 251–258, December 2011
32. Finkel, J., Grenager, T., Manning, C.: Incorporating non-local information into information extraction systems by gibbs sampling. In: Proceedings of the 43rd Annual Meeting of the Association for Computational Linguistics ACL 2005, pp. 363–370 (2005)
33. Twitter: Twitter Developers. https://dev.twitter.com

Fitness Landscape Analysis of the Permutation Flowshop Scheduling Problem with Total Flow Time Criterion

Marco Baioletti and Valentino Santucci(✉)

Department of Mathematics and Computer Science, University of Perugia, Perugia, Italy
{marco.baioletti,valentino.santucci}@unipg.it

Abstract. This paper provides a fitness landscape analysis of the Permutation Flowshop Scheduling Problem considering the Total Flow Time criterion (PFSP-TFT). Three different landscapes, based on three neighborhood relations, are considered. The experimental investigations analyze aspects such as the smoothness and the local optima structure of the landscapes. To the best of our knowledge, this is the first landscape analysis for PFSP-TFT.

1 Introduction and Related Work

The Permutation Flowshop Scheduling Problem (PFSP) is a scheduling problem widely encountered in areas such as manufacturing and large scale products fabrication [13]. The goal of PFSP is to determine the best permutation $\pi = \langle \pi[1], \ldots, \pi[n] \rangle$ of n jobs that have to be processed through a sequence of m machines. Preemption and job-passing are not allowed. Here we focus on the Total Flow Time (TFT) criterion which consists in minimizing the objective function

$$f(\pi) = \sum_{j=1}^{n} c(m, \pi[j]), \tag{1}$$

where $c(i, \pi[j])$ is the completion time of job $\pi[j]$ on machine i and is recursively calculated in terms of the processing times $p_{i,\pi[j]}$ as

$$c(i, \pi[j]) = \begin{cases} p_{i,\pi[j]} & \text{if} i = j = 1 \\ p_{i,\pi[j]} + c(i, \pi[j-1]) & \text{if} i = 1 \text{ and } j > 1 \\ p_{i,\pi[j]} + c(i-1, \pi[j]) & \text{if} i > 1 \text{ and } j = 1 \\ p_{i,\pi[j]} + \max\{c(i, \pi[j-1]), c(i-1, \pi[j])\} & \text{if} i > 1 \text{ and } j > 1 \end{cases} \tag{2}$$

The PFSP with the TFT criterion has been demonstrated to be NP hard for two or more machines [7,19]. Therefore, even due to its practical interest, many researches have been devoted to finding high quality and near optimal solutions by means of heuristic or meta-heuristic approaches, for instance with evolutionary algorithms [5,7,22].

O. Gervasi et al. (Eds.): ICCSA 2017, Part I, LNCS 10404, pp. 705–716, 2017.
DOI: 10.1007/978-3-319-62392-4_51

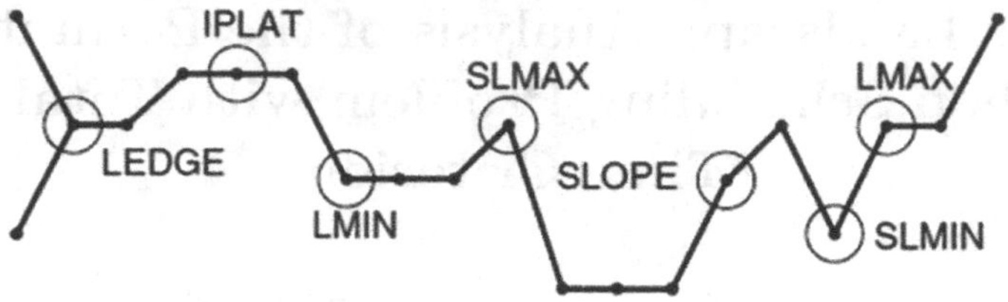

Fig. 1. Position types

Meta-heuristic techniques navigate the fitness landscape of the instance at hand. A fitness landscape is a triple $(S, \mathcal{N}, f)$ where: S is the set of solutions, $\mathcal{N}$ is a neighborhood relation among the solutions in S, and f is the objective/fitness function to optimize.

The fitness landscape can be analyzed by studying several features of interest [18,24,28]. These analyses are also important to understand the behavior of meta-heuristics and evolutionary algorithms [3,16] applied to combinatorial optimization problems [1,6].

A first aspect concerns the neutrality of the landscape and in general the classification of the solutions/points according to the fitness differences among the neighbors. There exists seven types of points: SLMIN, LMIN, IPLAT, SLOPE, LEDGE, LMAX and SLMAX. They are described in Fig. 1 [15]. A point P is of type IPLAT if all its neighbors have the same fitness values as P. A point is of type SLMAX (SLMIN) if it is a strictly local maximum (minimum), while the types LMAX and LMIN are non strict local maxima (or minima). A point P is a SLOPE if some of its neighbors have a greater fitness value and some other ones have a lower value than P. Finally, a LEGDE point P has also some neighbors with the same fitness values as P. The distribution of point types gives a quantitative analysis about the neutrality of the fitness landscape. In particular, a fitness landscape is more or less neutral according to the higher or lower percentage of IPLAT, LEGDE, LMAX and LMIN.

A second feature is to study the landscape correlation analysis [27]. Its main tool is the autocorrelation coefficient $\rho(1)$ which quantifies how much the fitness values of neighbor solutions are related to each other. Higher values for $\rho(1)$ mean that the landscape is smooth, while a small value means a rugged landscape.

A third analysis tries to check the presence of the so called "big-valley" hypothesis [14], i.e. if local minima of good quality are clustered and surround the global minimum point. Some of the combinatorial optimization problems have this structure and this is a positive characteristic for search algorithms, because it is easier to search the global optimum.

Another point that has been studied is the distribution of local minima and their distance from the global minimum [17]. Again, the distribution and the distance from local and global minima is strictly related to the performances of meta-heuristic algorithms.

Finally, another aspect of interest is the concept of fitness barrier [24], i.e. the number of steps necessary for exiting from a local minimum and finding a better point. A search space with a low fitness barrier can be better explored by a meta-heuristic algorithm because there are less chances that it can be trapped in a local minima.

In our case, S is the set of all the n-length permutations of jobs, and f is the total flow time objective of Eq. (1). Moreover, three fitness landscapes generated by three different neighborhood relations are considered, namely: the adjacent swap (ASW), interchange (INT) and insert (INS) neighborhoods [2,23]. ASW and INT neighbors are obtained by swapping two items in a permutation. While for INT neighbors no restriction is considered on the items to swap, the ASW neighborhood limits the swap to adjacent items. Finally, the INS neighborhood is obtained by removing an item and inserting back in a different position. To the best of our knowledge this is the first analysis conducted on PFSP-TFT.

The rest of the paper is organized as follows. Section 2 briefly describe the experimental setting used throughout the paper. A classification of the different type of solutions is provided in Sect. 3 together with an analysis of the landscape neutrality. Section 4 analyze the smoothness of the spaces. Sections 5 and 6 provide a study of the local minima distribution, while Sect. 7 discusses some aspects about the characteristics of the basins of attraction. Finally, conclusions are drawn in Sect. 8.

2 Experimental Setting

The experiments were held over a suite of 120 PFSP-TFT instances: ten problem instances for each $n \times m$ configuration, with $n \in \{10, 20, 50, 100\}$ and $m \in \{5, 10, 20\}$. The $n = 20, 50, 100$ instances were taken from the widely known benchmark suite of Taillard [25], while the 10 jobs instances were randomly generated using the same scheme of [25].

When the experimental analyses involve local minima, these have been exactly located through exhaustive enumeration for $n = 10$ instances, while for the larger instances they were collected by means of 2 000 local searches (each one starting from a randomly generated seed solution) performed using the "best improvement" strategy.

Finally, while the global optima of the 10 jobs instances were exactly identified, for the other instances we have considered the best known solutions taken from the state-of-the-art results reported in recent works [19,22].

3 Position Type Distributions and Neutrality Analysis

In this section we provide a classification of the position types [15] for the three search spaces.

The data in Table 1 have been obtained by a complete enumeration of the $n!$ solutions in the search space. Due to the cardinality, $n = 10$ has been considered.

Table 1. Position type distributions in percentage values

$n \times m$	Neigh.	SLMIN	LMIN	SLOPE	LEDGE	IPLAT	LMAX	SLMAX
10×5	ASW	0.41	0.05	89.85	9.43	0.00	0.06	0.20
	INT	<0.01	<0.01	80.28	19.71	0.00	<0.01	<0.01
	INS	<0.01	<0.01	76.87	23.13	0.00	<0.01	<0.01
10×10	ASW	0.55	0.05	93.91	5.07	0.00	0.04	0.38
	INT	<0.01	<0.01	87.93	12.07	0.00	<0.01	<0.01
	INS	<0.01	<0.01	81.27	18.72	0.00	<0.01	<0.01
10×20	ASW	0.48	0.03	94.26	4.76	0.00	0.04	0.42
	INT	<0.01	<0.01	89.72	10.28	0.00	<0.01	<0.01
	INS	<0.01	<0.01	84.38	15.62	0.00	<0.01	<0.01

Table 1 shows that the vast majority of the points are SLOPE and LEDGE points. Other types are very rare and, in particular, IPLAT points are not present in none of the three search spaces.

SLOPEs are more frequent in the ASW search space. Moreover, it looks that the quantity of SLOPEs increases with m. Therefore, the neutrality of the landscape is very likely to decrease when the number of machines increases.

The study of position types for $n > 10$ is possible by a random uniform sampling of the solutions in the search space. The results of this investigation are presented in Table 2. Note that, since the sampling have produced only SLOPE and LEDGE points, Table 2 only shows the percentage of SLOPEs. The remaining percentage are LEDGEs.

Table 2. Percentage of SLOPE points

$n \times m$	ASW	INT	INS
20×5	81.89	58.30	48.40
20×10	85.10	70.85	57.15
20×20	92.09	76.15	58.10
50×5	49.52	8.45	7.55
50×10	58.67	22.10	12.40
50×20	78.41	37.90	16.15
100×5	24.70	0.10	0.20
100×10	31.51	2.05	0.35
100×20	49.79	7.05	1.40

Under random sampling, Table 2 shows that the number of SLOPEs increases with m and decreases with n. For (relatively) small m and large n, LEDGE points are more frequent than SLOPEs, thus denoting that the neutrality degree is very likely to increase with n. This phenomenon is even stronger for the INT and INS neighborhoods. In these cases, as n increases, the number of SLOPEs drastically decreases. In particular, when $n = 100$, almost all the sampled points are LEDGEs.

In Table 3, we show the neutrality degrees [15], i.e. the number of the average percentage of neighbors with the same fitness. The percentage is computed with respect to the neighborhood size. For each configuration $n \times m$, we report the minimum and the maximum values. Data for $n = 10$ have been obtained by a complete enumeration, while for $n \in \{20, 50, 100\}$ a random uniform sampling of $N = 10\,000$ points has been drawn.

Table 3. Average neutrality degree per solution in percentage values

$n \times m$	ASW	INT	INS
10×5	$[1.07, 1.10]$	$[0.44, 0.48]$	$[0.31, 0.36]$
10×10	$[0.59, 0.62]$	$[0.27, 0.30]$	$[0.24, 0.28]$
10×20	$[0.55, 0.59]$	$[0.22, 0.26]$	$[0.20, 0.22]$
20×5	$[5.34, 5.41]$	$[0.14, 0.43]$	$[0.34, 0.37]$
20×10	$[5.31, 5.35]$	$[0.15, 0.36]$	$[0.32, 0.34]$
20×20	$[5.30, 5.32]$	$[0.11, 0.17]$	$[0.31, 0.33]$
50×5	$[2.12, 2.18]$	$[0.17, 0.29]$	$[0.08, 0.09]$
50×10	$[2.09, 2.12]$	$[0.10, 0.15]$	$[0.07, 0.08]$
50×20	$[2.07, 2.08]$	$[0.074, 0.10]$	$[0.06, 0.07]$
100×5	$[1.09, 1.13]$	$[0.14, 0.22]$	$[0.03, 0.04]$
100×10	$[1.04, 1.08]$	$[0.12, 0.19]$	$[0.02, 0.03]$
100×20	$[0.99, 1.04]$	$[0.10, 0.15]$	$[0.01, 0.02]$

These results clearly confirm the previous indications about the neutrality of the search spaces.

4 Landscape Correlation

Here, we study the autocorrelation coefficients and the correlation lengths for the three search spaces [27].

For each problem instance considered, a random walk of $N = 500\,000$ steps has been performed and the N visited solutions, together with their fitness values, have been registered. Then, the autocorrelation $\rho(1)$ and the correlation length $l = (\ln(|\rho(1)|))^{-1}$ have been computed. Intuitively, $\rho(1)$ measures the statistical correlation between neighboring solutions, while the correlation length gives an indication of how far it is possible to go without incurring in a steep descend or ascend.

The results of the experiment have been aggregated for every $n \times m$ configurations and shown in Table 4.

Table 4. Autocorrelations and correlation lengths

$n \times m$	Autocorr.			Corr. lengths		
	ASW	INT	INS	ASW	INT	INS
10×5	0.92	0.74	0.81	12.49	3.26	4.77
10×10	0.92	0.73	0.81	11.79	3.25	4.68
10×20	0.93	0.74	0.82	15.74	3.40	5.03
20×5	0.97	0.86	0.91	37.16	6.52	10.63
20×10	0.96	0.85	0.90	25.87	6.31	9.79
20×20	0.96	0.85	0.90	22.76	6.15	9.41
50×5	0.99	0.94	0.96	160.34	15.74	27.54
50×10	0.99	0.93	0.96	97.75	14.20	24.01
50×20	0.99	0.93	0.96	70.49	14.16	22.94
100×5	1.00	0.97	0.98	640.95	29.88	53.67
100×10	1.00	0.96	0.98	272.42	26.92	47.28
100×20	0.99	0.96	0.98	144.18	24.46	41.60

Table 4 shows that the autocorrelation is very large and becomes close to 1 when n increases. As described in [27], this is an evidence that the three spaces are very smooth, i.e., neighboring solutions tend to have similar fitness values [8]. This aspect is very important because search algorithms [10–12] are, in general, positively affected by the smoothness of the underlying fitness landscape.

5 Fitness-Distance Analysis of the Local Minima

As known in literature, evidence of the "big-valley" structure for a given fitness landscape can be inferred by showing that the fitness-distance correlation [26] of the local minima is large enough. A value larger than 0.15 is generally accepted as threshold [14,15].

For each $n \times m$ configuration and for each neighborhood relation, Table 5 provides the minimum and maximum correlation coefficients between the fitness and the distance of the local minima to the presumed global minimum. Moreover, also the number of instances presenting a correlation larger than 0.15 is shown.

These results show that the "big-valley" hypothesis is clearly more evident for the ASW neighborhood than for INT and INS. Moreover, in the latter cases, the number of correlated instances seems to decrease when n increases[1]. Finally, the number of machines m does not seem to influence the correlation coefficients.

[1] Note that the results of the INS neighborhood on the 10-jobs instances, due to the very small number of local minima, are not too much significant.

Table 5. Fitness-distance correlations

$n \times m$	ASW		INT		INS	
	fdc	#corr.inst	fdc	#corr.inst	fdc	#corr.inst
10×5	[0.30, 0.85]	10/10	[0.33, 0.70]	10/10	[0.33, 0.97]	10/10
10×10	[0.32, 0.75]	10/10	[0.29, 0.69]	10/10	[0.49, 0.88]	10/10
10×20	[0.46, 0.84]	10/10	[0.33, 0.73]	10/10	[0.57, 1.00]	10/10
20×5	[0.39, 0.83]	10/10	[0.25, 0.52]	10/10	[0.19, 0.50]	10/10
20×10	[0.48, 0.71]	10/10	[0.25, 0.52]	10/10	[0.03, 0.39]	**7/10**
20×20	[0.45, 0.69]	10/10	[0.20, 0.51]	10/10	[0.17, 0.51]	10/10
50×5	[0.47, 0.68]	10/10	[0.10, 0.30]	**7/10**	[0.07, 0.19]	**4/10**
50×10	[0.46, 0.59]	10/10	[0.06, 0.36]	**4/10**	[0.05, 0.18]	**1/10**
50×20	[0.44, 0.55]	10/10	[0.09, 0.34]	**5/10**	[0.06, 0.25]	**3/10**
100×5	[0.50, 0.64]	10/10	[0.04, 0.21]	**1/10**	[−0.01, 0.13]	**0/10**
100×10	[0.46, 0.59]	10/10	[−0.02, 0.20]	**3/10**	[−0.06, 0.14]	**0/10**
100×20	[0.37, 0.54]	10/10	[0.00, 0.13]	**0/10**	[−0.09, 0.20]	**1/10**

6 Distribution of the Local Minima

The distribution of the local minima has been investigated by means of two experiments. The first aims at estimate the centrality of the global minimum with respect to the other local minima, while the second goes in the opposite direction and analyzes the distances of the local minima from the global minimum.

The optimum centrality is defined as the percentage value *oc* defined as

$$oc = 100 \cdot \frac{d_{lmin} - d_{gmin}}{d_{gmin}}, \tag{3}$$

where d_{lmin} and d_{gmin} are, respectively, the average pairwise distance among the local minima and the average distance from the nearest global minimum. The measure *oc* estimates how much the global minimum is centrally located with respect to the other local minima [9]. A particular distinction has to be made between positive (i.e., $d_{lmin} > d_{gmin}$) and negative (i.e., $d_{lmin} < d_{gmin}$) values. For each $n \times m$ configuration and for each neighborhood relation, Table 6 reports the minimum and maximum *oc* values together with the number of instances where $oc > 0$.

Table 6 shows that, except the INS neighborhood in the 10-jobs instances, only few *oc* values are negative. Moreover negative values have a very low magnitude. This suggests us to conclude that the global optimum tends to be located inside the region of the local optima. Moreover, *oc* tends to decreases when n increases. Therefore, it is likely that the global optimum tends to "move" towards the borders of the local optima region when n increases.

The opposite analysis has been conducted by comparing the observed distribution of the local minima distances from the global minimum with respect to their expected distribution in the case that they were randomly sampled.

Table 6. Global optimum centrality

$n \times m$	ASW		INT		INS	
	oc	# $oc > 0$	oc	# $oc > 0$	oc	# $oc > 0$
10×5	$[5.55, 34.44]$	10/10	$[-28.97, -15.91]$	**0/10**	$[-6.92, 40.74]$	**9/10**
10×10	$[-5.44, 26.83]$	**9/10**	$[-34.89, -22.55]$	**0/10**	$[4.69, 23.38]$	10/10
10×20	$[15.21, 34.61]$	10/10	$[-37.89, -18.83]$	**0/10**	$[-0.79, 100.00]$	**9/10**
20×5	$[0.00, 7.20]$	10/10	$[1.18, 8.06]$	10/10	$[0.23, 10.10]$	10/10
20×10	$[-0.94, 4.41]$	**9/10**	$[0.30, 5.99]$	10/10	$[-1.28, 6.82]$	**6/10**
20×20	$[0.63, 4.83]$	10/10	$[1.92, 7.50]$	10/10	$[0.21, 5.70]$	10/10
50×5	$[0.64, 4.01]$	10/10	$[0.67, 2.03]$	10/10	$[-0.08, 1.59]$	**9/10**
50×10	$[-0.83, 1.89]$	**8/10**	$[0.48, 1.67]$	10/10	$[-0.18, 1.44]$	**9/10**
50×20	$[-0.28, 2.20]$	**8/10**	$[0.33, 1.61]$	10/10	$[-0.06, 1.26]$	**9/10**
100×5	$[0.83, 2.88]$	10/10	$[-0.01, 0.31]$	**9/10**	$[-0.58, 0.47]$	**7/10**
100×10	$[-0.76, 1.46]$	**6/10**	$[0.06, 0.74]$	10/10	$[-0.57, 0.65]$	**9/10**
100×20	$[-0.71, 0.79]$	**5/10**	$[-0.09, 1.05]$	**7/10**	$[-0.20, 0.52]$	**7/10**

Figures 2a, b and c report the observed (blue bars) and expected (red bars) distributions for the first instance of the 50×10 problems, respectively, for ASW, INT, and INS (the other instances seems to have a similar behaviour).

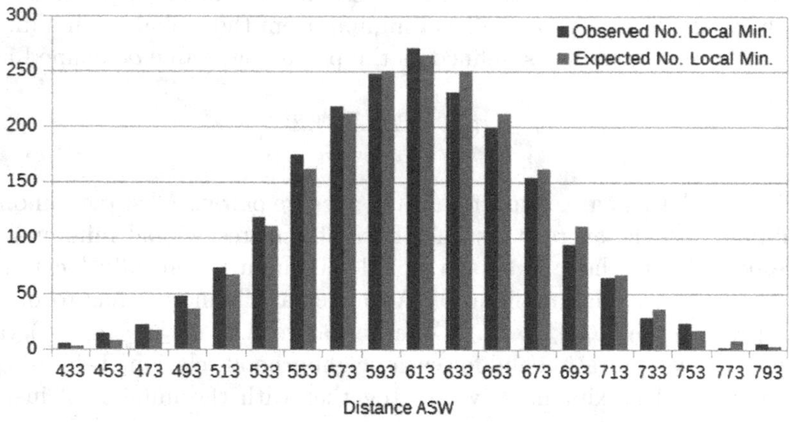

Fig. 2. Local minima distribution w.r.t. global min. for ASW (Color figure online)

These results show that the observed distribution is approximately similar to the expected one, thus local minima look like as they are randomly sampled from the set of all the permutations.

Furthermore, for ASW and INS the observed minima are a bit more than the estimated ones at the left of the distribution mode. The opposite happen for INT.

7 Fitness Barriers

In this section we study the fitness barriers [24], restricted to the ASW search space, which is the basis for the differential mutation operator described in [4, 20, 21] (Figs. 3 and 4).

For every instance ($n = 20, 50, 100$ and $m = 5, 10, 20$), we have found $N = 10\,000$ local minima by means of a simple local search based on adjacent swap moves. For each local minimum x, we have found, through a breadth-first search, the closest point y which has a smaller fitness, i.e., $f(y) < f(x)$. We have then considered two variables: the length L of the shortest path from x to y, and the largest fitness value H of the points in the path. Starting from H, we have computed the relative variation $R = \frac{H-f(x)}{f(x)}$.

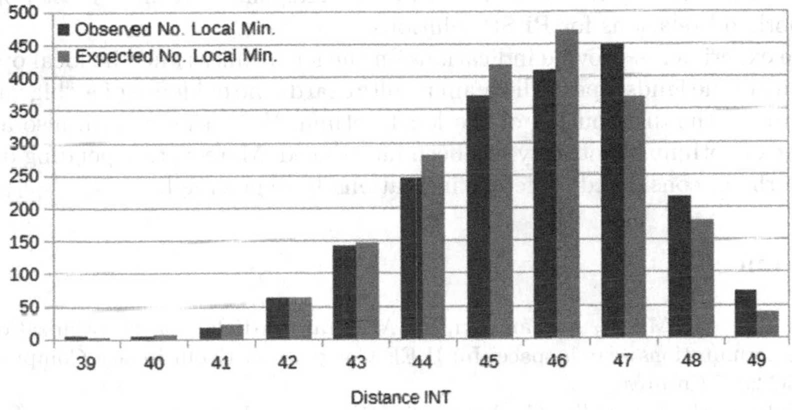

Fig. 3. Local minima distribution w.r.t. global min. for INT

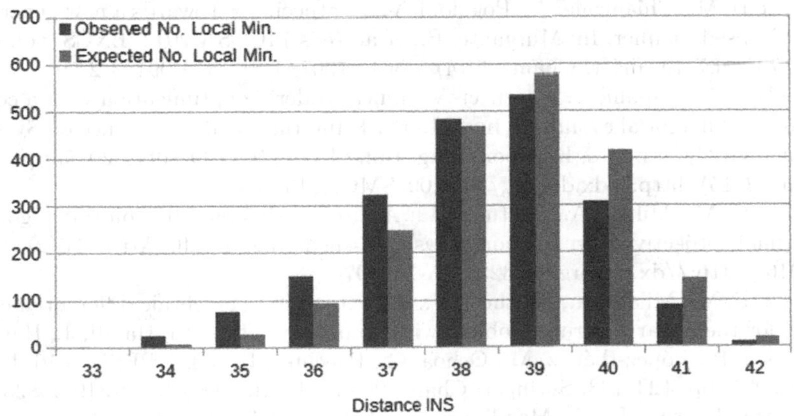

Fig. 4. Local minima distribution w.r.t. global min. for INS

L indicates how many steps one must go away from a local minimum in order to escape from its attraction basin. H is the barrier height and denotes which is the highest point one must pass through in order to find a better local minimum.

L has very low variability and, in all the experiments, presents an average value very close to 3. This means that very often it is sufficient to walk for three steps from a local minimum in order to find a better point.

The analysis of R shows that, in order to escape from a local minimum, it is sufficient to accept values which are around 1% (for $n = 20$), around 0.4% (for $n = 50$) and around 0.2% (for $n = 100$) worse than the minimum.

8 Conclusion

In this paper we have investigated the structure of PFSP-TFT instances. Three different landscapes have been considered by analyzing the three most popular neighborhood relations for PFSP solutions.

The experiments provide indications on the smoothness and the local optima structure of the landscapes. The main result regards the evidence of a "big valley" structure on the distribution of the local optima. With this regard, also a new measure of optimum centrality has been introduced. Moreover, depending on the neighborhood considered, different indications have emerged.

References

1. Baioletti, M., Milani, A., Santucci, V.: Algebraic particle swarm optimization for the permutations search space. In: IEEE Congress on Evolutionary Computation CEC 2017 (in press)
2. Baioletti, M., Busanello, G., Vantaggi, B.: Acyclic directed graphs representing independence models. Int. J. Approx. Reason. **52**(1), 2–18 (2011). http://dx.doi.org/10.1016/j.ijar.2010.09.005
3. Baioletti, M., Chiancone, A., Poggioni, V., Santucci, V.: Towards a new generation ACO-based planner. In: Murgante, B., et al. (eds.) ICCSA 2014. LNCS, vol. 8584, pp. 798–807. Springer, Cham (2014). doi:10.1007/978-3-319-09153-2_59
4. Baioletti, M., Milani, A., Santucci, V.: Linear ordering optimization with a combinatorial differential evolution. In: 2015 IEEE International Conference on Systems, Man, and Cybernetics, Kowloon Tong, Hong Kong, 9–12 October 2015, pp. 2135–2140 (2015). http://dx.doi.org/10.1109/SMC.2015.373
5. Baioletti, M., Milani, A., Santucci, V.: A discrete differential evolution algorithm for multi-objective permutation flowshop scheduling. Intell. Artif. **10**(2), 81–95 (2016). http://dx.doi.org/10.3233/IA-160097
6. Baioletti, M., Milani, A., Santucci, V.: An extension of algebraic differential evolution for the linear ordering problem with cumulative costs. In: Handl, J., Hart, E., Lewis, P.R., López-Ibáñez, M., Ochoa, G., Paechter, B. (eds.) PPSN 2016. LNCS, vol. 9921, pp. 123–133. Springer, Cham (2016). doi:10.1007/978-3-319-45823-6_12
7. Ceberio, J., Irurozki, E., Mendiburu, A., Lozano, J.A.: A distance-based ranking model estimation of distribution algorithm for the flowshop scheduling problem. IEEE Trans. Evol. Comput. **18**(2), 286–300 (2014). http://dx.doi.org/10.1109/TEVC.2013.2260548

8. Chiancone, A., Franzoni, V., Niyogi, R., Milani, A.: Improving link ranking quality by quasi-common neighbourhood. In: Proceedings of 15th International Conference on Computational Science and Its Applications, ICCSA 2015, pp. 21–26, June 2015. doi:10.1109/ICCSA.2015.19
9. Chiancone, A., et al.: A multistrain bacterial diffusion model for link prediction. Int. J. Pattern Recogn. Artif. Intell. **31**(8), 201–233 (2017). doi:10.1142/S0218001417590248
10. Franzoni, V., Mencacci, M., Mengoni, P., Milani, A.: Heuristics for semantic path search in wikipedia. In: Murgante, B., et al. (eds.) ICCSA 2014. LNCS, vol. 8584, pp. 327–340. Springer, Cham (2014). doi:10.1007/978-3-319-09153-2_25
11. Franzoni, V., Milani, A.: Semantic context extraction from collaborative networks. In: 2015 IEEE 19th International Conference on Computer Supported Cooperative Work in Design (CSCWD), pp. 131–136, May 2015. doi:10.1109/CSCWD.2015.7230946
12. Franzoni, V., Milani, A.: A semantic comparison of clustering algorithms for the evaluation of web-based similarity measures. In: Gervasi, O., et al. (eds.) ICCSA 2016. LNCS, vol. 9790, pp. 438–452. Springer, Cham (2016). doi:10.1007/978-3-319-42092-9_34
13. Gupta, J.N., Stafford Jr., E.F.: Flowshop scheduling research after five decades. Eur. J. Oper. Res. **169**(3), 699–711 (2006). http://dx.doi.org/10.1016/j.ejor.2005.02.001
14. Hains, D.R., Whitley, L.D., Howe, A.E.: Revisiting the big valley search space structure in the TSP. J. Oper. Res. Soc. **62**(2), 305–312 (2011). http://dx.doi.org/10.1057/jors.2010.116
15. Hoos, H.H., Stützle, T.: Stochastic Local Search: Foundations and Applications. Elsevier, Amsterdam (2004)
16. Milani, A., Santucci, V.: Community of scientist optimization: an autonomy oriented approach to distributed optimization. AI Commun. **25**(2), 157–172 (2012). http://dx.doi.org/10.3233/AIC-2012-0526
17. Ochoa, G., Verel, S., Daolio, F., Tomassini, M.: Local optima networks: a new model of combinatorial fitness landscapes. In: Richter, H., Engelbrecht, A. (eds.) Recent Advances in the Theory and Application of Fitness Landscapes. ECC, vol. 6, pp. 233–262. Springer, Heidelberg (2014). doi:10.1007/978-3-642-41888-4_9
18. Pitzer, E., Affenzeller, M.: A comprehensive survey on fitness landscape analysis. In: Fodor, J., Klempous, R., Araujo, C.P.S. (eds.) Recent Advances in Intelligent Engineering Systems. SCI, vol. 378, pp. 161–191. Springer, Heidelberg (2012). doi:10.1007/978-3-642-23229-9_8
19. Santucci, V., Baioletti, M., Milani, A.: Algebraic differential evolution algorithm for the permutation flowshop scheduling problem with total flowtime criterion. IEEE Trans. Evol. Comput. **20**(5), 682–694 (2016). http://dx.doi.org/10.1109/TEVC.2015.2507785
20. Santucci, V., Baioletti, M., Milani, A.: A differential evolution algorithm for the permutation flowshop scheduling problem with total flow time criterion. In: Bartz-Beielstein, T., Branke, J., Filipič, B., Smith, J. (eds.) PPSN 2014. LNCS, vol. 8672, pp. 161–170. Springer, Cham (2014). doi:10.1007/978-3-319-10762-2_16
21. Santucci, V., Baioletti, M., Milani, A.: An algebraic differential evolution for the linear ordering problem. In: Genetic and Evolutionary Computation Conference, GECCO 2015, Madrid, Spain, 11–15 July 2015, Companion Material Proceedings, pp. 1479–1480 (2015). http://doi.acm.org/10.1145/2739482.2764693

22. Santucci, V., Baioletti, M., Milani, A.: Solving permutation flowshop scheduling problems with a discrete differential evolution algorithm. AI Commun. **29**(2), 269–286 (2016). http://dx.doi.org/10.3233/AIC-2012-0526
23. Schiavinotto, T., Stützle, T.: A review of metrics on permutations for search landscape analysis. Comput. Oper. Res. **34**(10), 3143–3153 (2007). http://dx.doi.org/10.1016/j.cor.2005.11.022
24. Stadler, P.F.: Fitness landscapes. In: Lässig, M., Valleriani, A. (eds.) Biological Evolution and Statistical Physics. LNP, vol. 585, pp. 183–204. Springer, Heidelberg (2002). doi:10.1007/3-540-45692-9_10
25. Taillard, E.: Benchmarks for basic scheduling problems. Eur. J. Oper. Res. **64**(2), 278–285 (1993). http://dx.doi.org/10.1016/0377-2217(93)90182-M
26. Tomassini, M., Vanneschi, L., Collard, P., Clergue, M.: A study of fitness distance correlation as a difficulty measure in genetic programming. Evol. Comput. **13**(2), 213–239 (2005). http://dx.doi.org/10.1162/1063656054088549
27. Weinberger, E.: Correlated and uncorrelated fitness landscapes and how to tell the difference. Biol. Cybern. **63**(5), 325–336 (1990). http://dx.doi.org/10.1007/BF00202749
28. Wu, Y., McCall, J.A.W., Corne, D.: Fitness landscape analysis of bayesian network structure learning. In: Proceedings of the IEEE Congress on Evolutionary Computation, CEC 2011, New Orleans, LA, USA, 5–8 June 2011, pp. 981–988 (2011). http://dx.doi.org/10.1109/CEC.2011.5949724

Clustering Facebook for Biased Context Extraction

Valentina Franzoni[1,2(✉)], Yuanxi Li[3], Paolo Mengoni[1,4], and Alfredo Milani[1,3]

[1] Department of Mathematics and Computer Science, University of Perugia, Perugia, Italy
franzoni@dis.uniroma1.it, paolo.mengoni@unifi.it, milani@unipg.it

[2] Department of Computer Control and Management Engineering, University of Rome La Sapienza, Rome, Italy

[3] Department of Computer Science, Hong Kong Baptist University, Kowloon Tong, Hong Kong
csyxli@comp.hkbu.edu.hk

[4] Department of Mathematics and Computer Science, Florence University, Florence, Italy

Abstract. Facebook comments and shared posts often convey human biases, which play a pivotal role in information spreading and content consumption, where short information can be quickly consumed, and later ruminated. Such bias is nevertheless at the basis of human-generated content, and being able to extract contexts that does not amplify but represent such a bias can be relevant to data mining and artificial intelligence, because it is what shapes the opinion of users through social media. Starting from the observation that a separation in topic clusters, i.e. sub-contexts, spontaneously occur if evaluated by human common sense, especially in particular domains e.g. politics, technology, this work introduces a process for automated context extraction by means of a class of path-based semantic similarity measures which, using third party knowledge e.g. WordNet, Wikipedia, can create a bag of words relating to relevant concepts present in Facebook comments to topic-related posts, thus reflecting the collective knowledge of a community of users. It is thus easy to create human-readable views e.g. word clouds, or structured information to be readable by machines for further learning or content explanation, e.g. augmenting information with time stamps of posts and comments. Experimental evidence, obtained by the domain of information security and technology over a sample of 9M3k page users, where previous comments serve as a use case for forthcoming users, shows that a simple clustering on frequency-based bag of words can identify the main context words contained in Facebook comments identifiable by human common sense. Group similarity measures are also of great interest for many application domains, since they can be used to evaluate similarity of objects in term of the similarity of the associated sets, can then be calculated on the extracted context words to reflect the collective notion of semantic similarity, providing additional insights on which to reason, e.g. in terms of cognitive factors and behavioral patterns.

Keywords: Word similarity · Semantic distance · Artificial intelligence · Data mining · Collective knowledge · Knowledge discovery

O. Gervasi et al. (Eds.): ICCSA 2017, Part I, LNCS 10404, pp. 717–729, 2017.
DOI: 10.1007/978-3-319-62392-4_52

1 Introduction

Facebook comments can be elicited by the aggregation of users in homophily [23] communities, e.g. by interest or opinion. We start from the observation that users can become polarized comment after comment, where they comment expressing similar concepts or with respect to a similar level of abstraction. Besides the preferential attachment approach [23] in fact, users often comment the main topic using the same use cases. For example, in the domain of information security, where a previous comment asks how to solve a problem, other users will probably seek help and create questions about the same problem, because they trust the source (who can be a previous commenter, or the user/page posting the main post) and they think to have the same problem. In information technology it also happens, as every computer scientist knows, in the well-known "fix-my-PC" problem. In Facebook, previous commenters can reinforce, and then drive, the polarization on particular sub-topics. Such sub-topics, containing in most cases an information bias, will often be off-topic with respect to the main post topics. In our work, we propose a process to separate clusters of *in-topicness*, where concepts underlying the content of comments are grouped by similarity with the concepts underlying the main topic. Experimental evidence, evaluated by human common sense, shows that such sub-topics form sub-contexts. In this work, posts and comments are extracted from the Facebook graph and are preprocessed with basic Natural Language Processing techniques [13]. The obtained bag of words is considered a set of candidate topics for sub-contexts. Semantic path-based WordNet distance [12] Leacock-Chodorow similarity [22] and Wu-Palmer similarity [20] are calculated, by means of the hierarchy of an ontological knowledge base, e.g. WordNet [1] where experiments have been implemented using path-based distances between pairs of term pairs (word1 from the main topic, word2 from each comment) for computation simplicity, but can be exploited also on Web-based semantic measures. The proposed approach can be applied to different distances in a social or collaborative taxonomy (e.g. Wikipedia [6, 7] Linked Data [24]). Preprocessed data, augmented with the similarity values, are then submitted to a clustering algorithm (e.g. Expectation-Maximization or simple K-means [25]) to obtain the sub-context clusters, that we validate by human common sense as a preliminary analysis. Since clusters are linkable to the same third party knowledge base (in our case, WordNet), in which the content similarity is calculated, a further evaluation can be done by referring to word-to-word semantic distance, or validating already accepted tagged data sets, where clusters can be compared to class tags to which a word pertains, or not.

The exploration of social networks or Web content using their semantic meaning is a consolidated modern approach to information extraction. The similarity measurement between documents and text has been extensively studied for information retrieval and Web-based measures [11, 12].

Content Based Image Retrieval (CBIR) [3, 18] enables satisfactory similarity measurements of low level features. However, the semantic similarity of deep relationships among objects is not explored by CBIR or other state-of-the-art techniques in Concept Based Image Retrieval and artificial intelligence. A promising idea is that it is possible to generalize the semantic similarity, under the assumption that semantically similar terms behave similarly [4, 17, 19, 26]: the features of the main semantic proximity measures used in this work can be used in group similarity [27, 28] as a basis to extract semantic content, reflecting the collaborative change made on the web resources.

The example provided in Fig. 1 shows the different similarity recognition by humans and computers. Humans always have some bias [CIT SCIENCE FEB 2017] because of their cultural, educational or formation, besides the pure opinion that can be expressed in textual contributions to social or collaborative networks. Such a bias is a personal or community-based direction that will drive and shape opinions of other users participating to the same community, or potential ones. Such a bias is an important characteristic of the human being, and when politically wrong, it should be fixed by formation, not by filtering. With these premises, the most common problem of algorithms for automated tagging or context extraction is that they suffer to be domain-dependent. In particular machine learning approaches, which is the one with the best performance in many cases, suffers of the well-known problem of over-fitting. Now, it has been proved that semantics derived automatically from language corpora contain human-like biases: as quickly as it can learn, a machine learning process can amplify a bias. For instance, the pleasantness of a flower or unpleasantness of an insect can depend on cultural basis, but pushing too much the association between such accepted biases can lead over a racist threshold that, if generated by machines following human biases, is not acceptable by human politically-correct behavior.

In this point of view, is thus important that such biases are represented, being a content that will objectively shape opinions and cannot disappear in the analysis, but are not amplified, being considered a negative element. In other words, in this approach algorithms should not have opinions. The approach proposed in this paper is less domain-dependent, and does not pertain to that class of algorithms that can amplify the human bias, therefore can be preferred to machine learning, depending on the final goal, even when machine learning may have comparable or better results, which usually happens only in particular domains.

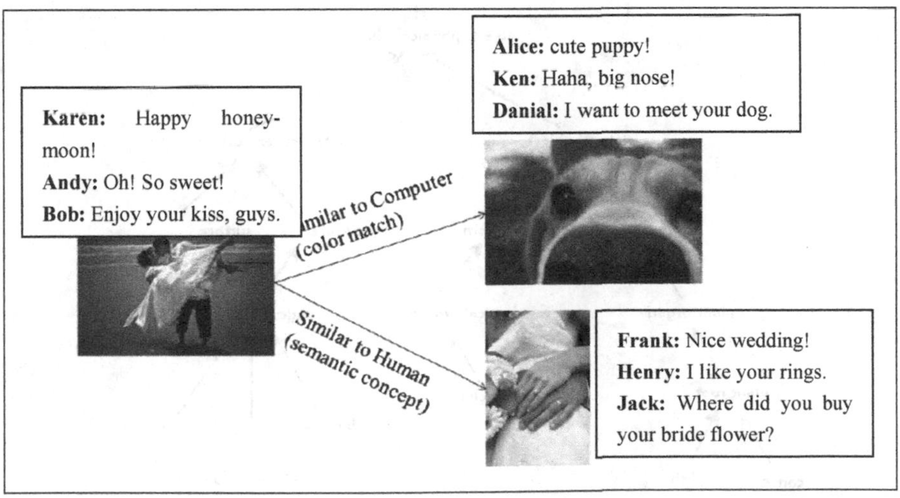

Fig. 1. Image similarity discovery comparison between computer and human

2 Related Work

2.1 WordNet Similarity

WordNet [1], is one of the applications of semantic lexicon propose for the English language and is a general knowledge base and common sense reasoning engine. Recent researches [2] on the topic in computational linguistics has emphasized the perspective of semantic relatedness of two lexemes in a lexical resource, or its opposite, semantic distance. The work in [12] brings together ontology and corpora, defining the similarity between two concepts *c1* and *c2* lexicalized in WordNet, named *WordNet Distance (WD)*, by the information content of the concepts that subsume them in the taxonomy. Then [27] proposes a similarity measure in WordNet between arbitrary objects where *lso* is the lowest super-ordinate (most specific common subsumer):

$$d(c_1, c_2) = \frac{2 \times \log p(lso(c_1, c_2))}{\log p(c_1) + \log p(c_2)} \tag{1}$$

The advantage of a WordNet similarity (where, results being normalized in a range [0, 1], similarity = 1 − distance) is to be based on a very mature and comprehensive lexical database, which provides measures of similarity and relatedness: WordNet, in fact, reflects universal knowledge because it is built by human experts; however, WordNet Distance is only for nouns and verbs in WordNet, but it is not dynamically updated. In Fig. 2, the "is a" relation example can be seen from [12].

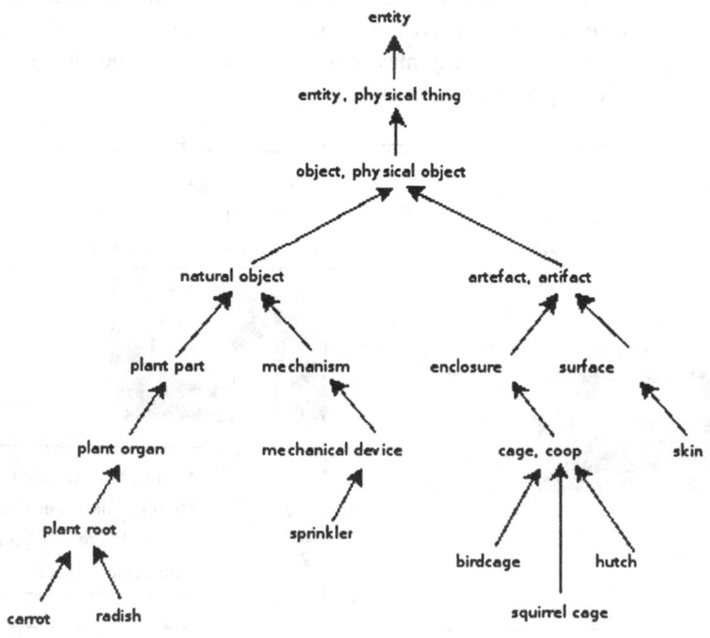

Fig. 2. WordNet "is a" relation example

2.2 Wikipedia Similarity

WikiRelate [5] was the first research to compute measures of semantic relatedness using Wikipedia. This approach takes familiar techniques that had previously been applied to WordNet and modified them to suit Wikipedia. The implementation of WikiRelate follows the hierarchical category structure of Wikipedia.

The *Wikipedia Link Vector Model (WLVM)* [6] uses Wikipedia to provide structured world knowledge about terms of interest. The probability of WLVM is defined by the total number of links to the target article over the total number of articles. Therefore, if t is the total number of articles within Wikipedia, the weighted value w for the link $a \rightarrow b$ is:

$$w(a \rightarrow b) = |a \rightarrow b| \times \log\left(\sum_{x=1}^{t} \frac{t}{|x \rightarrow b|}\right) \quad (2)$$

where a and b denote the search terms.

Among the approaches that use the hyperlink structure of Wikipedia rather than its category hierarchy or textual content, there is also the Heuristic Semantic Walk [26], that makes use of a search engine as a third-party knowledge base (e.g. Bing, Google) on which to calculate a Web-based similarity used as heuristic to drive a random walk. Wikipedia similarity reflects relationships as seen by the user community [7], which is dynamically changing as links and nodes are changed by the users collaborative effort. However, it only can apply to knowledge base organized as networks of concepts.

2.3 Flickr Similarity

Flickr distance (FD) [8] is another model for measuring the relationship between semantic concepts, in visual domains. For each concept, a collection of images is obtained from Flickr, based on which the improved latent topic-based visual language model is built to capture the visual characteristics of the concept. The Flickr distance between concepts *c1* and *c2* can be measured by the *square root of Jensen-Shannon divergence* [9, 15] between the corresponding visual language models, as follows:

$$\mathrm{D(C\_1,C\_2)} = \sqrt{\frac{\sum_{i=1}^{K}\sum_{j=1}^{K} \mathrm{D_{JS}}(\mathrm{P_{Zi}C_1}|\mathrm{P_{Zj}C_2})}{\mathrm{K}^2}} \quad (3)$$

where

$$\mathrm{D_{JS}} = (\mathrm{P_{Zi}C_1}|\mathrm{P_{Zj}C_2}) = \frac{1}{2}\mathrm{D_{KL}}(\mathrm{P_{Zi}C_1}|\mathrm{M}) + \frac{1}{2}\mathrm{D_{KL}}(\mathrm{P_{Zj}C_2}|\mathrm{M}) \quad (4)$$

K is the total number of latent topics, which is determined experimentally. $P_{Zi}\ C_1$ and $P_{Zj}\ C_2$ are the trigram distributions under latent topic $z_i c_1$ and $z_j c_2$ respectively, with M representing the mean of P_{Zi} C_1 and P_{Zj} C_2. The FD is based on Visual Language Models (VLM), which is a different concept relationship respect to WordNet Similarity and Wikipedia Similarity.

2.4 Context-Based Group Similarity

Set similarities in images [9, 10, 27], emotions [28] and, in general, web entities, can be calculated by means of underlying pair-based similarities with semantic proximity, based on user-provided concept clouds. A semantic concept cloud related to a Web object (e.g. image, video, post) includes all the semantic concepts associated to or extracted from the object. Typical sources for semantic concepts are tags, comments, descriptors, categories, or text surrounding an image. As shown in Fig. 3, Image I_i and Image I_j are a pair of images to be compared. T_{i1}, T_{i2},..., T_{im} are original user provided tags of image I_i, while T_{j1}; T_{j2},.., T_{jn} are original user provided tags of image *IJ*.

Given DI_{ij} as the distance (or equivalently, the similarity) of image I_i and image I_j, we define the ***Group Distance (GD)***:

$$DI_{ij} = AVG2\{AVG1[SEL(dT_{im \to jn})], AVG1[SEL(dT_{jn \to im})]\} \quad (5)$$

where *SEL* could be the maximum *MAX*, the average *AVG* or the minimum *MIN* of *d*, the similarity calculated by algorithm (*Confidence* or *NGD* [15] or *PMI* [14]), as in Eqs. (6–9).

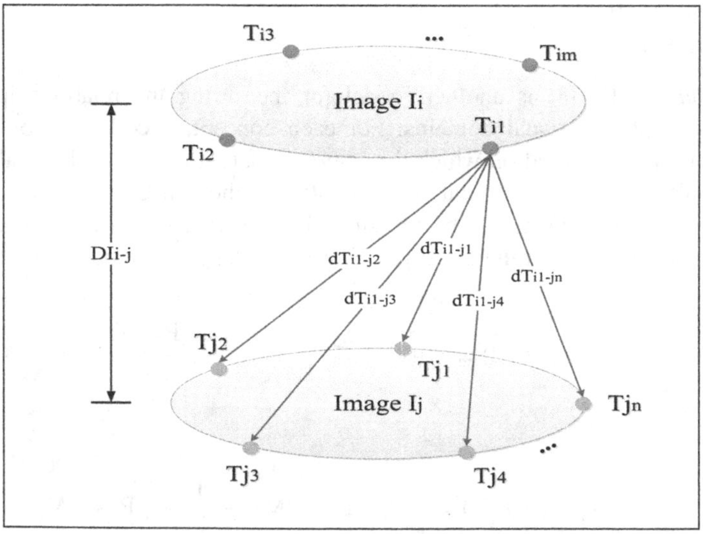

Fig. 3. Group similarity core algorithm

$$dT_{im} \rightarrow dT_{jn} = \begin{pmatrix} dT_{i1\rightarrow j1}, & dT_{i1\rightarrow j2}, & dT_{i1\rightarrow j3}, & \ldots & dT_{i1\rightarrow jn} \\ dT_{i2\rightarrow j1}, & dT_{i2\rightarrow j2}, & dT_{i2\rightarrow j3}, & \ldots & dT_{i2\rightarrow jn} \\ \ldots, & \ldots, & \ldots, & \ldots & \ldots \\ dT_{in\rightarrow j1}, & dT_{in\rightarrow j2}, & dT_{in\rightarrow j3}, & \ldots & dT_{in\rightarrow jn} \end{pmatrix}$$

$$dT_{im} \rightarrow dT_{jn} = \begin{pmatrix} dT_{j1\rightarrow i1}, & dT_{j1\rightarrow i2}, & dT_{j1\rightarrow i3}, & \ldots & dT_{j1\rightarrow im} \\ dT_{j2\rightarrow i1}, & dT_{j2\rightarrow i2}, & dT_{j2\rightarrow i3}, & \ldots & dT_{j2\rightarrow im} \\ \ldots, & \ldots, & \ldots, & \ldots & \ldots \\ dT_{jn\rightarrow i1}, & dT_{jn\rightarrow i2}, & dT_{jn\rightarrow i3}, & \ldots & dT_{jn\rightarrow im} \end{pmatrix} \quad (6)$$

$$AVG1\left[SEL(dT_{im\rightarrow jn})\right] = avg\left[SEL(dT_{i1\rightarrow jn}), SEL(dT_{i2\rightarrow jn}),, SEL(dT_{im\rightarrow jn})\right] \quad (7)$$

$$AVG1\left[SEL(dT_{jn\rightarrow im})\right] = avg\left[SEL(dT_{j1\rightarrow im}), SEL(dT_{j2\rightarrow im}),, SEL(dT_{jn\rightarrow im})\right] \quad (8)$$

$$AVG2 = AVGAVG1, AVG2 \quad (9)$$

3 Experiment

Information about Facebook post and comment similarity is extracted from raw data using a five-phases algorithm:

1. In the first phase Facebook post and comment data is harvested from public Facebook pages using an ad-hoc developed data pull app that have been registered on the social network.
2. Retrieved posts and comments are preprocessed to extract nouns.
3. Different ontology-based similarity measures are calculated on filtered nouns, where the distance between comments and the main topic are indagated.
4. Clustering is exploited on noun pairs augmented with similarity values.
5. Obtained clusters are visualized by a tag cloud and evaluated by means of human common sense.

Python, the Natural Language ToolKit [21], and TextBlob library are used to extract information, to analyze it using NLP techniques, and to compute word similarities.

3.1 Data Collection

Data are collected scraping the Facebook page *@Security*, which had (at time of experiments) over 9 million and 3 hundred thousand users. The access to Facebook data is allowed only to registered developers that write approved apps. The general policy on data access granted by Facebook include information from public Facebook pages or public posts written by normal users. To access private personal and post data the user should install an app on his Facebook account and grant it specific permissions: in this case, apps can access all the data, for a limited time.

Based on this premise, our data extraction algorithm from Facebook uses public posts from the page, and comments related to each post. These data are requested to Facebook using the Facebook Graph API, a low-level HTTP interface to *node*, *edge* and *field* information, where nodes are Facebook objects (users, photos, pages, posts, comments, et cetera) connected through *edges* (photos in the page, and their comments) while *fields* are the specific information contained in nodes, i.e. attributes.

3.2 Preprocessing Phase

The extracted Facebook post and comment data have undergone a Part Of Speech (POS) tagging. Such preprocessing phase is needed to identify nouns, verbs, adjectives and other phrase components (Table 1).

Table 1. @Security page example of raw data

Topic: "Being safe online often starts with the developers who create the products we use every day. Today we're sharing tips for developers to write more secure code and help avoid security risks. #SID2017"
SampleComment1: "It would be nice to control if my post can be shared by other people. My sister in law shares all my posts"
SampleComment2: "I'd like to report a problem. I clicked on a photo story about Tomatoes and had an Attack on my computer. My security stopped the attack, and I quickly shut down. I went back to the site later and took a screen shot of the site, but never again clicked on it. I would like to show you the site, but don't see any place to post the photo. Please contact me."
SampleComment3: "How can I stop my friends seeing comments I make on other friend's posts who are not also their friends?"

3.3 Word-Level and Set-Level Similarity

After identifying the nouns contained in the post/comment, similarity between post and comment nouns is computed using two different strategies, each using three measures. The third-party knowledge base used for experiments is the lexical resource WordNet. In WordNet we identify the set of synonyms of nouns (i.e. synset) to which the word pertains, then we extract the first term included in the synset (as a synset name). Then, similarity (i.e. distance, by its inverse) is calculated by means of relations linking words, traversing the taxonomy through the hypernym hierarchy, i.e. "IS-A" relations.

The two implemented strategies differ on how Facebook comment features are extracted. The first technique uses a comment tag, i.e. one tag per comment, where the tag is a word used in the comment, using a set similarity. The other technique is based on exploiting the inner set similarities, calculating the pair distances between each of the nouns used in each comment and each word of the main post, i.e. the commented topic.

An adjacency matrix is then built on similarities, pair by pair, where similarities are the path-based WordNet distance [12], Leacock-Chodorow similarity [22] and Wu-Palmer similarity [20].

Using path similarity, a measure on how two words are similar is calculated based on the shortest path distance between the two terms found analyzing the hypernym relationship tree.

Leacock-Chodorow combines a taxonomy shortest path (i.e. length) between two associated word senses and the maximum taxonomy depth (D) using the following formula:

$$Sim_{Lch=} - log \frac{lenght}{2 * D} \tag{10}$$

Wu-Palmer similarity measure uses the taxonomy depth of two associated concepts (a and b) and the depth of the least common subsumer LCS (i.e. the nearest common parent concept (Tables 2 and 3)).

$$Sim_{w\&p=} \frac{2 * \mathrm{depth(LCS)}}{\mathrm{depth(a)} + \mathrm{depth(b)}} \tag{11}$$

Table 2. @Security page example preprocessing: nouns extracted from text

Topic: "online, developers, products, day, Today, tips, developers, secure, code, security, risks, SID2017"
SampleComment1: "post, people, sister, law, shares, posts"
SampleComment2: "problem, photo, story, Tomatoes, Attack, computer, security, attack, site, screen, shot, site, site, place, photo, Please"
SampleComment3: "friends, comments, friend, posts, friends"

Table 3. @Security page example synset extraction from WordNet ontology

Topic: "developer, merchandise, day, today, tip, developer, code, security, hazard"
SampleComment1: "post, people, sister, law, share, post"
SampleComment2: "problem, photograph, narrative, tomato, attack, computer, security, attack, site, screen, shooting, site, site, topographic_point, photograph"
SampleComment3: "friend, remark, friend, post, friend"

3.4 Clustering Phase

Metrics of similarity provide data as distances in a Euclidean space. In general, any proximity measure can be used for clustering, even if it is not a metric, if the function following which the clustering algorithm will decide at each step where to include an evaluated point in a collection is defined (Table 4).

Table 4. @Security page example - similarity between each topic nouns' synset and each comment nouns' synset to be submitted for clusterization

Post noun	Post noun synset	Comment noun	Comment noun synset	Path sim.	LCH sim.	WUP sim.
Developers	Developer	Computer	Computer	0.091	1.240	0.444
Products	Merchandise	Computer	Computer	0.143	1.692	0.625
Day	Day	Computer	Computer	0.077	1.073	0.143
Today	Today	Computer	Computer	0.071	0.999	0.133
Tips	Tip	Computer	Computer	0.083	1.153	0.353
Developers	Developer	Computer	Computer	0.091	1.240	0.444
Code	Code	Computer	Computer	0.077	1.073	0.143
Security	Security	Computer	Computer	0.067	0.930	0.125
Risks	Hazard	Computer	Computer	0.091	1.240	0.286

The EM (Expectation-Maximization), as defined in [25] is an iterative algorithm for finding the maximum likelihood of estimated parameters, in statistical methods where the model depends on latent variables, e.g. from equations which cannot be resolved directly, or from data which were not observed, where the existence of such data can be assumed true. EM iteration rotates an expectation step (E), which iteratively calculates the expected likelihood on the current estimate of parameters, and a maximization step (M), which estimates which parameters maximize the expected likelihood, calculated in the E step until convergence, where updating the parameters does not increase anymore the likelihood.

K-means [25] is a clustering method to partition n observations into k clusters with the closest average (mean). The problem is computationally difficult (NP-hard), but algorithms exist, which make use of heuristics to converge quickly in a local optimum, similar to EM, through finishing steps.

3.5 Final Visualization

The human evaluation experiments have been held for the quality assessment of extracted sub-context, i.e. clusters. Experiments have been designed in a group of 12 experts, members of University of Perugia, from staff and students. Cloud tags related to sub-contexts clusters generated by the proposed algorithms from a pair of concept seeds extracted from Facebook comment pairs, have been submitted to the expert team. The experts have been asked to assess the relevance of the generated context on a range from 0 to 5 on a Linkert scale, by evaluating the context in the form of tags cloud where a term is shown in its cluster, with a size depending on its *in-topicness*. The clouds have been computed in three main of expertise for the different semantic proximity measures. In Fig. 4 we can see the tag cloud for the pair (Mars, Scientist) using a PMING-based HSW in (Wikipedia, Bing) from [11, 29] as an example. Tag clouds and dispersion graphs have been used, for visibility and readability issues. Tag clouds (see Fig. 4) basically show a parameter (i.e. similarity value) at a time, where bigger terms correspond to the most similar words. These representations are well suitable to human evaluation. In our case, tag clouds represent each concept to be compared to the main topic.

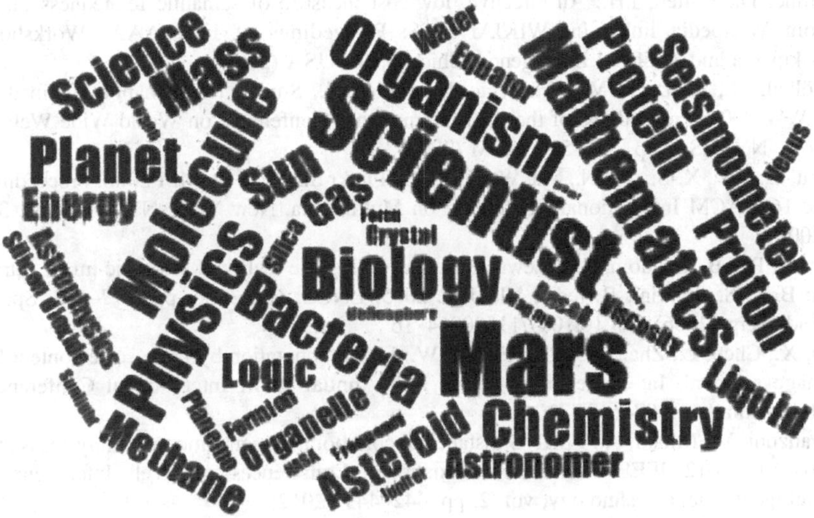

Fig. 4. Tag cloud for the pair (Mars, Scientist) using a PMING-based HSW in (Wikipedia, Bing)

4 Experimental Results and Discussion

In this work we introduced a method to investigate and identify the main context words obtained from Facebook posts and related user comments. The method is based on Natural Language Processing Part of Speech nouns extraction from sentences, similarity measurement using WordNet ontology, and clustering techniques.

Results show that clustering on frequency-based bag of words gives interesting results in the identification of topic contained in Facebook and it is more similar to human judgment than low level features comparison.

References

1. Miller, G.A.: Wordnet: a lexical database for english. Commun. ACM **38**(11), 39–41 (1995)
2. Budanitsky, A., Hirst, G.: Semantic distance in wordnet: an experimental, application-oriented evaluation of five measures. In: Proceedings of Workshop on WordNet and Other Lexical Resources, Pittsburgh, PA, USA, p. 641. North American Chapter of the Association for Computational Linguistics (2001)
3. Resnik, P.: Using information content to evaluate semantic similarity in a taxonomy. In: Proceedings of the 14th International Joint Conference on Artificial Intelligence, pp. 448–453 (1995)

4. Lin, D.: An information-theoretic definition of similarity. In: Proceedings of the 15th International Conference on Machine Learning, pp. 296–304. Morgan Kaufmann (1998)
5. Strube, M., Ponzetto, S.P.: WikiRelate! computing semantic relatedness using Wikipedia. In: Proceedings of the Twenty-First National Conference on Artificial Intelligence. AAAI Press, July 2006
6. Milne, D., Witten, I.H.: An effective, low-cost measure of semantic relatedness obtained from Wikipedia links. In: WIKIAI 2008: Proceedings of First AAAI Workshop on Wikipedia and Artificial Intelligence, Chicago, IL, USA (2008)
7. Völkel, M., Krötzsch, M., Vrandecic, D., Haller, H., Studer, R.: Semantic Wikipedia. In: WWW 2006: Proceedings of the 15th International Conference on World Wide Web, New York, NY, USA, pp. 585–594. ACM (2006)
8. Wu, L., Hua, X.-S., Yu, N., Ma, W.-Y., Li, S.: Flickr distance. In: MM 2008: Proceedings of the 16th ACM International Conference on Multimedia, New York, NY, USA, pp. 31–40 (2008)
9. Enser, P.G.B., Sandom, C.J., Lewis, P.H.: Surveying the reality of semantic image retrieval. In: Bres, S., Laurini, R. (eds.) VISUAL 2005. LNCS, vol. 3736, pp. 177–188. Springer, Heidelberg (2006). doi:10.1007/11590064_16
10. Li, X., Chen, L., Zhang, L., Lin, F., Ma, W.: Image annotation by large-scale content-based image retrieval. In: Proceedings of the 14th Annual ACM International Conference on Multimedia, pp. 607–610 (2006)
11. Franzoni, V., Milani, A.: PMING distance: a collaborative semantic proximity measure. In: WI-IAT, 2012 IEEE/WIC/ACM International Conferences on Web Intelligence and Intelligent Agent Technology, vol. 2, pp. 442–449 (2012)
12. Leung, C.H.C., Li, Y., Milani, A., Franzoni, V.: Collective evolutionary concept distance based query expansion for effective web document retrieval. In: Murgante, B., Misra, S., Carlini, M., Torre, C.M., Nguyen, H.-Q., Taniar, D., Apduhan, B.O., Gervasi, O. (eds.) ICCSA 2013. LNCS, vol. 7974, pp. 657–672. Springer, Heidelberg (2013). doi:10.1007/978-3-642-39649-6_47
13. Manning, D., Schutze, H.: Foundations of Statistical Natural Language Processing. The MIT Press, London (2002)
14. Turney, P.D.: Mining the web for synonyms: PMI-IR versus LSA on TOEFL. In: Raedt, L., Flach, P. (eds.) ECML 2001. LNCS, vol. 2167, pp. 491–502. Springer, Heidelberg (2001). doi:10.1007/3-540-44795-4_42
15. Cilibrasi, R.L., Vitanyi, P.M.: The Google similarity distance. IEEE Trans. Knowl. Data Eng. **19**(3), 370–383 (2007)
16. Li, Y.X.: Semantic image similarity based on deep knowledge for effective image retrieval. Research thesis (2014)
17. Franzoni, V., Mencacci, M., Mengoni, P., Milani, A.: Semantic heuristic search in collaborative networks: measures and contexts. In: WI-IAT (2), pp. 141–148 (2014)
18. Franzoni, V., Milani, A.: Heuristic semantic walk for concept chaining in collaborative networks. Int. J. Web Inf. Syst. **10**(1), 85–103 (2014). doi:10.1108/IJWIS-11-2013-0031
19. Franzoni, V., Mencacci, M., Mengoni, P., Milani, A.: Heuristics for semantic path search in Wikipedia. In: Murgante, B., et al. (eds.) ICCSA 2014. LNCS, vol. 8584, pp. 327–340. Springer, Cham (2014). doi:10.1007/978-3-319-09153-2_25
20. Wu, Z., Palmer, M.: Verbs semantics and lexical selection. In: Proceedings of the 32nd Annual Meeting on Association for Computational Linguistics. Association for Computational Linguistics (1994)
21. Bird, S., Klein, E., Loper, E.: Natural Language Processing with Python: Analyzing Text with the Natural Language Toolkit. O'Reilly Media Inc., Sebastopol (2009)

22. Leacock, C., Chodorow, M.: Combining local context and WordNet similarity for word sense identification. WordNet: Electron. Lexical Database **49**(2), 265–283 (1998)
23. Bakshy, E., Rosenn, I., Marlow, C., Adamic, L.: The role of social networks in information diffusion. In: Proceedings of the 21st International Conference on World Wide Web, pp. 519–528. ACM (2012)
24. Auer, S., Bizer, C., Kobilarov, G., Lehmann, J., Cyganiak, R., Ives, Z.: DBpedia: a nucleus for a web of open data. In: Aberer, K., et al. (eds.) SWC/ASWC 2007. LNCS, vol. 4825, pp. 722–735. Springer, Heidelberg (2007). doi:10.1007/978-3-540-76298-0_52
25. Franzoni, V., Milani, A.: A semantic comparison of clustering algorithms for the evaluation of web-based similarity measures. In: Gervasi, O., et al. (eds.) ICCSA 2016. LNCS, vol. 9790, pp. 438–452. Springer, Cham (2016). doi:10.1007/978-3-319-42092-9_34
26. Franzoni, V., Milani, A.: Heuristic semantic walk. In: Murgante, B., et al. (eds.) ICCSA 2013. LNCS, vol. 7974, pp. 643–656. Springer, Heidelberg (2013). doi:10.1007/978-3-642-39649-6_46
27. Franzoni, V., Leung, C.H.C., Li, Y., Mengoni, P., Milani, A.: Set similarity measures for images based on collective knowledge. In: Gervasi, O., et al. (eds.) ICCSA 2015. LNCS, vol. 9155, pp. 408–417. Springer, Cham (2015). doi:10.1007/978-3-319-21404-7_30
28. Biondi, G., Franzoni, V., Li, Y., Milani, A.: Web-based similarity for emotion recognition in web objects. In: UCC 2016, pp. 327–332 (2016)
29. Pallottelli, S., Franzoni, V., Milani, A.: Multi-path traces in semantic graphs for latent knowledge elicitation. In: ICNC 2015, pp. 281–288 (2015)

Workshop on Future Computing Systems, Technologies, and Applications (FiSTA 2017)

mruby – Rapid IoT Software Development

Kazuaki Tanaka[1(✉)] and Hirohito Higashi[2]

[1] Kyushu Institute of Technology, Kitakyushu, Japan
kazuaki@mse.kyutech.ac.jp
[2] Shimane IT Open-Innovation Center, Matsue, Japan

Abstract. Embedded systems are systems in which hardware and software are closely related. In the embedded system, the software controls the hardware.

We focused on "Ruby" which is an object-oriented programming language and scripting language to improve development efficiency of embedded systems. We developed mruby which applied Ruby so that it can be used in embedded systems, and this overview is described in this paper. mruby provides the real-time and hardware access features required in embedded systems development.

1 Introduction

Embedded systems are systems in which hardware and software are closely related. In the embedded system, the software controls the hardware.

Generally, since the value of the system is determined by the hardware and software, the cost of an embedded system is depending on the software development cost.

We focused on Ruby which is an object-oriented programming language and scripting language to improve development efficiency of embedded systems [1,2]. We developed mruby which applied Ruby so that it can be used in embedded systems, and this overview is described in this paper.

2 Internet of Things

In this section, we describe that the embedded system includes IoT and show the development way of the embedded system.

2.1 Embedded System

Recently, IoT (Internet of Things) that utilizes various devices connected to a network has become popular. According to a white paper from Cisco, 50 billion devices will be connected to internet [3]. The goal of IoT is used by information obtained from the device for service. The service provider is expected to implement and release their own ideas as a service in a short time.

O. Gervasi et al. (Eds.): ICCSA 2017, Part I, LNCS 10404, pp. 733–742, 2017.
DOI: 10.1007/978-3-319-62392-4_53

Since the IoT service is built by both hardware and software technologies are involved. Development of the system is sometimes difficult. In general, developers develop and release services through complicated program logic and its trial and error. Competitiveness in IoT is the quality of service, which is beneficial to users.

The development cost if the problem in releasing the IoT services. In IoT, since the area that realizes ideas greatly contributes to the quality of service, it is important how quickly software can be implemented. As embedded system developers are required to have a combination of hardware and software skills, development costs of embedded systems are high.

2.2 Software Prototyping

Development processes based on software prototyping are often used to develop IoT services. Software prototyping is a method to make it easier to grasp the image of the final product in the future by implementing incomplete software that operates in a short time.

IoT service has a large gap between ideas and implementations. The idea of IoT is conceived focusing on the service the user receives. On the other hand, the information acquired from the environment by the device is a sensor value, and implementation dependent on hardware is required. Furthermore, when analysis of sensor data is necessary, development of IoT service becomes complicated.

Software prototyping is an effective way of developing such an IoT service. IoT uses results obtained by blending various sensors for service. It is not often that program logic for processing sensor data has been decided from the beginning, and we will complete the IoT service by repeatedly testing with prototypes.

In software prototyping, it needs to be able to implement operating software "in a short time". Software prototyping develops prototypes based on ideas, tests prototypes, and iterates development using feedback. By using the feedback of the result of testing the prototype, the quality of the service improves. Therefore, it is essential to successful development to iterate this cycle quickly.

Software prototyping is widely known as agile development. Many IT startup companies are adopting this method in order to release software in a short time and continuously improve the product quality.

2.3 Embedded Systems Development

IoT is a system that combines hardware and software. Development of an embedded system has the following features.

- Coding in C Language
  Embedded system software controls hardware. Because hardware control requires direct access to memory and IO address space, many of them are coded in C language. Generally, software development in C language is expensive and development time tends to be long.

- Real-time Processing
  In the embedded system, the software operates according to a signal from the outside. The real-time processing ensures that a procedure starts in exact time after the signal from the outside.
  For example, consider an application that controls the motor at a constant rotation speed. Software starts to move at the output of the sensor that detects the rotation of the motor, and the rotation speed of the motor is obtained by measuring the time. Compares the rotation speed of the motor calculated by software with the target value and controls the output current driving the motor. In this application, time measurement is an important factor, and real time processing is used here.
- Limited Resources
  Processor power and memory available in devices is limited. Embedded systems are stored internally as part of the product. Embedded systems are developed with the smallest possible hardware to lower product costs and reduce the power consumed by products. That means, it is required that resources consumed at the time of execution are low.
- Device Dependent
  Embedded systems are used in a variety of environments and as various applications, so devices to be implemented vary. Furthermore, because embedded systems deal with hardware with software, description of software tends to depend on hardware.
  In one example, since the microprocessor used in the existing embedded system becomes obsolete and corresponds to a different processor, a correction that affects the entire software is necessary.

In the development of embedded systems, so far, development costs are high because of various restrictions and difficulties as showed above. In IoT, services are performed by software, and software scale is large. IPA (Information-technology Promotion Agency, Japan) reports that about half of the development cost of embedded products are the software development cost.

In order to reduce the cost of embedded systems, especially IoT products. It is expected to lower software development costs.

3 mruby

This section shows a method for efficiently developing software.

3.1 Programming Language Ruby

In developing Web applications, the programming language Ruby is typically used. Programming language Ruby is just an object-oriented scripting language, published by Matsumoto Yukihiro as open source software in 1995. In 2005, the Web application framework Ruby on Rails was released and used as a standard for Web applications in many commercial services. In 2012, Ruby was internationally standardized as ISO/IEC 30170.

Ruby is well known by its high development efficiency. High development efficiency of Ruby comes from the readability and flexibility of the code. The readability of code contributes not only to the development of software but also to maintenance and functional improvement. The reason why Ruby is adopted in a Web application is that development of software and subsequent addition of functions can be done smoothly.

The syntax of the Ruby program is not just flexible but also has many functions. Ruby has many built-in classes and methods, especially various abstract data supporting features such as Array and Hash are powerful. On the other hand, the Ruby interpreter needs a lot of resources to execute because of syntax and semantic analysis process. Many Ruby applications are Web applications, so there are no problems in consuming resources in the server machine.

3.2 mruby

We tried to apply Ruby's high development efficiency to the development of embedded software. This attempt was conducted as a project of the Ministry of Economy, Trade and Industry of Japan from 2010 to 2012. The programming language mruby is a deliverable of this project, which is published as open source software.

Ruby was an interpreter language, but mruby is a compiler language. The mruby compiler generates bytecode from the mruby program and executes the generated bytecode in mruby virtual machine (VM). In this mechanism, the compiler needs many resources, but the development environment has enough resources for the compilation, so there is no problem. The mrubyVM which executes bytecode is small and it works well even with small microprocessor.

The bytecode is a device-independent format. Various processor architectures and memory architectures are adopted in embedded systems. As mrubyVM absorbs differences between the architecture of the target devices, bytecode works the same way in all mrubyVM without any changes (Fig. 1).

3.3 mruby Compiler and VM

The mruby compiler creates bytecode from mruby's source code and mrubyVM executes this bytecode. As mrubyVM abstracts the hardware of the target device, bytecode does not depend on device architecture. In other words, by simply preparing the mruby VM of the target device, we can move the bytecode of the mruby program on that target device.

On the other hand, since the implementation of mrubyVM depends on hardware, it must be built for the target device. Generally, development of target device software requires a cross development environment. Since mrubyVM is implemented in C99, in which almost all cross development environments based. Portability of mrubyVM is high.

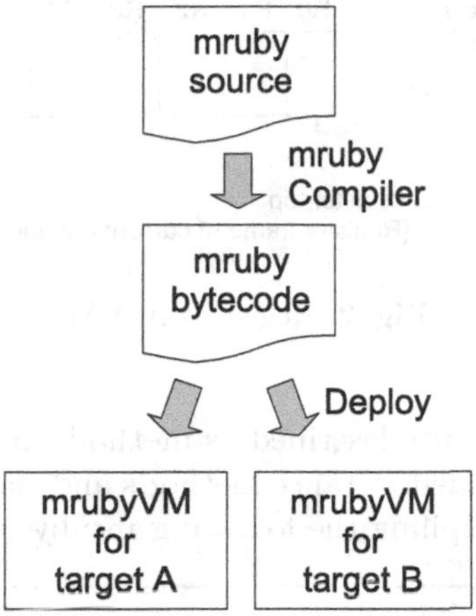

Fig. 1. Compile and Deploy

4 Implementation

In this section, we describe the implementation of mruby which satisfies the requirement of the embedded system mentioned in Sect. 2.3.

4.1 mruby VM

The biggest purpose of mruby is to abstract hardware and absorb differences between different devices. Embedded systems are composed of various architectures. The most prominent is a CPU, and the operation code handled by the CPU is an expression peculiar to each CPU. Also, the difference in memory architecture arises when word access is made to memory, which is called endian. Also, the difference in OS also gets a big influence. The smallest embedded system does not have a OS, and in a large embedded system, a general purpose OS having MMU (Memory Management Unit) function is used.

In order to absorb these various differences, mruby executes the program in the virtual machine (VM). This virtual machine is a register machine and has many registers (virtually infinite number of registers in a normal implementation), and the program performs arithmetic between registers. This differs from the office CPU in that all registers handle Ruby objects, which make it possible to efficiently execute Ruby programs.

Method invocation in Ruby is described by the register machine as showed in Fig. 2.

Many times of software execution of an object-oriented language are the time of method invocation on an object. In Ruby's method invocation, a method is searched and a matching method is called. Reducing the time of method invocation improves the performance of the program execution. Because allocation of registers is done by the compiler, it can be executed with a simple operation of registers at the time of execution. This point is the merit of mruby VM.

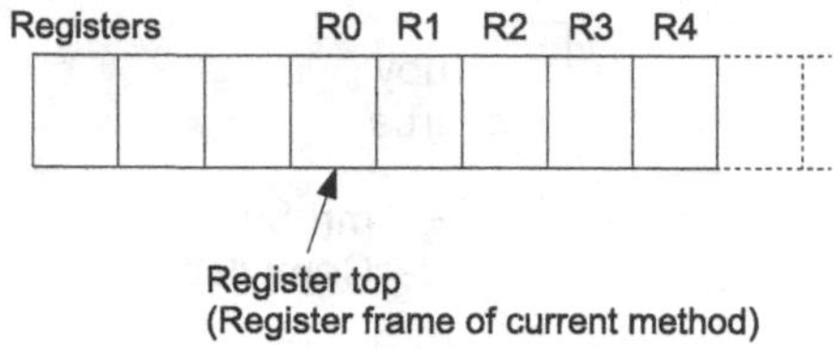

Fig. 2. Registers in VM

In mruby all programs are described as methods, methods belong to objects or class objects, and are called instance methods and class methods, respectively.

For example, after compiling the following mruby code.

```
a = Foo.new
b = a.bar
```

The following bytecodes are generated.

```
000 OP_GETCONST   R3    :Foo
001 OP_SEND       R3    :new   0
002 OP_MOVE       R1    R3           ; R1:a
003 OP_SEND       R3    :bar   0
004 OP_MOVE       R2    R3           ; R2:b
```

First, the class object of `Foo` is stored in `R3` and the `new` method is called (`OP_SEND` means a method call). Before establishing `OP_SEND`, the register top is changed to `R3` because the register frame begins `R3`. The return value of the new method of the `Foo` object is returned by `R3` which is the top of the register frame. And the return value is transferred into variable `a` which is register `R1`. The return value of `Foo.new` is instance of class `Foo`. Next, `bar` method is called using `R3`.

Final value which is the return value of `a.bar` is stored in `R3` and it is transferred into `R2`.

4.2 Incremental GC

Many object-oriented programming languages have a garbage collection(GC) feature for management of memory used by objects. GC can check the object (memory) that is no longer needed, manage the memory, and prevent memory leak at the same time.

GC algorithms of Ruby and Java are Mark&Sweep and Generation GC. These algorithms can check a lot of memory per unit time and have high performance. However, during GC execution, program execution by VM is stopped in these GC algorithms. When using Ruby or Java in the Web system, since the program stop by GC is at most several seconds, there are few problems in the operational phase.

Mark&Sweep and generational GC are not suitable for embedded systems that need real-time performance because VM execution stops during GC execution. Real-time of the embedded system is to guarantee the time from the occurrence of external input (or interrupt) until the start of execution of the target program. For example, periodic program execution every certain time by the timer interrupt.

Whereas ordinary GC stops VM execution and checks all objects, incremental GC guarantees VM downtime due to limitation of the number of objects checked at one time. Since the object to be checked is limited, overhead of accessing the object occurs, but real time can be guaranteed.

4.3 C Interface

In the embedded system, the software controls the hardware.

In order to describe software for embedded systems with mruby, it is necessary to have a mechanism for safely calling functions that are provided in the C language. mrubyVM has a function to create API calling C library. You can easily add a wrapper function that has to call the C library.

Hardware operation in the device is controlled over the IO port directly connected to the hardware. To directly operate the IO port, access to the physical address is indispensable, and a pointer to the C language is optimal for this purpose. Since the C language has the function of directly accessing the physical address of the device, many embedded systems are implemented in C language.

The following code is a simple C wrapper function for mruby. This wrapper function is for calling the `sin` function in C language from mruby's `sin` method. The implementation of the wrapper function is straightforward if there is a C language library.

```
static mrb_value
math_sin(mrb_state *mrb, mrb_value obj)
{
  mrb_float x;

  mrb_get_args(mrb, "f", &x);
  x = sin(x);

  return mrb_float_value(mrb, x);
}
```

5 Security

IoT devices are widely distributed in our daily life and they are connected to many equipment. The unauthorized use and information leakage cause serious problems. To prevent such problems, security features are necessary [4].

Since the available resources in an IoT device are limited, it is necessary lightweight security features. Although security is essential encryption, lightweight encryption algorithm in IoT is desired.

5.1 Cryptosystems

Cryptosystems that can be used for information encryption can be broadly classified into secret key cryptosystem and public key cryptosystem.

Common Key Cryptosystem. Common key algorithms are algorithms for cryptography that use the same cryptographic keys for both encryption of plain text and decryption of cipher text. Keys must be properly managed, but generally the amount of computation is small. It is implemented as DES, AES and so on.

Public Key Cryptosystem. Public key cryptography, or asymmetric cryptography, is any cryptosystem system that uses pairs of keys, public key and private key. The cipher text which is encrypted with one key can be decrypted only by another key. Public key is widely published. However, it is difficult to obtain a secret key from the public key, so key management is simple. It is only necessary to close and manage only the secret key in the system. It is implemented as RSA and so on.

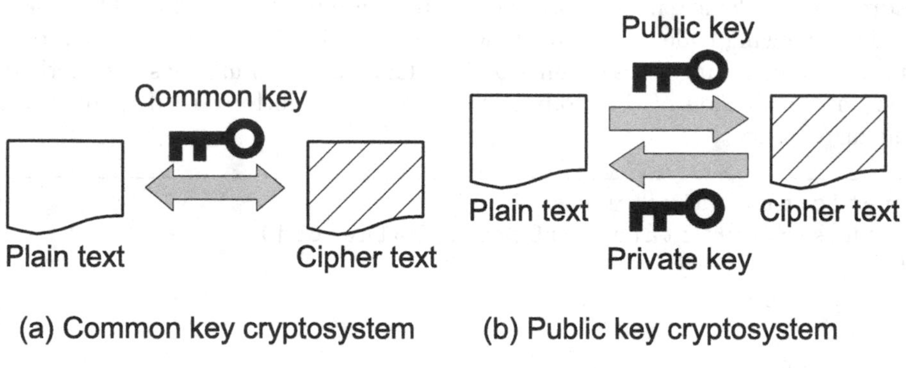

Fig. 3. Cryptosystems

An example of the use of encryption is shown in Fig. 3.

In the common key cryptosystem, a cipher text is generated from plain text using a common key. This cipher text is decrypted into plain text by the same common key. In the public key cryptosystem, a cipher text is generated from a plain text using a public key. This cipher text is decrypted only by the pair secret key. Conversely, a cipher text created using a secret key can be decrypted only by the public key.

Public key cryptography is used as a basic technology for information encryption and digital signature.

5.2 Security in IoT

In IoT, the information acquired in a variety of devices is utilized in conjunction with the service. Sensor data device is information having a value. It is necessary to encrypt the data when transmitting the data. Further, it operates embedded software in the IoT device is also required a mechanism which does not execute the illegal softwares. Encrypting data and avoiding illegal softwares must be realized in limited resources.

With the public key method, it is possible to decrypt the encrypted program using only the public key, and the encrypted program can be created only with the secret key. Thus, in general the software authentication, but the public key system is suitable, the resources IoT device can be used is limited, the implementation of a public key system is not practical.

5.3 mruby Security Options

In mruby, the application program is compiled into byte code, and mruby VM executes byte code sequentially as Fig. 1.

In order not to run unauthorized software, to implement a software authentication of the byte code in the mruby VM. Since byte code is continuously authenticated during execution, a mechanism of authentication with a small calculation amount is desired.

In addition to the implementation published in open source software, mruby offers several commercial options. As one of these options, a software

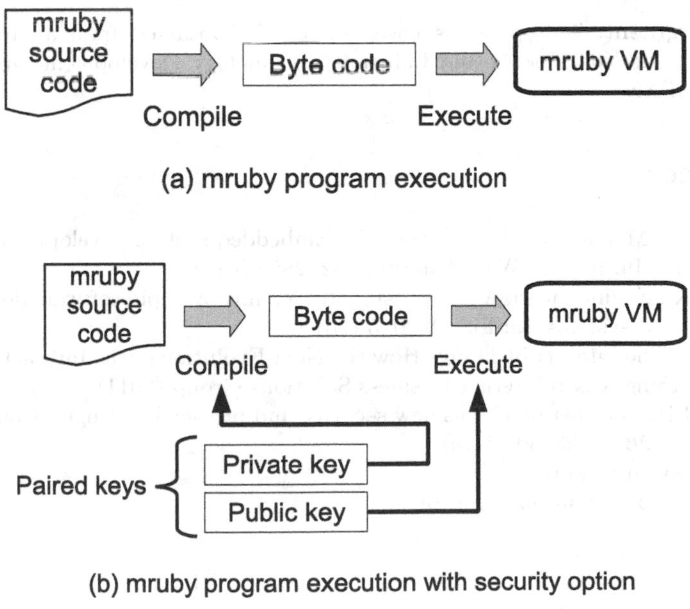

Fig. 4. mruby program execution without/with security option

authentication function is provided. In the software authentication function, it is checked whether or not the byte code has been compiled in the correct development environment.

Figure 4(a) shows execution of bytecode of open source version mruby, and Fig. 4(b) shows execution using that security option.

Security options provide compilers and VM, which hold private and public keys internally. The compiler generates encrypted byte code using the secret key when compiling mruby source code. Encrypted byte code can be executed only with decryptable mruby VM and mruby VM have a public key. Since the public key is widely published, and it cannot be forged by the public key. As a result, it is possible to run the only byte code that is encrypted created using a secret key in the pair. This procedure, avoids the execution of unauthorized software.

6 Conclusion

The mruby has been released as open source software [5]. mruby is a MIT license, and it can be freely used regardless of commercial or non-commercial use. We expect mruby to contribute to the development of embedded software engineers. Although mruby has not yet adopted its record yet and is under verification, it is a technology to improve the development efficiency of embedded software.

As future research and development, mruby/c with miniaturized mruby is being developed [6]. mruby/c is an implementation assuming execution with an one-chip microprocessor, aiming for cost reduction and power savings as well as development efficiency improvement.

Acknowledgment. This paper is based on results obtained from a project commissioned by the New Energy and Industrial Technology Development Organization (NEDO), JAPAN.

References

1. Tanaka, K., Matsumoto, Y., Arimori, H.: Embedded system development by lightweight ruby. In: ICCSA Workshop, pp. 282–285 (2011)
2. Tanaka, K., Nagumanthri, A.D., Matsumoto, Y.: mruby-rapid software development for embedded systems. In: FiSTA 2015 (2015)
3. Evans, D.: The Internet of Things-How the Next Evolution of the Internet Is Changing Everything. Cisco Internet Business Solutions Group (2011)
4. Weber, R.H.: Internet of Things new security and privacy challenges. Comput. Law Secur. Rep. **26**(1), 23–30 (2010)
5. http://www.mruby.org/
6. http://github.com/mrubyc/mrubyc

Design Approaches Emerging in Developing an Agricultural Ontology for Sri Lankan Farmers

Anusha Indika Walisadeera[1,3(✉)], Athula Ginige[2], and Gihan Nilendra Wikramanayake[3]

[1] Department of Computer Science, University of Ruhuna, Matara, Sri Lanka
waindika@cc.ruh.ac.lk
[2] School of Computing, Engineering and Mathematics, Western Sydney University, Penrith, NSW 2751, Australia
A.Ginige@westernsydney.edu.au
[3] University of Colombo School of Computing, Colombo 07, Sri Lanka
gnw@ucsc.cmb.ac.lk

Abstract. Building an ontology is not a simple task, because ontology development is a craft rather than an engineering activity. Ontology development depends on many factors, for examples, background of the domain experts, engineering techniques, need much time to investigate the domain in detail, and also it is a creative task. Then it is very easy to make mistakes due to lots of various new concepts for many users. Therefore ontology developers and researchers need best practices as well as ontology design patterns to construct good quality ontologies. However, the absence of structured guidelines, methods, and good practices hinders for the development of ontologies. We have developed a large user centered ontology to represent agricultural information and relevant knowledge in user context of Sri Lankan farmers. Through this development we have come across various design models and techniques. In this paper, we have highlighted those models and techniques that will be helpful for the users in the field of ontology development. This discussion is mainly based on three different scenarios. Scenario one mainly concerns for identifying the basic ontology components and their representations. Other scenarios such as event handling for complex real-world situations and conceptualization of overlapping concepts for the farmer location are discussed in detail and represented by using well-known Protégé tool.

Keywords: Ontology design · Design issues · Ontology design patterns · Modeling solutions

1 Introduction

Ontology development is a craft rather than an engineering activity [1]. Because building an ontology is not a simple task. It is very difficult task because it depends on many factors, for examples, background of the domain experts, engineering techniques, need much time to investigate the domain, and also it is a creative task. Furthermore,

O. Gervasi et al. (Eds.): ICCSA 2017, Part I, LNCS 10404, pp. 743–758, 2017.
DOI: 10.1007/978-3-319-62392-4_54

the absence of structured guidelines, methods, techniques and good practices hinders for the development of ontologies. Many users such as ontology developers and researchers in this field use different methods, approaches, and their own set of principles to build ontologies. However, there are lots of various new concepts for many users in the designing and implementing an ontology. Then it is very easy to make mistakes. Therefore ontology developers and researchers need best practices as well as ontology design patterns to construct good quality ontologies.

We have developed a large user centered ontology for Sri Lankan farmers in representing necessary agricultural information and relevant knowledge within the farmers' context. Through this development, we have come across various design techniques and design models. In this paper, we have discussed those in detail that will help for many users in the field of ontology development. The developed user centered ontology for Sri Lankan farmers to represent agricultural information and knowledge that can be queried based on the farmer context. From this ontology, specific farmer can get the specific answers to the specific questions such as "what crops will grow in 'my farm'?"; "what fertilizer to use for 'my crops' and when to apply to them?"; "which pest is destroying 'my crop'?"; and "what are the best control methods to apply 'my farm' based on 'my preferences'?"; etc. More details about the design of the ontology, ontology development, evaluation, and maintenance procedures are available in [2–4].

The remainder of the paper is organized as follows. Section 2 presents related research in this field. In Sect. 3, the design approaches (design models) are discussed in detail by using three different scenarios. Scenario one mainly concerns to identify basic ontology components and explain how those are represented in OWL 2 using the Protégé environment. In this study, the decidability is very important since agricultural information in user's context needs to be retrieved. Therefore the DL based (OWL 2 - DL) approach was selected to implement this ontology. In scenario two and three, handling of the events and overlapping concepts are mainly discussed respectively with implementation details. Validation and evaluation methods are described briefly in Sect. 4. Section 5 summarizes the design approaches used in this study and concludes the paper.

2 Related Work

In this section, we try to discover number of challenges we face when developing ontologies, some recommendations to overcome these challenges, good practices and design patterns for designing ontologies from the literature. Here we have cited some of the studies in this field that will help us to clearly identify what types of design issues available and what kind of modeling solutions are used for the ontology design.

Presently, ontology development has faced number of challenges such as knowledge acquisition, and lack of sufficient validated and generalized development and evaluation techniques [5]. Some of the major issues are listed below:

- Knowledge acquisition is an independent activity within ontology development, but it coincides with other activities. Ontology modeling is traditionally carried out by knowledge engineers and/or domain experts. Domain experts are not easily accessible and the experts' knowledge is likely to be incomplete, subjective and even outdated.
- Absence of standard methodologies for building ontologies and none of the approaches presented is fully mature. Developers apply their own approaches to develop ontologies then great effort is required for creating a consensual methodology for ontology construction.
- Ontology representation is the most fundamental issue in ontology development. Ontology representing languages should provide representation adequacy and inference efficiencies [5]. Some languages incorporate description logics to enhance the expressiveness of reasoning systems. Then special attention requires the selection of a suitable language to serve as medium for ontology expression.
- Since ontologies are used for several different purposes, different kinds of evaluations are required. The challenge is to evaluate in a quantitative manner how useful or accurate the developed ontology. Some general questions arise with evaluation such as why, what, when, how and where to evaluate; who evaluates; and what to evaluate against. Ontology can be evaluated from a variety of perspectives, ranging from content to technology, methodology and application [6] using objective measures such as completeness, consistency, and conciseness (3Cs) [7]. People have to use "ad hoc" techniques for evaluation because of the lack of methods and standard techniques for evaluating ontologies.
- Continuous growth and intensive maintenance of ontologies are serious challenges in the ontology development process due to conceptual changes, scope extensions, mistakes and quality improvements, etc. It becomes more complex in the case of large-scale ontologies. Manual ontology maintenance is a difficult task and automated solutions should be explored.

Some recommendations are identified to overcome above issues. They are:

- Before construction of ontology, an important step for a designer is to determine whether similar ontologies already have been created by others so that knowledge can be reused or extended. This minimizes the development effort and promotes interoperability with other applications that use the same ontology. When developing a domain specific ontology, it is an opportunity to extend based on more generic ontologies that exist within the same domain and to reuse previously defined class structure and axioms (e.g. thesauri, terminology and vocabulary).
- During knowledge acquisition, a set of knowledge-acquisition techniques can be used in an integrated manner [8]. They are: structured/non-structured interviews; informal/formal text analysis to study main concepts in books/handbooks; domain table and domain-graph analysis; and formula analysis. Practically, people can turn to other sources, e.g. dictionaries, Web documents, and database schemas to keep up with the requirements for the content of ontologies [9].

According to the literature in ontology engineering, Ontology Design Patterns (ODPs) are now a hot topic and can be considered as modeling solutions to the problems in ontology designing. These patterns are very helpful to ontology developers as well as researchers when modeling an ontology. Since it provides a development guideline that helps to solve design problems then they can improve the quality of the resulting ontologies.

Poveda et al. [10] have identified a set of pitfalls in solving design problems for which there are no available ODPs. Common pitfalls identified during ontology development process were classified by them as *Annotation pitfall*, *Reasoning pitfall*, *Naming pitfall*, *Logical pitfall*, and *Content pitfall* [10]. Annotation pitfall refers to the ontology usability from the user's point of view and improves the user's understanding of the ontology and its elements (e.g. Label vs. Comment). Reasoning pitfall refers to the implicit knowledge derived from ontology when reasoning procedures are applied to such an ontology (e.g. Defining wrong inverse relationships). Naming pitfall refers to the ontology usability from the user's point of view and specifically, to the naming of the ontology (e.g. Polysemy). Logical pitfall refers to the solution to design problems in which the primitives of the representation language used do not provide support (e.g. Part-of vs. Subclass). Content pitfall refers to the solution to the design problems related to the ontology domain (e.g. Incomplete information, Miscellaneous class). Pitfalls appear very frequently in ontology development process. Sometimes some pitfalls are not problems in practice. Using methodological guidelines it will help to avoid these pitfalls during the development process.

W3C work group has established "*Semantic Web Best Practices and Development (SWBPD)* [11]" to provide support to developers and the users of Semantic Web. It proposes a set of good practices and patterns with respect to logical patterns that solve the design problems in the OWL. Some other libraries are available, for example, "*NeOn Modelling Components*" under NeOn project [12]. They do not provide the pattern code but give descriptions of number of ODPs.

3 Design Approaches

We developed a large user centered ontology for Sri Lankan farmers in representing necessary agricultural information and knowledge within the farmers' context. Based on the experience we obtained from a long journey to develop our ontology, we have discussed here the modeling solutions to design the complex real world scenarios in the domain of agriculture. The following three sub sections (Sects. 3.1 to 3.3) are presented ontology design models by using different motivation scenarios.

3.1 Scenario I: Identifying Main Ontology Components

According to the field of computer science, many definitions for the term 'ontology' have been proposed. Different definitions provide different views on the same reality. In the literature, a popular and well-accepted definition of the ontology was proposed by Thomas Gruber as *"An ontology is an explicit specification of a conceptualization"* [13]. According to this definition, there are three main components of what ontology should consist of: concepts, relationships, and constraints. These components can be represented in many ways; concepts can be represented as classes, entities, sets, collections, etc. relationships between concepts can be categorized into mainly two groups: hierarchical relationships (taxonomies) and associative relationships (connects concepts by creating the semantic meaning of concepts). Additional knowledge about the domain is captured as constraints among concepts and relationships and it is represented by axioms (for reasoning).

For the design of our ontology, we selected Grüninger and Fox methodology, a first-order logic (FOL) based approach to design an ontology while providing a framework for evaluating the adequacy of the developed ontology. Being a formal ontology it is structurally and functionally rich enough for the description of the domain knowledge in context. Using this method, the ontology design is started from motivation scenario (problem) and the competency questions based on the scenario. Competency questions (CQs) determine the scope of the ontology and use to identify the contents of the ontology. However, the technique used for formulating CQs is out of the scope of this paper. Formulating competency questions related to this study is available in [2]. Then we used the middle-out strategy to identify main concepts. The main advantage of this approach is that it starts with most important concepts first. Once the higher level concepts are defined then specialized and generalized hierarchies get identified. Thus, these concepts are more likely to be stable. This results in less re-work and less overall effort.

There are few concepts which we can directly elicit, for instance, Crop as a main concept of this ontology based on scope of the ontology. Other major concepts need to be identified by analyzing each CQ. For example, the main concept in the CQ: *which crops are suitable to grow in the 'LowCountryDryZone' agro-climatic region?* is Zone. Once the concept is identified, then the specialized and generalized (if necessary) hierarchies need to be defined based on the concept properties, nature of the instances (instances are used for denoting specific members of a concept and represented by constants or variables), and other relevant constraints. The concept Zone (i.e. Climatic Zone) has properties such as maximum rainfall and minimum rainfall. By specializing Zone concept, the concept AgroZone (i.e. Agro-climatic Zone) is defined as a subclass of Zone, because there are several additional properties specific to AgroZone such as maximum and minimum temperature, and maximum and minimum elevation. The properties of concept Zone can be inherited by the AgroZone concept because of the taxonomic hierarchy (is_a relationship). AgroZone is a subclass of Zone if and only if every instance of AgroZone is also an instance of Zone (specialization). Then, AgroZone is also a subclass of Location (see Fig. 1). The Location concept is introduced to represent farmer location. Further, AgroZone can be divided into 46 zones based on the climatic (rainfall, elevation and temperature), soil, physical features with diversities of land usage. It was named as AgroEcologyZone. Then, AgroEcologyZone can inherit all the properties of super classes of it (see Fig. 1). In Sri Lankan agriculture domain, the farmers' location can be represented in many different ways. In this study, the farmer's location has been identified as zones, agro zones, agro ecological zones, provinces, districts, regional areas and elevation (i.e. elevation based locations). This knowledge is represented as shown in Fig. 1. Here we use the OWL terminology to represent this, (e.g. object property, data property, etc.).

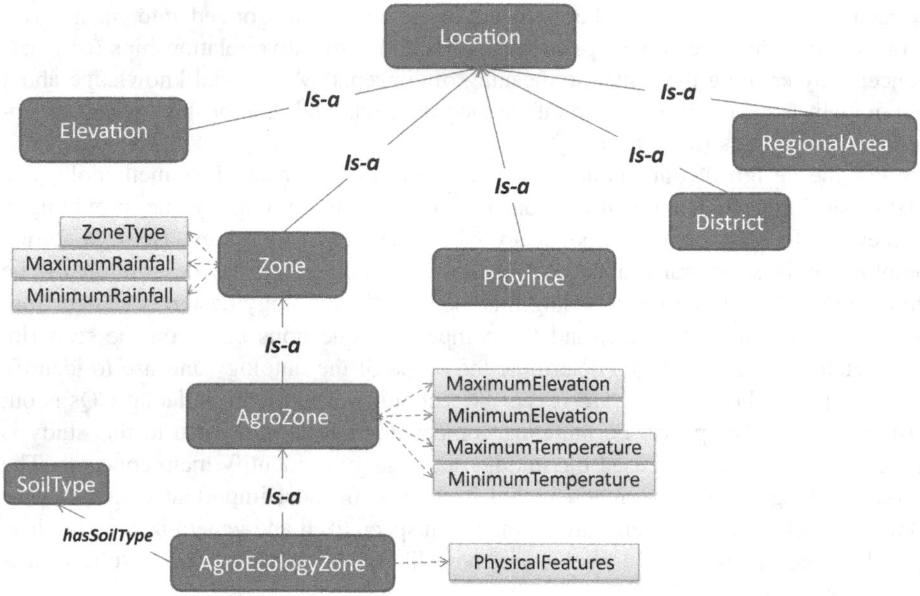

Fig. 1. Representation of Location concept.

Based on the definition of the concept Zone, the instances can be categorized as DryZone, IntermediateZone and WetZone. If there is no further categorization (has only first level categorization) and/or need to restrict conditions then this categorization can be restricted as a property of a concept (e.g. ConceptType) to reduce complexity of the design of the ontology. This decision of data property identification depends on the application, for example, on the type of information which we want to retrieve from the application (i.e. by restricting the values of the Data Property in Protégé) and kind of relationships between them (for example, no need to define inverse relationship). In this study, the Zone concept has three properties such as ZoneType, maximum and minimum rainfall. If there is a more than one level categorization, this has been organized as a taxonomic hierarchy. We now looked at more examples about data properties; Crop has main properties (Data Property) such as Crop Family, Hardiness, Nutrition, etc. Since a variety is a group of crops that share common qualities of crops of the same species, a variety has been identified as a subset of a crop (i.e. Variety is a sub-concept of Crop) (see Fig. 2). Then these properties can be inherited by Variety. Other than these properties Variety has own properties such as Length, Color, Shape, etc. (Data property).

The associative relationships (non-taxonomic hierarchy – Object Property in Protégé) are specified by identifying the concepts and relationships with meaningful relations. It needs to be defined between concepts as relationships and their inverse relationships (if applicable). For examples, there are associative relationships with inverse between Crop and Variety: *Crop hasVariety Variety*, *Variety isVarietyOf Crop* (see Fig. 2); Crop and SoilFactor: *Crop hasSoilFactor SoilFactor*, *SoilFactor isSoilFactorOf Crop*. A set of relationships describes the semantics of the domain.

The environmental factors have been defined to be Sunlight, Wind, Humidity, WaterSource, etc., which are subclasses of the EnvironmentalFactor (superclass). In this study, the union of these subclasses form the environmental factors which need to be specified for instances of crops as well as for farms. In here, the subclasses are used as *a set of mutually-disjoint classes* which covers EnvironmentalFactor (see Fig. 2). Every instance of EnvironmentalFactor is an instance of exactly one of the subclasses in the union. In similar way SoilFactor concept is defined with *a set of mutually-disjoint classes* such as PhValue, SoilDrainage, SoilNutrition, SoilTexture, SoilType, and SoilSpecialCharacteristics (see Fig. 2).

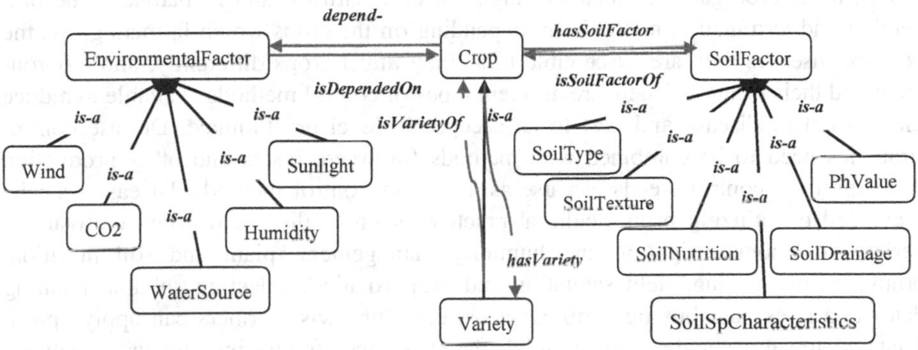

Fig. 2. Relationships among concepts. (Color figure online)

The definitions of the terms, constrains and their interpretation related to the query are specified using a set of axioms in FOL. The axioms have been defined to express these definitions and constraints. For example, the definition of one of the main climatic zones in Sri Lanka based on annual rainfall (in mm) is expressed below (e.g. DryZone is defined based on annual rainfall in between 1750 (Maximum Rainfall) and 0 (Minimum Rainfall)). Figure 3 shows this definition in Protégé environment.

$\forall x \quad (Zone(x) \wedge (\exists y \quad NonNegativeInteger(y) \wedge hasMaximumRainfall(x, \quad y) \wedge (y \quad 1750)) \wedge (\exists z \quad NonNegativeInteger(z) \wedge hasMinimumRainfall(x, \quad z) \wedge (z \geq 0)) \leftrightarrow DryZone(x));$

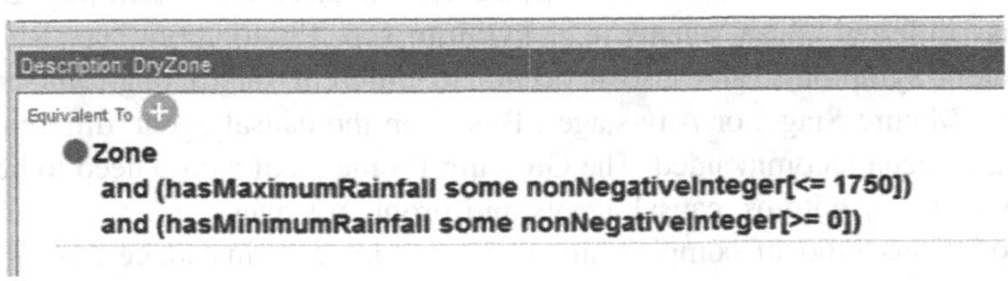

Fig. 3. The definition of "DryZone" represented in Protégé.

3.2 Scenario II: Event Handling

When modeling complex real world domains within the user context the representation of the entities and their relationships at different time intervals and in different locations is a challenge. In this research, this challenge is achieved by modeling the complex real-world scenario in the domain of agriculture.

Consider now we have to represent information and knowledge related to common growing problems and their management. Let's take this CQ: *What are the suitable cultural control methods to control Bacterial Wilt for Tomato?* By analyzing collected information from reliable sources we have identified that there are different types of growing problems especially for the vegetable cultivation. Mostly, crop diseases are caused by microorganisms such as fungi, bacteria, viruses, and nematodes. The prevention and termination methods are depending on the crops which farmers grow, the pests or diseases they are susceptible to as they affect crops differently, and environment and their location. There are different types of control methods available to reduce the amount of disease and pest to an acceptable level or eliminate. Disease control strategies need to be combined with methods for weed, insect and other production concerns (e.g. control weeds can use as a disease control method). Diseases can be prevented by utilizing proper cultural practices such as disease resistant cultivars or variety selections, irrigation and humidity management, plant and soil nutrition, pruning, row spacing, field sanitation and crop rotation, selecting suitable planting dates and rates, and buying certified seeds, etc. Otherwise farmers can apply appropriate chemical pesticides such as fungicides (to control fungi), insecticides (to control pest) and weedicides (to control weed) which are available in the market to control pest and diseases. Farmers can also apply a suitable biological control method to prevent or control the attack, for an example, Cytobagus salviniae to control Salvinia (Salvinia molesta). Normally, several methods are adapted at the same time in same place. Thus there are multiple factors to consider when selecting a suitable control method. Based on the above findings a control method selection criterion has been defined, such as *Environment* (soil, location, water), *Farming Stage* (application stage), and *Farmer Preferences* (types of control methods). More details related this is available in [3]. According to the data analysis, the examples show that Tomato and Brinjal face same diseases such as Bacterial Wilt and Damping-off. However, their symptoms and recommended control methods are different based on the crop (i.e. same diseases but different symptoms and different control methods, sometimes it may be the same). Therefore a disease control method depends on the disease related to the specific crop. In addition to that, it shows the crops have same disease problem (e.g. Damping-off) but has different causal agents (e.g. Pythium spp, Phytopthora spp, Rhizoctonia spp), different symptoms, and also it occurs in different stages (e.g. Nursery stage, Early stage, Mature Stage, or Any stage). Based on the causal agent, different control methods have been recommended. The Growing Problems of a crop need to be defined clearly based on symptoms, causal agents and problem stages.

To model this kind of complex situations, we have to introduce *Events* because events describe the behavior of the properties over time. Events describe the related information for a given time and location (space). The introduction of the concept of "events" allows all related information for the event to be linked together.

The binary relationships are most frequent in information models [14]. However, in some cases, the natural and convenient way to represent certain concepts is to use relationships to connect an individual to more than just one individual or value [15]. These relationships are called N-ary relations. Thus we have to define an event as *Growing Problem Event* to model the real world representation of the growing problem of a crop (see Fig. 4). The concepts are defined such as GrowingProblem, Symptom, Cause, CausalAgent, etc. (see Fig. 4). The GrowingProblem concept can be further categorized as Disease, Weed, InsectPest, and NematodePest. The data properties are defined (if available), for example, Disease has DiseaseType (e.g. Parasitic or Non-Parasitic). Next, the associative relationships and their inverse (when available) among concepts have explicitly been specified by maintaining the semantics (e.g. hasSymptom, isSymptomOf). The individual, GrowingProblemEvent_1 (of GrowingProblemEvent concept) represents a single object. It contains all the information held in the above relationships. The cardinality restrictions on the relationships are defined by specifying that each instance of GrowingProblemEvent has exactly one value for GrowingProblem, Symptom, CausalAgent, and so on. It is specified as *F* in Fig. 4 (*F* refers the *functional property* in OWL). A crop has many GrowingProblemEvent (i.e. non-functional) but a GrowingProblemEvent has exactly one value for Crop (i.e. functional).

Next, the suitable control methods need to be identified with respective to the events (i.e. Growing Problem Events). When selecting the control methods, it depends on many factors specially types of the control methods (e.g. cultural, chemical, or bio), soil type, location, time of application, and application stage. If the control method is chemical it involves the quantity as well (i.e. the amount of the fungicide, pesticide, or insecticide) then the exact amount of it needs to be given. Its unit, application method, special information, etc. also need to be provided. This is also complex real world situation and need to represent all the information by describing their relationships. We therefore have defined an event as *Control Method Event* to handle this complex situation. The Fig. 4 represents these events.

The concepts are defined related to this such as ControlMethod, SoilType, Quantity, Pesticide, Unit, ApplicationMethod, TimeOfApplication, Location, etc. ControMethod can also be further categorized. The data properties are defined such as *ControlMethod has ApplicationStage* (e.g. After Infestation or Before Infestation). Then associative relationships and their inverse have been explicitly specified by maintaining the semantics (e.g. hasControlMethod, isControlMethodOf). The individuals of GrowingProblemEvent or ControlMethodEvent represent a single object encapsulating all the information related to specified event. The cardinality restrictions on the relationships are also defined. A GrowingProblemEvent has many ControlMethodEvent (i.e. non-functional) but a ControlMethodEvent has exactly one value for the GrowingProblemEvent (i.e. functional).

Based on the existing information, the additional knowledge can be inferred using the composition of relations (e.g. the relation GRANDFATHEROF is composed by the relationships FATHEROF and PARENTOF). This property is used to infer the additional knowledge (see Fig. 4). For example, relation isAffectedBy defines as *Crop isAffectedBy GrowingProblem* and its inverse as *GrowingProblem affects Crop*. This has been implemented by using *object property chain* in OWL 2. These relationships specify as a combination of existing object properties. Figure 5 shows this representation in Protégé.

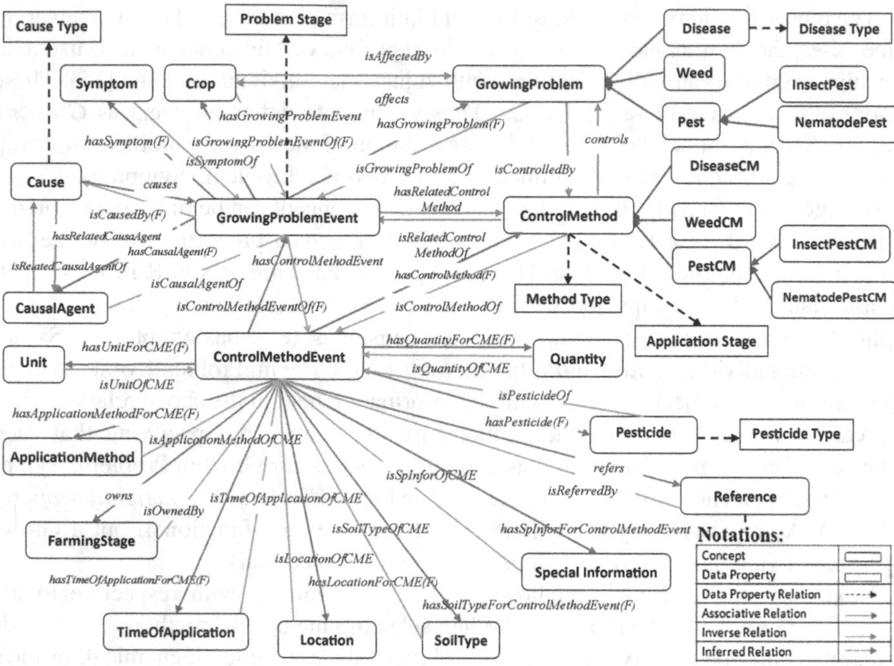

Fig. 4. Representation of GrowingProblemEvent and ControlMethodEvent.

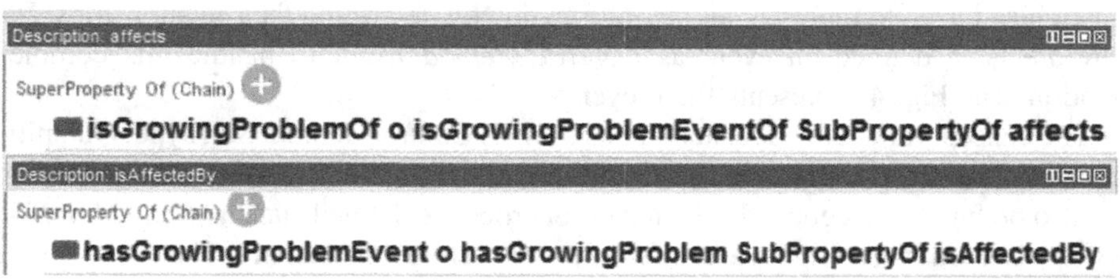

Fig. 5. Inferred relations represented in Protégé.

3.3 Scenario III: Handling Overlapping

In Sri Lankan agriculture domain, the farmers' location can be represented in many different ways (see Fig. 1). The Location concept is introduced to represent farmer location such as Zone, Province, District, RegionalArea, and Elevation (i.e. generalization). Then, concept Location is a super concept of Zone, District, Province, RegionalArea, and Elevation concepts (i.e. taxonomic hierarchy). Since AgroZone is a subclass of Zone then AgroZone is also a subclass of Location. The maps for Agro-climatic zones and Climatic zones in Sri Lanka are analyzed. Based on this analysis, zone can be further divided into 46 zones (Agro-Ecological Zones) (Fig. 1). Then the properties of Zone and AgroZone can be inherited by AgroEcologyZone concept because of taxonomic hierarchy.

The information about regional areas (330 Divisional Secretariat Divisions), districts (25 Districts), provinces (9 Provinces), zones (3 Zones), agro-zones (7 Agro-Zones), agro-ecology zones (46 AgroEcologyZones) and elevation based locations (3 Elevation) are also collected. Based on the analyzed data, it is clear that there is a conceptual overlapping among these concepts. For example, Southern province covers Galle, Matara, and Hambantota districts; Matara, Monaragala, and Kurunegala districts cover the Low Country Intermediate Zone. This conception needs to be handled semantically. Therefore this has been modeled by defining semantic relationships among the concepts. The relationships are defined as *belongsTo* and its inverse as *isBelonged*, and then the transitive property is specified on relationships (Fig. 6). This has been implemented by using *transitive property* in OWL 2.

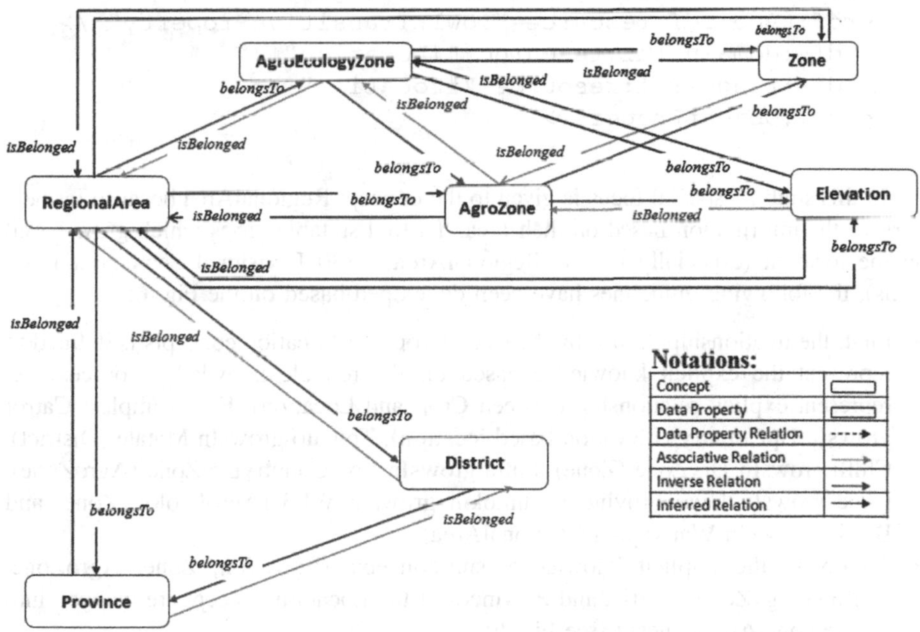

Fig. 6. Illustration of overlapping.

According to the Fig. 6, the RegionalAreas belong to Districts as well as AgroEcologyZone; the Districts belong to Provinces; AgroEcologyZones belong to Agro-Zones. AgroZones belong to Elevation as well as Zones. Then the knowledge can be inferred as showed in Table 1 (using inference engine in Protégé). As an example now farmers can find (i.e. infer) the regional areas where each crop will grow.

Table 1. Inferred knowledge

Transitive property	Inferred knowledge
RegionalArea belongsTo District and District belongsTo Province	RegionalArea belongsTo Province
RegionalArea belongsTo AgroEcologyZone and AgroEcologyZone belongsTo AgroZone	RegionalArea belongsTo AgroZone
AgroEcologyZone belongsTo AgroZone and AgroZone belongsTo Zone	AgroEcologyZone belongsTo Zone
AgroEcologyZone belongsTo AgroZone and AgroZone belongsTo Elevation	AgroEcologyZone belongsTo Elevation

This semantic can be represented as follows:
$\forall$ x,y,z $\in$ Location and x belongsTo y and y belongsTo z ==> x belongsTo z
This transitive property is symbolized using OWL:

```
<owl:ObjectProperty rdf:ID="belongsTo">
<rdf:type rdf:resource="&owl;TransitiveProperty" />
<rdfs:domain rdf:resource="#Location" />
<rdfs:range rdf:resource="#Location" />
</owl:ObjectProperty>
```

In this study, a special focus is given to the concept RegionalArea because farmers retrieve the information based on their area. To find suitable crops which grow based on the location (especially for the RegionalArea – 330 Divisional Secretariat Divisions), the following guidelines have been developed based on the Fig. 6:

- First, the relationship "growsIn" between Crop and Location concepts is defined to represent the explicit knowledge based on the reliable knowledge sources (i.e. represent explicit relationship between Crop and Location). For examples: Carrot growsIn UpCountry (Elevation based location), Tomato growsIn Matale (District), Chilli growsIn DryZone (Zone), Luffa growsIn LowCountryDryZone (AgroZone), Lime growsIn Uva (Province), Pumpkin growsIn WL3 (AgroEcologyZone) and Brinjal growsIn Wariyapola (RegionalArea).
- To obtain the implicit knowledge, sub-concepts (Elevation, Zone, AgroZone, AgroEcologyZone, District and Province) of the Location concept are mapped into the RegionalArea concept (see Fig. 6).
- Then following knowledge can be inferred based on the above mappings:
  - *If Crop grownsIn AgroEcologyZone then*
    If Crop grownsIn AgroEcologyZone and AgroEcologyZone isBelonged to RegionalArea then *Crop grownsIn RegionalArea*;
  - *If Crop grownsIn District then*
    If Crop grownsIn District and District isBelonged RegionalArea then *Crop grownsIn RegionalArea*;
  - *If Crop grownsIn Zone then*

If Crop grownsIn Zone and Zone isBelonged AgroZone then *Crop grownsIn AgroZone*;

- *If Crop grownsIn Elevation then*
  If Crop grownsIn Elevation and Elevation isBelonged AgroZone then *Crop grownsIn AgroZone*;
- *If Crop grownsIn AgroZone then*
  If Crop grownsIn AgroZone and AgroZone isBelonged RegionalArea then *Crop grownsIn RegionalArea*;
- *If Crop grownsIn Province then*
  If Crop grownsIn Province and Province isBelonged District then *Crop grownsIn District*;
  If Crop grownsIn Province and Province isBelonged RegionalArea then *Crop grownsIn RegionalArea*;

• This design has been implemented by using *object property chain* in OWL 2 (see Fig. 7). For the representation based on the property chain in OWL 2, the additional two relationships have been defined as *grows* and *growsOn* (see Fig. 7(a)). The relationship *growsIn* and *growsOn* are the sub-relationships of "grows" relationship. The definitions of relationships of *growsIn* (explicit relationship, i.e. original representation between Crop and Location) and *growsOn* are defined below and also shown in Fig. 7(b) and (c) respectively. The hierarchy of inverse relationship of "grows" relationship is depicted in Fig. 7(d).

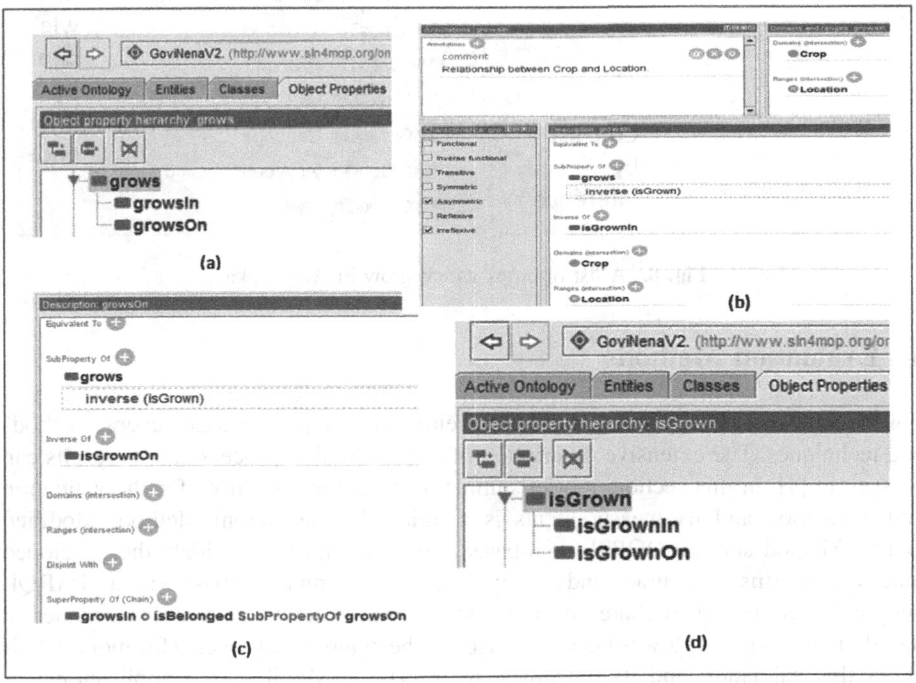

Fig. 7. Protégé screen views for representation of grows, growIn and grownOn relationships.

– *growsIn:* explicit relationship between Crop and Location concepts
  `Crop growsIn Location` (Fig. 7(b))
- *growsOn:* For example, if Crop grownsIn AgroEcologyZone and AgroEcologyZone isBelonged RegionalArea then "growsOn" (a new relationship) can be defined as Crop growsOn AgroEcologyZone. This can be implemented as follows:
  `growsIn o isBelonged ➔ growsOn` (Fig. 7(c))
- So that all the crops grown in RegionalArea, District, Province, AgroEcologyZone, AgroZone, Zone, Elevation based location are mapped into RegionalArea.
- Since the "grows" relationship is the super relationship of *growsIn* and *growsOn* relationships, all the relationships defined (implicit) with respect to the above two relationships (sub-relations) are considered as the relations of the "grows" relationship (super-relation).

Now farmers can find (i.e. infer) the crops which are grown in the regional areas using the implicit knowledge. For instance, CQ: *which crops are suitable to grow in 'Aranayaka' area?,* then it infers a list of crops which grow in Aranayaka (see Fig. 8).

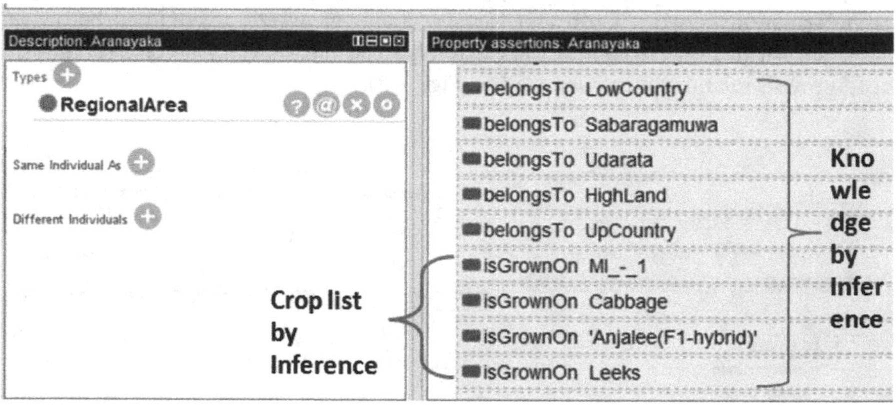

Fig. 8. A list of crops which grow in Aranayaka.

4 Evaluation Methods

For the validation and evaluation of the developed ontology, we used several methods and techniques. The extensive description of this evaluation process and its results can be seen in [4]. In this section, a brief summary of the methods used for the validation and evaluation and its measurements is mentioned. The Delphi Method, Modified Delphi Method and the OOPS! (web-based tool) were used to validate the developed ontology in terms of accuracy and quality. The online knowledge base with a SPARQL endpoint was created to share and reuse the domain knowledge that can be queried based on user context. It was also made use for the evaluation process (for more details about this validation and its outcomes, refer [4]). A Mobile-based application was developed and used to provide the information by using this ontology. This mobile application was trialed with a group of farmers in Sri Lanka. From this application farmer satisfaction (i.e. utility) on the knowledge provided by the ontology was examined (more details, refer [4]). There were three field trials. The initial system (working mobile prototype) was trialed with a group of 32 farmers in Sri Lanka. The second field trial was done using the real mobile application in Dambulla and Polonnaruwa area in Sri Lanka with 30 farmers. Since the user feedback especially depends on the service provided by an application, the third field trial was carried out to check the limitations in the design of the mobile application that can affect to usefulness of the system. The results of this field trial showed that the design of the application does not affect to the usability of the knowledge in the ontology. Because of the page limit we could not include more details about this evaluation, for examples, how the application was developed, how we used this application for the evaluation, how the results were obtained, how we analyze and a summary of the results, etc. But, all the details related to this evaluation are available in [4]. Very valuable feedback and comments were received from the field trials. Based on that, the designs of the ontology and subsequence developments were refined.

5 Conclusions

Currently, most of the research works are based on developing ontologies. However, ontology development is not a simple task. Based on the experience we obtained from a long journey to develop an ontology for the complex real word domain, we have identified several design approaches and how those approaches are implemented by using Protégé. In this paper, we mainly discussed these approaches by focusing on three different scenarios. Scenario one describes how we identify the basic ontology components and their representation in Protégé environment. How to design and implement a complex real word situation is described under scenario two by handling Events. In scenario three, conceptualization of the farmer location concept is modeled by considering the overlapping among the sub-concepts related to the location. This is implemented by using transitive property in OWL 2. The models and techniques we highlighted in this paper can be used as solutions to the similar problems in the different domains and fields.

Acknowledgements. We acknowledge the financial assistance provided to carry out this research work by the HRD Program of the HETC project of the Ministry of Higher Education, Sri Lanka. Assistance from the National Science Foundation to carry out the field visits is also acknowledged.

References

1. Fernández, M., Gómez-Pérez, A., Juristo, N.: METHONTOLOGY: from ontological art towards ontological engineering. In: Proceedings of the Ontological Engineering AAAI 1997 Spring Symposium Series, pp. 33–40. AAAI Press, Menlo Park, California (1997)

2. Walisadeera, A.I., Ginige, A., Wikramanayake, G.N.: User centered ontology for Sri Lankan farmers. Ecol. Inform. **26**(2), 140–150 (2015)
3. Walisadeera, A.I., Ginige, A., Wikramanayake, G.N.: Conceptualizing crop life cycle events to create a user centered ontology for farmers. In: Murgante, B., et al. (eds.) ICCSA 2014. LNCS, vol. 8583, pp. 791–806. Springer, Cham (2014). doi:10.1007/978-3-319-09156-3_55
4. Walisadeera, A.I., Ginige, A., Wikramanayake, G.N.: Ontology evaluation approaches: a case study from agriculture domain. In: Gervasi, O., et al. (eds.) ICCSA 2016. LNCS, vol. 9789, pp. 318–333. Springer, Cham (2016). doi:10.1007/978-3-319-42089-9_23
5. Zhou, L.: Ontology learning: state of the art and open issues. Inf. Technol. Manag. **8**, 241–252 (2007)
6. Staab, S., Gomez-Perez, A., Daelemana, W., Reinberger, M.L., Noy, N.F.: Why evaluate ontology technologies? Because it works! Intell. Syst. IEEE **19**(4), 74–81 (2004)
7. Gomez-Perez, A.: Some ideas and examples to evaluate ontologies. In: Proceedings of the 11th Conference on Artificial Intelligence for Applications, pp. 299–305 (1995)
8. Fernández-López, M.: Overview of methodologies for building ontologies. In: Proceedings of the IJCAI 1999 Workshop on Ontologies and Problem-Solving Methods (KRR5), Stockholm, Sweden (1999)
9. Ding, Y.: Ontology research and development part1 – a review of ontology generation. J. Inf. Sci. **28**(2), 123–136 (2002)
10. Poveda, M., Suárez-Figueroa, M.C., Gómez-Pérez, A.: Common pitfalls in ontology development. In: Meseguer, P., Mandow, L., Gasca, R.M. (eds.) CAEPIA 2009. LNCS (LNAI), vol. 5988, pp. 91–100. Springer, Heidelberg (2010). doi:10.1007/978-3-642-14264-2_10
11. Semantic web best practices and development. W3C Recommendation. http://www.w3.org/2001/sw/BestPrctices/
12. Suárez-Figueroa, M.C., Brockmans, S., Gangemi, A.,Gómez-Pérez, A., Lehmann, J., Lewen, H., Presutti, V., Sabou, M.: NeOn modeling components (2007). http://www.neon-project.org/deliverables/WP5/NeOn_2007_D5.1.1v3.pdf
13. Gruber, T.R.: A translation approach to portable ontology specifications. Technical report KSL 92-71, Knowledge System Laboratory, Stanford University, California (1993)
14. Dahchour, M., Pirotte, A.: The semantics of reifying n-ary relationships as classes. In: Proceedings of the 4th International Conference on Enterprise Information System (ICEIS 2002), Ciudad Real, Spain, pp. 580–586 (2002)
15. Noy, N., Rector, A.: Defining N-ary Relations on the Semantic Web (2006). http://www.w3.org/TR/swbp-n-aryRelations

Workshop on Data-driven modelling for Sustainability Assessment (DAMOST 2017)

Landslide Risk Analysis Along Strategic Touristic Roads in Basilicata (Southern Italy) Using the Modified RHRS 2.0 Method

Lucia Losasso(✉), Carmela Rinaldi, Domenico Alberico, and Francesco Sdao

University of Basilicata – School of Engineer, Viale dell'Ateneo Lucano 10, 85100 Potenza, Italy
lucia.losasso@unibas.it

Abstract. This work consists of the application of a modified method for the landslide risk assessment along a strategic touristic road. The proposed qualitative method is a modification of the original method Rockfall Hazard Rating System (RHRS) developed by Pierson et al. [26] at the Oregon State Highway Division (subsequently modified by other authors) and based on the exponential scoring functions.

The proposed application involves a careful analysis of environmental factors that influence the type of the mass movement as the slope, the use of the soil, the climatic conditions and the lithology, as such as parameters related to the structural characteristics of roads and traffic for example the road width, the number of lanes in each direction and the Average Vehicle Risk. The use of the different technical approaches, like double entry matrices and the implementation, for a few steps, of an Artificial Neural Network (ANN) allows to assess the analyzed landslide intensity, as well as the probability that it will occur in a given area along a transportation corridor. The application of such method involves a first phase of the data retrieval followed by a subsequent implementation and processing in a GIS softwares. The analysis carried out has been characterized by a step aimed at obtaining different layers, essential for the classification of the landsliding predisposing factors cataloged according to a final score by identifying the five risk threshold classes. So, in order to prepare appropriate interventions of protection and monitoring (if necessary, e.g. evacuation plans), underling the most dangerous areas has been fundamental.

Keywords: Basilicata region · Landslide risk assessment · Modified RHRS 2.0 · Qualitative method · Strategic transportation corridor

1 Introduction

In the last few decades the national territory has been increasingly plagued by natural disasters which prompted the scientific community and the territorially competent institutions to increase the knowledge about the risk and about the terms that contribute to its definition. The experiences gained in the past, in fact, have shown that the defense against natural hazards and the protection of environmental resources and public safety

O. Gervasi et al. (Eds.): ICCSA 2017, Part I, LNCS 10404, pp. 761–776, 2017.
DOI: 10.1007/978-3-319-62392-4_55

should be based primarily on a systematic work of prevention [29] that is the set of activities intended to avoid or minimize the possible occurrence of damages. The communication lines, in fact, with particular reference to the road network, assume an extremely important role as the inefficiency of the system can cause significant social and economic damages, as such as delays in the possible rescue operations. For this reason, a prior assessment of the risks on existing infrastructure is essential in order to plan appropriate interventions of the road network adjustment and the preparation of a priority lists of intervention. It is possible, therefore, to provide for an adequate risk management related to the operation of the road infrastructure by adopting methods of landslide risk analysis, prediction and prevention along roads allowing to prepare measures that can be assumed even in the construction phases of the new roads. The study presented in this work involved the analysis of representative landslides located along the Provincial Road 3 - SP 3 - and the State Road - SS 585 (Fondovalle del Noce) - arriving to the definition of the risk levels to which they are subject. The roads under analysis assume an important role because they are transportation corridors of primary importance for tourism that allows the connection with the Tyrrhenian coast, allowing to reach Maratea, a town known especially from the touristic point of view for its landscape in many points still intact and characterized by rocks, caves and sandy beaches. This area is also well known and the colossal statue of the Redeemer, 22 m high, is the second in the world after that of Rio de Janeiro.

2 Mass Movements Risk Assessment Along Roads

The transportation system may be subject to various types of natural risks that can damage it. They range from ordinary risks, that may occur daily, to extraordinary ones, that are exceptional. Moreover, the consequences of the events vary widely from case to case; it is possible to identify events (typically of low intensity) whose consequences are not serious and events that create serious damages to people and properties. If the risk definitions expressed in the literature are numerous, in general they concern the combination of two important elements: the probability and consequences. The risk of the road network can be expressed as the product of the probability that an event will occur and the consequences caused by this event. The reliability of a transportation corridor is the ability to keep offering, for a given period of time, the services it has been designed for.

Clearly, greater is the risk to which the transportation network is exposed, lower is the level of reliability that it is able to guarantee. It is therefore of fundamental importance to be able not only to define the concept of risk, but also to quantify it in some way.

2.1 State of Art

The instability phenomena caused by the mass movement along the roads represent a significant problem, which greatly undermines the travelers safety, even creating excessive obstacles to the traffic. In the literature several studies are promoted and financed aimed to identify valid methodological approaches in order to assess the risk of slope instability, allowing the implementation of important measures to prevent the landslide phenomenon. The approaches used in the literature for the landslides risk assessment along the road networks are based on the RHRS method (Rockfall Hazard Rating System, [26, 27]: between the second half of the 1980s and early 1990s, the Oregon Department of Transportation developed this system evaluating the geological dangers of rockfall in relation to the characteristics of the road on which they occur. This method allows to assign a score that takes into account the severity of the mass movements instability and also the characteristics of the infrastructure. The comparison between the scores for each analyzed site highlights the potentially most dangerous areas in order to address any repair measures. Because of the multiplicity of the risk assessment systems along roads, it is possible to identify two main evaluation categories, depending on the quantitative or qualitative analysis of landslide risk. The quantitative risk analysis is aimed to estimate the risk considering the probability of a loss. In particular, the probability of death of the population exposed at the risk for each year, due to landslides, is the annual probability that an accident involves the death of one or more occupants of a vehicle along a road [2]. This methodology has been applied by Bunce et al. [4] and Budetta [2] to assess landslide risk along the Highway 99 (British Columbia) and the A3 Napoli-Salerno motorway (Southern Italy) respectively, highlighting that, despite the quantitative analysis is the most used for the management of landslide risk (in the specific case of risk along highways) the methodology has some disadvantages. In fact a number of simplified assumptions about the traffic and the landslides have been adopted. For example, the traffic is evenly distributed in the time and in the space, without considering the landslide phenomenon, and an uniform average length for all vehicles is considered. The methodology also takes into account only the loss of human lives without considering the socio-economic consequences as the result of traffic interruptions, etc. and also the geological structure and the condition of the slopes are completely ignored. Furthermore, there is a considerable scarcity of historical data about the old roads, while, for the new roads, the data are not available. Finally, the methods based on quantitative analysis refer to the risk along stretches of road and not to individual slopes not allowing to identify the potentially dangerous ones. Despite the quantitative methods are more simple and immediate to use, for the calculation of geological risk along roads, it is very usual to use methods based on a qualitative analysis. In fact, the qualitative methods existing nowadays for the hydrogeological risk assessment have been developed, especially in the USA by various Departments of Transportation. Among the best known methods, it is possible to remember: the RHRS method (Rockfall Hazard Rating System) and SSRS method (Slope Stability Rating System, proposed by the Federal Highway Administration FHWA in the 1989), which allow the assessment of the risk as a sum of factors related to the failure hazard of the slopes, to the ditch effectiveness and to the consequences of a possible failure; the USRS method (Unstable Slopes Rating System) proposed by the Oregon Department of Transportation in 2001 which provides the sum of factors related to the estimated magnitude of failure, to the ditch effectiveness, to the cost of rehabilitation and to the highway class or even the HiSIMS method (Highway Slope instability Management System, [23]), which allows to assess the risk considering the factors related to the ditch effectiveness and to the consequences of a possible failure. These approaches allow to differentiate numerically the risks related to the identified sites, quantifying the level of risk to which the roads can be exposed. Most of these systems refers to the same criterion, considering an exponential score system with a base of 3 (3, 9, 27, 81 or 1, 3, 9, 27, 81) as originally proposed by Wyllie [33] and by Pierson and Van Viekle [27]. In this case, the total risk is given by the sum of the scores of a set of factors relating to the different categories. Only two system, the NPCs (New Priority Classification Systems – [32]) and the RCSA (Rock Cut Stability Assessment – [30]), assess the risk by multiplying the score of failure hazard by the consequences.

From the study of the literature, it has been possible to reveal that most of these systems have adopted or modified the original system RHRS of Pierson. In this study the RHRS 2.0 system [20] has been applied; it is a qualitative model that allows to assess the risk along the roads due to different types of mass movements, not only rockfalls, considering various factors described following.

3 The RHRS 2.0 Method

The RHRS 2.0 method derives from the necessity to make easier the score system to quantify the level of risk along the roads considering the geometric characteristics of the road and traffic, as well as the geological characteristics of the area (Table 1).

Table 1. Categories and scores of the RHRS 2.0.

Category	Rating criteria by score			
	Points 3	Points 9	Points 27	Points 81
Average slope (%)	<20	20–30	30–50	>50
Land use	Full vegetation, woodland area	Moderate vegetation, cultivated area	Sparse vegetation with a few scattered shrub	Absent vegetation
Average vehicle risk (%)	25%	50%	75%	100%
Roadway width (m)	13	10	8	6
Number of lines in each direction	3	2	2	1
Lithology	Calcareous, sandstones, intrusive and metamorphic rocks	weakly cemented sandstones and limestones, sands and conglomerates	Argillites, siltstones, conglomerates, flysch	slope debris, material subject to landslides, unstructured and/or mainly clay flysch
Climate	Low or moderate rainfall, no freezing periods	Moderate rainfall, short freezing periods	High rainfall, long freezing periods	Very high rainfall, long freezing periods
I/H	$(I/H)_1$	$(I/H)_2$	$(I/H)_3$	$(I/H)_4$
Total score	**24**			**648**

3.1 Average Slope

Through the morphological analysis of the digital elevation model, a parameterization of the surface can be carried out, the purpose of which is the numerical description of the continuous shape of the same surface [19]. Among the topographical attributes commonly considered from a DEM, in addition to its structure, the slope assumes a basic role, used for the evaluation of various parameters involved in many geological and geomorphological processes [18]; in fact, the slope is one of the most important causes of landsliding [16, 17]. The slope angle is a parameter easily correlated to the movement of a slope, because it is extremely tied to the acting forces and it is one of the most important morphological factors (greater is the slope, lower is the stability).

3.2 Land Use

This parameter refers to the different land cover; it is used to distinguish the different types of the soil in relation to the type of use and it is useful to identify the drainage network and infiltration, because a greater contribution to landsliding is also offered by the exploitation of areas named as fragile. It is important, in this regard, the lack of vegetation that exposes the slopes to the erosive action of rain water in proportion to their steepness. The presence of the woods, or the widespread vegetation, in addition to the action of the increase of the soil resistance by the roots, favors the interception of large quantities of rain, avoiding phenomena of erosion and degradation.

3.3 Average Vehicle Risk

The percentage of the exposure to the risk of vehicles (AVR = Average Vehicle Risk) is assessed taking into account the length of the road section at risk subject to falls, the average daily traffic and the speed limit (Eq. 1).

The AVR value is given by [26]:

$$AVR = (ADT * SL * 100\%)/PSL \tag{1}$$

Where:

AVR = exposure to risk of vehicles
ADT = average daily traffic (vehicles/day)
SL = length of the road section at risk (m)
PSL = posed speed limit (km/h)

The average risk per vehicle is one of the most important parameters to take into account because it quantifies the spatial probability of occurrence of a vehicle in the road section during a rock fall [3] and varies as a function of the hazard road length, the average daily traffic and the speed limit.

3.4 Roadway Width and Number of Lanes in Each Directions

The roadway width and, consequently, the number of lines in each directions provides the dimensional analysis of the lanes without including the shoulders.

The importance of this parameter is linked to the possibility to avoid obstacles on the road due to landslides of material (rock, earth or debris) from the slope.

3.5 Lithology

Among the different factors that cause instability, the lithology is one of the most important parameter that influences the slope movements. The lithology, together with other factors such as the angle of inclination and the slope profile, generally defines the type of movement and the breaking mechanism, since it is linked to predisposing factors that are influenced by the ground itself. Numerous studies, in fact, indicate that landslides are strongly influenced by the lithological properties of the Earth surface [1, 5, 7, 8, 12, 21, 22, 34] and therefore, the lithological conditions are the most frequently considered in the international literature.

3.6 Climate

The climatic factors are decisive in landslides triggering; this is particularly evident in climates characterized by long dry seasons and periods of intense and/or prolonged rainfall. This situation typically involves both the variation in flow rate of the drainage network, resulting in increase of erosive actions, both the raising of the free surfaces of the underground aquifers.

Regarding the rock fall, in particular, the freezing and thawing cycles determine the presence of the ice in the fractures and the water passage from the liquid state to the solid one with the increasing of the volume, creating favorable conditions to the detachment.

Finally, in the proposed method, the rainfall is assessed using the average values of precipitation in the study area.

3.7 Intensity/Hazard

Another parameter considered in the landslide risk assessment involves the use of a double-entry matrix (Table 2). It provides the analysis of the magnitude of a landslide, assessed by a scoring matrix (Table 3) taking into account the parameters related to the type of the phenomenon (earth flow, roto-translational slide, rock slide, rapid earth flow, rock fall, topples and debris flow) and to the descriptive characteristics (such as volume of involved material, velocity, run out, depth, affected area, deformation). The obtained value after assessing the magnitude class is intersected with the level of probability that the phenomenon occurs in a given area. This latter information is obtained by applying a mathematical model, the Artificial Neural Network, implemented by a software whose process has been described following.

Table 2. Score to assign for the H/I parameter assessment [20]

Magnitude				
Spatial hazard				
	Low	Medium	High	Very high
1–4	3 (I1)	3 (I1)	9 (I2)	9 (I2)
4–6	9 (I2)	9 (I2)	9 (I2)	27 (I3)
6–8	27 (I3)	27 (I3)	27 (I3)	81 (I4)
8–10	27 (I3)	81 (I4)	81 (I4)	81 (I4)

Landslide Magnitude. The magnitude is defined as a function of the parameters that characterize the landslides (Table 3), but it is not possible to define with precision a functional relationship between them, for this reason it is more appropriate to assess the magnitude in a relative scale rather than with a mathematical expression.

The method allows to consider each parameter in ranges of values characterized by a score (Table 3). Subsequently, the scores of each parameter relative to a landslide are summed, identifying the appropriate magnitude class. Each interval is associated to a magnitude value variable from 1 to 10 and to an intensity value variable from very low (I) to extremely high (X) (Table 3). To assess the landslide magnitude it is necessary to evaluate its size.

Table 3. Double-entry matrix to assess the landslide magnitude [25]

PARAMETER	RANGE OF VALUES						
Volume (m^3)	10	10-10^2	10^2-10^3	10^3-10^4	10^4-10^5	>10^5	
Rating	1	2	3	4	5	6	
Velocity (m/s)	$5x10^{-10}$	$5x10^{-10}$-$5x10^{-8}$	$5x10^{-8}$-$5x10^{-6}$	$5x10^{-6}$-$5x10^{-4}$	$5x10^{-4}$-$5x10^{-2}$	$5x10^{-2}$-5	>5
Rating	1	2	3	4	5	6	7
Run out (Km)	10^{-3}	10^{-3}-10^{-2}	10^{-2}-10^{-1}	10^{-1}-10^{-0}	>10		
Rating	1	2	3	4	5		
Depth (m)	1	1-12	12-25	25-35	35-50	>50	
Rating	1	2	3	4	5	6	
Affected area (Km^2)	0.01	0.01-0.25	0.25-0.50	0.50-0.75	0.75-1	>1	
Rating	1	2	3	4	5	6	
Deformation	heterogeneous	homogeneous	continuous	discontinuous			
Rating	4	2	4	2			
Typology	(Slowly) Earth flow	Roto-translational slides	Rock slides	Rapid earth flow	Rockfalls	Topples	Debris flow
Rating	2	4	8	8	10	10	10

TOTAL RATING	11	11-14	14-18	18-22	22-26	26-30	30-34	34-38	38-42	42
MAGNITUDE	I	II	III	IV	V	VI	VII	VIII	IX	X
INTENSITY	Very slow									Extremely rapid

The zonation of the territory in intensity classes is realized through the critical interpretation of the data collected in the inventory map, integrated with the informations derived from other basic documents.

Assessment of the Landslide Spatial Hazard (H) Using the Artificial Neural Networks. In the machine learning field, an Artificial Neural Network (ANN) is a mathematical model consisting of artificial "neurons".

In particular, similar to the human brain, a neural network consists of a number of neurons connected to each other by "weighed" connections, which receive stimuli and elaborate them. It can, therefore, be seen as a system able to give a response (output) to a question (input). In fact it is characterized by neurons working in parallel. A very common configuration used in an Artificial Neural Network consists of three layers: input layer, hidden layer and output layer.

In the recent years the neural networks (ANNs), frequently applied for the study of complex systems to solve engineering problems, are used to obtain thematic information to a sub-scale–pixel [11] to classify images [13], for the risk analysis [15], and finally, for the prediction, in the short period, of the environmental dynamics [10, 24, 31]. The most important peculiarity of these systems is the ability to learn mathematical-statistical models through the experience, or through the reading of the experimental data, without determining the mathematical relationships that bind the solutions to the problem. The artificial neural network is therefore not planned, but "trained" through a learning process based on empirical data. The result is a "black box" model, which is opposed to the "white box" because the internal components of the system are not known and it isn't possible to explain in clear terms the output. There are two different learning/training typologies [14]: supervised learning (used in this work) and unsupervised one. The supervised learning is the most used method. It is based on a training phase, constructing a data set with which to "train" the network, formed by a series of experimental pairs (real-input, real-output). If the training is successful, the network learns to recognize the implicit relationship that ties the input variables to the output and it is also able to respond correctly to stimuli that were not present in the training set. A neural network can be seen, therefore, as a system able to give a response (output) to a question (input). The training involves a series of operations:

- Choice of input parameters;
- Random allocation of weights to connections between neurons;
- Output Calculation;
- Comparison of the output with the expected value and error calculation.

A neural network possesses, as it is well known for all the networks, an algorithm that adjusts the weights (attenuations) of the links, so that it is suitable to provide a certain output in response to a given input. This process proceeds in an iterative manner until it reaches the convergence between the calculated value and the expected one, that is in this application, until the minimum squares error (RMSE) reaches the desired value (a RMSE value of zero indicates a good relationship between the desired value and the predicted one).

4 Study Area: The Provincial Road SP 3 and State Route SS585 (Fondovalle del Noce)

For the Basilicata Region the tourism industry is an economic sector of great importance, thanks to the beauty of its landscapes and to its historical-cultural heritage.

The Istat data (National Institute of Statistical Data) allow to estimate, on the overall regional touristic movement, an incidence of 55–60% of seaside tourism in the areas of Metaponto and Maratea. This location, as well as for its artistic and cultural heritage (watch towers, squares and monuments that make the city an open-air museum), is also known for the purity of the sea and the beauty of the coast, making the provincial road SP3 one of the most important transportation corridors to achieve such a relevant place in Basilicata. In this touristic context, in fact, plays a significant role the itinerary Lauria - Maratea (the two towns are connected by the Thirrenan Provincial Road SP3 that is linked for a short distance of about 32 km to the State Road 585). The road runs along the Noce River and crosses the picturesque landscape ranging from the mountains of Lagonegro to the *Riviera dei Cedri*, representing a fundamental road junction for those who want to reach Maratea from the highway A3. Unfortunately these transportation corridors are characterized by an intense state of instability (Fig. 1b) and are affected by active mass movements of various type (Fig. 1c and d). The observed area is geographically placed in the South West of Basilicata. The analyzed roads falls into the Sheet 521 Sect. 2 (Lauria) and into the Sheet 533 Sect. 4 (Maratea) of the cartography 1: 25.000 of the I.G.M. (Military Geographical Institute).

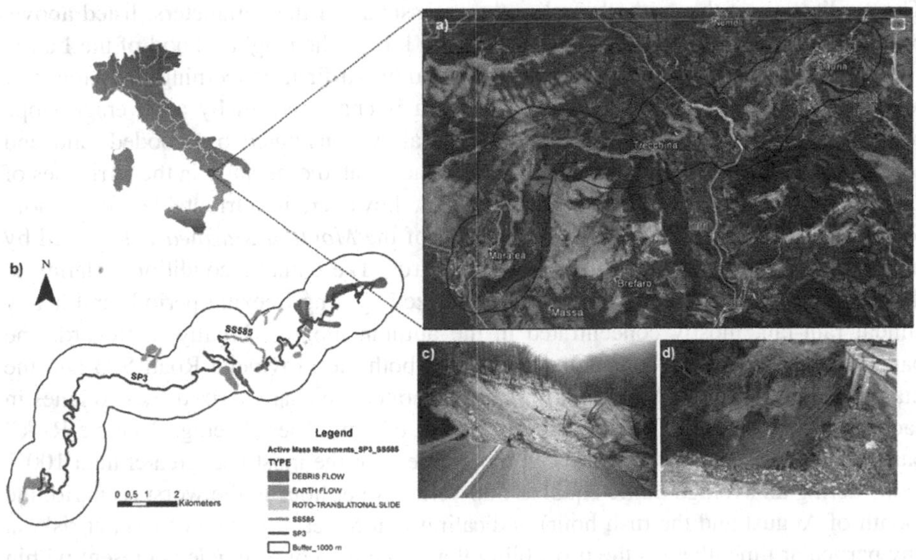

Fig. 1. (a) Study area; (b) Mass movements location in the study area; (c) Landslide occurred in location Panoramica (October 2016) by the side that overlooks the Seaport of Maratea: several tons of mud arrived to the port, invading homes and roads reaching the sea; (d) landslide occurred in the area of *Piano Menta* (Lauria) (March 2013).

4.1 Geomorphology of the Study Area

The landscape is mostly mountainous, with steep morphology, especially along the sides of the internal and coastal massifs. The major population centers including Lauria, Lagonegro and Trecchina, are mostly occupied by soils with predominantly pelitic component. From the morphological point of view it is possible to differentiate two areas: the north-east area shows a discontinuous relief, interrupted only by river valleys systems dissecting the morphostructures controlled by folds and thrust faults related to mio-Pliocene compressive tettogenesis. The south-western sector, instead, is marked by forms of fluvial dissection associated with horst-graben type morphostructures. The morphogenetic action of landslides occurring in the area is carried out through the slow or sudden displacement of a mass of rock and/or earth under the effect of gravity.

In particular, in correspondence of the reliefs of the *Monti di Lauria* and of the *Massiccio del Pollino* is possible to take over the carbonate succession referring respectively to the Unity of the *Monte Foraporta* and to the Unity of *Monte Pollino* [6, 28], characterized by limestone, dolomitic limestone, dolomite, carbonatic breccias, in layers and benches, sometimes intensely fractured. The areas crossed by the route are morphologically conditioned by the presence of the *Fiumicello* torrent. The same area has several watersheds and ridges that follow a morphology with strongly steep stretches.

4.2 Application of the RHRS 2.0 Method to the Study Area

The application of the method implies the assessment of the parameters, listed above, contextualized within the area of influence of 1 km, in the neighborhood of the Lauria – Maratea itinerary, that is the object of this study. At first, concerning the parameter related to the "Average Slope", the study area is characterized by an average slope between 10–25°. The zone, furthermore, is mainly constituted by wooded land and semi-natural areas, with few cultivated areas concentrated especially in the territories of Lauria and Trecchina. The lithological aspect, however, is attributable to Flyshoid formations. In particular, the morphostructure of the *Monti di Maratea* is bounded by important faults, characterized by notable discards. The climatic conditions relating to the examined territory, however, are characterized by long freezing periods and heavy annual rain-fall, mostly concentrated in the autumn months. Finally, as regards the parameters linked to the considered road path, both the Provincial Road SP3 and the State Road SS585 along the itinerary Lauria-Maratea are characterized by two lanes in each direction, each one constituted by a width of 3 m. The "Average Vehicle Risk" parameter, instead, assumes relatively high values, for the most part greater than 100% (considering an average traffic equal to 500 vehicles per hour in the worst scenario: the month of August and the rush hour), indicating that for each stretch of road at risk, at any particular time, there is the probability that more than one vehicle is present within the hazard zone [26]. Finally, the last considered parameter is I/H. For each landslide body, a value of intensity has been assessed using the matrix shown in the Table 3. Therefore the obtained magnitude class has been intersected with the hazard class obtained using the Artificial Neural Network (ANN). The ANN is implemented considering the landsliding predisposing parameters (Fig. 2), a training phase, represented by a portion of landslides, and a testing phase used to test the performance of the network. The choice of parameters has been conducted considering the characteristics of the study area in relation to the influence that they have on the mechanisms of landslides. Some factors are nominal variables, such as the lithology and the land use,

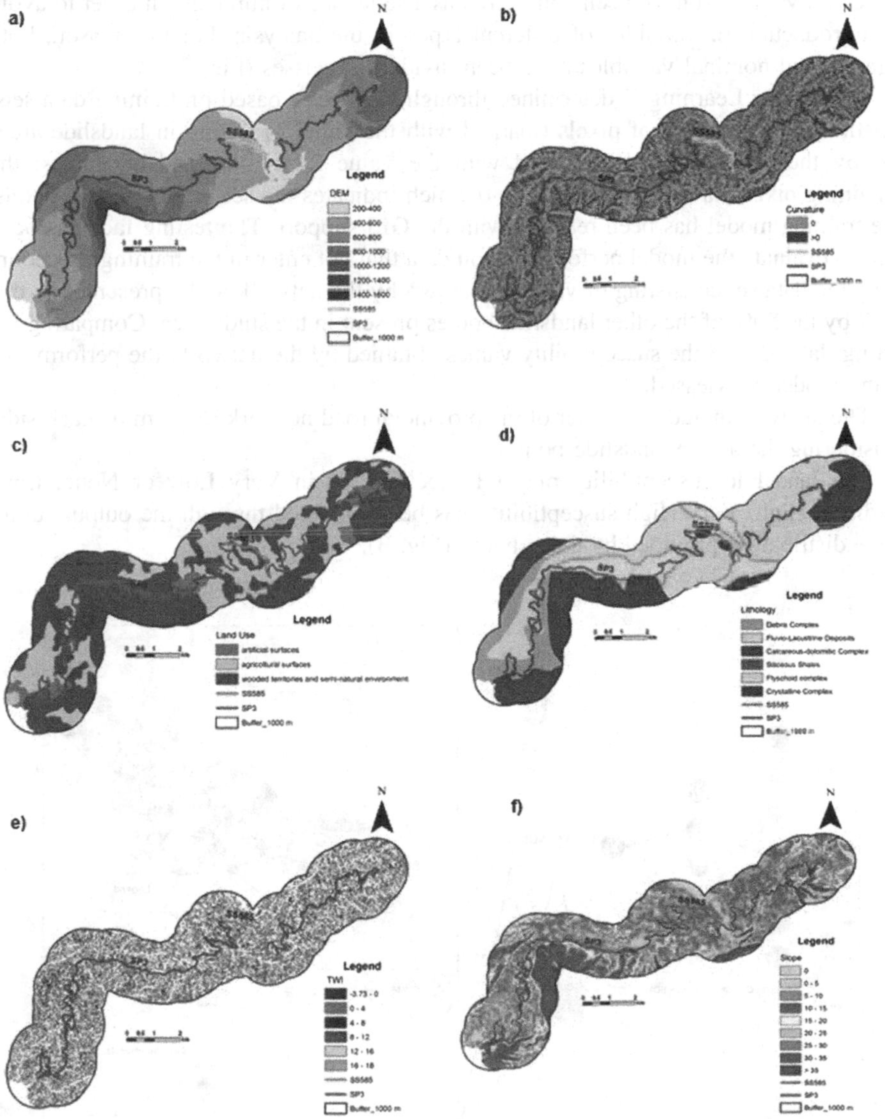

Fig. 2. Landslide predisposing factor maps used for the landslide susceptibility analysis: (a) DEM; (b) curvature; (c) land use; (d) lithology; (e) TWI; (f) slope

other are morphometric (slope, aspect, etc...) and they have been obtained from the DEM (Digital Elevation Model) having a cell resolution of 20 m × 20 m. After rasterized all parameters and converted to ASCII data, they have been processed and classified using the IDRISI Taiga software [9] in order to obtain input maps to insert later in the MLP (MultyLayerPerceptron) module, used to generate the Artificial Neural Network. The parameters considered for this work as input for the assessment of landslide susceptibility are: Aspect, Lithology, Slope, Curvature, DEM, TWI (Topographic Wetness Index), Land Use (Fig. 2). The used input parameters (nominal and numerical variables) have been considered as a sequence of numbers, in order to avoid the introduction of variables of different types in the analysis. For this reason, both numeric and nominal variables have been divided in classes (Fig. 2).

Training or Learning is determined through a process, based on training data sets, constituted by the 50% of pixels (marked with the value 1) falling in landslide areas and by the 50% of pixels (marked with the value 2) in non-landslide areas; the remaining pixels have a value equal to 0 which indicates the set of NO DATA pixels. The training model has been realized with the GIS support. The testing face has been used to evaluate the model performance on data that not enter in the training procedure using the data set consisting of values unknown by the network and represented in this work by the 50% of the other landslide bodies present in the study area. Comparing the testing data set and the susceptibility values obtained by the network, the performance of the model is assessed.

The analysis includes a buffer of the provincial road network of 1 km on each side, considering the active landslide bodies.

The landslide susceptibility map (H), reclassified in Very Low or None, Low, Medium, High, Very High susceptibility has been obtained through the output cumulative distribution provided by the network (Fig. 3).

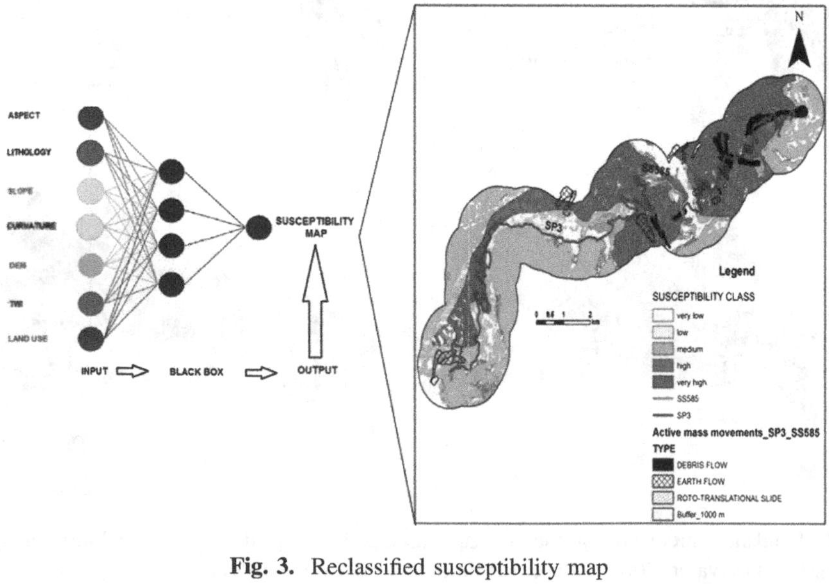

Fig. 3. Reclassified susceptibility map

4.3 Results and Conclusion

The application of the RHRS 2.0 method along the Provincial Road SP3 and along the State Road SS585, allowed to identify different risk classes due to several active mass movements. In particular, the risk along roads is assessed considering both the parameters linked to the characteristics of the infrastructures and the geological ones.

The sum of the scores of the analyzed parameters allowed to define different risk levels with increasing severity, according to the classification proposed in the Italian Legislation (DPCM 29/09/1998): R1 indicates the moderate risk for which the economic, social and environmental damages are marginal; R2 indicates the average risk for which minor damages to buildings, infrastructures and environmental assets are possible, so it doesn't affect the people safety, the building practicability and the functionality of the economic activities; R3 indicates a high risk level and possible problems for people safety, functional damage to buildings and infrastructure, disruption of socio-economic activities and significant damages to the environmental heritage; finally R4 indicates the very high risk level for which the loss of life and serious injuries, serious damages to buildings, infrastructures and environmental heritage and the destruction of socio-economic activities are possible.

The total length of the analyzed section is equal to about 32 km; of these the 3.5% is subject to risk R1, 14.2% is subject to risk R2, 13.4% at risk R3 and 2% at risk R4, while to the road sections not subject to risk, accounting for 67% of the total, have been assigned a risk class R0 corresponding to zero risk (Fig. 4). The results of this zoning allows to observe that between the four different risk classes defined by the DPCM 29/9/98, the predominant class is the R3.

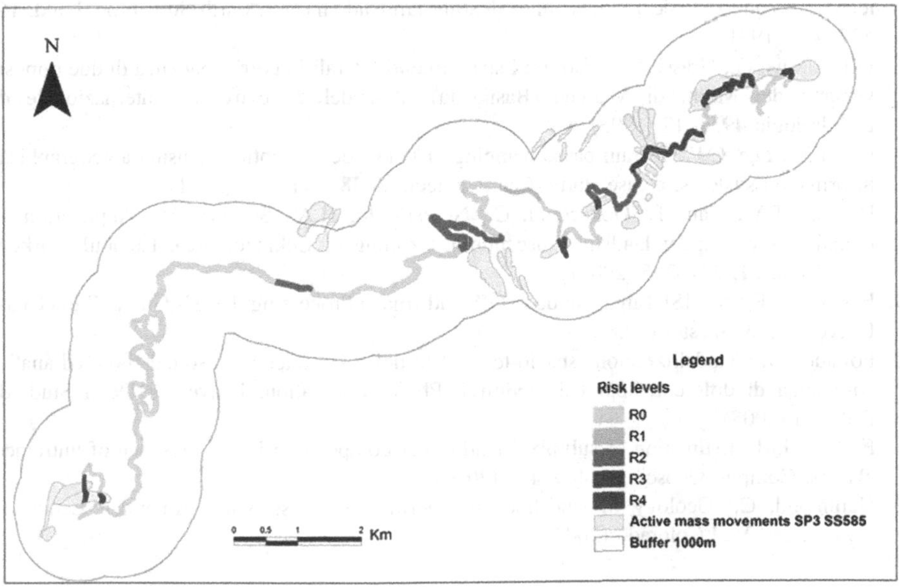

Fig. 4. Risk levels map along the strategic touristic road SS585 and SP3

In conclusion, it is possible to say that the 30 landslide bodies identified along the analyzed road significantly affect the road functionality. Furthermore, considering the tourist importance that characterizes the Lauria–Maratea itinerary, the presence of the active mass movements represents certainly a relevant problem. The RHRS 2.0 method results to be, therefore, a fundamental preliminary tool in order to mapping the road risk levels.

The innovation of the proposed method is primarily concerned with the possibility of assessing the risk along roads due not only to the rockfall but also to other types of mass movements (earth flow, roto-translational slide, rock slide, rapid earth flow, debris flow). In fact it is therefore considered an important tool for the competent authorities to timely adopt the maintenance and the management measures of the road infrastructures.

References

1. Anbalagan, R.: Landslide hazard evaluation and zonation mapping in mountainous terrain. Eng. Geol. **32**, 269–277 (1992)
2. Budetta, P.: Assessment of rockfall along roads. Bull. Nat. Hazards Earth Syst. Sci. **4**, 71–81 (2004)
3. Budetta, P., Nappi, M.: Comparison between qualitative rockfall risk rating systems for a road affected by high traffic intensity. Nat. Hazards Earth Syst. Sci. **13**, 1643–1653 (2013). doi:10.5194/nhess-13-1643-2013
4. Bunce, C.M., Cruden, D.M., Morgenstern, N.R.: Assessment of the hazard from rockfall on a highway. Can. Geotech. J. **34**, 344–356 (1997)
5. Carrara, A., Cardinali, M., Detti, R., Guzzetti, F., Pasqui, V., Reichenbach, P.: GIS techniques and statistical models in evaluating landslide hazard. Earth Surf. Proc. Land. **16**, 427–445 (1991)
6. D'Ecclesiis, G., Grassi, D., Sdao, F.: Espandimenti laterali in corrispondenza di due opposti versanti dei Monti di Maratea (Basilicata). Atti del 2° convegno internazionale di geoidrologia **49**, 1–17 (1993)
7. Dai, F.C., Lee, C.F.: Terrain-based mapping of landslide susceptibility using a geographical information systems: a case study. Can. Geotech. J. **38**, 911–923 (2001)
8. Duman, T.Y., Can, T., Gokceoglu, C., Nefeslioglu, H.A., Sonmez, H.: Application of logistic regression for landslide susceptibility zoning of Cekmece area, Istanbul, Turkey. Eng. Geol. **51**, 241–256 (2006)
9. Eastman, J.R.: IDRISI Taiga, Guide to GIS and Image Processing, User's Guide. Press Clark University, Worcester (2009)
10. Follador, M.: Modellizzazione spazio-temporale delle dinamiche di uso del suolo ed analisi comparata di differenti approcci predittivi. Ph.D. dissertation, Università Degli Studi di Bologna (2008)
11. Foody, G.M.: Estimation of sub-pixel land cover composition in the presence of untrained classes. Comput. Geosci. **26**, 469–478 (2000)
12. Hammond, C.: Geology in landslide engineering. In: First North American Landslide Conference, Vail Colorado (2007)

13. Joshi, M., Buchanan, K.T., Shroff, S., Orenic, T.V.: Delta and Hairy establish a periodic prepattern that positions sensory bristles in Drosophila legs. Dev. Biol. **293**(1), 64–76 (2006). doi:10.1016/j.ydbio.2006.01.005. (Export to RIS)
14. Kanevski, M., Maignan, M.: Analysis and Modelling of Spatial Environmental Data. EPFL Press, Lausanne (2004)
15. Kanungo, D.P., Arora, M.K., Sarkar, S., Gupta, R.P.: A comparative study of conventional, ANN black box, fuzzy and combined neural and fuzzy weighting procedures for landslide susceptibility zonation in Darjeeling Himalayas. Eng. Geol. **85**(3–4), 347–366 (2006)
16. Lee, S., Min, K.: Statistical analysis of landslide susceptibility at Yongin, Korea. Environ. Geol. **40**, 1095–1113 (2001)
17. Lee, S., Ryu, J.H., Won, J.S., Park, H.J.: Determination and application of the weights for landslide susceptibility mapping using an artificial neural network. Eng. Geol. **71**, 289–302 (2004)
18. Losasso, L., Derron, M.-H., Horton, P., Jaboyedoff, M., Sdao, F.: Definition and mapping of potential rockfall source and propagation areas at a regional scale in Basilicata region (Southern Italy). Rend. Online Soc. Geol. Ital. **41**, 175–178 (2016). doi:10.3301/ROL.2016.122
19. Losasso, L., Jaboyedoff, M., Sdao, F.: Potential rock fall source areas identification and rock fall propagation in the Province of Potenza territory using an empirically distributed approach. Landslides (2017a). doi:10.1007/s10346-017-0807-x
20. Losasso, L., Pascale, S., Sdao, F.: Landslides risk assessment along roads: the transportation corridors of the "Dolomiti Lucane" (Basilicata). In: 4th World Landslide Forum – Ljubljana. Advancing Culture of Living with Landslides: vol. 4 Diversity of Landslide Forms (2017b, in press)
21. Mejia-Navarro, M., Wohl, E.E.: Geological hazard and risk evaluation using GIS: methodology and model applied to Medellin, Colombia. Bull. Assoc. Eng. Geol. **31**, 459–481 (1994)
22. Mejia-Navarro, M., Garcia, L.A.: Natural hazard and risk assessment using decision support systems, application Glenwood Springs, Colorado. Environ. Eng. Geosci. **2**(3), 299–324 (1996)
23. Miller, S.M.: Development and Implementation of the Idaho Highway Slope Instability and Management System (HiSIMS). Idaho Transportation Department. report N03–07 (2003)
24. Nemmour, H., Chibani, Y.: Multiple support vector machines for land cover change detection: an application for mapping urban extension. ISPRS J. Photogram. Remote Sens. **61**, 125–133 (2006)
25. Pascale, S., Sdao, F., Sole, A.: A model for assessing the systemic vulnerability in landslide prone areas. Nat. Hazards Earth Syst. Sci. **10**, 1575–1590 (2010)
26. Pierson, L.A., Davis, S.A., Van Vickle, R.: Rockfall Hazard Rating System Implementation Manual: Oregon Department of Transportation, FHWA-OR-EG-90-01. FHWA, U.S. Department of Transportation (1990)
27. Pierson, L.A., Van Vickle, R.: Rockfall Hazard Rating System – Participant's manual, Federal Highway Administration, U.S. Department of Transportation Report FHWA-SA-93-057, 104 p. (1993)
28. Sansone, M.T.C., Rizzo, G.: Pumpellyite veins in the metadolerite of the Frido Unit (southern Appennines-Italy). Periodico di Mineralogia **81**, 75–92 (2012). doi:10.2451/2012PM0005
29. Sdao, F., Simeone, V.: Mass movements affecting goddess Mefitis sanctuary in Rossano di Vaglio (Basilicata, southern Italy). J. Cult. Herit. **8**, 77–80 (2007). doi:10.1016/j.culher.2006.10.004

30. Uribe-Extebarria, G., Morales, T., Uriarte, J.A., Ibarra, V.: Rock cut stability assessment in mountainous regions. Environ. Geol. **48**, 1002–1013 (2005)
31. Villa, N., Paegelow, M., Camacho, O.M.T., Cornez, L., Ferraty, F., Ferré, L., Sarda, P.: Various approaches for predicting land cover in mountain areas. Commun. Stat.-Simul. Comput. **36**, 73–86 (2007)
32. Wong, C.K.L.: New Priority Classification for Slopes and Retaining Walls (GEO Report No. 68), Geotechnical Engineering Office, Hong Kong (1998)
33. Wyllie, D.: Rock Slope Inventory/Maintenance Programs, FHWA Rockfall Mitigation Seminar, 13th Northwest Geotechnical Workshop, Portland, Oregon (1987)
34. Yalcin, A.: GIS-based landslide susceptibility mapping using analytical hierarchy process and bivariate statistics in Anderson (Turkey): comparison of results and confirmations. CATENA **1**, 1–12 (2008)

Author Index

Printed in the United States
By Bookmasters